Geotechnical, Geological and Earthquake Engineering

Volume 46

Series Editor
Atilla Ansal, School of Engineering, Özyeğin University, Istanbul, Turkey

Editorial Advisory Board
Julian Bommer, Imperial College London, U.K.
Jonathan D. Bray, University of California, Berkeley, U.S.A.
Kyriazis Pitilakis, Aristotle University of Thessaloniki, Greece
Susumu Yasuda, Tokyo Denki University, Japan

More information about this series at http://www.springer.com/series/6011

Kyriazis Pitilakis
Editor

Recent Advances in Earthquake Engineering in Europe

16th European Conference on Earthquake Engineering-Thessaloniki 2018

Springer

Editor
Kyriazis Pitilakis
Department of Civil Engineering
Aristotle University
Thessaloniki, Greece

ISSN 1573-6059 ISSN 1872-4671 (electronic)
Geotechnical, Geological and Earthquake Engineering
ISBN 978-3-319-75740-7 ISBN 978-3-319-75741-4 (eBook)
https://doi.org/10.1007/978-3-319-75741-4

Library of Congress Control Number: 2018937086

© Springer International Publishing AG, part of Springer Nature 2018
Chapter 1 is licensed under the terms of the Creative Commons Attribution 4.0 International License (http://creativecommons.org/licenses/by/4.0/). For further details see license information in the chapter.
This work is subject to copyright. All rights are reserved by the Publisher, whether the whole or part of the material is concerned, specifically the rights of translation, reprinting, reuse of illustrations, recitation, broadcasting, reproduction on microfilms or in any other physical way, and transmission or information storage and retrieval, electronic adaptation, computer software, or by similar or dissimilar methodology now known or hereafter developed.
The use of general descriptive names, registered names, trademarks, service marks, etc. in this publication does not imply, even in the absence of a specific statement, that such names are exempt from the relevant protective laws and regulations and therefore free for general use.
The publisher, the authors and the editors are safe to assume that the advice and information in this book are believed to be true and accurate at the date of publication. Neither the publisher nor the authors or the editors give a warranty, express or implied, with respect to the material contained herein or for any errors or omissions that may have been made. The publisher remains neutral with regard to jurisdictional claims in published maps and institutional affiliations.

Printed on acid-free paper

This Springer imprint is published by the registered company Springer International Publishing AG part of Springer Nature.
The registered company address is: Gewerbestrasse 11, 6330 Cham, Switzerland

Preface

Earthquake Engineering has made significant progress over the past years. Lessons learned from numerous devastating earthquakes all over the world, often associated with great physical, financial and societal losses, brought valuable new data, experiences and finally knowledge and fresh ideas on how science and engineering could effectively cope and reduce risk.

The intensified deployment and operation of dense networks providing high-quality ground motion records, combined with the observation of recorded damages and the important progress in numerical modelling and computation, offered to well-trained scientists and engineers better conception and understanding of the physical process of earthquakes, allowing to develop more accurate and efficient tools to assess seismic hazard and design ground motions including complex site effects.

Considerable progress has also been made in structural and geotechnical earthquake engineering based on numerous earthquake recordings and field observations, on high-quality monitoring of structural behaviour and damages, both in real instrumented structures and in laboratory model tests performed in large-scale facilities like shaking tables, reaction walls and centrifuges, and certainly based on the progress made in numerical modelling and computation capacity.

The improvement of the resilience of the important European cultural heritage omnipresent in the built environment and the conservation of historical monuments are of high priority in Europe, attracting special attention of the earthquake engineering community and public authorities.

Risk management policies and strategies also gained important insight from recent earthquakes, proving that effective coping of the impact of large earthquakes in modern societies is difficult and complicated even in highly developed countries, a fact that leads to the emerging need to develop more efficient methods, technologies, policies and strategies.

New scientific, societal and financial challenges also emerged from recorded induced seismicity, which changes drastically the seismic perception in several low- to negligible-seismicity countries, forcing them to strengthen considerably their seismic design codes and construction practices with obvious economic and safety impact.

In this respect, the philosophy underlying security and safety issues and the design methods are continuously improved based on the work of knowledgeable scientists and engineers, and certainly under the pressure of constantly evolving needs expressed by booming economies, exploration of new energies, construction of demanding buildings, infrastructures, critical facilities and industrial complexes, as well as under the stress of the population increase and concentration. Modern societies became inevitably more vulnerable to natural and anthropogenic hazards, and the only way to face these threats is to improve scientific knowledge and technology.

All this progress, knowledge and scientific-technical achievements are reflected quaternary in the European Conference on Earthquake Engineering, which gives the floor to young and mature scientists, researchers, engineers and policymakers to address the present state of the art and to discuss emerging future developments and needs in earthquake engineering and related topics. The lectures given every 4 years by distinguished academics, scientists, engineers, policymakers and seismic code developers provide a comprehensive panorama in Earthquake Engineering and a guidance to future developments.

To this end, this handbook is an outstanding collection of invited lectures written by internationally recognized academics, scientists, engineers, researchers, policymakers and code developers in Europe, presented at the 16th European Conference on Earthquake Engineering, organized by the European Association of Earthquake Engineering and the Hellenic Society of Earthquake Engineering and held in Thessaloniki, Greece, in June 2018.

It contains the 5th Nicholas Ambraseys distinguished lecture given by Peter Fajfar; four keynote lectures given by Atilla Ansal, Gian Michele Calvi, Michael Fardis and Sergio Lagomarsino; and twenty-two thematic lectures given by John Douglas, Sinan Akkar, Roberto Paolucci, Aspasia Zerva, Stefano Parolai, Amir Kaynia, Nikos Makris, George Gazetas, Oreste Bursi, Gobal Madabhushi, Dimitris Beskos, Raffaele Landolfo, Elizabeth Vintzileou, Andreas Kappos, Paulo Lourenco, Alper Ilki, Tiziana Rossetto, Ioannis Psycharis, Iunio Iervolino, Mauro Dolce, Philippe Bisch, Paolo Franchin and their co-workers. The lectures are put together as twenty-eight chapters, addressing a comprehensive collection of state-of-the-art and cutting-edge topics in earthquake engineering, engineering seismology and seismic risk assessment and management.

The book is of interest to civil engineers, engineering seismologists and seismic risk managers, covering a wide spectrum of fields from geotechnical and structural earthquake engineering to engineering seismology and seismic risk assessment and management. Scientists, professional engineers, researchers, civil protection policymakers and students interested in the seismic design of civil engineering structures and infrastructures, hazard, risk assessment and mitigation policies and strategies will find in this book not only the state-of-the-art advances in earthquake engineering, but also novel ideas on what will be the near future in earthquake engineering and resilient design of structures.

The editor would like to express his gratitude to all authors for their interest and effort in preparing their manuscripts and to his young colleagues Anastasios Anastasiadis, Dimitris Pitilakis and Sotiris Argyroudis. Without their enthusiasm, support and devotion, it would not have been possible to complete this book.

Thessaloniki, Greece   Kyriazis Pitilakis

Contents

1 **Analysis in Seismic Provisions for Buildings: Past, Present and Future** 1
Peter Fajfar

2 **Implications of Site Specific Response Analysis** 51
Atilla Ansal, Gökçe Tönük, and Aslı Kurtuluş

3 **A Redefinition of Seismic Input for Design and Assessment** 69
G. Michele Calvi, Daniela Rodrigues, and Vitor Silva

4 **From Force- to Displacement-Based Seismic Design of Concrete Structures and Beyond** 101
Michael N. Fardis

5 **Seismic Assessment of Existing Irregular Masonry Buildings by Nonlinear Static and Dynamic Analyses** 123
Sergio Lagomarsino, Daniela Camilletti, Serena Cattari, and Salvatore Marino

6 **Capturing Geographically-Varying Uncertainty in Earthquake Ground Motion Models or What We Think We Know May Change** 153
John Douglas

7 **Implementation of Near-Fault Forward Directivity Effects in Seismic Design Codes** 183
Sinan Akkar and Saed Moghimi

8 **3D Physics-Based Numerical Simulations: Advantages and Current Limitations of a New Frontier to Earthquake Ground Motion Prediction. The Istanbul Case Study** 203
Roberto Paolucci, Maria Infantino, Ilario Mazzieri, Ali Güney Özcebe, Chiara Smerzini, and Marco Stupazzini

9	**Issues with the Use of Spatially Variable Seismic Ground Motions in Engineering Applications**...................... Aspasia Zerva, Mohammad Reza Falamarz-Sheikhabadi, and Masoud Khazaei Poul	225
10	**Bridging the Gap Between Seismology and Engineering: Towards Real-Time Damage Assessment**................ Stefano Parolai, Michael Haas, Massimiliano Pittore, and Kevin Fleming	253
11	**Earthquake Geotechnics in Offshore Engineering**.............. Amir M. Kaynia	263
12	**The Dynamics of Rocking Isolation**........................ Nicos Makris	289
13	**Multistory Building Frames and Shear Walls Founded on "Rocking" Spread Footings**............................ G. Gazetas, D. Dais, F. Gelagoti, and R. Kourkoulis	309
14	**Seismic Design of Foundations in Difficult Soil Conditions: Examples of Solutions**..................................... Alain Pecker	323
15	**Structural Health Monitoring for Seismic Protection of Structure and Infrastructure Systems**.................... Oreste S. Bursi, Daniele Zonta, Emiliano Debiasi, and Davide Trapani	339
16	**Large Scale Testing Facilities – Use of High Gravity Centrifuge Tests to Investigate Soil Liquefaction Phenomena**.............. Gopal S. P. Madabhushi	359
17	**Seismic Analysis and Design of Composite Steel/Concrete Building Structures Involving Concrete-Filled Steel Tubular Columns**..... Konstantinos A. Skalomenos, George D. Hatzigeorgiou, and Dimitri E. Beskos	387
18	**Seismic Design of Steel Structures: New Trends of Research and Updates of Eurocode 8**.............................. Raffaele Landolfo	413
19	**Unreinforced Masonry Walls Subjected to In-Plane Shear: From Tests to Codes and Vice Versa**........................ Elizabeth Vintzileou	439
20	**Seismic Design of Bridges: Present and Future**................ Andreas J. Kappos	459

Contents

21 Technologies for Seismic Retrofitting and Strengthening of Earthen and Masonry Structures: Assessment and Application 501
Paulo B. Lourenço

22 Seismic Performance of a Full-Scale FRP Retrofitted Sub-standard RC Building 519
Alper Ilki, Erkan Tore, Cem Demir, and Mustafa Comert

23 Advances in the Assessment of Buildings Subjected to Earthquakes and Tsunami 545
Tiziana Rossetto, Crescenzo Petrone, Ian Eames, Camilo De La Barra, Andrew Foster, and Joshua Macabuag

24 Seismic Vulnerability of Classical Monuments 563
Ioannis N. Psycharis

25 What Seismic Risk Do We Design for When We Design Buildings? 583
Iunio Iervolino

26 The 2016–2017 Central Apennines Seismic Sequence: Analogies and Differences with Recent Italian Earthquakes 603
Mauro Dolce and Daniela Di Bucci

27 Eurocode 8. Evolution or Revolution? 639
Philippe Bisch

28 Research Needs Towards a Resilient Community 661
Paolo Franchin

About the Editor

Kyriazis Pitilakis is Professor in Geotechnical Earthquake Engineering at Aristotle University in Thessaloniki, Greece. He did his undergraduate degree in Civil Engineering at the same University and received his PhD from Ecole Centrale de Paris. He has more than 35 years of intensive academic, research and professional experience in earthquake and geotechnical engineering. Currently Vice President of the European Association of Earthquake Engineering (EAEE), he has been Chairman and is now Vice Chairman of the Technical Committee "Geotechnical Earthquake Engineering and Associated Problems" (TC203) of the International Society of Soil Mechanics and Geotechnical Engineering (ISSMGE), and past President of the Greek Society of Earthquake Engineering and of the Institute of Earthquake Engineering and Engineering Seismology in Greece (ITSAK). He served as Head of the Civil Engineering Department of the Aristotle University of Thessaloniki, and he currently acts as the Director of the Laboratory of Soil Mechanics, Foundations and Geotechnical Earthquake Engineering of the Civil Engineering Department of Aristotle University (http://sdgee.civil.auth.gr/). Professor Pitilakis has a vast experience in European research activities, acting as the coordinator of large European projects like EUROSEISTEST (http://euroseis.civil.auth.gr) and SYNER-G (www.syner-g.eu), and as participant in more than 30 European and national research projects. His main research and professional interests are in geotechnical and earthquake engineering, seismic design of foundations and infrastructures, soil dynamics, site effects and microzonation, vulnerability and risk assessment of civil engineering structures, infrastructures and monuments and lifeline earthquake engineering. He is an author of more than 550 scientific papers published in peer-reviewed scientific journals and proceedings, and he is also an author of several book chapters, an editor of three books published by Springer and a reviewer for numerous high-impact scientific journals in his field of interest. He has been an invited keynote lecturer in many international and national congresses and conferences, while being an active member of national, European and international committees for seismic standards. The French Republic has honoured him for his scientific and academic merits as Chevalier of the Order of Academic Palms (Ordre des Palmes Academiques).

Professor Pitilakis is the chairman of the 16th European Conference of Earthquake Engineering, held in Thessaloniki, Greece, in June 2018 (www.16ecee.org).

Chapter 1
Analysis in Seismic Provisions for Buildings: Past, Present and Future

Peter Fajfar

Abstract The analysis of structures is a fundamental part of seismic design and assessment. It began more than a hundred years ago, when static analysis with lateral loads of about 10% of the weight of the structure was adopted in seismic regulations. For a long time seismic loads of this size remained in the majority of seismic codes worldwide. In the course of time, more advanced analysis procedures were implemented, taking into account the dynamics and nonlinear response of structures. In the future, methods with explicit probabilistic considerations may be adopted as an option. In this paper, the development of seismic provisions as related to analysis is summarized, the present state is discussed, and possible further developments are envisaged.

1.1 Introduction: As Simple As Possible But Not Simpler

Seismic analysis is a tool for the estimation of structural response in the process of designing earthquake resistant structures and/or retrofitting vulnerable existing structures. In principle, the problem is difficult because the structural response to strong earthquakes is dynamic, nonlinear and random. All three characteristics are unusual in structural engineering, where the great majority of problems are (or at least can be adequately approximated as) static, linear and deterministic. Consequently, special skills and data are needed for seismic design, which an average designer does not necessarily have.

Methods for seismic analysis, intended for practical applications, are provided in seismic codes. (Note that in this paper the term "code" is used broadly to include codes, standards, guidelines, and specifications.) Whereas the most advanced analytical, numerical and experimental methods should be used in research aimed at the

---

Reprinted from Fajfar, P. Bull Earthquake Eng (2017). https://doi.org/10.1007/s10518-017-0290-8.

P. Fajfar (✉)
University of Ljubljana, Ljubljana, Slovenia
e-mail: Peter.Fajfar@fgg.uni-lj.si

© The Author(s) 2018
K. Pitilakis (ed.), *Recent Advances in Earthquake Engineering in Europe*,
Geotechnical, Geological and Earthquake Engineering 46,
https://doi.org/10.1007/978-3-319-75741-4_1

development of new knowledge, the methods used in codes should, as Albert Einstein said, be "as simple as possible, but not simpler". A balance between required accuracy and complexity of analysis should be found, depending on the importance of a structure and on the aim of the analysis. It should not be forgotten that the details of the ground motion during future earthquakes are unpredictable, whereas the details of the dynamic structural response, especially in the inelastic range, are highly uncertain. According to Aristotle, "it is the mark of an educated mind to rest satisfied with the degree of precision which the nature of the subject admits and not to seek exactness where only an approximation is possible" (Nicomachean Ethics, Book One, Chap. 3).

After computers became widely available, i.e. in the late 1960s and in 1970s, a rapid development of procedures for seismic analysis and supporting software was witnessed. Nowadays, due to tremendous development in computing power, numerical methods, and software, there are almost no limits related to computation. Unfortunately, knowledge about ground motion and structural behaviour, especially in the inelastic range, has not kept up the same speed. Also, we cannot expect that, in general, the basic capabilities of engineers will be better than in the past. So, there is a danger, as M. Sozen wrote already in 2002: "Today, ready access to versatile and powerful software enables the engineer to do more and think less." (M. Sozen, A Way of Thinking, EERI Newsletter, April 2002). Two other giants in earthquake engineering also made observations which have remained valid up to now. R. Clough, one of the fathers of the finite element method, stated: "Depending on the validity of the assumptions made in reducing the physical problem to a numerical algorithm, the computer output may provide a detailed picture of the true physical behavior or it may not even remotely resemble it" (Clough 1980, p.369). V. Bertero (2009, p.80) warned: "There are some negative aspects to the reliance on computers that we should be concerned about. It is unfortunate that there has been a trend among the young practicing engineers who are conducting structural analysis, design, and detailing using computers to think that the computer automatically provides reliability". Today it is lack of reliable data and the limited capabilities of designers which represent the weak link in the chain representing the design process, rather than computational tools, as was the case in the past.

An indication of the restricted ability of the profession (on average) to adequately predict the seismic structural response was presented by the results of a blind prediction contest of a simple full-scale reinforced concrete bridge column with a concentrated mass at the top, subjected to six consecutive unidirectional ground motions. A description of the contest, and of the results obtained, described in the following text, has been summarized from (Terzic et al. 2015). The column was not straightened or repaired between the tests. During the first ground motion, the column displaced within its elastic range. The second test initiated a nonlinear response of the column, whereas significant nonlinearity of the column response was observed during the third test. Each contestant/team had to predict peak response for global (displacement, acceleration, and residual displacement), intermediate (bending moment, shear, and axial force), and local (axial strain and curvature) response quantities for each earthquake. Predictions were submitted by 41 teams from 14 different countries. The contestants had either MSc or PhD degrees. They were supplied with data about the ground motions and structural details, including the complete dimensions of the test specimen, and the mechanical one-dimensional properties of the steel and concrete. In this way the largest sources of uncertainties, i.e. the characteristics of the ground motion and the material characteristics, were eliminated. The only remaining uncertainty was related to the modelling and analysis of the structural response. In spite of this fact, the results showed a very wide scatter in the blind predictions of the basic engineering response parameters. For example, the average coefficient of variation in predicting the maximum displacement and acceleration over the six ground motions was 39% and 48%, respectively. Biases in median predicted responses were significant, varying for the different tests from 5 to 35% for displacement, and from 25 to 118% for acceleration. More detailed results for the maximum displacements at the top of the column and the maximum shear forces at the base of the column are presented in Fig. 1.1. A large dispersion of the results can be observed even in the case of the elastic (Eq. 1.1) and nearly elastic (Eq. 1.2) structural behaviour.

The results of the blind prediction contest clearly demonstrate that the most advanced and sophisticated models and methods do not necessarily lead to adequate results. For example, it was observed that a comparable level of accuracy could be achieved if the column was modelled either with complex force-based fibre beam-column elements or with simpler beam-column elements with concentrated plastic hinges. Predictions of structural response greatly depended on the analyst's experience and modelling skills. Some of the results were completely invalid and could lead to gross errors if used in design. A simple check, e.g. with the response spectrum approach applied for a single-degree-of-freedom system, would indicate that the results were nonsensical.

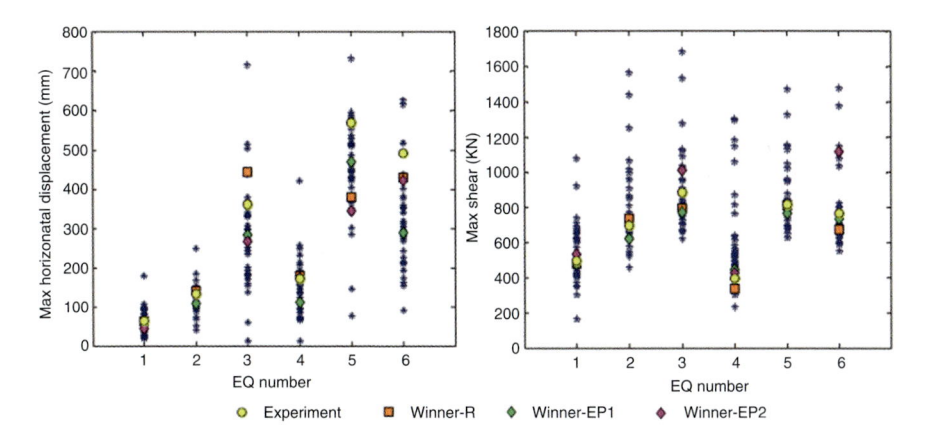

Fig. 1.1 Predictions of maximum horizontal displacements at the top of the column and maximum base shears versus measured values (Terzic et al. 2015)

This paper deals with analysis procedures used in seismic provisions. The development of seismic provisions related to the analysis of building structures is summarized, the present state is discussed, and possible further developments are envisaged. Although, in general, the situation in the whole world is discussed, in some cases emphasis is placed on the situation in Europe and on the European standard for the design of structures for earthquake resistance Eurocode 8 (CEN 2004), denoted in this paper as EC8. The discussion represents the views of the author and is based on his experience in teaching, research, consulting, and code development work.

1.2 History: Domination of the Equivalent Static Procedure and Introduction of Dynamics

1.2.1 Introduction

Earthquake engineering is a relatively young discipline, "it is a 20th century development" (Housner 1984). Although some types of old buildings have, for centuries, proved remarkably resistant to earthquake forces, their seismic resistance has been achieved by good conceptual design without any seismic analysis. Early provisions related to the earthquake resistance of buildings, e.g. in Lima, Peru (Krause 2014) and Lisbon, Portugal (Cardoso et al. 2004), which were adopted after the disastrous earthquakes of 1699 and 1755, respectively, were restricted to construction rules and height limitations. It appears that the first engineering recommendations for seismic analysis were made in 1909 in Italy. Apparently Housner considered this date as the starting date of Earthquake Engineering.

The period up to 1978 was dominated by the equivalent static procedure. "The equivalent static method gradually spread to seismic countries around the world. First it was used by progressive engineers and later was adopted by building codes. Until the 1940s it was the standard method of design required by building codes" (Housner 1984), and still today it is widely used for simple regular structures, with updated values for the seismic coefficients. "This basic method has stood the test of time as an adequate way to proportion the earthquake resistance of most buildings. Better methods would evolve, but the development of an adequate seismic force analysis method stands out in history as the first major saltation or jump in the state of the art." (Reitherman 2012, p.174). From the three basic features of seismic structural response, dynamics was the first to be introduced. Later, inelastic behaviour was approximately taken into account by the gradation of seismic loads for different structural systems, whereas randomness was considered implicitly by using various safety factors.

In the following sections we will summarize the development of seismic analysis procedures in different codes (see also Table 1.1). It will be shown that, initially, the equivalent static approach was used. With some exceptions, for several decades the seismic coefficient mostly amounted to about 0.1.

1 Analysis in Seismic Provisions for Buildings: Past, Present and Future

Table 1.1 The evolution of analysis provisions in seismic codes

1909 Italy	The first seismic regulations for buildings worldwide, with provisions for equivalent static analysis. In the first storey, the horizontal force was equal to 1/12th of the weight above, and in the second and third storeys, 1/8th of the weight above.
1924 Japan	The first seismic code in Japan. The seismic coefficient was equal to 10%.
1927 USA	First edition of the uniform building code (UBC) with optional seismic provisions. The seismic coefficient varied between 7.5% and 10% of the total dead load plus the live load of the building, depending on soil conditions.
1933 USA	First mandatory seismic codes in the United States (the Field and Riley acts in California). The seismic coefficient varied from 2% to 10%.
1943 USA	Los Angeles enacted the first code, which related the seismic coefficient to the flexibility of the building.
1956 USA	San Francisco enacted a code with explicit dependence of the seismic loads on the building period.
1957 USSR	Implementation of the modal response spectrum method, which later became the main analysis procedure in Europe.
1959 USA	The SEAOC model code took into account the impact of the energy dissipation capacity of structures in the inelastic range.
1977 Italy/ Slovenia	A very simple pushover procedure for masonry buildings was implemented in a regional code in Friuli, Italy.
1978 USA	The start of modern codes with ATC 3-06 guidelines (probabilistic seismic maps, force reduction R-factors).
1981 Yugoslavia	Adoption of linear and nonlinear response history analysis for very important buildings and prototypes of prefabricated buildings in the seismic code.
1986 USA	The pushover-based Capacity Spectrum Method was implemented in the "Triservices" guidelines.
2010 USA	Explicit probabilistic analysis permitted in ASCE 7-10.

Dynamic considerations were introduced by relating the seismic coefficient to the period of the building, indirectly via the number of storeys in 1943, and directly in 1956. The modal response spectrum method appeared for the first time in 1957. The impact of the energy dissipation capacity of structures in the inelastic range (although this was not explicitly stated in the code) was taken into account in 1959. Modern codes can be considered to have begun with the ATC 3-06 document "Tentative provisions for the development of seismic regulations for buildings", which was released in 1978 (ATC 1978). This document formed the basis upon which most of the subsequent guidelines and regulations were developed both in the United States and elsewhere in the world.

When discussing seismic code developments, the capacity design approach developed in the early 1970s in New Zealand, should not be ignored. It might be one of the most ingenious solutions in earthquake engineering. Structures designed by means of the capacity design approach are expected to possess adequate ductility both at the local and global level. In the case of such structures, it is completely legitimate to apply linear analysis with a force reduction factor which takes into account the energy dissipation capacity. Of course, a quantification of the inelastic behaviour is not possible. Since capacity design is not a direct part of the analysis process, it will not be further discussed in this paper.

1.2.2 Italy

After the earthquake in Messina in 1908, a committee of Italian experts (nine practicing engineers and five university professors) was formed. The committee prepared a quantitative recommendation for seismic analysis which was published in 1909. The committee proposed, mainly based on the results of a study of three timber-framed buildings which had survived the earthquake with little or no damage, that, "in future, structures must be so designed that they would resist, in the first story, a horizontal force equivalent to 1/12th of the weight above, and in the second and third story, $1/8^{th}$ of the weight above" (Freeman 1932, p.577). The procedure became mandatory through a Royal Decree in the same year (Sorrentino 2007). At that time, three-story buildings were the tallest permitted. In the committee report it was stated "that it was reasonable to add 50% to the seismic force to be provided against in upper stories, because of the observed greater amplitude of oscillation in tall as compared with low buildings and also in adherence to the universally admitted principle that the center of gravity of buildings should be as low as possible, and hence the upper stories should be lighter than the lower" (Freeman 1932, p.577). According to Reitherman (2012, p.193), the above-described technical regulation was later adjusted to provide factors of 1/8 and 1/6, respectively, after the disastrous 1915 Avezzano Earthquake, to the north of the Strait of Messina.

The committee's proposal for the determination of lateral forces actually introduced the seismic coefficient, which is used to multiply the weight of the building in order to obtain the total seismic force (the base shear force). The seismic coefficient enables performing a seismic analysis by means of the equivalent static method. It has a theoretical basis. According to D'Alembert's principle, a fictitious inertia force, proportional to the acceleration and mass of a particle, and acting in the direction opposite to the acceleration, can be used in a static analysis, in order to simulate a dynamic problem. The seismic coefficient represents the ratio between the acceleration of the mass and the acceleration of gravity. The acceleration of the mass depends on the acceleration of the ground and on the dynamic characteristics of the structure. At the beginning of the twentieth century there were no data about ground accelerations and the use of structural dynamics was, at that time, therefore infeasible. M. Panetti, who appears to be the main author of the recommendation of seismic coefficient in the report after the Messina earthquake, recognized "that the problem was really one of dynamics or kinetics, and that to solve this would involve such complication that one must have recourse to the assumption that a problem of statics could be so framed as to provide for safety" (Freeman 1932, p.577). So, according to Freeman, the proposed seismic coefficient, as well as the seismic coefficients used during the following decades in California and Japan, were "largely a guess tempered by judgement". Nevertheless, it is interesting to note that the order of magnitude of the seismic coefficient proposed in 1909 remained in seismic codes in different countries for many decades (see Sect. 1.2.6).

The most advanced of the engineers who studied the effects of the 1908 Messina earthquake was A. Danusso. He "was probably the first to propose a dynamic analysis method rather than static lateral force analysis method and, possibly for the first time in earthquake engineering, he stated that seismic demand does not depend upon the ground motion characteristics alone" (Sorrentino 2007). Danusso not only stated this relationship verbally, but presented his thinking mathematically. Danusso's approach was considered too much ahead of its time for practical implementation by the committee formed after the Messina earthquake, which found the static approach more practical for widespread application (Reitherman 2012, p.193). As a matter of historical fact, dynamic analysis has not been implemented in seismic provisions for the next half of the century.

The committee's proposal to conventionally substitute dynamic actions with purely static ones in representing seismic effects had a great impact on the subsequent development of early earthquake engineering in Italy, since it simplified the design procedures but ruled out from the code any dynamic considerations until the mid-seventies (1974), when a design spectrum was introduced (Sorrentino 2007) and an up-to-date seismic code was adopted.

The Italians were the first to propose, in 1909, the equivalent static procedure for seismic analysis, and to implement it in a code. The procedure has been a constituent part of practically all seismic codes up until present times. Also, apparently the first dynamic seismic computations stem from Italy. However, their achievements did not have a widespread effect worldwide. It was the Japanese achievements, as described in the next section, which became more popular.

1.2.3 Japan

In Japan, in 1916 R. Sano proposed the use of seismic coefficients in earthquake resistant building design. "He assumed a building to be rigid and directly connected to the ground surface, and suggested a seismic coefficient equal to the maximum ground acceleration normalized to gravity acceleration. Although he noted that the lateral acceleration response might be amplified from the ground acceleration with lateral deformation of the structure, he ignored the effect in determining the seismic coefficient. He estimated the maximum ground acceleration in the Honjo and Fukagawa areas on alluvial soil in Tokyo to be 0.30 g and above on the basis of the damage to houses in the 1855 Ansei-Edo (Tokyo) earthquake, and that in the Yamanote area on diluvial hard soil to be 0.15 g" (Otani 2004). T. Naito, one of Sano's students at the University of Tokyo, became, like Sano, not just a prominent earthquake engineer but also one of the nation's most prominent structural engineers. He followed Sano's seismic coefficient design approach, usually using a coefficient of 1/15 (7%) to design buildings before the 1923 Kanto Earthquake. The coefficient was uniformly applied to masses up the height of the building (Reitherman 2012, p.168).

The first Japanese building code was adopted in 1919. The 1923 Great Kanto earthquake led to the addition of an article in this code (adopted in 1924) to require the seismic design of all buildings for a seismic coefficient of 0.1. "From the incomplete measurement of ground displacement at the University of Tokyo, the maximum ground acceleration was estimated to be 0.3 g. The allowable stress in design was one third to one half of the material strength in the design regulations. Therefore, the design seismic coefficient 0.1 was determined by dividing the estimated maximum ground acceleration of 0.3 g by the safety factor of 3 of allowable stress" (Otani 2004).

The Japanese code, which applied to certain urban districts, was a landmark in the development of seismic codes worldwide. In the Building Standard Law, adopted in 1950, applicable to all buildings throughout Japan, the seismic coefficient was increased from 0.1 to 0.2 for buildings with heights of 16 m and less, increasing by 0.01 for every 4 m above (Otani 2004). Allowable stresses under temporary loads were set at twice the allowable stresses under permanent loads. From this year on, the seismic coefficients in Japan remained larger than elsewhere in the rest of the world. Later significant changes included the abolishment of the height limit of 100 ft. in 1963. In 1981 the Building Standard Law changed. Its main new features included a seismic coefficient which varied with period, and two-level design. The first phase design was similar to the design method used in earlier codes. It was intended as a strength check for frequent moderate events. The second phase design was new. It was intended as a check for strength and ductility for a maximum expected event (Whittaker et al. 1998). It is interesting to note that, in Japan, the seismic coefficient varied with the structural vibration period only after 1981.

1.2.4 United States

Surprisingly, the strong earthquake which hit San Francisco in 1906 did not trigger the development of seismic provisions in the United States. Wind force was intended to protect buildings against both wind and earthquake damage. Design recommendations were intended only for buildings taller than 100 feet (30.5 m), or taller than three times the building's least dimension, and consisted of applying a 30 psf (1.44 kPa) wind load to the building's elevation. Later, the recommended wind load was reduced to 20 psf, and then to 15 psf (Diebold et al. 2008). At that time, however, no building code provisions existed for the design of shorter structures to resist wind or earthquake loads.

The first regulations on earthquake-resistant structures appeared in the United States only in 1927, after the earthquake in Santa Barbara in 1925. The provisions were contained in an appendix to the Uniform Building Code, and were not mandatory. The equivalent static procedure was used. The seismic coefficient was varied between 7.5 and 10% of the total dead load plus the live load of the building, depending on soil conditions. This was the first time that a building code had recognized the likelihood of the amplification of ground motion by soft soil (Otani 2004). After the 1925 Santa Barbara earthquake, some California municipalities did adopt mandatory seismic provisions in their building codes.

The first mandatory seismic codes used in the United States were published in 1933, after the 1933 Long Beach earthquake, which caused extensive damage to school buildings. Two California State Laws were enacted. The Field Act required the earthquake-resistant design and construction of all new public schools in California. The adopted regulations required that masonry buildings without frames be designed to resist a lateral force equal to 10% of the sum of the dead load and a fraction of the design live load. For other buildings the seismic coefficient was set at 2–5%. Shortly after this, the Riley Act was adopted with a mandatory seismic design coefficient of 2% of the sum of the dead load and live load for most buildings. At about the same time, in Los Angeles a lateral force requirement of 8% of the total weight and half of the design live load was imposed. This requirement was also adopted by the 1935 Uniform Building Code for the zone of highest seismicity (Berg 1983, p.25–26), whereas, recognizing the different levels of seismic risk in different regions, the seismic coefficients in the other two zones were lower.

Although the first regulations were based on the static concept, and did not take into account the fact that the structural acceleration (and thus the seismic load) depends on the vibration period of the building, the developers of the early codes were aware of the dynamic nature of the structural response and of the importance of the periods of vibration of buildings. Freeman wrote: "Those who have given the matter careful study recognize that as a means of lessening the amplitude of oscillation due to synchronism [i.e. resonance], it is extremely important to make the oscillation period of the building as small as practicable, or smaller than the oscillation period of the ground ordinarily found in great earthquakes." (Freeman 1932, p.799). However, since the details of the ground motion, including the frequency content, were not known, any realistic quantitative considerations were impossible. The first generation of analysis methods evolved before being able to take into account response spectrum analysis and data from strong motion instruments, although the basic concept was known. Some of the most educated guesses of the frequency content of ground shaking were completely wrong considering today's knowledge. For example, Freeman (1930, p.37) stated that "In Japan it has been noted that the destructive oscillations of the ground in an earthquake are chiefly those having a period from 1 second to 1.5 seconds; therefore, some of the foremost Japanese engineers take great care in their designs to reduce the oscillation period of a building as nearly as practicable to from 0.5 to 0.6 second, or to less than the period of the most destructive quake, and also strive to increase the rigidity of the building as a whole in every practical way." Today, we know that, generally, in the 0.5–0.6 s range the spectral acceleration is usually larger than in the range from 1 to 1.5 s.

The first code, which related the seismic coefficient to the flexibility of the building, was enacted by the City of Los Angeles, in 1943. The level of the seismic coefficient was decreasing with the number of stories, reflecting the decrease of structural acceleration by increasing the flexibility and the natural period of the building. So, the building period was considered indirectly. This code also took into account the fact that the lateral force varies over the height of the structure. The building period became an explicit factor in the determination of seismic design forces only in 1956, when the City of San Francisco enacted a building code based on the recommendations produced by a joint committee of the Structural Engineers Association of Northern California and the San Francisco Section of the American Society of Civil Engineers (Berg 1983, p.26–27), published in 1951. The recommendations included an inverted triangle distribution of design lateral loads along the height of the building, which is still a basic feature of equivalent static lateral force procedures.

A theoretical basis for the consideration of the dependence of seismic load on the natural period of the building was made possible by the development of response spectra. Although the initial idea of the presentation of earthquake ground motion with spectra had appeared already in the early 1930s, the practical use of spectra has not been possible for over 30 years due to the lack of data about the ground motion during earthquakes, and because of the very extensive computational work required for the determination of spectra, which was virtually impossible without computers. For details on the historical development of response spectra, see, e.g., (Chopra 2007) and (Trifunac 2008).

The first code which took into account the impact of the energy dissipation capacity of structures in the inelastic range (although this was not explicitly stated in the code) was the SEAOC model code (the first "Blue Book"), in 1959. In order to distinguish between the inherent ductility and energy dissipation capacities of different structures, a coefficient K was introduced in the base shear equation. Its values were specified for four types of building construction. The basic seismic coefficient of 0.10 increased the most (by a factor of 1.33) in the case of wall structures, and decreased the most (by a factor of 0.67) in the case of space frames. According to Blume et al. (1961): "The introduction of K was a major step forward in code writing to provide in some degree for the real substance of the problem – energy absorption – and for the first time to recognize that equivalent acceleration or base shear coefficient C alone is not necessarily a direct index of earthquake resistance and public safety."

Seismic regulations at that time were mainly limited to analysis, and did not contain provisions on dimensioning and detailing. It was not until the end of the 1960s that, in the United States, provisions related to the detailing of ductile reinforced concrete frames were adopted.

An early appearance of Performance-Based Seismic Design can be found in the 1967 edition of the SEAOC Blue Book, where the following criteria were defined:

1. Resist minor earthquakes without damage.
2. Resist moderate earthquakes without structural damage, but with some nonstructural damage.

3. Resist major earthquakes, of the intensity of severity of the strongest experienced in California, without collapse, but with some structural as well as nonstructural damage.

The first document, which incorporated most of the modern principles of seismic analysis, was ATC 3-06 (ATC 1978), which was released in 1978 (Fig. 1.2) as a result of years of work performed by many experts in the United States. As indicated by the title of the document, the document represented tentative provisions for the development of seismic regulations for buildings. Its primary purpose was to present the current state of knowledge in the fields of engineering seismology and earthquake engineering, in the form of a code. It contained a series of new concepts that significantly departed from existing regulations, so the authors explicitly discouraged the document's use as a code, until its usefulness, practicality, and impact on costs was checked. Time has shown that the new concepts have been generally accepted, and that the document was a basis for the subsequent guidelines and regulations in the United States and elsewhere in the world, with some exception in Japan.

In the document the philosophy of design in seismic regions, set forth in the 1967 SEAOC Blue Book, was followed, according to which the primary purpose of seismic regulations is the protection of human life, which is achieved by preventing

Fig. 1.2 The cover-page of ATC 3-06

the collapse of buildings or their parts, whereas in strong earthquakes damage and thus material losses are permitted. At the very beginning it was stated: "The design earthquake motions specified in these provisions are selected so that there is low probability of their being exceeded during the normal lifetime expectancy of the building. Buildings and their components and elements ... may suffer damage, but should have a low probability of collapse due to seismic-induced ground shaking."

The document contained several significant differences compared to earlier seismic provisions. Seismic hazard maps, which represent the basis for design seismic loads, were based on a 90% probability that the seismic hazard, represented by the effective peak ground acceleration and the effective peak ground velocity (which serve as factors for constructing smoothed elastic response spectra), would not be exceeded in 50 years. This probability corresponds to a return period of the earthquake of 475 years. Buildings were classified into seismic hazard exposure groups. Seismic performance categories for buildings with design and analysis requirements depended on the seismicity index and the building seismic hazard exposure group. The analysis and design procedures were based upon a limit state that was characterized by significant yield rather than upon allowable stresses as in earlier codes. This was an important change which influenced also the level of the seismic coefficient.

An empirical force reduction factor R, called the" Response modification factor", was also introduced in ATC 3-06. Experience has shown that the great majority of well designed and constructed buildings survive strong ground motions, even if they were in fact designed for only a fraction of the forces that would develop if the structure behaved entirely as linearly elastic. A reduction of seismic forces is possible thanks to the beneficial effects of energy dissipation in ductile structures and inherent overstrength. Although the influence of the structural system and its capacity for energy dissipation has been recognized already in late 1950s, the force reduction factor (or simply R factor) in the current format was first proposed in ATC-3-06. Since then, R factor has been present, in various forms, in all seismic regulations (in the European standard EC8 it is called the behaviour factor q).

R factor allows, in a standard linear analysis, an approximate consideration of the favourable effects of the nonlinear behaviour of the structure, and therefore presents a very simple and practical tool for seismic design. However, it is necessary to bear in mind that describing a complex phenomenon of inelastic behaviour for a particular structure, by means of a single average number, can be confusing and misleading. For this reason, the R factor approach, although is very convenient for practical applications and has served the professional community well over decades, is able to provide only very rough answers to the problems encountered in seismic analysis and design. Also, it should be noted that "the values of R must be chosen and used with judgement", as stated in the Commentary to the ATC 03-6 document in Sec. 3.1. According to ATC-19 (ATC 1995), "The R factors for the various framing systems included in the ATC-3-06 report were selected through committee consensus on the basis of (a) the general observed performance of like buildings during past earthquakes, (b) estimates of general system toughness, and (c) estimates of the amount of damping present during inelastic response. Thus, there is little technical basis for the values of R proposed in ATC-3-06." Nevertheless, the order of magnitude of R factors (1.5 to 8, related to design at the strength level) has been widely used in many codes and has remained more or less unchanged until nowadays.

The model code recognized the existence of several "standard procedures for the analysis of forces and deformations in buildings subjected to earthquake ground motion", including Inelastic Response History Analysis. However, only two analysis procedures were specified in the document: Equivalent Lateral Force Procedure and Modal Analysis Procedure with one degree of freedom per floor in the direction being considered. In relation to modal analysis, it was stated in the Commentary in Sec. 3.5: "In various forms, modal analysis has been widely used in the earthquake-resistant design of special structures such as very tall buildings, offshore drilling platforms, dams, and nuclear power plants, but this is the first time that modal analysis has been included in the design provisions for buildings". The last part of this statement was true for the United States, but not worldwide, since modal analysis was adopted already in 1957 in the USSR's seismic code, as explained in the next section. In the Commentary it was recognized that the simple model used for modal analysis was "likely to be inadequate if the lateral motions in two orthogonal directions and the torsional motion are strongly coupled". In such a case and in some other cases, where the simple model was not adequate, the possibility of a "suitable generalization of the concepts" was mentioned.

Despite the tremendous progress which has been made in the field of earthquake engineering in recent decades, it can be concluded that the existing regulations, which of course contain many new features and improvements, are essentially still based on the basic principles that were defined in the ATC 3-06 document, with the partial exception of the United States, where the updating of seismic provisions has been the fastest.

1.2.5 Other Countries

At the beginning of 1930s, seismic codes were instituted only in Italy and Japan, and in a few towns in California. As mentioned in the previous section, the first mandatory seismic codes used in the United States were adopted in 1933. By the end of that decade, seismic codes were enacted in three more countries, New Zealand in 1935 (draft code in 1931), India in 1935, and Chile in 1939 (provisional regulations in 1930). In all cases a damaging earthquake triggered the adoption of seismic regulations (Reitherman 2012 p.200 and p.216). Seismic codes followed later in Canada in 1941 (Appendix to the National Building Code), Romania in 1942, Mexico in 1942, and Turkey in 1944 (a post-event response code had been adopted already in 1940). The USSR's first seismic code was adopted in 1941 (Reitherman 2012, p.288). The Standards and Regulations for Building in Seismic Regions, which was adopted in the USSR in 1957, included, apparently as the first code, the modal response spectrum method as the main analysis procedure. (Korčinski

1964). This analysis procedure was later included in several European seismic codes and has remained the most popular procedure for seismic analysis in Europe, up until the present day. In EC8, too, it is the basic analysis method. In 1963 and 1964, respectively, Slovenia and the whole of the former Yugoslavia adopted seismic codes, which were with respect to the analysis method similar to the Soviet code. In China, the first code was drafted in 1955, but was not adopted (Reitherman 2012, p.288). The first official seismic design code (The Code for the Seismic Design of Civil and Industrial Buildings) was issued in 1974.

1.2.6 Level of the Design Lateral Seismic Load

Interestingly, the level of the design horizontal seismic load (about 10% of the weight of the building), which was proposed in 1909 in Italy, and also used in Japan in the first half of the twentieth century, has been maintained in the seismic regulations for the majority of buildings up to the latest generation of regulations, when, on average, the design horizontal seismic load increased. An exception was Japan, where the seismic design loads increased to 20% of the weight already in 1950. The value of about 10%, proposed in Italy, was based on studies of three buildings which survived the 1908 Messina Earthquake. However, this study was, apparently, not known in other parts of the world. The Japanese engineer Naito wrote: "In Japan, as in other seismic countries, it is required by the building code [from 1924 onward] to take into account a horizontal force of at least 0.1 of the gravity weight, acting on every part of the building. But this seismic coefficient of 0.1 of gravity has no scientific basis either from past experience or from possible occurrence in the future. There is no sound basis for this factor, except that the acceleration of the Kwanto earthquake for the first strong portion as established from the seismographic records obtained at the Tokyo Imperial University was of this order." (Reitherman 2012, p.172). Freeman (1930) expressed a similar opinion: "There is a current notion fostered by seismologists, and incorporated in the tentative building laws of several California cities, that the engineer should work to a seismic coefficient of 1/10 g. ... Traced back to the source this rule is found to be a matter of opinion, not of measurement; a product, not of the seismometer, but of the "guessometer"."

Explanations as to why 10% of the weight of a building is an adequate design horizontal seismic load have changed over time. The seismic coefficient, which represents the ratio between the seismic load and the weight (plus some portion of the live load) depends on the ground acceleration and the dynamic characteristics of the structure. Initially, structures were assumed to be rigid, having the same acceleration as the ground. At the beginning of the twentieth century instruments for the recording of strong ground motion were not available. Some estimates of the level of maximum ground accelerations were obtained from observations of the sliding and overturning of rigid bodies such as stone cemetery monuments. Freeman (1932, p.76) prepared a table with the relations between the Rossi – Forel intensity and peak ground acceleration, as proposed by six different authors. The values of accelerations in m/s^2 for intensities VIII, IX and X were within the ranges 0.5–1.0 (mean 0.58), 1.0–2.0 (mean 1.23) and 2.5–3.0 (mean 2.62), respectively. Note that the intensity IX on the Rossi-Forel scale ("Partial or total destruction of some buildings") approximately corresponds to intensity IX according to the EMS. The data clearly indicate that the values of the peak ground acceleration were grossly underestimated.

The first strong motion accelerograms were recorded only in 1933, during the Long Beach earthquake. Maximum acceleration values did not significantly deviate from 0.1 g until 1940, when the famous El Centro 1940 accelerogram, with a maximum acceleration of more than 0.3 g, was recorded. At the time of the 1971 San Fernando Earthquake, the peak ground acceleration produced by that earthquake tripled what most engineers of the time thought possible (Reitherman 2012, p.272). With the awareness that ground acceleration can be much higher than expected, and that very considerable dynamic amplification can occur if the period of the structure is in the range of the predominant periods of the ground motion, both resulting in accelerations of the structure much greater than 0.1 g, it was possible to justify a seismic coefficient of about 10% only due to the favourable influence of several factors, mainly energy dissipation in the inelastic range, and so-called overstrength, i.e. the strength capacity above that required by the code design loads. In the 1974 edition of the SEAOC code, for the first time the basis for the design load levels was made explicit in the Commentary: "The minimum design forces prescribed by the SEAOC Recommendations are not to be implied as the actual forces to be expected during an earthquake. The actual motions generated by an earthquake may be expected to be significantly greater than the motions used to generate the prescribed minimum design forces. The justification for permitting lower values for design are many-fold and include: increased strength beyond working stress levels, damping contributed by all the building elements, an increase in ductility by the ability of members to yield beyond elastic limits, and other redundant contributions. (SEAOC Seismology Committee 1974, p.7-C)" (Reitherman 2012, p.277). More recently, other influences have been studied, e.g. the shape of the uniform hazard spectrum used in seismic design.

In Japan, already Sano and Naito advocated making structures as stiff as possible. Designing stiff and strong structures has remained popular in Japan until the present time. Otani (2004) wrote: "The importance of limiting story drift during an earthquake by providing large stiffness and high lateral resistance should be emphasized in earthquake engineering." Consequently, in Japan the design seismic lateral loads have been, since 1950, generally larger than in the rest of the world. More recently, the level of the design seismic loads has gradually increased, on average, also in other countries. The reasons for this trend are increasing seismic hazard, estimated by new probabilistic seismic hazard analyses and new ground motion databases, and also better care for the limitation of damage.

When comparing seismic coefficients, it should be noted that the design based on allowable stresses used in early codes has been changed to limit stress design, so that the values of seismic coefficients may not be directly comparable.

1.3 Present: Gradual Implementation of Nonlinear Methods

1.3.1 Introduction

Most buildings experience significant inelastic deformations when affected by strong earthquakes. The gap between measured ground accelerations and the seismic design forces defined in codes was one of the factors which contributed to thinking in quantitative terms beyond the elastic response of structures. At the Second World Conference on Earthquake Engineering in 1960 several leading researchers presented some early significant papers about inelastic response. However, there was a long way to go before the explicit nonlinear analysis found its way into more advanced seismic codes. Initially the most popular approach was the use of force reduction factors, and this approach is still popular today. Although this concept for taking into account the influence of inelastic behaviour in linear analysis has served the profession well for several decades, a truly realistic assessment of structural behaviour in the inelastic range can be made only by means of nonlinear analysis. It should be noted, however, that "good nonlinear analysis will not trigger good designs but it will, hopefully, prevent bad designs." (Krawinkler 2006). Moreover: "In concept, the simplest method that achieves the intended objective is the best one. The more complex the nonlinear analysis method, the more ambiguous the decision and interpretation process is. ... Good and complex are not synonymous, and in many cases they are conflicting." (Krawinkler 2006).

The current developments of the analysis procedures in seismic codes are characterized by a gradual implementation of nonlinear analysis, which should be able to explicitly simulate the second basic feature of structural response to strong seismic ground motion, i.e. inelastic behaviour. For such nonlinear analyses, data about the structure have to be known, so they are very well suited for the analysis of existing structures. In the case of newly designed structures, a preliminary design has to be made before starting a nonlinear analysis. Typical structural response measures that form the output from such an analysis (also called "engineering demand parameters") are: the storey drifts, the deformations of the "deformation-controlled" components, and the force demands in "force-controlled" (i.e. brittle) components that, in contemporary buildings, are expected to remain elastic. Basically, a designer performing a nonlinear analysis should think in terms of deformations rather than in terms of forces. In principle, all displacement-based procedures require a nonlinear analysis.

There are two groups of methods for nonlinear seismic analysis: response-history analysis, and static (pushover-based) analysis, each with several options. They will be discussed in the next two sections. An excellent guide on nonlinear methods for practicing engineers was prepared by Deierlein et al. (2010).

1.3.2 *Nonlinear Response History Analysis (NRHA)*

Nonlinear response history analysis (NRHA) is the most advanced deterministic analysis method available today. It represents a rigorous approach, and is irreplaceable for the research and for the design or assessment of important buildings. However, due to its complexity, it has, in practice, been mostly used only for special structures. NRHA is not only computationally demanding (a problem whose importance has been gradually reduced due to the development of advanced hardware and software), but also requires additional data which are not needed in pushover-based nonlinear analysis: a suite of accelerograms, and data about the hysteretic behaviour of structural members. A consensus about the proper way to model viscous damping, in the case of inelastic structural response, has not yet been reached. A wide range of assumptions are needed in all steps of the process, from ground motion selection to nonlinear modelling. Many of these assumptions are based on judgement. Blind predictions of the results of a bridge column test (see Sect. 1.1) are a good example of potential problems which can arise when NRHA is applied. Moreover, the complete analysis procedure is less transparent than in the case of simpler methods. For this reason, the great majority of codes which permit the use of NRHA require an independent review of the results of such analyses.

The advantages and disadvantages of NRHA have been summarized in a recent paper by the authors representing both academia and practice in the United States (Haselton et al. 2017). The advantages include "the ability to model a wide variety of nonlinear material behaviors, geometric nonlinearities (including large displacement effects), gap opening and contact behavior, and non-classical damping, and to identify the likely spatial and temporal distributions of inelasticity. Nonlinear response history analysis also has several disadvantages, including increased effort to develop the analytical model, increased time to perform the analysis (which is often complicated by difficulties in obtaining converged solutions), sensitivity of computed response to system parameters, large amounts of analysis results to evaluate and the inapplicability of superposition to combine non-seismic and seismic load effects." Moreover, "seemingly minor differences in the way damping is included, hysteretic characteristics are modeled, or ground motions are scaled, can result in substantially different predictions of response. ... The provisions governing nonlinear response history analysis are generally non-prescriptive in nature and require significant judgment on the part of the engineer performing the work."

To the author's knowledge, the first code in which the response-history analysis was implemented was the seismic code adopted in the former Yugoslavia in 1981 (JUS 1981). In this code, a "dynamic analysis method", meant as a linear and nonlinear response-history analysis, was included. In the code, it was stated: "By means of such an analysis the stresses and deformations occurring in the structure for the design and maximum expected earthquake can be determined, as well as the acceptable level of damage which may occur to the structural and non-structural elements of the building in the case of the maximum expected earthquake." The application of this method was mandatory for all out-of-category, i.e. very important buildings, and for prototypes of prefabricated buildings or structures which are produced industrially in large series. Such a requirement was very advanced (and maybe premature) at that time. (Fischinger et al. 2015).

In the USA, the 1991 Uniform Building Code (UBC) was the first to include procedures for the use of NRHA in design work. In that code, response-history analysis was required for base-isolated buildings and buildings incorporating passive energy dissipation systems (Haselton et al. 2017). After that, several codes in different countries included NRHA, typically with few accompanying provisions, leaving the analyst with considerable freedom of choice. Reasonably comprehensive provisions/guidelines have been prepared only for the most recent version of ASCE 7 standard, i.e. ASCE 7-16 (ASCE 2017). An overview of the status of nonlinear analysis in selected countries is provided in Sect. 1.3.5. In this paper, NRHA will not be discussed in detail.

1.3.3 Nonlinear Static (Pushover) Analysis

A pushover-based analysis was first introduced in the 1970s as a rapid evaluation procedure (Freeman et al. 1975). In 1980s, it got the name Capacity Spectrum Method (CSM). The method was also developed into a design verification procedure for the Tri-services (Army, Navy, and Air Force) "Seismic Design Guidelines for Essential Buildings" manual (Army 1986). An important milestone was the paper by Mahaney et al. (1993) in which the acceleration-displacement response spectrum (ADRS, later also called AD) format was introduced, enabling visualization of the assessment procedure. In 1996, CSM was adopted in the guidelines for Seismic Evaluation and Retrofit of Concrete Buildings, ATC 40 (ATC 1996). In order to account for the nonlinear inelastic behaviour of a structural system, effective viscous damping values are applied to the linear-elastic response spectrum (i.e. an "overdamped spectrum") in all CSM formulations. In the N2 method, initially developed by Fajfar and Fischinger (1987, 1989), and later formulated in acceleration-displacement (AD) format (Fajfar 1999, 2000), inelastic spectra are used instead of overdamped elastic spectra. The N2 method has been adopted in EC8 in 2004. In FEMA 273 (1997), the target displacement was determined by the "Coefficient Method". This approach, which has remained in all following FEMA documents, and has also been adopted in the ASCE 41-13 standard (ASCE 2014), resembles the use of inelastic spectra. In the United States and elsewhere, the use of pushover-based procedures has accelerated since the publication of the ATC 40 and FEMA 273 documents. In this section, the discussion will be limited to the variant of the pushover-based method using inelastic spectra. A comprehensive summary of pushover analysis procedures is provided in (Aydınoğlu and Önem 2010).

A simple pushover approach, which could be applied at the storey level and used for the analysis of the seismic resistance of low-rise masonry buildings, was developed in the late 1970s by M.Tomaževič (1978). This approach was adopted also in a regional code for the retrofitting of masonry buildings after the 1976 Friuli earthquake in the Italian region of Friuli-Venezia Guilia (Regione Autonoma Friuli-Venezia Giulia 1977).

Pushover-based methods combine nonlinear static (i.e. pushover) analysis with the response spectrum approach. Seismic demand can be determined for an equivalent single-degree-of-freedom (SDOF) system from an inelastic response spectrum (or an overdamped elastic response spectrum). A transformation of the multi-degree-of-freedom (MDOF) system to an equivalent SDOF system is needed. This transformation, which represents the main limitation of the applicability of pushover-based methods, would be exact only in the case that the analysed structure vibrated in a single mode with a deformation shape that did not change over time. This condition is, however, fulfilled only in the case of a linear elastic structure with the negligible influence of higher modes. Nevertheless, the assumption of a single time-invariant mode is used in pushover-based methods for inelastic structures, as an approximation.

Pushover-based analyses can be used as a rational practice-oriented tool for the seismic analysis. Compared to traditional elastic analyses, this kind of analysis provides a wealth of additional important information about the expected structural response, as well as a helpful insight into the structural aspects which control performance during severe earthquakes. Pushover-based analyses provide data on the strength and ductility of structures which cannot be obtained by elastic analysis. Furthermore, they are able to expose design weaknesses that could remain hidden in an elastic analysis. This means that in most cases they are able to detect the most critical parts of a structure. However, special attention should be paid to potential brittle failures, which are usually not simulated in the structural models. The results of pushover analysis must be checked in order to find out if a brittle failure controls the capacity of the structure.

For practical applications and educational purposes, graphical displays of the procedure are extremely important, even when all the results can be obtained numerically. Pushover-based methods experienced a breakthrough when the acceleration-displacement (AD) format was implemented. Presented graphically in AD format, pushover-based analyses can help designers and researchers to better understand the basic relations between seismic demand and capacity, and between the main structural parameters determining the seismic performance, i.e. stiffness, strength, deformation and ductility. They are a very useful educational tool for the familiarizing of students and practising engineers with general nonlinear seismic behaviour, and with the seismic demand and capacity concepts.

Pushover-based methods are usually applied for the performance evaluation of a known structure, i.e. an existing structure or a newly designed one. However, other types of analysis can, in principle, also be applied and visualised in the AD format. Basically, four quantities define the seismic performance: strength, displacement, ductility and stiffness. Design and/or performance evaluation begins by fixing one or two of them; the others are then determined by calculations. Different approaches differ in the quantities that are chosen at the beginning of the design or evaluation. Let's assume that the

Fig. 1.3 Comparison of demand and capacity in the acceleration – displacement (AD) format

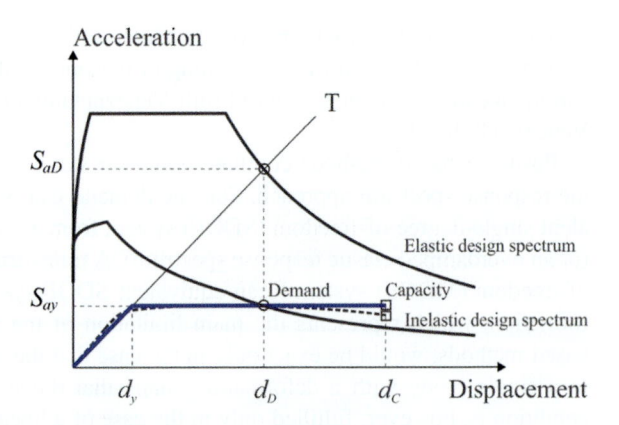

approximate mass is known. In the case of seismic performance evaluation, stiffness (period) and strength of the structure have to be known; the displacement and ductility demands are calculated. In direct displacement-based design, the starting points are typically the target displacement and/or ductility demands. The quantities to be determined are stiffness and strength. The conventional force-based design typically starts from the stiffness (which defines the period) and the approximate global ductility capacity. The seismic forces (defining the strength) are then determined, and finally the displacement demand is calculated. Capacity in terms of spectral acceleration can be determined from the capacity in terms of displacements. All these approaches can be easily visualised with the help of Fig. 1.3.

Note that, in all cases, the strength is the actual strength and not the design base shear according to seismic codes, which is in all practical cases less than the actual strength. Note also that stiffness and strength are usually related quantities.

Compared to NRHA, pushover-based methods are a much simpler and more transparent tool, requiring simpler input data. The amount of computation time is only a fraction of that required by NRHA and the use of the results obtained is straightforward. Of course, the above-listed advantages of pushover-based methods have to be weighed against their lower accuracy compared to NRHA, and against their limitations. It should be noted that pushover analyses are approximate in nature, and based on static loading. They have no strict theoretical background. In spite of extensions like those discussed in the next section, they may not provide acceptable results in the case of some building structures with important influences of higher modes, including torsion. For example, they may detect only the first local mechanism that will form, while not exposing other weaknesses that will be generated when the structure's dynamic characteristics change after formation of the first local mechanism. Pushover-based analysis is an excellent tool for understanding inelastic structural behaviour. When used for quantification purposes, the appropriate limitations should be observed. Additional discussion on the advantages, disadvantages and limitations of pushover analysis is available in, for instance (Krawinkler and Seneviratna 1998; Fajfar 2000; Krawinkler 2006).

1.3.4 The Influence of Higher Modes in Elevation and in Plan (Torsion)

The main assumption in basic pushover-based methods is that the structure vibrates predominantly in a single mode. This assumption is sometimes not fulfilled, especially in high-rise buildings, where higher mode effects may be important along the height of the building, and/or in plan-asymmetric buildings, where substantial torsional influences can occur. Several approaches have been proposed for taking the higher modes (including torsion) into account. The most popular is Modal pushover analysis, developed by Chopra and Goel (2002). Some of the proposed approaches require quite complex analyses. Baros and Anagnostopoulos (2008) wrote: "The nonlinear static pushover analyses were introduced as simple methods ... Refining them to a degree that may not be justified by their underlying assumptions and making them more complicated to apply than even the nonlinear response-history analysis ... is certainly not justified and defeats the purpose of using such procedures."

In this section, the extended N2 method (Kreslin and Fajfar 2012), which combines two earlier approaches, taking into account torsion (Fajfar et al. 2005) and higher mode effects in elevation (Kreslin and Fajfar 2011), into a single procedure enabling the analysis of plan-asymmetric medium- and high-rise buildings, will be discussed. The method is based on the assumption that the structure remains in the elastic range in higher modes. The seismic demand in terms of displacements and storey drifts can be obtained by combining together the deformation quantities obtained by the basic pushover analysis and those obtained by the elastic modal response spectrum analysis. Both analyses are standard procedures which have been implemented in most commercial computer programs. Thus the approach is conceptually simple, straightforward, and transparent.

In the elastic range, the vibration in different modes can be decoupled, with the analysis performed for each mode and for each component of the seismic ground motion separately. The results obtained for different modes using design spectra are then combined together by means of approximate combination rules, like the "Square Root Sum of Squares" (SRSS) or "Complete Quadratic Combination" (CQC) rule. The SRSS rule can be used to combine the results for different components of the ground motion. This approach has been widely accepted and used in practice, in spite of the approximations involved in the combination rules. In the inelastic range, the superposition theoretically does not apply. However, the coupling between vibrations in different modes is usually weak (Chopra 2017, p.819). Thus, for the majority of structures, some kind of superposition can be applied as an approximation in the inelastic range, too.

It has been observed that higher mode effects depend to a considerable extent on the magnitude of the plastic deformations. In general, higher mode effects in plan and in elevation decrease with increasing ground motion intensity. Thus, conservative estimates of amplification due to higher mode effects can usually be obtained by elastic modal response spectrum analysis. The results of elastic analysis, properly normalised, usually represent an upper bound to the results obtained for different intensities of the ground motion in those parts of the structure where higher mode effects are important, i.e. in the upper part of medium- or high-rise buildings, and at the flexible sides of plan-asymmetric buildings. An exception are the stiff sides of plan-asymmetric buildings. If the building is torsionally stiff, usually a de-amplification occurs at the stiff side, which mostly decreases with increasing plastic deformations. If the building is torsionally flexible, amplifications may also occur at the stiff side, which may be larger in the case of inelastic behaviour than in the case of elastic response.

The extended N2 method has been developed taking into account most of the above observations. It is assumed that the higher mode effects in the inelastic range are the same as in the elastic range, and that a (in most cases conservative) estimate of the distribution of seismic demand throughout the structure can be obtained by enveloping the seismic demand in terms of the deformations obtained by the basic pushover analysis, which neglects higher mode effects, and the normalized (the same roof displacement as in pushover analysis) results of elastic modal analysis, which includes higher mode effects. De-amplification due to torsion is not taken into account. The target displacements are determined as in the basic N2 method, or by any other procedure. Higher mode effects in plan (torsion) and in elevation are taken into account simultaneously. The analysis has to be performed independently for two horizontal directions of the seismic action. Combination of the results for the two horizontal components of the seismic action is performed only for the results obtained by elastic analysis. The combined internal forces, e.g. the bending moments and shear forces, should be consistent with the deformations. For more details, see (Kreslin and Fajfar 2012) and (Fardis et al. 2015). Note that a variant of Modal pushover analysis is also based on the assumption that the vibration of the building in higher modes is linearly elastic (Chopra et al. 2004).

Two examples of the results obtained by the extended N2 method are presented in Figs. 1.4 and 1.5. Figure 1.4 shows the storey drifts along the elevation of the 9-story high steel "Los Angeles building" which was investigated within the scope of the SAC project in the United States. Shown are results of the NRHA analysis, of the basic pushover analysis, of the elastic modal response spectrum analysis, and of the extended N2 method. Figure 1.5 shows the normalized displacements, i.e. the displacements at different locations in the plan divided by the displacement of the mass centre CM, obtained by NRHA for different intensities of ground motion, by the basic pushover analysis, by the elastic modal response spectrum analysis, and by the extended N2 method. The structural model corresponds to the SPEAR building which was tested within the scope of an European project.

The extended N2 method will be most probably implemented in the expected revised version of EC8. At the time when the current Part 1 of EC8 was finalised, the extended version of the N2 method for plan-asymmetric buildings had not yet been fully developed. Nevertheless, based on the preliminary results, the clause "Procedure for the estimation of torsional effects" was introduced, in which it was stated that "pushover analysis may significantly underestimate deformations at the stiff/ strong side of a torsionally flexible structure". It was also stated that "for such

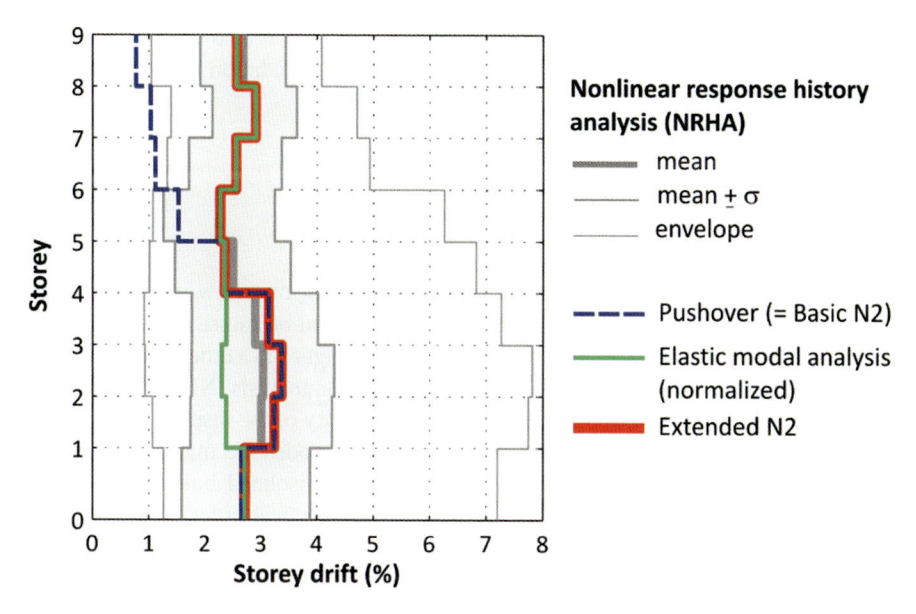

Fig. 1.4 Comparison of storey drifts obtained by different procedures for the 9-storey LA building (Adapted from Kreslin and Fajfar 2011)

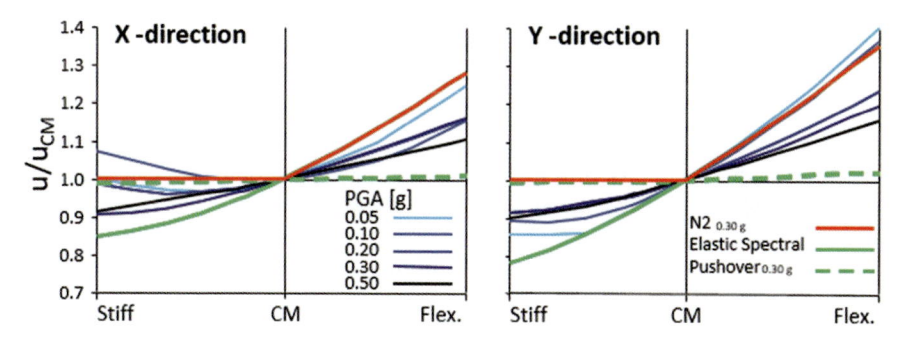

Fig. 1.5 Comparison of normalized roof displacement in plan obtained by NRHA analysis (mean values) for different intensities, elastic response spectrum analysis, pushover analysis, and the extended N2 method for the SPEAR building (Adapted from Fajfar et al. 2005)

structures, displacements at the stiff/strong side shall be increased, compared to those in the corresponding torsionally balanced structure" and that "this requirement is deemed to be satisfied if the amplification factor to be applied to the displacements of the stiff/strong side is based on the results of an elastic modal analysis of the spatial model".

ASCE 41-13 basically uses the same idea of enveloping the results of the two analysis procedures in order to take into account the higher modes in elevation. In C7.3.2.1 it is stated "Where the NSP [Nonlinear Static Procedure] is used on a structure that has significant higher mode response, the LDP [Linear Dynamic

Procedure, typically the modal response spectrum analysis] is also used to verify the adequacy of the evaluation or retrofit." The same recommendation is included in the very recent New Zealand guidelines (NZSEE 2017).

1.3.5 Nonlinear Analysis in Current Codes

In this section, an attempt has been made to present an overview of the status of nonlinear analysis in the codes of different countries and of its use in practice. The material is based on the responses of the author's colleagues to a quick survey and on the available literature. It can be observed that nonlinear analysis (NHRA more often than pushover analysis) has been adopted in the majority of the respective countries as an optional procedure which, however, is in some codes the mandatory for tall buildings and for special structural systems, e.g. base-isolated buildings. In next sections, these countries and regions are listed in alphabetical order.

1.3.5.1 Canada

The latest approved version of the National Building Code of Canada is NBCC 2015. The code applies only to new buildings; there is no code for existing building. The Code is accompanied by a Commentary, which explains the intent of the code and how best to meet it. The provisions or specifications given in the Commentary are not mandatory, but act as guidelines. According to the Code, dynamic analysis is the preferred method of analysis. The dynamic analysis procedure may be response spectrum analysis, or linear time history analysis, or NRHA. In the last case a special study is required. The Commentary gives the specifics of the nonlinear analysis procedure and the conditions for a special study. Inter alia, an independent design review is required when NRHA is used. Consultants (particularly in British Columbia) often use response spectrum analysis for the design of high-rise irregular buildings, and for important buildings such as hospitals. Linear time history analysis is used only infrequently, whereas NRHA is not used in practice. Nonlinear analysis is used mainly for the evaluation of existing buildings. It is considered to be a very complicated task, being hardly ever justified for the design of a new building. Pushover analysis is not directly referred to in the NBCC.

1.3.5.2 Chile

The present design building code, (NCh 433) does not consider and does not allow nonlinear analysis (static or dynamic). Nevertheless, all buildings taller than four storeys are required to have a peer review. This review is generally done using parallel modelling and design. Seismic designs of some of the tallest buildings have been reviewed using pushover analysis. The code for base isolated buildings

(NCh2745) requires NRHA in all irregular or tall isolated structures. Nonlinear properties are considered, at least for isolators. The code for industrial structures (NCh2369) has always allowed for nonlinear analysis (static and dynamic) but it has been used only on rare occasions, for structures with energy dissipation devices and complex structures. The new design building code, which has been drafted and is currently undergoing an evaluation process, does, however, take into account nonlinear analysis and performance-based design. Nonlinear structural analysis is taught at the undergraduate level at the universities. So young designers are able to use standard software that includes nonlinear static and dynamic analysis. In advanced design offices, nonlinear analysis is regularly used.

1.3.5.3 China

According to the 2010 version of the Chinese Code for Seismic Design of Buildings (GB 50011-2010), linear analysis (static or dynamic) is the main procedure. "For irregular building structures with obvious weak positions that may result in serious seismic damage, the elasto-plastic deformation analysis under the action of rare earthquake shall be carried out ... In this analysis, the elasto-plastic static analysing method or elasto-plastic time history analysing method may be adopted depending on the structural characteristics". In some cases simplified nonlinear analysis can be also used. The code has a separate section for performance-based design, where it is stated that either linear analysis with increased damping, or nonlinear static or dynamic analysis can be used for performance states where nonlinear behaviour is expected. In the case of tall buildings, linear analysis is performed for the frequent earthquake level, whereas nonlinear analysis, including pushover analysis and NHRA, is used for the major earthquake level. The type of the required nonlinear analysis depends on the height and complexity of the structure. NRHA should be performed for buildings with heights of more than 200 m or with severe irregularities. Buildings higher than 300 m have to be analyzed using two or more different computer programs in order to validate the results (Jiang et al. 2012).

1.3.5.4 Europe

Most countries in Europe use the European standard EN 1998 Design of structures for earthquake resistance (called Eurocode 8 or EC8), which consists of six parts. Part 1, enforced in 2004, applies to new buildings, and Part 3, enforced in 2005, applies to existing buildings. The main analysis procedure is linear modal response spectrum analysis. The equivalent static lateral force procedure can be used under some restrictions. Nonlinear analysis is permitted. It can be static (pushover) or NRHA. The basic requirements for pushover analysis (the N2 method) are provided in the main text of Part 1, whereas more details are given in an informative annex. NRHA is regulated very deficiently by only three clauses. In the revised version, which is currently under development, pushover analysis will play a more prominent role, whereas, in the draft of the revision, NRHA is still not adequately regulated. In practice, the use of NRHA is extremely rare, whereas the use of pushover analysis varies from country to country.

1.3.5.5 India

Nonlinear analysis has not been included in the building code, and is not used in design. No changes are expected in the near future.

1.3.5.6 Iran

The Iranian Seismic Code was updated in 2014 to the 4th Edition. The analysis procedures include linear procedures, i.e. linear static, response spectrum, and linear response history analyses, and nonlinear procedures, i.e. pushover and NRHA. The linear static procedure is permitted for all buildings with heights of up to 50 m with the exception of some irregular cases, whereas response spectrum and linear response history analyses are permitted for all cases. Pushover analysis and NRHA can be used for all kinds of buildings. However, one of the linear procedures should also be performed in addition to nonlinear analysis. In the 50% draft version of the code, the use of nonlinear analyses was encouraged by a 20% reduction in force and drift limitations. In the final version, such a reduction is not permitted. The pushover method is based on the EC8 and NEHRP 2009 approaches. A standard for the seismic rehabilitation of existing buildings (Standard No. 360) exists. The first edition was mainly based on FEMA 356, and the second edition is based mainly on ASCE 41-06. This standard has been widely used in retrofitting projects, and also sometimes in design projects. Pushover analysis is the most frequently used analysis method in seismic vulnerability studies of existing public and governmental buildings. It is used for all building types, including masonry buildings. The most popular is the coefficient method.

1.3.5.7 Japan

The seismic design requirements are specified in the Building Standard Law. In 1981 a two-phase seismic design procedure was introduced, which is still normally used in design offices. The performance-based approach was implemented in 2000. Most engineered steel and concrete buildings are designed with the aid of nonlinear analysis (pushover or NRHA) in the second phase design. For buildings with heights of less than 60 m, pushover analysis is usually conducted in order to check the ultimate resistance of members and of the building. For high-rise buildings (with heights of more than 60 m), NRHA is required. Usually, first a pushover analysis of a realistic 3D model is conducted in order to define the relationship between the storey-shear and the storey-drift for each storey. These relationships are modelled by means of trilinear models. NRHA is performed for a simplified structural model where stick element models (shear or shear-flexure model) are used. Very rarely, a complex 3D model is directly used for NRHA. Designs are subject to technical peer review. The software needs to be evaluated by the Minister of the Land, Infrastructure, Transport and Tourism.

1.3.5.8 Mexico

The present code (MOC-2008) permits NRHA. The design base shear determined in dynamic analysis should not be less than 80% of the base shear determined by the static force procedure. Over the last 10 years NRHA has been used in Mexico City for very tall buildings (having from 30 to 60 storeys), mainly because this was required by international insurance companies. Static pushover analysis has been seldom used in practice. The new code, which is expected to be published at the end of 2017, will require dynamic spectral analysis for the design of the majority of structures, and the results will need to be verified by a NRHA in the case of very tall or large buildings non-complying with regularity requirements.

1.3.5.9 New Zealand

The NZ seismic standard (NZS1170.5) requires either a response spectrum analysis or a response-history analysis (linear or nonlinear) in order to obtain local member actions in tall or torsionally sensitive buildings. Even in the case of other buildings (which are permitted to be analysed by equivalent static linear analysis), the designers can opt to use more advanced analysis methods (so using pushover analysis in such cases may perhaps be argued to be acceptable). In practice, majority of structures are still designed using the linear static approach, but NRHA is becoming more and more common. The use of the linear static method is also the most common method used when assessing existing buildings, although pushover analysis has also been used. Very recently, technical guidelines for the engineering assessment of existing buildings were published (NZSEE 2017). They recommend the use of SLaMA (Simple Lateral Mechanism Analysis) as a starting point for any detailed seismic assessment. SLaMA represents a form of static nonlinear (pushover) analysis method, which is a hand calculation upper-bound approach defined as "an analysis involving the combination of simple strength to deformation representations of identified mechanisms to determine the strength to deformation (push-over) relationship for the building as a whole". In addition to standard linear analyses and SLaMA, the standard nonlinear pushover procedure and NRHA are also included in the guidelines.

1.3.5.10 Taiwan

The seismic analysis procedures have remained practically unchanged since the 2005 version of the provisions. In the case of new buildings, linear analysis using the response spectrum method is a very common practice, regardless of the building's regularity or height. Pushover analysis is also very popular as a means to determine the peak roof displacement, the load factor to design the basement, the ground floor diaphragm, and the foundation or piling. Linear response history analysis is not common. NRHA is sometimes conducted for buildings using oil or metallic dampers. In the case of tall buildings with or without dampers, some engineers use general-purpose computer programs to evaluate the peak story drift demands, the deformation demands on structural members, and the force demands on columns, the 1st floor diaphragm, and the foundation or basement. Base-isolated buildings have become more popular, and engineers like to use nonlinear springs for base isolators only. Existing government buildings have to be retrofitted up to the current standard for new constructions. Pushover analysis is very common for the evaluation of existing condition of buildings and verifying the retrofit design.

1.3.5.11 Turkey

In the Specification for buildings to be built in seismic zones (which was adopted in 2007) nonlinear analysis is mentioned but not emphasized. A special chapter was added to the Code in 2007, which is devoted to the seismic assessment and retrofitting of existing buildings; it specifies the use of both the single-mode and the multi-mode pushover procedures as well as the use of NRHA. However, these methods have not been widely used in practice for assessment and retrofitting purposes. The new code (the Turkish Seismic Code for Building Structures) will be published by the end of 2017 and will be enforced after a one-year transition period. In addition to the traditional force-based design, which is based entirely on linear elastic analysis, the code contains a new section on displacement (performance) based design, where the entire analysis is nonlinear, either pushover (several different types of pushover analysis can be used) or response-history. Tall buildings, the isolation level of the base isolated buildings and all other buildings in seismic zones 1 and 2 (the most severe categories, out of 4), existing or new, which require advanced seismic performance, have to be modelled by considering the nonlinear response. For existing buildings, in addition to nonlinear procedures, a new displacement-based equivalent linear procedure can also be used for buildings where a soft storey mechanism is not expected. NRHA has been occasionally used for tall buildings (since 2007) based on a draft code prepared for the seismic design of tall buildings upon the request of the Istanbul Metropolitan Municipality. Except in the case of tall and base isolated buildings, there have so far been almost no practical applications of nonlinear analysis. For seismic rehabilitation, single mode pushover analysis has been used in the past.

1.3.5.12 United States

The structure of the guidelines, standards, and codes is complicated. The most representative seem to be the ASCE standards. According to ASCE 7-16, which applies to new structures, NRHA is permitted, but is not required. It is regulated in Chapter 16, which has been completely rewritten in the last version of the standard. A linear analysis has to be performed in addition to NRHA. An independent structural design review is required. Existing structures are regulated by ASCE 41-13. Nonlinear procedures are permitted for all buildings. They are used for the analysis of buildings where linear procedures are not permitted. There are some limitations on the use of nonlinear static (pushover) analysis, whereas NRHA can be used for all structures. However, when NRHA is used, "the authority having jurisdiction shall consider the requirement of review and approval by an independent third-party engineer with experience in seismic design and nonlinear procedures". Special Seismic Design Guidelines for Tall Buildings exist (PEER 2017) which present a recommended alternative to the prescriptive procedures for the seismic design of buildings contained in standards such as ASCE 7. NRHA is required for MCE_R (Maximum Considered Earthquake) evaluation. It should be performed in accordance with the provisions of ASCE 7-16, Chap. 16. Nonlinear static analysis may be used to supplement nonlinear dynamic analysis as an aid to understanding the yield hierarchy, overstrength, and effective values of R factors when this is deemed desirable. In general, nonlinear analysis has been well established in the United States.

1.4 Future: Risk Assessment. Is It a Feasible Option for Codes?

1.4.1 Introduction

Historically, among the three features of seismic analysis, dynamics has been implemented first, followed by inelasticity. The remaining feature is probability, which has not yet found its way into the practice, with the exception of some very important structures, although the theoretical background and computational procedures have been already developed. The evolution of reliability concepts and theory in structural design are described in *fib* (2012).

The seismic response of structures is characterized by large uncertainties, especially with respect to the ground motion, but also in the structural modelling, so that, in principle, a probabilistic approach would be appropriate for seismic performance assessment. However, an average engineer is not familiar with probabilistic methods and is very hesitant to use them. Also, a large part of the research community is skeptical about explicit probabilistic approaches other than those used in seismic hazard analysis.

The results of a quick survey which was performed among some colleagues from different countries indicate that almost all of them doubt that an explicit probabilistic approach will be adopted in future building codes in their countries, at least in the near future, with the exception of the United States, where this has already occurred to some extent in ASCE 7, but has been very rarely used in practice. A more realistic possibility is the adoption of an explicit quantitative consideration of risk in codes for critical infrastructure.

Nevertheless, in the long term, it will be difficult to completely avoid quantitative determination of risk. Also due to the public pressure on loss minimisation in addition to life safety in most developed countries with high seismicity, the profession will be sooner or later forced to accept some kind of risk-based design and assessment, at least for a better calibration of different safety factors and force reduction factors used in codes. Information on seismic risk would also facilitate discussions of design options between designers, building owners and other stakeholders. However, the mandatory use of explicit probabilistic approaches in seismic building codes, if it will ever happen, is still very distant. "This time lag, however, should be regarded as an opportunity to familiarize with the approaches before actual application." (*fib* 2012, Preface). The prerequisite for possible implementation of quantitative risk assessment in the codes are reasonably reliable input data and highly simplified procedures, which are presented in a format that is familiar to engineers, and which require only a small amount of additional effort and competence. Inclusion of optional reliability-based material in the seismic codes would help due to its educational role. The very first step has been very recently already taken in Europe with a draft informative annex to EC8, as described in the next section.

1.4.2 Current State of Probabilistic Analysis in Seismic Codes

Probabilistic seismic hazard analysis (PSHA) is already well established. It is, as a rule, performed by specialists and not by average designers. Since the early 1970s, it has been used for the preparation of seismic hazard maps and for the determination of the seismic design parameters of important structures. In the majority of codes, including EC8, the design ground motion corresponds to a return period of 475 years, which corresponds to a 10% probability of exceedance in 50 years. Some codes use the return period of 2500 years (a 2% probability of exceedance in 50 years). The analysis of structures is typically performed by deterministic analysis, using the ground motion parameters corresponding to a prescribed return period of the ground motion. In this analysis, the uncertainties are implicitly taken into account by means of various safety factors. An explicit probabilistic approach, which allows for the explicit quantification of the probability of exceedance of different limit states, has not yet been implemented in building seismic codes (with the exception of the ASCE-7 standard, as explained later in this section). When using current seismic codes, "at the end of the design process there is no way of evaluating the reliability actually achieved. One can only safely state that the adoption of all the prescribed design and detailing rules should lead to rates of failure substantially lower than the rates of exceedance of the design action" (*fib* 2012, p.3).

"Risk-targeted maximum considered earthquake" (MCE$_R$) ground motion maps are a fairly new development in the United States. When using the previous uniform hazard maps, corresponding to a 2% probability of the ground motion being exceeded in 50 years, buildings were designed to resist uniform hazard *ground shaking* levels, whereas buildings designed by using the MCE$_R$ maps are intended to provide the same level of seismic performance, meaning that they have the same *risk* level. According to Kircher et al. (2012), "The risk-targeted MCE$_R$ ground motions are defined by two essential collapse objectives. First objective answers the question – what is the acceptable collapse risk, that is, what is the acceptable probability of collapse of a structure in 50 years? This probability was selected to be 1 percent (in 50 years) ... The second collapse objective answers the question – what is the acceptable probability of collapse of a structure given the MCE$_R$ ground motions occur at the site of interest? This conditional probability was selected to be 10 percent (given MCE$_R$ ground motions). .... The 1 percent in 50-year collapse risk objective is the result of integrating the hazard function (which is different for each site) and the derivative of the "hypothetical" collapse fragility defined by the 10 percent conditional probability (and an appropriate amount of collapse uncertainty)." The new maps have been adopted in the 2010 edition of the ASCE 7 standard, and have remained in this standard also in the latest (ASCE 7-16) edition. However, many researchers and practicing engineers in United States are not happy with the new concept. There are several reasons for this. According to Hamburger, who is one of the key persons in the development of seismic codes in the United States, clients think they understand probabilistic ground motion (e.g. a 500-year or a 2500-year event), but do not understand designing for a ground motion that produces a 1% risk of collapse in 50 years. Most engineers do not understand this well either, and have a difficult time explaining it. Moreover, the 10% probability of collapse conditioned on MCE$_R$, which is the basis for design, seems to be too pessimistic given the very few collapses that have historically been experienced, mostly in a population of buildings that does not conform to the present code (RO Hamburger 2017, personal communication, 2. November 2017). The author and his colleagues have also some doubts about the concept of risk-targeted seismic maps. Whereas seismic hazard depends only on ground motion, risk depends both on ground motion and structural behaviour. The usual hazard maps are based solely on ground motion considerations, whereas the risk-targeted maps inevitably also involve structural considerations, which makes the process more complex and less transparent. The whole seismic design or assessment process is facilitated if the ground motion problems can be separated from the structural problems.

Performance-based earthquake engineering seeks to improve seismic risk decision-making through new design and assessment methods that have a probabilistic basis, as well as trying to take into account properly the inherent uncertainty. The goals of performance-based earthquake engineering can be achieved by using methods which allow for the explicit quantification of risk of collapse and of different damage states. Such procedures have already been developed. The most influential has been the work of Cornell and colleagues at Stanford University. In 1996, Cornell (1996a) published "a convenient explicit probabilistic basis for new seismic design and evaluation guidelines and codes". The proposed procedure avoided using "often very esoteric concepts of random vibration theory ... or simulation of (practically) prohibitively large number of samples of events". Instead, "the proposed procedure makes use only of traditional and available tools". In the same year, Cornell (1996b) also presented his "personal view of the needs and possible future path of project-specific earthquake resistant design and re-assessment based on explicit reliability analysis" which "cannot be done without effective estimates of the likelihoods of loss- and injury-inducing structural behaviour". The procedure, proposed in (Cornell 1996a), he classified as "future, but currently feasible, practice". In fact, the procedure became very popular among researchers. The well-known SAC-FEMA approach (Cornell et al. 2002) is based on this work. However, a procedure which requires a large number of NRHA analyses encounters difficulties when searching for a way into the codes.

In 2012, the Applied Technology Council (ATC) developed a methodology for the seismic performance assessment of buildings (FEMA P-58, 2012). The technical basis of the methodology was the framework for performance based earthquake engineering developed by researchers at the Pacific Earthquake Engineering Research Center (PEER) between 1997 and 2010. As acknowledged in the Dedication of the report FEMA P-58, A. Cornell and H. Krawinkler were the leading forces in the development of this framework. The methodology is intended for use in a performance-based seismic design process. It is applicable to the assessment of new or existing buildings. The procedures are probabilistic, uncertainties are considered explicitly, and performance is expressed as the probable consequences, in terms of human losses (deaths and serious injuries), direct economic losses (building repair or replacement costs), and indirect losses (repair time and unsafe placarding) resulting from building damage due to earthquake shaking. The implementation of the methodology is described in companion volumes. An electronic Performance Assessment Calculation Tool (PACT) was also provided. Nevertheless, the comprehensive methodology has not yet been included in the codes.

A rigorous explicit probabilistic approach for structural analysis, which utilises recent research results, is, for example, the procedure provided in Appendix F to the FEMA P-695 document (2009). This methodology requires: (a) a structural model that can simulate collapse, (b) use of many (perhaps hundreds) of nonlinear response history analyses, and (c) explicit treatment of many types of uncertainties. "This process is too complex and lengthy for routine use in design." (Haselton et al. 2017). The explicit approach is nonetheless permitted by Section 1.3.1.3 (Performance-Based Procedures) of ASCE 7-10 and ASCE 7-16. Tables with target (acceptable, tolerable) reliabilities for failure were previously provided in the Commentary of

ASCE 7-10, but have moved to the main part of the standard in ASCE 7-16. If using performance-based design, designers are supposed to demonstrate through testing and analysis that the design is capable of achieving these reliabilities. The analysis and design shall be "subject to the approval of the authority having jurisdiction for individual projects". However, the standard permits implicit demonstration that the target reliability can be achieved, therefore, in practice, explicit demonstration through calculation of a failure probability will almost certainly never be done (RO Hamburger 2017, personal communication, 2. November 2017). Moreover, Hamburger states: "I would not look to rapid adoption of these [explicit probabilistic] approaches procedures by the profession. Most engineers are not familiar with probability and statistics, let alone structural reliability theory. Most have a qualitative understanding of the basis for Load Resistance Factor Design, but are more than happy to assume that the load and resistance factors do what is necessary to give them safe designs."

In Europe, a very comprehensive document dealing with explicit probabilistic analysis is the Italian Guide for the Probabilistic Assessment of the Seismic Safety of Existing Buildings (CNR 2014), developed by P.E.Pinto and P.Franchin. The Guide is intended to "be of help in future revisions of the current standards". Three verification methods having different levels of complexity are presented.

In order to facilitate a gradual introduction of probabilistic considerations into practice and codes, highly simplified practice-oriented approaches for the determination of seismic risk are needed, at least as a first step. Very recently, an informative Annex to EC8, Part 1, entitled Simplified reliability-based verification format (CEN 2017), has been drafted by Dolšek et al. (2017b). It provides a basis for a simplified verification of the performance of a structure in probabilistic terms. It gives information about (a) a simplified reliability-based verification format; (b) a procedure for the estimation of the relationship between performance factors and the degree of reliability; and (c) a procedure for the estimation of behaviour (i.e. force reduction) factors with respect to target reliability for the NC limit state.

A very simple method called the Pushover-based Risk Assessment Method (PRA), which is in line with the Annex and which requires only a very minor effort in addition to a standard pushover-based analysis, is summarized in the next section. For more details, see Kosič et al. (2017), Dolšek and Fajfar (2007), and Fajfar and Dolšek (2012). By combining Cornell's closed form solution (Cornell 1996a) and the pushover-based N2 method (Fajfar 2000), which is used for the determination of the capacity of the structure, the annual probability of "failure" of a structure can be easily estimated, provided that predetermined default values for dispersions are available. Compared to Cornell's original approach, in the PRA method a large number of nonlinear response history analyses is replaced by a few pushover analyses (in most cases just a single one). Of course, like other simplified methods, the PRA method has limitations, which are basically the same as those which apply to Cornell's closed form solution and to the basic N2 method.

1.4.3 Summary of the Pushover-Based Risk Assessment (PRA) Method

The "failure" probability of building structures, i.e. the annual probability of exceeding the near collapse limit state (NC), which is assumed to be related to a complete economic failure of a structure, can be estimated (Cornell 1996a; Fajfar and Dolšek 2012; Kosič et al. 2017) as:

$$P_f = P_{NC} = \exp\left[0.5 \ k^2 \beta_{NC}^2\right] H(S_{aNC}) = \exp\left[0.5 \ k^2 \beta_{NC}^2\right] k_0 \ S_{aNC}^{-k}, \quad (1.1)$$

where S_{aNC} is the median NC limit state spectral acceleration at the fundamental period of the structure (i.e. the capacity at failure), and β_{NC} is the dispersion measure, expressed as the standard deviation of the natural logarithm of S_{aNC} due to record-to-record variability and modelling uncertainty. The parameters k and k_0 are related to the hazard curve $H(S_a)$ which is assumed to be linear in the logarithmic domain

$$H(S_a) = k_0 \ S_a^{-k}. \quad (1.2)$$

A seismic hazard curve shows the annual rate or probability at which a specific ground motion level will be exceeded at the site of interest. The reciprocal of the annual probability of exceedence is the return period $T_R = 1/H$.

In principle, dispersion β_{NC} and hazard function should be consistent with the intensity measure S_a, since they depend on it.

The capacity at failure S_{aNC} is estimated using the pushover-based N2 method (Fajfar 2000), whereas predetermined dispersion values are used for β_{NC}. Note that, in principle, Eq. (1.1) can be applied for any limit state, provided that the median value and the dispersion of the selected intensity measure are related to the selected limit state. A widely accepted definition of the NC limit state, referred to in this paper also as "failure", is still not available. One possibility is to relate the global NC limit state to the NC limit state of the most critical vertical element. Another option is to assume that the NC limit state corresponds to 80% strength at the softening branch of the pushover curve. The NC limit state was selected instead of the C (collapse) limit state, since it is much easier to estimate capacities for the NC limit state than for the C limit state. It should be noted, however, that the tolerable probabilities of exceedance are higher for the NC than for the C limit state.

The determination of the spectral acceleration at the NC limit state S_{aNC} can be visualized in Fig. 1.6, where the equal displacement rule is assumed. In the acceleration – displacement (AD) format, S_{aNC} is defined by the crossing point of the vertical line, representing the displacement at the NC limit state, i.e. the displacement capacity of the equivalent single-degree-of-freedom system representing the structure, and the diagonal line, representing the period of the structure. The crossing point is a point on the acceleration spectrum, which corresponds to the NC level ground motion.

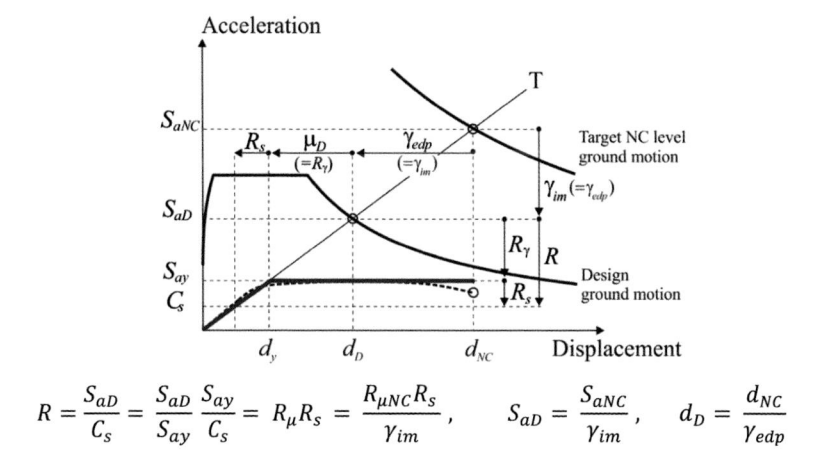

Fig. 1.6 Illustration of parameters relevant for the determination of the capacities and force reduction factors. In the plot the validity of the equal displacement rule was assumed. The presented parameters apply to a general case, with the exception of the equalities in brackets which apply only in the case if the equal displacement rule is valid

$$R = \frac{S_{aD}}{C_s} = \frac{S_{aD}}{S_{ay}}\frac{S_{ay}}{C_s} = R_\mu R_s = \frac{R_{\mu NC} R_s}{\gamma_{im}}, \quad S_{aD} = \frac{S_{aNC}}{\gamma_{im}}, \quad d_D = \frac{d_{NC}}{\gamma_{edp}}$$

According to Eq. (1.1), the failure probability is equal to the hazard curve evaluated at the median capacity S_{aNC}, multiplied by an exponential magnification factor which depends on the product of the variability of the S_{aNC} (expressed by β_{NC}) and the slope (in log-log terms) k of the hazard curve. For frequently used values ($\beta_{NC} = 0.5$ and $k = 3.0$) the correction factor amounts to 3.1. In such a case the probability of failure is about three times larger than the probability of the exceedance of the ground motion corresponding to the median capacity at failure in terms of the spectral acceleration S_{aNC}. If there was no variability ($\beta_{NC} = 0$), both probabilities would be equal.

Extensive studies have been performed in order to determine typical dispersions β_{NC} of the capacity at failure for reinforced concrete (RC) building structures using $S_a(T)$ as the intensity measure (Kosič et al. 2014, 2016). The results of these studies showed that the values depend on the structural system and on the period of the structure T. However, in a simplified approach, it may be reasonable to assume $\beta_{NC} = 0.5$ as an appropriate estimate for RC building structures, with the exception of very stiff structures, where β_{NC} is smaller. This value takes into account both aleatory (related to ground motion) and epistemic (related to structural modelling) uncertainty.

Several options are available for the estimation of the parameters k and k_0. Best k and k_0 estimates can be obtained by fitting the actual hazard curve by a linear function in logarithmic domain. In absence of an appropriate hazard curve, k can be estimated from seismic hazard maps for two return periods. If hazard maps for two different return periods are not available, the only (very approximate) option is to assume a value of k depending on the geographical location of the structure. Appropriate values of k are usually within the range from 1 to 3 (exceptionally to 4). If the value of k, specific for the region, is not known, a value of $k = 3.0$ has often been used as an option in high seismicity regions. In low seismicity regions the k values are usually smaller. Note that k also depends on the intensity measure used in Eq. (1.1). In the case of the spectral acceleration S_a, it depends on the period of the structure. This dependence should be taken into account when a more accurate analysis is being sought (Dolšek et al. 2017a).

For the determination of the parameter k_0 at least one value of S_a, corresponding to a specific return period needs to be known for the location under consideration, e.g. S_{aD} that corresponds to the return period of the design ground motion T_D and represents the spectral acceleration in the elastic design spectrum. Knowing the value S_{aD} and the corresponding annual frequency of exceedance ($H = 1/T_D$), the parameter k_0 can be obtained from Eq. (1.2) as follows:

$$k_0 = 1/\left(T_D \cdot S_{aD}^{-k}\right). \tag{1.3}$$

Considering Eq. (1.3), Eq. (1.1) can be written in the form

$$P_{NC} = \exp\left[0.5\ k^2 \beta_{NC}^2\right] \frac{1}{T_D}\left(\frac{S_{aNC}}{S_{aD}}\right)^{-k}. \tag{1.4}$$

1.4.4 Applications of the PRA Method

The results of extensive studies have demonstrated that the PRA method has the potential to predict the seismic risk of low- to medium-rise building structures with reasonable accuracy, so it could become a practical tool for engineers. Typical values of probabilities of exceedance of the NC limit state in a life-span of 50 years are, in the case of buildings designed according to modern codes, about 1%. In the case of older buildings, not designed for seismic resistance, the probabilities are usually at least one order of magnitude higher (see, e.g., Kosič et al. 2014, 2016). It should be noted, however, that the absolute values of the estimated failure probability are highly sensitive to the input data and simplifying assumptions, especially those related to the seismic hazard. Comparisons between different structures are more reliable. Comparative probabilistic analyses can provide valuable additional data necessary for decision-making. Due to its simplicity, the PRA method can also serve as a tool for the introduction of explicit probabilistic considerations into structural engineering practice.

The basic equation for the probability of failure can also be used in risk-targeted design. In recent years this issue has been intensively studied by M. Dolšek and his PhD students. Already in 2014, Žižmond and Dolšek (2014) prepared a manuscript proposing "risk-targeted behaviour factor". In that manuscript the concept of force reduction factors based on acceptable risk was developed and formulated in a form very similar to Eq. (1.12). The idea and the formulation, originally presented in that manuscript, are of great help for understanding the empirical values of force reduction factors and allow a scientifically based quantification of these factors. Unfortunately, that manuscript was not accepted for publication in the international journal to which it was submitted. A very substantial part of the text which follows in this section is based on the work, ideas and formulations of M. Dolšek and his former students, especially J. Žižmond and N. Lazar Sinković.

For design purposes, Eq. (1.4) has to be inverted in order to express the ratio between the spectral acceleration at failure S_{aNC} and the design spectral acceleration S_{aD} as a function of the return period corresponding to the target probability of failure $T_{NC} = 1/P_{NC}$ and the parameters k and β_{NC}

$$\frac{S_{aNC}}{S_{aD}} = \left(\frac{T_{NC}}{T_D}\right)^{\frac{1}{k}} \exp\left[0.5\ k\ \beta_{NC}^2\right] = \gamma_{im}, \qquad (1.5)$$

where index im stands for Intensity Measure (i.e. spectral acceleration at the fundamental period of the structure).

The factor γ_{im}, initially proposed by Žižmond and Dolšek (2014, 2017), further elaborated in (Dolšek et al. 2017a), and used also in the draft Annex to revised EC8 (Dolšek et al. 2017b; CEN 2017), is the product of two factors.

The first factor $((T_{NC}/T_D)^{1/k})$ takes into account the fact that the target probability of failure (the NC limit state) is smaller than the probability of the design ground motion or, expressed in terms of return periods, the target return period of failure is larger than the return period of design ground motion. For example, in EC8 the design ground motion for common structures corresponds to a return period of 475 years and is related to the significant damage (SD) limit state. Although the SD limit state is not well defined, it certainly corresponds to a much smaller damage than the NC limit state. It is attained at a much smaller deformation. The value of this factor is equal to 1.0 if $T_{NC} = T_D$.

The second factor takes into account the uncertainties both in the ground motion (record to record dispersion) and in the modelling. In codes, safety factors, which take into account possible less favourable conditions, should be used. Safety factors are typically based on probabilistic considerations and on experience. In a standard deterministic analysis, the fixed values prescribed in codes are used. If a probabilistic analysis is used, the uncertainty can be explicitly included in the calculations. The value of this term is 1.0 if $\beta_{NC} = 0$, i.e. if there is no uncertainty. Note that the second factor is similar to the magnification factor in Eq. (1.1), the difference is only in the exponent of k (1 versus 2).

The factor γ_{im} depends on the target and design return period, on the slope of the hazard curve k, and on the dispersion β_{NC}. For some frequently used input data ($k = 3$, $\beta_{NC} = 0.5$, $T_{NC} = 5000$ years, $T_D = 475$ years), $\gamma_{im} = 2.19 \cdot 1.46 = 3.20$.

By analogy to Eq. (1.5), the ratio of displacements d_{NC}/d_D, can be written as:

$$\frac{d_{NC}}{d_D} = \gamma_{edp}, \tag{1.6}$$

where the index *edp* stands for Engineering Demand Parameter (Dolšek et al. 2017a).

If the equal displacement rule applies, the ratio of accelerations is equal to the ratio of displacements (Fig. 1.6) so that both γ values are equal:

$$\gamma_{edp} = \gamma_{im}. \tag{1.7}$$

If the equal displacement rule does not apply, e.g. in the case of buildings with fundamental period in the short period range, a more general relation between the two γ factors can be derived, provided that the relationship between the spectral acceleration and displacement, i.e. between *im* and *edp*, is defined in closed form. Using the simple R-μ-T relationship (Fajfar 2000), implemented in EC8, Dolšek et al. (2017a) derived a relation for structures with short periods, i.e. for structures with the fundamental period T less than the characteristic period of the ground motion T_C:

$$\gamma_{edp} = \frac{\left(\gamma_{im} \frac{S_{aD}}{S_{ay}} - 1\right) \frac{T_C}{T} + 1}{\left(\frac{S_{aD}}{S_{ay}} - 1\right) \frac{T_C}{T} + 1}, \tag{1.8}$$

where S_{ay} represents the acceleration at the yield point of the structure (Fig. 1.6).

Dolšek et al. (2017a) called both γ factors "risk-targeted safety factors", since they depend, inter alia, on the selected acceptable collapse risk. γ_{edp} can be used, for example, in displacement-based design for determining the displacement capacity, whereas γ_{im} can be applied when determining the force reduction factors used in force-based design, as explained in the text which follows. In both cases, the γ factors should be applied to "best estimate" values, i.e. values determined without any partial safety factors.

In the case of the nonlinear analysis in EC8, the design ground motion corresponds to a return period of 475 years, whereas the target return period for failure (the NC limit state) is much larger. The deformation demand, determined for a 475 year return period, cannot be directly compared with the median deformation capacity corresponding to the NC limit state. In order to allow a direct comparison of demand and capacity with proper consideration of safety, the median NC capacity in terms of displacements has to be reduced by dividing the NC displacement capacity with the risk-targeted safety factor γ_{edp}, which takes into account both the difference in return periods and dispersion. Earlier in this section it was shown that, for some frequently used data, γ_{im} is equal to about 3. For structures with the fundamental period in the moderate or long period range, for which the equal displacement rule applies, γ_{edp} is equal to γ_{im}. Thus, in such a case the displacement capacity to be compared with the displacement demand, i.e. the design displacement capacity, should be about three times smaller than the NC displacement capacity. For more details see (Dolšek et al. 2017a).

1 Analysis in Seismic Provisions for Buildings: Past, Present and Future

Using the idea and formulation of Žižmond and Dolšek (2014, 2017), the risk-targeted safety factor γ_{im} can be used for the determination of the risk-targeted force reduction factor R, as shown in the text which follows.

The force reduction factor R can be presented as a product of two factors:

$$R = R_\mu \cdot R_s, \qquad (1.9)$$

where R_s is the so-called overstrength factor, and R_μ is the reduction factor component which takes into account the deformation capacity and the inelastic dissipation energy capacity. R_μ is also called ductility dependent part of the R factor. The overstrength factor, which is an inherent property of properly designed, detailed, constructed and maintained highly redundant structures, is defined as the ratio between the actual strength of the structure and the strength used in code design. R_μ is usually expressed in terms of a ductility factor, i.e. the ratio between the deformation capacity (usually displacements) and the deformation at yield. In general, it depends on the period of the structure T. If the equal displacement rule applies, the ductility dependent R_μ factor is equal to the ductility factor:

$$R_\mu = \mu. \qquad (1.10)$$

In other cases, e.g. in the case of stiff structures with short periods, other relations are used, e.g. the simple relation (Fajfar 2000), implemented in EC8:

$$R_\mu = (\mu - 1)\frac{T}{T_C} + 1, \quad T < T_C \qquad (1.11)$$

where T_C is the characteristic period of ground motion.

When determining the value of a force reduction factor to be used in a code, it is of paramount importance to take into account an appropriate value of the displacement and ductility, which control the ductility dependent part of the reduction factor R_μ. The difference between the return period of the design ground motion and the target return period of failure has to be taken into account, as well as uncertainties. This can be achieved by using the design displacement capacity, as defined in the previous example of application, i.e. the median displacement capacity at failure (the NC limit state) divided by the factor γ_{edp}. Alternatively, if the median NC displacement capacity is used for the determination of the factor R_μ (denoted as $R_{\mu NC}$), the force reduction factor according to Eq. (1.9) should be reduced by the factor γ_{im} (Žižmond and Dolšek 2014, 2017; Dolšek et al. 2017b; CEN 2017):

$$R = \frac{R_{\mu NC} \cdot R_S}{\gamma_{im}} \qquad (1.12)$$

The different parameters which are related to the force reduction factor are illustrated in Fig. 1.6. The presentation of the parameters is similar to that in Fig. 1-2 presented in FEMA P-695 (FEMA 2009). The parameters (with the exception of accelerations and displacements at both axes which have dimensions) are dimensionless ratios of accelerations or displacements. They are depicted as the differences between two related parameters rather than as ratios of the parameters. In order to facilitate presentation, in the plot the validity of the equal displacement rule was assumed. Nevertheless, the presented parameters apply to a general case, with the exception of the equalities in brackets which apply only in the case if the equal displacement rule is valid. C_s is the seismic coefficient, i.e. the reduced acceleration used for the determination of design seismic action.

Equation (1.9) has often been used for the determination of force reduction factors from the results of experimental and numerical investigations. Recently, many studies have been aimed at determining appropriate values of reduction factors, e.g. FEMA P-695 (2009). This study, which took into account probabilistic considerations, is an example of a correct procedure. On the other hand, there have been some studies aimed at determining force reduction factors for specific structural systems, in which Eq. (1.9) was used incorrectly, i.e. too large values of the ductility dependent R_μ factor were applied.

The principle of the reduction of forces and the derivation of relevant equations, shown in the previous text, is based on a single-degree-of-freedom (SDOF) system. Nevertheless, the force reduction factor approach has been widely used in codes also for multi-storey buildings modelled as multi-degree-of-freedom (MDOF) systems which can be approximately represented by an equivalent SDOF system.

1.4.5 Tolerable Probability of Failure

1.4.5.1 General

When determining the tolerable (called also acceptable or target) probability of failure, the possible consequences in terms of risk to life and injury, the potential economic losses, and the likely social and environmental effects need be taken into account. The choice of the target level of reliability should also take into account the amount of expense and effort required to reduce the risk of failure. The acceptable risk is, of course, a reflection of personal and societal value judgements, as well as disaster-based experience, and differs from one cultural environment to another. It is therefore no wonder that generally accepted quantitative values for target structural reliability, which could possibly be used in seismic design, do not exist. Nevertheless, some data are provided in the existing seismic regulations, as discussed in the text which follows.

The standard measures for structural reliability are the reliability index β and the probability of failure P_f, both of which are expressed either at the annual level or for the reference period of the expected working life of the structure (50 years in the case of ordinary buildings). The relationship between the probability of failure and the reliability index is given as:

Table 1.2 The relationship between P_f and β (CEN 2002)

P_f	10^{-1}	10^{-2}	10^{-3}	10^{-4}	10^{-5}	10^{-6}	10^{-7}
β	1.28	2.33	3.09	3.72	4.26	4.75	5.20

$$P_f = \Phi(-\beta) \tag{1.13}$$

where $\Phi(.)$ is the standard normal cumulative probability distribution function. Numerical values of the $\beta - P_f$ relationship are given in Table 1.2.

The reliability index β should not be confused with the dispersion measure β which is used in Sects. 1.4.3 and 1.4.4.

In EN 1990 (also referred to as Eurocode 0, CEN 2002) the target reliability depends on the consequence class, i.e. on the consequences of the potential failure of buildings. For ordinary buildings, the target reliability index for the "ultimate limit state" related to one-year reference period is equal to $\beta = 4.7$ (the same value as that in the ISO 13822 standard). The corresponding probability of failure is about $P_f \approx 10^{-6}$. In Sect. 1.2.2 of EN 1990 it is stated that the choice of the levels of reliability should take into account, in addition to the consequences of failure, also "the expense and procedures necessary to reduce the risk of failure". However, the provided target reliability indices do not depend on cost issues. In the draft of the revised EN 1990, which is at the development stage, the target annual probability of failure $P_f \approx 10^{-6}$ remained unchanged. However, it is explicitly stated that seismic limit states are excluded.

In the *fib* model code (2010), Section 3.3.3 is related to reliability. As in EN 1990, it is stated that the target level of reliability should take into account the possible consequences as well as the amount of expense and effort required to reduce the risk of failure. It is also stated that due attention should be paid to differentiating the reliability level of new and existing structures, since the costs of increasing the safety of existing structures are usually high compared to the costs for new structures. Such a differentiation of the reliability level can be performed on the basis of a "well founded analysis". If such an analysis is omitted, recommended target reliability indices can be used. The target reliabilities for ordinary buildings are the same as those given in EN 1990. In the case of recommended values, no distinction is made with regard the costs of safety measures.

The Probabilistic model code (JCCS 2000) provides target reliability indices which depend not only on the consequences of failure but also on the relative cost of safety measures. In the code (p.18), it is stated: "A large uncertainty in either loading or resistance (coefficients of variation larger than 40%), as for instance the case of many accidental and seismic situations, a lower reliability class should be used. The point is that for these large uncertainties the additional costs to achieve a high reliability are prohibitive." For ordinary buildings and normal relative costs of safety measures, a reliability index for 1 year $\beta = 4.2$ ($P_f \approx 10^{-5}$) is used, whereas in the case of large costs (typical for seismic protection) $\beta = 3.3$ ($P_f \approx 5 \cdot 10^{-4}$). In the model code, it is also stated that failure consequences also depend on the type of failure, which can be ductile or brittle. "A structural element which would be likely to collapse suddenly without warning should be designed for a higher level of reliability than one for which a collapse is preceded by some kind of warning which enables measures to be taken to avoid severe consequences." (JCCS 2000).

The target probabilities of failure in all three codes, which do not take into account the difference in cost needed for increasing safety in a seismic design situation compared to standard design, are much lower than that which could possibly be achieved by complying with the current seismic codes. Only if the differentiation of the reliability index according to the cost of safety measures is made, as suggested in all three codes, but realized in the form of reduced target reliability indices only in the case of the Probabilistic model code, the target probabilities are increased to more realistic values. The annul probability $P_f \approx 5 \cdot 10^{-4}$ (2.5% in 50 years), provided in the Probabilistic model code for ordinary buildings, is even substantially higher than the currently most popular value $P_f = 2 \cdot 10^{-4}$ (1% in 50 years, $\beta \approx 3.5$), suggested also in the draft Annex to the revised EC8 (CEN 2017; Dolšek et al. 2017b), and confirmed in a discussion among European code developers. The latter value is comparable to the probabilities of failure estimated for buildings compliant with current seismic codes.

In ASCE 7-16 reliability indices for "load conditions that do not include earthquake, tsunami, or extraordinary events" are provided. For "structural stability caused by earthquake", target reliabilities are given in terms of "conditional probability of failure caused by the MCE_R shaking hazard". For ordinary buildings the target probability amounts to 10%. (For discussion on the MCE_R concept, see Sect. 1.4.2.). It should be noted that the conditional probability of failure due an assumed earthquake cannot be directly compared with the probability of failure due to earthquakes in the working life of a building.

When discussing acceptable risk of failure, a distinction should be made between physical collapse (i.e. the collapse or C limit state) and economic failure without physical collapse (i.e. the near collapse or NC limit state). Of course, the tolerable probability for collapse, which would very likely result in human casualties, is smaller. Apparently, the definition of failure in most codes corresponds to the NC limit state. In EN 1990 it is stated: "States prior to structural collapse, which, for simplicity, are considered in place of the collapse itself, may be treated as ultimate limit states." Also the estimation of seismic risk in probabilistic analyses is mostly related to the NC limit state, since it is extremely difficult to numerically simulate the complete physical collapse of a building.

1.4.5.2 Acceptable Risk in Slovenia

The Republic of Slovenia is a country with a moderate seismic hazard, where earthquakes up to a magnitude of about 7 and an EMS intensity of IX are possible throughout its territory. An internet-based survey was conducted in 2013 in order to gather data about the perception of seismic risk in Slovenia. In view of the expected differences of opinion between experts and the lay public, the survey was conducted in two groups, where respondents were differentiated according to their expertise in the field of project design and building construction. The first group of respondents were members of the Slovenian Chamber of Engineers (denoted as "experts").

217 responses were received. Most of the respondents were working in the field of project design (55%) or construction (41%). Their answers were compared to the answers of the lay public sample (502 respondents) which was located using snowball sampling. It should be noted that the sample of lay people was not representative for all inhabitants of Slovenia, and was to a large extent limited to people with higher levels of education. The results of the survey did not, in most cases, show significant differences between the two samples. High agreement between the answers was observed also in the answers regarding the tolerable probability of collapse of buildings built according to the current seismic regulations. Both groups were asked how many buildings, on average, can be tolerated to collapse as a direct consequence of an earthquake during their expected working life (i.e. 50 years). The results are shown in Fig. 1.7 (left). In a similar question respondents were asked about the tolerable probability of economic failure (i.e. the building does not collapse physically, but repair is not economically justified, corresponding to the NC limit state) for buildings built according to the current seismic regulations (Fig. 1.7, right)). A significantly higher tolerable probability than in the case of building collapse was expected. However, surprisingly, there was only a small difference between the tolerable probability of collapse and the tolerable probability of economic failure. For both groups of respondents, a large scatter of results can be observed. The mean values of the tolerable probabilities of collapse and economic failure in a working life of 50 years are presented in Table 1.3. More details of the survey are available in (Fajfar et al. 2014).

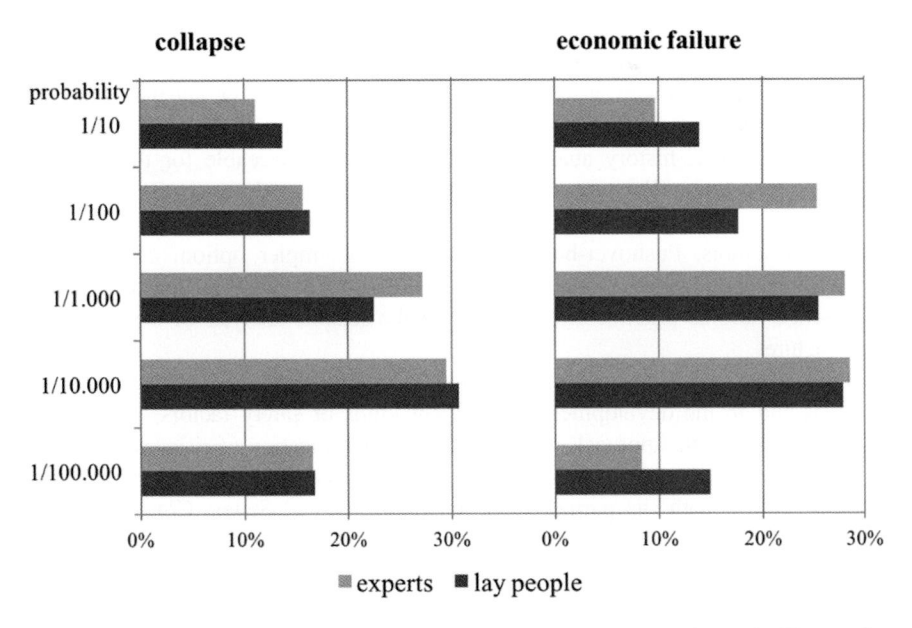

Fig. 1.7 Distribution of tolerable probabilities of collapse and economic failure in 50 years for ordinary buildings, built according to the current seismic regulations

Table 1.3 Mean values of the tolerable probabilities of collapse and economic failure in 50 years for ordinary buildings, built according to the current seismic regulations

	Prob. of collapse	Prob. of econ. failure
Experts	$1/1780 = 5.62 \cdot 10^{-4}$	$1/1000 = 10.0 \cdot 10^{-4}$
Lay people	$1/1740 = 5.75 \cdot 10^{-4}$	$1/1320 = 7.58 \cdot 10^{-4}$

The results shown in Table 1.3 suggest that both experts and lay people expect, on average, a lower probability of failure than that which has been estimated for buildings complying with the current seismic regulations.

1.5 Conclusions: Analysis Procedures Available for All Needs

The seismic response of structures subjected to strong ground motion is dynamic, inelastic, and random. Historically, the three major response characteristics of seismic structural response have only gradually penetrated into the analysis of structures, which is one of the most important parts of seismic provisions. Such analysis provides estimates of seismic demands, which are then compared with corresponding capacities.

After static analysis, which was used in early codes, linear dynamic methods were developed and are currently well established. Explicit nonlinear analysis, taking into account inelastic structural behaviour, is a relatively new achievement. In recent seismic codes, explicit nonlinear analysis has been typically adopted as an optional procedure, which, however, is in some codes mandatory for special structural systems. The most advanced deterministic analysis method available today is nonlinear response history analysis (NRHA). It is irreplaceable for the research and for the design and/or assessment of very important buildings. However, due to its complexity, it has, in practice, only rarely been used, with the exception of some special structures. Pushover-based methods are a simpler option, combining a nonlinear static (pushover) analysis with the response spectrum approach. They represent a rational practice-oriented tool for the seismic analysis of many but not all structures.

Probabilistic considerations have been taken into account in seismic hazard analysis and in the development of different kinds of safety factors, whereas an explicit probabilistic approach, which allows a quantification of risk, as needed in performance-based earthquake engineering, has not yet been implemented in seismic building codes, with the partial exception of the United States. Considering the reluctance of the great majority of engineers and also a large part of researchers to accept probabilistic methods, it is extremely unlikely that these methods will be adopted in seismic codes in the foreseeable future as a mandatory option. They are indispensable, however, for a better calibration of different safety factors and force reduction factors in current codes.

A rigorous seismic analysis that could take into account all three major response characteristics (dynamics, inelasticity and randomness) is too demanding for most practical applications. As well as this, the large uncertainties which are related to both ground motion and the modelling of structures do not permit accurate predictions of the structural response during future earthquakes. Simplified approaches have been developed in different historical periods, depending on the knowledge and computational tools available at the time. When a new procedure was implemented, the existing analysis procedures mostly remained in the codes. For example, dynamic analysis has not replaced static analysis, and nonlinear analysis has not replaced linear analysis; the new procedures just complemented the old ones. Currently, several analysis procedures with different complexity levels are available for different needs. They have been developed to a stage in which, if properly applied, permit, together with appropriate conceptual design, dimensioning and detailing, the construction of building structures which, with a very high probability, would not collapse due to earthquakes. This has been often demonstrated in recent earthquakes. Unfortunately, the limitation of structural and non-structural damage is still an issue. Further improvements are needed, inter alia, in the reliability of input data and structural modelling. In practice, the competence of designers, as well as the financial and time restrictions imposed on the design projects, also have to be taken into account.

Acknowledgements The author is greatly indebted to the Awards Committee of the European Association for Earthquake Engineering, who granted him the Prof. Nicholas Ambraseys Distinguished Lecture Award. Much appreciation is expressed to the author's former PhD students who were involved in the development of the N2 method. Among them, Maja Kreslin, Damjan Marušić and Iztok Peruš greatly contributed to its extension. An important contribution to the development of the PRA approach was made by Mirko Kosič. He also helped with the technical editing of this paper. Great thanks go to the author's colleagues who kindly responded to a survey about the use of analysis methods in their own countries: Alireza Azarbakht, Mehmet Nuray Aydinoğlu, Ruben Boroschek, Rajesh Dhakal, Mohsen Ghafory-Ashtiany, Ron O. Hamburger, Jag Humar, Sudhir K. Jain, Masahiro Kurata, Koichi Kusunoki, Gregory MacRea, Roberto Meli, Eduardo Miranda, Khalid Mosalam, Taichiro Okazaki, Abdolreza Sarvghad-Moghadam, Haluk Sucuoğlu, and Keh-Chyuan Tsai. Very useful was the interaction with Edmund Booth and Michael N. Fardis. The perceptive comments of two reviewers, Matej Fischinger and Miha Tomaževič, are also much appreciated. The research work of the author and his colleagues has been continuously supported by the Slovenian Research Agency and its predecessors. This support is hereby gratefully acknowledged.

References

Army (1986) Seismic design guidelines for essential buildings. Departments of the Army (TM 5-809-10-1), Navy (NAVFAC P355.1), and the Air Force (AFM 88-3, Chapter 13, Section A), Washington, DC

ASCE (2014) Seismic evaluation and retrofit of existing buildings, ASCE/SEI 41-13. American Society of Civil Engineers, Reston

ASCE (2017) Minimum design loads and associated criteria for buildings and other structures, ASCE/SEI 7-16. American Society of Civil Engineers, Reston

ATC (1978) Tentative provisions for the development of seismic regulations for buildings, ATC 3-06. National Bureau of Standards, Washington, DC

ATC (1995) Structural response modification factors, ATC-19. Applied Technology Council, Redwood City

ATC (1996) Seismic evaluation and retrofit of concrete buildings (report SSC 96-01 of California seismic safety commission), Report ATC-40. Applied Technology Council, Redwood City

Aydinoğlu MN, Önem G (2010) Evaluation of analysis procedures for seismic assessment and retrofit design. In: Garevski M, Ansal A (eds) Earthquake engineering in Europe, geotechnical, geological and earthquake engineering, vol 17. Springer, Dordrecht, pp 171–198

Baros D, Anagnostopoulos SA (2008) An assessment of static non-linear pushover analyses in 2-D and 3-D applications. In: Bento R, Pinho R (eds) 3Dpushover 2008 – nonlinear static methods for design/assessment of 3D structures. IST Press, Lisbon

Berg GV (1983) Seismic design codes and procedures. Earthquake Engineering Research Institute, Berkeley

Bertero VV (2009) Vitelmo V. Bertero. In Connections: EERI oral history series, Reitherman R (interviewer). Earthquake Engineering Research Institute, Oakland

Blume JA, Newmark NM, Corning LH (1961) Design of multistory reinforced concrete buildings for earthquake motions. Portland Cement Association, Chicago

Cardoso R, Lopes M, Bento R (2004) Earthquake resistant structures of Portuguese old 'Pombalino' buildings. In: Proceedings of the 13th world conference on earthquake engineering, Paper no. 918, Vancouver

CEN (2002) European standard EN 1990:2002. Eurocode: Basis of structural design. European Committee for Standardization, Brussels

CEN (2004) European standard EN 1998–1:2004, Eurocode 8: design of structures for earthquake resistance, Part 1: general rules, seismic action and rules for buildings. European Committee for Standardization, Brussels

CEN (2017) Draft Annex F (Informative): simplified reliability-based verification format, Annex to European standard EN 1998–1. Eurocode 8: Design of structures for earthquake resistance. Part 1: General rules, seismic action and rules for buildings. European Committee for Standardization, Brussels (Prepared by Dolšek M, Žižmond J, Kosič M, Lazar Sinković N)

Chopra AK (2007) Elastic response spectrum: a historical note. Earthq Eng Struct Dyn 36:3–12

Chopra AK (2017) Dynamics of structures: theory and applications to earthquake engineering, 5th edn. Pearson, Hoboken

Chopra AK, Goel RK (2002) A modal pushover analysis procedure for estimating seismic demands for buildings. Earthq Eng Struct Dyn 31(3):561–582

Chopra AK, Goel RK, Chintanapakdee C (2004) Evaluation of a modified MPA procedure assuming higher modes as elastic to estimate seismic demands. Earthquake Spectra 20(3):757–778

Clough R (1980) The finite element method after twenty-five years: a personal view. Comput Struct 12(4):361–370

CNR (2014) Guide for the probabilistic assessment of the seismic safety of existing buildings, Report CNR-DT 212/2013. CNR – Advisory Committee on Technical Recommendations for Construction, Rome

Cornell CA (1996a) Calculating building seismic performance reliability: a basis for multi-level design norms. In: Proceedings of the 11th world conference on earthquake engineering, Mexico City, Mexico, Paper no. 2122

Cornell CA (1996b) Reliability-based earthquake-resistant design: the future. In: Proceedings of the 11th world conference on earthquake engineering, Mexico City, Mexico, Paper no. 2166

Cornell CA, Jalayar F, Hamburger RO, Foutch DA (2002) Probabilistic basis for 2000 SAC Federal Emergency Management Agency steel moment frame guidelines. J Struct Eng ASCE 128(4):526–533

Deierlein GG, Reinhorn AM, Wilford MR (2010) Nonlinear structural analysis for seismic design, a guide for practicing engineers, NEHRP seismic design technical brief no. 4, NIST GRC 10-917-5. National Institute of Standards and Technology, Gaithersburg

Diebold J, Moore K, Hale T, Mochizuki G (2008) SEAOC Blue Book: seismic design recommendations 1959 to 2008. In: Proceedings of the 14th world conference on earthquake engineering, Beijing, China

Dolšek M, Fajfar P (2007) Simplified probabilistic seismic performance assessment of plan-asymmetric buildings. Earthq Eng Struct Dyn 36(13):2021–2041

Dolšek M, Lazar Sinković N, Žižmond J (2017a) IM-based and EDP-based decision models for the verification of the seismic collapse safety of buildings. Earthq Eng Struct Dyn 36 (15):2665–2682

Dolšek M, Žižmond J, Kosič M, Lazar Sinković N (2017b) Simplified reliability-based verification format, working material for annex F to revised EN 1998-1. University of Ljubljana, Ljubljana

Fajfar P (1999) Capacity spectrum method based on inelastic demand spectra. Earthq Eng Struct Dyn 28:979–993

Fajfar P (2000) A nonlinear analysis method for performance-based seismic design. Earthquake Spectra 16(3):573–592

Fajfar P, Dolšek M (2012) A practice-oriented estimation of the failure probability of building structures. Earthq Eng Struct Dyn 41(3):531–547

Fajfar P, Fischinger M (1987) Non-linear seismic analysis of RC buildings: implications of a case study. Eur Earthq Eng 1:31–43

Fajfar P, Fischinger M (1989) N2 – a method for non-linear seismic analysis of regular buildings. In: Proceedings of the 9th world conference on earthquake engineering, Tokyo, Kyoto, 1988, Maruzen, Tokyo, vol V:111–116

Fajfar P, Marušić D, Peruš I (2005) Torsional effects in the pushover-based seismic analysis of buildings. J Earthq Eng 9(6):831–854

Fajfar P, Polič M, Klinc R (2014) Perception of seismic risk by experts and lay people (in Slovenian). Gradbeni vestnik 63:111–118

Fardis MN, Carvalho EC, Fajfar P, Pecker A (2015) Seismic design of concrete buildings to Eurocode 8. CRC Press, Boca Raton

FEMA (1997) NEHRP guidelines for the seismic rehabilitation of buildings, FEMA 273. Federal Emergency Management Agency, Washington, DC

FEMA (2009) Quantification of building seismic performance factors, FEMA P-695. Federal Emergency Management Agency, Washington, DC

FEMA (2012) Next-generation methodology for seismic performance assessment of buildings, prepared by ATC for FEMA, FEMA P-58. Federal Emergency Management Agency, Washington, DC

fib (2010) Model code, International Federation for Structural Concrete, Lausanne

fib (2012) Probabilistic performance-based seismic design, Technical report prepared by Task Group 7.7. International Federation for Structural Concrete, Lausanne

Fischinger M, Isaković T, Fajfar P (2015) Yugoslav seismic code JUS 31/81 and Eurocode 8 – a tribute to the IZIIS contribution towards development of the modern EU seismic standards. In: Proceedings of the international conference on earthquake engineering and seismology IZIIS50. Institute of Earthquake Engineering and Engineering Seismology, Skopje

Freeman JR (1930) Engineering data needed on earthquake motion for use in the design of earthquake-resisting structures. In: Proceedings of the 1930 meeting of the Eastern section of the seismological Society of America. U.S. Bureau of Standards, Washington, DC

Freeman JR (1932) Earthquake damage and earthquake insurance. McGraw-Hill, New York

Freeman SA, Nicoletti JP, Tyrell JV (1975) Evaluations of existing buildings for seismic risk – a case study of Puget sound naval shipyard, Bremerton, Washington. In: Proceedings of the U.S. national conference on earthquake engineering, Berkeley, pp 113–122

Haselton CB, Baker JW, Stewart JP, Whittaker AS, Luco N, Fry A, Hamburger RO, Zimmerman RB, Hooper JD, Charney FA, Pekelnicky RG (2017) Response history analysis for the design of new buildings in the NEHRP provisions and ASCE/SEI 7 standard: Part I – overview and specification of ground motions. Earthquake Spectra 33(2):373–395

Housner GW (1984) Historical view of earthquake engineering. In: Proceedings of the 8th world conference on earthquake engineering, post-conference volume: 25–39. San Francisco, CA

JCSS (2000). Probabilistic model code – Part 1: basis of design. Joint Committee on Structural Safety

Jiang H, Lu X, Zhu J (2012) Performance-based seismic analysis and design of code-exceeding tall buildings in Mainland China. Struct Eng Mech 43(4):545–560

JUS (1981) Code of technical regulations for the design and construction of buildings in seismic regions. Official Gazette of SFR Yugoslavia, 31/81 (English translation in Earthquake resistant regulations, a world list – 1992, IAEE)

Kircher CA, Harris JL, Heintz JA, Hortacsu A (2012) ATC-84 project: improved seismic performance factors for design of new buildings. In: Proceedings of the 10th U.S. national conference on earthquake engineering. Anchorage, AK

Korčinski IL (1964) Fundamentals of design of buildings in seismic (in Serbian). Gradjevinska knjiga, Beograd (translation, original in Russian 1961)

Kosič M, Fajfar P, Dolšek M (2014) Approximate seismic risk assessment of building structures with explicit consideration of uncertainties. Earthq Eng Struct Dyn 43(10):1483–1502

Kosič M, Dolšek M, Fajfar P (2016) Dispersions for pushover-based risk assessment of reinforced concrete frames and cantilever walls. Earthq Eng Struct Dyn 45(13):2163–2183

Kosič M, Dolšek M, Fajfar P (2017) Pushover-based risk assessment method: a practical tool for risk assessment of building structures. In: Proceedings of the 16th world conference on earthquake engineering. Santiago, Chile

Krause G (2014) Erdbebensicheres Bauen im Spanischen Weltreich, DGG- Mitteilungen 2/2014: 14–15, Deutsche Geophysikalische Gesellschaft e.V. ISSN 0934 – 6554 DGG

Krawinkler H (2006) Importance of good nonlinear analysis. Struct Des Tall Special Build 15 (5):515–531

Krawinkler H, Seneviratna GDPK (1998) Pros and cons of a pushover analysis for seismic performance evaluation. Eng Struct 20(4–6):452–464

Kreslin M, Fajfar P (2011) The extended N2 method taking into account higher mode effects in elevation. Earthq Eng Struct Dyn 40(14):1571–1589

Kreslin M, Fajfar P (2012) The extended N2 method considering higher mode effects in both plan and elevation. Bull Earthq Eng 10(2):695–715

Mahaney JA, Paret TF, Kehoe BE, Freeman SA (1993) The capacity spectrum method for evaluating structural response during the Loma Prieta Earthquake. In: Earthquake hazard reduction in the Central and Eastern United States: a time for examination and action, Proceedings of 1993 national earthquake conference. Earthquake Engineering Research Institute, Oakland, pp 501–510

NZSEE (2017) The seismic assessment of existing buildings, technical guidelines for engineering assessments, Part A: assessment objectives and principles. Ministry of Business, Innovation and Employment. Earthquake Commission, New Zealand Society for Earthquake Engineering, Structural Engineering Society and New Zealand Geotechnical Society

Otani S (2004) Earthquake resistant design of reinforced concrete buildings: past and future. J Adv Concr Technol 2(1):3–24

PEER (2017) Guidelines for performance-based seismic design of tall buildings. PEER report 2017/06. Pacific Earthquake Engineering Research Center, Univrsity of California, Berkeley

Regione Autonoma Friuli-Venezia Giulia (1977) Raccomandazioni per la riparazione strutturale degli edifici in muratura. DT 2, Udine, Italy

Reitherman RK (2012) Earthquakes and engineers, An international history. ASCE, American Society of Civil Engineers, Reston

Sorrentino L (2007) The early entrance of dynamics in earthquake engineering: Arturo Danusso's contribution. ISET J Earthq Technol 44(1):1–24

Terzic V, Schoettler MJ, Restrepo JI, Mahin SA (2015) Concrete column blind prediction contest 2010: outcomes and observations. PEER Report 2015/01. Pacific Earthquake Engineering Research Center, Berkeley

Tomaževič M (1978) Improvement of computer program POR (in Slovenian). Report ZRMK-IK. Institute for Testing and Research in Materials and Structures, Ljubljana

Trifunac MD (2008) Early history of the response spectrum method, historical review. Soil Dyn Earthq Eng 28:676–685

Whittaker A, Moehle J, Higashino M (1998) Evolution of seismic building design practice in Japan. Struct Des Tall Build 7:93–111

Žižmond J, Dolšek M (2014) Risk-targeted seismic intensities and the behaviour factor for the force-based design of structures. Unpublished manuscript. University of Ljubljana, Faculty of Civil and Geodetic Engineering, Ljubljana

Žižmond J, Dolšek M (2017) The formulation of risk-targeted behaviour factor and its application to reinforced concrete buildings. In: Proceedings of the 16th world conference on earthquake engineering, Santiago, Chile, Paper N° 1659

Open Access This chapter is licensed under the terms of the Creative Commons Attribution 4.0 International License (http://creativecommons.org/licenses/by/4.0/), which permits use, sharing, adaptation, distribution and reproduction in any medium or format, as long as you give appropriate credit to the original author(s) and the source, provide a link to the Creative Commons license and indicate if changes were made.

The images or other third party material in this chapter are included in the chapter's Creative Commons license, unless indicated otherwise in a credit line to the material. If material is not included in the chapter's Creative Commons license and your intended use is not permitted by statutory regulation or exceeds the permitted use, you will need to obtain permission directly from the copyright holder.

Chapter 2
Implications of Site Specific Response Analysis

Atilla Ansal, Gökçe Tönük, and Aslı Kurtuluş

Abstract Definition of design earthquake characteristics, more specifically uniform hazard acceleration response spectrum, on the ground surface is the primary component for performance based design of structures and assessment of seismic vulnerabilities in urban environments. The adopted approach for this purpose requires a probabilistic local seismic hazard assessment, definition of representative site profiles down to the engineering bedrock, and 1D or 2D equivalent or nonlinear, total or effective stress site response analyses depending on the complexity and importance of the structures to be built. Thus, a site-specific response analysis starts with the probabilistic estimation of regional seismicity and earthquake source characteristics, soil stratification, engineering properties of encountered soil layers in the soil profile. The local seismic hazard analysis would yield probabilistic uniform hazard acceleration response spectrum on the bedrock outcrop. Thus, site specific response analyses also need to produce a probabilistic uniform hazard acceleration response spectrum on the ground surface. A general review will be presented based on the previous studies conducted by the author and his co-workers in comparison to major observations and methodologies to demonstrate the implications of site-specific response analysis.

2.1 Introduction

The major uncertainties in site-specific response analysis arises from the variability of (a) local seismic hazard assessment, (b) selection and scaling of the hazard compatible input earthquake time histories, (c) soil stratification and corresponding engineering properties of encountered soil and rock layers, and (d) method of site response analysis.

A. Ansal (✉) · A. Kurtuluş
Özyeğin University, Istanbul, Turkey
e-mail: atilla.ansal@ozyegin.edu.tr

G. Tönük
MEF University, Istanbul, Turkey

The uncertainties related to local seismic hazard assessment (Erdik 2017), even though it has primary importance on the outcome of the site-specific response analyses, will not be considered here and the discussion in this article will start with the second source of uncertainties related to selection and scaling of the hazard compatible input earthquake time histories.

In general, procedures for the selection of design ground motions are specified in codes. For example, ASCE 7-05 (2016) requires at least five recorded or simulated rock outcrop horizontal ground motion acceleration time histories are selected from events having magnitudes and fault distances that are compatible with those that control the MCE (Maximum Considered Earthquake).

One option may be to select, as much as possible, large number of acceleration records compatible with the local earthquake hazard in terms of fault mechanism, magnitude and distance range recorded on stiff site conditions to account for the variability in earthquake source and path effects. In addition, these records would require scaling with respect to uniform hazard acceleration response spectrum estimated on the bedrock outcrop.

Likewise, the observed variability and level of uncertainty in site conditions and engineering properties of soil layers was observed to play in important role in modelling site response (Thompson et al. 2009; Li and Assimaki 2010). Thus, similar to the selection of large number hazard compatible acceleration records, one option may be conducting site response analyses for large number of soil profiles for the investigated site to assess design acceleration spectra with respect to different performance levels. At this stage, one may also consider using Monte Carlo Simulations to increase the number of soil profiles (Ansal et al. 2015a). The goal of a site-specific response analysis is to develop a uniform hazard acceleration response spectrum on the ground surface.

For standard engineering projects, the seismic design criteria may be taken from the appropriate earthquake design codes. However, for important engineering projects where the consequences of failure are more serious, and the cost would be very high, it is preferable to determine the earthquake design criteria from a site specific probabilistic seismic hazard analysis.

The first option for specifying site specific acceleration spectrum on the ground surface is to use contemporary ground motion prediction equations (Abrahamson et al. 2014; Boore et al. 2014; Campbell and Bozorgnia 2014). These formulations would yield only generic assessment of the earthquake characteristics on the ground surface. Baturay and Stewart (2003) observed that relative effectiveness of the site response analyses is illustrated by the reduced standard deviation of residuals for sites at short periods and predictions relative to generic site terms improved significantly for soft sites having large impedance contrasts within the profiles.

The second option is to use empirical amplification factors such as those suggested by Borcherdt (1994) and NEHRP (2015) based on equivalent average shear wave velocity. However, it was observed based on recorded ground motion data (Ansal et al. 2014, 2015a) and based on parametric studies (Baturay and Stewart 2003; Haase et al. 2011) that use of average shear wave velocity (V_{s30}) may not always yield spectral accelerations on the conservative side on the ground surface. In addition, these procedures are deterministic procedures that would lead to surface ground motion levels with unknown exceedance probabilities, that may be non-uniform, non-conservative, and inconsistent across frequency range (Bazzurro and Cornell 2004a, b). This approach may also be considered as a hybrid approach where the mean amplifications are used.

Rathje and Ozbey (2006) developed a probabilistic site response analysis model based on random vibration theory. The proposed approach is an attractive alternative to other site response methods, since it does not require input acceleration records. However, the results obtained were systematically different in the order of 20–50% larger than time-series analysis. The authors confirm that site response analysis based on random vibration theory yields the largest overprediction occurring for sites with smaller natural frequencies and sites underlain by hard rock. The overprediction is caused by an increase in duration generated by the site response (Kottke and Rathje 2013). Correcting for this change in duration brings the random vibration theory site response results within 20% of the results obtained by site response analysis based on acceleration records.

There has been significant amount of work done related to the sources of variability and bias in site response analysis. Kaklamanos et al. (2013) conducted a detailed study based on the data obtained in the Kiban-Kyoshin network (KIK-net) to determine the critical parameters that contribute to the uncertainty in site response analysis. They observed that 1D equivalent-linear site-response method generally yields underprediction of ground motions, except in the range of 0.5–2 s, where the bias is slightly negative. Relative to empirical site amplification factors, site specific ground response analyses offer a reduction in the total standard deviation at short spectral periods. In addition, Kaklamanos et al. (2015) in the comparison of 1D linear, equivalent-linear, and nonlinear site response models observed that strains greater than approximately 0.05% nonlinear models exhibit slight improvements over equivalent-linear models.

Site-specific response analyses are deterministic computations of site response given certain input parameters. The results of these calculations need to be merged with the probabilistically derived ground motion hazard for rock outcrop site conditions. Bazzurro and Cornell (2004a) recommended a convolution method for combining a nonlinear site amplification function with a rock hazard curve to estimate a soil hazard curve. The principal advantage of this approach relative to the hybrid approach is that uncertainties in the site amplification function are directly incorporated into the analysis. Another more simplified approach may be to evaluate site response analyses output adopting a probabilistic approach as suggested by Ansal et al. (2011, 2015b).

The principle option is to develop completely probabilistic site specific seismic hazard analysis. Site-specific probabilistic ground-motion estimates should be based on the full site-amplification distribution instead of a single deterministic median value. A probabilistic methodology using site-amplification distributions to modify rock ground-motion attenuation relations into site specific relations prior to calculating seismic hazard need to be considered (Cramer 2003). The use of a completely probabilistic approach can make about a 10% difference in ground motion estimates over simply multiplying a bedrock probabilistic ground motion by a median site-amplification factor even larger differences at smaller probabilities of exceedance. The value of this approach is that a probabilistic answer incorporating the uncertainty in our knowledge of site amplification of ground motions can be calculated.

The purpose is to improve the accuracy of predicted ground motions relative to the use of median predictions from empirical ground motion prediction equations (GMPEs) with their site terms (Stewart et al. 2014). However, site response is considerably more complex including surface waves, basin effects (including focusing and basin edge-generated surface waves), and topographic effects. Thus, 1D site-specific response analyses may not always be effective for accurately modelling and/or predicting site effects, however, in comparison to alternatives of GMPEs and empirical amplification factors, site response analysis would model the probable surface spectrum more adequately. If needed, 2D analysis may also be performed (Ansal et al. 2017a) to account for 2D effects.

2.2 Selection of Input Acceleration Records

Bommer and Acevedo (2004) and Bommer et al. (2000) recommended to use recorded acceleration time histories in dynamic response analysis. It is also preferable to use recorded strong ground motion records for site response analysis as observed by Ansal and Tönük (2007) since the use of simulated acceleration records, in general may yield overconservative results.

The selection and scaling of input acceleration time histories is an important stage in site response analysis (Ansal et al. 2015a, b; Tönük and Ansal 2010; Tönük et al. 2014; Bradley 2010, 2012; Haase et al. 2011; Kottke and Rathje 2011; Malhotra 2003; Rathje et al., 2010). One of the primary concern is the selection of acceleration time histories recorded under similar tectonic conditions within the range of the expected magnitude, fault distance, and fault type recorded on stiff site conditions with shear wave velocity of $v_s \geq 750$ m/s. In addition, recorded PGA and spectral accelerations should be in the similar range in comparison to those estimated based on uniform hazard acceleration response spectrum on the rock outcrop.

One important issue is related to number of acceleration records that need to be used for site response analysis. It was observed that if the number of selected acceleration records are in the range of 22–23, the calculated mean response spectrum is consistent with only minor changes with additional input records as shown in Fig. 2.1. In the case of using limited number of acceleration time histories for example 5 as suggested by ASCE 7-05 (2016) the mean spectra and standard deviation of all input records may be significantly different.

In the case of site response analysis, different number of input acceleration records would yield different mean acceleration spectra on the ground surface as shown in Fig. 2.2. The comparison of calculated mean acceleration spectra for 21 and 23 input acceleration records are very similar indicating that for this case study 21–23 acceleration inputs would be sufficient to calculate the mean spectral

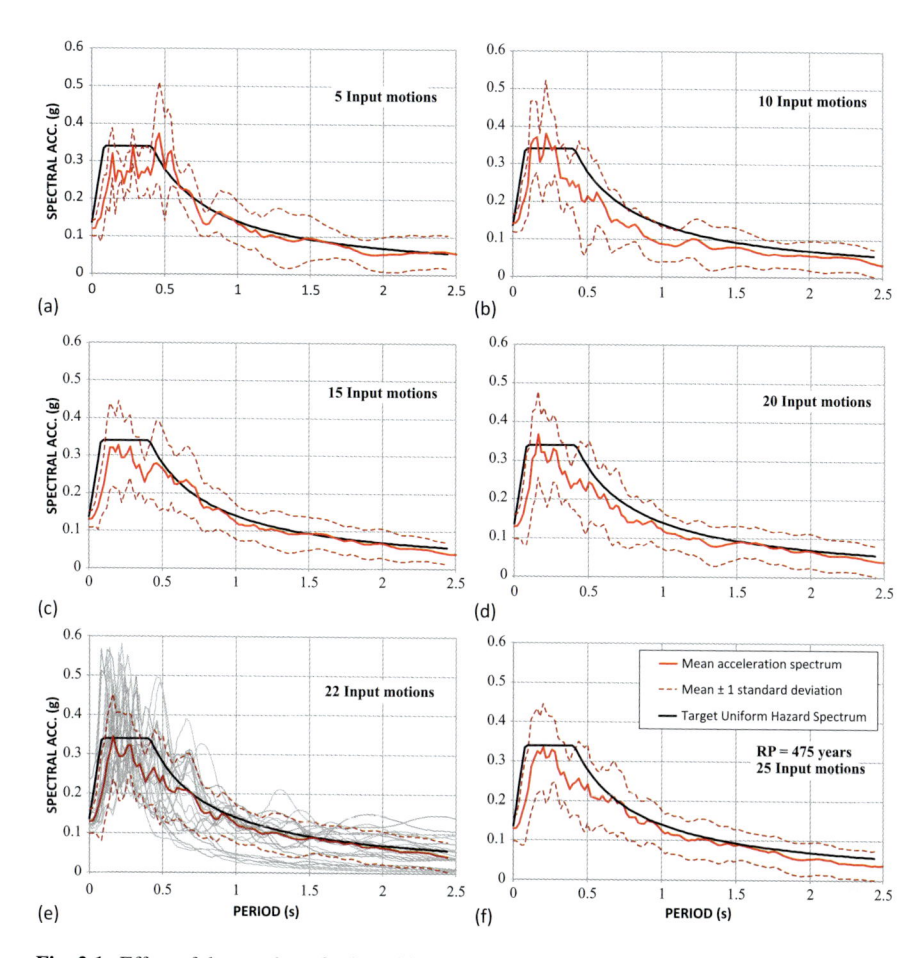

Fig. 2.1 Effect of the number of selected input acceleration time histories

acceleration response spectrum on the ground surface. The other issue is depending on the properties of the selected limited number of acceleration time histories, the site response analyses would yield different results.

2.3 Scaling of Input Acceleration Records

The adopted scaling procedure have three goals, (a) to obtain the best fits with respect to rock outcrop target uniform hazard acceleration spectrum, (b) to match the target acceleration spectrum within the considered period range (c) to decrease the scatter in the acceleration spectra after scaling. Using 1D equivalent linear site response model at sites with pre-determined hazard, the response variability is investigated for selected 22 acceleration records compatible with the site specific

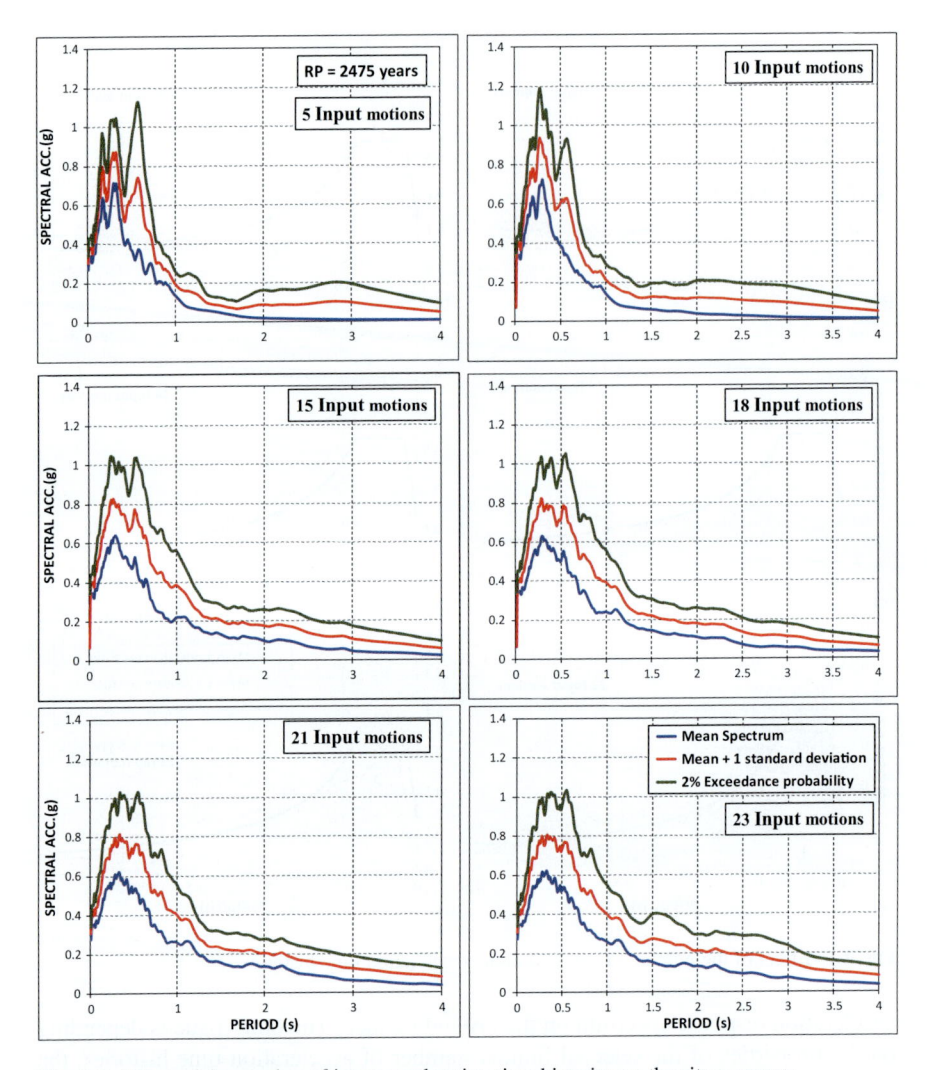

Fig. 2.2 Effect of the number of input acceleration time histories on the site response

earthquake hazard in terms of expected fault mechanism, magnitude, and distance. The selected records are scaled to different intensity measures, namely; peak ground acceleration (PGA), peak ground velocity (PGV), and Arias intensity (AI), the analyses are repeated, and the variability introduced by scaling is evaluated (Durukal et al. 2006). Scaling of input time histories was carried out in time-domain that involves changing only the amplitude of the time series. The scaled records were applied as outcrop motion on the engineering bedrock.

The distribution of the calculated PGAs on the ground surface for all selected borings and the PGA corresponding to 10% exceedance level based on distribution function for the three scaling options are shown in Fig. 2.3. Considering all three

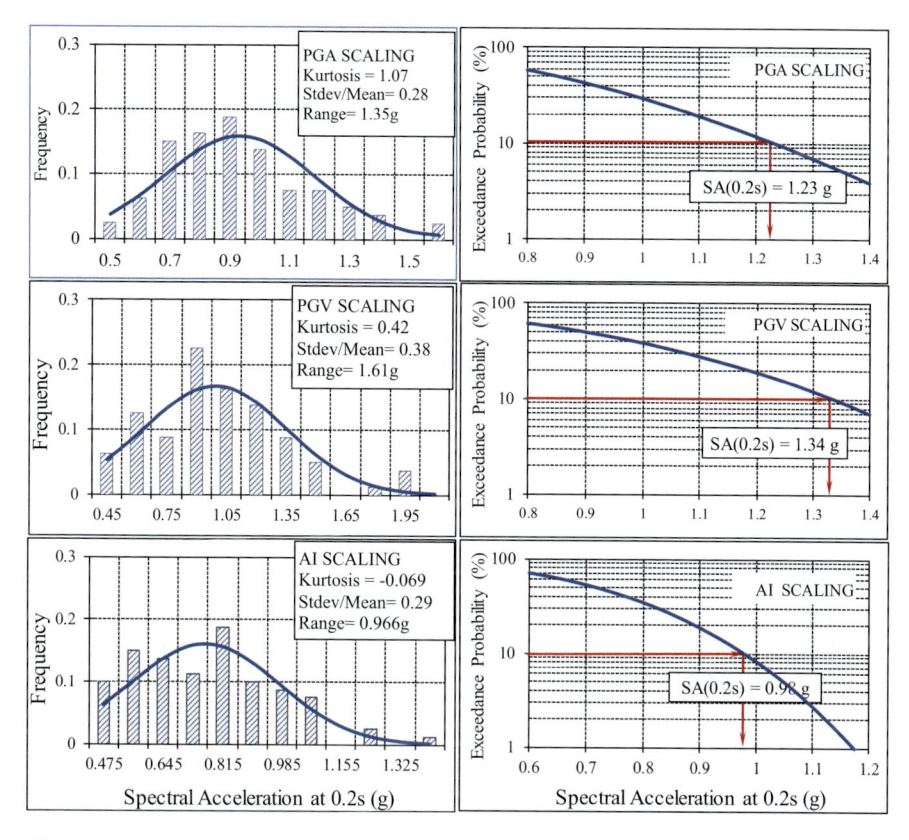

Fig. 2.3 Histograms of Sa(0.2 s) on the ground surface for PGA, PGV, and AI scaled records (Ansal et al. 2006)

parameters calculated to determine the range of variability in each set (kurtosis and normalized standard deviation being minimum and range being the smallest), the PGA scaling is observed to be the most suitable scaling parameter in terms of calculated peak ground accelerations on the ground surface (Ansal et al. 2006).

At the second stage, the distance compatibility criteria are evaluated by conducting site response analysis using different sets of earthquake time histories recorded at different fault distances scaled in a similar manner (Ansal et al. 2006). As observed in Fig. 2.4, in the case PGA scaling, fault distance becomes an important factor while in the case of Arias Intensity scaling distance appears as an insignificant factor.

Assuming the suitability of PGA as a primary scaling parameter, a parametric study was conducted to evaluate the effects of input motion scaling with respect to target outcrop uniform hazard acceleration spectrum based on PGA scaling. The most widely adopted is PGA scaling of the selected hazard compatible set of SM acceleration records (Ansal et al. 2006). The acceleration spectra of these PGA scaled acceleration records with respect to uniform outcrop hazard acceleration spectra for return periods of 2475 and 475 years are shown in Fig. 2.5a.

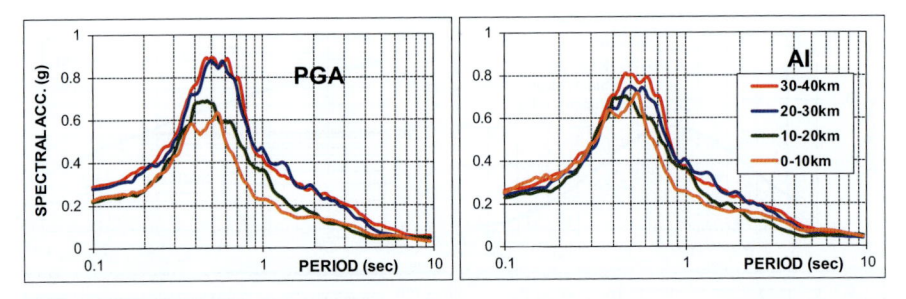

Fig. 2.4 Calculated elastic acceleration response spectrum for four distance ranges (**a**) 0–10 km; (**b**) 10–20 km; (**c**) 20–30 km (**d**) 30–40 km using PGA and Arias intensity scaled time histories (Ansal et al. 2006)

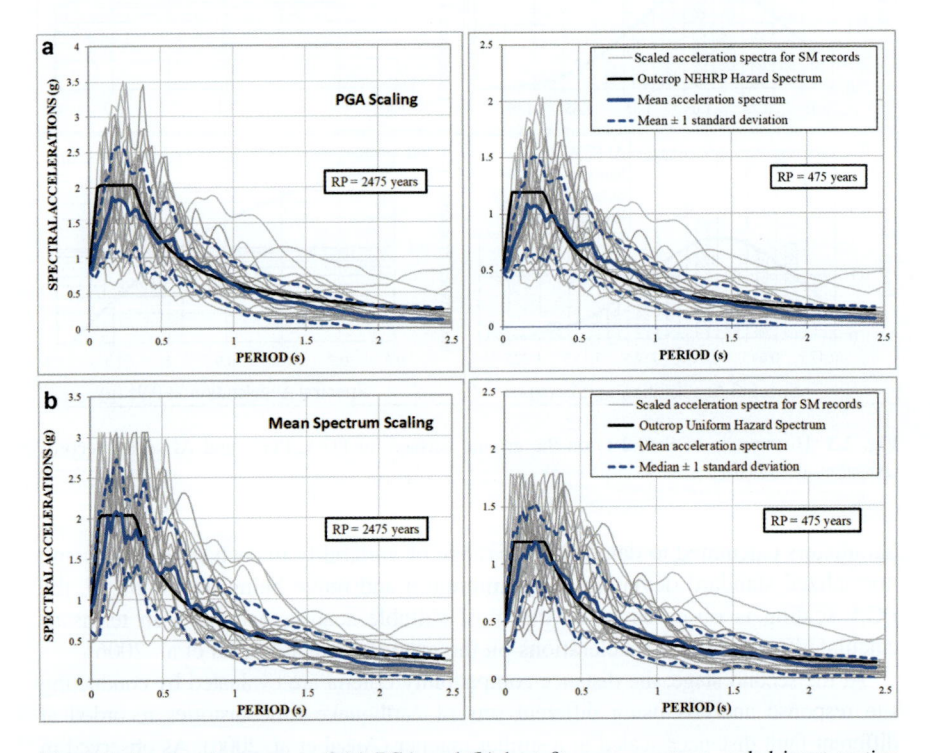

Fig. 2.5 Acceleration spectra of (**a**) PGA and (**b**) best fit mean spectrum scaled input motion records

The second option is to modify the input motion set to have better outcrop uniform hazard spectra compatibility by using a simple optimization scheme by PGA scaling to have the best fit of the mean acceleration spectra with respect to the outcrop uniform hazard spectra. The spectra of all scaled SM records and the mean spectrum with respect to outcrop hazard spectrum are shown in Fig. 2.5b. It was observed that in the case of mean spectrum scaling, a better fit is obtained with

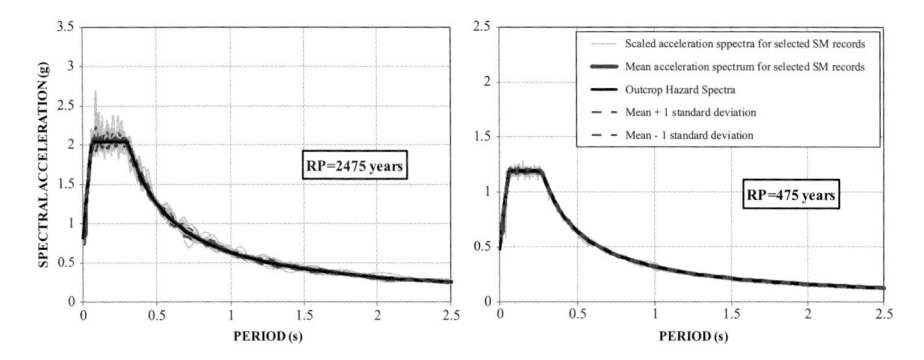

Fig. 2.6 Acceleration spectra of spectra based scaling (Abrahamson 1993)

respect to target uniform hazard acceleration spectrum and the scatter of calculated acceleration spectra for all input records have decreased significantly.

The other possible scaling option is also to modify the frequency content of the selected acceleration records to have better fit to the outcrop uniform hazard spectrum. The spectra of all scaled strong motion records based on Abrahamson (1992, 1993) methodology and the mean spectrum are shown in Fig. 2.6 in comparison to uniform hazard spectrum. This approach gave the best fits with respect to spectra scaling for the outcrop uniform hazard spectra with very limited scatter in the individual acceleration spectra of the 22 scaled strong motion records and with very low standard deviation (Ansal et al. 2012).

2.4 Site Response on the Ground Surface

Site response analyses were carried out for 25 borings conducted as a part of the geotechnical investigation with shear wave velocity profiles determined based on SPT blow counts, seismic surface and in-hole tests. The input strong motion acceleration records are scaled with respect to three scaling options (PGA, mean spectrum, spectral) are used for site response analyses. The acceleration response spectra on the ground surface are shown in Fig. 2.7 corresponding to return periods of 2475 and 475 years.

The mean acceleration response spectra calculated on the ground surface based on three scaling alternatives as shown in Fig. 2.7, indicate slight to significant deamplification with respect to outcrop uniform hazard spectrum. One of the reasons for deamplification in the case of spectra scaling is most likely due to frequency changes applied to acceleration time histories to match the target acceleration spectrum. The spectra compatibility for the selected SM records that was improved by mean spectrum scaling to have better fit with the outcrop uniform hazard spectrum gave slightly higher spectral accelerations with respect to conventional PGA scaling approach.

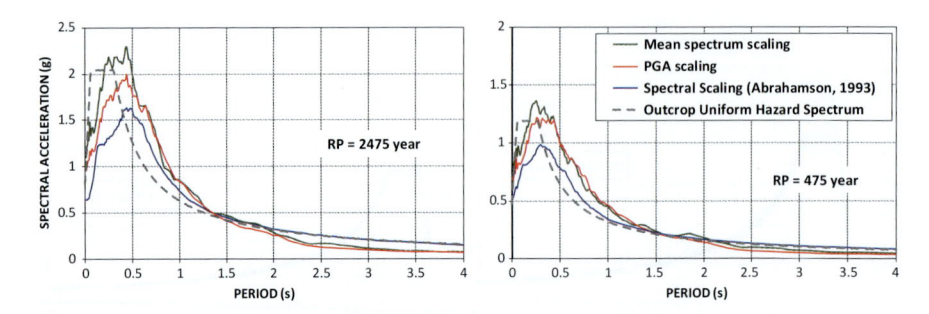

Fig. 2.7 Comparison of mean acceleration spectra calculated on the ground surface using three different approaches for scaling the hazard compatible input SM records

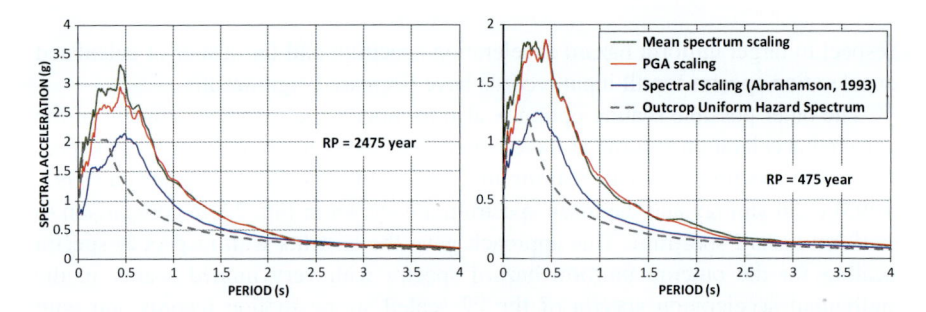

Fig. 2.8 Comparison of mean + 1 standard deviation acceleration spectra on the ground surface using three different scaling procedures for the hazard compatible input SM records

One possible other observation from Fig. 2.7, that it would be unconservative to adopt mean acceleration spectra for design purposes since in all three scaling options, the calculated ground surface mean spectral accelerations are in the range of outcrop uniform hazard spectrum, this would be on the unconservative side. The option recommended by the author was to adopt mean + 1 standard deviation as possible design uniform hazard spectrum on the ground surface for the return periods of 475 and 2475 years as shown in Fig. 2.8 (Ansal and Tonuk 2009).

The geotechnical data comprised of 25 soil profiles indicating possible variability at one site was adopted to study the site-specific design acceleration spectra for corresponding to return periods of 475 and 2475 years. The highest level of spectral accelerations was calculated by mean spectrum scaling approach. The conventional PGA scaling yielded similar or slightly lower spectral accelerations. In the case of spectra scaling (Abrahamson 1993), the spectral accelerations on the ground surface were less than to those corresponding to PGA scaling, interestingly for the case studied; this approach did not lead to spectral amplifications. Considering the additional time for spectral scaling and in the light of these preliminary results, it would not be feasible and conservative to adopt spectral scaling site specific response analyses.

2 Implications of Site Specific Response Analysis

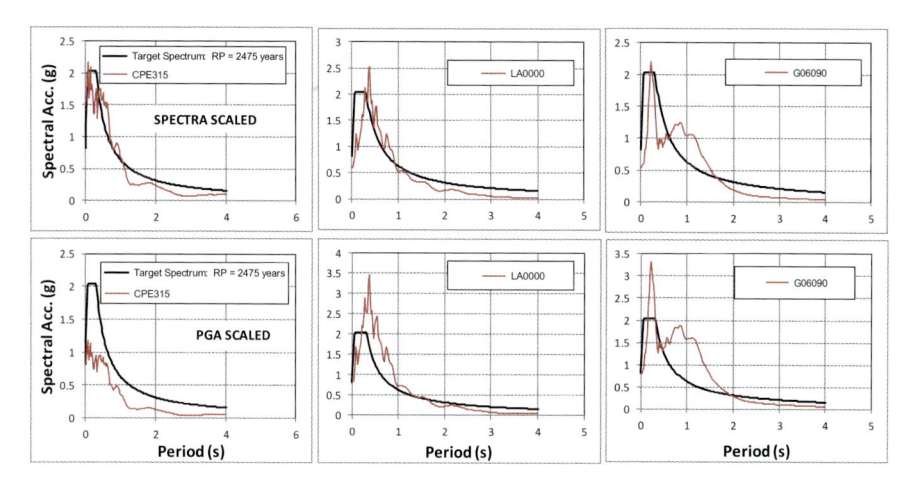

Fig. 2.9 Comparison of some worst case PGA and spectra scaled input acceleration records

Ansal and Tönük (2007) and Ansal et al. (2006) observed that PGA scaling generally would yield conservative results in comparison to other scaling alternatives. Since in most cases structural designs are based on acceleration response spectra rather than PGA, as an alternative option, each input hazard compatible acceleration record is scaled with respect to the target uniform hazard acceleration spectrum (defined as spectra scaling) by minimizing the difference between the calculated and the target spectrum utilizing a simple nonlinear optimization routine by modifying the scaling factor, PGA for each record. The comparison of the three worst cases is shown in Fig. 2.9 where PGA and spectra scaling yield significantly different input acceleration spectrum that demonstrates the main advantage of the spectra scaling. In addition, as can be observed in Fig. 2.10, the scatter of the acceleration spectra for input acceleration records are slightly reduced in the case of spectra scaling approach. This fitting option was also suggested by ASCE 7-05 (2016).

2.5 Site Specific Design Spectrum

The possible differences in soil profiles within relatively short distances and observations in previous earthquakes (Hartzell et al. 1997) have indicated necessity to adopt detailed site response analysis for the estimation of site-specific design earthquake acceleration spectra. Baturay and Stewart (2003) observed that site response analyses are needed for spectral acceleration predictions for especially soft sites.

A parametric study was conducted for a site based on 40 available soil borings where shear wave velocity profiles were modelled by averaging the measured or calculated values for each soil layer. The variations of shear wave velocities with

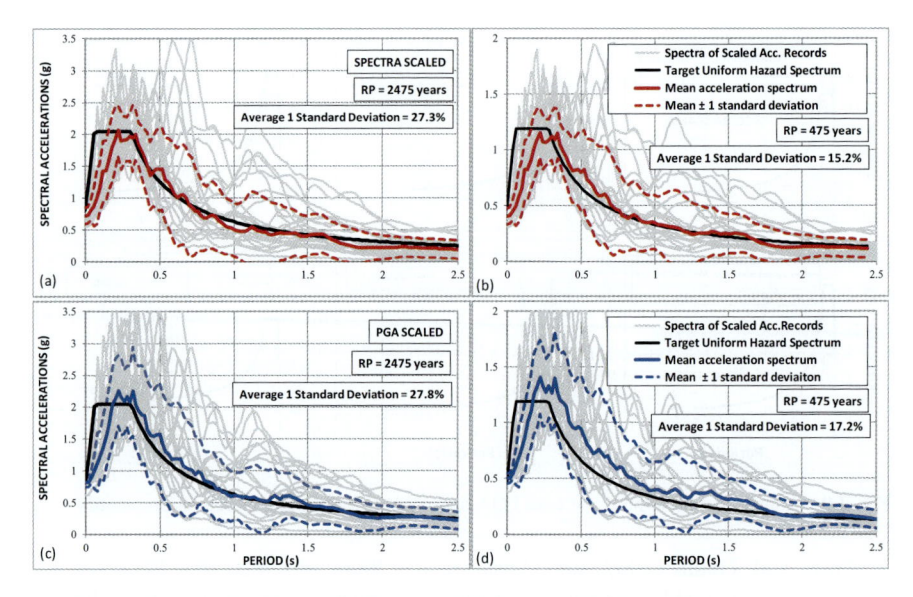

Fig. 2.10 Comparison of PGA and spectra scaled input acceleration records

depth were determined from SPT blow counts using the empirical relationship proposed by Iyisan (1996) and from shear wave velocity measurements with depth by in-hole seismic wave velocity measurements (PS Logging Method).

The modified version of the site response analysis code, Shake91 updated by Idriss and Sun (1992) to account for frequency and stress dependence of dynamic shear moduli and damping similar to what was suggested by Assimaki and Kausel (2002) was used to evaluate the effects of local soil stratification and to calculate acceleration response spectra on the ground surface (Ansal et al. 2009, 2015a, b). Site response analysis were conducted for 40 shear wave velocity profiles using the 22 input acceleration time histories (Peer 2014) yielding 880 site response analyses for exceedance levels corresponding to 475 and 2475 years return period.

One possible approach to determine the uniform hazard acceleration response spectrum on the ground surface is to assume that it is possible to carry out a probability analyses for peak ground and spectral accelerations separately. The effect of input earthquake record, in terms of peak ground acceleration and acceleration response spectrum can be modelled by adopting suitable probability distribution functions (Ansal et al. 2011, 2015a, b). Assuming the statistical distribution of calculated peak ground accelerations (Fig. 2.11) can best be modelled by Beta distribution. PGA values corresponding to return periods of 475 and 2475 years are calculated based on Beta distribution using the identical exceedance probabilities adopted in the seismic hazard study and based on mean + 1 standard deviation. The difference between PGA calculated by the probabilistic approach and as the mean + 1 standard deviation may be considered insignificant for practical engineering applications for the case study presented in Fig. 2.11 (Ansal et al. 2017b).

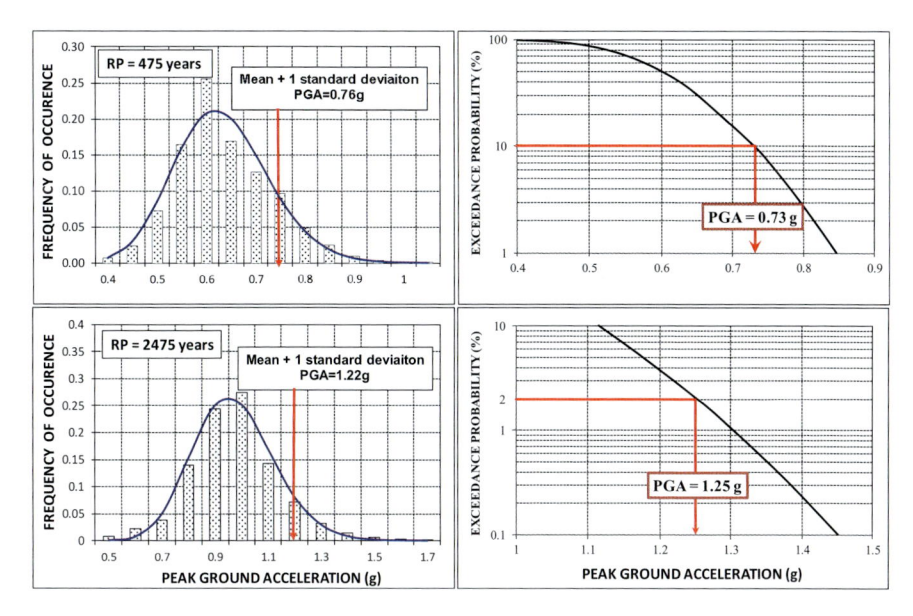

Fig. 2.11 Distribution and probability of PGA for 475 and 2475 years return period

The proposed approach is very similar to developing a probabilistic amplification distribution function considering major parameters affecting site response (soil stratification, shear wave velocity profiles, input acceleration records) at the first stage and in the second stage by multiplying it with the probabilistic rock outcrop distribution to obtain the probabilistic ground surface uniform hazard acceleration spectrum (Cramer 2003, 2006). The effect of the soil layers on the ground motion characteristics at the surface was expressed as frequency-dependent amplification function, AF(f), as suggested by Bazzurro and Cornell (2004a);

$$AF(f) = S_a^s(f)/S_a^r(f) \qquad (2.1)$$

where f is a generic oscillator frequency, $S_a^s(f)$ and $S_a^r(f)$ are the 5%-damped spectral-acceleration values at the soil surface and at the bedrock, respectively.

The acceleration spectra for RP = 475 and 2475 years for 40 soil profiles and 22 spectra scaled input acceleration records are calculated by assuming normal distribution for each period level based on 880 calculated acceleration response spectra as shown in Fig. 2.12. The spectral accelerations calculated for 10% exceedance probability based on probabilistic approach (Fig. 2.12a) are similar or slightly above those calculated based on mean + 1 standard deviation. However, in the case of RP = 2475 years, the probabilistic spectrum corresponding to 2% exceedance (Fig. 2.12b) is higher, thus more conservative compared to spectrum calculated based on mean + 1 standard deviation. In the case of 475-year return period, it may be justified to adopt mean + 1 standard deviation spectrum as the design spectrum since the results are very similar with the probabilistic approach where

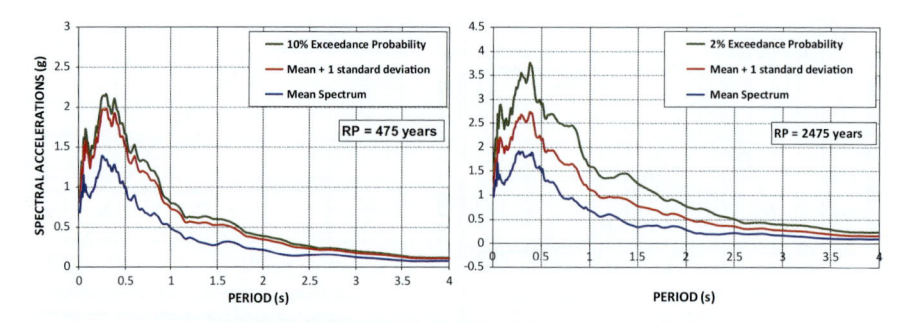

Fig. 2.12 Acceleration spectra on the ground surface for RP = 475 and 2475 years

the average difference ratio with respect to probabilistic approach is 9%. However, in the case of 2475-year return period, the difference between probabilistic interpretation and mean + 1 standard deviation spectrum is different with the average difference ratio is 28%. Thus, for the case of 2475 year return period, even though it will be a conservative choice, it may be more reliable and safer to design with respect to 2% exceedance probability.

The main purpose is to develop the uniform hazard acceleration spectrum on the ground surface based on the uniform hazard acceleration spectrum on the rock outcrop that was calculated by the probabilistic hazard analyses. As summarized above the variability due to source and path effects was considered by conducting site response analysis for large number of hazard compatible acceleration time histories. The other important causes of uncertainty are arising from the variability of soil stratification and engineering properties of encountered soil layers. One option is to conduct as many site response analyses as possible, as in the approach adopted for the case of hazard compatible input earthquake acceleration records. The main question at this point is, what would be the sufficient number of soil profiles to account for the variability of the site conditions. In large engineering projects, there can be large number of borings.

In one case study as shown in Fig. 2.13a, a parametric study was conducted at a site with relatively large number of borings. In this case study, it is very visible that up to 15 soil profiles, site response analysis would yield relatively overconservative high spectral accelerations. The sufficient number of borings to determine a reliable uniform hazard design acceleration spectrum is observed to be approximately 30 since the average difference between two consecutive calculated spectral accelerations are insignificant (less than 0.01 g) as shown in Fig. 2.13b. The increase in the number of soil profiles to 40 leads to a further slight decrease in the difference indicating the suitability of this approach in evaluating the variability in site conditions and in the applicability of the probabilistic approach as suggested previously in Figs. 2.11 and 2.12.

In the case of using Monte Carlo simulation for generating additional soil profiles based on available soil profiles, the number of necessary soil profiles is observed to be approximately same as 30 based on the variability on the acceleration spectra with number of soil profiles as observed in Fig. 2.14a and in terms of mean and maximum spectral accelerations as shown in Fig. 2.14b.

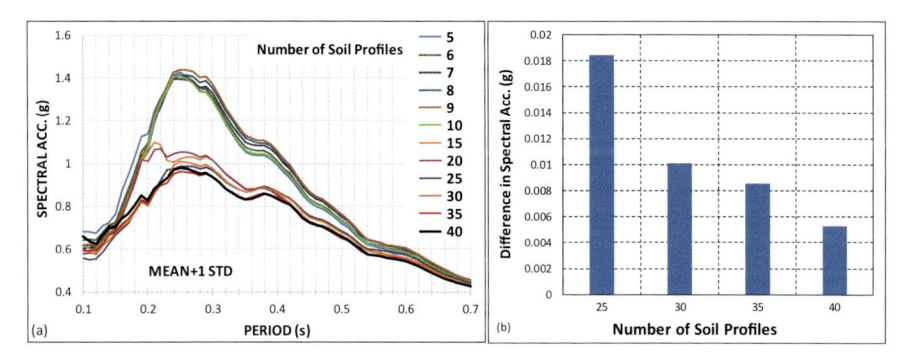

Fig. 2.13 The effect of number of soil profiles based on (**a**) a case study and (**b**) Difference in spectral accelerations in consecutive analysis for increasing number of soil profiles

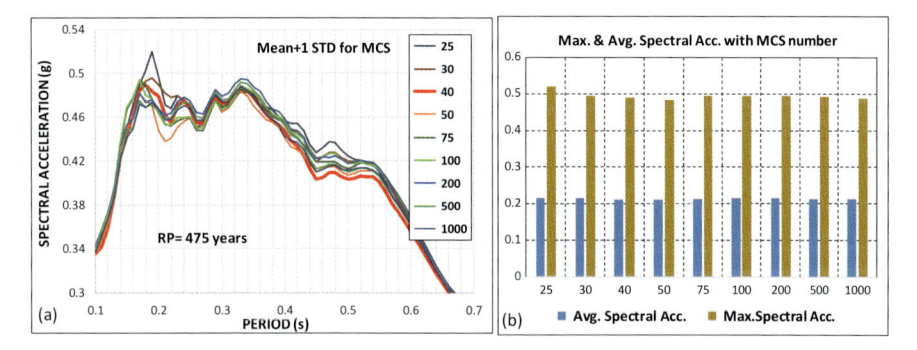

Fig. 2.14 The effect of number of soil profiles based on (**a**) Monte Carlo simulations and (**b**) Max. and mean spectral accelerations with respect to number of soil profiles

2.6 Conclusions

Based on the results of case studies summarized, it is possible to develop a simplified procedure to determine acceleration design response spectrum on the ground surface for two mostly adopted return periods. The procedure is not fully a probabilistic procedure, however, based on the observations the proposed approach may be assumed to yield uniform hazard acceleration design spectrum on the ground surface. The approach is based on the determination of amplification function based on major factors affecting site response analysis. The case studies composed of different number of soil profiles from different site investigations were used to conduct parametric studies. The approach is based on performing multiple number of site response analysis using hazard compatible acceleration records, that are scaled with respect to target acceleration response spectrum. It was observed that reliable results would be calculated on the ground surface if the number of input acceleration records are in the order of 22–23. It was also observed that the response acceleration spectra calculated on the ground surface, similarly to account for site variability, the number soil profiles are estimated based on a case study to be in the order of 30 soil profiles.

It is possible to define the design uniform hazard acceleration response spectrum for RP = 475 and 2475 years based on sufficient number of soil profiles using sufficient number of hazard compatible acceleration time histories adopting a distribution function to evaluate the statistical variation and to calculate the exceedance levels as foreseen in the regional hazard study.

Acknowledgement Faculty Members and Staff of Kandilli Observatory and Earthquake Research Institute of Bogazici University for all their support and contributions during different phases of this research.

References

Abrahamson NA (1992) Non-stationary spectral matching. Seismol Res Lett 63(1):30
Abrahamson NA (1993) Non-stationary spectral matching program RSPMATCH, User manual
Abrahamson NA, Silva WJ, Kamai R (2014) Summary of the ASK14 ground motion relation for active crustal regions. Earthq Spectra 30:1025–1055
Ansal A, Durukal E, Tönük G (2006) Selection and scaling of real acceleration time histories for site response analyses. In: Proceedings of ETC12 workshop, Athens, Greece, pp 93–98
Ansal A, Tönük G (2007) Chapter 4: source and site effects for microzonation. In: Pitilakis K (ed) Earthquake geotechnical engineering. Springer, Berlin, pp 73–92
Ansal A, Kurtuluş A, Tönük G (2009) Earthquake damage scenario software for urban areas computational. Struct Dyn Earthq Eng 2:377–391. Book series: Structures and Infrastructures Series, Editor(s): Papadrakakis, M; Charmpis, DC; Lagaros, ND; Tsompanakis, Y
Ansal A, Tönük G (2009) Site specific earthquake characteristics for performance based design. In: Keynote Lecture, Proceedings of international conference on earthquake engineering, seismology and engineering seismology volume, pp 59–63, Banja Luka, Republic of Srpska, Bosna Herzegnovia
Ansal A, Tönük G, Kurtuluş A (2011) Site specific earthquake characteristics for performance based design. In: Proceedings of the 5th international conference on geotechnical earthquake engineering, Santiago, Chile
Ansal A, Tönük G, Kurtuluş A, Çetiner B (2012) Effect of spectra scaling on site specific design earthquake characteristics based on 1D site response analysis. In: Proceedings of 15WCEE, Lisbon, Portugal
Ansal A, Tönük G, Kurtuluş A (2014) Site response from Istanbul vertical arrays and strong motion network. In: Proceedings of 10NCEE, nees.org/resources/12584
Ansal A, Tönük G, Kurtuluş A (2015a) A probabilistic procedure for site specific design earthquake. In: Theme lecture, 6th international conference on earthquake geotechnical engineering, Christchurch, New Zealand
Ansal A, Tönük, Kurtuluş A (2015b) A methodology for site specific design earthquake. In: SECED 2015 conference: earthquake risk and engineering towards a Resilient World. Cambridge UK, July 2015
Ansal A, Fercan Ö, Kurtuluş A, Tönük G (2017a) 2D site response analysis of the Istanbul rapid response network. Theme Lecture, PBD III, Vancouver
Ansal A, Tönük G, Kurtuluş A (2017b) A simplified approach for site specific design spectrum. In: Keynote lecture, international conference on earthquake engineering and structural dynamics, in Honor of Prof. Ragnar Sigbjörnsson, GEE
ASCE Standard (2016) Minimum design loads for buildings and other structures, ASCE/SEI 7-05

Assimaki D, Kausel E (2002) An equivalent linear algorithm with frequency- and pressure-dependent moduli and damping for the seismic analysis of deep sites. Soil Dyn Earthq Eng 22:959–965

Baturay M, Stewart P (2003) Uncertainty and bias in ground motion estimates from ground response analyses. Bull Seismol Soc Am 93(5):2025–2042

Bazzurro P, Cornell AC (2004a) Nonlinear soil-site effects in probabilistic seismic-hazard analysis. Bull Seismol Soc Am 94(6):2110–2123

Bazzurro P, Cornell CA (2004b) Ground-motion amplification in nonlinear soil sites with uncertain properties. Bull Seismol Soc Am 94(6):2090–2109

Bommer JJ, Acevedo AB (2004) The use real earthquake accelerograms as input to dynamic analysis. J Earthq Eng 8(1):43–91

Bommer JJ, Scott SG, Sarma SK (2000) Hazard-consistent earthquake scenarios. Soil Dyn Earthq Eng 19:219–231

Boore DM, Stewart JP, Seyhan E, Atkinson GA (2014) NGA-West2 equations for predicting PGA, PGV, and 5% damped PSA for shallow crustal earthquakes. Earthq Spectra 30:1057–1085

Borcherdt RD (1994) Estimates of site dependent response spectra for design (Methodology and justification). Earthq Spectra 10(4):617–654

Bradley BA (2010) A generalized conditional intensity measure approach and holistic ground-motion selection. Earthq Eng Struct Dyn 39:1321

Bradley BA (2012) A ground motion selection algorithm based on the generalized conditional intensity measure approach. Soil Dyn Earthq Eng 40:48–61

Campbell KW, Bozorgnia Y (2014) NGA-West2 ground motion model for the average horizontal components of PGA, PGV, and 5% damped linear acceleration response spectra. Earthq Spectra 30:1087–1115

Cramer CH (2003) Site specific seismic hazard analysis that is completely probabilistic. Bull Seismol Soc Am 93(4):1841–1846

Cramer CH (2006) Quantifying the uncertainty in site amplification modeling and its effects on site-specific seismic hazard estimation in the upper Mississippi embankment and adjacent areas. Bull Seismol Soc Am 96(6):2008–2020

Durukal E, Ansal A, Tonuk G (2006) Effect of ground motion scaling and uncertainties in site characterisation on site response analyses. In: Proceedings of eighth US national conference on earthquake engineering, San Francisco, California

Erdik M (2017) Earthquake risk assessment, 5th Prof. Ambraseys lecture. Bull Earthq Eng (15):12 5055–5092

Haase JS, Choi YS, Bowling T, Nowack RL (2011) Probabilistic seismic-hazard assessment including site effects. Bull Seismol Soc Am 101(3):1039–1054

Hartzell S, Cranswick E, Frankel A, Carver D, Meremonte M (1997) Variability of site response in the Los Angeles urban area for Evansville, Indiana, and the surrounding region. Bull Seismol Soc Am 87:1377–1400

Idriss IM, Sun JI (1992) Shake91, A computer program for conducting equivalent linear seismic response analysis of horizontally layered soil eposits. University of California, Berkeley

Iyisan R (1996) Correlations between shear wave velocity and In-situ penetration test results. Tech J Turkish Chamb Civil Eng 7(2):1187–1199. (in Turkish)

Kaklamanos J, Bradley BA, Thompson EM, Baise LG (2013) Critical parameters affecting bias and variability in site-response analyses using Kik-net downhole Array data. Bull Seismol Soc Am 103(3):1733–1749

Kaklamanos J, Baise LG, Thompson EM, Dorfmann L (2015) Comparison of 1D linear, equivalent-linear, and nonlinear site response models at six KiK-net validation sites. Soil Dyn Earthq Eng 69:207–219

Kottke AR, Rathje EM (2011) A semi-automated procedure for selecting and scaling recorded earthquake motions for dynamic analysis. Earthq Spectra 24(4):911–932

Kottke AR, Rathje EM (2013) Comparison of time series and random-vibration theory site-response methods. Bull Seismol Soc Am 103(3):2111–2127

Li W, Assimaki D (2010) Site- and motion-dependent parametric uncertainty of site-response analyses in earthquake simulations. Bull Seismol Soc Am 100:954–968

Malhotra PK (2003) Strong-motion records for site-specific analysis. Earthq Spectra 19(3):557–578

NEHRP (2015) Recommended seismic provisions for new buildings and other structures (FEMA P-1050-1)

PEER 2014 Pacific Earthquake Engineering Research Center, Strong motion Database. http://peer.berkeley.edu/2014

Rathje EM, Ozbey MC (2006) Site-specific validation of random vibration theory-based seismic site response analysis. J Geotech Geoenviron Eng 132(7):911–922

Rathje EM, Kottke AR, Trent WL (2010) Influence of input motion and site property variabilities on seismic site response analysis. J Geotech Geoenviron Eng 136:607–619

Stewart JP, Afshari, K, and Hashash, YMA (2014) Guidelines for performing hazard-consistent one-dimensional ground response analysis for ground motion prediction. PEER Report 2014/16, Pacific Earthquake Engineering Research Center, Headquarters, University of California at Berkeley

Thompson EM, Baise LG, Kayen RE, Guzina BB (2009) Impediments to predicting site response: seismic property estimation and modeling simplifications. Bull Seismol Soc Am 99(5):2927–2949

Tönük G, Ansal A (2010) Selection and scaling of ground motion records for site response analysis. In: 14th European conference of earthquake engineering, Ohrid, paper no. 1386

Tönük G, Ansal A, Kurtuluş A, Çetiner B (2014) Site specific response analysis for performance based design earthquake characteristics. Bull Earthq Eng 12(3):1091–1105

Chapter 3
A Redefinition of Seismic Input for Design and Assessment

G. Michele Calvi, Daniela Rodrigues, and Vitor Silva

Abstract For several decades, seismologists and engineers have struggled to perfect the shape of design spectra, analyzing recorded signals and speculating on probabilities. In this process, several solutions have been proposed, including considering more than one period to define a spectral shape, or proposing different spectral shapes as a function of the return period of the design ground motion.

However, the basic assumption of adopting essentially three fundamental criteria, i.e.: constant acceleration at low periods, constant displacement at long periods, constant velocity in an intermediate period range, has never been thoroughly questioned.

In this contribution, the grounds of a constant velocity assumption is discussed and shown to be disputable and not physically based. Spectral shapes based on different logics are shown to be consistent with the experimental evidence of several hundred recorded ground motions and to lead to significant differences in terms of displacement and acceleration demand.

The main parameters considered to define the seismic input are magnitude and epicenter distance. The possible influence of other parameters – such as focal depth and fault distance, duration and number of significant cycles, local amplification – will be the subject of future studies.

Novel forms of ground motion prediction equations and of hazard maps may result from this approach.

Specific points of interest include the generation and adaptation of acceleration and displacement time histories for design, the possibility of including the effects of energy dissipation on the side of capacity rather than on that of demand, the consistent generation of floor spectra for design and assessment of non-structural elements.

G. Michele Calvi (✉) · D. Rodrigues
Eucentre Foundation and IUSS, Pavia, Italy
e-mail: gm.calvi@eucentre.it

V. Silva
GEM Foundation, Pavia, Italy

3.1 Origin of Response Spectra in Earthquake Engineering

The origin and development of response spectra in earthquake engineering is masterfully described by A. Chopra (2007). Readers will learn that the concept of elastic response spectrum – which summarizes the peak response of all possible linear single-degree-of-freedom systems to a particular component of ground motion – was first intuitively conceived in Japan by K. Suyehiro in 1926. He actually built a *Vibration Analyzer* that could be exposed to an earthquake's strong motion and record the maximum amplitudes of deflection of a set of rods varying in their periods in the range 0.2–2.0 s.

Significant advancement based on instrumental measurements followed until the fifties, while in parallel theoretical development took place mainly at Caltech, with contributions by Von Karman, Biot, Hudson, Popov and others (e.g.: Housner (1941)).

What appears clearly when exploring those extraordinary advancements, is the obtrusive dichotomy between physical and mathematical understanding and data availability.

Not at all surprising is thus the extensive use of the only strong motion record available at the time – recorded at El Centro during the Imperial Valley earthquake of 18 May 1940 – and its transformation into the paradigm of a strong earthquake epicentral ground motion.

Chopra's story ends with the seventies, when the concept of elastic response spectrum had become fully established with fundamental contributions by Veletsos and Newmark. However, in the eighties a widely used summary was contained in a monograph published by EERI (Newmark and Hall 1982). In this booklet, the spectra

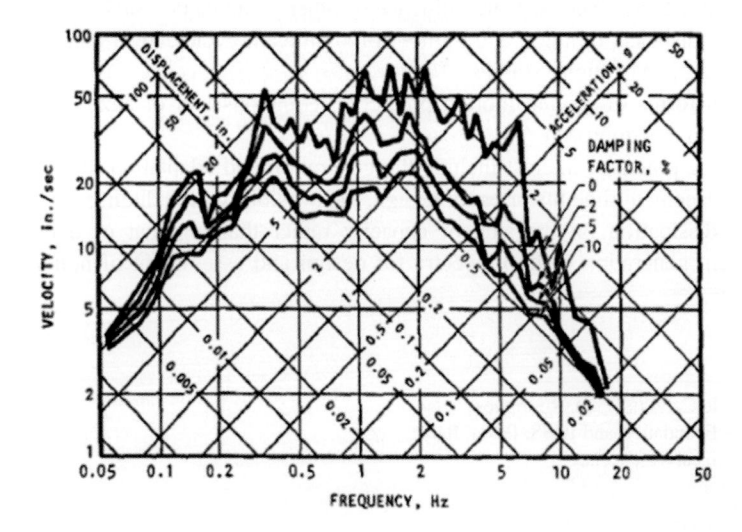

Fig. 3.1 Response spectra from the NS El Centro record, as reported in (Newmark and Hall 1982)

3 A Redefinition of Seismic Input for Design and Assessment

derived from the El Centro record were reported as shown in Fig. 3.1 and a design spectrum was recommended as depicted in Fig. 3.2, based on a peak ground acceleration equal to 0.5 g, which was assumed to correspond to one sigma cumulative probability.

According to the description provided, and referring to periods rather than frequencies, a design spectrum should show a constant maximum displacement demand at period approximately longer than 3 s, a constant maximum acceleration demand at periods between about 0.15 and 0.5 s and a constant maximum velocity demand between 0.5 and 3 s.

3.2 Modern Design Spectra

It is interesting to observe that the evolution of design spectra, based on progressively available strong motion data, has implied all sorts of proposed variation of period ranges and acceleration values, as a function of magnitude, epicenter or fault distance, local soil amplification, orographic effects and forward or backward directivity. However, the general shape of the spectrum, based on constant displacement, constant velocity and constant acceleration, has never been questioned.

Newmark's design spectrum of Fig. 3.2, is redrawn in Fig. 3.3, as a combined acceleration – displacement spectrum. This spectrum is more commonly plotted as an acceleration spectrum in codes of practice, while in displacement based

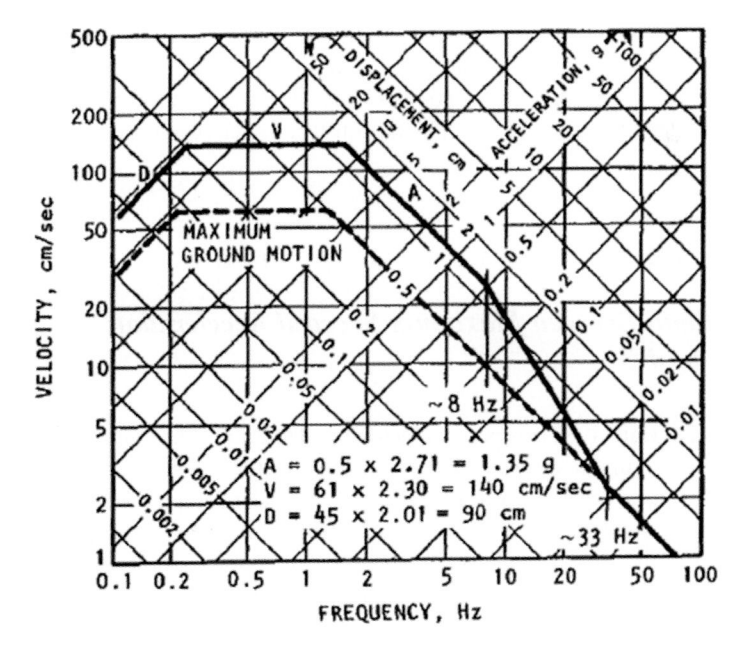

Fig. 3.2 Elastic design spectrum as defined by Newmark and Hall (1982), for 0.5 g PGA, 5% damping and one sigma cumulative probability

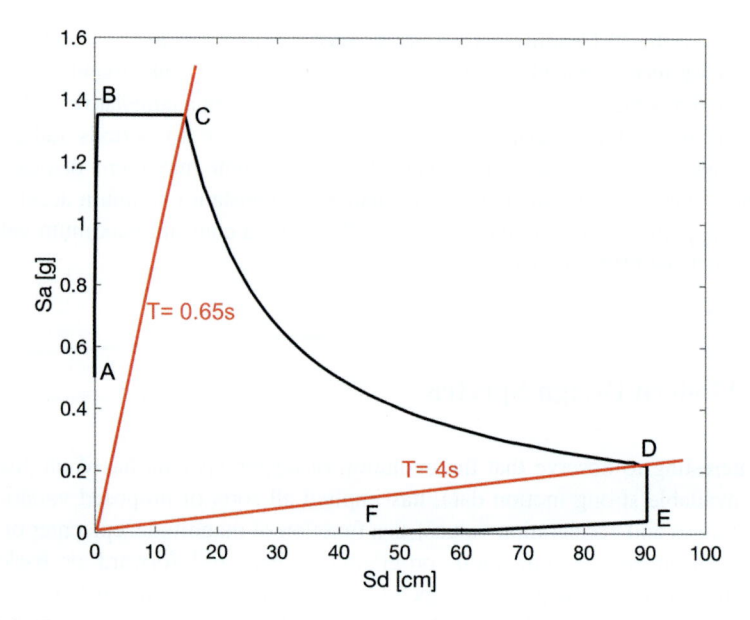

Fig. 3.3 Newmark's design spectrum of Fig. 3.2, shown as a combined acceleration – displacement spectrum

approaches, a displacement spectrum is often characterized simply by its "corner point" (D in Fig. 3.3), assuming a spectral shape made by two straight lines.

Considering the combined representation in Fig. 3.3, it is evident that the segments A-B and E-F are of no practical interest and consequently a full description of a design spectrum could be obtained from points C and D, provided that the assumption of constant acceleration, constant velocity and constant displacement is accepted.

3.3 Considerations on Elastic Design Spectra

3.3.1 Ratio Between Maximum Spectral Acceleration and PGA

Newmark and Hall (1982) proposed a factor equal to 2.71 as the multiplier of PGA to estimate the maximum spectral acceleration ($S_{a,max}$) for one sigma probability.

Similar values have been adopted by most codes of practice, with a basic parameter ranging between 2.5 and 3.0, according to the local soil amplification.

Analysis of modern data indicates a large dispersion of the ratio between $S_{a,max}$ and PGA (e.g.: Booth 2007). Considering, in addition, that the value of PGA is irrelevant for any practical purpose (and possibly misleading, due to its low correlation with $S_{a,max}$ and with other relevant response parameters), the obvious conclusion would be to simply adopt point C in Fig. 3.3 as the main acceleration-related parameter.

This would imply defining functions correlating magnitude, fault distance and local soil with period T_C and spectral acceleration S_{aC} at point C.

3.3.2 Maximum Spectral Displacement and Corner Period

Correlations between magnitude/distance and displacement spectra shapes have been recommended within the framework of displacement–based design (Priestley et al. 2007). In this context, the displacement spectrum is based on one point only, i.e. point D in Fig. 3.3.

The displacement spectrum is thus represented by a straight line from the origin to point D and remains constant afterwards. According to the recommendations given Priestley et al. (2007, following the indications of Tolis and Faccioli (1999)), point D could be defined according to the following equations:

$$T_D = 1.0 + 25(M_w - 5.7) \quad \text{(s)} \tag{3.1}$$

$$\Delta_D = C_S \cdot \frac{10^{(M_w - 3.2)}}{r} \quad \text{(mm)} \tag{3.2}$$

Where M_w is the moment Magnitude, **r** the fault rupture distance in km, C_s is a local soil factor, equal to 0.7 for rock, 1.0 for firm ground, 1.4 for intermediate soil, 1.8 for very soft soil.

3.3.3 Combination of the Two Horizontal Components

Equations of the sort of (3.1) and (3.2) assumes that peak values of acceleration and displacement are the same for the same distance and magnitude. However, the two recorded horizontal components of a ground motion, clearly characterized by the same magnitude, distance, soil, etc. usually show different values in terms of peak spectral acceleration and displacement, in many cases even mixing the larger demand in acceleration with the smaller displacement and vice versa.

Several solutions are possible, for example adopting the spectrum resulting from the envelope of the two, or calculating for each abscissa an ordinate resulting from the square root of the sum of the squares of the ordinates of the two spectra.

The preferable solution could possibly be to derive one single horizontal acceleration signal, combining the two components, instant by instant. This will imply to produce spectra that do not correspond to any specific direction, being actually sort of "rotating" spectra, without any specification of the actual direction for each maximum value. The resulting spectrum will be similar to the envelope of the spectral components, being normally dominated by the largest acceleration vector instant by instant, and it seems quite consistent with the concept of spectrum.

It has to be noted that this suggestion has nothing to do with the combination of actions on buildings, resulting from their response in different directions. As discussed by Stewart et al. (2011), the need of considering the different responses in different directions of *azimuth-dependent* structures remains untouched.

3.3.4 Why Constant Velocity?

As previously mentioned, the shape of a design spectrum has been traditionally defined adopting three criteria in different period ranges, i.e.: constant acceleration at low periods, constant displacement at long periods, constant velocity in an intermediate range.

While the first two indications are based on physical considerations, the choice of keeping a constant velocity in the intermediate range is only apparently sound, though never questioned in the past.

As an example, assume to adopt a straight line between point C and D of Fig. 3.3, as shown in Fig. 3.4. The resulting design spectrum shows possible differences of 100% in terms of displacement and acceleration demand in the wide intermediate region that characterizes the vast majority of structures.

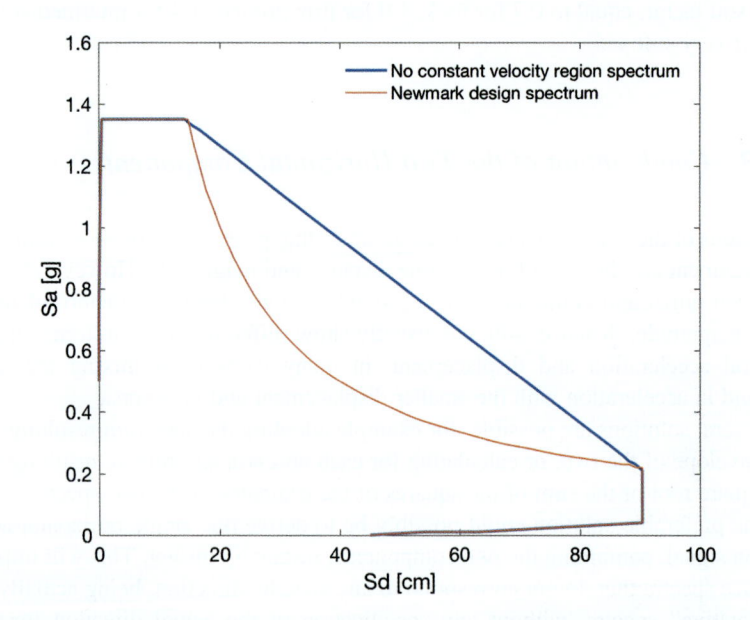

Fig. 3.4 The Newmark design spectrum compared with a spectrum with no constant velocity region, where S_a varies linearly with S_d in the intermediate periods region

However, such major differences are less evident if the data are plotted adopting a tri-logarithmic scale, as in Figs. 3.1 and 3.2 and as shown in Fig. 3.5. The variation of about 30% in the peak velocity (from about 140 to about 200 cm/s) does not appear to imply evident mismatching with the experimental data.

The constant velocity rule has never been questioned in so many years: this may be an effect of the great respect felt for the researchers who developed the concept of spectra.

3.3.5 El Centro Response Spectra

The ground motions recorded at El Centro on May 18 1940, used as a main reference until the eighties, are plotted in Figs. 3.8 and 3.9, using the digital data provided by Irvine, and assuming 5% damping. The median and one sigma design spectra have been produced following the indications by Newmark and Hall (1982), adopting the following peak ground values: PGA = 0.35 g, PGV = 26.4 cm/s, PGD = 18 cm and the following multipliers: A = 2.12 V = 1.65 D = 1.39 (median design spectrum) and A = 2.71 V = 2.30 D = 2.01 (one sigma spectrum). The corresponding spectrum with a linear variation of S_a vs. S_d in lieu of the constant velocity region has also been derived.

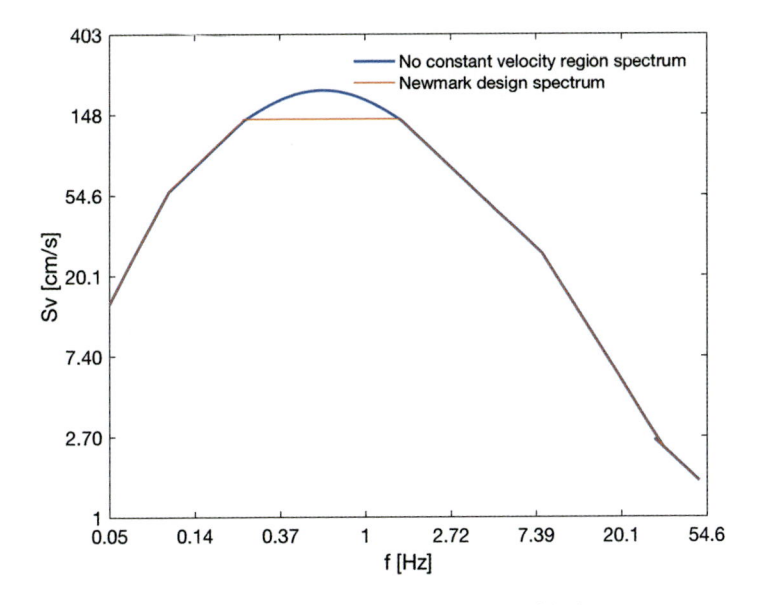

Fig. 3.5 The same spectra shown in Fig. 3.4, drawn on a tri-logarithmic paper

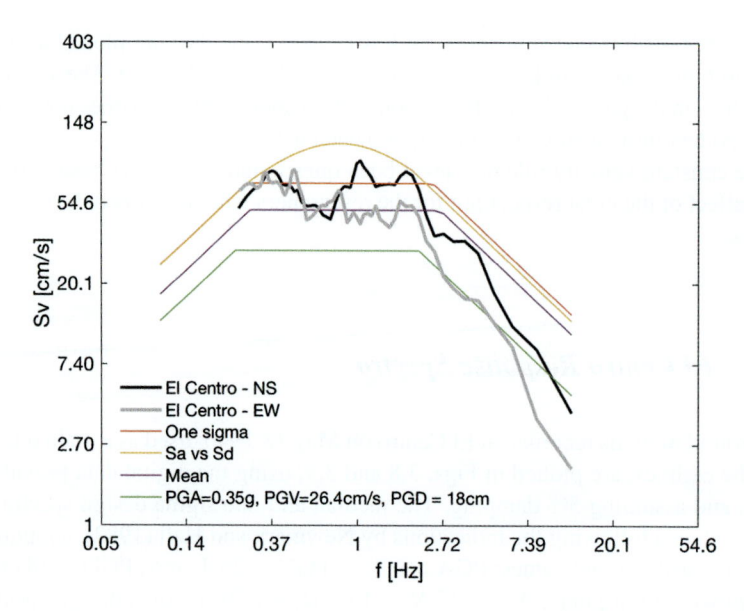

Fig. 3.6 Response spectra of the components of the El Centro record (data derived from Irvine), compared with the peak ground values and with the mean and one sigma design spectra suggested by Newmark and Hall (1982). The yellow spectrum has been produced assuming a linear variation of spectral acceleration versus spectral displacement in the intermediate period region (see Fig. 3.7)

In Fig. 3.6 all spectra are plotted on a tri-logarithmic plot paper, while in Fig. 3.7 the spectra are represented as displacement, velocity and acceleration vs. period and in the combined S_a–S_d form.

While in Fig. 3.6 the logarithmic scale squeezes the curves into similar shapes, in Fig. 3.7 it is evident how the adoption of a hyperbolic or linear descending branch results in the assessment of completely different demands in terms of acceleration and displacement.

The hyperbolic descending branch implied by a constant velocity assumption is not evidently superior to the linear hypothesis in matching the recorded data; more ground motions should be analyzed to assess the best approximation for a design spectrum.

In Fig. 3.7, the spectra resulting from an accelerogram generated by the vectorial combination, instant by instant, of the NS and ES acceleration components are also reported. These spectra differ significantly from those obtained from the NS component only at long periods of vibration, where the displacement demand is larger and the EW component seems to be dominant.

The design spectrum suggested by Newmark and Hall (1982) seems to take into consideration the displacement demand of the ES component, even if this was not explicitly stated. This extended use of common sense and intelligence is the base of the longevity of the spectral shape defined in the central decades of the previous century.

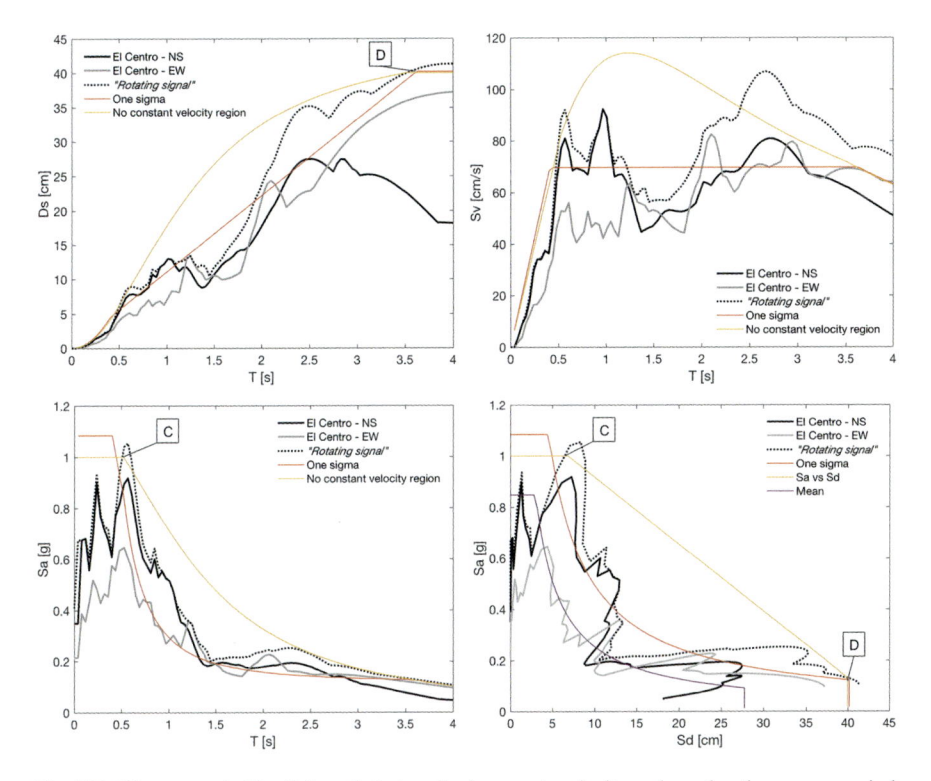

Fig. 3.7 The spectra in Fig. 3.6 re-plotted as displacement, velocity and acceleration versus period and in the combined S_a–S_v representation. The spectra derived from the *"rotating signal"* are added

3.4 Dealing with Non-linear Response

3.4.1 Early Approaches

As thoroughly described by R. Riddell (2008), the first attempts to determine the maximum response of nonlinear systems by means of inelastic response spectrum, date back to the fifties.

The basic idea to derive inelastic spectra from their elastic counterpart was based on defining ranges of the spectrum where acceleration, velocity and displacement were assumed to be "conserved" or modified by some correction factor. The key figure reported by Newmark and Hall (1982) is reproduced in Fig. 3.8. Here, an elastic-perfectly-plastic system is assumed, characterized by a ratio between displacement capacity and yield displacement equal to the displacement ductility value µ. Under these assumptions, the rules applied to derive inelastic spectra were:

- Divide the ordinate of the elastic spectrum by µ for frequencies up to about 2 Hz (regions D and V in Fig. 3.8) to obtain the acceleration inelastic spectrum.

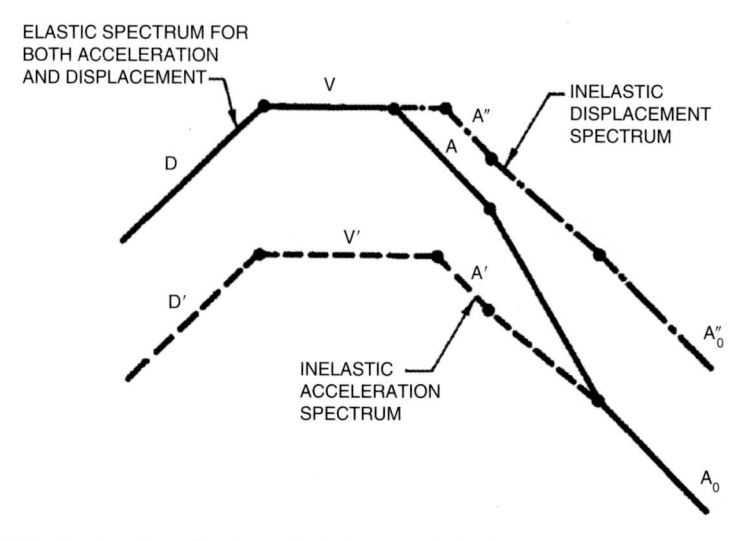

Fig. 3.8 Derivation of acceleration and displacement inelastic spectra from an elastic spectrum (Newmark and Hall 1982 and Riddel 2008)

- Do the same in the frequency range between 2 and 8 Hz (region A), dividing by $(2\,\mu-1)^{0.5}$ instead of μ.
- Keep the same acceleration in the elastic and inelastic spectrum for frequencies higher than 33 Hz.
- Link linearly the ordinates at 8 and 33 Hz in the logarithmic plot.
- To obtain the inelastic displacement spectrum multiply all the ordinates of the inelastic acceleration spectrum by μ.

This approach has been discussed, corrected, and modified in minor aspects for decades, but essentially used in its basic structure in all force-based approaches (and as such in all codes of practice) until today.

The procedure is different if a displacement-based approach is applied (Priestley et al. 2007). In this case it is suggested to take into consideration a correction of the spectral shape only due to the energy dissipated in the hysteretic cycles, and to use an elastic equivalent system with a stiffness (and thus a period) secant to the intersection between design displacement and corrected displacement spectrum.

As discussed by Priestley et al. (2007) and Priestley (2003), the required strength resulting from a displacement-based approach can be higher or lower with respect to that resulting from a traditional force-based approach. The result depends on the characteristics of the structural system, but as well on the site seismicity, since a displacement-based approach induces more dramatic variations in the required strength.

Possible outcomes from the applications of these different approaches, and of alternative capacity curves methods (Freeman 1975; Fajfar and Fishinger (1988) are discussed in Calvi (2018) and Calvi et al. (2018), where it is shown that the results can vary in the order of three times and seem to be a rather casual matter.

3.4.2 Correction of the Elastic Design Spectrum to Account for Energy Dissipation

Today it is indisputably accepted that demand and capacity should be kept as much as possible on different sides of an inequality and that the non-linear response should be captured as much as possible on the capacity side.

This can be accomplished using various forms of push-over analysis and applying one of the many existing methods to compare demand and capacity. One option is to apply a formulation of the capacity spectrum method, originally proposed by Freeman (1975), a second one is to resort to the N2 method, originally proposed by Fajfar and Fishinger (1988) and currently recommended by EC8. A further possibility is to consider a linear equivalent approaches as discussed by Priestley et al. (2007), but originally proposed by Shibata and Sozen (1976). A detailed discussion of this subject falls outside the scope of this paper, the interested reader is addressed to the critical summaries provided by Pinho et al. (2013) and by Silva et al. (2014).

Considering instead the side of the demand, it is still preferred to correct the elastic spectra to account for energy dissipation rather than modifying the capacity curve. This can be accomplished (Priestley et al. (2007)) by applying to the displacement spectrum ordinates a *displacement reduction factor* η_ξ as a function of an *equivalent hysteretic damping* ξ_h, estimated by some energy equivalence (Jacobsen (1960)). While several formulations are available, consider as an example the equations recommended by Priestley et al. (2007):

$$\eta_\xi = \left(\frac{0.07}{0.02 + \xi}\right)^{0.5} \quad (3.3)$$

$$\xi = \xi_0 + \xi_h = \xi_0 + C\left(\frac{\mu - 1}{2\pi\mu}\right) \quad (3.4)$$

Where ξ is the total equivalent damping of the system, ξ_0 is the inherent damping (usually taken as 0.05), C is a function of the shape of the hysteresis loop (generally in the range of 0.3–0.6) and μ is the ductility of the system.

An open problem is an appropriate tuning of η, expressed as a function of magnitude, source-to-site distance, site conditions, faulting style, etc. (e.g., Priestley et al. (2007) recommended to reduce the exponent from 0.5 to 0.25 in case of near field ground motion). Other parameters have actually been recommended, for example, based on analyses of specific Turkish ground motions, Akkar (2015) recommends to reduce η with period elongation, however, a different shape of the "constant velocity" branch will have an effect on demand likely more significant than a correction based on period of vibration.

A reformulation of design spectra will pave the way for consequent re-visitations of non-linear floor spectra, which are the most viable tool to design non-structural elements and are today treated by most codes of practice in an improper way, as thoroughly discussed by Calvi (2014) and Calvi and Sullivan (2014).

3.5 Definition of Corner Periods Based on Recorded Data

3.5.1 Extension of the Study and Available Data

A proposal for an alternative definition of design spectra will be based on conceptual speculation and on a large database of recorded ground motions, geographically limited to the Italian territory. The database includes 360 couples of records, derived from 24 events, recorded between 1972 and 2017. The location of the epicenters and of the recording stations are shown in Fig. 3.9, while in Table 3.1 the date, the number of records (each of them being actually made by two separated records in orthogonal directions) and the minimum and maximum distance between epicenter and recording stations are indicated for each event.

The selection of events and records has been simply based on the quality of the signals, considering all available cases, however, the complete set is evidently rather compact in time, geographical area, and source mechanism. Most of the records actually pertains to sequences recorded in the Apennines in 1997, 2009, 2012 and 2016.

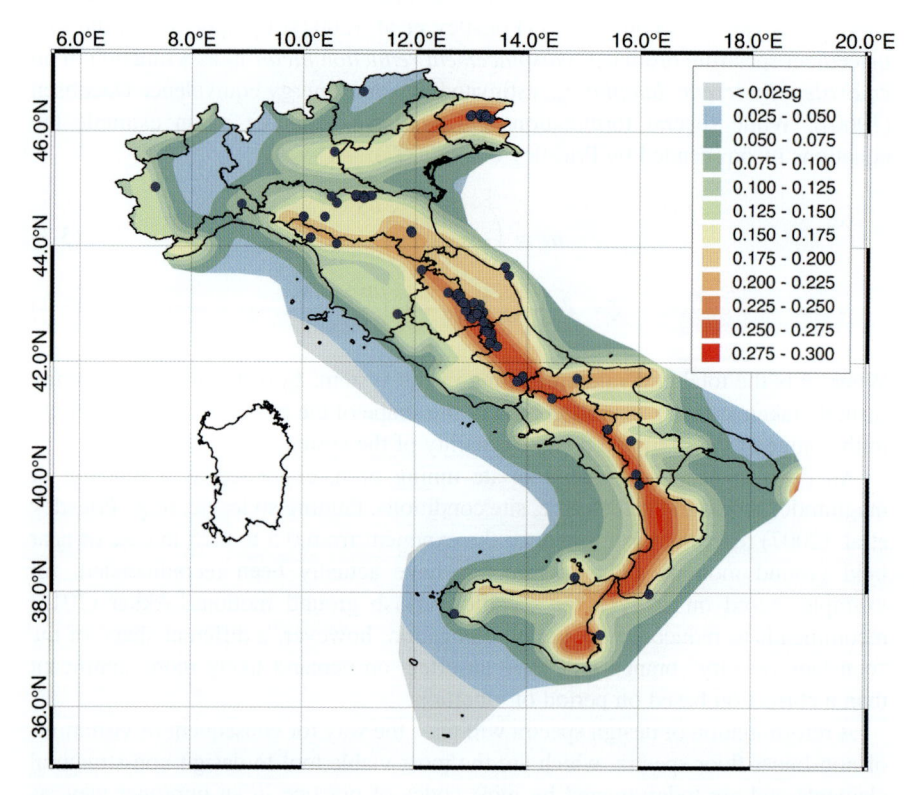

Fig. 3.9 Location of the epicenters of all events and records considered, reported on the Italian hazard map showing the 10% probability PGA values in 50 years (Stucchi et al. 2011)

3 A Redefinition of Seismic Input for Design and Assessment

For the purpose of this study, the available records have been divided into 20 bins characterized by appropriate ranges of magnitude and distance, as shown in Table 3.2.

3.5.2 Derivation of Single "Rotating" Signals

As previously discussed, the derivation of a single design spectra for a given distance and magnitude implies that one signal should be used for each record, not two orthogonal components. Consistently, each couple of acceleration records has been combined, deriving a single acceleration vector at each instant, simply characterized by its module. The resulting single signal has been used for all further elaborations. Examples of component and combined records are reported in Fig. 3.10.

Table 3.1 Records used in this study: date, number of different couples of record, maximum and minimum distance between epicenter and recording stations

Year	Magnitude	No. of records	Min distance (km)	Max distance (km)
1972	4.8	2	7.69	11.44
1976	4.5–6.4	29	1.91	85.41
1977	5.3	4	6.15	11.43
1978	5.2–6.1	5	9.15	46.64
1979	5.9	3	4.59	43.51
1980	4.6–5.0	4	5.83	16.65
1981	4.9–5.2	9	9.45	21.38
1982	4.6	1	8.07	8.07
1984	4.7–5.9	18	5.34	68.11
1990	5.6–5.8	5	26.65	65.26
1996	5.4	2	13.25	16.45
1997	4.5–6.0	52	0.96	79.50
1998	4.8–5.6	17	5.14	66.04
2000	4.5–4.8	3	1.71	7.56
2001	4.7–4.8	2	2.54	18.62
2002	5.7	1	41.06	41.06
2003	4.8	1	17.06	17.06
2004	5.3	1	14.37	14.37
2008	4.9	1	9.13	9.13
2009	4.6–6.3	38	0.73	49.17
2012	4.7–5.9	42	1.72	67.35
2013	4.9–5.2	8	4.37	81.30
2016	4.5–6.5	99	2.58	94.27
2017	4.6–5.5	13	5.24	28.85

Table 3.2 Magnitude and distance bins and number of records available for each one of them

Number of available records	r < 10 km	10 < r < 20 km	20 < r < 30 km	30 < r < 50 km	50 km < r
6.0 < M < 6.5	8	3	5	17	14
5.5 < M < 6.0	15	38	27	24	11
5.0 < M < 5.5	34	46	11	7	4
4.5 < M < 5.0	54	42	0	0	0

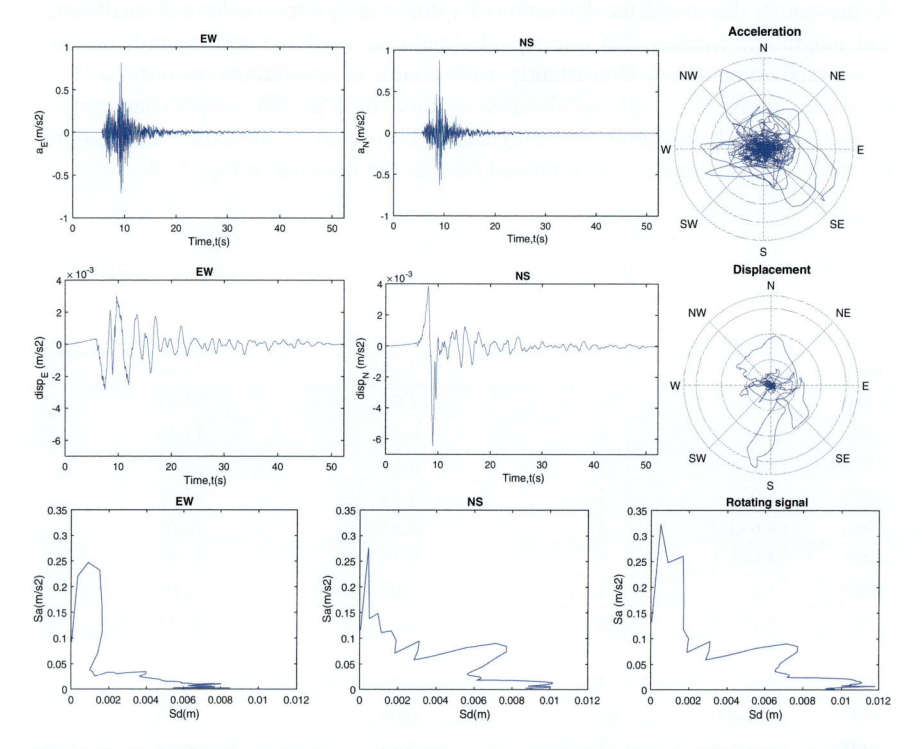

Fig. 3.10 Example of acceleration and displacement versus time, for the EW, NS and rotating signal. Corresponding S_a versus S_d spectra. The case considered for this example has a magnitude M = 5.4; the distance between recording station and epicenter is 11.69 km

3.5.3 Anchoring Points

As previously discussed, design spectra corresponding to a given ground motion could be anchored to two points, identified as the longest period (T_C) at which the spectral acceleration will be at the peak amplification (S_{aC}) and the shortest period (T_D) at which the spectral displacement will reach its maximum value (S_{dD}).

For each available record, the values of T_C and T_D and of the corresponding spectral acceleration (S_{aC}) and displacement (S_{dD}) have been calculated, as shown in

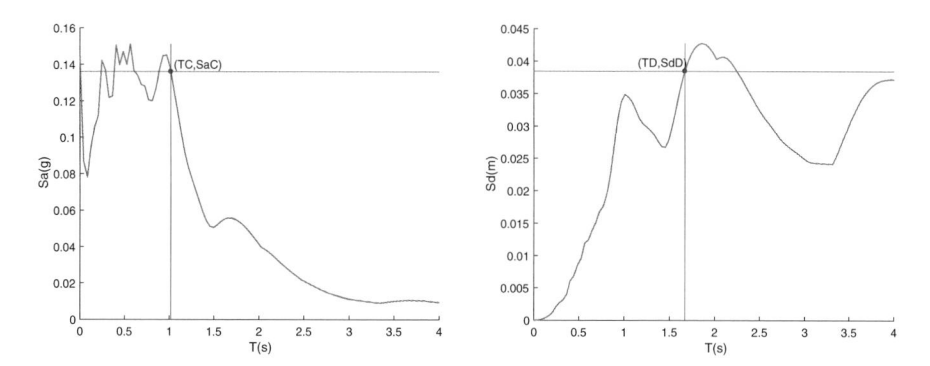

3 A Redefinition of Seismic Input for Design and Assessment   83

Fig. 3.11 Definition of the two corner period values. The horizontal lines on the acceleration and displacement spectra are taken at 90% of the respective maximum values. The points adopted are at the first (displacement) and last (acceleration) intersection between spectra and horizontal lines

Fig. 3.11. T_C and S_{aC} have been defined at the intersection between the acceleration response spectrum and a horizontal line marking 90% of the maximum spectral acceleration, considering the last intersection to the right of the plot. T_D and S_{dD} have been defined at the intersection between the displacement response spectrum and a horizontal line marking 90% of the maximum spectral displacement, considering the first intersection to the left of the plot. In all cases, a critical damping of 5% has been adopted.

For each one of the bins defined in Table 3.2, the mean and the $+1\sigma$ values of each parameter have been calculated. In Fig. 3.12 for each bin the mean and the $+1\sigma$ values of S_{aC} and S_{dD} are shown and compared with all available data. In Fig. 3.13 similar plots related to T_c and T_D, are shown, but in these cases reference is made to the mean values only. In fact, while it is felt appropriate to embed in the procedure some protection with respect to the values adopted for spectral acceleration and displacement, it is believed that the period values should be kept at their best fit. The reason for this choice will become clearer in the following discussion.

3.6 Definition of Interpolating Equations for the Corner Values

3.6.1 Construction of 3D Response Surfaces

Based on the empirical data introduced in the previous section, a number of interpolating equations have been developed, to estimate the longest period (T_C) at which the spectral acceleration will be at the peak amplification (S_{aC}) and the shortest period (T_D) at which the spectral displacement will reach its maximum value (S_{dD}).

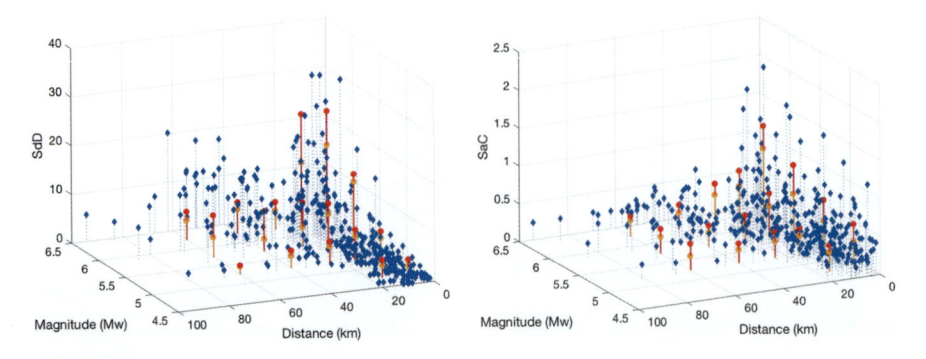

Fig. 3.12 Mean (orange) and +1σ (red) values of S_{dD} (left) and S_{aC} (right) for each magnitude-distance bin, compared with all available data

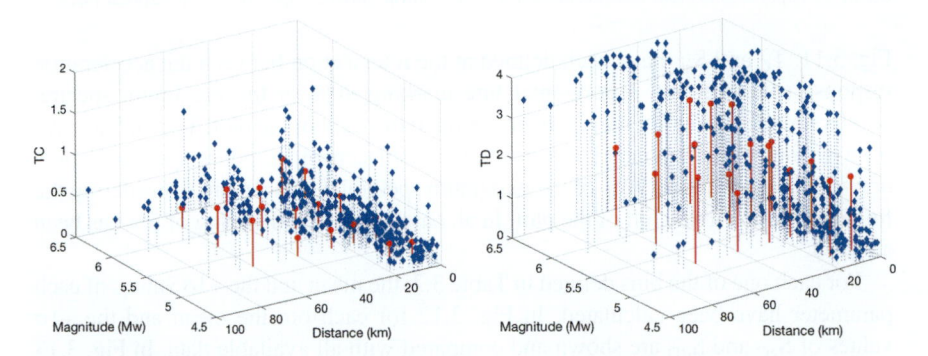

Fig. 3.13 Mean values of T_C (left) and T_D (right) for each magnitude-distance bin, compared with all available data

The response surface parametric equations employed to estimate periods and values have been optimized as a function of all available recorded data, being based on the combination of curves defined for specific values of magnitude and distance, as described in the following sections.

3.6.2 Definition of the Interpolation Function of T_C

It is assumed that T_c can be interpolated by a plane surface resulting by the combination of two lines. The first line is assumed at the maximum considered distance (r = 70 km) and it is thus function of the magnitude only:

$$T_{C(r=0)} = k_{TCo} - k_{TC1}(6.5 - M) \quad (s) \qquad (3.5)$$

3 A Redefinition of Seismic Input for Design and Assessment

Two parameters are here used: k_{TC0} is the value of T_C at M = 6.5 and r = 70 and k_{TC1} is the rate of change of T_C with M.

The second line expresses T_C at M = 6.5, as a function of the distance only:

$$T_{C(M=6.5)} = k_{TCo} - k_{TC2}(70 - r) \quad (s) \tag{3.6}$$

Where k_{TC0} has been already defined and k_{TC2} is the rate of change of T_C with r.

The resulting interpolating plane is then expressed as:

$$T_C = \frac{(k_{TCo} - k_{TC1}(6.5 - M))(k_{TCo} - k_{TC2}(70 - r))}{k_{TCo}} \quad (s) \tag{3.7}$$

3.6.3 Definition of the Interpolation Function of T_D

Similarly, T_D is first expressed at long distance (r = 70 km) as a function of the magnitude only:

$$T_{D(r=0)} = k_{TDo} - k_{TD1}(6.5 - M) \quad (s) \tag{3.8}$$

Where k_{TD0} is value of T_D at M = 6.5 and r = 70 and k_{TD1} is the rate of change of T_D with M.

T_D is then expressed at M = 6.5 as a function of the distance only:

$$T_{D(M=6.5)} = k_{TDo} - k_{TD2}(70 - r) \quad (s) \tag{3.9}$$

Where k_{TD0} has already been defined and k_{TD2} is the rate of change of T_D with r.

The final combined equation is thus:

$$T_D = \frac{(k_{TDo} - k_{TD1}(6.5 - M))(k_{TDo} - k_{TD2}(70 - r))}{k_{TDo}} \quad (s) \tag{3.10}$$

3.6.4 Definition of a General Equation for the Interpolation of S_{aC} and S_{dD}

While the functions to interpolate the period data have been derived from a combination of linear functions, the attenuation of acceleration and displacement with magnitude and distance are evidently more complex. To allow a proper consideration of the effects of distance and magnitude, the general form of equation considered is an over damped sinusoidal equation, expressed in terms of four parameters (k_0 to k_3), as shown below:

$$S = k_0 + k_1 e^{-k_2(f(x))} \cos k_3(f(x)) \tag{3.11}$$

3.6.5 Definition of the Interpolation Function of S_{aC}

S_{aC} is first expressed at r = 0, as a function of the magnitude only:

$$S_{aC(r=0)} = k_{a0} + k_{a1} e^{-k_{aM2}(6.5-M)} \cos \frac{2}{\pi k_{aM3}} (6.5 - M) \tag{3.12}$$

The four parameters appearing in the equations are defined as follows:

k_{a0} = a minimum threshold value of S_{aC}, at minimum magnitude, at the epicenter (r = 0)
k_{a1} = a value that summed to k_{a0} will give the value of S_{ac} at r = 0 and M = 6.5, i.e. the maximum spectral acceleration
k_{aM2} = a factor that increases or reduces damping
k_{aM3} = the magnitude value at which it is assumed that S_{aC} is minimum (i.e. equal to k_{a0}) for r = 0

S_{aC} is then defined at M = 6.5, as a function of the distance only:

$$S_{aC(M=6.5)} = k_{a0} + k_{a1} e^{-k_{ar2} r} \cos \frac{2}{\pi k_{ar3}} r \tag{3.13}$$

In this equation, the parameters k_{a0} and k_{a1} have a different meaning, but will maintain the same numerical values. Their sum must remain the same for physical reasons, k_{a0} will be forced to the same value selecting an appropriate distance from the epicenter:

k_{a0} = a minimum threshold value of S_{aC}, at maximum magnitude (M = 6.5), at a maximum distance (possibly r = 70 km, to be defined)
k_{a1} = a value that summed to k_{a0} will give the value of Sac at r = 0 and M = 6.5, i.e. the maximum spectral acceleration

While two additional parameters are used:

k_{ar2} = a factor that makes the damping stronger or weaker
k_{ar3} = the distance at which it is assumed that SaC is minimum (i.e. equal to k_{a0}) for M = 6.5

The resulting combined equation for the S_{aC} surface is:

$$S_{aC} = \frac{S_{aC}(M)_{(r=0)} S_{aC}(r)_{(M=6.5)}}{S_{aC(M=6.5,r=0)}} = \frac{S_{aC}(M)_{(r=0)} S_{aC}(r)_{(M=6.5)}}{k_{a0} + k_{a1}} \qquad (3.14)$$

3.6.6 Definition of the Interpolation Function of S_{dD}

The procedure is analogous to the previous one. S_{dD} is first defined at r = 0 as a function of the magnitude only:

$$S_{dD(r=0)} = k_{d0} + k_{d1} e^{-k_{dM2}(6.5-M)} \cos\frac{2}{\pi k_{dM3}}(6.5 - M) \qquad (3.15)$$

With the following parameters:

k_{d0} = a minimum threshold value of S_{dD}, at minimum magnitude, at the epicenter (r = 0)
k_{d1} = a value that summed to k_{d0} will give the value of SdD at r = 0 and M = 6.5, i.e. the maximum spectral displacement
k_{dM2} = a factor that increases or reduces damping
k_{dM3} = the magnitude value at which it is assumed that S_{dD} is minimum (i.e. equal to k_{d0}) for r = 0

S_{dD} is then expressed at M = 6.5 as a function of the distance only:

$$S_{dD(M=6.5)} = k_{d0} + k_{d1} e^{-k_{dr2} r} \cos\frac{2}{\pi k_{dr3}} r \qquad (3.16)$$

As above, k_{d0} and k_{d1} are forced to maintain the same numerical values with a different meaning:

k_{d0} = a minimum threshold value of SdD, at maximum magnitude (M = 6.5), at a maximum distance (possibly r = 70 km, to be defined)
k_{d1} = a value that summed to k_{d0} will give an approximate value for S_{dD} at r = 0 and M = 6.5, i.e. the maximum spectral displacement

While k_{dr2} is a factor that increases or reduces damping and k_{dr3} is the distance at which it is assumed that S_{dD} is minimum (i.e. equal to k_{d0}) for M = 6.5.
The resulting global equation is:

$$S_{dD} = \frac{S_{dD}(M)_{(r=0)} S_{dD}(r)_{(M=6.5)}}{S_{dD(M=6.5,r=0)}} = \frac{S_{dD}(M)_{(r=0)} S_{dD}(r)_{(M=6.5)}}{k_{d0} + k_{d1}} \qquad (3.17)$$

Table 3.3 k values resulting from an optimization procedure described in the text (column 2 and 3) and k values used in the recommended equations

Parameter	Mean	+1σ	Used
k_{TC0} (value of T_C at M = 6.5 and r = 70) [s]	0.46	0.61	0.46
k_{TC1} (rate of change of T_C with M)	0.042	0.053	0.042
k_{TC2} (rate of change of T_C with r)	0.0028	0.0034	0.003
k_{TD0} (value of T_D at M = 6.5 and r = 70) [s]	2.27	2.69	2.27
k_{TD1} (rate of change of T_D with M)	0.13	0.11	0.13
k_{TD2} (rate of change of T_D with r)	0.0053	0.0043	0.004
k_{a0} (value of S_{aC}, at minimum magnitude and zero distance) [g]	0.2	0.27	0.27
k_{a1} (value to be summed to k_{a0} to obtain S_{ac} at r = 0 and M = 6.5) [g]	1.15	1.35	1.35
k_{aM2} (magnitude-acceleration damping correction factor)	0.5	0.48	0.48
k_{aM3} (magnitude at which S_{aC} is assumed minimum (i.e. = k_{a0}) for r = 0)	4.5	4.5	4.5
k_{ar2} (distance-acceleration damping correction factor)	0.11	0.09	0.09
k_{ar3} (distance at which S_{aC} is assumed = k_{a0}, for M = 6.5) [km]	70	70	70
k_{d0} (value of S_{dD}, at minimum magnitude and zero distance) [cm]	4	5.9	5.9
k_{d1} (value to be summed to k_{d0} to obtain S_{dD} at r = 0 and M = 6.5) [cm]	25	29.5	29.5
k_{dM2} (magnitude-displacement damping correction factor)	1.39	1.55	1.55
k_{dM3} (magnitude at which S_{dD} is assumed minimum (i.e. = k_{d0}) for r = 0)	4.5	4.5	4.5
k_{dr2} (distance-displacement damping factor)	0.1	0.1	0.1
k_{dr3} (distance at which S_{dD} is assumed = k_{d0}) for M = 6.5) [km]	70	70	70

3.6.7 Calculation of Best Fit Parameters

Due to the large number of parameters that had to be estimated, a nonlinear optimization procedure was employed. In this process, the algorithm proposes a combination of parameters that minimizes the residuals (expressed as a percentage) between the actual empirical data (see Figs. 3.12 and 3.13), and the associated 3D surfaces. This procedure produced the k values summarized in the second and third columns of Table 3.3; the final values used in the following analyses are reported on the fourth column.

The surfaces resulting from the adoption of the k values reported in the fourth column of Table 3.3 are depicted in Figs. 3.14 and 3.15, and compared with +1σ data (for spectral acceleration and displacement) and with mean data (for periods).

Comparing the values recommended with the ones resulting from the optimization procedure in Table 3.3, it is immediate to observe that the recommended values are approximating the mean values for the definition of periods of vibration, while

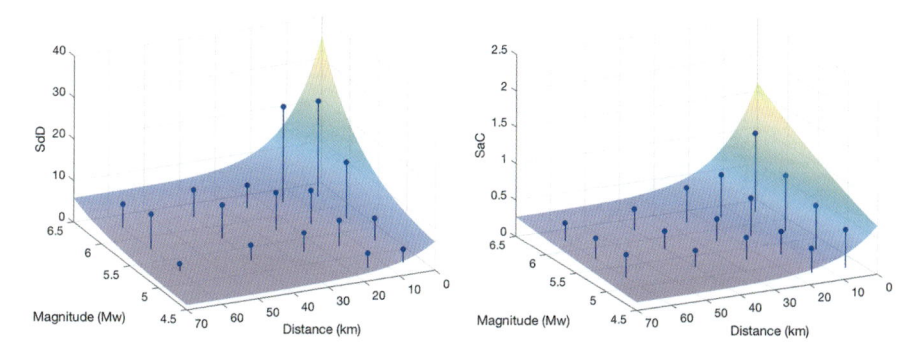

3 A Redefinition of Seismic Input for Design and Assessment   89

Fig. 3.14 Comparison of best fit interpolating surface and +1σ values of S_{dD} (left) and S_{aC} (right)

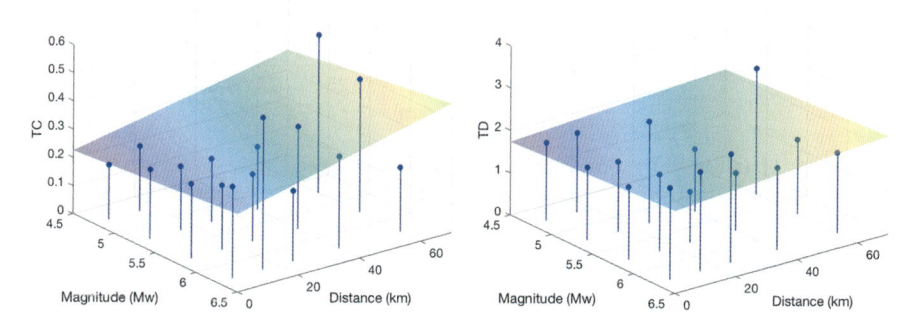

Fig. 3.15 Comparison of best fit interpolating surface and mean values of T_C (left) and T_D (right)

they are approximating the +1sigma values in the case of spectral acceleration and displacement. This decision had been anticipated and is due to the wish of amplifying the combined S_a–S_d spectrum to a + 1σ probability of attainment, but to maintain a homothetic spectral shape. An example is illustrated in Fig. 3.16, with reference to the case of M = 6.5 and r = 0.

Note, as well, that the results are suggesting that T_D depends mainly on magnitude, not on the distance from the source (i.e., k_{TD2} could be set equal to zero).

Finally, it seems interesting to observe that the multiplier coefficients to pass from mean to +1σ spectral acceleration and displacement are here estimated around 1.26 for both S_a and S_d. This value imply that the dispersion of peak spectral acceleration and peak spectral displacement are similar, somehow unexpectedly. Actually, the corresponding values according to Newmark and Hall (1982) were 1.28 and 1.45, indicating a greater dispersion in displacement values.

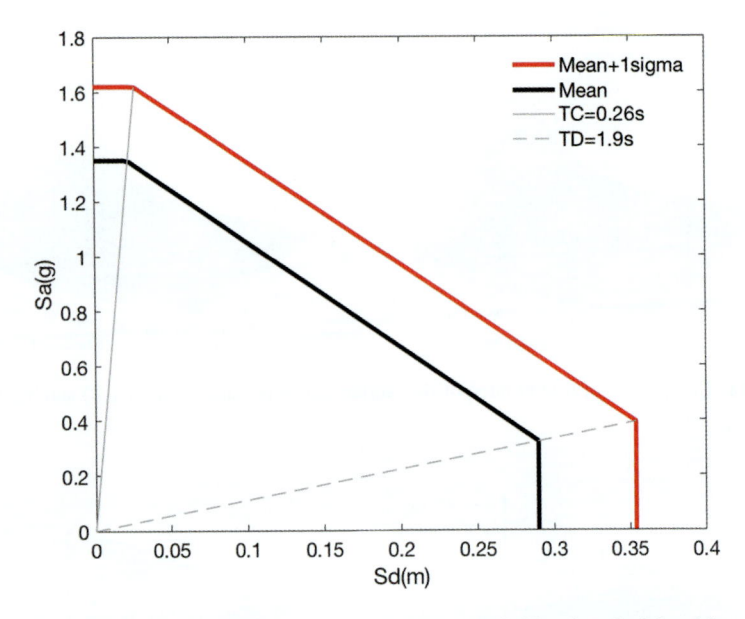

Fig. 3.16 Mean and +1σ spectra resulting from the recommended k values for M = 6.5 and r = 0. The two corner periods are kept equal to the mean values to conserve the shape of the spectrum

3.7 Spectral Shape in the Intermediate Period Region

3.7.1 Definition of an Interpolating Function

In the intermediate region between the corner points, an equation based on a single parameter α has been derived to shape the connecting curve, again as a function of magnitude and distance. The equation similar to the function that allows modifying the shape of a force-displacement curve of a viscous damper, which is based on the combination of two sinusoidal functions, as follows:

$$y = \sin^\alpha t \tag{3.18}$$

$$z = \cos^\alpha t \tag{3.19}$$

$$y = \sin^\alpha\left(\cos^{-1}\left(z^{1/\alpha}\right)\right) \tag{3.20}$$

Varying α, this equation can take any form between points (0; 1) and (1; 0), as shown in Fig. 3.17.

Applying a coordinate transformation, from point (0; 1) to point (S_{dC}; S_{aC}) and from point (1; 0) to point (S_{dD}; S_{aD}), the following equation is derived:

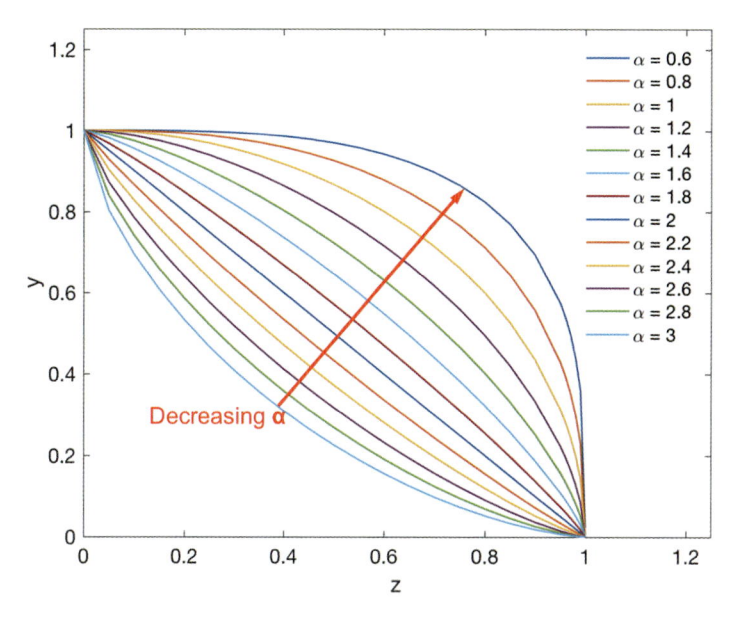

3 A Redefinition of Seismic Input for Design and Assessment

Fig. 3.17 Graphical representation of Eq. (3.20) with varying α

$$S_a = S_{aD} + (S_{aC} - S_{aD}) \cdot \sin^\alpha \left(\cos^{-1} \left(\frac{S_d - S_{dC}}{S_{dD} - S_{dC}} \right)^{\frac{1}{\alpha}} \right) \quad (3.21)$$

Where two derived parameters are used:

$$S_{dC} = S_{aC} \frac{T_C^2}{4\pi^2} \quad (3.22a)$$

$$S_{aD} = S_{dD} \frac{4\pi^2}{T_D^2} \quad (3.22b)$$

3.7.2 Calculation of the α Parameter

The optimal value of α has been again calculated assuming a surface shape in the space defined by magnitude and distance, and estimating the best interpolating values of the parameters to fit the recorded data. The following general equation of a plane to define α has been assumed:

$$\alpha = k_{\alpha 0} - k_{\alpha 1}(M - 4.5) - k_{\alpha 2} r \quad (3.23)$$

Where $k_{\alpha 0}$ is the value of α at $M = 4.5$ and $r = 0$, $k_{\alpha 1}$ is the rate of change of α with M and $k_{\alpha 2}$ is rate of change of α with r.

Table 3.4 Best fit α value obtained for each magnitude and distance bin

Best α value	r < 10 km	10 < r < 20 km	20 < r < 30 km	30 < r < 50 km	50 km < r
6.0 < M < 6.5	2.5	2.4	2.3	2.1	1.8
5.5 < M < 6.0	3.2	2.7	2.2	2.0	–
5.0 < M < 5.5	3.4	3.1	2.5	–	–
4.5 < M < 5.0	3.5	3.2	–	–	–

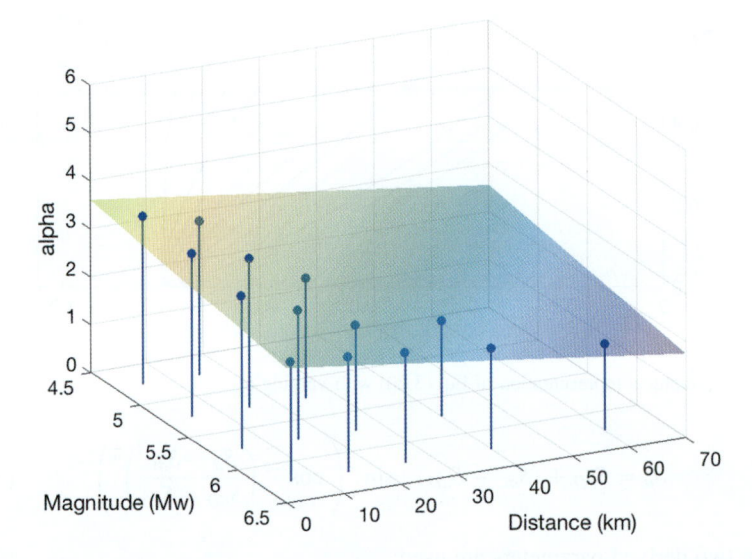

Fig. 3.18 The best fitting surface of α, as resulting from Eq. 3.16, compared with the bin values reported in Table 3.4.

A best fit analysis of the data has been performed calculating the value of α that minimizes the area between the curved line connecting points C and D and the actual response spectrum line, for each one of the bins previously defined. The best fit α values obtained for each bin are reported in Table 3.4.

Based on these values, the optimal k_α parameters inserted in the following equation have been obtained:

$$\alpha = 3.6 - 0.4(M - 4.5) - 0.015r \tag{3.24}$$

The single bin values reported in Table 3.4 and the surface produced by Eq. (3.24) are depicted and compared in Fig. 3.18.

It is interesting to note that the adopted α values are varying between 3.6 and 1.8. As a consequence, the acceleration and displacement design values corresponding to a given period of vibration may be considerably larger than the values resulting from the classical assumption of a linear variation of the displacement as a function of period (i.e., a variation of the spectral acceleration with 1/T, or a variation of the ratio S_a/S_d with $1/T^2$)).

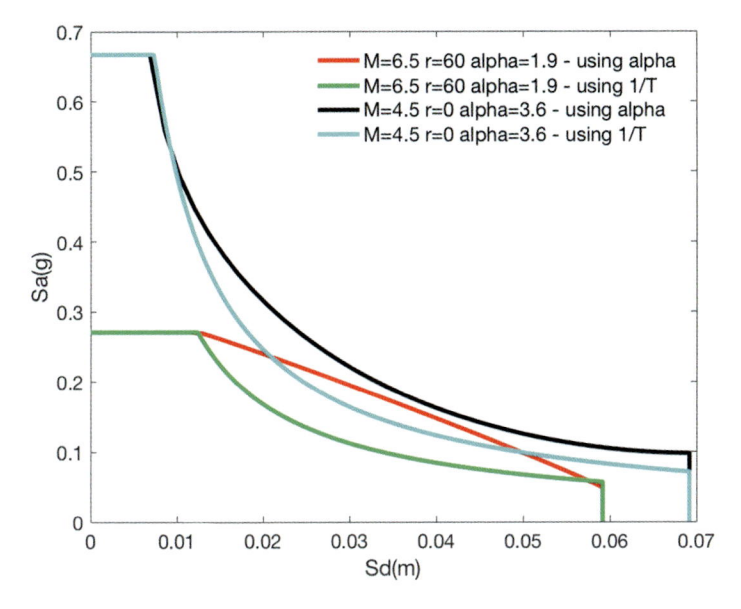

3   A Redefinition of Seismic Input for Design and Assessment   93

Fig. 3.19 Range of variation of the spectral shape derived in this work (red) compared with the corresponding spectral shapes obtained assuming a linear variation of the displacement with period

This effect is evident in Fig. 3.19, where the range of variation of the spectral shapes derived in this work are compared with spectra resulting from the assumption mentioned above, assuming the same maximum spectral acceleration and corresponding maximum period of vibration.

Even in the extreme case of the epicenter location (r < 10 km) of a small magnitude earthquake (M = 4.5) the predicted demands are different, but the variation becomes particularly pronounced for large magnitude events at a significant distance.

Note that the period region in which the differences are more relevant is in the range of 0.8–1.5 s, i.e. a range related to a large fraction of mid to high-rise buildings.

3.8 Resulting Spectra

The final design spectra resulting for each one of the bins are depicted in Figs. 3.20, 3.21, 3.22 and 3.23. In each plot, the mean and +1σ spectra resulting from the ground motions pertaining to the bin are also reported, as well as all spectra resulting from any record with lower magnitude and larger distance than those characterizing the bin limits. In general, the data resulting from the lower distance and the higher magnitude characterizing each bin have been adopted, with the exception of the case with distance between 0 and 10 km, in which case the data related to a distance equal to 5 km have been used, assuming some sort of saturation at lower distances.

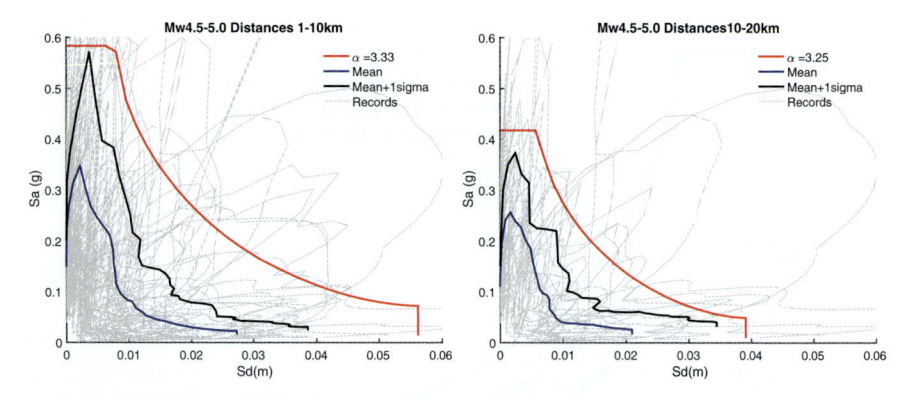

Fig. 3.20 Design spectra (red line) for 4.5 < M < 5.0, at distances r < 10 km and 10 km < r < 20 km, compared with the mean (blue) and +1σ (black) spectra resulting from the corresponding records. All spectra from any record with lower magnitude and larger distance are also depicted (grey)

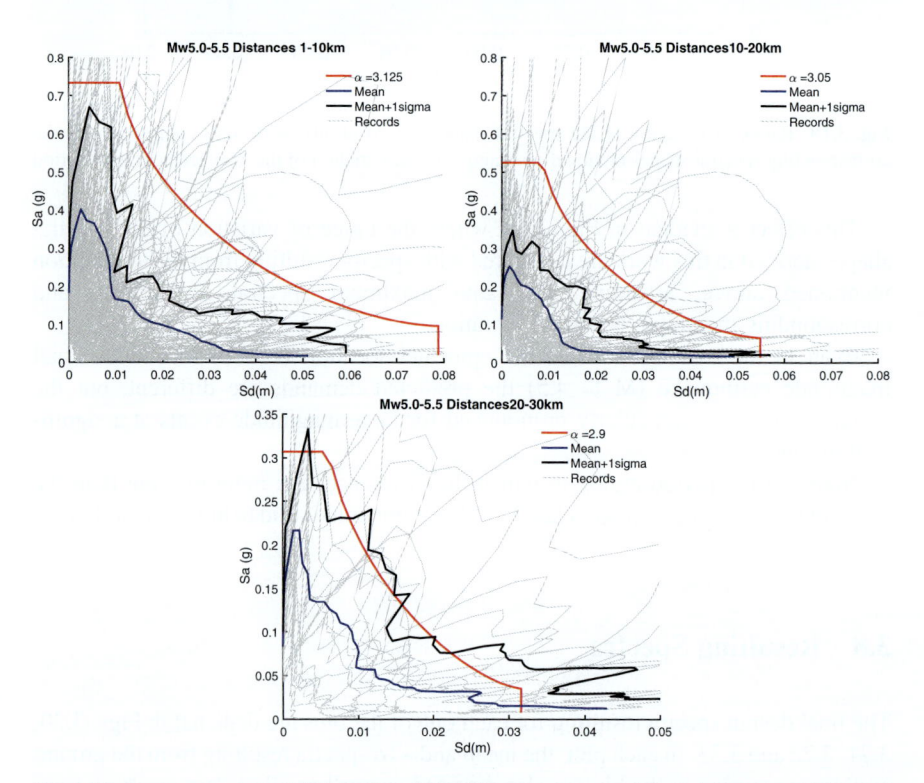

Fig. 3.21 Design spectra (red line) for 5.0 < M < 5.5, at distances r < 10 km, 10 km < r < 20 km and 20 km < r < 30 km, compared with the mean (blue) and +1σ (black) spectra resulting from the corresponding records. All spectra from any record with lower magnitude and larger distance are also depicted (grey)

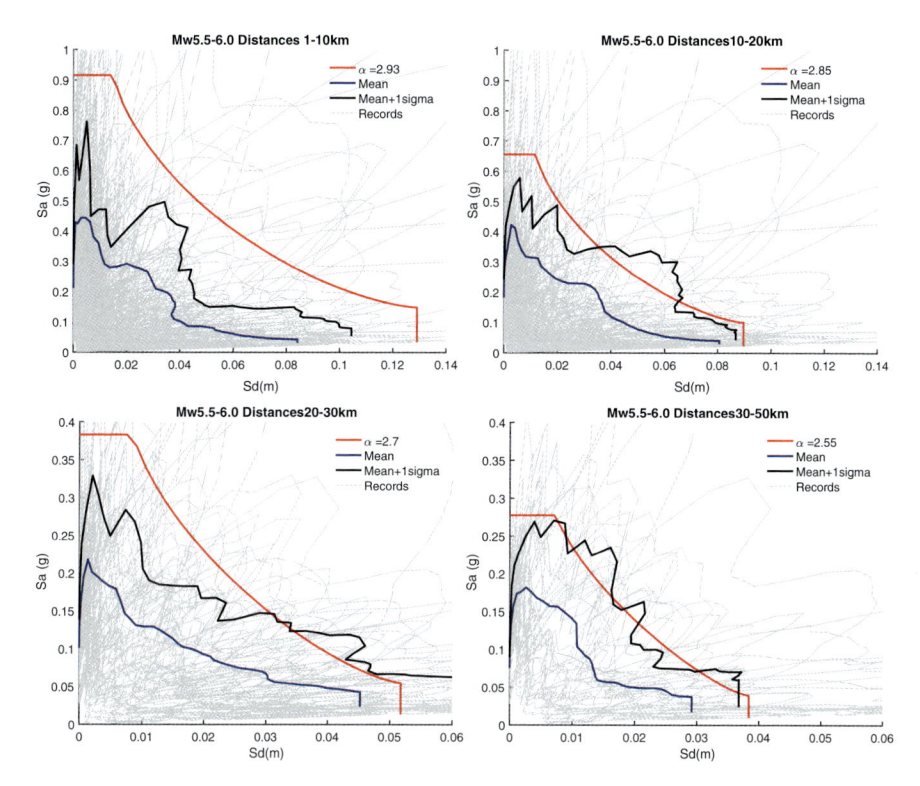

3 A Redefinition of Seismic Input for Design and Assessment

Fig. 3.22 Design spectra (red line) for 5.5 < M < 6.0, at distances r < 10 km, 10 km < r < 20 km, 20 km < 30 km and 30 km < r < 50 km, compared with the mean (blue) and +1σ (black) spectra resulting from the corresponding records. All spectra from any record with lower magnitude and larger distance are also depicted (grey)

Observing the design spectra obtained for each bin of distance and magnitude, it may be noted that there is in general a good correlation with the recorded data, considering that a single set of equations with a relatively low number of parameters has been consistently applied in all cases.

All design spectra in each point of the curves are producing acceleration and displacement demand larger than the corresponding mean values, and only in few cases the +1σ spectral curves are exceeding the values resulting from the design equations. The desired correspondence between design and corresponding +1σ spectra seems thus to have been achieved at a satisfactory level, particularly observing that:

- The performance of the method appears to be slightly worse for cases of very low magnitude or very long distance. This is considered acceptable.
- The performance also seems worse when bins comprise a smaller number of records. Additional data may improve the approximation.
- Assuming an acceleration amplification factor equal to 2.5, the peak values of spectral acceleration at a distance of 5 km from the epicenter corresponds to

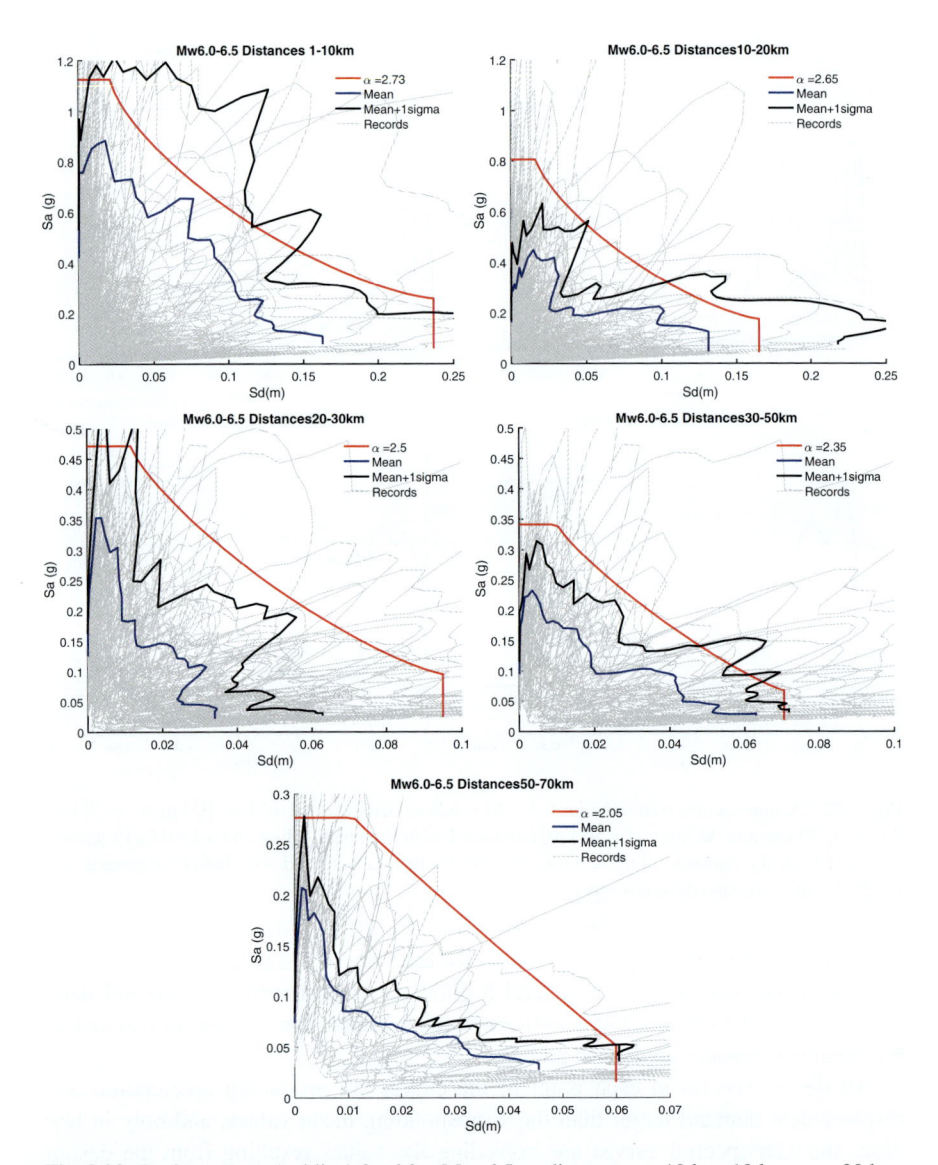

Fig. 3.23 Design spectra (red line) for $6.0 < M < 6.5$, at distances $r < 10$ km, 10 km $< r < 20$ km, 20 km < 30 km, 30 km $< r < 50$ km and 50 km < 70 km compared with the mean (blue) and $+1\sigma$ (black) spectra resulting from the corresponding bin records. All spectra from any record with lower magnitude and larger distance are also depicted (grey)

PGAs varying between 0.2 and 0.45 g for magnitudes varying between 4.5 and 6.5. These values seem quite reasonable and in line with the current state of the art. Clearly, the spectral acceleration plateau levels do not capture the single maximum recorded peaks (Suzuki and Iervolino (2017)), as expected for design spectra and shown by some extreme single spectra depicted in the figures.

- The predicted maximum spectral displacements are quite in line with those predicted by Eq. 3.2 for firm soil at large magnitudes and small distances, i.e. when the equation is applicable.

3.9 Conclusions

Design spectra should be instrumental to design, and as such not necessarily derived as a weighted combination or an average of response spectra.

The work presented in this paper focuses on the derivation of design spectra assuming that magnitude of the event and distance from the source are sufficient parameters to produce a combined spectral acceleration versus spectral displacement spectrum.

It is suggested that peak ground acceleration should be abandoned as a key parameter of demand and hazard. Spectral shape and its anchoring should be based on the definition of two points (C and D in this paper), i.e. maximum spectral acceleration and displacement with the corresponding periods of vibration.

The spectra shape in the intermediate region, roughly ranging from 0.5 to 3 s, has been investigated, abandoning the concept of constant velocity, which has no theoretical nor experimental base. An equation to define the shape of the intermediate region as a function of a single parameter (α, Eq. (3.5)) is proposed. Such a spectral shape will be equally applicable to force-based and displacement-based methods, as well as to design and assessment procedures.

It is suggested to derive one single horizontal acceleration signal from recorded ground motions, combining the two components instant by instant. The resulting signal, and consequently the corresponding generated spectrum, will be "rotating", indicating in each instant the magnitude of the demand vector. This approach seems quite consistent with the spectrum concept and will allow defining single values for point C and D for each ground motion.

The definition of such single spectrum has nothing to do with the combination of actions on buildings, resulting from their response in different directions (Stewart et al. (2011).

Considering 360 records originated by 24 different events in a relatively small window of time and location (i.e. all the available signal with an adequate quality recorded in Italy between 1972 and 2017) it has been possible to show that a limited number of equations and empirically derived parameters can produce reasonably accurate spectral shapes.

These results seem very promising and may result in more reliable design input representations.

The objective of producing spectra with a rather uniform probability of exceedance seems to have been reasonably attained, with reference to specific ground motions.

In any case, the procedure discussed in this paper is not conditioned by the desired protection level and allows in principle to increase or reduce the spectral values in a consistent way.

In addition, all parameters used to define the equations to derive the spectra have an intuitive and physical meaning, bringing a refreshing clarity to the process of estimating ground shaking based on magnitude and distance.

Though very promising, the proposal described in this work requires significant further research on a number of subjects to introduce these ideas and procedures into practice, as summarized below.

(a) The actual dependency of acceleration and displacement demand and spectral shape on magnitude and distance only should be investigated considering a much more extensive database (e.g.: Akkar et al. (2014).

- In the case of a large earthquakes the distance to be considered is from epicenter, fault or focus (Bommer and Akkar (2012); Monelli et al. (2014); Silva (2016))? The focal depth has some influence on the results?
- The source mechanism and type of fault have an effect on the results or will these be included in the produced magnitude (Chiou and Youngs (2008))?

(b) It is unquestionable that local soil effect will have an impact on the spectra shape and a lot of past studies have been focusing on this aspect (e.g.: Borcherdt (1994); Pitilakis et al. (2012)). All the findings from the past and possibly some novel experimental and numerical evidence should be considered and re-visited to provide the appropriate correction factors to the spectra here derived.

(c) Magnitude and distance have certainly an effect on ground motion duration and number of relevant cycles (e.g.: Novikova and Trifunac (1994)). This matter has to be included in the prediction equations, possibly together with effects of forward directivity.

(d) Within the framework of design based on spectral demand and structure capacity, the effects of energy dissipation have been traditionally included in the demand side, correcting the spectra. If this should be the case, it will be necessary to consider the possible effects of different periods of vibration (Akkar 2015). However, it seems in principle more rational to define alternative rules to switch the consideration of dissipation on the side of capacity, consistently with its nature related to structure response, not to ground motion demand.

(e) The equations derived can be regarded as a different form of ground motion prediction equations. This may allow the derivation of innovative seismic hazard maps (e.g.: Stucchi et al. (2011)), passing directly from a probability assessment of potential events to the combination of spectra associated to each event at different locations to produce probability based design spectra.

(f) The representation of design spectra in the combined $S_a - S_d$ form calls for the derivation or selection of consistent ground motion time histories (Bradley (2010); Mukherjee and Gupta (2002)). Any correction to better fit a specific spectral region may be defined by an interval in both key parameters.

(g) The implications of the input definition presented in this paper are certainly significant and may foster a radical change in design philosophy, though still within a general displacement-based approach. Considering the progressive orientation of design towards damage limitations, it appears particularly intriguing to explore the possibility of a direct derivation of combined floor spectra (Calvi 2014; Calvi and Sullivan 2014).

Considering the results discussed in this paper and the potential extensions mentioned above, it is expected that the new input definition will result in noticeable implications on design and assessment. Within the framework of displacement base design, the preliminary definition of a design displacement compatible with a desired damage level, possibly referred to specific classes of non-structural elements, may consent the immediate estimate of the corresponding energy dissipation and required strength.

Within the framework of displacement based assessment, the analysis of damage conditions resulting from known spectral demands may enlighten the actual correlations between demand, equivalent stiffness and damping and corresponding damage level. This may specifically apply to back analysis of recent events used in this study, such as the earthquake sequence of Central Italy in 2016 (Luzi et al. (2017)), or to successive events, such as the earthquake in Ischia, Italy, of August 2017.

Finally, the possible application of the concepts derived in this work for natural events to cases of induced seismicity could be of extreme interest: different correlations between magnitude and distance and displacement and acceleration demand may characterize short duration induced events with relatively low magnitude and very superficial focus.

References

Akkar S (2015) Revised probabilistic hazard map of Turkey and its implications on seismic design. EU workshop "elaboration of maps for climatic and seismic actions for structural design in the Balkan region", Zagreb

Akkar S, Sandıkkaya MA, Şenyurt M, Azari Sisi A, Ay BÖ, Traversa P, Douglas J, Cotton F, Luzi L, Hernandez B, Godey S (2014) Reference database for seismic ground-motion in Europe (RESORCE). Bull Earthq Eng 12(1):311–339

Bommer JJ, Akkar S (2012) Consistent source-to-site distance metrics in ground-motion prediction equations and seismic source models for PSHA. Earthq Spectra 28(1):1–15

Booth E (2007) The estimation of peak ground-motion parameters from spectral ordinates. J Earthq Eng 11(1):13–32

Borcherdt RD (1994) Estimates of site-dependent response spectra for design (methodology and justification). Earthq Spectra 10(4):617–653

Bradley BA (2010) A generalized conditional intensity measure approach and holistic ground-motion selection. Earthq Eng Struct Dyn 39:1321–1342

Calvi PM (2014) Relative displacement floor spectra for seismic design of non structural elements. J Earthq Eng 18(7):1037–1059

Calvi GM (2018) A re-visitation of design earthquake spectra (earthquakes and unicorns). Earthq Eng Struct Dyn, submitted for publication

Calvi PM, Sullivan TJ (2014) Estimating floor spectra in multiple degree of freedom structures. Earthq Struct 7(1):17–38

Calvi GM, Rodrigues D, Silva V (2018) Response and design spectra from Italian earthquakes 2012–2017. Earthq Eng Struct Dyn, submitted for publication

Chiou B, Youngs R (2008) An NGA model for the average horizontal component of peak ground motion and response spectra. Earthq Spectra 24(1):173–215

Chopra AK (2007) Elastic response spectrum; a historical note. Earthq Eng Struct Dyn 36:3–12

Fajfar P, Fishinger M (1988) N2 – a method for nonlinear analysis of regular buildings. In: Proceedings of the 9th WCEE, vol V. Tokyo and Kyoto, pp 111–116

Freeman SA (1975) Evaluation of existing buildings for seismic risk: a case study of Puget sound naval shipyard, Bremerton, Washington. In: Proceedings of the 1st US national conference on earthquake engineering, EERI, Oakland, CA, pp 133–122

Housner GW (1941) Calculating response of an oscillator to arbitrary ground motion. Bull Seismol Soc Am 31:143–149

Irvine T. http://www.vibrationdata.com/elcentro.htm, Vibrationdata, Madison, AL

Jacobsen LS (1960) Damping in composite structures. In: Proceedings of the 2nd WCEE, vol II. Tokyo and Kyoto, pp 1029–1044

Luzi L, Pacor F, Puglia R, Lanzano G, Felicetta C, D'Amico M, Michelini A, Faenza L, Lauciani V, Iervolino I, Baltzopoulos G, Chioccarelli E (2017) The Central Italy seismic sequence between august and December 2016: analysis of strong-motion observations. Seismol Res Lett 88(5):1219–1231

Monelli D, Pagani M, Weatherill G, Danciu L, Garcia J (2014) Modeling distributed seismicity for probabilistic seismic-hazard analysis: implementation and insights with the OpenQuake-engine. Bull Seismol Soc Am 104:1636–1649

Mukherjee S, Gupta VK (2002) Wavelet-based generation of spectrum-compatible time-histories. Soil Dyn Earthq Eng 22(9–12):799–804

Newmark NM, Hall WJ (1982) Earthquake spectra and design. Engineering monographs, EERI, Oakland, CA

Novikova EI, Trifunac MD (1994) Duration of strong ground motion in terms of earthquake magnitude, epicentral distance, site conditions and site geometry. Earthq Eng Struct Dyn 23:1023–1043

Pinho R, Marques M, Monteiro R, Casarotti C, Delgado R (2013) Evaluation of nonlinear static procedures in the assessment of building frames. Earthq Spectra 29(4):1459–1476

Pitilakis K, Riga E, Anastasiadis A (2012) Design spectra and amplification factors for Eurocode 8. Bull Earthq Eng 10(5):1377–1400

Priestley MJN (2003) Myths and fallacies in earthquake engineering. IUSS Press, Pavia

Priestley MJN, Calvi GM, Kowalsky MJ (2007) Displacement based seismic design of structures. IUSS Press, Pavia

Riddell R (2008) Inelastic response spectrum; early history. Earthq Eng Struct Dyn 37:1175–1183

Shibata A, Sozen M (1976) Substitute structure method for seismic design in reinforced concrete. ASCE J Struct Eng 102(1):1–18

Silva V (2016) Critical issues in earthquake scenario loss modeling. J Earthq Eng 20(8):1322–1341

Silva V, Crowley H, Pinho R, Varum H, Sousa R (2014) Evaluation of analytical methodologies to derive vulnerability functions. Earthq Eng Struct Dyn 43(2):181–204

Stewart JP, Abrahansom NA, Atkinson GM, Baker JW, Boore DM, Bozorgnia Y, Campbell KW, Comartin CD, Idriss IM, Lew M, Mehrain M, Mohele JP, Naeim F, Sabol TA (2011) Representation of bidirectional ground motions for design spectra in building codes. Earthq Spectra 27(3):927–937

Stucchi M, Meletti C, Montaldo V, Crowley H, Calvi GM, Boschi E (2011) Seismic hazard assessment (2003–2009) for the Italian building code. Bull Seismol Soc Am 101(4):1885–1911

Suzuki A, Iervolino I (2017) Italian vs worldwide history of largest PGA and PGV. Ann Geophys 60(5):S0551

Tolis SV, Faccioli F (1999) Displacement design spectra. J Earthq Eng 3(1):107–125

Chapter 4
From Force- to Displacement-Based Seismic Design of Concrete Structures and Beyond

Michael N. Fardis

Abstract Earthquakes impart to structures energy and produce displacements, both of which depend on the structure's pre-yielding natural period but not on its strength. The resulting seismic force is normally equal to the structure's lateral resistance. Nevertheless, seismic design is still carried out for empirically specified lateral forces, proportional to the ground motion intensity. Displacement-based seismic design (DBD) requires realistic estimation of seismic deformation demands and of the corresponding deformation capacities. A comprehensive and seamless portfolio of models for the secant-to-yield-point stiffness (which is essential for the calculation of displacements and deformations by linear or nonlinear analysis) and the ultimate deformation under cyclic loading has been developed, covering all types of concrete members, with continuous or lap-spliced bars, ribbed or smooth. The effect of wrapping the member in Fiber Reinforced Polymers is also considered. DBD is now making an entry into European standards, sidelining the earlier, more promising idea of energy-based seismic design, although energy lends itself better than displacements as a basis for seismic design: (a) being a scalar, it relates best to the 3D seismic response and damage; (b) it has a solid basis: energy balance; (c) its evolution during the computed response flags numerical problems. The initial enthusiasm for seismic energy 25 years ago led to a boom in activity on energy demand, but ran out of steam without touching on the more challenging issue of energy capacity of components. This is a fertile field for seismic engineering research.

M. N. Fardis (✉)
University of Patras, Patras, Greece
e-mail: fardis@upatras.gr

4.1 Introduction

Structural design boils down to the verification of the Safety Inequality:

$$\text{Demand} \leq \text{Capacity} \tag{4.1}$$

If both sides of Eq. (4.1) are internal forces or moments, the design is force-based (FBD); if they are displacements or deformations (deflections, rotations, curvatures, strains), the design is displacement-based (DBD); if both sides are energies, it is energy-based (EBD). The Vision 2000 report on future seismic design codes (SEAOC 1995) identified DBD and EBD as the promising approaches to be developed within a conceptual framework for performance-based seismic design.

To the present day seismic design is force-based: like wind, earthquakes are considered to impose lateral forces. An elastic analysis of the structure gives the seismic internal force/moment demands in order to verify or adjust the capacities (i.e., resistances in terms of internal forces/moments) against them. This time-honored practice survives thanks to the solid foundation provided by equilibrium, and to its convenience: internal forces/moments due to gravity, which is a force-based loading, are combined with those from a linear analysis for the earthquake.

Contrary to the presumption that an earthquake exerts specified forces on a structure, in reality it imparts to it (through the foundation) seismic energy. From another viewpoint, it produces displacements relative to the ground. The forces are a by-product of the displacements, not the other way around. The magnitude of the internal forces is controlled by the members' resistance at a number of locations sufficient to turn the structure into a mechanism, swaying back and forth up to the peak lateral displacement produced by the earthquake. Fortunately, this displacement may be estimated through an empirical observation: the "equal displacement" approximation (often called "rule") holds that the peak seismic displacement is about equal to that of the elastic structure, for any value of the ratio, R, between the resultant of the lateral forces produced by the earthquake on the elastic structure to the global lateral force resistance equilibrated by member resistances. If the global lateral force-displacement relationship is elastic-perfectly plastic, the ratio, μ, of the peak seismic displacement to the displacement at the instant the structure turns from elastic to a plastic mechanism (i.e., the ductility factor) is equal to the force reduction factor R (in Europe "behavior factor", q). In systems with fundamental natural period T less than the "control" or "predominant" period of the earthquake, T_C, a value of μ larger than R is required (Vidic et al. 1994).

In FBD member deformations are indirectly addressed at the end of the design process via the ductility factor μ: members which are part of the plastic mechanism and have their resistance checked via Eq. (4.1) against the internal moment/force demands from the elastic seismic analysis divided by R, should have deformation capacity not less than μ times the deformation at yielding. By contrast, in DBD, proposed first in (Moehle 1992), then by (Priestley 1993), Eq. (4.1) should be directly checked in terms of seismic deformation demands and capacities.

DBD saw independent versions being conceived and developing into two fiercely competing schools of thought and practice. As a matter of fact, design of members to match seismic displacement or deformation demands with their own capacities is not central to either of these versions of DBD. One version converts top displacement demands to internal forces and moments for which members are designed (as in FBD); in the other version, members provisionally designed for other types of loadings (gravity, wind, etc.) have their design updated to match their capacities to the seismic deformation demands. The rationality of the new paradigm notwithstanding, DBD has not made a serious dent yet in seismic design codes or practice: it is used only for seismic assessment and retrofitting of older structures (CEN 2005; ASCE 2007). The new generation of Eurocodes (CEN/TC250 2013) tentatively considers DBD as an alternative to FBD for new designs.

The author's research team identified early on the chord rotation at the member ends as the appropriate deformation measure for DBD, and has proposed DBD methodologies (Fardis and Panagiotakos 1997; Panagiotakos and Fardis 1998, 1999a; 2001a; Bardakis and Fardis 2011a) in which the reinforcement and the detailing of members already designed for non-seismic actions are tailored to the inelastic chord rotation demands estimated through linear elastic analysis – equivalent static or of the modal response spectrum type. To support these proposals, inelastic chord rotation demands from nonlinear response-history analyses were compared to their counterparts from linear elastic analysis, and rules were proposed for the estimation of the former from the latter depending on the location in the building (Economou et al. 1993; Panagiotakos and Fardis 1999b; Kosmopoulos and Fardis 2007). On average, estimates from the two analysis approaches were found to be equal, at least in concrete structures, which normally have fundamental period longer than the dominant period of the motion. This finding was the basis of the Eurocode 8 rules (CEN 2005), which allow such a simple estimation under certain conditions, without the multiplicative factors of the US coefficient approach (ASCE 2007). The most important contribution of the author's team to DBD, though, concerns the capacity side: a large and diverse database of cyclic tests on all types of concrete members was gradually assembled and provided the ground for models for the chord rotation at yielding (which, in addition to serving as a limit deformation for serviceability, gives the effective stiffness of the member, see Eq. (4.2)) and at ultimate conditions (Panagiotakos and Fardis 2001b). The second version of these models, based on an enriched database and covering members with lap-spliced bars or wrapped in Fiber Reinforced Polymer (FRP) over their end regions, and members having smooth bars as longitudinal reinforcement, were adopted in the first generation of Eurocode 8 as a European Standard (EN) (Biskinis and Fardis 2007, 2010a, 2010b, 2013b). The third version, highlighted in Sect. 4.2 below, builds on over 4000 member tests. It is a comprehensive and seamless portfolio that covers the full spectrum of concrete members encountered in new designs or in the old substandard building stock, as well as in seismic retrofitting applications (Grammatikou et al. 2015, 2016, 2018a, b, c). The core of the third version has tentatively been adopted in the draft-new-Eurocode 8, part of the upcoming second generation of Eurocodes (CEN/TC250 2013).

4.2 Member Properties for Displacement-Based Design

4.2.1 Which Properties Do We Need to Know?

DBD needs simple means to estimate the member deformation capacities and seismic deformation demands, which appear in Eq. (4.1). Deformation measures to be used in Eq. (4.1) should be easy to extract from analysis results, as well as meaningful indicators of member failure. Strains are not a good measure of loss of member resistance; moreover, their prediction by analysis is model-dependent. In contrast, the chord rotation capacity – total or plastic – correlates well with local damage and loss of resistance. On the demand side, inelastic chord rotations may be estimated even by linear analysis (Economou et al. 1993; Panagiotakos and Fardis 1999b; Kosmopoulos and Fardis 2007; Bardakis and Fardis 2011b).

The elastic stiffness of members holds the key to a relatively accurate estimation of member chord rotation demands. The value used for this stiffness is that of the fully cracked member. The reason is that an earthquake strong enough to drive members to yielding and beyond is rare; so, it gives the necessary time to the variable loads and ambient temperature, to restrained thermal and drying shrinkage and even to small earthquakes to occur earlier, at magnitudes large enough to thoroughly crack every member. Hence, a seismic response analysis should use as elastic stiffness the secant value to the apparent yield point in a bilinear approximation of the envelope to the cyclic moment-deformation response.

Section 4.2 summarizes a comprehensive portfolio of rules for the estimation of a member's "effective stiffness", EI_{eff}, and ultimate chord rotation, θ_u – the latter identified with a drop in moment resistance of at least 20% of its maximum possible value. For details see (Grammatikou et al. 2016, 2018a, b, c).

4.2.2 Effective Stiffness of Members with Ribbed or Smooth Bars

A common practice in nonlinear seismic response analysis is to model every discrete member – a beam or a column between adjacent beam-column joints, the part of a wall between successive floors, etc. – as a single nonlinear element. Moreover, all sources of flexibility between its end nodes, namely flexure, shear and slippage of longitudinal bars from their anchorage zone(s) outside the physical member, are lumped into an effective flexibility of the member in bending. Its inverse (i.e., the member's secant-to-yield-point effective stiffness) relates the yield moment of the member's end section to the yield value, θ_y, of an apparent chord rotation, in which all deformations between the theoretical node and the end of the shear span are lumped, as if they were all due to bending. Chord rotation refers to the shear span, L_s, of the end of interest, which is defined as the moment-to-shear ratio there. Then, at that end of the member, we have:

$$EI_{eff} = \frac{M_y L_s}{3\theta_y} \quad (4.2)$$

The yield moment, M_y, may be computed from section analysis, lumping all bars that are near the extreme tension or compression fibers into concentrated tension and compression reinforcement, and considering the cross-sectional area of all other bars in the section as uniformly distributed between those lumped near the extreme fibers. Elastic σ-ε laws may be used; yielding of the tension reinforcement (or over three-eighths of the tension zone's perimeter in circular sections) may serve as a section's yield criterion (Grammatikou et al. 2016). If the end region is wrapped in FRP, the concrete Modulus is estimated from the strength of the FRP-confined concrete, f_{cc}, e.g., in MPa as $E_c = 10,000(f_{cc})^{1/3}$ (fib 2012). Then, estimation of EI_{eff} boils down to estimation of θ_y and use of Eq. (4.2).

A single-valued elastic stiffness is obtained for the member by averaging the values from Eq. (4.2) for positive and negative bending at the two member ends.

4.2.2.1 Members with Continuous Ribbed (Deformed) Bars

The plane sections hypothesis is taken to apply in members with ribbed bars. If ϕ_y is the yield curvature of the end section (computed alongside M_y), flexural deformations along L_s contribute to θ_y the first term in Eq. (4.3), which takes into account that yielding of tension bars spreads from the end section till the point where they cross a 45° crack (if a crack forms due to shear before the end section yields):

$$\theta_y = \phi_y \frac{L_s + a_V z}{3} + \frac{\phi_y d_b f_y}{8\sqrt{f_c}} + \theta_{shear,y} \quad (4.3)$$

The internal lever arm, z, appears in the first term if $V_y = M_y/L_s$ exceeds the shear resistance without shear reinforcement; then $a_V = 1$; if it doesn't, then $a_V = 0$. The second term is the fixed-end-rotation of the end section due to slippage of the member's tension bars – having diameter d_b and yield stress f_y (MPa) – from their anchorage zone past the member end; it is based on an assumed bond stress (in MPa) equal to $\sqrt{f_c}$(MPa) and is confirmed by experimental measurements of the fixed-end-rotation. The last term in Eq. (4.3) is the shear deformation, given by the following empirical fits to experimental results (Grammatikou et al. 2018a, b):

$$\text{Rectangular columns or beams}: \quad \theta_{shear} = 0.0019\left(1 + \frac{h}{1.6L_s}\right) \quad (4.4a)$$

$$\text{Walls(rectangular or not), box sections:} \quad \theta_{shear} = 0.0011\left(1 + \frac{h}{3L_s}\right) \quad (4.4b)$$

$$\text{Circular columns:} \quad \theta_{shear} = 0.0025\left(1 - \max\left(1; \frac{L_s}{8D}\right)\right) \quad (4.4c)$$

where h is the depth of the section and D the diameter of a circular section.

4.2.2.2 Members with Lap-Spliced Ribbed Bars – Effect of FRP-Wrapping

The rules in this section apply to members of any cross-section, if the longitudinal bars have straight ends lap-spliced over a length l_o, starting at the end section.

Rule 1: Both bars in a pair of lapped compression bars count as compression reinforcement (Biskinis and Fardis 2007, 2010a).

Rule 2: In a section analysis for M_y and ϕ_y the maximum stress, f_{sm}, that can build up in the lap-spliced tension bars is (Grammatikou et al. 2018a):

$$f_{sm} = \min\left(\sqrt{\frac{l_o}{l_{oy,min}}};\ 1\right) f_y \quad (4.5)$$

$$\text{where:} \quad l_{oy,min} = \frac{d_b f_y}{4 f_{ct} \max\left(\frac{c_{min}}{d_b};\ 0.7\right)\left(1 + 4\frac{E_f t_f}{E_c R_c}\right)^2} \quad (4.6)$$

with c_{min}: minimum concrete cover of lapped bars, or half the clear distance to the closest lap-spliced bar, if smaller; t_f: thickness of FRP (if any), R_c: radius of circular section or at chamfered corners of rectangular one; E_f, E_c: elastic modulus of FRP or concrete; $f_{ct}(MPa) = 0.3(f_c(MPa))^{2/3}$ concrete tensile strength.

Rule 3: Rules 1 and 2 apply to the value of ϕ_y for Eq. (4.3). The second term in Eq. (4.3) is multiplied by the ratio of M_y, as modified by Rules 1 and 2 for the splicing, to M_y without a splice. Besides, in order to decide if $a_V = 1$ in the first term, the end moment at diagonal cracking, $L_s V_{Rc}$, is compared to the value of M_y that accounts for the splices via Rules 1 and 2 (Biskinis and Fardis 2010a).

4.2.2.3 Members with Smooth Bars, Including Columns with Lap-Splices at Floor Levels

Although section analysis estimates well the yield moment at the end section of a member with smooth (plain) bars, the plane sections hypothesis cannot be taken to apply throughout the shear span. A non-flexural physical model, of the strut-and-tie type, assumes that tension dominates throughout the length of a smooth bar, linearly varying from the yield stress at the extreme tension fibers of an end section which has yielded, to the maximum possible stress that can develop at its anchorage. According to (*fib* 2012) ahead of a standard 180° hook that stress is equal to 60 times the bar's bond strength, which, for vertical smooth bars and stresses in MPa, was given in the ENV version of Eurocode 2 as equal to $0.36\sqrt{f_c}$. So, the stress in a vertical smooth bar at a standard 180° hook may be taken as $f_o = 22\sqrt{f_c}$.

The vertical bars of multi-story building columns are fixed at floor levels; they are lap-spliced there with the top end of the bars of the underlying column, as in Fig. 4.1a. Figure 4.1a depicts the assumed distribution of bar stresses at the instant the columns reach their yield moments at the top and bottom of two successive stories. This distribution is similar to that assumed in (Grammatikou et al. 2018c) for double-fixity columns and verified by comparing the predicted moments and chord rotations at yielding with the results of tests on single-story columns. Figure 4.1b depicts the distribution of steel stresses at the upper- and lowermost stories and in starter bars embedded in the foundation. For generality, the yield stress and the stresses at the top and bottom ends of the bars, indexed by y, ot and ob, respectively, are taken to be different at each story and are indexed accordingly; index 0 is used for the starter bars. H_i is the total height of story i, $h_{b,i}$ the beam depth at the top of story i, and z_i the internal lever arm of a column in that story; $l_{o,i}$ is the lapping of vertical bars at the base of column i.

The yield chord rotation at the top and bottom end of a column in story i at yielding of the corresponding end section is estimated from a strut-and-tie model (Grammatikou et al. 2018c). With the distribution of steel stresses in Fig. 4.1 and neglecting the effect of shortening of the concrete strut, we get:

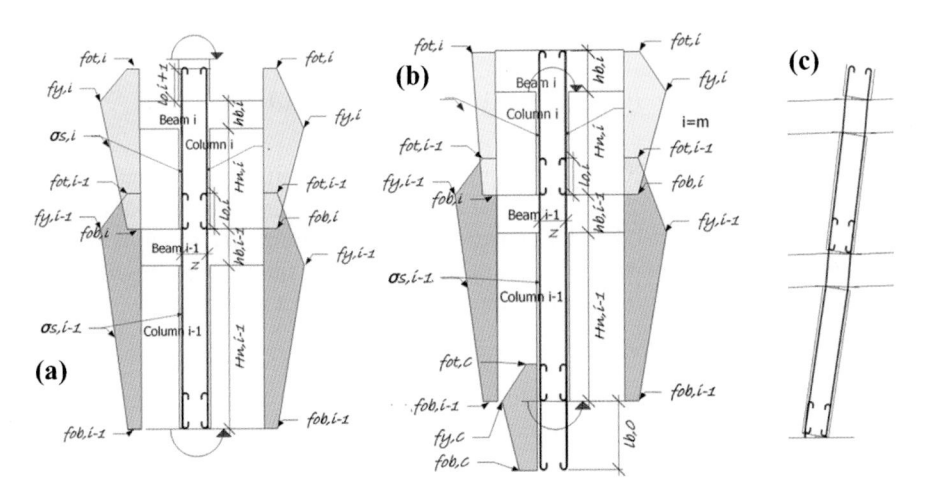

Fig. 4.1 Multistory column with smooth bars lap-spliced at floor levels at incipient yielding: (**a**) assumed distribution of bar tensile stress in two adjacent stories; (**b**) as in (**a**) but at the lower- and upper-most story; (**c**) deformed column

$$\theta_{y,top,i} = \frac{H_i\left(f_{y,i} + f_{ob,i}\right) + h_{b,i}\left(f_{ot,i} - f_{ob,i}\right) + l_{o,i+1}\left(f_{y,i} + f_{ot,i}\right)}{2E_s z_i} \quad (4.7a)$$

$$\theta_{y,bot,i} = \frac{H_{i-1}\left(f_{y,i-1} + f_{ob,i-1}\right) + l_{o,i}\left(f_{y,i-1} + f_{ot,i-1}\right)}{2E_s z_i} \quad (4.7b)$$

At the top story (Fig. 4.1b), indexed by $i = m$, the value $l_{o,i+1} = 0$ should be used in Eq. (4.7a). At the connection to the foundation (Fig. 4.1b), indexed by 0 if $i = 1$ at the lowermost story, the embedment depth of starter bars in the foundation, denoted by $l_{b,0}$ in Fig. 4.1b, should be used in Eq. (4.7b) instead of H_{i-1}.

4.2.3 Physical Models for the Ultimate Chord Rotation

4.2.3.1 Members with Continuous Ribbed Bars

Inelastic flexural deformations are lumped in a plastic hinge length, L_{pl}, measured from the end which has yielded. The inelastic part of the curvature, ϕ-ϕ_y, is taken as constant over L_{pl} and as zero outside. A post-elastic fixed-end rotation, $\Delta\theta_{slip}$, develops after the end section yields, owing to penetration of yielding in the anchorage zone of the tension bars beyond the member end. The ultimate chord rotation of the shear span, θ_u, and the ultimate curvature of the end section, ϕ_u, are taken to occur when the end section's moment resistance drops to less than 80% of its peak value. By the time ϕ_u is reached, $\Delta\theta_{slip}$ has increased to $\Delta\theta_{u,slip}$. So:

$$\theta_u = \theta_y + (\phi_u - \phi_y) L_{pl}\left(1 - \frac{L_{pl}}{2L_s}\right) + \Delta\theta_{u,slip} \quad (4.8)$$

Biskinis and Fardis (2010b) give equations and flowcharts for the calculation of the ultimate curvature, ϕ_u, for all possible failure modes of a section in flexure. For a parabolic-rectangular σ-ε law for concrete and an elastic-linearly strain-hardening one for steel, the following ultimate strains give optimal fitting of the resulting ϕ_u values to measured ones in cyclic loading (Grammatikou et al. 2016):

(i) Before spalling of the concrete cover:

- for tension bars having uniform elongation $\varepsilon_{su,nom}$ at tensile strength in a standard coupon test, in a section without FRP wrapping:

$$\varepsilon_{su} = 0.4\varepsilon_{su,nom} \quad (4.9)$$

- for the $N_{b,\,tension}$ bars in the tension zone of a section wrapped in FRP:

$$\varepsilon_{su,FRP} = 0.6\varepsilon_{su,nom}(1-0.15\ln(N_{b,tension}))^{1/2} \quad (4.10)$$

- for unconfined concrete:

$$0.0035 \leq \varepsilon_{cu} = \left(\frac{18.5}{h(mm)}\right)^2 \leq 0.01 \quad (4.11)$$

- for FRP-confined concrete:

$$\varepsilon_{cu,FRP} = \varepsilon_{cu,Eq.(4.11)}$$
$$+ \alpha_f \beta_f \min\left(0.5; \frac{0.6\varepsilon_{u,f}E_f}{f_c}\rho_f\right)\left(1 - \min\left(0.5; \frac{0.6\varepsilon_{u,f}E_f}{f_c}\rho_f\right)\right) \quad (4.12)$$

where:

$$- \quad \alpha_f = 1 - \frac{(b_y - 2R)^2 + (b_x - 2R)^2}{3b_x b_y} \quad (4.13)$$

with b_y, b_x: sides of rectangle circumscribed around section, and R: radius of the corners of the section ($b_y = b_x = 2R = D$ in circular sections);
- $\beta_f = 0.115$ for Carbon or Glass FRP (CFRP, GFRP), $\beta_f = 0.1$ for Aramid FRP (AFRP);
- $\varepsilon_{u,f}$: failure strain, equal to 1.5% for CFRP or ARFP, 2% for GFRP;
- E_f, ρ_f: Elastic Modulus and geometric ratio of FRP in loading direction.

(ii) After spalling of the concrete cover (including FRP-wrapped sections, if the ultimate strain of Eq. (4.12) has been reached) (Grammatikou et al. 2018a):

- for the tension bars:

$$\varepsilon_{su} = \frac{4}{15}\left(1 + 3\frac{d_b}{s_h}\right)\left(1 - 0.75e^{-0.4N_{b.compr}}\right)\varepsilon_{su,nom} \quad (4.14)$$

ε_{su} drops with increasing stirrup-spacing-to-bar-diameter ratio, s_h/d_b, because in cyclic loading bars break in tension after they buckle in a previous half-cycle; it also increases with increasing number of bars at the extreme compression fibers, $N_{b,compr}$ (taken equal to 2 in a circular section), because these bars share with the member its curvature in the longitudinal direction (ie, their convex side faces inwards). To buckle outwards, their curvature must be reversed, which is unlikely for intermediate bars before the confined core crumbles; this is not an issue for the corner backs, which buckle sideways. The larger $N_{b,compr}$ is, the more the intermediate bars are.

- for the confined concrete core inside the steel ties:
  - in a rectangular compression zone:

$$\varepsilon_{cu,c} = \varepsilon_{cu,Eq.(4.11)} + 0.04\sqrt{\frac{a\rho_w f_{yw}}{f_c}} \qquad (4.15a)$$

  - in a circular section:

$$\varepsilon_{cu,c} = \varepsilon_{cu,Eq.(4.11)} + 0.07\sqrt{\frac{a\rho_w f_{yw}}{f_c}} \qquad (4.15b)$$

The depth of the confined core, h_o, is used in Eq. (4.11) as h; ρ_w and f_{yw} are the volumetric ratio and the yield stress of transverse steel; the confinement effectiveness factor a is:
  - in rectangular sections:

$$a = \left(1 - \frac{s_h}{2b_o}\right)\left(1 - \frac{s_h}{2h_o}\right)\left(1 - \frac{\sum b_i^2}{6b_o h_o}\right) \geq 0 \qquad (4.16a)$$

where b_o, h_o: confined core dimensions to the centerline of the outer tie, b_i: spacing of centers of adjacent bars (indexed by i) which are restrained by a tie corner or a hook; and s_h: centerline spacing of ties.
  - for circular sections:

$$a = \left(1 - \frac{s_h}{2D_o}\right)^n \qquad (4.16b)$$

with $n = 2$ for individual circular hoops. $n = 1$ for spiral reinforcement.

The strength of concrete confined by steel ties, f_{cc}, and the associated strain are given by Eq. (4.17) (*fib* 2012), with concrete strain at f_c, ε_{co}, equal to 0.002 and a from Eq. (4.16a, b).

$$f_{cc} = f_c(1 + K), \varepsilon_{co,c} = \varepsilon_{co}(1 + 5K), \text{ with } K = 3.5\left(\frac{a\rho_w f_{yw}}{f_c}\right)^{3/4} \qquad (4.17)$$

If ϕ_u is computed according to the above and ϕ_y from section analysis, the post-yield cyclic fixed-end-rotation due to bar slippage is (Grammatikou et al. 2018a):

$$\Delta\theta_{u,slip} = 4.5d_b\phi_u, \text{ or } \Delta\theta_{u,slip} = 4.25d_b(\phi_u + \phi_y) \qquad (4.18)$$

For so-determined ϕ_u, ϕ_y and $\Delta\theta_{u,slip}$, and with $\nu = N/A_c f_c$ (A_c: cross-sectional area, N axial force, positive for compression), expressions for L_{pl} were derived for cyclic-flexure-controlled members conforming to recent seismic design codes:

- For columns with circular section (Grammatikou et al. 2018a):

$$\frac{L_{pl}}{D} = 0.7(1 - \min(0.7; \nu))\left(1 + \frac{1}{7}\min\left(9; \frac{L_s}{h}\right)\right) \quad (4.19a)$$

- For sections made of rectangular parts with web width b_w, depth h:

$$\begin{aligned}\frac{L_{pl}}{h} =\ & 0.3(1 - 0.45\min(0.7;\ \nu))\left(1 + 0.4\min\left(9;\ \frac{L_s}{h}\right)\right) \\ & \times \left(1 - \frac{1}{3}\sqrt{\min\left(2.5;\ \max\left(0.05; \frac{b_w}{h}\right)\right)}\right)\end{aligned} \quad (4.19b)$$

Poor details reduce the cyclic flexure-controlled ultimate curvature much more than the ultimate chord rotation. So, the plastic hinge length to be used in Eq. (4.8) for non-conforming members alongside the smaller ultimate curvature, is:

$$L_{pl,nc} = 1.3 L_{pl, Eqs.(4.19)} \quad (4.20)$$

Equation (4.20) does not apply after wrapping the plastic hinge region with FRP, as the wrapping heals poor detailing due to nonconformity with seismic design codes.

4.2.3.2 Members with Lap-Spliced Ribbed Bars – Effect of FRP-Wrapping

The ultimate chord rotation is unaffected by the splicing l_o if the latter exceeds:

$$l_{ou,\min} = \frac{d_b f_y}{f_{ct}\left(1 + a_c a_n a_s \sqrt{\frac{d_b}{2R_c}} \min\left(3; \frac{p_c}{f_{ct}}\right)\left(1 - \frac{1}{6}\min\left(3; \frac{p_c}{f_{ct}}\right)\right)\right)} \quad (4.21)$$

where (Grammatikou et al. 2018a):

- $a_n = 1$ in circular sections, $a_n = n_{restr}/n_{tot}$ in rectangular ones with n_{restr} lapped bar pairs restrained at corners or hooks of ties or inside a chamfered corner of an FRP jacket, out of a total of n_{tot} spliced bar pairs;
- $a_s = 1$ for confinement by FRP; for confinement by ties a_s is the product of the first two terms in Eq. (4.16a); in circular columns the diameter of the circular hoop, D_o replaces the centerline dimensions of the rectangular outer tie, b_o, h_o;

- $a_c = 7.5$ for confinement by steel ties, $a_c = 9.5$ for confinement by CFRP, $a_c = 10.5$ for GFRP, and $a_c = 12$ for confinement by AFRP;
- R_c is equal to the bending radius of the steel tie or of the FRP jacket;
- $p_c = A_{sh}f_{yw}/(s_h R_c)$ for confinement by steel ties; $p_c = t_f f_{u,f}/R_c$ for confinement by FRP of thickness t_f.

The minimum of the values given by Eq. (4.21) for confinement by ties or FRP is used as $l_{ou,min}$.

If the available lapping, l_o, is less than $l_{ou,min}$, the ultimate chord rotation decreases: with θ_y and ϕ_y determined according to Rules 1 to 3 in Sect. 4.2.2.2, θ_u is computed from Eqs. (4.8), (4.9), (4.10), (4.11), (4.12), (4.13), (4.14), (4.15a, 4.15b), (4.16a, 4.16b), (4.17), (4.18, 4.19a), (4.19b), with Eqs. (4.10), (4.12) applying to members with FRP wrapping and (4.9), (4.11) to those without. The ultimate curvature, ϕ_u, is computed by applying Rule 1 to the lapped compression bars and by reducing the maximum elongation of the extreme tension bars at ultimate conditions due to steel failure (Grammatikou et al. 2018a):

$$\varepsilon_{su,laps} = \min\left(1; \frac{l_o}{l_{ou,min}}\right)\varepsilon_{su} \geq \min\left(1; \frac{l_o}{l_{oy,min}}\right)\frac{f_y}{E_s} \quad (4.22)$$

Equations (4.9), (4.14) give ε_{su} in members without FRPs and Eqs. (4.10) and (4.14) for FRP-wrapped ones; Eqs. (4.6) and (4.21) give $l_{oy,min}$ and $l_{ou,min}$, respectively.

Equation (4.20) doesn't apply if the plastic hinge region is wrapped in FRP; such wrapping heals poor detailing due to nonconformity with seismic design codes. However, Eq. (4.20) applies to nonconforming members without FRP wrapping.

4.2.3.3 Multi-Story Columns with Smooth Bars Lap-Spliced at Floor Levels

The strut-and-tie model gives the ultimate chord rotation at column ends in story i:

$$\theta_{u,top,i} = \theta_{y,top,i} + (\phi_{u,i} - \phi_{y,i})\left(0.08\frac{H_i - h_{b,i}}{2} + 0.6(l_{o,i+1} + h_{b,i})\right) + \frac{\phi_{u,i}\xi_{u,i}d_i}{2}\left(\frac{H_i - h_{b,i}}{z_i} + \frac{z_i}{H_i - h_{b,i}}\right) \quad (4.23)$$

$$\theta_{u,bot,i} = \theta_{y,bot,i} + a_{lap,i}\left((\phi_{u,i-1} - \phi_{y,i-1})\left(0.08\frac{H_{i-1} + h_{b,i-1}}{2} + 0.6 l_{o,i}\right) + \frac{\phi_{u,i-1}\xi_{u,i-1}d_i}{2}\left(\frac{H_i - h_{b,i}}{z_i} + \frac{z_i}{H_i - h_{b,i}}\right)\right) \quad (4.24)$$

With $d_{b,i-1}$ the diameter of the bars in column i-1, the reduction factor due to lap-splicing of hooked bars is (Grammatikou et al. 2018c):

$$a_{lap,i} = \min\left(1; \frac{l_{o,i}}{50d_{b,i-1}}\right) \quad (4.25)$$

If there are FRP jackets of length L_f around each column end in story i, then:

$$\theta_{u,top,i,FRP} = \theta_{y,top,i} + (\phi_{u,i} - \phi_{y,i})\left(0.06\frac{H_i - h_{b,i}}{2} + 0.8(l_{o,i+1} + h_{b,i})\right)$$
$$+ \left(\phi_{u,i}\xi_{u,i}\left(\frac{1}{2} - \frac{L_f}{H_i - h_{b,i}}\right) + \phi_{u,c,i}\xi_{u,c,i}\frac{L_f}{H_i - h_{b,i}}\right)d_i\left(\frac{H_i - h_{b,i}}{z_i} + \frac{z_i}{H_i - h_{b,i}}\right) \quad (4.26)$$

$$\theta_{u,bot,i,FRP} = \theta_{y,bot,i}$$
$$+ a_{lap,FRP,i}\begin{pmatrix}(\phi_{u,i-1} - \phi_{y,i-1})\left(0.08\frac{H_{i-1} + h_{b,i-1}}{2} + 0.6l_{o,i}\right)\\ + \left(\phi_{u,i-1}\xi_{u,i-1}\left(\frac{1}{2} - \frac{L_f}{H_i - h_{b,i}}\right)\right.\\ \left.+\phi_{u,c,i-1}\xi_{u,c,i-1}\frac{L_f}{H_i - h_{b,i}}\right)d_i\left(\frac{H_i - h_{b,i}}{z_i} + \frac{z_i}{H_i - h_{b,i}}\right)\end{pmatrix} \quad (4.27)$$

If $d_{b,i-1}$ is the diameter of the column bars in story $i-1$, lap-spliced with 180° hooks to those of the column in story i, then in Eq. (4.27) (Grammatikou et al. 2018c):

$$a_{lap,FRP,i}$$
$$= \min\left(1; \frac{l_{o,i}}{50d_{b,i-1}}\left(1 + 300\min\left(0.05; \frac{\rho_f E_f}{E_c}\right)_i\left(0.1 - \min\left(0.05; \frac{\rho_f E_f}{E_c}\right)\right)\right)_i\right) \quad (4.28)$$

The geometry, the reinforcement and the axial force of the column in story i are used to compute $\phi_{y,i}$, $\phi_{u,i}$ and the neutral axis depth at ultimate conditions, $\xi_{u,i}$, in Eq. (4.23) and (4.26). In Eq. (4.24) and (4.27) $\phi_{y,i-1}$, $\phi_{u,i-1}$ and $\xi_{u,i-1}$ are calculated using the geometry and the axial force of the column in story i and the reinforcement of that of story i-1, taking into account the effects of the lap splice.

At the top of the uppermost story (indexed by $i = m$), $l_{o,i+1} = 0$ in Eqs. (4.23), (4.26). At the bottom of the lowermost story ($i = 1$), Eq. (4.24) is replaced by:

$$\theta_{u,bot,1} = \theta_{y,bot,1}$$
$$+ a_{lap,1} \begin{pmatrix} (\phi_{u,0} - \phi_{y,0})(0.08\max(l_{o,1};l_{b,0}) + 0.6\min(l_{o,1};l_{b,0})) \\ + \dfrac{\phi_{u,0}\xi_{u,0}d_1}{2}\left(\dfrac{H_1 - h_{b,1}}{z_1} + \dfrac{z_1}{H_1 - h_{b,1}}\right) \end{pmatrix}$$
(4.29)

Equation (4.27) is replaced similarly to Eq. (4.29) (Grammatikou et al. 2018c).

4.3 Empirical Ultimate Chord Rotation of Members with Section Consisting of Rectangular Parts

Of the five empirical ultimate chord rotation models proposed in (Grammatikou et al. 2018b) for members with section consisting of rectangular parts, the following was chosen in the revision of Eurocode 8 (CEN/TC250 2013):

$$\theta_u = \theta_y + \theta_u^{pl} = \theta_{y,Eqs.(4.3),(4.4a),(4.4b)} + 0.0206(1 - 0.41a_{w,r})(1 - 0.31a_{nr})$$
$$(1 - 0.22a_{nc})(0.2^{\min(0.7;\nu)})$$

$$\left[\dfrac{\max(0.01;\omega')}{\max(0.01;\omega_{tot} - \omega')}\right]^{\frac{1}{4}} \left(\min(50; f_c(MPa))\right)^{0.1} \left[\min\left(9; \dfrac{L_s}{h}\right)\right]^{0.35} 24^{\max\left(\frac{a\rho_s f_{yw}}{f_c}; \left(\frac{\alpha\rho f_u}{f_c}\right)_f\right)} 1.225^{100\rho_d}$$
(4.30)

where:

- $a_{w,r} = 1$ for rectangular walls, $a_{w,r} = 0$ for all other types of members;
- $a_{nr} = 1$ for T-, C- or box sections, $a_{nr} = 0$ for rectangular walls or columns;
- $a_{nc} = 0$ for conformity to recent seismic codes, $a_{nc} = 1$ for nonconformity;
- $\nu = N/bhf_c$, with b: width of compression zone and N: axial force, positive for compression;
- $\omega_{tot} = \Sigma\rho f_y/f_c$, $\omega' = \rho' f_y'/f_c$: mechanical ratio of all longitudinal bars and of the compression reinforcement, respectively;
- $L_s/h = M/Vh$: shear-span-to-depth ratio at the section of maximum moment;
- ρ_d: steel ratio of diagonal bars per diagonal direction of the member's lateral view.
- f_{yw}, $\rho_s = A_{sh}/b_w s_h$: yield stress and ratio of transverse steel parallel to applied shear;
- a: confinement effectiveness factor for steel ties from Eq.(4.16a);
- $(\alpha\rho f_u/f_c)_f$: confinement term due to FRP wrapping (if there is one):

$$- \left(\dfrac{\alpha\rho f_u}{f_c}\right)_f = a_f c_f \min\left[0.4; \dfrac{0.6\varepsilon_{u,f} E_f \rho_f}{f_c}\right]\left(1 - 0.5\min\left[0.4; \dfrac{0.6\varepsilon_{u,f} E_f \rho_f}{f_c}\right]\right) \quad (4.31)$$

where the new variable beyond those defined in connection with Eq. (4.12) is:

- $c_f = 1.9$ for CFRP, $c_f = 1.15$ for GFRP, and $c_f = 0.9$ for AFRP.

Lap splicing is taken to reduce the plastic part of the ultimate chord rotation as:

$$\theta^{pl}_{u,laps} = \min\left(1.4\sqrt{\frac{l_o}{l_{ou,\min}}} - 0.4;\ 1\right)\theta^{pl}_{u,Eq.(4.30)} \qquad (4.32)$$

The penultimate term in Eq. (4.30) becomes $24^{\max\left(\frac{a\rho_s f_{yw}}{f'c};\left(\frac{a\rho f_u}{f_c}\right)_{f,eff}\right)}$ if the lap-splice region is wrapped in FRP. The term in the exponent due to the FRP is taken from Eq. (4.31), but is ignored if the bars around the perimeter are eight or more. Correction for nonconformity with recent seismic design codes is not needed then (i.e., $a_{nc} = 0$), as wrapping the member end in FRP heals the effects of poor detailing.

4.4 The Case for Energy as the Basis of Seismic Design

Energy provides a better ground for seismic design than forces or displacements:

- Forces, the incumbent, are a phony basis: they are not the product of the earthquake but of the resistance of the structure; very different earthquakes will exert on the structure essentially the same forces.
- Displacements, the challenger, have a twofold basis:
  - an empirical observation: the equal displacement "rule" and its variant for structures with short period;
  - a physical reality, albeit not so straightforward to use as energy conservation or equilibrium: geometric continuity in the deformed structure.
- Energy holds many advantages over both forces and displacements:
  - Energy balance (or conservation), the basis of the energy approach, is a law of nature, as solid, familiar to engineers and easy to apply as the foundation of the force-based approach, i.e., equilibrium.
  - Given an earthquake, the input energy per unit mass depends almost exclusively on the structure's fundamental period and is roughly independent of the viscous damping ratio, the post-yield hardening ratio, the degree of inelastic action (as measured by the ductility factor) and the number of degrees of freedom (Zahrah and Hall 1984; Uang and Bertero 1990; Fajfar et al. 1992; Fajfar and Vidic 1994; Sucuoglu and Nurtug 1995; Surahman 2007; Cheng et al. 2014). This is the equivalent of the "equal displacement rule", matching the prime advantage of displacements over forces.

- Forces and displacements are vectors; their components in two (often arbitrarily chosen) orthogonal horizontal directions are normally considered separately in design, despite the fact that the most critical response to concurrent shaking in both horizontal directions and the associated damage occur in an intermediate direction. The true response is better described by a scalar measure, such as energy.
- Energy embodies more damage-related-information than displacements, such as the number of equivalent cycles (Zahrah and Hall 1984; Teran-Gilmore and Jirsa 2007).
- Numerical instabilities or lack of convergence during a nonlinear response-history analysis clearly show up in the evolution of the components of energy; so, a positive side-effect of tracing the energy components is the awareness of any numerical problems.

4.5 State-of-the-Art in Energy-Based Seismic Design

Shortly before the emergence of DBD, a few seminal publications co-authored by bigwigs of the S/T community of Earthquake Engineering (Zahrah and Hall 1984; Akiyama 1988, 1992; Uang and Bertero 1990) drew its attention to the seismic energy imparted to the structure through the foundation. Initial enthusiasm for seismic energy was great, as evidenced by the dozens of papers that followed in the short- to medium-term and are overviewed below.

Early research addressed mainly the total seismic energy input and its flow within the structure during the response, i.e., it focused on the demand (.Fajfar and Vidic 1994; Sucuoglu and Nurtug 1995; Bruneau and Wang 1996; Decanini and Mollaioli 1998, 2001; Safak 2000; Chai and Fajfar 2000; Manfredi 2001; Riddell and Garcia 2001; Chou and Uang 2000, 2003; Benavent-Climent et al. 2002, 2010; Kalkan and Kunnath 2007; Arroyo and Ordaz 2007; Teran-Gilmore and Jirsa 2007; Amiri et al. 2008; Cheng et al. 2014, 2015; Alici and Sucuoglu 2016). Standard shapes were proposed for the input energy spectra or for the energy-equivalent velocity (Decanini and Mollaioli 1998, 2001; Chai and Fajfar 2000; Manfredi 2001; Riddell and Garcia 2001; Benavent-Climent et al. 2002, 2010); these shapes peak at the dominant period of the motion, T_C. The most complete proposal seems to be that of Arroyo and Ordaz (2007): it gives the ratio of hysteretic-energy-equivalent velocity to peak ground velocity (PGV) in terms of the ratio of spectral pseudo-acceleration to peak ground acceleration (PGA), of spectral pseudo-velocity to PGV, of motion duration normalized to the ratio of the maximum spectral displacement to the maximum spectral pseudo-velocity, of soil type and of the ductility factor; it also includes a shift of the dominant period with increasing ductility.

The State-of-the-art on attenuation of the energy content of the motion with distance from the source is represented by (Cheng et al. 2014). It used 1550 pairs of horizontal records from 63 earthquakes to express the geometric mean of the energy-equivalent velocity in the two horizontal directions (maximum during the motion of absolute or relative values) in terms of the natural period of a single-degree-of-freedom (SDOF) system, the Magnitude of the earthquake and the distance to its source, the average shear wave velocity in the top 30 m of soil and the fault mechanism (normal, reverse or strike-slip). Cheng et al. (2015) established also the cross-correlation function of equivalent velocities along the two horizontal directions at the same or different periods, and the autocorrelation function along the same direction at different periods. Other works, which do not have such a general scope (Chapman 1999; Chou and Uang 2000; Safak 2000; Ordaz et al. 2003; Alici and Sucuoglu 2016), contain valuable alternative ideas.

Unlike the concerted and fruitful research on input energy spectra of SDOF systems, energy-based analytical studies of multi-degree-of-freedom (MDOF) ones were sporadic, uncoordinated and inconclusive. Outcomes so far suggest that the total input energy is nearly independent of eccentricities and asymmetries in plan (Goel 1997) and about the same in a multi-story building as in its equivalent SDOF system (Chou and Uang 2003; Surahman 2007). The feasibility of estimating the height- or plan-wise distribution of inelastic energy demand via analysis of a 3-DOF modal system, or (modal) pushover analysis and modal ductility factors instead of nonlinear response-history analysis, has been explored (Surahman 2007; Prasanth et al. 2008; Lin and Tsai 2011; Rathore et al. 2011). Benavent-Climent and Zahran (2010a, b) generalised the procedure in (Chou and Uang 2003) for the height-wise distribution of inelastic energy demand and developed an integrated approach involving a composite damage index.

Research efforts to consider the capacity along with the energy demand were few and inconclusive: Leelataviwat et al. (2002) set the input energy as equal to the work done at plastic hinges at all beam ends and all column bases, assuming an inverted triangular mode and force distribution and a beam-sway mechanism, with all plastic rotations equal to the top drift ratio, and solved for the base shear and the top drift. Leelataviwat et al. (2009) extended the MDOF approach by introducing energy demand and energy capacity curves, to be plotted against top drift; the intersection of these curves is the balance point. Along similar lines D'Ambrisi and Mezzi (2015) expressed the capacity curve approach in energy terms.

Mollaioli et al. (2011) carried out nonlinear response-history analyses of SDOF systems with various hysteresis models, using about 900 ground motions (including pulse-like ones) to establish a relationship between energy demand and displacement demand, applicable to MDOF systems with various height-wise stiffness distributions. The so-established displacement demands are meant to be checked against the corresponding displacement capacities. That work, despite being the most advanced so far in the direction of bringing the capacities into play, may lay claim to the title of an energy-based approach only as far as the demands are concerned; capacities are still in terms of displacements, not of energy.

In an effort to cast energy-based design in a probabilistic framework, Ghosh and Collins (2006) simulated 1300 motions for a Los Angeles site to convert the probability distribution of normalized energy at given period of the SDOF system to a probability distribution of the design base shear coefficient. Member energy capacities were considered fixed and pushover analysis of an equivalent SDOF system was used to derive from them the energy capacity of the system.

With time, research interest shifted to seismic design with energy dissipation devices or other supplemental protection techniques, where energy concepts is the natural approach to follow (Benavent-Climent 2011; Habibi et al. 2013; Paolacci 2013; Reggio and De Angelis 2015; Sorace et al. 2016). By contrast, research output concerning energy-based seismic design or evaluation and retrofitting of conventional structures without seismic protection devices reduced to a trickle.

In summary, the State-of-the-art may be considered as satisfactory only as far as the total energy demand is concerned. Its flow and distribution within the structure during the inelastic response is an open issue, to be answered on a case-by-case basis through nonlinear response-history analysis. Energy is converted during the response to fluctuating kinetic and potential energy and to heat, which is dissipated through hysteretic and viscous damping. The peaks of potential energy are equal to the energy input up to that point in time, minus the energy dissipated hitherto by hysteresis and viscous damping. As potential energy is mostly or exclusively deformation energy, its peaks during the response correspond to peaks of deformation, i.e., of damage. The amount of hysteretic and viscous energy dissipation is normally large and holds the key to the magnitude of the potential energy, hence to the predicted severity of damage or proximity to failure. This raises questions about the distribution of energy demand, due to the recent doubts cast upon the use of viscous damping in nonlinear response history analyses (Hall 2006; Charney 2008; Chopra and McKenna 2016; Carr et al. 2017). The recently acknowledged, yet unresolved, problems attributed to the viscous damping model are an important hurdle to the distribution of energy demand within the structure; this is an issue that the energy approach's research community has not faced yet. Another issue for the analysis is the potential energy of weights supported on rocking vertical elements of large horizontal size. Despite its potential importance, this component of the energy balance is normally ignored, as it reflects a special type of geometric non-linearity, not included in typical P-Δ effects.

The largest gap of knowledge, though, concerns the energy capacity of structural elements and systems, which is more challenging (Fajfar et al. 1991) and has not been addressed at all so far.

4.6 Concluding Remarks

To date, the promising energy approach to seismic design has fallen short of its true potential. Instead of developing all the way, to bear fruits for engineering practice, it has remained an academic exercise, and indeed a quite imbalanced one, as it has focused on the facet of the problem that is easier to tackle, i.e., the energy demand, ignoring the energy capacity aspect. So, the energy approach, incomplete and out of the limelight despite its merits, missed the opportunity to influence the first generation of Eurocode 8. The prospects are not better for the second generation of Eurocodes, currently being drafted (CEN/TC250 2013).

Major achievements of the past concerning the seismic input energy and the progress so far regarding other aspects of the demand side will all be wasted, and an opportunity for a new road to seismic design will be missed, unless:

- A concerted effort is undertaken in analysis to resolve issues in modeling the energy dissipation and to find an easy way to account for the variation in the potential energy of weights supported on large rocking elements, such as concrete walls of significant length (a geometrically nonlinear problem).
- The capacity of various types of elements to dissipate energy by hysteresis and to safely store deformation energy is quantified in terms of their geometric features and material properties.
- Energy-based design procedures are devised and applied on a pilot basis, leading to a new, energy-based conceptual design thinking.

If earthquake engineering research re-engages with the very promising and appealing concept of seismic energy, it is feasible to formulate a full-fledged new paradigm of seismic design in time for the third generation of Eurocodes, after 2030.

References

Akiyama H (1988) Earthquake-resistant design based on the energy concept. In: 9th world conference on earthquake engineering, vol V. Tokyo-Kyoto, pp 905–910
Akiyama H (1992) Damage-controlled earthquake resistant design method based on energy concept. In: Fajfar P, Krawinkler H (eds) Nonlinear seismic analysis and design of reinforced concrete buildings. Elsevier, London, pp 236–242
Alici FS, Sucuoglu H (2016) Prediction of input energy spectrum: attenuation models and velocity spectrum scaling. Earthq Eng Struct Dyn 45:2137–2161
Amiri GG, Darzi GA, Amiri JV (2008) Design elastic input energy spectra based on Iranian earthquakes. Can J Civ Eng 35(6):635–646
Arroyo D, Ordaz M (2007) On the estimation of hysteretic energy demands for SDOF systems. Earthq Eng Struct Dyn 36:2365–2382
ASCE (2007) Seismic rehabilitation of existing buildings. ASCE Standard ASCE/SEI 41-06 American Society of Civil Engineers, Reston, VA
Bardakis VG, Fardis MN (2011a) Nonlinear dynamic v elastic analysis for seismic deformation demands in concrete bridges having deck integral with the piers. Bull Earthq Eng 9:519–536
Bardakis VG, Fardis MN (2011b) A displacement-based seismic design procedure for concrete bridges having deck integral with the piers. Bull Earthq Eng 9:537–560
Benavent-Climent A (2011) An energy-based method for seismic retrofit of existing frames using hysteretic dampers. Soil Dyn Earthq Eng 31:1385–1396
Benavent-Climent A, Zahran R (2010a) Seismic evaluation of existing RC frames with wide beams using an energy-based approach. Earthq Struct 1(1):93–108
Benavent-Climent A, Zahran R (2010b) An energy-based procedure for the assessment of seismic capacity of existing frames: application to RC wide beam systems in Spain. Soil Dyn Earthq Eng 30(5):354–367
Benavent-Climent A, Pujades LG, Lopez-Almansa F (2002) Design energy input spectra for moderate seismicity regions. Earthq Eng Struct Dyn 31:1151–1172

Benavent-Climent A, Lopez-Almansa F, Bravo-Gonzalez DA (2010) Design energy input spectra for moderate-to-high seismicity regions based on Colombian earthquakes. Soil Dyn Earthq Eng 30:1129–1148

Biskinis D, Fardis MN (2007) Effect of lap splices on flexural resistance and cyclic deformation capacity of RC members. Beton- Stahlbetonbau Sond Englisch 102:51–59

Biskinis D, Fardis MN (2010a) Deformations at flexural yielding of members with continuous or lap-spliced bars. Struct Concr 11(3):127–138

Biskinis D, Fardis MN (2010b) Flexure-controlled ultimate deformations of members with continuous or lap-spliced bars. Struct Concr 11(2):93–108

Biskinis D, Fardis MN (2013a) Stiffness and cyclic deformation capacity of circular RC columns with or without lap-splices and FRP wrapping. Bull Earthq Eng 11(5):1447–1466

Biskinis D, Fardis MN (2013b) Models for FRP-wrapped rectangular RC columns with continuous or lap-spliced bars under cyclic lateral loading. Eng Struct 57:199–212

Bruneau M, Wang N (1996) Some aspects of energy methods for the inelastic seismic response of ductile SDOF structures. Eng Struct 18(1):1–12

Carr AJ, Puthanpurayil AM, Lavan O, Dhakal RP (2017) Damping models for inelastic time history analysis: a proposed modelling approach. In: 16th world conference in earthquake engineering, Santiago

CEN (2005) European Standard EN 1998-3:2005: Eurocode 8: design of structures for earthquake resistance. Assessment and retrofitting of buildings CEN, Brussels

CEN/TC 250(2013) Response to Mandate M/515 "Towards a second generation of EN Eurocodes' CEN/TC 250 Document - N 993, CEN, Brussels

Chai YH, Fajfar P (2000) A procedure for estimating input energy spectra for seismic design. J Earthq Eng 4(4):539–561

Chapman MC (1999) On the use of elastic input energy for seismic hazard analysis. Earthq Spectra 15(4):607–635

Charney F (2008) Unintended consequences of modeling damping in structures J Struct Eng ASCE 134(4):581-592

Cheng Y, Lucchini A, Mollaioli F (2014) Proposal of new ground-motion prediction equations for elastic input energy spectra. Earthq Struct 7(4):485–510

Cheng Y, Lucchini A, Mollaioli F (2015) Correlation of elastic input energy equivalent velocity spectral values. Earthq Struct 8(5):957–976

Chopra AK, McKenna F (2016) Modeling viscous damping in nonlinear response history analysis of buildings for earthquake excitation. Earthq Eng Struct Dyn 45:193–211

Chou CC, Uang CM (2000) Establishing absorbed energy spectra – an attenuation approach. Earthq Eng Struct Dyn 29:1441–1455

Chou CC, Uang CM (2003) A procedure for evaluating seismic energy demand of framed structures. Earthq Eng Struct Dyn 32:229–244

D'Ambrisi A, Mezzi M (2015) An energy-based approach for nonlinear static analysis of structures. Bull Earthq Eng 13:1513–1530

Decanini LD, Mollaioli F (1998) Formulation of elastic earthquake input energy spectra. Earthq Eng Struct Dyn 27:1503–1522

Decanini LD, Mollaioli F (2001) An energy-based methodology for the seismic assessment of seismic demand. Soil Dyn Earthq Eng 21:113–137

Economou SN, Fardis MN, Harisis A (1993) Linear elastic v nonlinear dynamic seismic response analysis of RC buildings EURODYN'93, 2nd European conference on structural dynamics, Trondheim, pp 63–70

Fajfar P, Vidic T (1994) Consistent inelastic design spectra: hysteretic and input energy. Earthq Eng Struct Dyn 23(5):523–537

Fajfar P, Vidic T, Fischinger M (1991) On the energy input into structures. In: Proceedings of the Pacific conference on earthquake engineering, New Zealand, pp 199–210

Fajfar P, Vidic T, Fischinger M (1992) On energy demand and supply in SDOF systems. In: Fajfar P, Krawinkler H (eds) Nonlinear seismic analysis and design of reinforced concrete buildings. Elsevier, pp 41–62

Fardis MN, Panagiotakos TB (1997) Displacement-based design of RC buildings: proposed approach and application. In: Fajfar P, Krawinkler H (eds) Seismic design methodologies for the next generation of codes. Balkema, Rotterdam, pp 195–206

fib (2012) Model Code 2010 Bull. 65/66, Federation Internationale du Beton, Lausanne

Ghosh S, Collins KR (2006) Merging energy-based design criteria and reliability base methods: exploring a new concept. Earthq Eng Struct Dyn 35:1677–1698

Goel RK (1997) Seismic response of asymmetric systems: energy-based approach. ASCE J Struct Eng 123(11):1444–1453

Grammatikou S, Biskinis D, Fardis MN (2015) Strength, deformation capacity and failure modes of RC walls under cyclic loading. Bull Earthq Eng 13(11):3277–3300

Grammatikou S, Biskinis D, Fardis MN (2016) Ultimate strain criteria for RC members in monotonic or cyclic flexure. ASCE J Struct Eng 142(9):04016046

Grammatikou S, Biskinis D, Fardis MN (2018) Effects of load cycling, FRP jackets and lap-splicing of longitudinal bars on effective stiffness and ultimate deformation of flexure-controlled RC members. ASCE J Struct Eng 144. (accepted)

Grammatikou S, Biskinis D, Fardis MN (2018) Flexural rotation capacity models fitted to test results using different statistical approaches. Struct Concr 19. https://doi.org/10.1002/suco.201600238

Grammatikou S, Biskinis D, Fardis MN (2018) Models for the flexure-controlled strength, stiffness and cyclic deformation capacity of concrete columns with smooth bars, including lap-splicing and FRP jackets. Bull Earthq Eng 16(1):341–375. https://doi.org/10.1007/s10518-017-0202-y

Habibi A, Chan RWK, Albermani F (2013) Energy-based design method for seismic retrofitting with passive energy dissipation systems. Eng Struct 46:77–86

Hall JF (2006) Problems encountered from the use (or misuse) of Rayleigh damping. Earthq Eng Struct Dyn 35:525–545

Kalkan E, Kunnath S (2007) Effective cyclic energy as a measure of seismic demand. J Earthq Eng 11(5):725–751

Kosmopoulos A, Fardis MN (2007) Estimation of inelastic seismic deformations in asymmetric multistory RC buildings. Earthq Eng Struct Dyn 36(9):1209–1234

Leelataviwat S, Goel SC, Stojadinović B (2002) Energy-based seismic design of structures using yield mechanism and target drift. ASCE J Struct Eng 128(8):1046–1054

Leelataviwat S, Saewon W, Goel SC (2009) Applications of energy balance concept in seismic evaluation of structures. ASCE J Struct Eng 135(2):113–121

Lin J-L, Tsai K-C (2011) Estimation of the seismic energy demands of two-way asymmetric-plan building systems. Bull Earthq Eng 9(8):603–621

Manfredi G (2001) Evaluation of seismic energy demand. Earthq Eng Struct Dyn 30:485–499

Moehle JP (1992) Displacement-based design of RC structures subjected to earthquakes. Earthq Spectra 8(3):403–428

Mollaioli F, Bruno S, Decanini L, Saragoni R (2011) Correlations between energy and displacement demands for performance-based seismic engineering. Pure Appl Geophys 168(1-2):237–259

Ordaz M, Huerta B, Reinoso E (2003) Exact computation of input-energy spectra from Fourier amplitude spectra. Earthq Eng Struct Dyn 32:597–605

Panagiotakos TB, Fardis MN (1998) Deformation-controlled seismic design of RC structures. In: Proceedings of the 11th European conference on earthquake engineering, Paris

Panagiotakos TB, Fardis MN (1999a) Deformation-controlled earthquake resistant design of RC buildings. J Earthq Eng 3:495–518

Panagiotakos TB, Fardis MN (1999b) Estimation of inelastic deformation demands in multistory RC buildings. Earthq Eng Struct Dyn 28:501–528

Panagiotakos TB, Fardis MN (2001a) A Displacement-based seismic design procedure of RC buildings and comparison with EC8. Earthq Eng Struct Dyn 30:1439–1462

Panagiotakos TB, Fardis MN (2001b) Deformations of RC members at yielding and ultimate. ACI Struct J 98(2):135–148

Paolacci F (2013) An energy-based design for seismic resistant structures with viscoelastic dampers. Earthq Struct **4**(2):219–239

Prasanth T, Ghosh S, Collins KR (2008) Estimation of hysteretic energy demand using concepts of modal pushover analysis. Earthq Eng Struct Dyn 37:975–990

Priestley MJN (1993) Myths and fallacies in earthquake engineering – conflicts between design and reality. In: Proceedings of the Tom Paulay symposium: recent developments in lateral force transfer in buildings La Jolla, CA

Rathore M, Chowdhury AR, Ghosh S (2011) Approximate methods for estimating hysteretic energy demand on plan-asymmetric buildings. J Earthq Eng 15(1):99–123

Reggio A, De Angelis M (2015) Optimal energy-based seismic design of non-conventional Tuned Mass Damper (Tmd) implemented via interstory isolation. Earthq Eng Struct Dyn 44 (10):1623–1642

Riddell R, Garcia JE (2001) Hysteretic energy spectrum and damage control. Earthq Eng Struct Dyn 30:1791–1816

Safak E (2000) Characterization of seismic hazard and structural response by energy flux. Soil Dyn Earthq Eng 20:39–43

SEAOC (1995) Performance based seismic engineering of buildings VISION 2000 committee, Structural Engineers Association of California, Sacramento, CA

Sorace S, Terenzi G, Mori C (2016) Passive energy dissipation-based retrofit strategies for R/C frame water towers. Eng Struct 106:385–398

Sucuoglu H, Nurtug A (1995) Earthquake ground motion characteristics and seismic energy dissipation. Earthq Eng Struct Dyn 24(9):1195–1213

Surahman A (2007) Earthquake-resistant structural design through energy demand and capacity. Earthq Eng Struct Dyn 36(14):2099–2117

Teran-Gilmore A, Jirsa JO (2007) Energy demands for seismic design against low-cycle fatigue. Earthq Eng Struct Dyn 36:383–404

Uang CM, Bertero VV (1990) Evaluation of seismic energy in structures. Earthq Eng Struct Dyn 19:77–90

Vidic T, Fajfar P, Fischinger M (1994) Consistent inelastic design spectra: strength and displacement. Earthq Eng Struct Dyn 23:502–521

Zahrah TF, Hall WJ (1984) Earthquake energy absorption in SDOF structures. ASCE J Struct Eng 110(8):1757–1772

Chapter 5
Seismic Assessment of Existing Irregular Masonry Buildings by Nonlinear Static and Dynamic Analyses

Sergio Lagomarsino, Daniela Camilletti, Serena Cattari, and Salvatore Marino

Abstract The use of nonlinear static (pushover) analysis in the case of existing irregular masonry buildings is validated through a comparison with results from nonlinear dynamic analyses, assumed as reference because considered as able to represent the actual seismic behavior. After the selection of a regular prototype case study building, different irregular configurations are defined (in terms of plan irregularity and finite stiffness of horizontal diaphragms). Specific proposals are considered for the selection of load patterns to be used in pushover analysis and the definition of limit states on the capacity curve. A general overview of possible approaches for modelling and analysis of masonry buildings is presented in the introduction.

5.1 Introduction

The seismic assessment of existing masonry buildings is a relevant issue both from the social and cultural point of view. Recent earthquakes have confirmed the high seismic vulnerability of this type of structures, which suffered damage even for low intensity earthquakes and underwent collapses for levels of intensity for which the life safety should be guaranteed. However, some masonry buildings have shown a good behavior, when specific constructive details were adopted, both at the moment of the construction and after preventive strengthening interventions.

Therefore, the conservation of pre-modern masonry buildings is possible, also assuring their use under safety conditions, after an accurate seismic assessment, which is also the starting point for the design of an effective seismic retrofitting, under the wish of a minimum intervention, both for economic and conservative reasons.

S. Lagomarsino (✉) · D. Camilletti · S. Cattari · S. Marino
Department of Civil, Chemical and Environmental Engineering, University of Genoa, Genoa, Italy
e-mail: sergio.lagomarsino@unige.it

It is worth noting that masonry buildings, in particular pre-modern ones, have been realized through an empirical approach, that is a trial and error process leading, along the time, to the use of "rules of thumb" mainly consisting of geometric proportions among structural elements, without the use of mechanical-based models. Moreover, besides material deterioration, these buildings have often undergone transformations, not always appropriate from the structural point of view. Hence, the identification of a structural model for the assessment is not straightforward, because in principle a clear distinction between structural and nonstructural elements is not possible. In addition, material properties are inhomogeneous and difficult to be measured by in-situ nondestructive tests.

The seismic assessment of existing masonry buildings is based on three steps: modelling, analysis and verification. The seismic response of masonry buildings depends on the behavior of masonry walls, both in-plane and out-of-plane, on the connection between walls, and on the interaction with horizontal diaphragms.

When out-of-plane mechanisms are prevented, the seismic capacity of a masonry building should be evaluated by considering the in-plane behavior of masonry walls only. In many cases an equivalent frame model can be adopted by identifying masonry vertical (piers) and horizontal (spandrels) structural elements, for which proper resistance models (in terms of generalized forces) and deformation capacities (in terms of drift limits) are formulated.

For piers, different failure criteria are present in the literature (see for a state of the art Magenes and Calvi (1997) and Calderini et al. (2009)). Flexural failure, with partialization at the end sections and crushing at the tip, due to normal force and bending, should be always considered. Diagonal cracking at the center of the panel is related to the diagonal tensile strength (isotropic behavior), in the case of irregular masonry, while it depends on the mortar joint strength, the local friction and the interlocking between units, in the case of regular masonry. In the latter case, also shear sliding on a horizontal mortar joint should be considered. Drift limits for irregular masonry have been recently derived from experimental tests by Vanin et al. (2017).

For spandrels, less experimental tests are available (Gattesco et al. 2008; Beyer and Dazio 2012; Graziotti et al. 2012; Parisi et al. 2014) and failure criteria (Beyer 2012; Beyer and Mangalathu 2013) should consider also the horizontal tensile strength of masonry (due to interlocking and vertical compressive stress) and the characteristics of lintels and other coupled horizontal elements (tie rods, ring beams). It is worth noting that the normal force in spandrels is usually very low, as horizontal seismic actions are distributed to each node in proportion to the tributary mass; normal forces are generated only when a redistribution of shear forces between masonry piers occurs or if the lengthening of the spandrel is resisted by elements, such as ring beams or tie rods. Moreover, in this cases the 3D equivalent frame model may be not accurate.

Piecewise linear force-deformation relationships should be defined at element level, in terms of shear force and element drift ratio. The elastic stiffness should correspond to cracked conditions, or a bilinear model may be used until the maximum strength. After that, the progressive strength degradation should be included at element level, in order to evaluate the ultimate displacement capacity at global level by considering the progressive damage and failure under seismic (horizontal) and gravity (vertical) loads of all resisting walls; in this case safety verification should be made in global terms. It is worth noting that this near collapse condition is still compatible with the stability under gravity loads, for piers, and for the prevention of local collapse in spandrels. On the contrary, if the progressive strength degradation is not included in the model, the nonlinear analysis should consider verification in local terms.

For masonry walls in which it is not possible to define in a reliable manner the equivalent frame (piers and spandrels), two-dimensional or three-dimensional continuous or discrete models may be adopted. However, in all cases the safety verification should be carried out in terms of generalized forces, in specific relevant sections, and in terms of the generalized shear deformation of properly singled out masonry panels. It is worth noting that continuous models are usually feasible for linear analysis. For nonlinear analysis, constitutive models should be able to reproduce the local behavior of masonry and should be validated with failure criteria on ex-post defined panels.

Horizontal diaphragms in masonry buildings should be classified as rigid, stiff or flexible. If horizontal diaphragms are rigid or stiff, the building should be analyzed by a global 3D model wherein the in-plane behavior of masonry walls is considered. If horizontal diaphragms are flexible, each single wall may be analyzed and verified independently, being subjected to its own effects of seismic actions (including those transferred by supported floors) and to those related to connected out-of-plane loaded walls.

The seismic assessment of pre-modern masonry buildings should also consider the verification of possible partial mechanisms, mainly characterized by out-of-plane response of a wall portion that is not well connected to the adjacent walls loaded in-plane. These mechanisms are usually not captured by the global model and may be analyzed using sub-structuring models, considering a portion (macro-element) that may be assumed as behaving independently from the rest of the building. Two alternative approaches, of increasing accuracy should be used (Lagomarsino 2015): (a) equilibrium limit analysis of a kinematic chain of rigid blocks (evaluation of the seismic horizontal force multiplier at the activation of the mechanism); (b) incremental equilibrium limit analysis with geometric nonlinearity (evaluation of the pushover curve and the displacement capacity; either a rigid or an elastic initial behavior may be assumed, depending on the mechanism features).

Regarding the seismic analysis, the use of linear methods is problematic for many reasons (Camilletti et al. 2017): (i) the difficulty of assuming a value for the behavior factor (q-factor approach) and the overstrength coefficient, which may range from 1 to values also greater than 2 (Magenes 2000); (ii) the lack of reliability of the lateral force analysis, with unreduced response spectrum and verification in terms of deformation, because masonry buildings have shorter periods than reinforce concrete ones (deformation demand in nonlinear range is higher than the elastic one); (iii) the lack of reliability of the equivalent frame model in the elastic range, due to the strong simplification related to the assumption of rigid nodes of finite dimension.

In the ambit of nonlinear analyses, the static incremental approach (pushover analysis) is very convenient in the engineering practice, because of the limited computational effort, but its reliability in the case of irregular buildings (in plan and elevation) without rigid diaphragms needs to be proved.

A distinctive feature of masonry buildings is the limited ductility of elements (in particular piers, under some of the possible failure mechanisms). For this reason, it is useful to model within the analysis the strength degradation at element level, in order evaluate the global behavior of the building even after this local limit conditions. Modeling the strength degradation at element level allows to evaluate the gradual strength degradation at global level. The verification is then possible at this scale, by identifying the limit states on the capacity curve, provided that specific checks are performed in each wall if diaphragms are not rigid.

The displacement demand at global level may be evaluated by different formulations (N2 method – Fajfar 2000; Capacity Spectrum Method – Freeman 1998, 2004; Coefficient method - ASCE-SEI 2014), after the derivation of the equivalent SDOF system from the pushover curve. The safety verification is made by checking at global level that the displacement demand is lower than the displacement capacity. The verification of local failure, related to the condition wherein one masonry element is not able to bear gravity loads, is considered by a proper limitation of the displacement capacity of each single pier. It is worth noting that the verification of the damage level in each masonry element is implicitly made during the nonlinear static analysis, by considering force-deformation relationships with strength degradation and the consequence at global level may be directly detected on the capacity curve.

Within this more general context, in the next sections the reliability of nonlinear static (pushover) analysis is investigated in the case of irregular masonry buildings, by considering a proper set of different prototype buildings. These buildings were also studied by nonlinear dynamic analysis, in order to have a reference solution.

5.2 Distinctive Features of Nonlinear Static and Dynamic Analyses

Since as aforementioned masonry has a strongly nonlinear behavior even for low seismic actions, the use of nonlinear methods of analysis is suggested, more than in the case of other structural typologies. Among the available nonlinear methods, the Nonlinear Dynamic Analysis (NLDA) is the most accurate tool, but its use is still limited, especially at engineering practice, since it is time-consuming and requires a significant computational effort, the availability of proper constitutive laws effective also in cyclic field and an expert judgment for the choice of the time histories to use. Moreover, the interpretation of the results is not a simple task (Lagomarsino and Cattari 2015b).

Nonlinear Static Analysis (NLSA), that is widely adopted in international standards (e.g. ASCE-SEI 2014; CEN 2004; G.U. 2008), is a valid alternative for performing nonlinear analyses of masonry structures since it requires a lower computational effort, thus being more practice-oriented.

As known, the procedure is based on the following steps: (1) execution of a pushover analysis, with a proper load pattern; (2) identification of displacements related to the attainment of different Performance Levels (PLs) or Limit States; (3) derivation of the capacity curve of an equivalent nonlinear Single Degree of Freedom (SDOF) system; (4) for the seismic input motion to be considered for the verification of each PL, evaluation of the displacement demand by a properly reduced spectrum; (5) comparison between displacement demand and capacity.

This method was originally developed for reinforced concrete (RC) or steel framed structures, under the hypothesis of rigid horizontal diaphragms and, possibly, in the case of regular configurations. Then, several proposals have been formulated in literature to improve the reliability of such procedure in case of structures strongly irregular or for which the contribution of higher modes is not negligible. They follow different approaches based on the execution of multi-modal or adaptive analyses (Aydinoglu and Onem 2010) or the introduction of corrective factors to amplify the displacement demand or corrective eccentricities to reproduce torsional effects induced by irregularities; a state of the art of such modified procedures for RC buildings is in De Stefano and Mariani (2014). However, these proposals are specifically meant for steel or RC structures, and so they deal with the hypothesis of rigid floors and with the irregularities that are frequent in these structural typologies (thus mainly related plan and/or elevation irregularity).

Critical issues arise in the case of masonry buildings. Here the sources of irregularity are different and involve not only plan and/or elevation irregularity, but also the stiffness of the horizontal diaphragms. These latter are often constituted by timber floors or masonry vaults, that are far from the idealization of *rigid* diaphragm and this strongly influences the global seismic behavior.

The main critical issues related to the application of the nonlinear static approach in case of irregular masonry buildings can be summarized as follows, referring to the aforementioned steps of the procedure.

First of all, the implementation of a pushover analysis (step 1) requires the availability of proper nonlinear models. Subsequently, it is necessary to choose the load pattern to use, and this is a critical step, since it is impossible to know in advance the inertial forces that a seismic event may activate on a given building. Seismic codes propose to use one (ASCE-SEI 2014) or two different load patterns (G.U. 2008; CEN 2004), generally assumed as the uniform (i.e. proportional to the masses of the building) and the modal (the only one recommended in ASCE-SEI 2004) or the inverted-triangular (i.e. proportional to the product of the masses times their heights) distributions. However, these load patterns are not always reliable if applied to irregular masonry buildings with flexible floors, in fact they might not be representative to describe the actual seismic behavior of these buildings. Indeed, for these cases the participant mass on the first mode of vibrations can be very low, and the influence of the higher modes can be significant. More details are given in Sect. 5.3.2.

The definition of PLs (step 2), that in codes involve economic, structural and safety issues, obviously presupposes a correlation to the seismic response of the structure that is monitored in structural models by one or more engineering demand parameters able to identify Damage Levels (DL) on the pushover curve. In general, codes do not distinguish between PLs and DLs and the definition of PLs is treated by checking the corresponding DLs in each single element (ASCE-SEI 2014) or by considering interstorey drift thresholds and/or heuristic criteria on the stiffness and strength degradation of the pushover curve (G.U. 2008; CEN 2005). However, in case of complex masonry buildings, often with plan/elevation irregularities and with flexible diaphragms, the first method turns out to be too conservative, the latter often inadequate since it cannot take into account the localization of damage in single walls. To overcome this issue, the multiscale approach (Lagomarsino and Cattari 2015a) has been adopted and described in Sect. 5.3.3.

For regular buildings with rigid diaphragms the conversion into capacity curve (step 3) is almost insensitive to the selection of the control node for the pushover analysis, whereas in the case of flexible floors it can be strongly affected by this choice. More details on this issue are given in Sect. 5.3.2.

Therefore, although NLSA represents a very suitable tool for the seismic analyses of masonry buildings, an extensive validation of this procedure in case of irregular structures and/or with flexible diaphragms is still lacking, and it is the object of this paper. Specific directions are also provided on the most appropriate load patterns to use in these cases in the pushover analysis to obtain reliable results.

To reach this goal, ten prototype building configurations have been developed and described in Sect. 5.4, assumed as representative of unreinforced masonry (URM) buildings with different types and levels of irregularity. In Sect. 5.5, for each case study, the results obtained by NLSA, performed using different load patterns, have been compared with those considered as the reference actual behavior, obtained by NLDA through the Incremental Dynamic Analyses (IDA), as proposed by Vamvatsikos and Cornell (2002).

5.3 Methodology Adopted for the Comparison Between Nonlinear Static and Dynamic Analyses

The procedure adopted to study the accuracy of the nonlinear static approach when applied to irregular masonry buildings is summarized by the following steps:

- definition of configurations of masonry buildings characterized by different types of irregularities (plan irregularity, diaphragms with different stiffness);
- execution of nonlinear static analyses with different load patterns;
- execution of nonlinear incremental dynamic analyses, considered as the reference solution;
- definition of reference limit states, in terms of damage level – DLs;
- comparison between the results of static and dynamic analyses.

As mentioned also in Sect.5.2, codes adopt various and in some case different recommendations on all such issues. In the following, for sake of brevity, reference is made only to the European and Italian codes, while a more comprehensive overview of other codes and literature proposals is discussed in Marino et al. (2018).

The following subparagraphs explain in detail the procedure used for performing both the NLSA and the NLDA: the modeling technique and the constitutive laws applied (Sect. 5.3.1), the load pattern used in NLSA and seismic input adopted for executing the NLDA (Sect. 5.3.2), the approach adopted for the definition of the performance levels (Sect. 5.3.3) and the comparisons between the obtained results (Sect. 5.3.4).

5.3.1 Modelling Technique and Constitutive Laws Used

The execution of both NLSAs and NLDAs requires nonlinear models able to properly describe the building response until near collapse conditions.

Among the available modeling techniques for masonry structures, the equivalent frame approach represents a feasible tool for modeling 3D complex buildings, which has been validated by several case studies and is the most effective in terms of computational effort (Cattari et al. 2014; Penna et al. 2016). According to this approach each masonry wall has to be idealized in structural elements (piers and spandrels) in which the nonlinear response is concentrated, connected by rigid areas (nodes). It is worth noting that usually only the in-plane response of masonry walls may be considered.

The numerical parametric analyses have been carried out with the software TREMURI (Lagomarsino et al. 2013), which is based on the use of the equivalent frame modelling technique. TREMURI program implements various constitute laws among which a multilinear cyclic hysteretic constitutive law, with failure criteria for piers and spandrels that allow performing both static and dynamic nonlinear analyses (Cattari and Lagomarsino 2013a). The elastic response phase is described according to the beam theory by defining the initial Young's (E) and shear (G) moduli of masonry, and then the progressive degradation is approximated using a secant stiffness (the values adopted for the numerical models are defined in Table 5.1). The elastic values are defined by multiplying the secant stiffness by a coefficient, in this case equal to 2 (as suggested in CEN 2004). The maximum shear strength is defined on the basis of common criteria proposed in literature as a function of

Table 5.1 Summary of the threshold used for piers and spandrels

[%]	δ_{E3}[a]	δ_{E4}[a]	δ_{E5}[a]	β_{E3}[a]	β_{E4}[a]
Piers[b]	0.6 – 0.3	1.0–0.5	1.5–0.7	0–30	15–60
Spandrels	0.2	0.6	2	40	40

[a]δ_{Ei} drift limits, β_{Ei} strength decays as shown in Fig. 5.1
[b]The first value is assumed in case of prevailing flexural behaviour, the second in case of shear

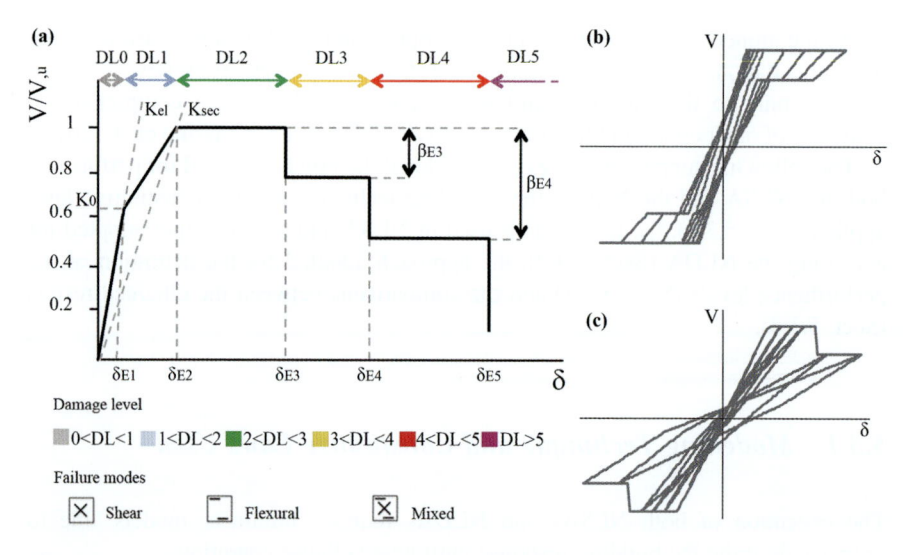

Fig. 5.1 Backbone response of a masonry panel (adapted from Cattari and Lagomarsino 2013a) (**a**); examples of hysteretic response for the case of shear (**b**) and flexural (**c**) failure modes

different failure modes examined (either flexural or shear). In particular, for the diagonal shear cracking the criterion proposed by Mann and Müller (1980) has been adopted; for the flexural failure the criterion proposed in the Eurocode 8, Part 3 (CEN 2005) has been used in the case of masonry piers, while for the spandrels the criterion described in Cattari and Lagomarsino (2008) has been applied. These constitutive laws allow also to describe the nonlinear response until very severe damage levels (from 1 to 5) through progressive strength degradation (β_{Ei}) in correspondence of assigned values of drift (δ_{Ei}), which are associated with the achievement of reference damage levels (DL_i with i = 1,5, where DL5 is associated with "collapse" of the panel, representing as the state when the panel has lost the capacity to support horizontal loads) as shown in Fig. 5.1a. Moreover, an accurate description of the hysteretic response is also included, based on a phenomenological approach that is able to account for the typical response of panels having a prevailing shear or flexural behavior (Fig. 5.1b/c). Mixed failure modes are also taken into account and parameters are calibrated in order to differentiate between the behavior of piers and spandrels. The values adopted for the constitutive laws are summarized in Table 5.1.

In the software TREMURI, diaphragms are modelled as 3- or 4-nodes orthotropic membrane finite (plane stress) elements with an equivalent thickness (t). They are identified by a principal direction (floor spanning direction), with two values of Young modulus along the two orthogonal directions (parallel, E_1, and perpendicular, E_2, to the spanning direction), Poisson ratio (ν) and the in-plane shear modulus (G_{eq}). This latter represents the shear stiffness of the floor and influences the horizontal force transferred among the walls, both in linear and nonlinear phases. Different values of G_{eq} have been adopted for the numerical analyses and they are defined in Table 5.2.

Table 5.2 Mechanical properties of masonry, properties of the three different types of diaphragms considered (rigid, stiff and flexible) and values of damping (elastic and hysteretic) assumed for the prototype buildings analyzed

Masonry properties		Diaphragms properties				Damping			
E_m [MPa]	750		Rigid	Stiff	Flexible	$\xi_{elastic}$ [%]	5		
G_m [MPa]	250	E_1 [MPa]	58800	12000	12000			Mean	σ
f_m [MPa]	2.8	E_2 [MPa]	30000	1000	1000	$\xi_{hist, DL2,u}$[%]		5.93	1.96
$\tilde{c}$ [MPa][a]	0.11	G_{eq}[MPa]	12500	100	10	$\xi_{hist, DL2,t}$ [%]		6.45	1.28
$\hat{\mu}$ [−][*]	0.34	t [m]	0.04	0.04	0.04	$\xi_{hist, DL3,u}$ [%]		8.93	1.20
ρ [kN/m³]	18	ν [−]	0.2	0.2	0.2	$\xi_{hist, DL3,t}$ [%]		7.26	1.42

Notes:
f_m masonry compression strength; $\hat{\mu}$ and $\tilde{c}$ equivalent fiction and cohesion, E_m and G_m Young modulus and shear modulus of masonry; ρ specific weight of masonry; E_1 Young modulus of the diaphragms in the direction of the joists; E_2 Young modulus orthogonal to E_2; ν Poisson's ratio; t thickness of the diaphragms.
[a]the values of $\hat{\mu}$ and $\tilde{c}$ are those modified as proposed in Mann e Muller (1980) and obtained from:
$\tilde{c} = \dfrac{c}{1+\varphi\mu}$; $\tilde{\mu} = \dfrac{\mu}{1+\varphi\mu}$ where c and μ represent the local values of cohesion and friction of the joint and φ is a parameter which describes the texture of masonry (here assumed equal to 1).

5.3.2 Load Pattern Used in NLSA and Seismic Input Adopted

Codes usually allow the use of nonlinear static procedure if the participant mass on the first mode is greater than a given percentage (around 75%), a condition that is met for regular buildings with rigid floors. However, in the case of masonry buildings nonlinear static analysis should be used also for irregular buildings, taking into consideration also the aforementioned problematic use of alternative methods as the linear one (Magenes 2000) and the still unsustainable computational effort at engineering practice level of NLDA. For this reason, the Italian Building Code (G.U. 2008) and its Commentary (G.U. 2009) allow the use of this method for the seismic analysis of masonry structures with limitations that are less strict than those for the other structural typologies. In particular its applicability for the design of masonry buildings is allowed at first in case of participant mass higher than 60% instead of 75%, and then in any case for the assessment existing buildings.

In the present research, the pushover analyses were carried out by using different load patterns (LPs), kept invariant during the analysis. Both the European (CEN 2004) and the Italian (G.U. 2008) codes propose to use two different load patterns for the pushover analysis. According to the Italian code, they have to be selected within two groups. The first group (principal group) contains load patterns aimed to activate inertial forces proportional to the first mode (assumed in a simplified or rigorous form), while in the second one (secondary group) the uniform and the adaptive distributions are explicitly indicated. Regarding the first group, an alternative to the first mode distribution is represented by the one derived from the inverted-triangular deformation profile. However, in the case of irregular masonry buildings with flexible floors these load patterns are not always reliable. On one hand, the first mode is not representative because the related participant mass can be very low, in particular when diaphragms are flexible. On the other hand, the triangular load pattern may provide unreliable results if: (i) the walls have not the same stiffness and in the absence of rigid floors they are not forced to deform in the same way, (ii) the deformed shape of the building under analysis is far from the linear one.

In order to overcome these critical issues, two additional load patterns have been here investigated, obtained by a combination of load patterns derived from all modal shapes which do not present the inversion of sign in displacement, called *first modes* (in other words, in the presence of flexible diaphragms there are many similar modes, which may be associated to the first mode of each single wall, considered as independent, see Table 5.3):

- SRSS, using the Square Root of Sum of Squares rule applied to the above mentioned modes;
- CQC, using a Complete Quadratic Combination of the same modes.

Hence a total of five load patterns have been applied and the results compared.

The execution of the pushover analysis presupposes then the choice of a control node and the control displacement to be plotted in the pushover curve, aiming the latter to be more representative of the displacements of the building during the earthquake. In presence of flexible diaphragms, the definition of both is not straightforward. Indeed, if the diaphragms are rigid, the control node and the control displacement should be assumed in the center of mass of top level of the building, but when diaphragms are flexible this does not correspond to a physically significant point of the building. For this reason, in this research: a) the control node (i.e. the node whose displacements are incremented in order to conduct a nonlinear static analysis) was located on a node at top floor of the wall that collapses first (see Lagomarsino and Cattari 2015b); b) the control displacement, which is the one plotted in pushover curves, was assumed as the average displacement of all nodes at top floor weighted by their tributary masses.

The nonlinear dynamic analyses have been performed in the form of Incremental Dynamic Analysis (IDA), which consists in scaling each time history in order to obtain increasing intensity values of the seismic action (Vamvatsikos and Cornell 2002).

Since each seismic event has specific features and can produce different effects on the same structure (record to record variability), 10 different records have been used. The reference Intensity Measure (IM) has been used here as the spectral acceleration $Sa(T_1)$, assuming the period $T_1=0.36$ s as representative of the main modes of vibration of the considered buildings (in Sect. 5.4.1 the dynamic properties of the case study buildings are discussed). The records have been selected to be conditioned to the spectral shape that corresponds to a return period of 475 years for the seismic hazard in L'Aquila, according to the Italian hazard map.

Figure 5.2a shows the acceleration response spectra of the 10 records, normalized to the value $Sa(T_1) = 2.089$ m/s^2, in this way the median spectrum has unitary PGA.

Table 5.3 Summary of the main results from modal analysis, properties of first modes, as defined in Sect. 5.3.2

Diaphragms properties	Model	Mode	T[s]	M_X [%]	M_Y [%]	Γ_X	Γ_Y	Γ_X/Γ_Y [%]
Rigid	Ar	1	0.296	0.00	82.8	0.010	1.460	0.66
	Br	1	0.301	0.00	82.9	0.007	1.457	0.51
	Airr	1	0.335	0.12	71.4	−0.071	1.713	−4.14
		3	0.211	3.39	11.1	−0.361	0.652	−55.3
	Birr	1	0.341	0.12	71.4	−0.071	1.698	−4.16
		3	0.216	3.33	11.4	−0.355	0.658	54.0
Intermediate (stiff)	Ar	1	0.324	0.04	70.6	0.042	1.757	2.41
		4	0.215	0.00	11.9	0.010	0.697	1.47
	Br	1	0.329	0.03	70.6	0.040	1.761	2.25
		4	0.215	0.00	12.0	0.009	0.705	1.34
	Airr	1	0.359	0.03	66.1	−0.037	1.708	−2.17
		3	0.273	7.93	4.42	0.511	−0.381	−134
		4	0.208	0.00	8.81	−0.011	0.551	−1.93
		6	0.172	0.10	3.16	−0.070	0.401	−17.5
	Birr	1	0.364	0.03	66.1	−0.037	1.689	−2.21
		3	0.178	9.77	4.32	0.546	−0.364	−150
		4	0.212	0.00	9.04	−0.012	0.560	−2.13
		6	0.176	0.08	3.13	−0.062	0.390	−15.9
Flexible	Ar	1	0.381	0.04	36.4	0.050	1.585	3.15
		3	0.303	0.01	10.4	−0.025	0.954	−2.66
		5	0.264	1.62	18.5	0.350	1.180	29.7
		6	0.237	1.18	17.4	−0.331	1.272	−26.01
	Airr	1	0.384	0.01	50.6	−0.020	1.736	−1.13
		2	0.360	2.96	3.65	−0.451	0.501	−90.1
		4	0.302	0.00	11.7	0.002	1.017	0.22
		6	0.218	7.58	9.00	−0.593	0.646	−91.8
		7	0.201	13.3	7.21	0.760	0.558	73.5

Figure 5.2b shows the correspondent median spectrum, which shape is very similar to the spectrum provided by the Italian Building Code (G.U. 2008), as well as the spectra associated to the 16% and the 84% percentiles.

For each case study, the results obtained by the nonlinear Incremental Dynamic Analyses (IDA) are considered as the reference actual behavior, to be compared with the seismic demand provided by NLSA, using different possible procedures.

5.3.3 *Definition of the Limit States*

The identification of limit states from the results of a pushover analysis is a critical issue, in particular when the near collapse conditions are considered. The easiest way

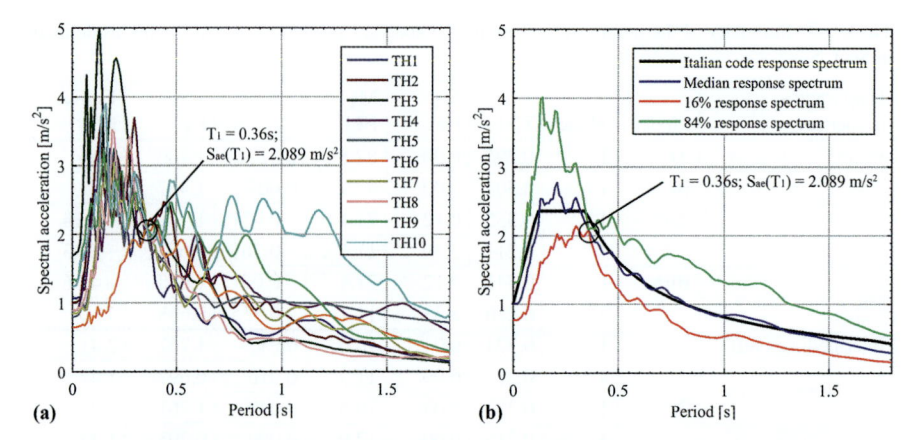

Fig. 5.2 Acceleration response spectra of the 10 Time Histories (TH) used for the Incremental Dynamic Analyses (**a**); correspondent median spectrum, together with the spectra associated to the 16% and 84% percentiles and Italian code response spectrum (**b**). The point ($T_1 = 0.36$ s; $S_a(T_1) = 2.089$ m/s^2), to which all the spectra are conditioned, is underlined

is to refer to the control displacement for which the first masonry element attains the correspondent damage level (ASCE-SEI 2014), but this approach is very conservative, because in irregular masonry buildings a single structural elements may show an high level of damage without precluding the safety of occupants. When strength degradation in single masonry elements is modelled, a possible alternative is to identify the limit state as the control displacement for which the capacity curve has a drop down of 20% of the maximum base shear (CEN 2005); however, this approach is very easy to be implemented but not conservative when diaphragms are not rigid, because single walls or local portions of the building could undergo collapse conditions before the global capacity curve shows a significant strength degradation.

In order to overcome these problems, the multiscale approach was developed within the PERPETUATE research project Lagomarsino and Cattari 2015a) and has been adopted here. This approach is proposed also in CNR-DT212 (2014) and it takes into account the behavior of single elements (E), macroelements (M) and of the global building (G). For each scale, proper variables are introduced and their evolution in nonlinear phase is monitored: the cumulative rate of panels (piers and spandrels as identified in the equivalent frame idealization of URM walls) that reach a certain damage level (E); drift in masonry walls and horizontal diaphragms (M); normalized total base shear, from global pushover curve (G). The reaching of assigned thresholds for such variables allows to define in the case of NLSA the displacements on the pushover curve corresponding to the attainment of each limit state (or Performance Level – PL) at these different scales ($u_{E,PLk}$, $u_{M,PLk}$ and $u_{G,PLk}$), being the minimum value among the three limit states in the building. In case of NLDA, the results of each single analysis have to be properly processed. To this aim and coherently with the multiscale approach adopted in NLSA, a scalar variable Y_{DLk} ($=Y_{PLk}$) is adopted (Fig. 5.3b): it derives from the maximum among proper

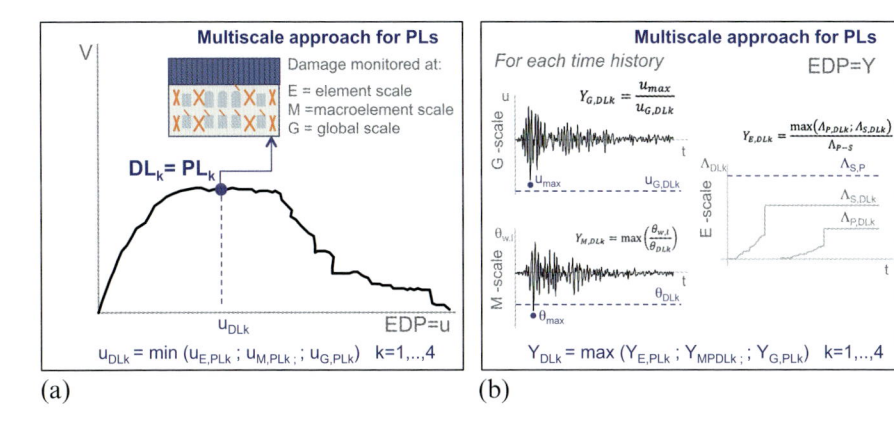

Fig. 5.3 Multiscale approach for the definition of the Performance Levels (PLs) in nonlinear static (**a**) and nonlinear dynamic (**b**) analysis (adapted from Lagomarsino and Cattari 2015b)

ratios between the maximum value of the variables monitored at three different scales, reached through the application of the selected record, and the corresponding threshold. It is assumed that the attainment of $Y_{PLk}=1$ indicates the reaching of the examined PL.

The adoption of this multiscale approach turns out very useful, in particular when a damage concentration is expected on single walls that however could not correspond to a significant decay of the overall base shear. The multiscale approach may be applied with analogous principles in case of both static and dynamic nonlinear analyses, guaranteeing a consistent comparison between the results provided by these two methods, as discussed in Lagomarsino and Cattari (2015b) and summarized in Fig. 5.3.

In general, 4 Limit States are considered in the codes (1 – operational; 2 – damage limitation or immediate occupancy; 3 – significant damage or life safety; 4 – near collapse or collapse prevention), which may be related to the 4 Damage Levels (DLs) related to damage in structural and nonstructural elements (i.e. the first 4 damage grades of the EMS-98 macroseismic scale – Grunthal 1998). In this paper two of the four PLs defined by the multiscale approach have been considered (PL$_k$, k = 2,3), those related to the damage limitation and the significant damage, related to the most important performances usually assumed also in codes (that is immediate occupancy and life safety).

5.3.4 Comparisons Between the Results of the Static and Dynamic Analyses

In order to check the reliability of the static approach depending on the different load patterns used in the pushover analysis, a series of comparisons have been undertaken between the results of the static and dynamic analyses.

To this end, the most significant parameter is the Intensity Measure that produces the attainment of the different Performance Levels, indicated herein with the acronym IM_{DLk} (with k = 2,3).

For NLDAs, the values of IM_{DLk} have been evaluated by using, for each one of the ten records, the multiscale approach described in Sect. 5.3.3 by checking the attainment of condition $Y_{PLk} = 1$. Then, for both the two PLs, the median value and the 16% and 84% percentiles of IM have been evaluated, under the hypothesis IM is log-normally distributed.

For NLSAs the calculation of IM_{DL} requires to preliminarily define the displacement on the pushover curve in which the given DL is attained; to this aim the multiscale approach has been used. Then, as aforementioned, it is necessary to operate the conversion of the pushover curve (representative of the original Multi Degree of Freedom System) into the capacity curve of an equivalent SDOF system. Such conversion is performed herein by the participation factor Γ, as adopted in G.U. (2008) and CEN (2004) and proposed by Fajfar (2000), which is computed as indicated in Eq. (5.1):

$$\Gamma = \frac{\sum m_i \phi_i}{\sum m_i \phi_i^2} = \frac{m^*}{\sum m_i \phi_i^2} \qquad (5.1)$$

where m^* is the mass of the equivalent SDOF system, m_i is the mass of the i-th node of the EF model (or story if a stick model is assumed) and Φ_i represents the normalized displacement of the i-th node. For the calculation of Φ_i the Italian code Commentary (G.U. 2009) suggests to refer to the displacement pattern of the fundamental modal shape, independently from the load pattern chosen. However, this factor is very sensitive to the applied load pattern and the related deformed shape, which furthermore varies in the nonlinear response. Therefore, the reference deformed shape adopted for the conversion is the displacement profile obtained in the elastic phase by the application of each specific load pattern. Since the building is in its elastic phase only at the very initial steps of the pushover analysis it is important to subtract the deformation caused by the application of the gravity loads. Subsequently, the displacement profile is normalized to the control displacement, as defined in Sect. 5.3.2.

Once the pushover curve is converted into the capacity curve, the next step is the evaluation of the displacement demand. To this aim, as stated in Sect. 5.1, different methods are available in the general framework of the DBA: the Capacity Spectrum Method (Freeman 1998, 2004); the N2 method (Fajfar 2000); the Coefficient Method (ASCE-SEI 2014).

Herein, IM_{DL} has been evaluated by the Capacity Spectrum Method (CSM), that was preferred among the others because it does not require the transformation of the capacity curve into an equivalent bilinear and the assumption on soil conditions. This method is based on the use of overdamped spectra, made according to the damping correction factor η, proposed in CEN (2004) and G.U. (2008), which is a function of the equivalent viscous damping (ξ_{eq}). This latter is the sum of the viscous and hysteretic contribution, and it has been calibrated through cyclic pushover analyses performed on each prototype building examined by applying as maximum displacement the one corresponding to the attainment of each DL (u_{DL}). For each DL two full cycles of loading have been performed and then the hysteretic damping (ξ_{hist}) has been calculated as indicated in Eq. (5.2):

$$\xi_{hist} = \frac{E_d}{2\pi(E_{S0+} + E_{S0-})} \quad (5.2)$$

where E_d is the energy dissipated during the cycle considered and E_{S0} is the elastic energy produced (+ for positive direction of loading, − for the negative). For more information see Lagomarsino and Cattari (2015a).

In order to have consistent results with those from the NLDAs, the spectra correspondent to the records used in the IDA have been used to determine the IM_{DL} (in particular the median spectrum and the spectra related to the 16% and the 84% percentiles).

Another useful comparison is that between the ISA (Incremental Static Analysis) and the IDA (Incremental Dynamic Analysis) curves. The IDA curves can be simply obtained by plotting in a graph the displacement demand deriving from each performed IDA (represented by the maximum control displacement occurred during the analysis) as a function of IM applied for that analysis; considering the displacement demand values deriving from the different records as log-normally distributed, it is possible to obtain only three curves, associated to the median value, the 16% and the 84% percentiles. The ISA curves derive from the nonlinear static analyses; the procedure explained before for the computation of earthquake intensity associated to the specific DL may be applied for any current point of the pushover curve, thus leading to the ISA curve. The latter shows in the abscissa the control displacement adopted for the NLSA and in the ordinate the IM.

Again, in order to have consistent results between static and dynamic analyses, the spectra correspondent to the records used in the NLDA have been applied for the calculation of the IM_{PL} of the static procedure. Therefore, for each given configuration, three ISA curves have been evaluated, obtained by using the median spectrum and the spectra related to the 16% and the 84% percentiles. By comparing ISA and IDA curves it is possible to provide a more comprehensive comparison of the results between static and dynamic analyses (Fig. 5.4).

In particular, this comparison is useful to highlight if the possible differences in the values of IM_{DL} are mainly related to discrepancies in the attainment of DLs or to intrinsic limits of the static method (e.g. on the conversion into equivalent SDOF, the approximate evaluation of damping, etc.). In the latter case, the IDA and ISA curve are expected to be different, whereas in the first case although the curves could also be very similar, the IMs may have different values for example due to a different damage distribution that may influence checks at the different scales (E, M, G).

It is evident that dynamic analyses produce a considerable amount of data (i.e. the displacement and the acceleration time histories of all nodes of the model) that go far beyond the strict computation of IM_{DL} and could be very useful to make further comparisons with the results obtained from the static analyses. In particular, it is

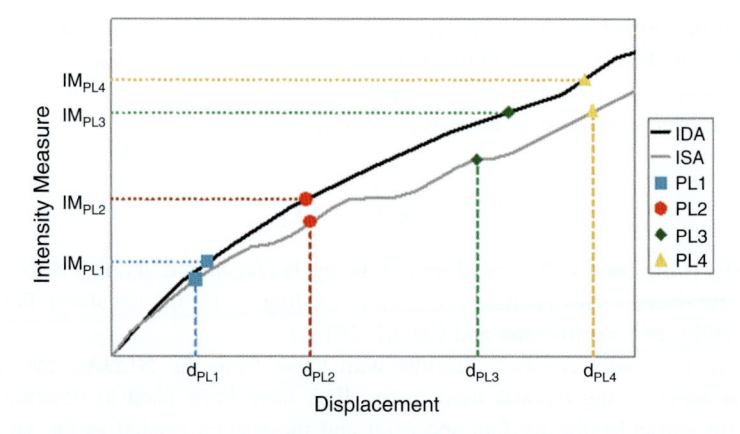

Fig. 5.4 Example of a comparison of ISA-IDA curves. The horizontal and vertical lines indicate the reaching of 4 Performance Levels (PL) in the IDA curves (defined in terms of Intensity Measure – IM_{DL}) and in the ISA curves (defined in terms of displacement – d_{DL}), respectively

interesting to compare the results in terms of deformation modes and load profiles associated to the reaching of each given DL.

In the case of NLSA, for each one of the considered load profiles, it is possible to draw the deformed shape correspondent to a given DL.

In the case of NLDA, the use of the Proper Orthogonal Decomposition (POD, Lumley 1970) method has been adopted to process the output of the analyses (time histories of displacements and accelerations in each node of the building). The method basically consists in the eigenvalue decomposition of the covariance matrix estimated from the acceleration or displacement time histories resulting from the nonlinear dynamic analysis (Lagomarsino and Cattari 2015b). This technique turned out to be particularly effective to interpret the seismic response in terms of dominant behaviors (Cattari et al. 2014), instead of referring to single instants of the response history (e.g. the maximum displacement occurred in a point at the top level of the building). In particular, herein it is used to capture the dominant displacement profiles of each masonry wall (and so the dominant deformed shape of a given structure) as well as the principal inertial load distributions activated by the seismic action in correspondence of the attainment of each given DL.

5.4 Prototype Buildings for the Parametric Analyses

The parametric analysis for validating the use of NLSA in the case of irregular masonry buildings has been performed by considering different prototype buildings obtained from the same architectural base configuration, which may be considered representative of the typical existing buildings in Italy and, more in general, in many European countries. The base building configuration, which was analyzed in Cattari

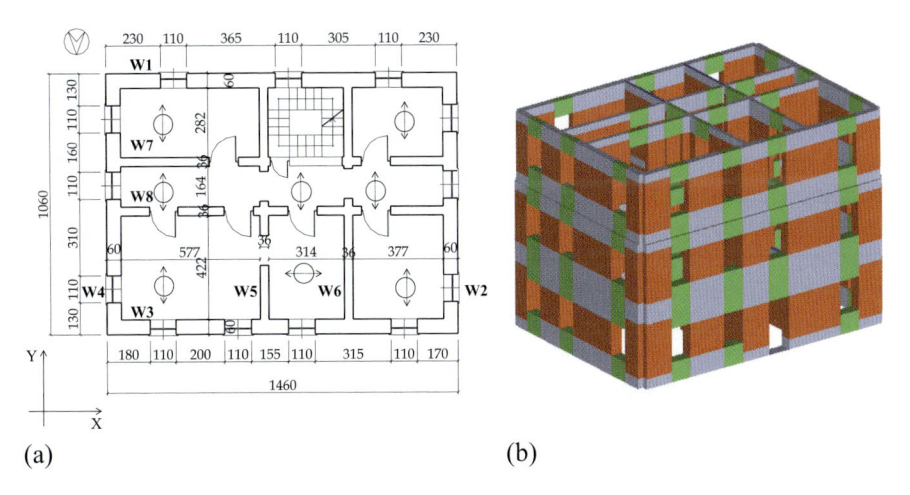

Fig. 5.5 Basic configuration of the prototype buildings: (**a**) plan view and (**b**) 3D view showing the equivalent frame idealization (orange = piers; green = spandrels; blue = rigid nodes)

and Lagomarsino (2013b), is a three-story full clay masonry building with lime mortar and steel tie rods at each level (Fig. 5.5). The thickness of the external walls is 0,48 m at the ground floor and at the first floor, while it is 0,36 m at the top level; the internal walls have a constant thickness (0,24 m) for all the levels. For all the defined configurations the same type of masonry has been used, and its mechanical properties are summarized in Table 5.2.

Starting from this base configuration, which has a regular openings distribution and is characterized by rigid diaphragms (RC slabs), different variants have been defined in order to examine the effects related to horizontal diaphragms stiffness and to plan irregularity.

First of all, the structural details variation has been explored, by replacing the tie rods with RC ring beams, coupled with the spandrels at each level, thus leading to 2 different configurations: building A (with steel tie rods) and building B (with RC ring beams). The ring beams had rectangular cross-section, 0.25 m high, with longitudinal reinforcement made by four 16 mm diameter bars both on top and bottom side and shear reinforcement consisted of 10 mm bars spaced 150 mm.

As far as the floor stiffness is concerned, two additional configurations have been defined, representative of a stiff and a flexible condition, respectively. In fact, in the case of URM ancient and existing constructions, timber floors or vaults are often present and they are far from the idealization of rigid diaphragms. In particular, in the case of timber floors the shear stiffness depends principally on: (i) the sheathing (single or double); (ii) the quality of the connection between by the joists and the board; (iii) the presence of steel dowels embedded in masonry walls. Reference values for the shear stiffness are provided in Brignola et al. (2012), ASCE-SEI (2014) and NZSEE recommendation (2015). In the case of vaults, the shear stiffness depends mainly on: (i) the thickness, (ii) the properties of the materials used and (iii) the shape and the geometric proportions (i.e. the rise-to-span ratio, as discussed in Cattari et al. 2008). Therefore, in the examined case studies, the *flexible* condition was meant to be representative of a single straight sheathing, for timber floors, and of barrel and cross vaults with a high rise-to-span ratio; the *stiff* condition was meant to be representative of double straight sheathing with good connection provided by joists, for timber floors, and cloister vaults or barrel and cross vaults with a low rise-to-span ratio. In Table 5.2 the values assumed for E_1, E_2 and G_{eq} in the three different typologies of diaphragms considered (rigid, stiff and flexible) are summarized.

In order to introduce plan irregularity, the stiffness of two outer walls has been changed by closing six of the nine openings of one of these walls (wall 2, Fig. 5.6) and enlarging the correspondent six windows in the other one (wall 4, Fig. 5.6). In such a way it was possible to obtain a plan irregular configuration, that is characterized by a significant distance between the center of the mass and the center of the stiffness than the base configuration (Fig. 5.6).

By combining the various types of irregularity, 10 different configurations have been defined (summarized in Table 5.3): six with tie rods (models A) and four with ring beams (models B). Among models the A and B, half of them have regular plane configuration, the other half irregular; for the A models three different stiffness of the diaphragms were adopted, for the B models only the rigid and stiff ones.

Regarding the values of the equivalent damping adopted for the calculation of IM_{DL}, Table 5.2 shows the mean values obtained for the 10 models (being the equivalent damping calculated for each configuration through the execution of cyclic pushover analyses, as mentioned in Sect. 5.2.4). In case of load pattern proportional to the masses the damping values $\xi_{hist, DLk,u}$ were used, while in other cases the ones named as $\xi_{hist, DLk,t}$ resulting from the cyclic pushover with load pattern proportional to the inverted triangular.

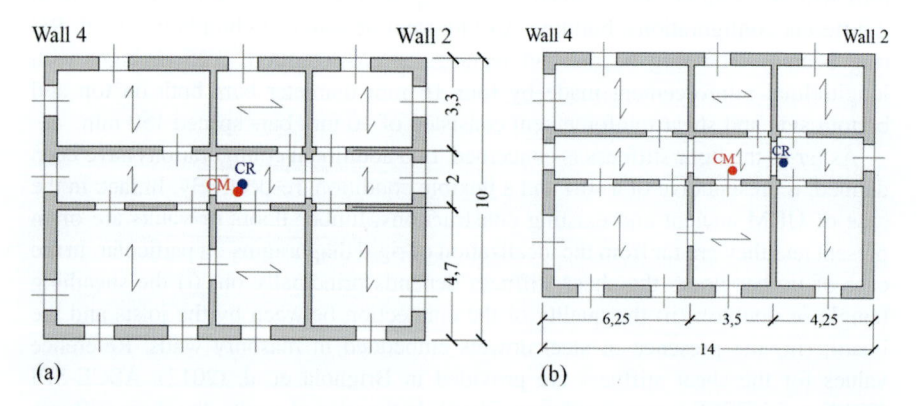

Fig. 5.6 Plan configuration of the basic model (**a**) and of the plan irregular model (**b**); CM = centre of the masses; CS = centre of the stiffness

5.4.1 Results of the Modal Analyses for Verifying the Effectiveness of Conceived Prototype Buildings

Modal analysis allows to interpret the influence of three different features: (i) - in-plane diaphragms stiffness; (ii) plan irregularity; (iii) effect of ring beams.

First of all, it is worth noting that all the configurations exhibit in each wall a linear modal shape. Therefore, it is expected that the inverted triangular load distribution would be appropriate for these buildings.

The modal analyses confirmed that the dynamic behavior of the case study models is strongly influenced by the in-plane diaphragm stiffness. If they are rigid it is easy to observe a global response (Fig. 5.7a, b, for regular and irregular configuration, respectively). In the case of stiff diaphragms, each wall moves independently from the others and it becomes difficult to identify a global behavior of the building (Fig. 5.7c, d, for the regular building). In the models that present flexible diaphragms the modal shapes associated to each mode are equivalent to the mode of each wall (Fig. 5.7e, f, g, h, for the regular configuration only).

Moreover, by decreasing the diaphragm stiffness in the first fundamental mode: (i) the period increases; (ii) the effective modal mass decreases. Indeed, when the diaphragms are rigid all the walls are connected and forced to move together, and more mass participates. It is also possible to observe that the period of the 1st mode of model A_{r_rig} is equivalent to the average of the periods of the four modes that activate the four walls of model A_{r_flex} (see Fig. 5.7 and Table 5.3). This observation is not valid for the model A_{irr_rig} because in this case also the torsional mode activates a significant mass, whereas when the diaphragms are flexible (A_{irr_flex}) there is no torsional effect.

The introduction of the plan irregularity leads to similar effects of the diaphragm stiffness (i.e. increase of the 1st mode period and decrease of the effective mass);

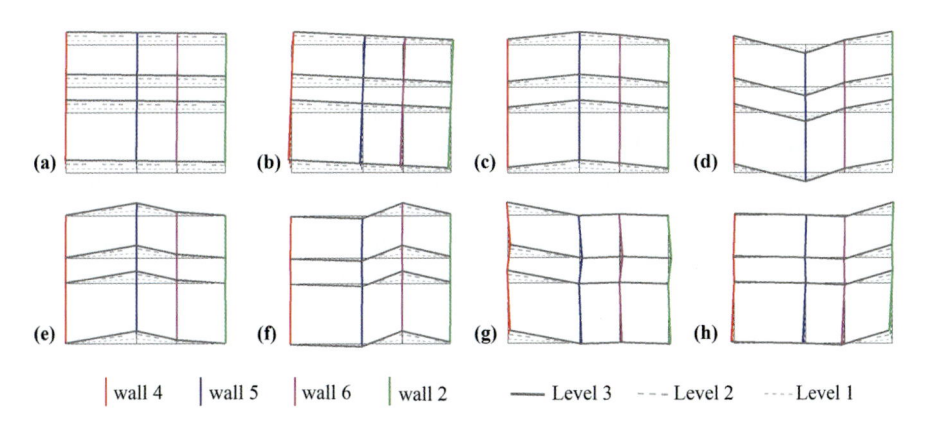

Fig. 5.7 Modal plan deformed shapes: A_{r_rig} 1st mode (**a**), A_{irr_rig} 1st mode (**b**), A_{r_int} 1st mode (**c**), A_{r_int} 4th mode (**d**), A_{r_flex} 1st mode (**e**), A_{r_flex} 3rd mode (**f**), A_{r_flex} 5th mode (**g**), A_{r_flex} 6th mode (**h**)

another consequence is that the 1st mode is not anymore purely translational but there is also a significant torsional component (Fig. 5.7 a, b).

The introduction of ring beams does not cause evident variations in the modal response. The modal shapes are similar to those for the models with tie rods and also the most significant modes are the same. A small difference is the slight increase of the modal periods, due to the slight increase of the mass, which turns out to be more relevant than the stiffness increase of the spandrels.

The modal properties of the analyzed configurations are summarized in Table 5.3. It is worth noting that although the introduction of plane irregularity caused the presence of torsional effects, they were not so relevant to require the consideration of the bidirectional effect of the ground motion. Indeed, in CNR-DT212 (2014) it is suggested that significant effects due to the bidirectional ground motion are present when the ratio between the minimum and maximum participation factors is greater than 10%. This ratio is reported in the last column of Table 5.3 and it is possible to observe that for the 1st mode it is always smaller than 10%.

The main conclusions from modal analysis are: (i) when the diaphragms are not rigid the 1st mode is not enough to represent the dynamic behavior of the building but also higher modes should be taken into account; (ii) the introduction of the plane irregularity caused torsional effects that however were not significant enough to require a study of the bidirectional effect of the earthquake motion. The latter allowed to undertake both NLSA (and NLDA) applying the forces (or the accelerations) only in one direction, in particular in the Y direction, which is the one influenced by the eccentricity of mass/stiffness.

5.5 NLSAs and NLDAs Results Comparison

The first comparison is made in terms of load patterns, in order to check which is the more accurate among the five different load patterns that were considered for the NLSAs. To this end, the shape of these load patterns are compared with the main inertia forces activated during the NLDAs through the use of POD technique (described in Sect. 5.3.4) and shown in Fig. 5.8 for Wall 2 in two different models. It is worth noting that due to both the plan irregularity and the flexibility of the diaphragms, each wall is loaded differently. It is possible to observe that the uniform distribution always results very different from the others, while among the load patterns aimed to simulate the "modal shape" the SRSS and CQC distributions turned out to be more reliable. On the contrary, the 1st modal load pattern (proposed in many international codes) turns out to be not accurate enough, in particular in case of flexible floors, where it leads to a situation in which some walls are almost not loaded (Fig. 5.8b).

The comparison of the results in terms of deformed shapes associated to the PL attainment is shown in Fig. 5.9, where in the abscissa the average displacements of all nodes at that floor, weighted by their masses, is reported. The solid lines refer to NLSAs with load patterns applied in direction Y+ (see Fig. 5.5a), while the dashed

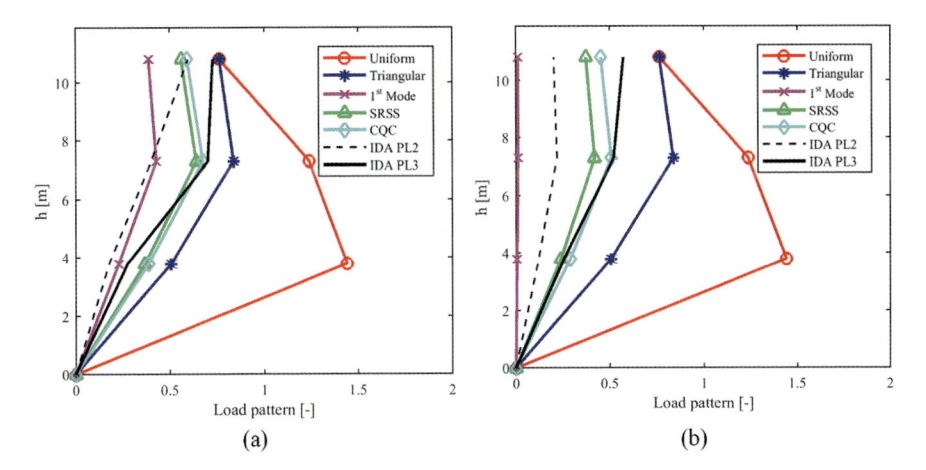

Fig. 5.8 Comparison between the median load patterns deriving from Incremental Dynamic Analyses (IDA), obtained through the application of the Proper Orthogonal Decomposition (POD), and the 5 different load patterns used for the pushover analyses (kept invariant during the analysis), considering a specific wall (wall 2) of the Airr_rig model (**a**) and Airr_flex (**b**)

ones are for the direction Y-. For the NLDAs, the median deformed shape together with those associated to the 16% and 84% percentiles are shown, assuming a lognormal distribution; the deformed shape is plotted considering the maximum mean displacement of the nodes at each storey throughout all the time history. It is possible to observe that if the floors are flexible (Fig. 5.9b1 and b2) the displacement capacities related to the attainment of the DLs are more disperse than the rigid case (Fig. 5.9 a1 and a2). Furthermore, for the prototype buildings under analysis, if the flexibility of the diaphragms increases, the DLs are reached for a smaller displacement capacity. Similarly, if plan irregularity is introduced, the displacement capacity reduces. A graphical representation of this trend is depicted in Fig. 5.13.

With regard to the definition of the proper LPs to use, as suggested in CEN (2004) and G.U. (2008), in order to get safe results in terms of IM_{PL}, it is suggested to use at least two different load patterns. For the choice of them, at this stage of the research some recommendations could be made: (i) the use of the 1st modal load pattern is not correct when diaphragms are not rigid; (ii) the uniform load pattern always behaved differently than the others and allowed to detect cases in which a soft story failure mechanism occurs in the NLDA; (iii) as second load pattern to be used, for the model under analysis both the triangular and SRSS provide reliable results; (iv) no relevant difference were detected in the results among the SRSS and CQC load patterns (this justifies the use of the former that is easier to be calculated).

Although in most cases also the load pattern proportional to the first modal shape provided safe results, its use, in presence of flexible diaphragms is discouraged for two main reasons: (a) the walls that have their own period of vibration far from the 1st mode one could not receive any force; (b) it activates a small percentage of mass. The issues of point (a) was already highlighted in Fig. 5.8b, indeed in model A_{irr_flex}, wall 2 receives almost zero forces, since its displacements are close to zero in the 1st

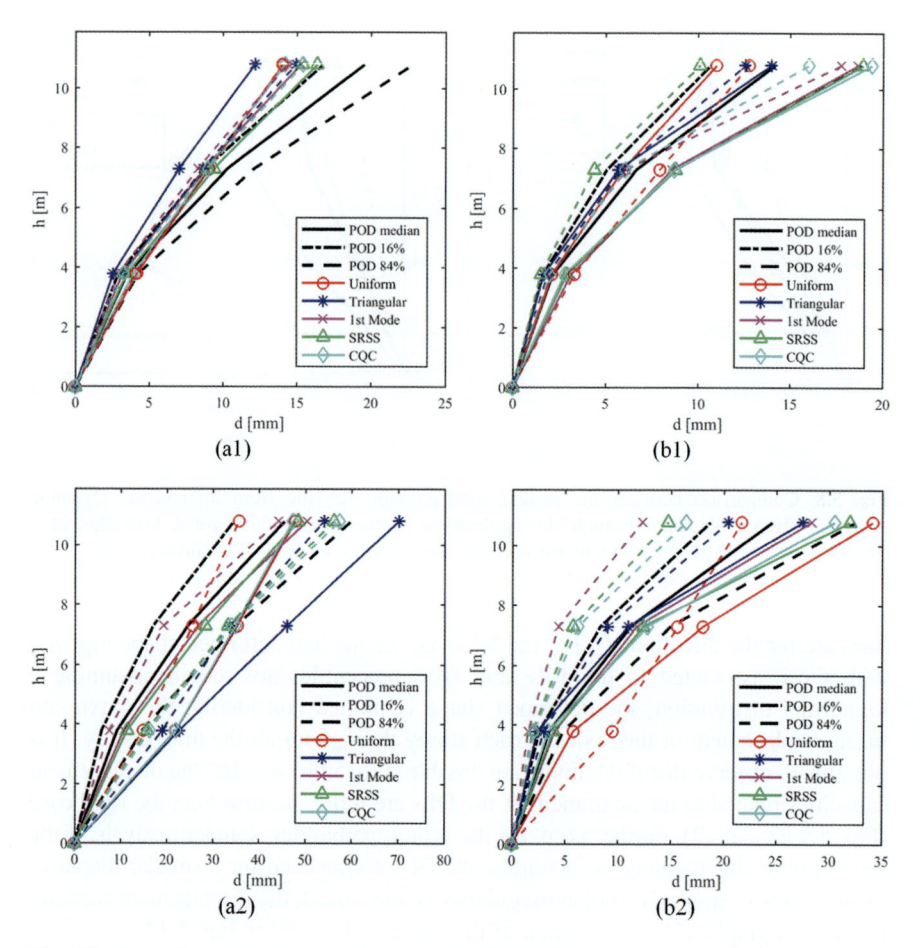

Fig. 5.9 Comparison between the deformed shapes deriving from Incremental Dynamic Analyses (IDA), obtained through the application of the POD technique, and those deriving from pushover analyses in correspondence of DL2 (1) and DL3 (2) for models Airr_rig (**a**) and Airr_flex (**b**)

mode of the model (see Fig. 5.7e). Another consequence of this issue is that the prediction of the damage occurred is not correct. Indeed, at the attainment of each DL after NLDAs all the walls showed similar damage, because all of them were activated by the input ground acceleration. On the contrary, as shown in Fig. 5.11, when the same DL was reached with the NLSAs performed applied the load pattern proportional to 1st mode, only one or two walls were damaged.

Despite these issues, the load pattern proportional to the first mode provided safe results only due to the use of the multiscale approach. Indeed, as written above, the application of this load pattern caused a concentration of damage in a single wall (Fig. 5.10) that brought to the attainment of a DL although it does not cause a decrease of the global base shear. Indeed, both the DLs were reached even before the attainment of the maximum base shear. On the contrary, if the multiscale approach is

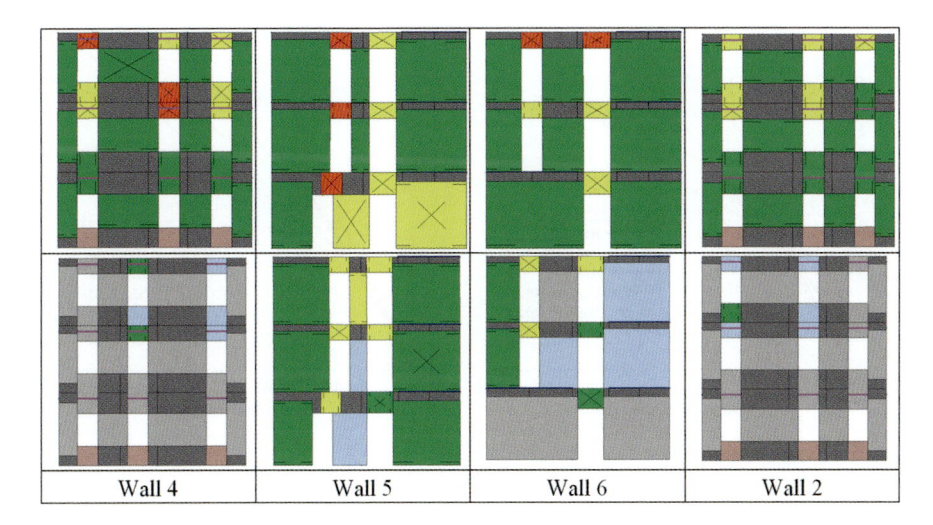

5 Seismic Assessment of Existing Irregular Masonry Buildings by...

| Wall 4 | Wall 5 | Wall 6 | Wall 2 |

Fig. 5.10 Ar_flex model: damage in the walls at the attainment of DL3: up from one NLDA performed, down from NLSA applying a load pattern proportional to 1st mode. Refer to Fig. 5.1a for the meaning of colours and symbols

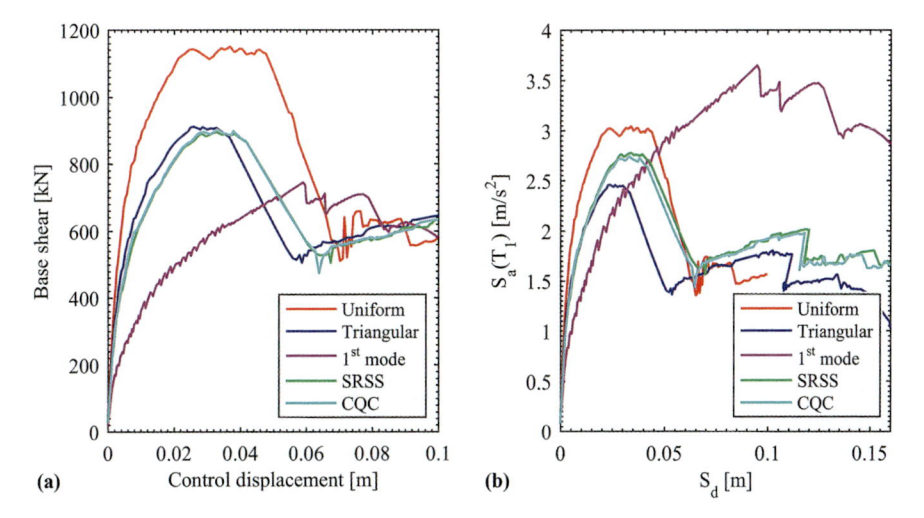

Fig. 5.11 Ar_flex model: pushover (**a**) and capacity (**b**) curves associated to the five different load patterns used for the NLSAs

not used (DL_3 in correspondence of a base shear decay of 20%) the use of this load pattern would have brought to strongly unsafe results.

The consequences of point (b) are shown in Fig. 5.11. Although the use of the LP proportional to 1st mode provides the pushover curve with the smallest base shear, it becomes the curve with the highest capacity when transformed into the capacity curve (representative of the equivalent SDOF). This happened because in order to

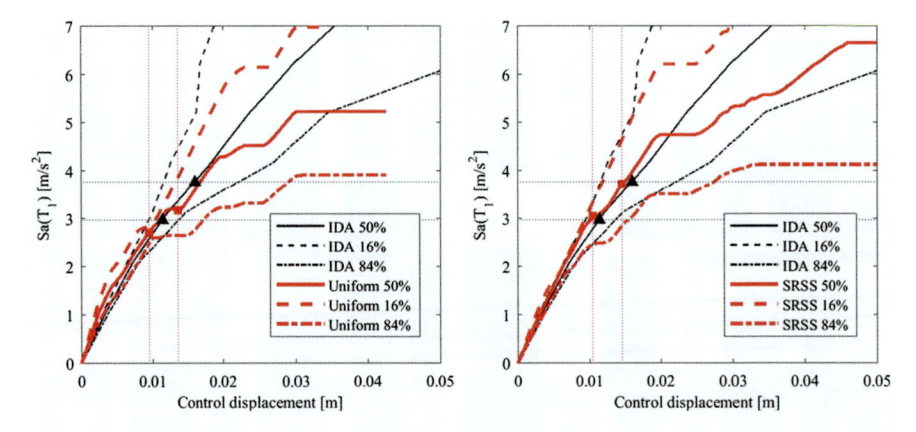

Fig. 5.12 IDA – ISA curves comparison for the Birr_int model. Left the uniform load pattern is applied, right the SRSS

get the capacity the base shear need to be divided for the mass of the equivalent SDOF that is, in this case, very small if compared with the ones obtained from the others load patterns. Another consequence of this issue is that for high values of displacement, the ISA curve diverges from the IDA curve and leads to non-conservative results.

The nonlinear static method results more or less conservative due to two different aspects: (1) the accuracy of the ISA curve, in comparison with the IDA curve; (2) the capacity to detect the correct displacement capacity for a given limit state. Both of these aspects are highlighted in Fig. 5.12. With continuous lines the ISA and IDA curves correspondent to the 50% percentile are plotted, while with the dashed and dashed-dotted lines represent the 16% and the 84% percentiles respectively; in particular, the curves obtained with the uniform and the SRSS distributions applied to model B_{irr_int} are considered. The positions of the DLs derived from NLSAs are in terms of displacements, and shown by dotted vertical lines; the ones derived from NLDAs are in terms of accelerations and shown by dotted horizontal lines. It can be observed that in both cases the static method provided conservative results, but for different reasons. The ISA curve derived from uniform load pattern is always below the IDA curve, apart for the very initial phase. Also the displacement correspondent to the attainment of the two DLs calculated with the static method is smaller than the one correspondent to the dynamic one. Therefore it is obvious that for this LP the static method provides conservative results. On the contrary, when applied the SRSS load pattern, for a longer section the ISA curve is above the IDA curve (however this overestimation is small) and a slight anticipation in the definition of the position of DLs with the static approach is enough to provide safe results.

In Fig. 5.13 the results of the comparison in terms of the displacements correspondent to the attainment of the two DLs are shown. All the analyzed configurations are represented (each one of them is identified with a specific abbreviation used in Table 5.3) and the control displacement at the attainment of the two DLs is reported. It is possible to observe that for DL2 the NLSAs give results very close to

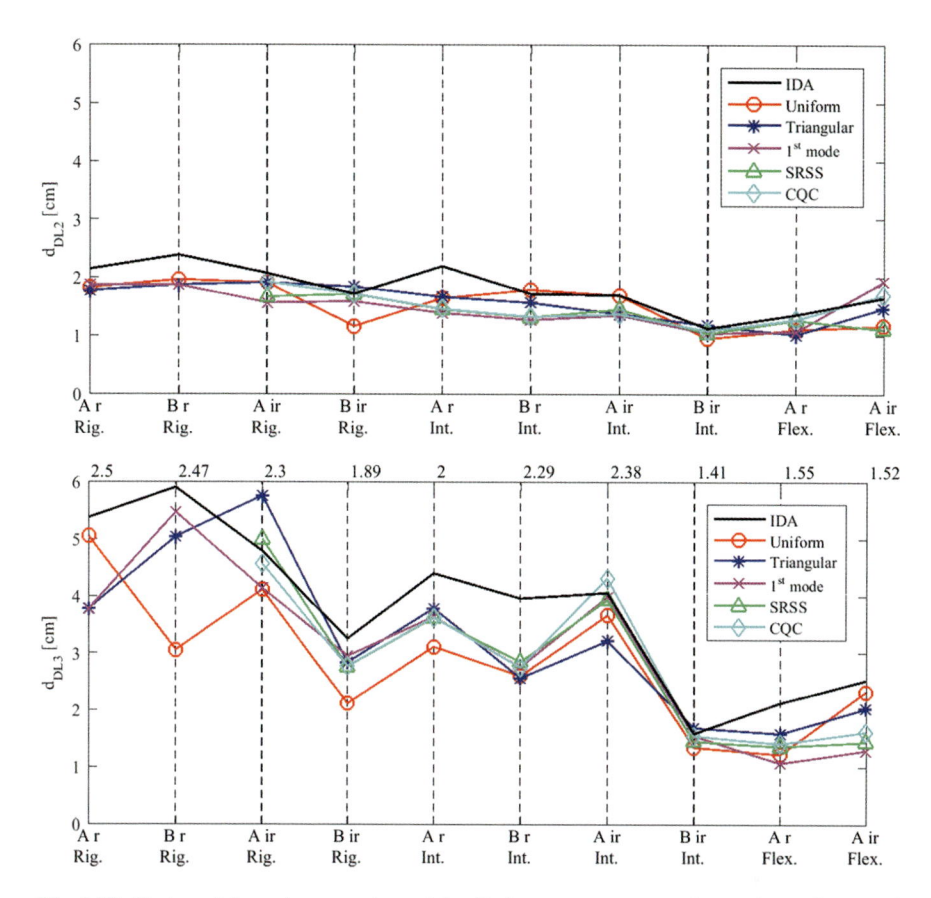

Fig. 5.13 Static and dynamic comparison of the displacements correspondent to the attainment of DL2 (above) and to the DL3 (below)

the NLDA, and it is almost insensitive to the choose of the load pattern. Major differences among the load patterns emerge for DL3, but also in this case, with the exceptions of few analyses, the results are always below the dynamic one.

Above in the graph, for each case study, the dynamic ductility is reported, the latter is calculated as the ratio of DL3 over DL2 dynamic displacement. As a general trend, the ductility reduces with the increase of irregularities. For both DLs, but mainly for DL3, the increase of irregularities reduces also the displacement correspondent to the attainment of DLs for both static and dynamic analyses.

In Fig. 5.14 the comparison in terms of IM_{DL} is shown. As for the displacements, the static method results on the safe side (i.e. it provides smaller values than the dynamic) for almost all the models, and the few exceptions are represented by the configurations that have a more significant level of irregularity. Indeed, in these cases the static results are closer to the dynamic. However, considering at least two load patterns (the uniform and one of the others) the static method always provides conservative results and they are included in the range of 60–80% of the dynamic. It

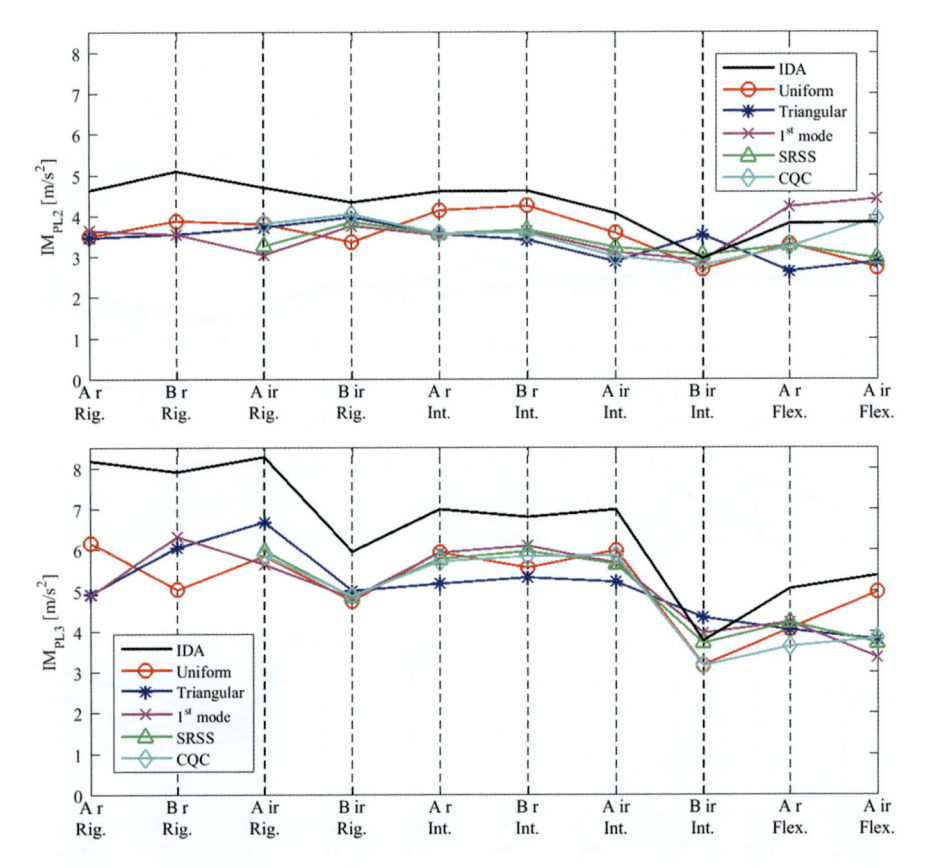

Fig. 5.14 Static and dynamic IMs comparison correspondent to the Damage Level 2 (above) and to the Damage Level 3 (below)

is worth reminding that the IM results derive from the use of the CSM, while other methods may not provide safe results as well. Indeed, Marino et al. (2018) proved that the CSM resulted a very conservative method if compared with other literature and code proposals, such as the N2 (Fajfar 2000) or the Coefficient Method (ASCE-SEI 2014).

As final remarks, Fig. 5.14 justifies statements (iii) and (iv) above mentioned, indeed: (iii) although there are not strong differences among the Triangular and SRSS load patterns, the use of the latter is suggested because the first, especially for DL3, provides worse results if compared with the dynamic (and this error should increase in the case of irregularity in elevation, a condition that was not analyzed here); (iv) there are not significant differences among SRSS and CQC.

5.6 Conclusions

The applicability of NLSA for the seismic assessment of irregular masonry buildings has been investigated by a systematic comparison with the results provided by NLDA, assumed as actual reference solution.

Ten configurations have been considered, defined in order to be representative of URM buildings, starting from a regular building with rigid floors and introducing plan irregularity and stiff or flexible diaphragms. The nonlinear static approach turns out to be reliable and also able to provide conservative results when, as prescribed by the current codes, at least two different load patterns are used in the pushover analysis. With regard to this aspect, from the comparison with nonlinear dynamic analyses, the most reliable choice has revealed to be the use of the uniform distribution combined with another load pattern representative of the dynamic behavior of the structure in the initial phase. Among the tested load patterns, the one obtained through the SRSS combination applied to the modal shapes which do not present the inversion of sign in displacement (modes of 1st type) seems to be the most accurate. Indeed, it leads to reliable and stable results also for irregular buildings with stiff or flexible diaphragms. However, it is worth noting that also the triangular load pattern provided conservative results, at least when buildings are regular in elevation. Moreover, it was possible to observe that the use of a LP proportional to first modal shape should be discouraged, especially in presence of not rigid diaphragms.

The applicability of NLSA also in the case of irregular masonry buildings is an important result for the engineering practice because, at present, the use of NLDA is still problematic, due to the lack of practice-oriented computational models able to consider the cyclic behavior of masonry elements with strength degradation. Anyhow, the use of NLDA at code level is far to be well defined, mainly for three critical issues: (i) the need to select a proper set of records (which is not simply a problem of spectral fitting); (ii) the definition of limit states for the results of a time history analysis; (iii) the computational effort.

However, these conclusions about the reliability of nonlinear static approach should be supported by the application to other irregular configurations, in particular related to the stiffness and strength in elevation. Moreover, other reference buildings should be considered, in terms of plan configuration and number of floors. Finally, an accurate comparison of different available procedures for the evaluation of the displacement demand should be interesting, because all of them were primarily validated in the case of structures characterized by longer periods, ductile behavior and rigid floors.

Acknowledgments The research was funded by the Italian Civil Protection Department through the Italian Network of Seismic Laboratories (ReLUIS), within the framework of the ReLUIS-DPC 2014-2018 Project.

References

ASCE/SEI 41–13 (2014) Seismic evaluation and retrofit of existing buildings. American Society of Civil Engineers, Reston. ISBN 978-0-7844-7791-5

Aydinoglu MN, Onem G (2010) Evaluation of analysis procedures for seismic assessment and retrofit design. In: Garevsky M, Ansal A (eds) Earthquake Engineering in Europe. Springer, Berlin, pp 171–198

Beyer K (2012) Peak and residual strengths of brick masonry spandrels. Eng Struct 41:533–547

Beyer K, Dazio A (2012) Quasi-static cyclic tests on masonry spandrels. Earthq Spectra 28(3):907–929

Beyer K, Mangalathu S (2013) Review of strength models for masonry spandrels. Bull Earth Eng 11:521–542

Brignola A, Pampanin S, Podestà S (2012) Experimental evaluation of the in-plane stiffness of timber diaphragms. Earthq Spectra 28(4):1687–1909

Calderini C, Cattari S, Lagomarsino S (2009) In-plane strength of unreinforced masonry piers. Earthq Eng Struct Dyn 38(2):243–267

Camilletti D, Cattari S, Lagomarsino S, Bonaldo D, Guidi G, Bracchi S, Galasco A, Magenes G, Manzini CF, Penna A, Rota M (2017) RINTC Project: nonlinear dynamic analyses of Italian code-conforming URM buildings for collapse risk assessment. In: Proceedings of 6th international conference on computational methods in structural dynamics and earthquake engineering (COMPDYN 2017), Rhodes Island (Greece), June 15th to 17th 2017

Cattari S, Lagomarsino S (2008) A strength criterion for the flexural behaviour of spandrel in unreinforced masonry walls. In: Proceedings of the 14th world conference on earthquake engineering, Beijing, 12–17 October 2008

Cattari S, Resemini S, Lagomarsino S (2008) Modelling of vaults as equivalent diaphragms in 3D seismic analysis of masonry buildings. In: Proceedings of the 6th SAHC conference, Bath, 2–4 July 2008

Cattari S, Lagomarsino S (2013a). Masonry structures, pp 151–200. In: Sullivan T, Calvi GM (ed) Developments in the field of displacement based seismic assessment. IUSS Press (PAVIA) and EUCENTRE, pp 524, ISBN: 978-88-6198-090-7

Cattari S, Lagomarsino S (2013b) Assessment of mixed masonry-reinforced concrete buildings by non linear static analyses. Earthq Struct 4(3):241–264

Cattari S, Chioccariello A, Degeè H, Donoux C, Lagomarsino S, Mordant C (2014) Seismic assessment of masonry buildings from shaking tables tests and nonlinear dynamic simulations by the proper orthogonal decomposition (POD). In: Proceedings of the 2nd European conference on earthquake engineering and seismology, Istanbul, Turkey, 24–29 August 2014

CEN (2004) Eurocode 8: design of structures for earthquake resistance - Part 1: General rules, seismic actions and rules for buildings. EN1998 – 1:2004. Comité Européen de Normalisation, Brussels

CEN (2005) Eurocode 8: design of structures for earthquake resistance – Part 3: Assessment and retrofitting of buildings. EN1998 – 3:2005. Comité Européen de Normalisation, Brussels

CNR-DT212 (2014) Guide for the probabilistic assessment of the seismic safety of existing buildings. National Research Council of Italy, Rome

De Stefano M, Mariani V (2014) Pushover analysis for plan irregular structures. In: Ansal A (ed) Perspectives on European earthquake engineering and seismology, geotechnical, geological and earthquake engineering, vol 34. Springer, Berlin. https://doi.org/10.1007/978-3-319-07118-3_3

Fajfar P (2000) A nonlinear analysis method for performance-based seismic design. Earthq Spectra 16(3):573–591

Freeman SA (1998) The capacity spectrum method as a tool for seismic design. In: Proceedings of 11th European conference of earthquake engineering. Paris, France

Freeman SA (2004) Review of the development of the capacity spectrum method. ISET J Earthq Technol 41(1):1–13

Gattesco N, Clemente I, Macorini L, Noè S (2008) Experimental investigation of the behavior of spandrels in ancient masonry buildings. In: Proceedings of the 14th world conference on earthquake engineering, Beijing, China

Graziotti F, Magenes G, Penna A (2012) Experimental cyclic behaviour of stone masonry spandrels. In: Proceedings of the 15th world conference on earthquake engineering, Lisboa, PT

Grunthal G (1998) European Macroseismic scale 1998: EMS-98. In: Grunthal G (ed) Chaiers du Centre Européen de Géodynamique et de Séismologie, vol 15, Luxembourg. ISBN: 2-87977-008-4

G.U. (2008) Decreto Ministeriale 14/1/2008: Norme tecniche per le costruzioni. Ministry of Infrastructure, Rome (in Italian)

G.U. (2009) Circolare C.S.Ll.Pp. No. 617 of 2/2/2009: Istruzioni per l'applicazione delle nuove norme tecniche per le costruzioni di cui al Decreto Ministeriale 14 gennaio 2008. G.U. S.O. n. 27 of 26/2/2009, No. 47. Ministry of Infrastructure, Rome, Italy (in Italian)

Lagomarsino S, Penna A, Galasco A, Cattari S (2013) TREMURI program: an equivalent frame model for the nonlinear seismic analysis of masonry buildings. Eng Struct 56:1787–1799

Lagomarsino S (2015) Seismic assessment of rocking masonry structures. Bull Earth Eng 13 (1):97–28

Lagomarsino S, Cattari S (2015a) PERPETUATE guidelines for seismic performance-based assessment of cultural heritage masonry structures. Bull Earthq Eng 13(1):13–47

Lagomarsino S, Cattari S (2015b) Performance of historical masonry structures through pushover and nonlinear dynamic analyses. In: Ansal A (ed) Perspectives on European earthquake engineering and seismology, Geotechnical, Geological and Earthquake Engineering, vol 39, pp 265–292. https://doi.org/10.1007/978-3-319-16964-4_11

Lumley JL (1970) Stochastic tools in turbulence. Academic Press, New York

Magenes G (2000) A method for pushover analysis in seismic assessment of masonry buildings. In: Proceedings of the 12th WCEE, Auckland, New Zealand

Magenes G, Calvi GM (1997) In-plane seismic response of brick masonry walls. Earthq Eng Struct Dyn 26:1091–1112

Mann W, Müller H (1980) Failure of shear-stressed masonry – An enlarged theory, tests and application to shear-walls. In: Proceedings of the 7th international symposium on load-bearing brickwork, London

Marino S, Cattari S, Lagomarsino S (2018) Use of nonlinear static procedures for irregular URM buildings in literature and codes. In: Proceedings of 16th European conference on earthquake engineering, Thessaloniki, 18–21 June 2018

NZSEE (2015) Assessment and improvement of the structural performance of buildings in earthquakes - section 10 revision. Seismic Assessment of Unreinforced Masonry Buildings. New Zealand Society for Earthquake Engineering, Corrigendum n° 4, 22nd April 2015, ISBN 978-0-473-26634-9

Parisi F, Augenti N, Prota A (2014) Implications of the spandrel type on the lateral behavior of unreinforced masonry walls. Earthq Eng Struct Dyn 43:1867–1887

Penna A, Senaldi IE, Galasco A, Magenes G (2016) Numerical Simulation of Shaking table test on full-scale stone masonry buildings. Int J Archit Herit 10(2–3):146–163

Vamvatsikos D, Cornell CA (2002) Incremental dynamic analysis. Earthq Eng Struct Dyn 31 (3):491–514

Vanin F, Zaganelli D, Penna A, Beyer K (2017) Estimates for the stiffness, strength and drift capacity of stone masonry walls based on 123 quasi-cyclic tests reported in the literature. Bull Earthq Eng. https://doi.org/10.1007/s10518-017-0188-5

Chapter 6
Capturing Geographically-Varying Uncertainty in Earthquake Ground Motion Models or What We Think We Know May Change

John Douglas

Abstract Our knowledge of earthquake ground motions of engineering significance varies geographically. The prediction of earthquake shaking in parts of the globe with high seismicity and a long history of observations from dense strong-motion networks, such as coastal California, much of Japan and central Italy, should be associated with lower uncertainty than ground-motion models for use in much of the rest of the world, where moderate and large earthquakes occur infrequently and monitoring networks are sparse or only recently installed. This variation in uncertainty, however, is not often captured in the models currently used for seismic hazard assessments, particularly for national or continental-scale studies.

In this theme lecture, firstly I review recent proposals for developing ground-motion logic trees and then I develop and test a new approach for application in Europe. The proposed procedure is based on the backbone approach with scale factors that are derived to account for potential differences between regions. Weights are proposed for each of the logic-tree branches to model large epistemic uncertainty in the absence of local data. When local data are available these weights are updated so that the epistemic uncertainty captured by the logic tree reduces. I argue that this approach is more defensible than a logic tree populated by previously published ground-motion models. It should lead to more stable and robust seismic hazard assessments that capture our doubt over future earthquake shaking.

6.1 Introduction

Capturing epistemic uncertainty within probabilistic seismic hazard assessments (PSHAs) has become a topic of increasing interest over the past couple of decades, especially since the publication of the SSHAC approach (Budnitz et al. 1997). Put

J. Douglas (✉)
Department of Civil and Environmental Engineering, University of Strathclyde, Glasgow, UK
e-mail: john.douglas@strath.ac.uk

simply this means that the seismic hazard model, comprising a characterisation of the seismic sources (locations, magnitude-frequency relations, maximum magnitudes) and the ground-motion model (median ground motion for a given magnitude and distance and its aleatory variability, characterising the probability distribution around this median), need to capture our knowledge *and also* our doubt about earthquakes and their associated shaking in the region of interest.

Epistemic uncertainty is generally quantified by constructing a logic tree with weighted branches modelling our degrees of belief in different inputs (Kulkarni et al. 1984), e.g.: what is the largest earthquake that could occur along a fault (maximum magnitude)? For this article I am using the terms "aleatory variability" and "epistemic uncertainty" in the way they are commonly used in the engineering seismology community, i.e. aleatory variability is accounted for in the hazard integral whereas epistemic uncertainty is capture within a logic tree. Stafford (2015) proposes a different framework and terminology. Because a given structure may only need to withstand a single potentially-damaging earthquake during its lifetime, Atkinson (2011) proposes that all between-event uncertainties (i.e. including some of those currently modelled as aleatory variability) should be considered epistemic. This viewpoint is not considered in the following as it is not (yet) standard practice.

The comparisons of the uncertainties captured in various recent PSHAs shown by Douglas et al. (2014) suggest potential inconsistencies in some logic trees. Douglas et al. (2014) report, for various studies and locations, a measure of the width of the fractiles/percentiles of the seismic hazard curves equal to $100[\log(y_{84})-\log(y_{16})]$ where y_{84} and y_{16} are the 84th and 16th percentiles of peak ground acceleration (PGA) or response spectral acceleration (SA) for a natural period of 1s. This measure indicates how much uncertainty there is in the assessed hazard. Douglas et al. (2014) argue that this measure should vary geographically with the level of knowledge of engineering-significant ground motions, specifically areas with limited data (generally stable regions) showing high values (large uncertainty) and areas with considerable data showing lower uncertainties. This behaviour was generally seen in Europe [e.g. within the European Seismic Hazard Model (ESHM, Woessner et al. 2015)] but not in all studies or for all locations. Comparing the ESHM with site-specific studies for the same locations suggests that the overall uncertainty modelled in the ESHM is too low or alternatively the fractiles of the site-specific studies too wide. The key driver of the modelled uncertainty in PSHAs is often the ground-motion logic tree (e.g. Toro 2006) and hence this is the first place to start when seeking a method to construct PSHAs that reflect the underlying level of knowledge.

The next section of this article assesses the level of uncertainty captured within some typical ground-motion logic trees. The following section discusses the various approaches applied in the past decade to construct ground-motion logic trees. The main focus of this article is to propose a new approach, which is presented in the subsequent section and then applied in the penultimate section for three European countries. The article ends with some conclusions.

6.2 Uncertainties Captured in Logic Trees

Toro (2006) presents approximate results to quantify the impact of epistemic uncertainty in the median ground motion on the mean hazard curve. He states that the ratio, R, of the mean to median ground motion (e.g. PGA) for a given annual frequency of exceedance roughly equals:

$$R = \exp\left(0.5 \, k \, \sigma_\mu^2\right), \quad (6.1)$$

where k is the slope of the hazard curve in log-log space and σ_μ is the lognormal epistemic uncertainty in the median ground motion (in terms of natural logarithms[1]). He notes that this increase in ground motion is the same as would be caused by an increase in the aleatory variability from σ_0 to σ_1, where $\sigma_1^2 = \sigma_0^2 + \sigma_\mu^2$.

The slope of the hazard curve, k, generally is between 1 (generally, areas of low seismicity) and 4 (generally, areas of high seismicity) but k depends on the range of annual frequencies of exceedance considered (e.g. Weatherill et al. 2013; H. Bungum, written communication, 2017). Douglas (2010b) reports values of σ_μ between 0.23 (for well-studied areas such as western North America) and 0.69 (for the largest subduction events) from comparisons of median PGAs from many published ground motion prediction equations (GMPEs) for a few scenarios. Toro (2006) reports similar values from logic trees used in various US site-specific studies. Using Eq. (6.1) leads to the percentage increases in the median hazard given in Table 6.1. These calculations show that the effect of large epistemic uncertainties in the ground-motion model can be dramatic, e.g. more than doubling of the median ground motion (105% increase) for $\sigma_\mu = 0.69$ and k = 3 and the effect is highly sensitive to the exact level of uncertainty modelled. Consequently there is a need for a rigorous method to assess what uncertainty should be captured.

Rearranging Eq. (6.1) and assuming that the epistemic uncertainty in the hazard results is entirely due to the ground-motion logic tree, allows σ_μ to be estimated for published PSHAs. Table 6.2 reports the σ_μ obtained by this approach for a representative selection of the PSHAs considered by Douglas et al. (2014). The k value for each site and PSHA is estimated from the slope of the hazard curve computed using the ground-motion amplitudes for annual frequency of exceedance of 1/475 and 1/2475. This table suggests that the epistemic uncertainty captured in the

Table 6.1 Percentage increase when going from median to mean ground motion due to epistemic uncertainty in the ground-motion model of σ_μ for hazard curves with a slope k

	Ln	Log$_{10}$	1	2	3	4
	0.23	0.1	3	5	8	11
σ_μ	0.46	0.2	11	24	37	53
	0.69	0.3	27	61	105	160
	0.92	0.4	53	134	257	446

---

[1]Natural logarithms are used throughout for clarity.

Table 6.2 Estimated epistemic uncertainty in the ground-motion logic tree for some representative sites and PSHAs (for 475 year return period and PGA, except for Thyspunt where the spectral acceleration at 0.01 s is used)

Site	References	k	σ_μ (ln)
Edinburgh, UK	ESHM (Woessner et al. 2015)	1.68	0.23
Berlin, Germany	ESHM (Woessner et al. 2015)	1.55	0.34
Istanbul, Turkey	ESHM (Woessner et al. 2015)	2.51	0.13
Mühleberg, Switzerland	ESHM (Woessner et al. 2015)	2.09	0.18
"	**PEGASOS (Nagra** 2004)	**2.15**	**0.41**
Bruce, Canada	**AMEC Geomatrix Inc.** (2011)	**1.94**	**0.49**
Thyspunt	**Bommer et al.** (2015)	**1.19**	**0.51**
Yucca Mountain	**CRWMS M&O (Stepp et al.** 2001)	**1.80**	**0.45**

See Douglas et al. (2014) for details of studies. Site-specific studies are in bold. Only rock sites are considered

ground-motion logic tree of regional PSHAs such as ESHM are too low, particularly for stable areas, as they are far below those captured in site-specific studies (e.g. Toro 2006) and those reported by Douglas (2010b) based simply on comparing predictions from GMPEs for some scenarios.

A quick check that could be performed when developing ground-motion logic trees for regional applications is to compute the σ_μ for a few key scenarios (see below for guidance from disaggregation on what earthquakes are likely to dominate the hazard) and compare it to the values reported in Table 6.2 for site-specific studies. Values much lower than these values would need to be carefully justified. Also the σ_μ for ground-motion logic trees for application in active areas should generally be lower than the σ_μ for stable areas, because of the lack of ground-motion data of engineering significance from such areas to constrain the GMPEs.

As stated by USNRC (2012), the aim of any PSHA should be to capture the "centre, body and range of technical defensible interpretations", even if the PSHA is regional and not site-specific and even if it is being conducted at a low SSHAC level (1 or 2). The results of PSHAs should also be stable with time, i.e. if a new PSHA was conducted for the sample location in the future the results of the new study should not be greatly different to the original results. This requirement means that sufficient uncertainty has to be modelled so that the fractiles of the assessed hazard from the original and subsequent PSHAs broadly overlap.

Bommer and Scherbaum (2008) note that the different models on the branches of logic trees should be mutually exclusive and collectively exhaustive (MECE) so that the branch weights can be considered as probabilities, an implicit assumption when computing mean hazard curves and those for different fractiles. This means that one of the models (although we do not know which) is the true model and that all the models are independent. This criterion is likely not to hold for logic trees developed using GMPEs derived from overlapping datasets.

This study discusses how to populate logic trees for the ground-motion component of the seismic hazard model to capture epistemic uncertainty within national or continental-scale seismic hazard assessments, where our knowledge of ground motions in moderate and large earthquakes varies. The focus is not on site-specific studies (e.g. those conducted for critical infrastructure), which have different challenges (e.g. higher regulatory scrutiny) and opportunities (e.g. much smaller geographical range and more resources per km^2 covered). The approach proposed in this article may, however, be of interest for these studies, particularly in regions with limited data. The focus of this study is on models for the prediction of the median ground motions rather than the, equally important, models for the aleatory variability (sigma).

6.3 Current State of Practice

There are three principal ways in which geographically-varying logic trees could be constructed, if the approach of Savy et al. (2002) using point-based estimates from expert judgement for various magnitude-distance-period scenarios is excluded as being too cumbersome for regional-scale PSHA. These are summarised in this section, focussing on the relatively new approach of backbone models. Goulet et al. (2017, Chapter 2) provide a recent comprehensive review of the development of ground-motion logic trees to capture epistemic uncertainty, particularly within US projects.

6.3.1 Multiple Ground Motion Prediction Equations

The most common way of constructing ground-motion logic trees is to populate the branches with a selection of previously published GMPEs. A recent example of such a logic tree was that used in the ESHM and presented in Delavaud et al. (2012). A recent overview of this approach and ways of selecting and weighting the GMPEs is provided by Kale and Akkar (2017) in the context of the Earthquake Model for the Middle East (EMME, Danciu et al. 2018).

For areas with a set of recently-published and robust GMPEs (e.g. California, Japan and Italy) this approach appears relatively straightforward. Nevertheless, for these areas there are difficulties in deciding which of the many available models to choose and how many models should be included. For California two recent projects (NGA-West1 and 2) have developed a set of five GMPEs using a consistent database and independent and dependent variables (Power et al. 2008; Bozorgnia et al. 2014). Therefore, it would appear that a logic tree for California should comprise the most recent versions of these five GMPEs. However, there are doubts that such a logic tree would capture all the epistemic uncertainty concerning earthquake ground motions in future earthquakes because the models may be too similar. In this situation it has been proposed that additional logic-tree branches equivalent to a backbone approach are required (see below).

For many areas, however, there are no indigenous GMPEs, or those that are available are based on extrapolations from weak-motion data, often using the stochastic method (e.g. Rietbrock et al. 2013). When adopting a multiple GMPE approach for such areas, the problem is which of the models published for other regions should be included and will these GMPEs truly model ground motions in future earthquakes. Delavaud et al. (2012) were uncertain about whether ground motions in much of northern Europe (continental crust) were similar to those in the Mediterranean region, for which many indigenous models exist, or closer to those in the Scandinavian shield, for which they believed GMPEs developed for tectonically-similar eastern North America could be used. Therefore, their ground-motion logic tree included GMPEs for both tectonic regimes for that part of the continent. The logic tree proposed by Stewart et al. (2015) for the Global Earthquake Model discusses the considerable epistemic uncertainty in assessing ground motions for the majority of the world and they seek to propose a logic tree that captures this uncertainty by choosing robust models that displaying differing characteristics, e.g. decay rates in subduction earthquake ground motions. However, this is not an objective procedure nor was the resulting logic tree checked to see whether it models sufficient (or what appears to be sufficient) uncertainty. EMME sought to combine and improve on both these procedures for the construction of its ground-motion logic tree (Danciu et al. 2018).

Musson (2012) proposes that the weights on the logic tree are the probability of each GMPE being "the best model available". As pointed out by Bommer (2012), this implies that if there is only a single model available for a tectonic regime, this GMPE would automatically get a weight of unity, implying no epistemic uncertainty. In fact the uncertainty may be high, particularly as a single model often implies a lack of data from which to build more.

6.3.2 Hybrid Empirical Composite Ground-Motion Model

To create ground-motion models that are more regionally-specific, Campbell (2003) proposed the hybrid-empirical method where existing empirical GMPEs are adjusted based on the ratio of stochastic models for the target and host regions. In Campbell (2003) this method is applied to adjust GMPEs for California to make them applicable for eastern North America. Douglas et al. (2006) developed this method to account for uncertainties in developing stochastic models for the target region, where invariably there are fewer recorded data than in the host region (because otherwise robust GMPEs for the target region could have been proposed directly). These uncertainties in the various parameters of the stochastic model, e.g. stress (drop) parameter, are accounted for using a logic tree so that many stochastic models are created and applied when computing the ratios between target and host regions. In addition, Douglas et al. (2006) apply the technique to GMPEs from various host regions, again to capture uncertainty in the final logic tree of the adjusted models.

This approach appears more rigorous and transparent than the multiple GMPE approach using previously-published models but it is a time-consuming approach, particularly for a continent containing many target regions. In addition, there is subjectivity in deciding on the branches and weights for the stochastic models for the target region. Finally, although the uncertainties are propagated to the final adjusted GMPEs there is only a single (mean) model for each host GMPE and, consequently, a logic tree comprised simply of the adjusted GMPEs would not correctly model the uncertainty. However, the uncertainties could be tracked throughout the adjustment [e.g. Figure 15 of Douglas et al. 2006] to obtain a logic tree with branches modelling this uncertainty. For the site parameters V_s and kappa this is commonly done for site-specific studies and hence it could be extended to other parameters. This leads to the final main approach: backbone GMPEs, as discussed in the following section.

6.3.3 Backbone GMPEs

In the past decade and often in the context of site-specific PSHAs for critical facilities the backbone approach (Atkinson et al. 2014) has been used to construct ground-motion logic trees. In this approach a single or a handful of existing GMPEs are scaled up and down to account for uncertainty in the median motion. The scaling factors employed are generally related to uncertainty in the average stress (drop) parameter in the region, as well as other inputs to the stochastic model, e.g. anelastic attenuation. Using this approach leads to multiple GMPEs that are explicitly MECE. The level of uncertainty modelled is also made transparent.

Starting in 2008 and continuing in 2014, the US National Seismic Hazard Model applies a simple backbone approach to increase the modelled epistemic uncertainty in the ground-motion logic tree for shallow crustal seismicity in the western states (Petersen et al. 2014). Petersen et al. (2014) argue that the selected GMPEs for this tectonic regime (all five of the NGA-West2 models) show too much similarity because they were derived using similar data and approaches following considerable interaction between the GMPE developers. Therefore, the epistemic uncertainty captured by these models is too low. To overcome this, for each original GMPE branch in the logic tree they add a higher and lower branch equal to the original GMPE shifted up or down by a factor that varies in nine magnitude (**M** 5-6, 6-7 and 7+) and distance bins (<10 km, 10–30 km and >30 km). This factor is given by: $\exp[0.4\sqrt{n/N}]$, where n and N equal the number of earthquakes used to derive the GMPE within the **M** 7+ and R < 10 km bin and the number in the specific magnitude-distance bin, respectively. The **M** 7+ and R < 10 km bin is generally the one with the smallest number of earthquakes for which an epistemic uncertainty of 50% [i.e. exp.(0.4)] was assumed. The uncertainties for all other bins are scaled with respect to this bin.

The Petersen et al. (2014) approach recognizes that a lack of data is the reason for epistemic uncertainty and tries to capture this in a relatively simple manner. The basis of the function used to scale the uncertainty factor with the number of earthquakes is not given in Petersen et al. (2014) but it is likely related to the equation for the standard error of the mean, where the standard deviation is divided by $\sqrt{N}$ where N is the number of values used to compute the standard deviation. This is a reasonable basis for the function. Also reasonable is the use of the number of earthquakes in each bin rather than the number of records because uncertainty in what the average source characteristics (e.g. stress drop) are for a region (related to the number of earthquakes observed from which to assess this) is often more important than what the average path or site characteristics are (related to the number of records).

The principal criticism of the factor of Petersen et al. (2014) is the apparently arbitrary decision to assume a 50% uncertainty for the **M** 7+ and R < 10 km bin from which to scale all others. This value should be related to how well average ground motions are known for that bin, which itself should be controlled by the available data. If the value was explicitly defined by the available data, over time the epistemic uncertainties modelled by this approach would reduce (as data are collected). Also it would allow the approach to be transportable to other regions or for other GMPEs. Currently if a GMPE is based on a single event for **M** > 7 then the additional uncertainty is the same as if a GMPE is based on many hundreds of earthquakes for that magnitude range. A recent study by Douglas and Boore (2017) suggests that the epistemic uncertainty in ground motions in the **M** 7+ bins is lower than could have been thought given the limited data used to constrain GMPEs in that magnitude range.

A more sophisticated method has been proposed by Al Atik and Youngs (2014) to add additional branches to model the statistical uncertainty characterized by the confidence limits from regression analysis based on a finite dataset [see Douglas 2007, 2010a for estimates of these confidence limits for other GMPEs]. Generally this additional uncertainty is smaller than the uncertainty coming from model-to-model differences. As Al Atik and Youngs (2014) show it is also smaller than the additional uncertainty added to the US National Seismic Hazard Model using the approach discussed above.

Atkinson and Adams (2013) use a backbone approach to develop ground-motion logic trees for the PSHA underlying the Canadian National Building Code. For crustal seismicity in western Canada, they examine the spread of the NGA-West1 models for magnitudes and distances critical for the PSHA of this region and averages of data to define a representative GMPE [in this case the GMPE of Boore and Atkinson 2008] and an upper and lower GMPE to cover the observed range of median predictions. They find that weakly distance-dependent additive and subtractive terms of between 0.23 and 0.69 in terms of natural logarithms can envelope the observed spread in the models. [Atkinson 2011 also develops a simple ground-motion logic tree using a similar approach, although as noted above she adopts an unconventional split between epistemic and aleatory components]. Because this western region is the best studied area of Canada, and epistemic uncertainties for active crustal GMPEs should be the lowest, this uncertainty is assumed to be a lower limit when constructing logic trees for other Canadian regions. For example, for eastern Canada, which is a stable continental region, Atkinson and Adams (2013) found that applying a similar approach led to lower uncertainties for some magnitudes and distances than in the western Canadian logic tree and hence they added uncertainty to take account of the fewer ground-motion observations from eastern Canada.

Gehl (2017) applies this approach to produce a pan-European representative GMPE. Kale and Akkar (2017) propose a similar technique for the selection of multiple GMPEs that cover the centre, body and range but they also include calculation of the seismic hazard to check that no particular model dominates.

An example of a backbone model from a site-specific study is that developed by Bommer et al. (2015) for the Thyspunt (South Africa) nuclear power plant hazard assessment. In this approach, as well as adjusting for V_s and kappa (and accounting for uncertainty in these site parameters), they also add branches to scale the predictions from three existing GMPEs to account for uncertainty in the median stress drop for earthquakes in the surrounding region. Four branches are considered: one for the chance that average stress drops in the region are lower than average (because the tectonics are extensional), one that the average stress drops are the same as in the original GMPEs (i.e. a scale factor of unity) and two for higher stress drops due to the area being part of a stable continental region.

In the procedure proposed by Goulet et al. (2017) to develop ground-motion logic trees for central and eastern North America, the suite of seed ground-motion models are extended, making sure that certain physical criteria are met, to a continuous distribution of ground-motion models so that the set is then MECE. The ground-motion space modelled by this continuous distribution is visualized using the mapping approach of Sammon (1969), which approximates the high-dimensional (magnitude, distance, period) ground-motion space on a 2D map (Scherbaum et al. 2010). To obtain a continuous distribution the expected epistemic uncertainty in ground motions at different distances is imposed based on analogies to western North America and understanding where the seed GMPEs are best constrained. The continuous distribution of ground-motion models is then discretized to a representative set that is easier to handle computationally within PSHA. Weights are then assigned to this set based on prior knowledge and residuals between each models predictions and strong-motion records from the considered region. This rigorous approach, although scientifically appealing, requires considerable computational effort and choices to be made. Therefore, in this article I am seeking a more straightforward approach but using some of the ideas from the procedure of Goulet et al. (2017).

6.4 Retrospective Test of Logic Trees for California

Because it is not possible to test objectively whether a ground-motion logic tree developed today will correctly capture observations of future earthquakes, Douglas (2016) undertakes a retrospective analysis using the 35 years following 1981 as a basis. As the analysis of Douglas (2016) was only presented within an oral presentation at the 35th General Assembly of the European Seismological Commission, I include a summary here.

Douglas (2016) chose 1981 as a basis for his analysis because, if a PSHA for horizontal PGA had been conducted at that time following current practice[2] for shallow crustal seismicity (particularly in California), it is likely that the ground-motion logic tree would have included the three robust GMPEs that had only recently been published: Trifunac (1976), Campbell (1981) and Joyner and Boore (1981). Therefore, using the data in the NGA-West2 database (Ancheta et al. 2014) from 1981 onwards (the database ends in 2011) enables a comparison between the predictions from such a ground-motion logic tree and the observed PGAs. In particular, a check can be made of whether the right level of uncertainty was captured by this simple logic tree consisting of GMPEs that would no longer be considered state-of-the-art. The observation that inspired this approach is that made in Douglas (2010b, p. 1519), who notes that the scatter in predictions of median PGAs from many dozens of GMPEs (a proxy for the epistemic uncertainty) is wider than the confidence limits in the average observed PGA for narrow magnitude-distance bins from a large strong-motion database.

The ground-motion logic tree considered is one comprised of the three GMPEs each given a weight of one-third. The basis of the comparison is to compare, for all magnitude-distance bins, the predicted median PGAs from the logic tree, both in terms of the weighted average and the upper and lower PGAs branches (corresponding to the median PGA from one of the three GMPEs but which one will vary with M and R), with the median PGAs (and its 5 and 95% confidence limits, which are assumed proxies for the uncertainty due to the lack of data for that bin) computed from the PGAs observed in post-1981 earthquakes. Ideally the PGA from the lower branch of the logic tree should equal the 5% confidence limit from the observations, and weighted average from the logic tree should equal the median from the observations and the upper branch should equal the 95% confidence limit from the observations. If the logic tree's branches are wider than the 5–95% confidence limits from the observations then the logic tree is capturing too much uncertainty whereas if they are narrower sufficient uncertainty is not being modelled.

The 100 bins used for the analysis were constructed using 10 intervals 0.5 units wide between M 3 and 8 and 10 logarithmically-spaced Joyner-Boore distance intervals between 0 and 300 km. The variability in a single observation (i.e. the standard deviation) as well as the uncertainty in a median observation (i.e. the standard error) from the database are shown in Fig. 6.1. From this figure it can be seen that despite low apparent aleatory variability at large magnitudes (around 0.5) the uncertainty in the median PGA is still quite high (between 0.1 and 0.15, i.e. factors of 10–15%) and despite high apparent aleatory variability at low magnitudes and large distances (around 1.0) the uncertainty in the median PGA is low (between 0.04 and 0.08). For a much larger database the graph on the left (aleatory variability) would likely show a similar trend because it is related to ground-motion variability, which cannot be modelled simply using magnitude and distance

---

[2]The use of logic trees within PSHA were only proposed in 1984 (Kulkarni et al. 1984) so this is a truly hypothetical situation.

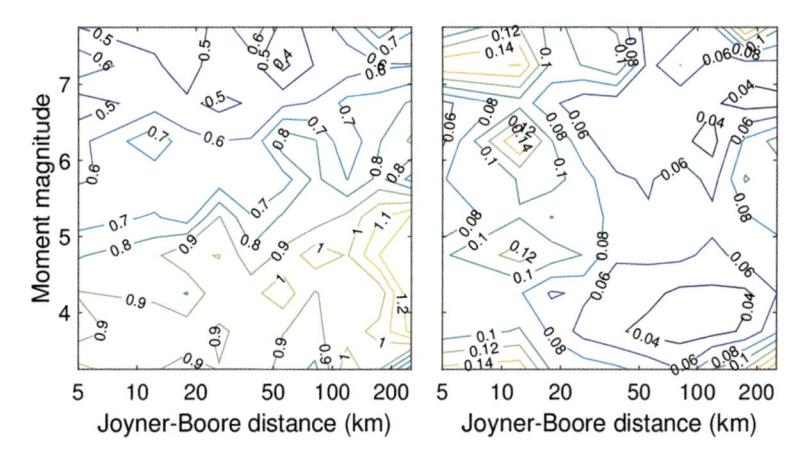

6 Capturing Geographically-Varying Uncertainty in Earthquake Ground... 163

Fig. 6.1 Standard deviations of observed PGAs in 10 × 10 bins (left) and standard errors of observed PGAs (right) from NGA-West2 database for earthquakes that occurred since 1981

(e.g. Douglas and Smit 2001), whereas the graph on the right would approach zero throughout as the true *median* PGA given a magnitude and distance would be known exactly.

Figure 6.2 shows the comparison between the predictions from the logic tree and the observations. Ideally the bottom row of graphs would show contours around unity, meaning that the correct amount of uncertainty is being captured. This is roughly the case for the upper branch of the logic tree (bottom middle and right graphs). Contours higher than unity mean that too much uncertainty is being captured. This is true for the lower branch of the logic tree (bottom left graph)

Figure 6.3 shows the standard deviation of the logic tree (cf. Toro 2006). This graph shows that the epistemic uncertainty captured by this logic tree is quite low. Referring to Table 6.1 indicates that the difference between the mean and median PGA for a given annual frequency of exceedance will hence also be small (less than 50% for typical values of k).

Finally, Fig. 6.4 shows the Sammon's map (Sammon 1969; Scherbaum et al. 2010) of the three original GMPEs, the GMPEs divided by two and multiplied by two (to simulate a simple backbone approach) and the binned observed PGAs. The observations are surrounded by the GMPEs, although this map suggests that the Joyner and Boore (1981) GMPE could be removed from the logic tree as it is further from the observations than the other two models.

In conclusion, this simple logic tree would have been appropriate for a seismic hazard assessment conducted in 1981 (at least until 2011, the end of the database) because the epistemic uncertainty captured is roughly the same as observed in the data, although it is slightly too wide at the lower end. This suggests that perhaps we do not need more sophisticated GMPEs or logic trees. We should be wary of the limitations of this analysis, however. Firstly, this analysis was only for simple

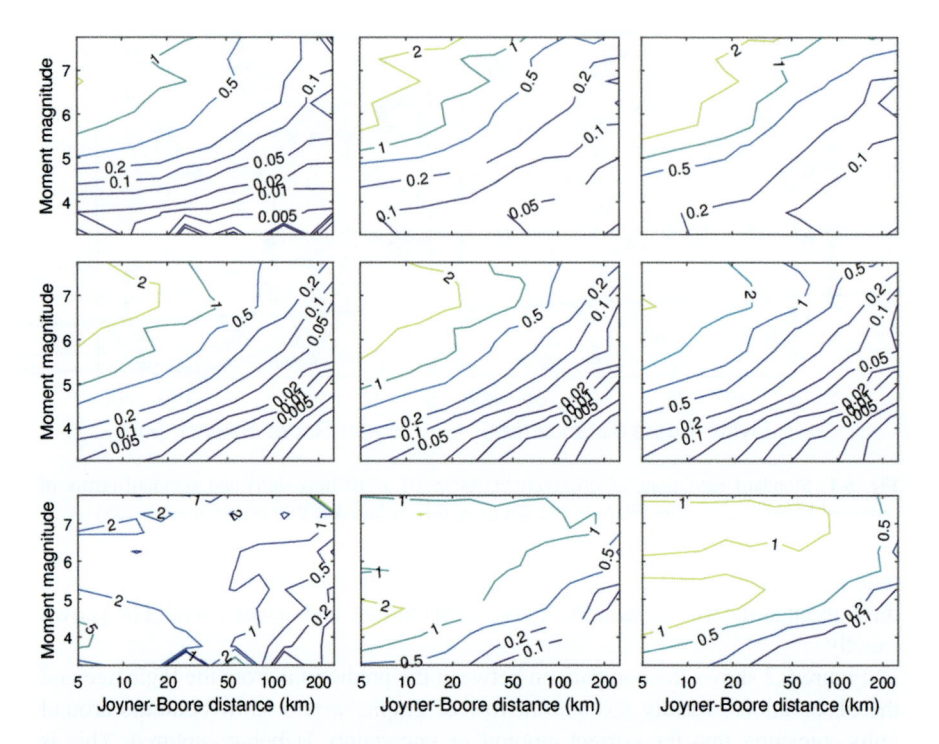

Fig. 6.2 Predicted median PGAs from the lower (left), median (middle) and upper (right) branches of a logic tree comprised of Trifunac (1976), Campbell (1981) and Joyner and Boore (1981) with equal weighting (top row); 5% confidence limit, median and 95% confidence limits for the median observed PGAs from the NGA-West2 database (middle row); and the ratio of the middle row to the top row (observations/logic tree) (bottom row)

GMPEs that did not account for site effects (or, if they did, only in a crude way) and only for PGA. Secondly, the analysis relies on making the strong assumption of no regional dependency in earthquake ground motions (i.e. we can combine all the strong-motion data together to assess the medians and confidence limits). Thirdly, it assumes that the data available from the period 1981–2011 are sufficient to obtain robust statistics and that data from future events will not significantly change the assessed medians and confidence limits. A few well-recorded **M** 7+ earthquakes with apparently 'abnormal' ground motions could significantly change the analysis for that magnitude range; although, as noted above, Douglas and Boore (2017) suggest that current predictions for this magnitude range appear robust. Fourthly, and probably most importantly, this type of analysis cannot be conducted for areas with little or no observations without invoking the assumption of no regional dependency.

6 Capturing Geographically-Varying Uncertainty in Earthquake Ground... 165

Fig. 6.3 Standard deviation of the logic tree in natural logarithm units

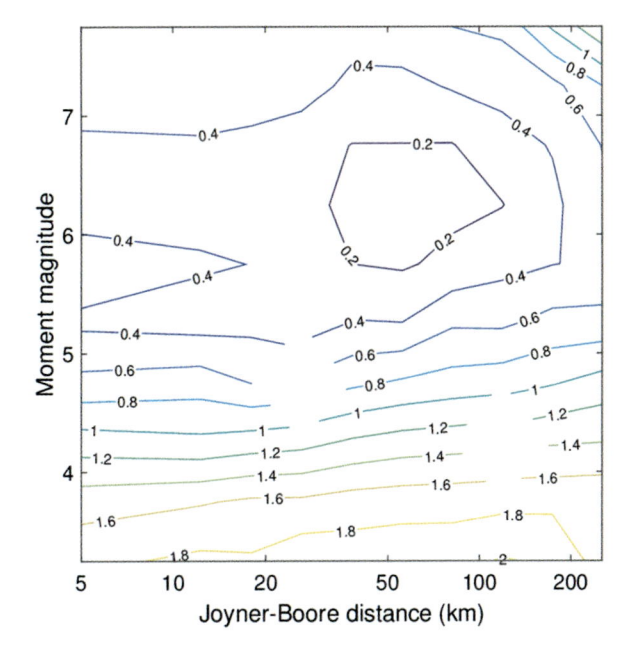

Fig. 6.4 Sammon's map of the three unadjusted GMPEs, the GMPEs multiplied by two, the GMPEs divided by two and the observed PGAs. Note that the absolute positions are arbitrary – only the relative positions are meaningful

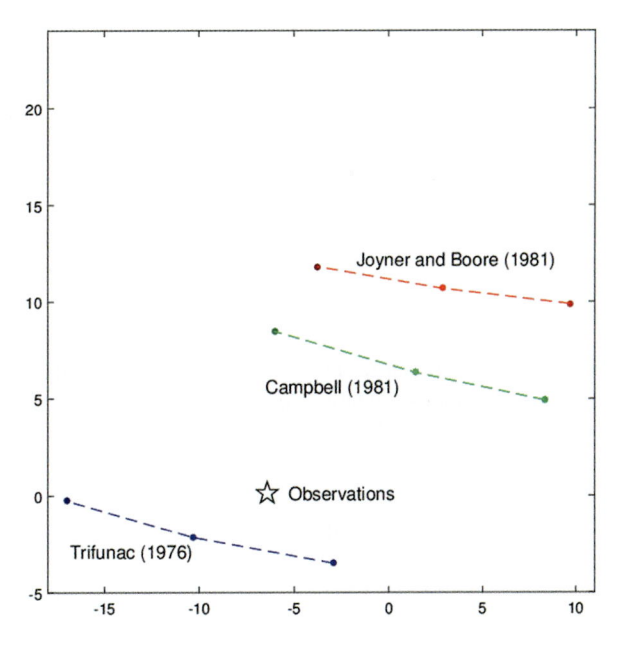

6.5 Proposed New Approach

When developing ground-motion models for induced seismicity Douglas et al. (2013) noted large differences in observed ground motions amongst the sites they considered. They relate this principally to differences in the average stress (drop) parameter for earthquakes near each site. Before a scheme that may induce seismicity is begun it is not possible to know what the average stress parameter in future earthquakes would be. Hence, Douglas et al. (2013) propose that the logic tree used for the initial hazard assessment for the scheme is populated by the 36 GMPEs they derive from stochastic models that cover the possible range of key parameters (stress parameter and attenuation modelled by Q and kappa). If information on what the average values of these parameters are for the site in question then the logic-tree weights can be tuned to reflect this. If and when ground-motion data are collected from the site then again the weights can be modified to emphasis more probable GMPEs and reduce the modelled uncertainty. Edwards and Douglas (2013) showed that this approach worked in practice by using data from the Cooper Basin (Australia) geothermal site, which were not used to develop the original 36 models.

It is proposed here that a similar approach could be applied for natural seismicity. The idea is the opposite of taking a set of GMPEs and then widening out the branches to capture uncertainty. Rather, many branches with default weights are considered and, when data are available, the weights adjusted to reflect our improved knowledge. This provides a framework where the reduction in epistemic uncertainty through the collection of new information is explicitly captured.

The philosophy of this approach is the same as employed by Douglas et al. (2009) to develop potential mean shear-wave velocity profiles *and their uncertainty* by starting with all possible profiles (generated using a Monte Carlo approach with underlying distributions based on an analysis of a large set of observed profiles) and then applying the available constraints to obtain a set of profiles that accounts for what you know *and what you do not know* about site conditions.

When developing this approach the simplest possible logic trees are sought because extra complexity is probably not justified for national or continental PSHAs given the large uncertainties and the need to make the hazard calculations for many locations tractable. As noted above, the intended use of this proposal is not site-specific seismic hazard assessments where time and resources would be available for analysis of all data. The focus is broad-brush PSHAs, which would be more typical of those with a wide geographical scope. Therefore, although the procedure of Goulet et al. (2017) is appealing, it is perhaps too complex for application beyond projects with considerable resources.

Disaggregation of PSHAs (Bazzurro and Cornell 1999) allows the earthquake scenarios that contribute most to the seismic hazard at a site to be determined. This information is useful for this study because it provides guidance on what magnitude-distance range needs to be the principal focus of the ground-motion model. Although it must be recalled that all magnitudes and distances that are not precluded by the seismic source model will influence the hazard and hence, even if some scenarios dominate, the ground-motion model should be accurate for all scenarios.

As an example of a low-to-moderate-seismicity European country, Table 6.3 reports the mean magnitude, distance and epsilon (i.e. number of standard deviations above the median ground-motion) of the disaggregated scenarios reported by Goda et al. (2013) for the UK. As an example for a moderate-to-high-seismicity European country, Fig. 6.5 shows the distribution of mean magnitude, distance and epsilons of the disaggregated scenarios reported by Barani et al. (2009) for Italy. These results indicate that generally the most important earthquake scenarios for return periods used for seismic hazard mapping are magnitudes between about 5.0 and 6.5 and distances up to about 60 km for PGA and SA(0.2 s) and between about 5.5 and 7.0 and distances up to about 100 km for SA(1.0 s). Epsilons are generally between 1.0 and 2.0 with higher values as the return period increases.

Table 6.3 Mean magnitude and distance of the dominant earthquake scenario and the value of epsilon for 2500 year return period for two UK cities (Goda et al. 2013)

Location	Period (s)	Mean M	Mean R (km)	Epsilon
Cardiff	PGA	5.0	19	1.2
"	1.0	5.3	31	1.2
Oban	PGA	5.2	25	1.0
"	1.0	5.5	39	1.1

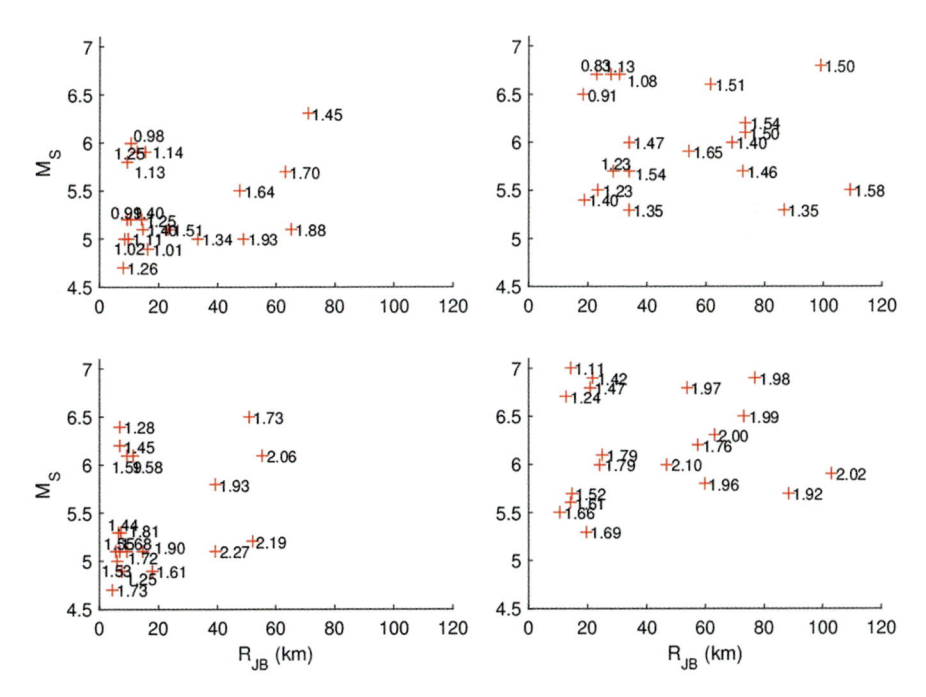

Fig. 6.5 Mean magnitude, distance and epsilon of the disaggregated scenarios for the 19 Italian cities considered by Barani et al. (2009). Top: 475 year return period, bottom: 2475 year return period, left: SA(0.2 s) and right: SA(1.0 s)

In the following subsection, a full ground-motion logic tree for shallow non-subduction earthquakes in Europe and the Middle East is proposed. This is followed by a subsection adjusting the weights as discussed previously for two countries with limited ground-motion data (Georgia and Iran) as well as one with much data (Italy).

6.5.1 Development of a Full Ground-Motion Logic Tree

The aim of this section is to generate a full population of possible GMPEs, one of which (although which one is currently unknown) is the correct ground-motion model for a given location in Europe and the Middle East. Following Bommer and Scherbaum (2008), I seek to create a family of MECE GMPEs. Only PGA and SA (1.0 s) are considered in the following due to space limitations. Also the analysis is conducted assuming $V_{S30} = 800$ m/s (Eurocode site class A/B boundary).

Considering just the principal inputs to a GMPE, i.e. magnitude and distance, the available strong-motion records for each magnitude-distance pair can be seen as samples from an underlying distribution for which we do not know the mean (when using logarithms of the intensity measure). (Its standard deviation is also unknown but as noted above this study is focused on the median ground motions not the variability). Using published GMPEs and well-recorded earthquakes it is possible to assess a possible range for the mean, which can then be related to the epistemic uncertainty that should be captured within the ground-motion logic tree. This part of the procedure is similar to the approach of Atkinson and Adams (2013) in developing a suite of models from a backbone model.

For active regions where strong-motion data have been collected over the past few decades, it is possible to assess how GMPEs are likely to change with the accumulation of new data. A study that provides guidance on this issue is by Bindi et al. (2009) who re-derive the Sabetta and Pugliese (1987) GMPE for Italy, which was derived using only 95 PGAs from 17 earthquakes (from before 1985), using the same functional form but 235 PGAs from 27 earthquakes from the period 1972–2002. Bindi et al. (2009) find that the 1987 model overpredicts median PGAs by less than 5% for magnitude 4.5 and short distances but by more than 50% for magnitude 7 and distances around 100 km (their Figure 5). Therefore, GMPEs based on sparse data are susceptible to significant change when updated. Hence, there is a need to recognize that what we think we know now may change. This doubt should be reflected in the epistemic uncertainty captured in the ground-motion logic tree.

The causes of epistemic uncertainty can be divided into two. Firstly the statistical uncertainty: even for regions with much strong-motion data (e.g. Italy) variability in the ground motions and a finite sample means that the median ground motion for a given magnitude and distance is not known precisely (cf. the formula for the standard error with $\sqrt{n}$ on the denominator). This is modelled by the confidence limits of the regression analysis, which are only rarely published (e.g. Douglas 2007, 2010a; Al Atik and

Youngs 2014; Bindi et al. 2017). Secondly, the 'regional' uncertainty: for regions with little strong-motion data from large earthquakes (e.g. much of northern Europe) it is not known whether ground motions show significant differences to those in well-observed regions (for which native GMPEs exist), e.g. because of differences in median stress drop. Calibrating the width of these additional branches is challenging as we need an assessment of our *unknown* knowledge (i.e. how much we do not know) – it is often hoped that selecting GMPEs from various regions covers this uncertainty.

6.5.2 Statistical Uncertainty

To estimate the statistical uncertainty due to regression analysis using relatively complex functional forms on finite datasets Figure 3 of Douglas (2007), Figure 6 of Douglas (2010a), results from Al Atik and Youngs (2014) and Figure 6 of Bindi et al. (2017) are used to assess the component of σ_μ coming only from this factor: $\sigma_{statistical}$. The 95% confidence limits shown on Figure 3 of Douglas (2007) for seven GMPEs and the ratios of the 95% to 50% confidence limits of Douglas (2010a) for six GMPEs can be converted to obtain $\sigma_{statistical}$. For the most poorly constrained models at the edges of their magnitude-distance range of applicability (e.g. Ambraseys et al. 2005; Sabetta and Pugliese 1987) $\sigma_{statistical}$ approaches 0.3 (natural logarithms). Better constrained models and within the magnitude-distance 'comfort zone' (Bommer et al. 2010) of GMPEs $\sigma_{statistical}$ from the models considered by Douglas (2007, 2010a), the NGA-West2 models considered by Al Atik and Youngs (2014) and the GMPE of Bindi et al. (2017) are similar with values around 0.1. For magnitudes larger than 7 there is an increase in $\sigma_{statistical}$, which is expected because of the sparsity of data from large earthquakes.

6.5.3 'Regional' Uncertainty

Here the potential that median ground motions in a region are different to those in well-observed regions is assessed by comparing median ground motions for scenarios and regions for which the median ground motions are well-known. We are using the spread in average ground motions in countries with extensive strong-motion databases as a proxy for what could be the spread for countries for which observations are currently sparse. Subsequently this information is used to assess the 'regional' uncertainty. Rather than potentially double count the statistical uncertainty, poorly-sampled areas are not considered when assessing this component of uncertainty.

As shown by the graphs of statistical uncertainty referred to in the previous subsection, GMPEs are best constrained for the distance range from roughly 20 to 60 km. This is also the range where anelastic attenuation is unlikely to be having a large influence and hence it is possible to identify the difference between average ground motions in various regions that is due predominantly to differences in average stress drop. Using the same assumption as Yenier and Atkinson (2015) that differences in stress drop are present in ground motions for all distances makes it possible to apply the factors derived from abundant data from this restricted distance range to all distances. As shown above, seismic hazard is often dominated by earthquakes within 100 km for the return periods of most interest for national mapping and, therefore, the range of distances used to estimate differences in average regional stress drop overlap. As shown by Figures 4 and 5 of Douglas and Jousset (2011), changing the stress drop does not significantly change the magnitude scaling so again it is an acceptable first-order solution to apply the derived adjustments for the limited magnitude range 5–6 to all magnitudes.

At distances beyond about 60 km (for short oscillator periods) regional differences in anelastic attenuation will make ground-motion predictions for different regions diverge (e.g. Kotha et al. 2016, Figure 5). As a first-order solution to account for this potential uncertainty within the population of possible GMPEs, the three regional models of Kotha et al. (2016) (i.e. the 'Italy', 'Turkey' and 'Other' models) are used as the backbone models that are then branched out to account for potential differences in the average stress drop. This makes the assumption that the anelastic attenuation in these three regions is an adequate sample of the population of all anelastic attenuation rates in Europe and the Middle East — this is probably untrue but it is assumed for convenience. As noted above seismic hazard is often dominated by earthquakes within 100 km and hence the effect of potential variations in anelastic attenuation is unlikely to be particularly important. The backbone models assumed here could be improved in future applications of the approach. Boore et al. (2014), for example, also provide terms to account for variations in anelastic attenuations between regions, which could be used instead.

An effect that could lead to regional dependency in ground motions, but which is neglected here because of a lack of a simple approach for its incorporation, is the influence of crustal structure. Previous studies (e.g. Dahle et al. 1990; Somerville et al. 1990; Douglas et al. 2004, 2007) have shown that effects such as wave reflections off the Moho can have a strong influence on ground motions at intermediate source-to-site distances (>50 km). This effect could potentially be incorporated into the approach proposed here by developing backbone models for different typical crustal structures, perhaps using simulations and the equivalent hypocentral distance technique of Douglas et al. (2004, 2007).

Any regional dependency at short distances (<10 km) is not possible to currently assess due to sparse datasets at such distances even when they are not separated by region (Fig. 6.6). Therefore, I have not attempted to account for any 'regional' uncertainty here. The statistical uncertainty discussed above, combined with the 'regional' uncertainty due to differences in average stress drop, is expected to account for sufficient uncertainty in this distance range. Although, again, this is a topic where additional work may be warranted.

For the calculations made in this section the Engineering Strong-Motion (ESM) flat-file 2017 (Lanzano et al. 2017) is used. The distribution of data in this flat-file with respect to magnitude, distance and various countries is shown in Fig. 6.6. Data

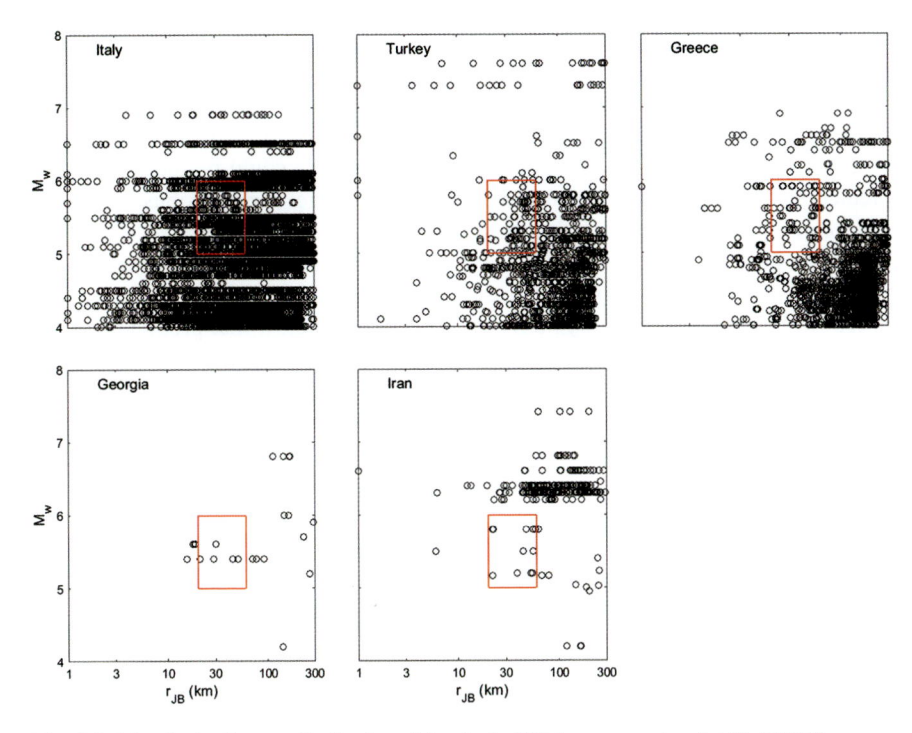

6 Capturing Geographically-Varying Uncertainty in Earthquake Ground... 171

Fig. 6.6 Magnitude-distance distribution of data in the ESM strong-motion flat-file 2017 (Lanzano et al. 2017) used for the analysis in this section. The red box indicates the magnitude-distance interval used to assess the 'regional' uncertainty

from three countries with many strong-motion records (Italy, Turkey and Greece) are used to assess empirically the possible size of regional dependency due to average stress drop differences. Data within the magnitude-distance range of $5 \leq M_w \leq 6$ and $20 \leq r_{JB} \leq 60$ km are used for these calculations for the reasons given above and to obtain statistically robust estimates of the 'regional' uncertainty without relying on weak-motion data ($M_w < 5$), whose relevance to the adjustment of empirical GMPEs is unclear.

The residuals with respect to the generic Kotha et al. (2016) GMPE, i.e. the model where the regional terms are turned off, for each record are computed. The equations for problem 1 of appendix of Spudich et al. (1999) are used to evaluate the average bias and its uncertainty (for all the data and for each of the three countries separately) to account for the correlations between data from the same earthquake. In addition, the average differences for $5 \leq M_w \leq 6$ and $20 \leq r_{JB} \leq 60$ km between the Kotha et al. (2016) GMPE and recent country-specific GMPEs that are robust, at least for this restricted magnitude-distance range, are computed as an additional constraint. The country-specific GMPEs are: Akkar and Çağnan (2010) (Turkey), Bindi et al. (2011) (Italy), Danciu and Tselentis (2007) (Greece), Sedaghati and Pezeshk (2017) (Iran) and Yenier and Atkinson (2015) (central and eastern North America). The model for central and eastern North America is included as an example of a GMPE for a stable continental region so as to include within the 'regional' uncertainty the possibility that ground motions in an area may show similarities to such regions. This is potentially important for much of northern Europe but for which robust native GMPEs are lacking. Both approaches to evaluate this 'regional' uncertainty should be equivalent but that based on country-specific GMPEs could be more robust as predictions for this restrictive magnitude-distance range borrow robustness from neighbouring magnitudes and distances.

The results of these residual analyses are shown in Fig. 6.7. From these average residuals it can be seen that ground motions for this magnitude-distance range from some countries (e.g. Italy) are on average below that predicted by the Kotha et al. (2016) GMPE whereas ground motions from some countries (e.g. Greece) are higher than predicted by this pan-European model. Using this set of averages as a sample from the population of average deviations for each country/region in Europe and the Middle East, a simple logic tree can be proposed to capture the 'regional' uncertainty that these averages seek to

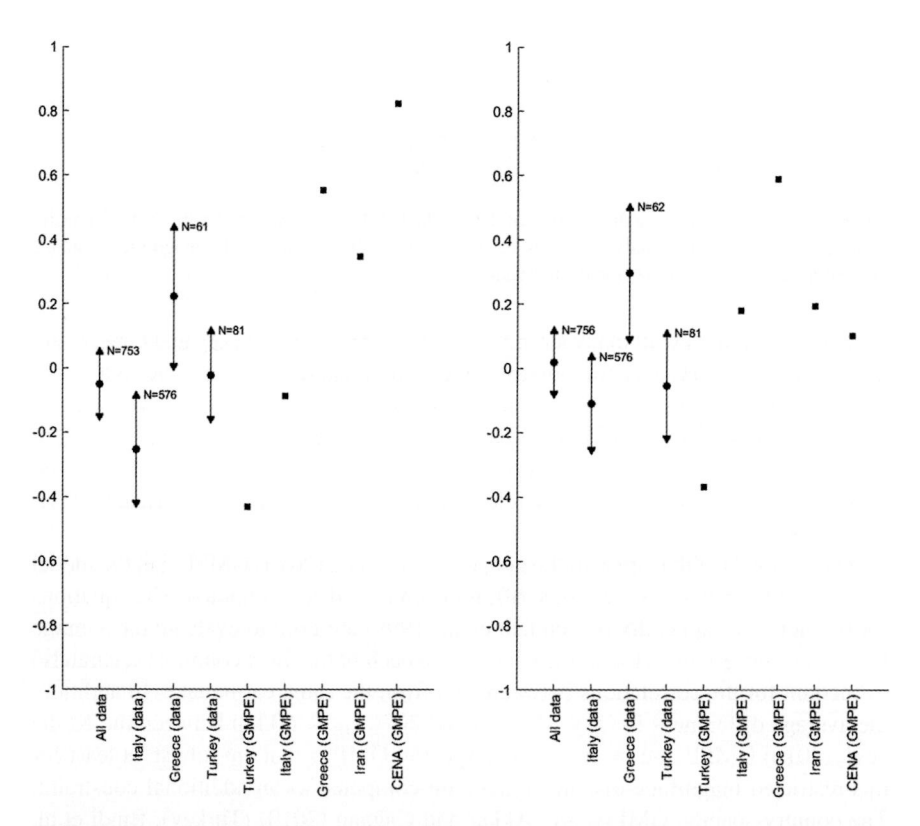

Fig. 6.7 Mean residuals and their 5–95% confidence limits for all ESM data and for data from three countries (the number of records used to compute the averages are indicated) as well as the average residuals between country-specific GMPEs and the GMPE of Kotha et al. (2016) for $5 \leq M_w \leq 6$ and $20 \leq r_{JB} \leq 60$ km and for PGA (left) and SA(1 s) (right)

model. This approach is similar to studies that develop a suite of stochastic models accounting for epistemic uncertainty in the average stress drop in a region (e.g. Douglas et al. 2013; Bommer et al. 2017).

For PGA, symmetrical lower, middle and upper branches equal to predictions from: the Kotha et al. (2016) model × 0.6 [i.e. exp(-0.5)], the Kotha et al. (2016) model × 1.2 [i.e. exp(0.2)], and the Kotha et al. (2016) model × 2.5 [i.e. exp(0.9)], with weights using a standard three-point distribution of 0.185, 0.63 and 0.185, respectively, would roughly capture the spread in these average residuals. For SA(1 s), symmetrical lower, middle and upper branches equal to predictions from: the Kotha et al. (2016) model × 0.7 [i.e. exp(-0.4)], the Kotha et al. (2016) model × 1.1 [i.e. exp(0.1)], and the Kotha et al. (2016) model × 1.8 [i.e. exp(0.6)], with the same weights again would roughly capture the spread in these average residuals. As discussed above, it is assumed that these adjustments apply for all magnitudes and distances.

6.5.4 Final Logic Tree

The first set of branches is the three regional models of Kotha et al. (2016) accounting for variations in anelastic attenuation, each with equal weights of 1/3. The second set of branches is the lower, middle and upper branches that model the effect of uncertainty in the average stress drop for a given region. The third set of branches of the logic tree is those proposed by Al Atik and Youngs (2014) to account for the statistical uncertainty component $\sigma_{statisical}$, where the upper and lower branches equal the 95% confidence limits using this standard deviation [although the values of Al Atik and Youngs 2014 for normal and reverse faulting are switched because Kotha et al. 2016's GMPE is better constrained for normal and strike-slip than reverse]. As noted above the model expressed in equations 9 to 11 of Al Atik and Youngs (2014) is adopted to account for this component as we are adopting well-constrained GMPEs, which are branched out to account for potential regional dependency. If less well-constrained models were used then a larger value for $\sigma_{statisical}$ should be used.

This ground-motion logic tree implies the values of σ_μ, characterising the level epistemic uncertainty, shown in the contour plot on Fig. 6.8. The epistemic uncertainty is independent of magnitude except for $M > 7$, where the statistical uncertainty from the model of Al Atik and Youngs (2014) increases slightly. The effect of the three models for anelastic attenuation is to increase the epistemic uncertainty at larger distances (>70 km). The overall epistemic uncertainty is similar to those implied by the site-specific logic trees listed in Table 6.2, giving confidence that roughly the right level of uncertainty is being captured. For SA(1 s) the graph of σ_μ is similar but the values are slightly lower (0.39 for short distances increasing to 0.66 at 300 km).

In total $3 \times 3 \times 3 = 27$ GMPEs make up the population that it is assumed to represent all possible ground-motion models for application in Europe and the

Fig. 6.8 σ_y implied by the proposed ground-motion logic tree for PGA

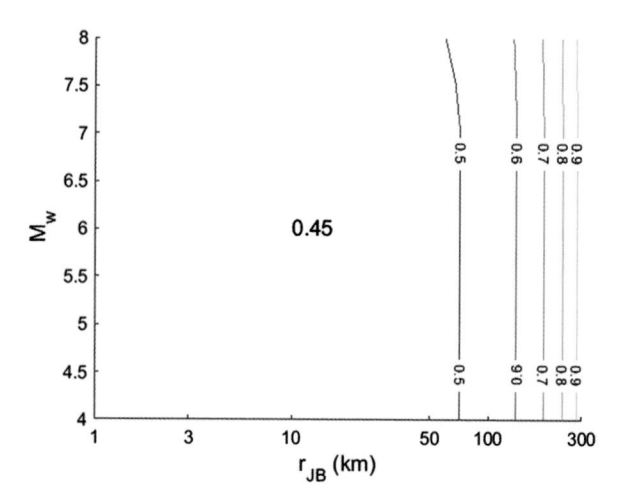

Middle East. When there is no additional information on earthquake ground motions in a region this complete population should be used with the weights noted above. When observations are available the default weights can be altered to reflect this additional information. This is demonstrated in the next section for Georgia, Iran and Italy.

6.5.5 Pruning the Ground-Motion Logic Tree for Three Countries

Georgia and Iran are chosen here as examples of two countries with considerably less strong-motion data in the ESM database (Fig. 6.6). All available data from each country in the interval $M_w \geq 4$ and $0 \leq r_{JB} \leq 300$ km are used to adjust the weights of the full logic tree derived in the previous section. Because the statistical uncertainty in the family of potential GMPEs remains, those branches of the logic tree are left unchanged and the testing only alters the weights of the first two sets of branches that account for potential regional dependency.

The log-likelihood approach of Scherbaum et al. (2009) and the data from Georgia and Iran are used to update the original weights of the 9 branches related to the regional dependence. Combining these branches with the set related to the statistical uncertainty leads to the predicted PGAs shown in Fig. 6.9. Also shown are predicted PGAs applying the same technique for Italy as a demonstration for a country with much data. The result of the adjusted weights is to decrease the width of the confidence limits particularly at moderate and long distances because there are many records to modify the weights of the three models of anelastic attenuation. However, the effect of the weighting on the level of epistemic uncertainty modelled is small even for Italy (σ_μ only decreases from about 0.45 to 0.41 at close distances although by larger amounts at greater distances) because of the large uncertainty in

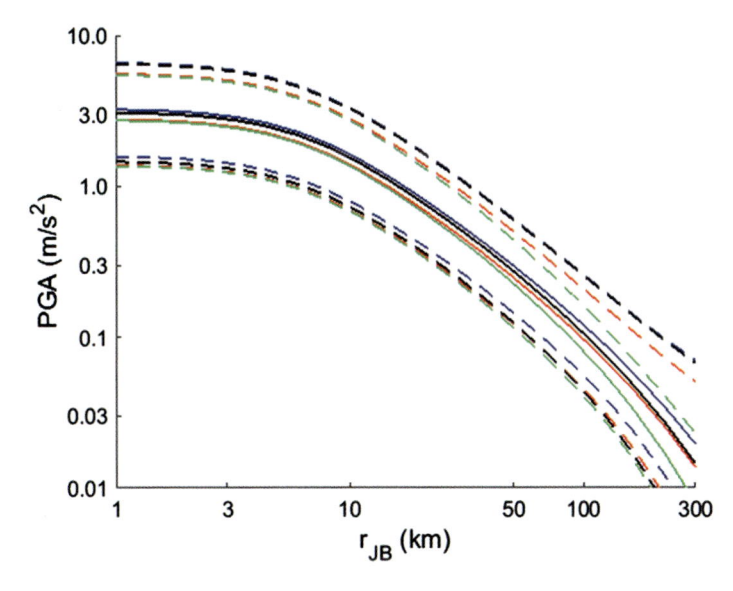

Fig. 6.9 Median predicted PGA and its 5–95% confidence limits (dashed lines) for an **M** 6 earthquake with respect to Joyner-Boore distance. Black curves correspond to the original logic tree and the other colours correspond to logic trees for specific countries: Georgia (red), Iran (blue) and Italy (green)

average bias even when there are many records (Fig. 6.7). Therefore, the changes to the weights for the lower, middle and upper branches of the 'regional' uncertainty are limited. Despite this there is a change in the predicted PGA from the original logic tree. Similar results are obtained for SA(1.0 s).

Rather than simply using all the strong-motion data available from a region to adjust the weights it may be more appropriate to use only those records from the magnitude-distance range likely to be relevant from the point of seismic hazard. Also other information, e.g. tectonic analogies, independent estimates of anelastic attenuation, stochastic models derived from weak-motion data, could be used to modify the weights. However, we should be humble about our knowledge of a region with little strong-motion data.

6.6 Conclusions

In this article I have reviewed previous approaches to develop ground-motion logic trees that account for geographically-varying epistemic uncertainty. As demonstrated by the relatively low epistemic uncertainty implied by some recent continental seismic hazard assessments, the classic approach of selecting a handful of ground-motion models from the literature can lead to inconsistencies when compared with site-specific studies. Therefore, the backbone approach is attractive as it allows epistemic uncertainty to be more easily and transparently modelled. However, the principal difficulty is calibrating this approach when lacking observations, e.g. how much do we *not* know about earthquake ground motions in country X where an **M** > 5 earthquake has never been recorded within 50 km?

To provide guidance on applying the backbone approach in national and continental scale hazard assessments I have proposed a relatively simple ground-motion logic tree to account for potential variations in average stress drop and anelastic attenuation between regions as well as the statistical uncertainty inherent in regression-based models. It is based on assuming that the variation amongst regions that are currently poorly-observed will be similar to the differences amongst regions with relatively large strong-motion databases. This is a potential weakness of the proposal because these regions are also generally those with the highest seismicity and consequently, for target areas that are tectonically stable, the range of average ground motions modelled could be too narrow. The final step in the proposed procedure is to modify the weights of the logic tree by making use of any available strong-motion data from the target region. This step slightly reduces the epistemic uncertainty and adjusts the predictions to make them more applicable to the region. When such data are not available, the full uncertainty implied by the ground-motion model is incorporated into seismic hazard assessments.

When developing ground-motion models for use within national or continental hazard assessments there is a balance to be struck about the resolution captured. At one extreme, a single logic tree could be used for all locations, while at the other, each individual seismic source (e.g. a given fault) could have its own model. In this study, each country was assumed to have its own model but as tectonics do not generally follow national boundaries this was only done for convenience. The higher the resolution the smaller the databases available to calibrate the models. These smaller databases lead to higher standard errors in the averages and the risk of modelling an event-specific (or sequence-specific) rather than a regional-specific effect. For example, do the large differences between observed ground motions between two areas of central Italy (Molise and Umbria-Mache) as evidenced by Douglas (2007), for example, mean that these areas require separate ground-motion models? Or are the relatively small sets of observations from these two regions from one or two earthquake sequences insufficient to draw conclusion?

In addition to the potential problems mentioned above concerning the calibration of the average stress drop branches as well as the spatial resolution of the models, other parts of the procedure require additional work, e.g. the use of the Kotha et al. (2016) models to capture regionally dependence in anelastic attenuation, the updating of the weights using the log-likelihood procedure and how to incorporate knowledge gained from weak-motion data. In conclusion, the approach proposed here aims to be a first-order procedure to develop ground-motion logic trees capturing geographically-varying uncertainty for use in seismic hazard assessments covering a large area (i.e. not site-specific studies).

Acknowledgements I thank the conference organizers for inviting me to deliver a Theme Lecture. I thank the developers of the ESM strong-motion flat-file 2017 for providing these data. Finally, I thank Dino Bindi, Hilmar Bungum, Fabrice Cotton, Laurentiu Danciu, Ben Edwards and Graeme Weatherill for their comments on a previous version of this study. In an effort not to increase the length of this article, I have chosen not to follow some of their suggestions, despite agreeing with them.

References

Akkar S, Çağnan Z (2010) A local ground-motion predictive model for Turkey and its comparison with other regional and global ground-motion models. Bull Seismol Soc Am 100 (6):2978–2995. https://doi.org/10.1785/0120090367

Al Atik L, Youngs RR (2014) Epistemic uncertainty for NGA-West2 models. Earthq Spectra 30 (3):1301–1318. https://doi.org/10.1193/062813EQS173M

Ambraseys NN, Douglas J, Sarma SK, Smit PM (2005) Equations for the estimation of strong ground motions from shallow crustal earthquakes using data from Europe and the Middle East: horizontal peak ground acceleration and spectral acceleration. Bull Earthq Eng 3(1):1–53. https://doi.org/10.1007/s10518-005-0183-0

AMEC Geomatrix, Inc (2011) Seismic hazard assessment, OPG's geologic repository for low and intermediate level waste, NWMO DGR-TR-2011-20, revision R000

Ancheta TD, Darragh RB, Stewart JP, Seyhan E, Silva WJ, Chiou BSJ, Wooddell KE, Graves RW, Kottke AR, Boore DM, Kishida T, Donahue JL (2014) NGA-West2 database. Earthq Spectra 30 (3):989–1005. https://doi.org/10.1193/070913EQS197M

Atkinson GM (2011) An empirical perspective on uncertainty in earthquake ground motion prediction. Can J Civ Eng 38(9):1002–1015. https://doi.org/10.1139/L10-120

Atkinson GM, Adams J (2013) Ground motion prediction equations for application to the 2015 Canadian national seismic hazard maps. Can J Civ Eng 40(10):988–998. https://doi.org/10.1139/cjce-2012-0544

Atkinson GM, Bommer JJ, Abrahamson NA (2014) Alternative approaches to modeling epistemic uncertainty in ground motions in probabilistic seismic-hazard analysis. Seismol Res Lett 85 (6):1141–1144. https://doi.org/10.1785/0220140120

Barani S, Spallarossa D, Bazzurro P (2009) Disaggregation of probabilistic ground-motion hazard in Italy. Bull Seismol Soc Am 99(5):2638–2661. https://doi.org/10.1785/0120080348

Bazzurro P, Cornell CA (1999) Disaggregation of seismic hazard. Bull Seismol Soc Am 89 (2):501–520

Bindi D, Luzi L, Pacor F, Sabetta F, Massa M (2009) Towards a new reference ground motion prediction equation for Italy: update of the Sabetta-Pugliese (1996). Bull Earthq Eng 7:591–608. https://doi.org/10.1007/s10518-009-9107-8

Bindi D, Pacor F, Luzi L, Puglia R, Massa M, Ameri G, Paolucci R (2011) Ground motion prediction equations derived from the Italian strong motion database. Bull Earthq Eng 9:1899–1920. https://doi.org/10.1007/s10518-011-9313-z

Bindi D, Cotton F, Kotha SR, Bosse C, Stromeyer D, Grünthal G (2017) Application-driven ground motion prediction equation for seismic hazard assessments in non-cratonic moderate-seismicity areas. J Seismol 21(5):1201–1218. https://doi.org/10.1007/s10950-017-9661-5

Bommer JJ (2012) Challenges of building logic trees for probabilistic seismic hazard analysis. Earthq Spectra 28(4):1723–1735. https://doi.org/10.1193/1.4000079

Bommer JJ, Scherbaum F (2008) The use and misuse of logic trees in probabilistic seismic hazard analysis. Earthq Spectra 24(4):997–1009. https://doi.org/10.1193/1.2977755

Bommer JJ, Douglas J, Scherbaum F, Cotton F, Bungum H, Fäh D (2010) On the selection of ground-motion prediction equations for seismic hazard analysis. Seismol Res Lett 81 (5):783–793. https://doi.org/10.1785/gssrl.81.5.783

Bommer JJ, Coppersmith KJ, Coppersmith RT, Hanson KL, Mangongolo A, Neveling J, Rathje EM, Rodriguez-Marek A, Scherbaum F, Shelembe R, Stafford PJ, Strasser FO (2015) A SSHAC level 3 probabilistic seismic hazard analysis for a new-build nuclear site in South Africa. Earthq Spectra 31(2):661–698. https://doi.org/10.1193/060913EQS145M

Bommer JJ, Stafford PJ, Edwards B, Dost B, van Dedem E, Rodriguez-Marek A, Kruiver P, van Elk J, Doornhof D, Ntinalexis M (2017) Framework for a ground-motion model for induced seismic hazard and risk analysis in the Groningen gas field, The Netherlands. Earthq Spectra 33 (2):481–498. https://doi.org/10.1193/082916EQS138M

Boore DM, Atkinson GM (2008) Ground-motion prediction equations for the average horizontal component of PGA, PGV, and 5%-damped PSA at spectral periods between 0.01s and 10.0s. Earthq Spectra 24(1):99–138. https://doi.org/10.1193/1.2830434

Boore DM, Stewart JP, Seyhan E, Atkinson GM (2014) NGA-West 2 equations for predicting PGA, PGV, and 5%-damped PSA for shallow crustal earthquakes. Earthq Spectra 30 (3):1057–1085. https://doi.org/10.1193/070113EQS184M

Bozorgnia Y et al (2014) NGA-West2 research project. Earthq Spectra 30(3):973–987. https://doi.org/10.1193/072113EQS209M

Budnitz RJ, Apostolakis G, Boore DM, Cluff LS, Coppersmith KJ, Cornell CA, Morris PA (1997) Recommendations for probabilistic seismic hazard analysis: guidance on uncertainty and use of experts, NUREG/CR-6372, U.S. Nuclear Regulatory Commission, Washington, DC

Campbell KW (1981) Near-source attenuation of peak horizontal acceleration. Bull Seismol Soc Am 71(6):2039–2070

Campbell KW (2003) Prediction of strong ground motion using the hybrid empirical method and its use in the development of ground-motion (attenuation) relations in eastern North America. Bull Seismol Soc Am 93(3):1012–1033

Dahle A, Bungum H, Kvamme LB (1990) Attenuation models inferred from intraplate earthquake recordings. Earthq Eng Struct Dyn 19(8):1125–1141

Danciu L, Tselentis G-A (2007) Engineering ground-motion parameters attenuation relationships for Greece. Bull Seismol Soc Am 97(1B):162–183. https://doi.org/10.1785/0120040087

Danciu L, Kale Ö, Akkar S (2018) The 2014 earthquake model of the Middle East: ground motion model and uncertainties. Bull Earthq Eng. https://doi.org/10.1007/s10518-016-9989-1

Delavaud E, Cotton F, Akkar S, Scherbaum F, Danciu L, Beauval C, Drouet S, Douglas J, Basili R, Sandikkaya MA, Segou M, Faccioli E, Theodoulidis N (2012) Toward a ground-motion logic tree for probabilistic seismic hazard assessment in Europe. J Seismol 16(3):451–473. https://doi.org/10.1007/s10950-012-9281-z

Douglas J (2007) On the regional dependence of earthquake response spectra. ISET J Earthq Technol 44(1):71–99

Douglas J (2010a) Assessing the epistemic uncertainty of ground-motion predictions. In: Proceedings of the 9th U.S. National and 10th Canadian conference on Earthquake Engineering, Paper No. 219. Earthquake Engineering Research Institute

Douglas J (2010b) Consistency of ground-motion predictions from the past four decades. Bull Earthq Eng 8(6):1515–1526. https://doi.org/10.1007/s10518-010-9195-5

Douglas J (2016) Retrospectively checking the epistemic uncertainty required in logic trees for ground-motion prediction. In: 35th General Assembly of the European Seismological Commission, Abstract ESC2016-63

Douglas J, Boore DM (2017) Peak ground accelerations from large (M≥7.2) shallow crustal earthquakes: a comparison with predictions from eight recent ground-motion models. Bull Earthq Eng 16(1):1–21. https://doi.org/10.1007/s10518-017-0194-7

Douglas J, Jousset P (2011) Modeling the difference in ground-motion magnitude-scaling in small and large earthquakes. Seismol Res Lett 82(4):504–508. https://doi.org/10.1785/gssrl.82.4.504

Douglas J, Smit PM (2001) How accurate can strong ground motion attenuation relations be? Bull Seismol Soc Am 91(6):1917–1923

Douglas J, Suhadolc P, Costa G (2004) On the incorporation of the effect of crustal structure into empirical strong ground motion estimation. Bull Earthq Eng 2(1):75–99. https://doi.org/10.1023/B:BEEE.0000038950.95341.74

Douglas J, Bungum H, Scherbaum F (2006) Ground-motion prediction equations for southern Spain and southern Norway obtained using the composite model perspective. J Earthq Eng 10(1):33–72

Douglas J, Aochi H, Suhadolc P, Costa G (2007) The importance of crustal structure in explaining the observed uncertainties in ground motion estimation. Bull Earthq Eng 5(1):17–26. https://doi.org/10.1007/s10518-006-9017-y

Douglas J, Gehl P, Bonilla LF, Scotti O, Régnier J, Duval A-M, Bertrand E (2009) Making the most of available site information for empirical ground-motion prediction. Bull Seismol Soc Am 99(3):1502–1520. https://doi.org/10.1785/0120080075

Douglas J, Edwards B, Convertito V, Sharma N, Tramelli A, Kraaijpoel D, Cabrera BM, Maercklin N, Troise C (2013) Predicting ground motion from induced earthquakes in geothermal areas. Bull Seismol Soc Am 103(3):1875–1897. https://doi.org/10.1785/0120120197

Douglas J, Ulrich T, Bertil D, Rey J (2014) Comparison of the ranges of uncertainty captured in different seismic-hazard studies. Seismol Res Lett 85(5):977–985. https://doi.org/10.1785/0220140084

Edwards B, Douglas J (2013) Selecting ground-motion models developed for induced seismicity in geothermal areas. Geophys J Int 195(2):1314–1322. https://doi.org/10.1093/gji/ggt310

Gehl P (2017) Bayesian networks for the multi-risk assessment of road infrastructure. PhD thesis, University of London (University College London)

Goda K, Aspinall W, Taylor CA (2013) Seismic hazard analysis for the U.K.: sensitivity to spatial seismicity modelling and ground motion prediction equations. Seismol Res Lett 84(1):112–129. https://doi.org/10.1785/0220120064

Goulet CA, Bozorgnia Y, Kuehn N, Al Atik L, Youngs RR, Graves RW, Atkinson GM (2017) NGA-East ground-motion models for the U.S. Geological Survey National Seismic Hazard Maps, PEER Report no. 2017/03. Pacific Earthquake Engineering Research Center, University of California, Berkeley

Joyner WB, Boore DM (1981) Peak horizontal acceleration and velocity from strong-motion records including records from the 1979 Imperial Valley, California, earthquake. Bull Seismol Soc Am 71(6):2011–2038

Kale Ö, Akkar S (2017) A ground-motion logic-tree scheme for regional seismic hazard studies. Earthq Spectra 33(3):837–856. https://doi.org/10.1193/051316EQS080M

Kotha SR, Bindi D, Cotton F (2016) Partially non-ergodic region specific GMPE for Europe and Middle-East. Bull Earthq Eng 14(4):1245–1263. https://doi.org/10.1007/s10518-016-9875-x

Kulkarni RB, Youngs RR, Coppersmith KJ (1984) Assessment of confidence intervals for results of seismic hazard analysis. In: Proceedings of eighth world conference on earthquake engineering, San Francisco, 21–28 July 1984, 1, pp 263–270

Lanzano G, Puglia R, Russo E, Luzi L, Bindi D, Cotton F, D'Amico M, Felicetta C, Pacor F, ORFEUS WG5 (2017) ESM strong-motion flat-file 2017. Istituto Nazionale di Geofisica e Vulcanologia (INGV), Helmholtz-Zentrum Potsdam Deutsches GeoForschungsZentrum (GFZ), Observatories & Research Facilities for European Seismology (ORFEUS). PID: 11099/ESM_6269e409-ea78-4a00-bbee-14d0e3c39e41_flatfile_2017

Musson RMW (2012) On the nature of logic trees in probabilistic seismic hazard assessment. Earthq Spectra 28(3):1291–1296

National Cooperative for the Disposal of RadioactiveWaste (NAGRA) (2004) Probabilistic seismic hazard analysis for Swiss nuclear power plant sites (PEGASOS Project) prepared for Unterausschuss Kernenergie der Überlandwerke (UAK), Final report Vols. 1/6, 2557 pp., to be obtained on request at swissnuclear by writing to info@swissnuclear.ch

Petersen MD, Moschetti MP, Powers PM, Mueller CS, Haller KM, Frankel AD, Zeng Y, Rezaeian S, Harmsen SC, Boyd OS, Field N, Chen R, Rukstales KS, Luco N, Wheeler RL, Williams RA, Olsen AH (2014). Documentation for the 2014 update of the United States national seismic hazard maps, U.S. Geological Survey Open-File Report 2014–1091, 243 p. https://dx.doi.org/10.333/ofr20141091

Power M, Chiou B, Abrahamson N, Bozorgnia Y, Shantz T, Roblee C (2008) An overview of the NGA project. Earthq Spectra 24(1):3–21. https://doi.org/10.1193/1.2894833

Rietbrock A, Strasser F, Edwards B (2013) A stochastic earthquake ground-motion prediction model for the United Kingdom. Bull Seismol Soc Am 103(1):57–77. https://doi.org/10.1785/0120110231

Sabetta F, Pugliese A (1987) Attenuation of peak horizontal acceleration and velocity from Italian strong-motion records. Bull Seismol Soc Am 77:1491–1513

Sammon JW (1969) A nonlinear mapping for data structure analysis. IEEE Trans Comput C18 (5):401–409

Savy JB, Foxall W, Abrahamson N, Bernreuter D (2002) Guidance for performing probabilistic seismic hazard analysis for a nuclear plant site: example application to the southeastern United States, NUREG/CR-6607. Livermore, Lawrence Livermore National Laboratory

Scherbaum F, Delavaud E, Riggelsen C (2009) Model selection in seismic hazard analysis: an information-theoretic perspective. Bull Seismol Soc Am 99(6):3234–3247. https://doi.org/10.1785/0120080347

Scherbaum F, Kuehn NM, Ohrnberger M, Koehler A (2010) Exploring the proximity of ground-motion models using high-dimensional visualization techniques. Earthq Spectra 26 (4):1117–1138. https://doi.org/10.1193/1.3478697

Sedaghati F, Pezeshk S (2017) Partially nonergodic empirical ground-motion models for predicting horizontal and vertical PGV, PGA, and 5% damped linear acceleration response spectra using data from the Iranian plateau. Bull Seismol Soc Am 107(2):934–948. https://doi.org/10.1785/0120160205

Somerville PG, McLaren JP, Saikia CK, Helmberger DV (1990) The 25 November 1988 Saguenay, Quebec, earthquake: source parameters and the attenuation of strong ground motions. Bull Seismol Soc Am 80(5):1118–1143

Spudich P, Joyner WB, Lindh AG, Boore DM, Margaris BM, Fletcher JB (1999) SEA99: a revised ground motion prediction relation for use in extensional tectonic regimes. Bull Seismol Soc Am 89(5):1156–1170

Stafford PJ (2015) Variability and uncertainty in empirical ground-motion prediction for probabilistic hazard and risk analyses. In: Perspectives on European earthquake engineering and seismology, Geotechnical, Geological and Earthquake Engineering, vol 39. Springer, Cham, p 97. https://doi.org/10.1007/978-3-319-16964-4_4

Stepp JC, Wong I, Whitney J, Quittmeyer R, Abrahamson N, Toro G, Youngs R, Coppersmith K, Savy J, Sullivan T, members YMPSHAP (2001) Probabilistic seismic hazard analyses for ground motions and fault displacement at Yucca Mountain, Nevada. Earthq Spectra 17 (1):113–151

Stewart JP, Douglas J, Javanbarg M, Abrahamson NA, Bozorgnia Y, Boore DM, Campbell KW, Delavaud E, Erdik M, Stafford PJ (2015) Selection of ground motion prediction equations for the global earthquake model. Earthq Spectra 31(1):19–45. https://doi.org/10.1193/013013EQS017M

Toro GR (2006) The effects of ground-motion uncertainty on seismic hazard results: examples and approximate results. In: Annual meeting of the seismological society of America. https://doi.org/10.13140/RG.2.1.1322.2007

Trifunac MD (1976) Preliminary analysis of the peaks of strong earthquake ground motion – dependence of peaks on earthquake magnitude, epicentral distance, and recording site conditions. Bull Seismol Soc Am 66(1):189–219

United States Nuclear Regulatory Commission (USNRC) (2012) Practical implementation guidelines for SSHAC level 3 and 4 hazard studies, NUREG 2117 (Rev 1), Division of Engineering Office, USA

Weatherill G, Danciu L, Crowley H (2013), Future directions for seismic input in European design codes in the context of the seismic hazard harmonisation in Europe (SHARE) project. In: Vienna Congress on Recent Advances in Earthquake Engineering and Structural Dynamics, Paper no 494

Woessner J, Laurentiu D, Giardini D, Crowley H, Cotton F, Grünthal G, Valensise G, Arvidsson R, Basili R, Demircioglu MB, Hiemer S, Meletti C, Musson RW, Rovida AN, Sesetyan K, Stucchi M, The SHARE Consortium (2015) The 2013 European seismic hazard model: key components and results. Bull Earthq Eng 13(12):3553–3596

Yenier E, Atkinson GM (2015) Regionally adjustable generic ground-motion prediction equation based on equivalent point-source simulations: application to central and eastern North America. Bull Seismol Soc Am 105(4):1989–2009. https://doi.org/10.1785/0120140332

Chapter 7
Implementation of Near-Fault Forward Directivity Effects in Seismic Design Codes

Sinan Akkar and Saed Moghimi

Abstract Near-fault ground motions exhibiting forward directivity effects are critical for seismic design because they impose very large seismic demands on buildings due to their large-amplitude pulselike waveforms. The current challenge in seismic design codes is to recommend simple (easy-to-apply) yet proper rules to explain the near-fault forward directivity (NFFD) phenomenon for seismic demands. This effort is not new and has been the subject of research for over two decades. This paper contributes to these efforts and proposes an alternative set of rules to modify the elastic design spectrum of 475-year and 2475-year return periods for NFFD effects. The directivity rules discussed here are evolved from a relatively large number of probabilistic earthquake scenarios (probabilistic seismic hazard assessment, PSHA) that employ two recent directivity models. The paper first gives the background of the probabilistic earthquake scenarios and then introduces the proposed NFFD rules for seismic design codes. We conclude the paper by presenting some cases with the proposed rules to see how spectral amplitudes modify due to directivity.

7.1 Introduction

In principle, when the fault rupture and the seismic waves originating from the rupture propagate towards a site at a speed close to the shear-wave velocity, the waveforms arriving at the site contain a high-amplitude pulse that is predominantly observed in the direction normal to the strike. This phenomenon is called as forward directivity and the forward-directivity pulse usually occurs at the beginning of the ground motion. Ground motions dominated by forward directivity are known for

---

S. Akkar (✉)
Department of Earthquake Engineering, Boğaziçi University, Istanbul, Turkey
e-mail: sinan.akkar@boun.edu.tr

S. Moghimi
Department of Civil Engineering, İstanbul Aydın University, Istanbul, Turkey
e-mail: saedmoghimi@aydin.edu.tr

© Springer International Publishing AG, part of Springer Nature 2018
K. Pitilakis (ed.), *Recent Advances in Earthquake Engineering in Europe*,
Geotechnical, Geological and Earthquake Engineering 46,
https://doi.org/10.1007/978-3-319-75741-4_7

their severe damage potential (e.g., Bertero et al. 1978; Hall et al. 1995; Gupta and Krawinkler 1999; Alavi and Krawinkler 2001, 2004; Chioccarelli and Iervolino 2010). Note that sites located opposite to the rupture propagation are subjected to smaller amplitude and rather longer duration waveforms and such ground motions are potentially less damaging. (This phenomenon is called as backward directivity). That's why seismic codes opt to place special emphasis in the seismic demand attributes subjected to forward directivity. The fling step is another important property of near-fault records featuring directivity. Unlike the pulselike signal, which is the result of dynamic action of the fault movement, the fling step (observed in the displacement waveforms) is the static feature of the ruptured fault. Fling step (permanent fault displacement) might be a critical design parameter for long-span bridges or pipelines crossing the ruptured fault segments but, until now, pulselike signals are of main concern in many building provisions as their effects on design spectrum (leading to larger spectral ordinates than commonly expected) requires immediate attention for proper design of many building classes.

Figure 7.1 (from Tothong et al. 2007) shows the median response spectra of fault-normal pulselike (directivity dominant) ground motions for two different pulse periods (T_p; period of pulselike signals). The mean pulse periods of the ground motions are $\overline{T}_p = 1.0$s (Fig. 7.1a) and $\overline{T}_p = 1.9$s (Fig. 7.1b). Also shown in these plots is the representative response spectrum estimates from a conventional ground-motion predictive model (Abrahamson and Silva 1997) as well as the broad-band directivity ground-motion model of Somerville et al. (1997); the pioneer ground-motion model to estimate spectral ordinates subjected to near-fault forward directivity. The comparative plots indicate that conventional ground-motion predictive models (GMPMs) fail to capture the amplified spectral ordinates in the vicinity of pulse period. The Somerville et al. broad-band directivity model, by definition, amplifies spectral ordinates after a certain period ($T = 0.6$s) towards longer periods. However, the monotonic and continuous spectral amplification in this model cannot fully capture the increase in spectral ordinates in the vicinity of pulse period. This

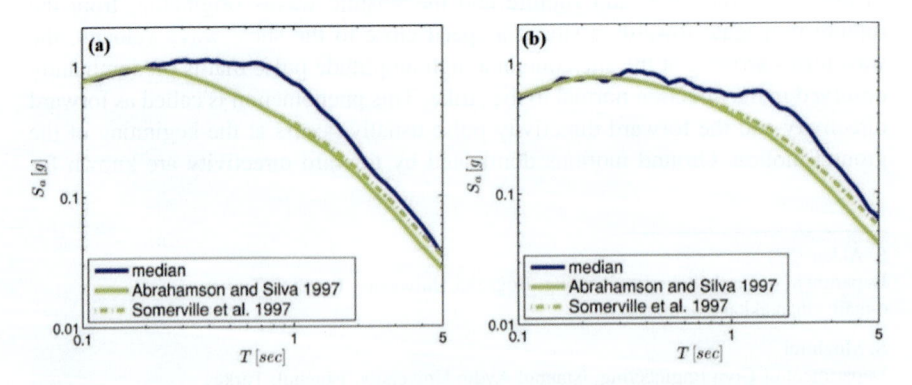

Fig. 7.1 Median response spectra of fault-normal directivity dominant horizontal ground motions having mean pulse periods of (**a**) $\overline{T}_p = 1.0$s and (**b**) $\overline{T}_p = 1.9$s (from Tothong et al. 2007). Each panel also shows the spectral ordinate estimates from the conventional Abrahamson and Silva (1997) GMPM and the broad-band forward directivity GMPM by Somerville et al. (1997)

example advocates the complexity and challenge in reflecting the forward directivity waveform characteristics on to seismic ground motion demands (i.e., response spectrum). Recent seismological studies propose improved broad-band as well as alternative narrow-band GMPMs that accentuate directivity dominated spectral amplifications in the neighborhood of pulse-period. These models are presented in Spudich et al. (2013, 2014) that are developed under the framework of NGA-West 2 project (Bozorgnia et al. 2014). It should be noted that pulse period is a function of earthquake magnitude (Mavroeidis and Papageorgiou 2003; Somerville 2003; Baker 2007): NFFD ground motions of small-magnitude events would exhibit short-period pulselike signals that result in spectral amplifications within short-to-medium period range. The directivity dominant recordings from large-magnitude events, on the other hand, would possess larger pulse periods and the spectral ordinates peak towards a longer period range.

The brief discussions in the previous paragraph emphasize the necessity to consider the dynamic NFFD effects on design spectrum. To the best knowledge of the authors, the first seismic design code that explicitly accounts for the dynamic directivity effects is the 1997 edition of the Uniform Building Code (UBC 1997) that amplifies the design spectrum for source-to-site distances less than 15 km when the site is located in the most seismic prone zone according to the US seismic zonation map at that time. Based on the seismic activity of the capable fault (described by the maximum possible magnitude and the slip rate) as well as the closest distance to the ruptured fault segment (R_{rup}), this code provides the so-called near-fault amplification factors (N_a and N_v) for acceleration- and velocity-sensitive spectral regions. It is believed that the proposed near-fault amplification factors are originated from the Somerville et al. (1997) broad-band directivity model. The approach implemented by UBC inspired some other seismic design codes such as those from Taiwan (Chai et al. 2000), Iran (Yaghmaei-Sabegh and Mohammad-Alizadeh 2012) and China (Li et al. 2007). The seismic design regulations in the New Zealand Standard (NZS 2004) impose forward directivity effects through a period-dependent function that also accounts for mean annual exceedance probability of seismic hazard (i.e., return period) and source-to-site distance. This continuous function modifies the design spectrum for periods $T > 1.5s$ and for distances up to $R_{rup} \leq 20$ km. The seismic design provisions by China, Taiwan, Iran and UBC (1997) amplify design spectrum for the whole period range and for distances $R_{rup} \leq 15$ km. The updated seismic design criteria of CALTRANS (2013) also provide near-fault adjustment factors that amplify the design spectrum for $T \geq 0.5s$ and $R_{rup} \leq 25$ km. These discussions suggest that the seismic provisions of New Zealand and CALTRANS follow a similar rationale in terms of near-fault spectral amplification. We should note that the directivity-based spectral amplifications vary significantly between the cited design codes. In some regulations, these amplification factors may reach to a value of 1.6 (or even higher) in the long-period range. Depending on the considered earthquake scenarios in these codes, the near-fault factors generally modify the design spectrum by 10–40% in the entire spectrum band. Needless to say an unbiased comparison between these codes in terms of NFFD factors require structuring well-established case studies in order not to yield biased conclusions about the consistency of proposed near-fault adjustments. This is not the scope of this study.

The objective of this paper is to introduce alternative period-dependent NFFD factors to modify the design spectrum at different return by considering the fault-site geometry, fault activity and maximum magnitude that can occur on the considered fault segment (referred to as characteristic magnitude, M_{ch}). The proposed NFFD factors are based on a comprehensive set of PSHA case studies (Akkar et al. 2018) that account for near-fault effects by two recent narrow-band directivity models (Shahi and Baker 2011; Chiou and Spudich 2013 as discussed in Spudich et al. 2013). The paper briefs the probabilistic case studies that form the bases of the proposed NFFD formulation and presents the expressions proposed for modifying the design spectrum for forward directivity. Simple case studies are given at the end of the paper to demonstrate the implementation of the proposed modification factors.

7.2 Background of Probabilistic Case Studies Used for Developing the Proposed Near-Fault Directivity Factors

Figure 7.2 shows the fault-site layout used in the PSHA cases that are used in the development of proposed NFFD factors. The sites are distributed symmetrically with respect to R_y (vertical axis) that crosses at the mid-length of the strike-slip fault[1]. There is a mirror image distribution of the sites on the left-hand side with

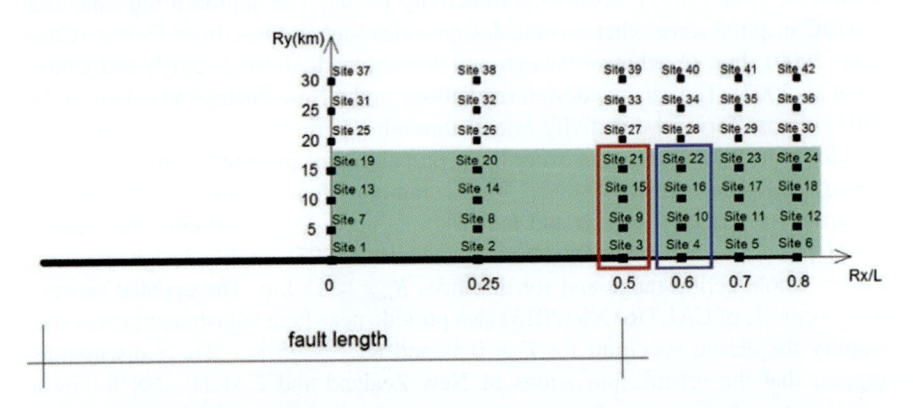

Fig. 7.2 Spatial distribution of sites and the fault for fictitious PSHA case studies. Shaded region and the blue and red rectangles encircling certain sites are used in the derivation of proposed near-fault spectral amplification functions (Modified from Akkar et al. 2018)

---

[1] The style-of-faulting is 90 degrees dipping strike-slip fault throughout this study. Thus, the results presented here are strictly valid for strike-slip faults although the estimated near-fault spectral amplifications may also be valid for dip-slip faults provided the controlling conditions are similar.

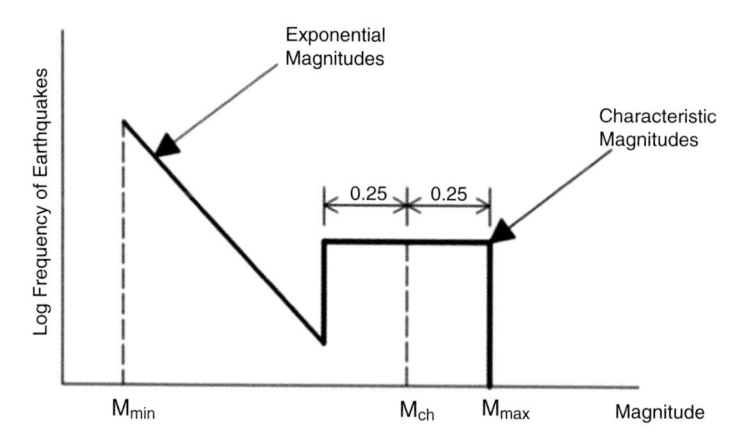

Fig. 7.3 Illustration of characteristic earthquake recurrence model (Youngs and Coppersmith 1985)

respect to the fault center (designated by $R_x/L = 0$ in Fig. 7.2). The fault length is designated by L and the fault-length normalized horizontal axis runs parallel to the fault strike. The site locations are equally spaced at every 5 km in the strike-normal direction and they are extended beyond the fault edges by 0.3 L in the strike parallel direction to capture the spatial variation of forward directivity along the horizontal plane. We consider a generic rock site definition ($V_{S30} = 760$ m/s) in all PSHA cases.

Five fictitious fault lengths ($L = 20$ km, 50 km, 100 km, 150 km and 300 km) are chosen for PSHA scenarios that can generate characteristic earthquakes of M_{ch} 6.25, 6.7, 7.0, 7.25 and 7.5, respectively under full rupture conditions. The fault length and M_{ch} matching is done from the empirical Wells and Coppersmith (1994) M_w vs. rupture area (RA) relations. The maximum fault width (w) is taken as 10 km and slip rates ($\dot{s}$) of 0.5 cm/year, 1.0 cm/year and 2.0 cm/year are used for the fictitious faults to represent low, moderate and high seismic activity, respectively. The stochastic earthquake recurrence is described by the characteristic earthquake model (Fig. 7.3) of Youngs and Coppersmith (1985). The exponential part in this model considers earthquake activities between $5.0 \leq M_w \leq M_{ch-0.25}$. The uncertainty in the characteristic earthquake magnitudes is represented as a uniform distribution within $M_{ch} \pm 0.25$ band. The rupture area for each discrete magnitude in PSHA scenarios are obtained from RA vs. M_w relationships of Wells and Coppersmith (1994). The ruptured fault length for each scenario is determined by dividing the square root of rupture area with constant fault width whenever $\sqrt{RA} > w$. Otherwise, the rupture length is computed as $\sqrt{RA}$. The hypocenter of the rupture is always taken at the center of the ruptured area in all PSHA scenarios.

For each M_{ch} and slip rate combination, we ran PSHA by considering the directivity models proposed in Shahi and Baker (SHB11; 2011) and Chiou and Spudich (CHS13; Chapter 6 in Spudich et al. 2013). The PSHA runs were repeated for no directivity cases that are optionally available for SHB11 and CHS13. The spectral ratios between directivity and no directivity PSHA runs are used in the proposed near-fault spectral amplification functions. We focus on two specific return periods for the proposed NFFD spectral amplifications: $T_R = 475$ year and $T_R = 2475$ year. These return periods represent the most frequently used hazard levels in seismic design and performance assessment in most modern seismic design codes. Although Akkar et al. (2018) and Moghimi (2017) gave detailed explanations about the SHB11 and CHS13 directivity models, we briefed them in the following paragraphs for the sake of clarity and completeness.

SHB11 is a probabilistic seismic hazard assessment model and considers spectral amplitude modifications at sites subjected to pulselike waveforms due to directivity. It considers the probability of pulse occurrence for a given fault-site geometry and the distribution of magnitude-dependent pulse period, T_p. This model improves its predecessor (Tothong et al. 2007) by including the probability of observing a pulse in a particular orientation given a pulse is observed at the site.

SHB11 estimates the amplification (and deamplification) of response spectrum ordinates for the existence (and absence) of pulse at sites located in the vicinity of fault. Equation (7.1) presents the basic theory behind SHB11 to predict ground-motion spectral amplitudes for directivity. $P^*(S_a > x \,|m, r, z)$ is the probability of spectral ordinate, S_a, exceeding x ($S_a > x$) given an earthquake of magnitude m at distance r under modified GMPM. z represents fault-site geometry and it is important to mark the directivity effects. It comprises of the parameters s and α where s is the distance along the rupture plane from the epicenter toward the site and α is the smallest angle between incident S-wave and the fault strike. (See Fig. 7.4 for the description of parameters). The first two probabilities on the right hand side of Eq. (7.1) are probabilities of observing a pulse and $S_a > x$ upon the occurrence of pulse. The last two probabilities consider observing no pulse and $S_a > x$ when no pulse is observed. Thus, the modification of spectral intensities, S_a, depend on the pulse occurrence or non-occurrence cases. Thus, SHB11 splits the probability of $S_a > x$ into two cases depending on whether or not the pulselike ground motion is observed.

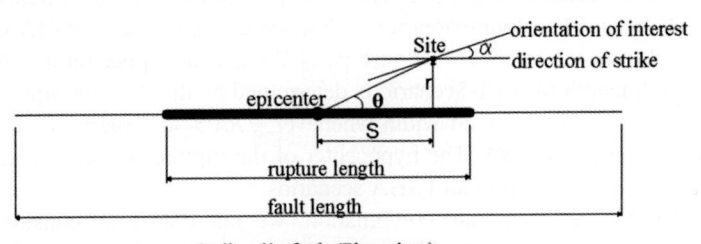

Fig. 7.4 The main strike-slip fault related parameters in SHB11 (Modified from Akkar et al. 2018)

$$\begin{aligned}P^*(S_a > x|m,r,z) \\ = P(pulse|\,m,r,z) \cdot P(S_a > x|m,r,z,pulse) \\ + [1 - P(pulse|m,r,z)] \cdot P(S_a > x|m,r,no\ pulse)\end{aligned} \quad (7.1)$$

The probabilities $P(S_a > x|m,r,z,pulse)$ and $P(S_a > x|m,r,no\ pulse)$ are obtained from modified GMPMs for directivity effects. In their model SHB11 modifies Boore and Atkinson (BA08; 2008) for directivity effects to compute above two probabilities. However, they indicate that the proposed expressions in their model are equally applicable to all conventional GMPMs. We used SHB11 model together with the modified and original BA08 to predict near-fault directivity and no-directivity spectral ordinates. We note that the horizontal spectral component definition of the original BA08 is GMRotI50 (Boore et al. 2006) that can be grossly defined as the geometric mean of two mutually perpendicular horizontal components. In our case studies SHB11 model is used to estimate the directivity effects on fault-normal component. Therefore, near-fault directivity amplifications computed for SHB11 represent the ratio between directivity-dominant fault-normal spectrum to no-directivity geometrical mean spectrum.

CHS13 defines the so-called Direct Point Parameter (*DPP*) to model directivity by considering the slip distribution and radiation pattern of a finite fault as well as the isochrone velocity (a quantity closely related to rupture velocity; high isochrone velocity is the indication of strong directivity effects). Given the relative location of the site with respect to the ruptured fault plane, *DPP* is the indicator about the strength of directivity (see Akkar et al. 2018). *DPP* can be computed at equidistant sites from the ruptured fault segment as illustrated in Fig. 7.5. The site depicting the maximum *DPP*, given the rupture and fault geometry, is subjected to the largest directivity effect. The *DPP* concept is used by the GMPM of Chiou and Youngs (CY14; 2014) as a predictor parameter to account for directivity in spectral amplitude estimates. CY14 centers DPP on its mean (*DPP$_{mean}$*) over a suite of equidistant sites (Fig. 7.5) and the dominancy of directivity at a specific site i along the same racetrack is determined by subtracting *DPP$_{mean}$* from *DPP$_i$*. The difference between *DPP$_i$* and *DPP$_{mean}$* is called *ΔDPP* and a large *ΔDPP* indicates stronger forward directivity effects in CY14. When *ΔDPP* is zero, the directivity does not dominate

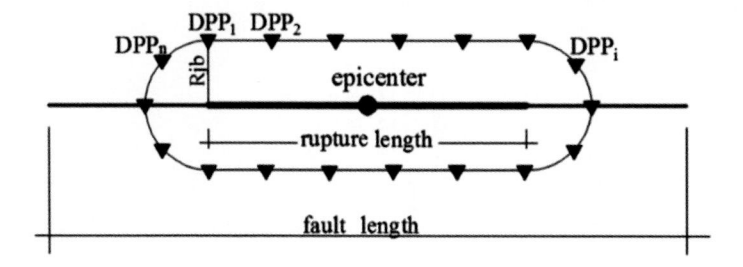

Fig. 7.5 The plan view of a fault and sites located at equal distances from the ruptured fault to illustrate the computation of Direct Point Parameter proposed by CHS13 and implemented by CY14 (Modified from Akkar et al. 2018)

the spectral amplitudes at the site of interest for CY14. We used CY14 in our PSHA scenarios by considering and disregarding ΔDPP (i.e., $\Delta DPP \neq 0$ and $\Delta DPP = 0$ for directivity and no-directivity cases, respectively) and the ratio between the spectral periods for these two independent cases yield the near-fault directivity amplifications for CHS13 model. The horizontal spectral component definition for CY14 is RotD50 (Boore 2010) and directivity to no-directivity spectral ratios for CHS13 is based on this horizontal component definition. The details and implementation of CHS13 as well as SHB11 directivity models to PSHA are discussed in Akkar et al. (2018) and Moghimi (2017).

7.3 A Short Discussion on the Near-Fault Directivity Amplifications Predicted by SHB11 and CHS13

Figure 7.6 shows the distribution of 475-year near-fault directivity amplifications at $T = 4$s by SHB11 and CHS13 for a fictitious perfect strike-slip fault rupture (dipping at 90°) that can generate a characteristic magnitude of M_{ch} 7 (the maximum ruptured fault length is $L = 100$ km). The fictitious fault is assumed to have a slip rate of 2 cm/year (high seismic activity). As depicted by Fig. 7.6 the spectral amplification contours of SHB11 (Fig. 7.6a) are larger than those of CHS13 (Fig. 7.6b) because SHB11 estimates the directivity amplifications for fault-normal component whereas CHS13 considers RotD50 horizontal component for directivity. Both models tend to estimate the largest directivity amplifications close to the ends of the fault whereas no amplification is computed by the two models at the center of the fault. This is consistent with the Somerville et al. (1997) model. The directivity amplifications of CHS13 are exclusively concentrated at the ends of the fault and they extend beyond the fault edges. This is the result of isochrone theory; the theoretical background of this model. The directivity amplifications by SHB11 are shaped by pulse occurrence probability that systematically increases towards the ends of the fault (see the details

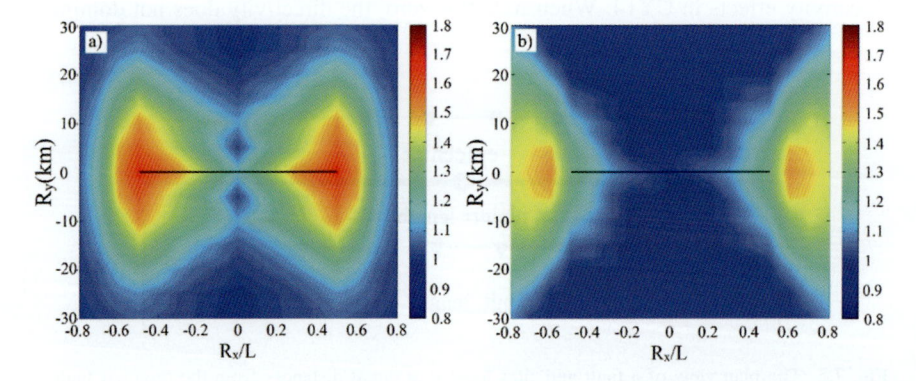

Fig. 7.6 475-year return period near-fault spectral amplification distributions at $T = 4$s for a fictitious pure strike-slip fault (solid black line) with M_{ch} 7 and $\dot{s} = 2$ cm/year (**a**) SHB11 and (**b**) CHS13 (Modified from Akkar et al. 2018)

of this model in Akkar et al. 2018; Moghimi 2017 and Shahi and Baker 2011). The SHB11 directivity amplifications significantly decrease for $R_x/L > 0.6$, which is, again, due to the decreased probability of observing pulse occurrence at remote sites that are far from the ends of the fault segment. Such modeling constraints do not exist in CHS13. This simple case clearly shows that there are differences in the directivity modeling between SHB11 and CHS13. The use of different GMPMs by the two models may also contribute to the observed differences in directivity spectral amplifications.

Figure 7.7 shows a sample case for the period-dependent variations of 2475-year directivity spectral amplifications for a seismically very active fault (total slip rate of $\dot{s}$ = 2.0 cm/year) that is capable of producing characteristic earthquakes of M_{ch} 7.3 (i.e., L = 300 km). The amplifications in each panel (light gray curves) are plotted for all the sites as shown in Fig. 7.2. The panels in the first column show the amplifications estimated by SHB11 whereas the curves in the second column are the amplifications by CHS13. Also shown in these panels are the median spectral amplifications for R_y (first row; R_y = 0 km, 5 km, 10 km, 15 km, 20 km, 25 km and 30 km) and for R_x/L (second row; R_x/L = 0, 0.25, 0.5, 0.6, 0.7, 0.8) to mark the dependency of directivity amplifications on fault-site geometry. The spectral amplifications at 7 sites located along the same R_x/L are used to compute the median spectral amplifications for a given R_x/L. Similarly, the median spectral amplification for a specific R_y is computed from the spectral amplifications at 6 sites located along the same R_y.

The plots in Fig. 7.7 show the considerable difference between the period-dependent directivity spectral amplification shapes predicted by SHB11 and CHS13. The possible sources of observed differences are discussed in the previous paragraph. The amplification curves of SHB11 show a fast increase until a peak (the corresponding period is called as T_{max}) and this is followed by a steep decrease. The CHS13 amplifications show a milder period-dependent increase and they show a slight reduction (barely visible) after reaching a certain maximum value (the corresponding period is called as T_{corner}). As it can be noted from the median amplifications the directivity amplifications are inversely proportional to R_y: maximum at the on-fault sites (R_y = 0 km) and they decrease as R_y increases. Dependency of directivity amplifications on R_y decreases for sites that are located remotely with respect to the fault strike. This observation is more noticeable in CHS13.

The directivity amplifications display a more complex picture for R_x/L. The median spectral amplifications along the fault ends (R_x/L = 0.5 and R_x/L = 0.6) are maximum for SHB11 because this model gives higher possibility of pulse occurrence at the ends of the ruptured fault. SHB11 also suggests R_x/L = 0.25 as another potential location for large directivity amplifications. Remote sites from the fault ends (represented by R_x/L = 0.8) are disregarded for directivity effects by SHB11. The directivity amplifications by CHS13 are significant at sites located along and beyond the fault ends (i.e., $0.5 \leq R_x/L \leq 0.6$). For the case study presented here, CHS13 does not predict (or barely predicts) directivity amplifications at sites remotely located from the fault ends ($R_x/L \geq 0.7$). In contrast to SHB11, this directivity model does yield any directivity amplification at sites located along R_x/L = 0.25.

We note that more comprehensive discussions about the influence of fault-site geometry as well as the contributions of characteristic magnitude, slip rate and

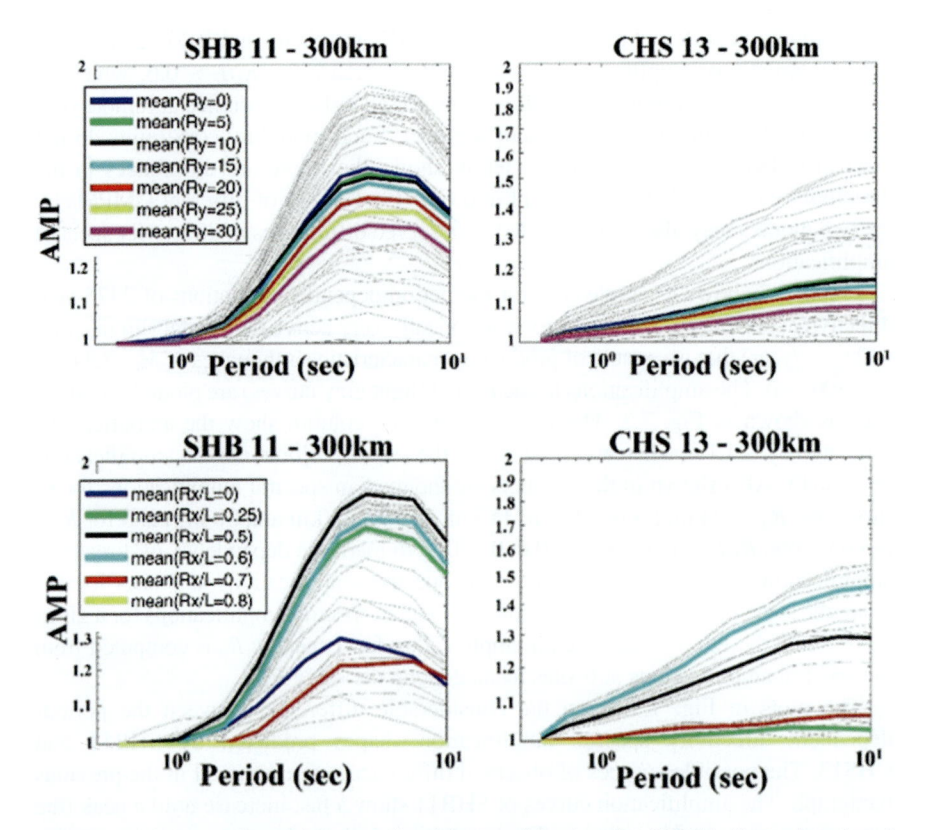

Fig. 7.7 Period-dependent 2475-year directivity amplifications by SHB11 (first column) and CHS13 (second column) for a fault length $L = 300$ km (M_{ch} 7.3) having a total slip rate of 2.0 cm/year. The first row shows median directivity amplifications for a given R_y. The second row panels show the same information for a constant R_x/L

ruptured fault length to directivity spectral amplifications are presented in Akkar et al. (2018) and Moghimi (2017). We give an overall picture about each one of these parameters via limited case studies as a background for the directivity functions that are presented in the following section. The discussions held here clearly suggest that different directivity models predict different spectral amplifications for reasons underlying in their theoretical context.

7.4 Directivity Functions for Their Implementation to Seismic Codes

The proposed functions to modify spectral demands for directivity effects are inferred from the observations of PSHA scenarios that are discussed in detail in Akkar et al. (2018) and Moghimi (2017). They are partially addressed in the previous section as well. Since SHB11 and CHS13 predict different patterns of directivity amplifications, separate expressions are developed for each model. Firstly, directivity amplification expressions are developed for sites where the highest directivity spectral amplifications are observed. This step is followed by defining source-site geometry scaling factors (referred to as geometry scale factors hereafter) to estimate the directivity spectral amplifications at the other locations around the fault. The details on the development of these functions are given in the following items.

1. Compute the median directivity amplifications at sites that experience the largest amplification ($R_x/L = 0.5$ for SHB11 and $R_x/L = 0.6$ for CHS13 that are encircled by red and blue rectangles in Fig. 7.2). The median directivity amplifications are computed for $R_y \leq 15$ km (sites 3, 9, 15 and 21 for SHB11 and sites 4, 10, 16 and 22 for CHS13 -encircled by red and blue rectangles in Fig. 7.2) because the directivity amplifications taper down very fast for $R_y > 15$ km. A similar distance capping is also implemented in CALTRANS (2013). These median directivity amplifications are called as "*base*" amplifications that are further detailed in items 2 and 3 below.

2. The median directivity amplifications are computed at spectral periods T_{max} for SHB11 ($Amp_{Tmax,base}$) and T_{corner} for CHS13 ($Amp_{Tcorner,base}$). The same computations are also repeated for $T = 10$s ($Amp_{T10,base}$) for SHB11. Equations (7.2a, 7.2b and 7.2c) display the expressions for SHB11 whereas Eqs. (7.3a and 7.3b) show the same relations for CHS13. Note that for magnitudes beyond M_{ch} 7.25, the directivity amplifications are assumed to follow a constant value. The characteristic magnitude dependent base directivity amplifications are assumed to be linearly increasing between $0.6s \leq T \leq T_{max}$ and $0.6s \leq T \leq T_{corner}$ for SHB11 and CHS13, respectively in Eqs. (7.2a, 7.2b and 7.2c) and (7.3a and 7.3b). They tend to decrease linearly for SHB11 between $T_{max} < T \leq 10$s whereas they remain constant after T_{corner} in the case of CHS13. Tables 7.1, 7.2, and 7.3 show the regression coefficients of Eqs. (7.2a, 7.2b and 7.2c) and (7.3a and 7.3b). The regression coefficients are functions of return period (2475-year and 475 year in this study) as well as slip rate ($\dot{s}$) for SHB11. CHS13 directivity amplifications are insensitive to slip rate as discussed in Moghimi (2017) and Akkar et al. (2018).

$$Amp_{Tmax,base} = \alpha_{Tmax} \cdot M_{ch} + \beta_{Tmax}; 6.25 < M_{ch} \leq 7.25 \quad (7.2a)$$
$$Amp_{Tmax,base} = 7.25 \cdot \alpha_{Tmax} + \beta_{Tmax}; M_{ch} > 7.25 \quad (7.2b)$$
$$Amp_{T10,base} = \alpha_{T10} \cdot M_{ch} + \beta_{T10} \quad (7.2c)$$
$$Amp_{Tcorner,base} = \alpha_{Tcorner} \cdot M_{ch} + \beta_{Tcorner}; 6.25 < M_{ch} \leq 7.25 \quad (7.3a)$$
$$Amp_{Tcorner,base} = 7.25 \cdot \alpha_{Tcorner} + \beta_{Tcorner}; M_{ch} > 7.25 \quad (7.3b)$$

3. Establish relationships between T_{max} vs. M_{ch} (or T_{corner} vs. M_{ch}) because Eqs. (7.2a, 7.2b and 7.2c) and (7.3a and 7.3b) are functions of characteristic magnitude that suggest a change in T_{max} (or T_{corner}) with M_{ch} (Akkar et al. 2018;

Table 7.1 α_{Tmax} and β_{Tmax} coefficients for $Amp_{Tmax,base}$ – SHB11

$\dot{s}$ (cm/year)	475 year α_{Tmax}	475 year β_{Tmax}	2475 year α_{Tmax}	2475 year β_{Tmax}
0.5	0.146	0.149	0.495	−1.9
1.0	0.241	−0.364	0.546	−2.168
2.0	0.454	−1.664	0.554	−2.167

Table 7.2 α_{T10} and β_{T10} coefficients for $Amp_{T10,base}$ – SHB11

$\dot{s}$ (cm/year)	475 year α_{T10}	475 year β_{T10}	2475 year α_{T10x}	2475 year β_{T10}
0.5	0.045	0.72	0.313	−0.95
1.0	0.167	−0.04	0.384	−1.4
2.0	0.229	−0.4	0.425	−1.65

Table 7.3 $\alpha_{Tcorner}$ and $\beta_{Tcorner}$ coefficients for $Amp_{Tcorner,base}$ – CHS13

475 year $\alpha_{Tcorner}$	475 year $\beta_{Tcorner}$	2475 year $\alpha_{Tcorner}$	2475 year $\beta_{Tcorner}$
0.4	−1.4931	0.464	−1.9

Moghimi 2017). Equation (7.4) presents this expression that is valid both for T_{max} and T_{corner} because PSHA scenarios indicate similar values for T_{max} and T_{corner} for a given M_{ch}. Note that T_{mc} stands for T_{max} or T_{corner} in Eq. (7.4). Note also that M_{ch} is dependent of ruptured fault length in our study to establish a physical relation between the size of the fault and the maximum possible earthquake that can be generated by that fault.

$$T_{mc} = 2.72 \cdot M_{ch} - 15.37 \qquad (7.4)$$

4. To estimate the directivity amplifications of sites at other locations around the fault, geometry scale factors (GSF_{Tmax} and GSF_{T10} for SHB11, and $GSF_{Tcorner}$ for CHS13) are developed. These factors are still used to estimate the directivity amplifications within $R_y \leq 15$ km (shaded area in Fig. 7.2) because for distances greater than $R_y > 15$ km, we use a distance capping as discussed later. GSF_{Tmax} and GSF_{T10} modify $Amp_{Tmax,base}$ and $Amp_{T10,base}$ to estimate Amp_{Tmax} and Amp_{T10} at locations different than $R_x/L = 0.5$ for SHB11. In a similar manner, $GSF_{Tcorner}$ modifies $Amp_{Tcorner,base}$ at locations different than $R_x/L = 0.6$ for CHS13.
5. Given a specific M_{ch}, GSF is the normalized directivity amplifications at $R_x/L \neq 0.5$ by those at $R_x/L = 0.5$ for SHB11 (and at $R_x/L \neq 0.6$ by those at $R_x/L = 0.6$ for CHS13). This item describes the expressions for computing GSF_{Tmax} and GSF_{T10} in SHB11 that are given in Eqs. (7.5a and 7.5b) and (7.6a and 7.6b). The parameters SF_{Tmax} and SF_{T10} that are used in these equations are given in Table 7.4.

ns of Near-Fault Forward Directivity Effects in Seismic Design Codes

Table 7.4 SF_{Tmax} and SF_{T10} values for SHB11 model for computing geometric scale factors at different R_x/L

	$R_x/L=0$	$R_x/L=0.25$	$R_x/L=0.5$	$R_x/L=0.6$	$R_x/L=0.7$	$R_x/L=0.8$
2475 year-SF_{Tmax}	0.67	0.89	1	0.93	0.7	0.6
2475 year-SF_{T10}	0.78	0.94	1	0.93	0.83	0.78
475 year-SF_{Tmax}	0.83	0.85	1	0.93	0.85	0.83
475 year-SF_{T10}	0.96	0.96	1	0.98	0.96	0.96

$$GSF_{Tmax} = [1 + (SF_{Tmax} - 1) \cdot (M_{ch} - 6.25)]; 6.25 \leq M_{ch} \leq 7.25 \quad (7.5a)$$
$$GSF_{Tmax} = SF_{Tmax}; M_{ch} > 7.25 \quad (7.5b)$$
$$GSF_{T10} = [1 + (SF_{T10} - 1) \cdot (M_{ch} - 6.25)]; 6.25 \leq M_{ch} \leq 7.25 \quad (7.6a)$$
$$GSF_{T10} = SF_{T10}; M_{ch} > 7.25 \quad (7.6b)$$

Equations (7.5a and 7.5b) and (7.6a and 7.6b) indicate that the characteristic magnitude dependent GSF_{Tmax} and GSF_{T10} start with unity at M_{ch} 6.25 and increases linearly up to M_{ch} 7.25. They attain a constant value after M_{ch} 7.25 that is equal to the geometric scale factor at M_{ch} 7.25. In brief, the directivity amplifications at T_{max} and $T = 10s$ (Amp_{Tmax} and Amp_{T10}, respectively) for site locations other than $R_x/L = 0.5$ are computed from $Amp_{Tmax,base}$, $Amp_{T=10,base}$, GSF_{Tmax} and GSF_{T10}. For convenience, Eqs. (7.7a and 7.7b) and (7.8a and 7.8b) are the compact forms of Amp_{Tmax} and Amp_{T10} by considering all the relevant expressions as discussed above.

$$Amp_{Tmax} = Amp_{Tmax,base} \cdot GSF_{Tmax}$$
$$= (\alpha_{Tmax} \cdot M_{ch} + \beta_{Tmax})[1 + (SF_{Tmax} - 1) \cdot (M_{ch} - 6.25)]; \quad (7.7a)$$
$$6.25 < M_{ch} \leq 7.25$$

$$Amp_{Tmax} = Amp_{Tmax,base} \cdot GSF_{Tmax}$$
$$= (\alpha_{Tmax} \cdot 7.25 + \beta_{Tmax}) \cdot SF_{Tmax}; M_{ch} > 7.25 \quad (7.7b)$$

$$Amp_{T10} = Amp_{T10,base} \cdot GSF_{T10}$$
$$= (\alpha_{T10} \cdot M_{ch} + \beta_{T10})[1 + (SF_{T10} - 1) \cdot (M_{ch} - 6.25)]; \quad (7.8a)$$
$$6.25 < M_{ch} \leq 7.25$$

$$Amp_{T10} = Amp_{T10,base} \cdot GSF_{T10}$$
$$= (\alpha_{T10} \cdot 7.25 + \beta_{T10}) \cdot SF_{T10}; M_{ch} > 7.25 \quad (7.8b)$$

As in the case of base directivity amplification expressions, the extension of directivity amplifications at periods other than T_{max} and $T = 10s$ is done by assuming a bilinear variation in directivity amplifications between $0.6s \leq T \leq T_{max}$ and $T_{max} < T \leq 10s$. Equations (7.9a and 7.9b) present this bilinear relation.

Table 7.5 $SF_{Tcorner}$ values for CHS13 model for computing geometric scale factors at different R_x/L values

	$R_x/L = 0$	$R_x/L = 0.25$	$R_x/L = 0.5$	$R_x/L = 0.6$	$R_x/L = 0.7$	$R_x/L = 0.8$
475 year-T_{corner}	0.73	0.74	0.93	1	0.98	0.89
2475 year-T_{corner}	0.69	0.70	0.86	1	0.98	0.88

$$Amp(T) = 1 + \left[(Amp_{T_{max}} - 1) \cdot \left(\frac{T - 0.6}{T_{max} - 0.6}\right)\right]; 0.6s \leq T \leq T_{max} \quad (7.9a)$$

$$Amp(T) = Amp_{T_{max}} + \left[(Amp_{T10} - Amp_{T_{max}}) \cdot \left(\frac{T - T_{max}}{10 - T_{max}}\right)\right]; T_{max} < T \leq 10s \quad (7.9b)$$

The directivity amplifications of SHB11 as presented above are valid for $R_y \leq 15$ km. For $R_y > 15$ km, the near-fault directivity amplifications taper down to unity at $R_y = 30$ km that is discussed separately after introducing the $GSF_{Tcorner}$ for CHS13 in the following item.

6. The expressions for the computation of $GSF_{Tcorner}$ are given in Eqs. 7.10a and 7.10b. $GSF_{Tcorner}$ varies linearly between $6.25 \leq M_{ch} \leq 7.25$ whereas it is kept constant after M_{ch} 7.25 with the corresponding value at M_{ch} 7.25. As inferred from the given expressions $GSF_{Tcorner}$ is unity for M_{ch} 6.25. The $SF_{Tcorner}$ values in Eqs. 7.10a and 7.10b are given in Table 7.5.

$$GSF_{Tcorner} = [1 + (SF_{Tcorner} - 1) \cdot (M_{ch} - 6.25)]; 6.25 < M_{ch} \leq 7.25 \quad (7.10a)$$

$$GSF_{Tcorner} = SF_{Tcorner}; M_{ch} > 7.25 \quad (7.10b)$$

After determining $AMP_{Tcorner,base}$ and geometric scale factor ($GSF_{Tcorner}$) the directivity amplification at any location around the fault for CHS13 narrow-band directivity model can be calculated from Eqs. 7.11a and 7.11b. For spectral periods larger than T_{corner} the directivity amplifications by CHS13 take constant values and they are equal to $AMP_{Tcorner}$.

$$\begin{aligned} Amp_{Tcorner} &= Amp_{Tcorner,base} \cdot GSF_{Tcorner} \\ &= (\alpha_{Tcorner} \cdot M_{ch} + \beta_{Tcorner})[1 + (SF_{Tcorner} - 1) \cdot (M_{ch} - 6.25)]; \\ & \quad 6.25 < M_{ch} \leq 7.25 \end{aligned} \quad (7.11a)$$

$$\begin{aligned} Amp_{Tcorner} &= Amp_{Tcorner,base} \cdot GSF_{Tcorner} \\ &= (\alpha_{Tcorner} \cdot 7.25 + \beta_{Tcorner}) \cdot SF_{Tcorner}; M_{ch} > 7.25 \end{aligned} \quad (7.11b)$$

7 Implementation of Near-Fault Forward Directivity Effects in Seismic Design Codes

The extension of directivity amplifications for periods other than T_{corner} is given in Equation 7.12a and 7.12b where a linear trend is assumed between $0.5s \leq T \leq T_{corner}$ and a constant value for $T > T_{corner}$.

$$Amp(T) = 1 + \left[(Amp_{T_{corner}} - 1) \cdot \left(\frac{T - 0.5}{T_{max} - 0.5}\right)\right]; 0.5s \leq T \leq T_{max} \quad (7.12a)$$

$$Amp(T) = Amp_{T_{corner}}; T_{max} < T \leq 10s \quad (7.12b)$$

A distance taper is also implemented for sites beyond $R_y = 15$ km for CHS13 that accounts for the decrease in directivity effects and it will be discussed in the next item.

7. The directivity amplification equations presented in the above lines assume an invariant directivity amplification both for SHB11 and CHS13 for distances up to $R_y = 15$ km. The directivity amplifications taper down linearly to unity between 15 km $< R_y \leq 30$ km (see detailed discussions in Moghimi 2017). Equation 7.13 show the expressions to implement this approach.

$$AMP^{(SHB11 or CHS13)}(T) = AMP_{(0km \leq R_{rup} \leq 15km)}^{(SHB11 or CHS13)}(T) +$$
$$\left[\left(1 - AMP_{0km \leq R_{rup} \leq 15km}^{SHB11\ or\ CHS13}(T)\right) \cdot \left(\frac{R_{rup} - 15}{15}\right)\right]; 15km < R_{rup} \leq 30km \quad (7.13)$$

7.5 Implementation of Proposed Directivity Amplification Rules

We explored the effect of directivity amplifications on spectral ordinates by considering some specific cases. Figure 7.8 illustrates the period-dependent variation of directivity amplifications at $R_x/L = 0.5$ for the rules developed from SHB11 directivity model in the cases of fictitious strike-slip faults having characteristic magnitudes M_{ch} 6.25 and M_{ch} 7.25. The computed amplifications are valid for $R_y \leq 15$ km. The plots include the slip rate effects for $\dot{s} = 0.5$ cm/year, $\dot{s} = 1.0$ cm/year and $\dot{s} = 2.0$ cm/year. As it is depicted the directivity rules developed from SHB11 imply larger directivity amplifications with the increase of slip rates as well as M_{ch}. Needless to say, the change in return periods from 475-year to 2475-year also lead to larger directivity amplifications. Note that the maximum directivity amplifications occur at $T_{max} = 1.6s$ and $T_{max} = 4.4s$ for M_{ch} 6.25 and M_{ch} 7.25, respectively.

Figure 7.9 also compares SHB11 based directivity amplifications for $R_x/L = 0$ and $R_x/L = 0.6$ when fictitious strike-slip fault has a slip rate of 1 cm/year and ruptures with a characteristic magnitude of M_{ch} 6.25 and M_{ch} 7.25. The computed amplifications are valid for $R_y \leq 15$ km. Regardless of fault-site geometry directivity

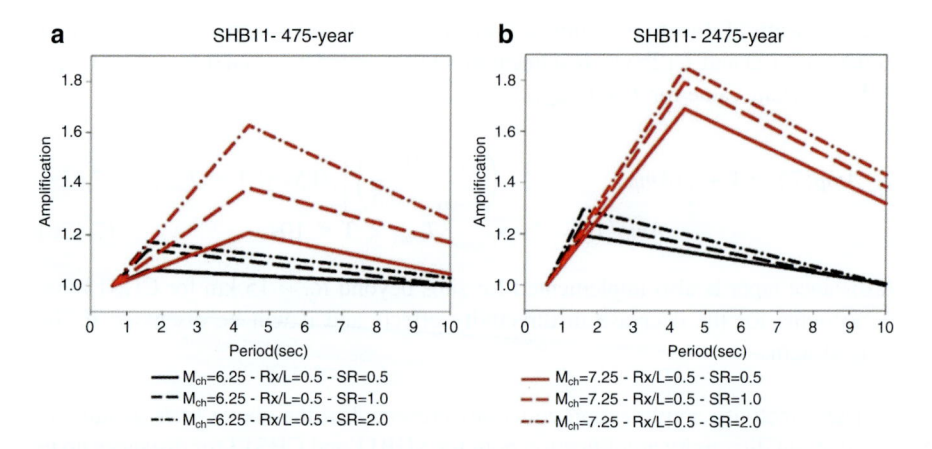

Fig. 7.8 475-year and 2475-year directivity spectral amplifications based on SHB11 model for fictitious strike-slip faults of M_{ch} 6.25 and M_{ch} 7.25 for $\dot{s} = 0.5$ cm/year, $\dot{s} = 1.0$ cm/year and $\dot{s} = 2.0$ cm/year (Modified from Moghimi 2017)

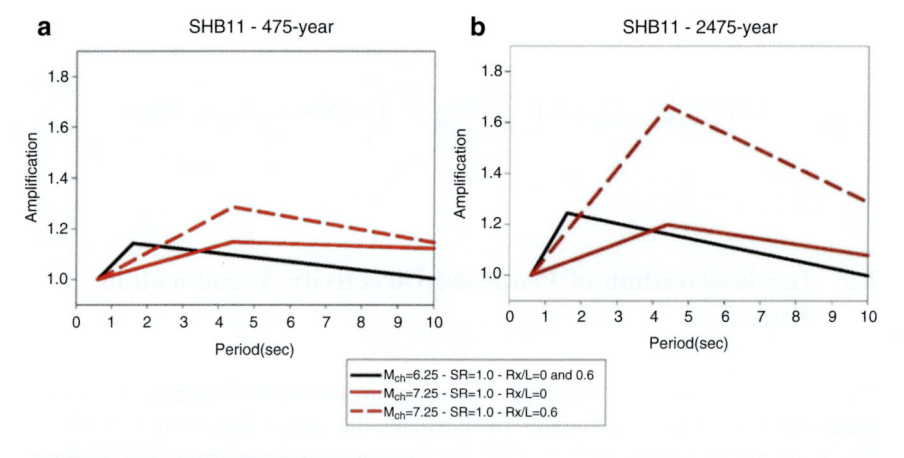

Fig. 7.9 475-year and 2475-year directivity spectral amplifications based on SHB11 model when $\dot{s} = 1.0$ cm/year fictitious strike-slip faults rupture with characteristic magnitudes of either M_{ch} 6.25 or M_{ch} 7.25 (Modified from Moghimi 2017)

amplifications do not change for M_{ch} 6.25 as our directivity rules assume indifferent spatial variation of directivity amplification at small magnitude events. (Side note: the smallest characteristic magnitude considered in our directivity rules is M_{ch} 6.25 that is approximately the lower bound limit of narrow-band directivity models used in this study). Complementary to the above discussion the directivity amplifications attain higher values for $R_x/L = 0.6$ as the characteristic magnitude and return periods attain larger values.

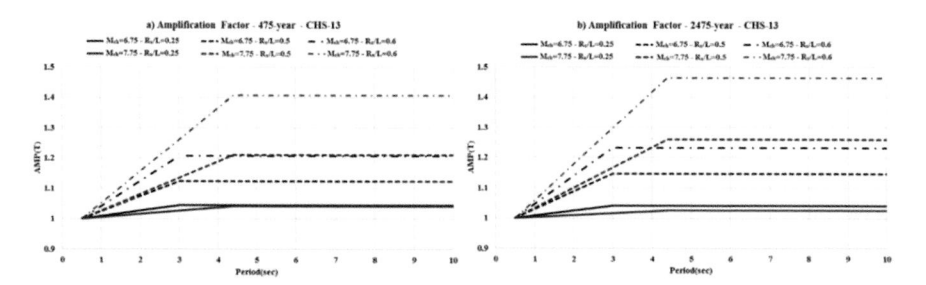

7 Implementation of Near-Fault Forward Directivity Effects in Seismic Design Codes

Fig. 7.10 Implementation of CHS13 based directivity amplification rules (**a**) 475-year return period and (**b**) 2475-year return period for fictitious strike-slip faults of M_{ch} 6.25 and M_{ch} 7.25 at $R_x/L = 0.25$, 0.5 and 0.6 ($\dot{s} = 2$ cm/year)

Figure 7.10 shows the 475-year and 2475-year return period directivity amplifications for the CHS13 based directivity rule plotted for M_{ch} 6.75 and M_{ch} 7.25 at three site locations ($R_x/L = 0.25$, 0.5 and 0.6). In all cases, the displayed directivity amplifications are valid for distances up to $R_y = 15$ km. As it is depicted from this figure, the site location and M_{ch} can significantly affect the directivity amplifications. The corner periods shift towards longer periods with increasing M_{ch} that also results in increased directivity amplifications. As R_x/L attains values closer to 0.6, the directivity amplifications increase. Inherently, the larger return periods (2475-year vs. 475-year return periods in this case) result in larger directivity amplifications. These observations are similar to those highlighted for the SHB11 based directivity rules. The difference is in the period-dependent directivity amplification trends as well as their amplitudes that originate from different horizontal component definitions as well as the modeling approaches between Shahi and Baker (2011) and Chiou and Spudich (Chapter 6 in Spudich et al. 2013).

7.6 Summary and Conclusions

This study describes the directivity amplification rules for strike-slip faults that are based on the observations of a comprehensive PSHA scenarios. The probabilistic scenarios considered the most recent narrow-band directivity models of Shahi and Baker (2011) and Chiou and Spudich (Chapter 6 in Spudich et al. 2013). The use of two different directivity models lead to a better understanding of modeling uncertainty while estimating the spectral amplifications due to near-fault directivity effects. The proposed directivity rules can be implemented to code design spectra for return periods of 475-year and 2475-year. These two return periods are frequently implemented by the current modern seismic codes for design and performance assessment of structures.

Acknowledgments This study is financially supported by the Scientific and Technical Council of Turkey with an award number 113 M308. The authors express their sincere gratitude for this support provided by the Scientific and Technical Research Council of Turkey (TUBITAK).

References

Abrahamson NA, Silva WJ (1997) Empirical response spectral attenuation relationships for shallow crustal earthquakes. Seismol Res Lett 68:94–127

Akkar S, Moghimi S, Arıcı Y (2018) A study on major seismological and fault-site parameters affecting near-fault directivity ground-motion demands due to strike-slip faulting for their possible inclusion in seismic design codes. Soil Dyn Earthq Eng 104:88–105

Alavi B, Krawinkler H (2001) Effects of near-fault ground motions on frame structures. The John A. Blume Earthquake Center Report No. 138, Department of Civil and Environmental Engineering, Stanford University

Alavi B, Krawinkler H (2004) Behavior of moment-resisting frame structures subjected to near-fault ground motions. Earthq Eng Struct Dyn 33:687–706. https://doi.org/10.1002/eqe.369

Baker JW (2007) Quantitative classification of near-fault ground motions using wavelet analysis. Bull Seismol Soc Am 97:1486–1501. https://doi.org/10.1785/0120060255

Bertero V, Mahin S, Herrera R (1978) A seismic design implications of near-fault San Fernando earthquake records. Earthq Eng Struct Dyn 6:31–42. https://doi.org/10.1002/eqe.4290060105

Boore DM (2010) Orientation-independent, nongeometric-mean measures of seismic intensity from two horizontal components of motion. Bull Seismol Soc Am 100:1830–1835. https://doi.org/10.1785/0120090400

Boore DM, Atkinson G (2008) Ground-motion prediction equations for the average horizontal component of PGA, PGV, and 5%-damped PSA at spectral periods between 0.01 s and 10.0 s. Earthq Spectra 24:99–138

Boore DM, Watson-Lamprey J, Abrahamson NA (2006) Orientation independent measures of ground motion. Bull Seismol Soc Am 96:1502–1511. https://doi.org/10.1785/0120050209

Bozorgnia Y, Abrahamson NA, Al-Atik L, Ancheta TD, Atkinson GM, Baker JW et al (2014) NGA-West2 research project. Earthq Spectra 30:973–987. https://doi.org/10.1193/072113EQS209M

CALTRANS (2013) Seismic design criteria. California Department of Transportation: Sacramento. Version 1.7

Chai JF, Loh CH, Chen CY (2000) Consideration of the near fault effect on seismic design code for sites near the Chelungpu fault. J Chin Inst Eng 23:447–454

Chioccarelli E, Iervolino I (2010) Near-source seismic demand and pulse-like records: a discussion for L'Aquila earthquake. Earthq Eng Struct Dyn 39:1039–1062. https://doi.org/10.1002/eqe.987

Chiou BSJ, Youngs RR (2014) Update of the Chiou and Youngs NGA model for the average horizontal component of peak ground motion and response spectra. Earthq Spectra 30:1117–1153. https://doi.org/10.1193/072813EQS219M

Gupta A, Krawinkler H (1999) Seismic demands for performance evaluation of steel moment resisting frame structures. Report No. 132, Stanford University, Stanford, CA, (132)

Hall JF et al (1995) Near-source ground motion and its effects on flexible buildings. Earthq Spectra 11:569–605

Li X et al (2007) Response spectrum of seismic design code for zones lack of near-fault strong earthquake records. Acta Seismologica Sismica Engl Ed 20:447–453

Mavroeidis GP, Papageorgiou AS (2003) A mathematical representation of near-fault ground motions. Bull Seismol Soc Am 93(3):1099–1131

Moghimi S (2017) Addressing the near-fault directivity effects for their implementation to design spectrum. PhD dissertation, Department of Civil Engineering Middle East Technical University, Ankara Turkey

New Zealand Standard (NZS) 1170.5:2004 (2004) Structural design actions part 5: earthquake actions. New Zealand – Commentary

Shahi SK, Baker JW (2011) An empirically calibrated framework for including the effects of near-fault directivity in probabilistic seismic hazard analysis. Bull Seismol Soc Am 101:742–755. https://doi.org/10.1785/0120100090

Somerville PG (2003) Magnitude scaling of the near fault rupture directivity pulse. Phys Earth Planet Inter 137:201–212. https://doi.org/10.1016/S0031-9201(03)00015-3

Somerville PG, Smith NF, Graves RW, Abrahamson NA (1997) Modification of empirical strong ground motion attenuation relations to include the amplitude and duration effects of rupture directivity. Seismol Res Lett 68:199–222. https://doi.org/10.1785/gssrl.68.1.199

Spudich P, Bayless JR, Baker J, Chiou BSJ, Rowshandel B, Shahi S, Somerville P (2013) Final report of the NGA-West2 directivity working group. Pacific Engineering Research Center Report. PEER Report 2013/09. NGA West 2 Database

Spudich P, Rowshandel B, Shahi SK, Baker JW, Chiou BSJ (2014) Comparison of NGA-West2 directivity models. Earthquake Spectra 30:1199–1221. https://doi.org/10.1193/080313EQS222M

Tothong P, Cornell CA, Baker JW (2007) Explicit directivity-pulse inclusion in probabilistic seismic hazard analysis. Earthquake Spectra 23:867–891. https://doi.org/10.1193/1.2790487

Uniform Building Code. (UBC) (1997) Uniform Building Code. International Conference Building Officials, Whittier

Wells DL, Coppersmith KJ (1994) New empirical relationships among magnitude, rupture length, rupture width, rupture area, and surface displacement. Bull Seismol Soc Am 84:974–1002

Yaghmaei-Sabegh S, Mohammad-Alizadeh H (2012) Improvement of Iranian seismic design code considering the near-fault effects. Int J Eng 25:147–158. https://doi.org/10.5829/idosi.ije.2012.25.02c.08

Youngs RR, Coppersmith KJ (1985) Implications of fault slip rates and earthquake recurrence mtpfdodels to probabilistic seismic hazard estimates. Bulletin of the Se

Chapter 8
3D Physics-Based Numerical Simulations: Advantages and Current Limitations of a New Frontier to Earthquake Ground Motion Prediction. The Istanbul Case Study

Roberto Paolucci, Maria Infantino, Ilario Mazzieri, Ali Güney Özcebe, Chiara Smerzini, and Marco Stupazzini

Abstract In this paper, an overview is presented to motivate the use of 3D physics-based numerical simulations of seismic wave propagation to support enhanced Probabilistic Seismic Hazard Assessment. With reference to the case study of Istanbul, we introduce the activities required to construct a numerical model of the surface geology and topography and to determine the input conditions to trigger future earthquakes in a physically sound way. Owing to the intrinsic frequency limitations of the numerical simulations, a post-processing technique to produce realistic broadband waveforms is introduced, allowing to correlate short-period to long-period spectral ordinates from an Artificial Neural Network. Finally, the results obtained in Istanbul from numerous physics-based ground motion scenarios of $M7+$ earthquakes allow us to throw light on the potential added value to PSHA of the 3D numerical simulations. Namely, to provide locally constrained probabilistic distributions of ground motion intensity measures, matching the actual footprint of a large earthquake in the specific area under study.

8.1 Introduction

Empirical ground motion prediction equations (GMPE) and 3D physics-based numerical simulations (3DPBNS) are generally presented as alternative tools for earthquake ground motion prediction and for its application to seismic hazard

---

R. Paolucci (✉) · M. Infantino · A. G. Özcebe · C. Smerzini
Department of Civil and Environmental Engineering, Politecnico di Milano, Milan, Italy
e-mail: roberto.paolucci@polimi.it

I. Mazzieri
Department of Mathematics, Politecnico di Milano, Milan, Italy

M. Stupazzini
MunichRe, Munich, Germany

Table 8.1 Advantages and limitations of GMPEs and 3DPBNSs

	PROs	CONs
GMPE	Ease-of-use	Lack of records to solve important conditions, such as near-source and complex geological environments
	Calibrated on records	Only peak values of motion
	Adapted to different tectonic environments and site conditions	Recalibration when new data are available
		No correlation of ground motion intensities among multiple sites and among different spectral periods
3DPBNS	Flexibility to produce synthetics in arbitrary site and source conditions	High-frequency computational and modelling limit
	Parametric analyses allowed	High computational costs
	Spatial correlation of simulated ground motion	Need of expert users
	Insight into the earthquake physics	Hardly available information to construct a reliable 3D model
		Large epistemic uncertainties
		Few well documented validation case studies on real earthquakes

assessment studies. While the use of GMPEs is well consolidated, in the framework of both probabilistic and deterministic studies, 3DPBNS seem to be still confined to a relatively restricted range of applications, where earthquake ground motion scenarios are produced in an almost deterministic way.

Such dichotomy is obstructive and does not allow to fully exploit those techniques, shading lights on their limitations on one hand and, on the other hand, on their potential advantages to produce more reliable results, as summarized in Table 8.1. More specifically, the limitation of GMPEs not to be sufficiently well calibrated for those conditions, such as large earthquake magnitude, near-source, soft soil sites, complex geological irregularities, that typically govern seismic hazard at a site, decreases the reliability of probabilistic seismic hazard assessment (PSHA) results, typically based on GMPE application as a tool for ground motion prediction. Moreover, 3DPBNS have not been sufficiently developed yet to yield a consensus on the engineering applicability of their broad-band results, also because, in spite of the ongoing progress in the recent years, there are still relatively few cases of fully worked out validation exercises on real earthquake case studies and comparison with records.

In Fig. 8.1, the typical conditions for which GMPEs and 3DPBNSs should be considered in a PSHA study are sketched: on one hand, GMPE provide reliable results when the source-to-site distance is sufficiently large, e.g., at least larger than the size of the fault, and no complex geological conditions are present, while, on the other hand, the vicinity to the source and the complex geological conditions should lead to the selection of 3DPBNS as the main tool for earthquake ground motion prediction.

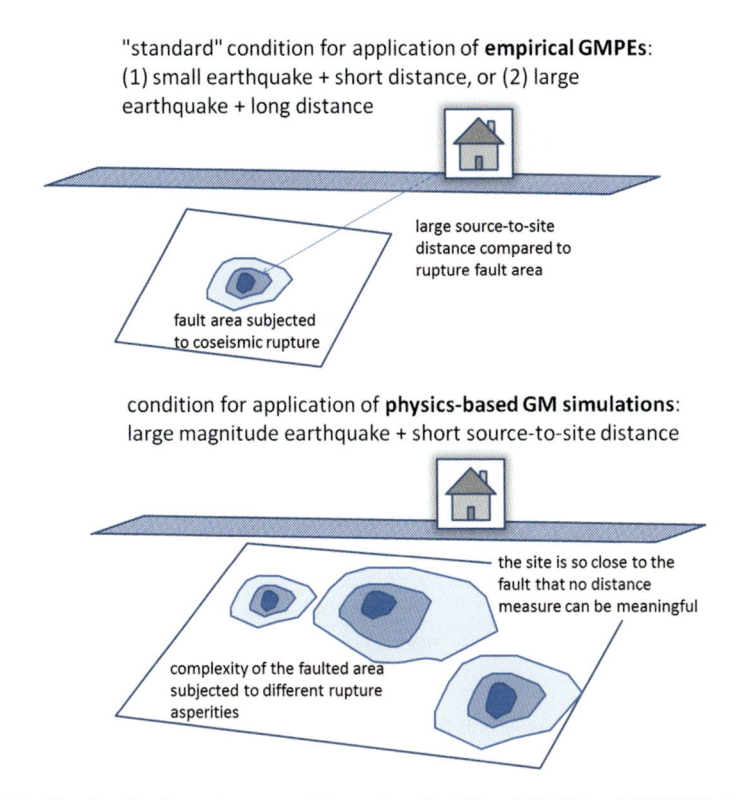

8 3D Physics-Based Numerical Simulations: Advantages and Current... 205

Fig. 8.1 Sketches for the optimum conditions of applicability of GMPE and 3DPBNS, depending on the distance of the site to the seismic source

To capture the potential drawbacks of fully relying on GMPEs for PSHA, typically expressed in terms of uniform hazard spectra at rock sites, it is worth considering Fig. 8.2, where the Peak Ground Acceleration (PGA) values from the NGA 2014 dataset (Ancheta et al. 2014) are extracted, for records with magnitude $M > 6$ and R_{JB} (i.e., distance from the surface projection of fault) < 20 km, and plotted as a function of $V_{S,30}$ (i.e., average shear wave velocity in the top 30 m according to seismic norms). The scarcity of suitable records for rock conditions (e.g., $V_{S,30} > 800$ m/s) is evident, as well as the consideration that some of such records are obtained in conditions (such as the Pacoima and Lexington dam in Fig. 8.2) far away from the ideal reference free-field rock.

The Marmara Sea region is an ideal area were the potential advantages and limitations of GMPEs and 3DPBNSs can be tested, because of coupling the high seismic hazard related to the major seismic gap on the North Anatolian Fault, expected to produce Magnitude 7+ earthquakes (for a comprehensive review see, among many others, Bohnhoff et al. 2013; Akinci et al. 2017; Aochi et al. 2017), with the huge seismic risk exposure of Istanbul.

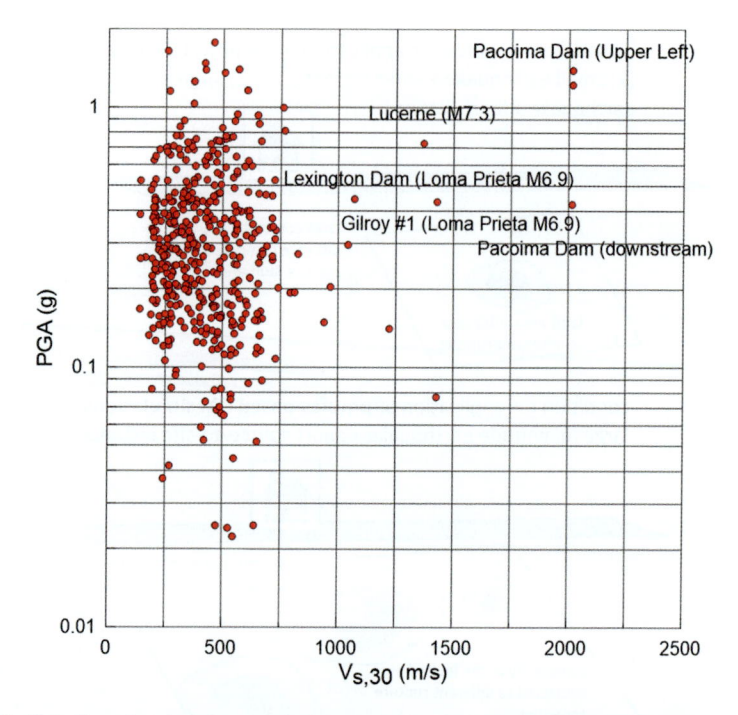

Fig. 8.2 Selection of earthquake records from the NGA 2014 dataset, with $M > 6$ and $R_{JB} < 20$ km, as a function of $V_{S,30}$. Some of the stations corresponding to "rock" conditions are denoted

In this paper, we aim at introducing the complete workflow of a PSHA study carried out in Istanbul, in the framework of updating the seismic hazard model for reinsurance purposes, where results of a comprehensive set of 3DPBNS were exploited to provide an enhanced seismic hazard assessment. This paper provides an overview of such study, regarding in particular:

1. construction of the numerical model for 3DPBNS;
2. simulations of different ground motion scenarios by generating realistic fault-slip distributions for different scenario earthquakes;
3. construction of broad-band synthetics through an Artificial Neural Network-based procedure;
4. input of the 3DPBNS results into a PSHA framework.

The simulations were carried out using the numerical code SPEED (http://speed.mox.polimi.it/), designed for the seismic wave propagation analysis in large areas, including the coupled effects of a seismic fault rupture, the propagation path through Earth's layers, localized geological irregularities, such as alluvial basins, and soil-structure interaction problems. Based on a discontinuous version of the classical spectral element method introduced by Faccioli et al. (1997), SPEED (Mazzieri et al. 2013) is naturally oriented to solve multi-scale numerical problems, allowing one to use non-conforming meshes (h-adaptivity) and different polynomial approximation degrees (*N*-adaptivity) in the numerical model. By taking advantage of the hybrid MPI-OpenMP parallel programming SPEED runs on multi-core computers and large clusters (e.g., Marconi at CINECA, https://www.cineca.it/en/content/marconi).

SPEED was successfully verified within the Grenoble benchmark (Stupazzini et al. 2009; Chaljub et al. 2010), validated by comparison with several among the most recent worldwide earthquakes, including L'Aquila 2009, $M_w6.3$ (Smerzini and Villani 2012; Evangelista et al. 2017), Chile 2010, $M_w5.2$ (Pilz et al. 2011), Christchurch 2011, $M_w6.2$ (Guidotti et al. 2011), Emilia 2012, $M_w6.0$ (Paolucci et al. 2015), and also applied to simulate devastating earthquakes of the past, such as the Marsica 1915, $M_w6.7$ (Paolucci et al. 2016). A repository of results of the 3DPBNS carried out by SPEED can be consulted at the web site http://speed.mox.polimi.it/.

8.2 From the Tectonic and Geological Framework to the 3D Spectral Element Model

Leaving details to Infantino (2016), we summarize here the main input data required and how they are cast into a 3D numerical model, with reference to the Istanbul case.

8.2.1 3D Geological Model

First, a geological model has been constructed, based on collection of the following data:

- digital elevation model and bathymetry, see Fig. 8.3 top;
- crustal structure, typically in form of a layered model of S and P wave velocity, V_S and V_P;
- local shallow geological structure, typically in the form of a spatial model of V_S and V_P, variable both in the horizontal and vertical direction, and possibly including the corresponding models for internal soil damping and local variation of shear modulus and damping as a function of shear strain (or, in 3D, of the second invariant of the strain tensor).

Information of the local shallow geology structure is by far the most difficult to gather in a format suitable for the 3D numerical modelling of the area. However, it should be kept in mind that the level of detail of input data should be balanced with the actual computer power limitations: on the one hand, it is useless to get small-scale details on the local geology, when the numerical mesh is bounded to resolve up to, say, 2 Hz, in order to make the number of degrees-of-freedom and the computer time affordable. On the other hand, even with unlimited computational resources, a detailed in-field survey on a vast area is very seldom available. For this reason,

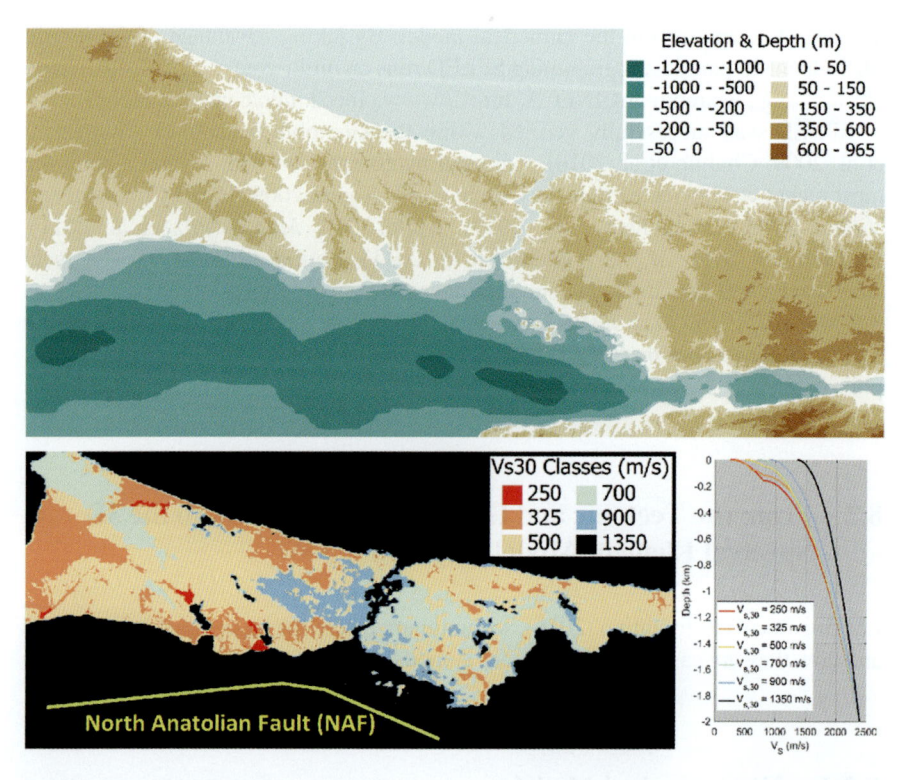

Fig. 8.3 Top: bathymetry and digital elevation model of the Marmara Sea area adopted in SPEED. Bottom: map of classes of $V_{S,30}$ based on Özgül (2011) and simplified velocity profiles for each $V_{S,30}$ class, adopted in SPEED

extending the frequency limit of 3DPBNSs beyond about 2 Hz is nowadays practically meaningless, unless focus is limited to predicting ground motions at rock or stiff soil sites.

For the Istanbul case, according to the geotechnical site characterization provided by Özgül (2011), the following procedure has been adopted to define the 3D soil model. First, the maps presented by Özgül (2011) have been digitized to obtain the distribution of $V_{S,30}$ and rock/soil classification for the whole Istanbul region. Second, by making use of three sets of data, namely, $V_{S,30}$, rock/soil map and slope information, six site classes have been assigned ranging from $V_{S,30} = 250$ m/s to $V_{S,30} = 1350$ m/s, see Fig. 8.3 bottom left and, for each class, a V_S profile has been considered, as shown in Fig. 8.3 bottom right.

Although a non-linear visco-elastic model is available in SPEED, as introduced by Stupazzini et al. (2009), this was not used in the present numerical simulations, and results presented in this work refer to the linear visco-elastic case, where quality factor correlation of ground motion intensities among multiple sites and among different spectral periods is not accounted for.

8.2.2 The Seismic Source Model

The second set of input data refers to the seismic source model. Two basic families of models exist: (1) dynamic source models, where rupture is initiated by specifying a stress perturbation within a given, more or less irregular, area of the fault to reach a yield condition and introducing suitable friction relationships between stress and the resulting fault slip (Madariaga and Olsen 2002); (2) kinematic source models, where a more or less heterogenous distribution of co-seismic slip is applied along the fault, together with a slip source function, typically in the form of a sigmoid function, with initiation time and length depending on the local rupture velocity and rise-time, respectively. As shown in Fig. 8.4, a kinematic source model, requires to input, on each node of the fault plane, both the mechanical properties of the fault material (e.g. shear modulus) and the kinematic properties for the characterization of the space-time evolution of seismic slip.

The kinematic approach is preferred in most engineering applications of 3DPBNS, because it complies with the following key features:

– it is cost-effective, i.e., it does not imply a significant increase of the computer time;
– it can be adapted to model effectively not only the low-frequency, but also the high-frequency seismic energy radiation.

The latter feature requires the spatial slip distribution along the fault, as well as the other fault parameters, such as the rise time, the peak time of the slip, the rupture velocity, to fulfill spatial correlation constraints derived from dynamic rupture simulations (see e.g., Mai and Beroza 2003; Gallovič and Brokešová 2004; Causse et al. 2009). In the SPEED code, the kinematic approaches proposed by Herrero and

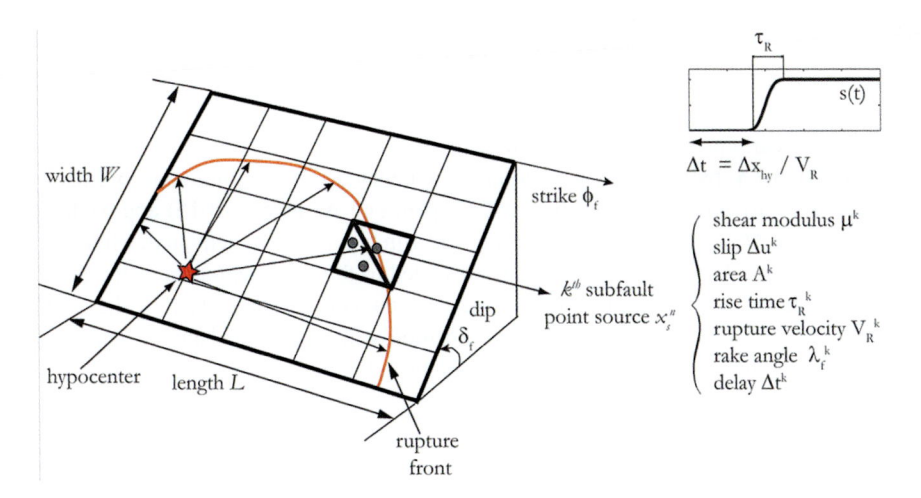

Fig. 8.4 Sketch of kinematic numerical modeling of an extended seismic source

Bernard (1994) and by Crempien and Archuleta (2015) were implemented. In the first one, the heterogeneities of the slip distribution are assumed to present a k^{-2} spectral decay in the wavenumber domain, leading to the Brune (1970) spectrum ω^{-2} fall-off in the frequency domain, while in the second one a comprehensive recipe was proposed for broadband seismograms generation based on correlation of fault parameters, complying with the SCEC validation criteria (Goulet et al. 2015).

Both kinematic approaches mentioned above are then suitable to generate broadband input motions for future earthquakes with prescribed Magnitude along a given fault, by setting random variability of the fault parameters. In Fig. 8.5, the sketch of the Marmara Sea area under study is provided, together with the North Anatolian Fault (NAF) segmentation and the slip distribution models for the sample M_w7 scenario earthquakes considered in the following. Three modes of fault rupture propagation are selected, with reference to directivity condition with respect to Istanbul: forward (scenario 1), neutral (scenario 2) and backward (scenario 3) directivity.

8.2.3 The SPEED Numerical Model of the Marmara Sea Region

Finally, the information above is condensed into a spectral element numerical model (Fig. 8.6), consisting of more than 2 million hexahedral elements and corresponding, with a spectral degree $N = 4$, to about 500 million degrees of freedom.

The spectral degree was selected in order for the maximum frequency to be accurately propagated by the numerical integration scheme to be about 1.5 Hz (using a rule of thumb of about five grid points per minimum wavelength in heterogeneous media modelled by spectral elements, according to Faccioli et al. 1997). The runs were carried out at the Fermi supercomputer (now replaced by Marconi) at CINECA, Italy, requiring about 24 h with 8192 cores for simulating $T = 60$ s of wave propagation with a time step $\Delta t = 0.001$ s.

Thirty earthquake ground motion scenarios were generated, with M_w ranging from 7.0 to 7.4, by randomly generating the kinematic slip distribution according to either Herrero and Bernard (1994) or Crempien and Archuleta (2015) approaches. For most of $M_w7.0$ earthquakes, only segment 1 in Fig. 8.5b was activated, while, for all $M_w7.2$ and 7.4 earthquakes, rupture extended to the three segments of the NAF facing Istanbul. As shown by Infantino (2016), the convex shape itself of the NAF along the Marmara Sea and its relative position with respect to Istanbul makes it more likely that a directive scenario may occur in the city.

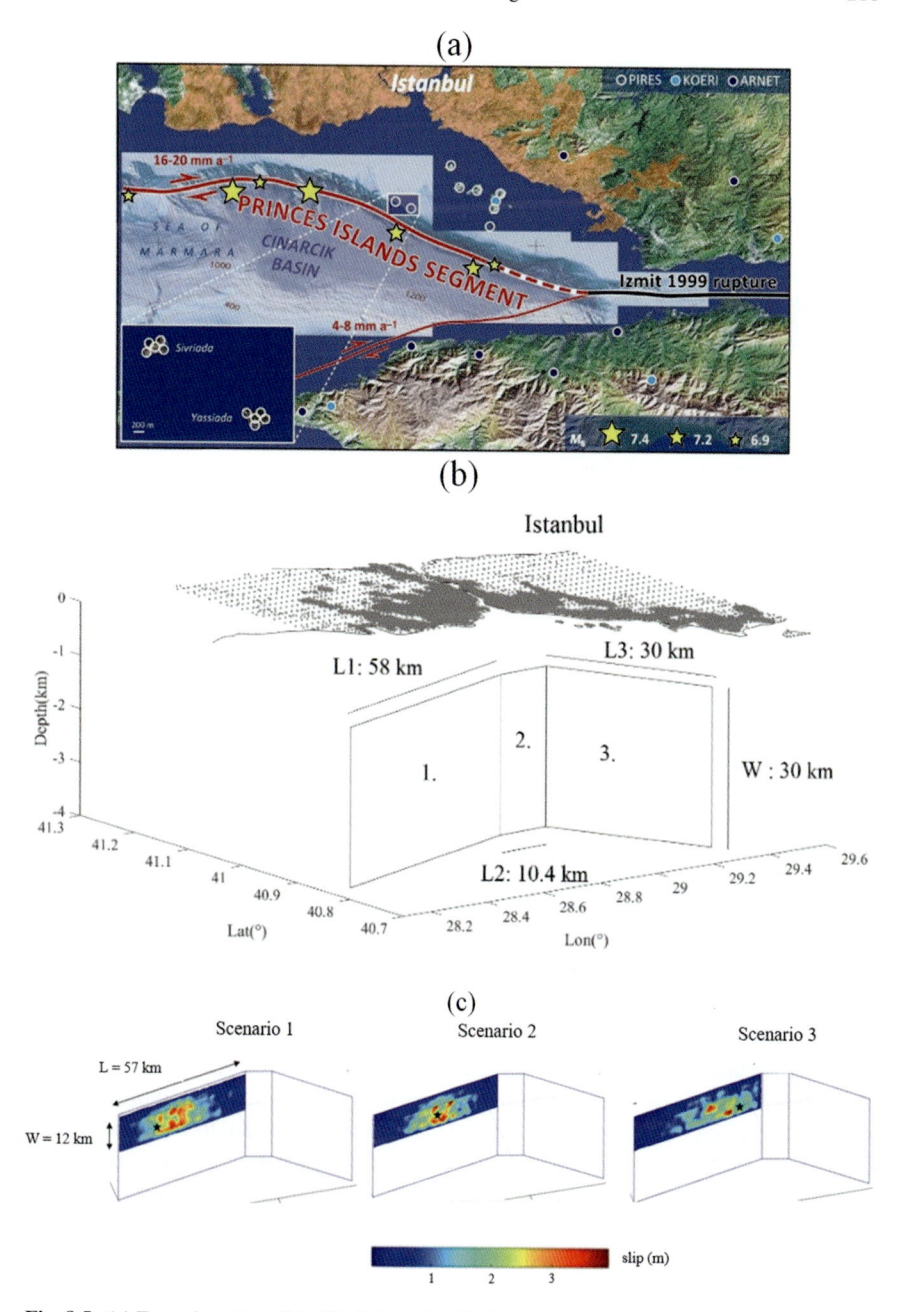

Fig. 8.5 (a) Tectonic setting of the North Anatolian Fault, reproduced from Bohnhoff et al. (2013). The fault segment investigated, the bold red line, is the eastern part of the current Marmara seismic gap. The black line marks the observed rupture zone of the 1999 Izmit event, while the dashed red line indicates its potential link with the Princes Islands segment. (b) Geometric scheme of the fault considered in this study and location of Istanbul identified by grey points. (c) Three fault slip distributions generated according to Herrero and Bernard (1994) approach for a $M_w7.0$ earthquake scenario and different directivity conditions with respect to Istanbul: forward (scenario 1), neutral (scenario 2) and backward (scenario 3) directivity. The epicenter locations are represented by black stars

Horizontally stratified crustal model		
Depth (km)	V_s (m/s)	$Q(-)$
0-5	Fig. 3	$V_s/10$
5-10	3490	350
10-20	3500	350
20-30	3920	400

Computational model	
Hexahedral elements	2.257.482
Degrees of freedom	~500 million
Frequency range	0 to 1.5 Hz
Element size range	180 m to 2 km

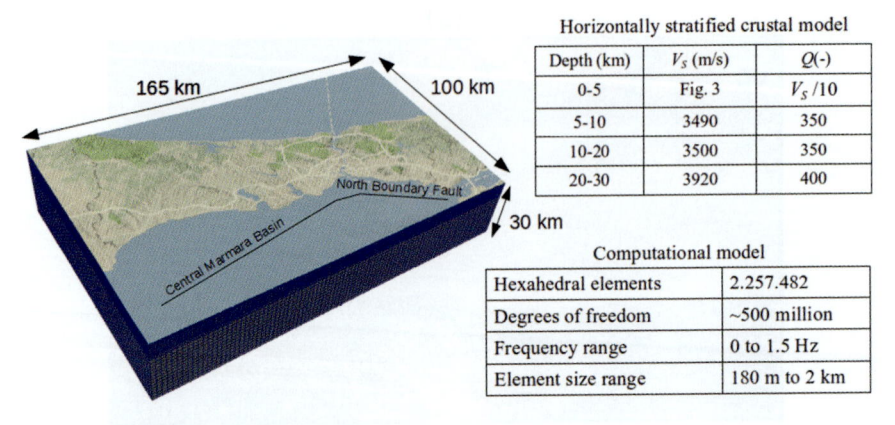

Fig. 8.6 Sketch of the SPEED numerical model

8.3 Producing Broad-Band Synthetics from 3DPBNS

One of the main drawbacks of 3DPBNS is that synthetics are reliable only in the long period range, typically above 0.75–1 s, owing to the limitations posed both by computational constraints as well as by lack of detailed data to constrain the soil model as well the source at high frequency (see overview in Paolucci et al. 2014). On the other hand, earthquake engineering applications need broad-band (referred to as BB hereafter) ground motion time histories with realistic features in a range of frequencies that is broad enough to cover the vibration characteristics of fundamental and higher modes of the structures, say from 0 to 25 Hz.

The most commonly used approach to produce BB waveforms relies on a hybrid modelling which combines in the frequency domain the low-frequency waveforms from 3DPBNS with high-frequency signals from stochastic approaches, based either or point- or finite- source modeling (e.g., Boore 2003; Motazedian and Atkinson 2005) or Empirical Green's function (e.g., Kamae et al. 1998; Mai et al. 2010). The main disadvantages of such an approach, especially when applied for hazard assessment at regional scale, are the lack of correlation between the low and high frequency parts of ground motion and the strong impact of the choice of the transition frequency over which the deterministic and stochastic parts of the Fourier spectrum are glued.

In this work, an approach based on Artificial Neural Networks (ANN), referred to as ANN2BB, has been adopted. Referring to Paolucci et al. (2018) for a thorough introduction of the method and verification tests against real case studies, we summarize herein its key points. Denoting by T^* the minimum period of the 3DPBNS-based synthetics, the ANN2BB approach consists of the following steps:

(a) an ANN is trained based on a strong motion records dataset (namely, SIMBAD, see Smerzini et al. 2014), separately for horizontal (geometric mean) and vertical components. The ANN allows to predict short period horizontal/vertical spectral ordinates ($T < T^*$) taking as input the long period ones obtained from the 3DPBNS ($T \geq T^*$);
(b) for each simulated waveform, a target ANN2BB response spectrum is generated by application of the previously trained ANN: therefore, the resulting spectrum is equal to that simulated by 3DPBNS at long periods, while at short periods it consists of the ANN outputs;
(c) a hybrid 3DPBNS-stochastic modeling is applied to inject high-frequency in the simulated low-frequency waveform and, hence, to make it usable for the following step;
(d) finally, the hybrid waveform is iteratively modified in the frequency domain, with no phase change, until its response spectrum matches the target ANN2BB spectrum.

In Paolucci et al. (2018) it has been demonstrated by comparison with earthquake observations that this approach allows: (1) to obtain realistic waveforms both in time and frequency domains, in line with earthquake observations; (2) to predict maps of short-period peak values of ground motion incorporating those effects reproduced by the physics-based simulations, such as source directivity/directionality and complex basin effects; (3) to preserve the spatial correlation features of ground motion, of particular interest for seismic hazard assessment at regional scale.

The ANN2BB approach was used to process all ground motion scenarios produced by SPEED. In Fig. 8.7 a sample of results is shown, consisting of the NS acceleration (top) and velocity (bottom) time histories at four sites in the Istanbul area, for the three selected scenarios (see Fig. 8.5). Realistic time histories can be appreciated, together with the consistency of arrival times and dependency of simulated ground motion on the directivity features of the rupture scenario, with largest values corresponding to Scenario 1, where the largest asperities are aligned along the pathway from the hypocenter to Istanbul.

In Fig. 8.8, the effect of ANN2BB post-processing is illustrated in detail, with reference to the Scenario 2 synthetics at the Ayasofya site. It is clear that processing enriches the high frequency portion of the waveform, while keeping the medium-to-low frequency portions practically invariant, as can be seen by the velocity and displacement time histories.

Finally, we show in Fig. 8.9 the maps of horizontal (geometric mean) peak ground velocity (PGV) for the three considered scenarios, together with the median PGV map for the whole set of $M_w 7$ scenarios, as well as the median PGV map

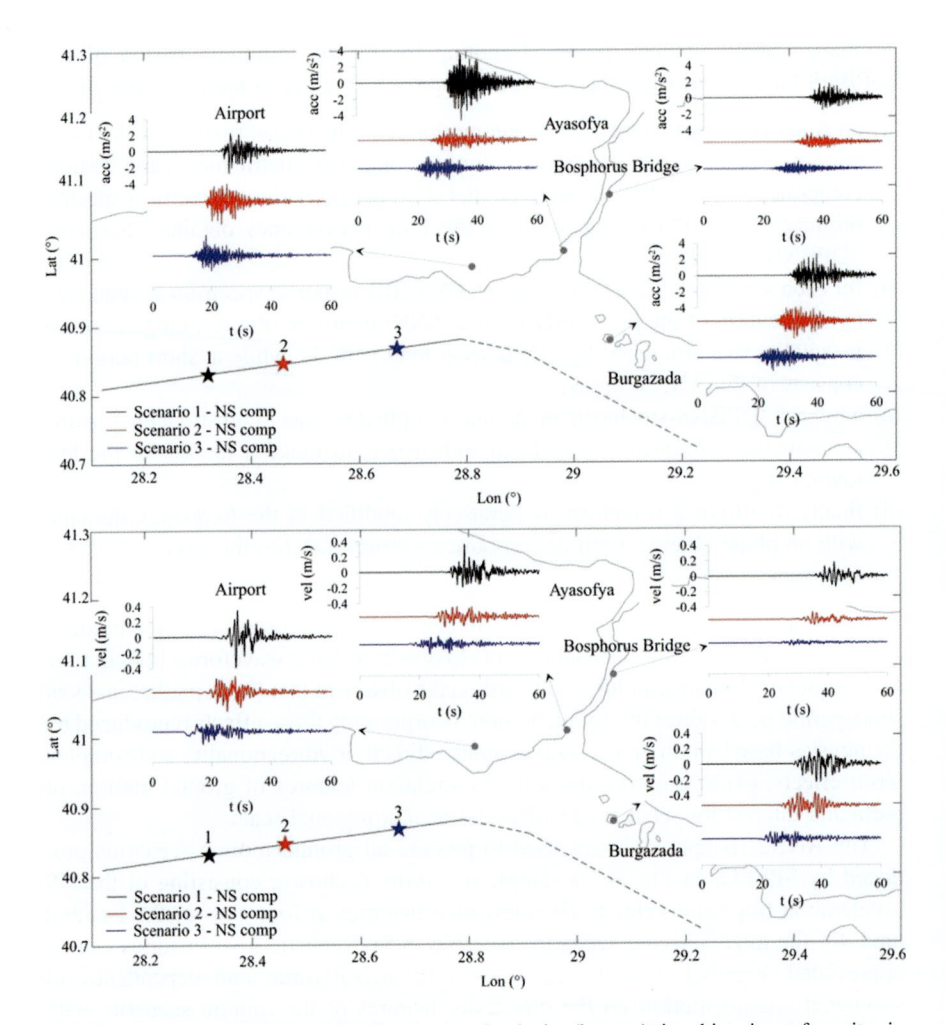

Fig. 8.7 NS components of acceleration (top) and velocity (bottom) time histories at four sites in Istanbul, obtained with three different earthquake scenarios (1, 2 and 3) of $M_w7.0$ identified by different colors. The fault is shown with a continuous black line while the epicenter positions are represented by stars. Slip distributions are illustrated in Fig. 8.4

according to the GMPE proposed by Cauzzi et al. (2015), referred to as CAEA15. It can be observed that there is a good agreement in terms of median values provided by the GMPE for the M_w7 scenarios, with the median map approaching the neutral directivity Scenario 2, especially in the western part of Istanbul. It should be noted that, as discussed by Infantino (2016), such agreement is not found any more for

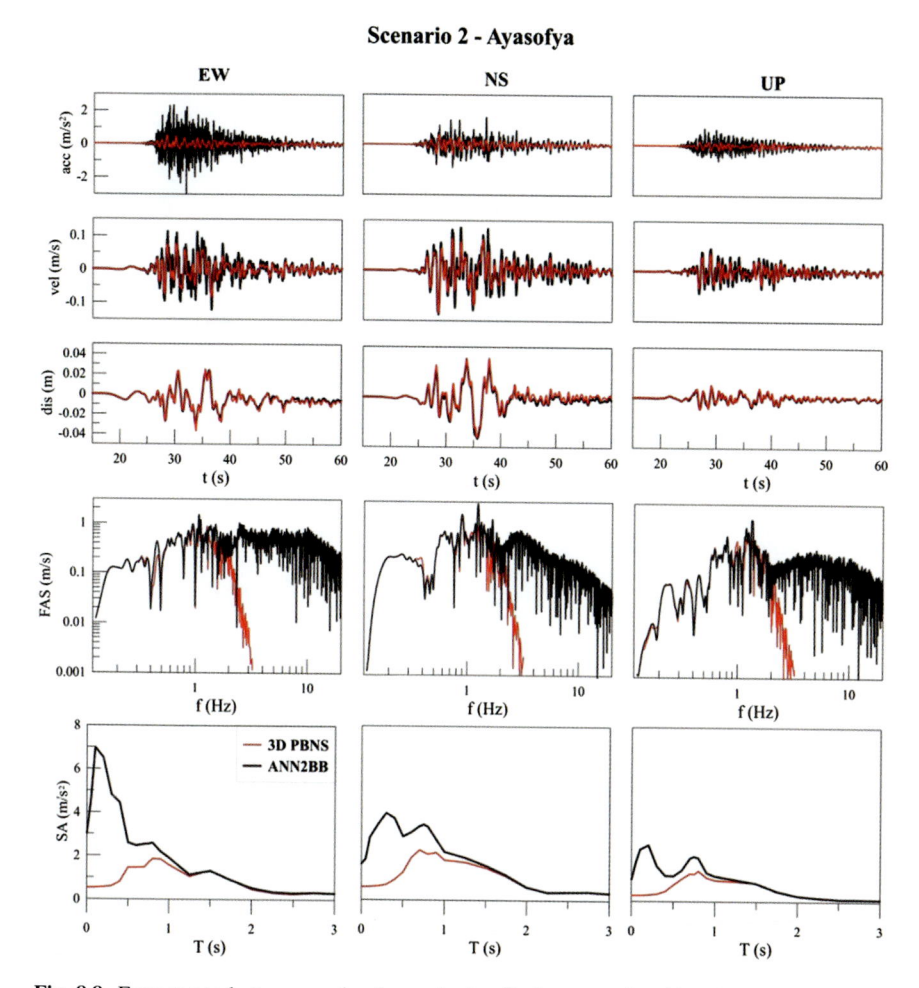

Fig. 8.8 From top to bottom: acceleration, velocity, displacement time histories, Fourier Amplitude Spectra and Response Spectra for the three components at the site of Ayasofya obtained with the 3DPBNS (red line) and ANN2BB (black line) for Scenario 2 (Fig. 8.5b)

those scenarios involving rupture along the three segments of the NAF (M_w7.2 and 7.4), for which only the incorporation of forward directivity effects in GMPEs (such as Bray and Rodriguez-Marek 2004) can explain the high ground motion values obtained in Istanbul area.

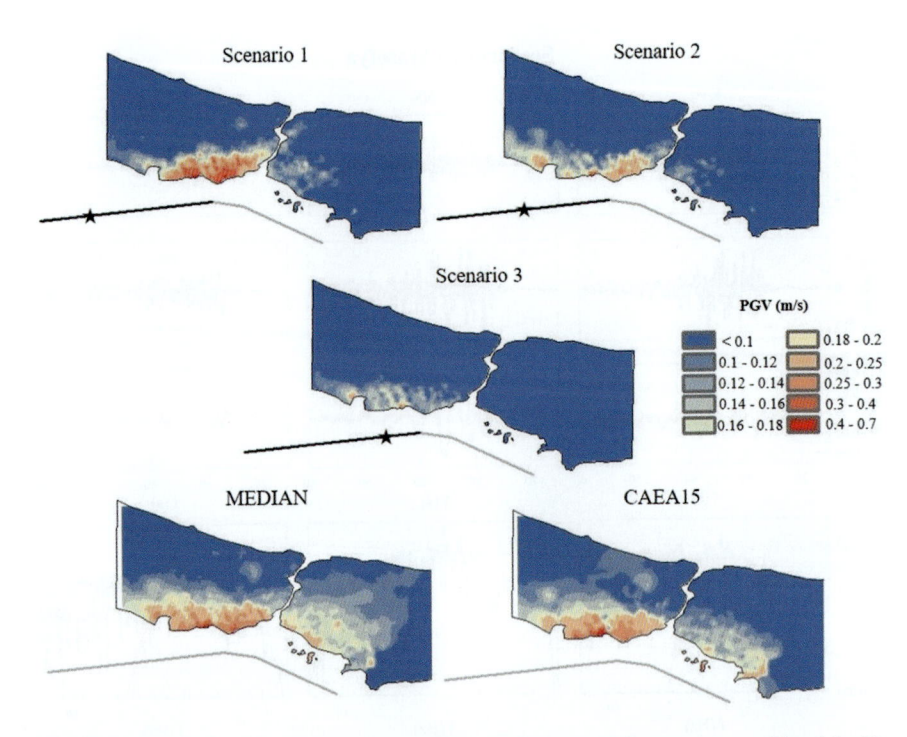

Fig. 8.9 Top: horizontal (geometric mean) PGV maps of the $M_w 7$ scenarios considered in this work. Bottom: median PGV map based on all the 30 $M_w 7$ scenarios (left) compared with the corresponding map based on the GMPE by Cauzzi et al. (2015) – CAEA15

8.4 How to Take Advantage of 3DPBNS in the Framework of Probabilistic Seismic Hazard Assessment

In a nutshell, probabilistic seismic hazard assessment (PSHA) at a site can be set in the form

$$[IM > x] = \sum_{j=1,N} Prob[IM > x|scenario_j] Prob[scenario_j] \quad (8.1)$$

where IM is a given ground motion intensity measure. In the classical PSHA, the N scenario earthquakes may occur throughout a seismic zone with annual probability $Prob[scenario_j]$ given by a frequency-Magnitude relationship set in the form of the classical Gutenberg-Richter relationship, or other forms, including e.g. the characteristic earthquake model.

The ground motion attenuation model is defined within the first factor at the right-hand side of eq. (8.1), and it is typically provided through a GMPE. Therefore, in the classical PSHA, eq. (8.1) is typically expressed in the form, referred for simplicity to a single seismic zone:

$$Prob[IM > x] = \iiint Prob[IM > x|m,r,\varepsilon] f_M(m) f_R(r) f_\varepsilon(\varepsilon) dm dr d\varepsilon \quad (8.2)$$

where $f(\cdot)$ denotes a probability density function, R is the distance from the site of a small area within the seismic zone where a magnitude M earthquake occurs, and ε is the spread of the GMPE.

In principle, moving from a GMPE-based to a 3DPBNS-based PSHA implies only the different evaluation of the term $Prob[IM > x|m,r,\varepsilon]$. In the first case, this term is computed based on the ergodic assumption that the probability distribution of *IM* can be obtained based on statistical evaluation of datasets of strong motion records from other regions of the world with similar tectonic framework. In the second case, it is computed by repetition of 3DPBNS of a given scenario earthquake at the site, in a number sufficient for a reliable evaluation of the probability distribution. For simplicity of notation, we will denote by *PSHAe* the *enhanced* seismic hazard assessment based on 3DPBNS.

Although *PSHAe* provides an obvious advantage to the standard GMPE, as already discussed in the Introduction, there are few examples in the literature of practical applications of such *PSHAe* (see e.g., Convertito et al. 2006; Graves et al. 2011; Villani et al. 2014), especially because of the large computational effort implied by computing the term *Prob [IM > x|scenario]*, extended to all potential scenarios, from low to large magnitude earthquakes.

For this reason, it is wise to limit application of the *PSHAe* to those contexts where seismic hazard is dominated by few near-source scenarios, for example a characteristic earthquake from a known seismic fault, with a relatively narrow range of possible magnitudes. The resulting framework for *PSHAe* may be set as in Fig. 8.10, where GMPE-based and 3DPBNS-based PSHA are combined and applied selectively in their suitable ranges of magnitude.

Two different approaches may be envisaged for application of the *PSHAe*. With reference to Fig. 8.11, the histograms at four sites in the Istanbul area are considered, showing the frequency distribution of computed PGV values according to 30 realizations through 3DPBNS of a $M_w 7$ event along the North Anatolian Fault segment considered in this work. Frequency histograms are compared with the lognormal distribution, fit on

Fig. 8.10 A frequency-magnitude relationship explaining the range of applicability of different ground motion prediction models in the framework of an enhanced PSHA

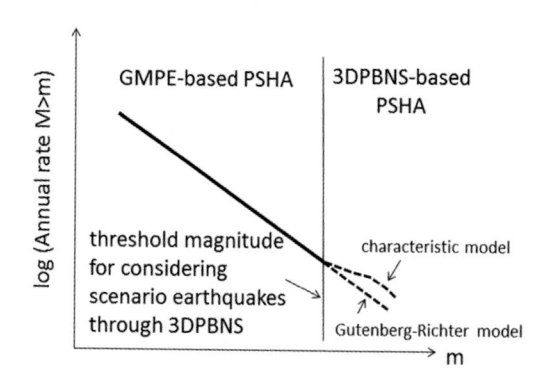

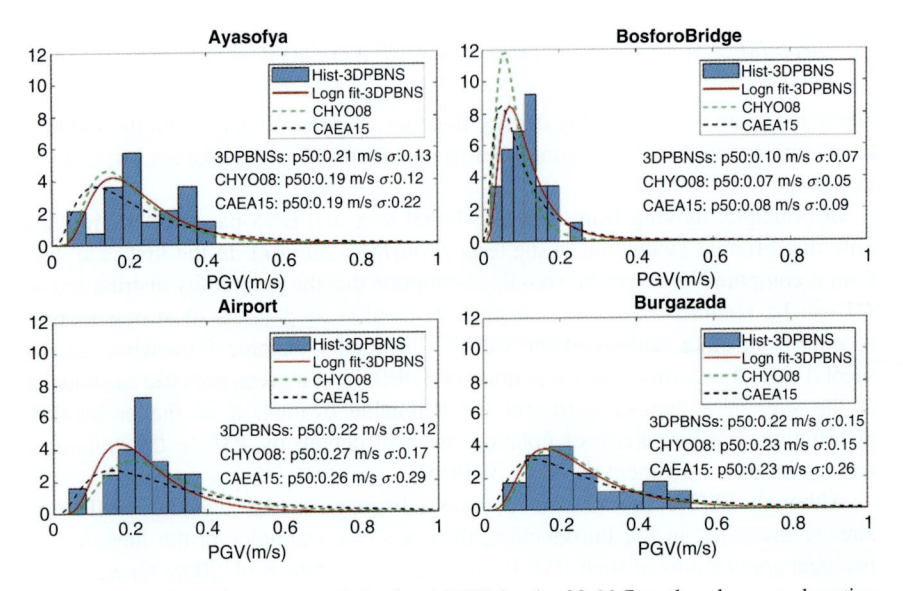

Fig. 8.11 Frequency histograms of simulated PGV for the 30 M_w7 earthquake ground motion scenarios simulated by SPEED at the four Istanbul sites considered in this study. Superimposed are the lognormal models based on the GMPEs by Cauzzi et al. (2015) and Chiou and Youngs (2008)

the numerical results, as well as the corresponding distribution from two GMPEs (Cauzzi et al. 2015 – CAEA15; Chiou and Youngs 2008 – CHYO08).

The first approach was called *high-resolution PSHA* by Villani et al. (2014) and consists, for the considered scenario earthquake, of replacing the moments of the lognormal distribution from the GMPE with those obtained from the 3DPBNS at each site of interest. This approach was implemented in the CRISIS software (Ordaz et al. 2013) through the so-called *generalized attenuation functions* (*GAF*). Of course, this *GAF-based PSHAe* is expected to provide different results from the GMPE-based PSHA, if the resulting probability distributions are significantly different.

However, comparing the frequency histograms with the probability density functions, it turns out that maximum PGVs from 3DPBNS are bounded, while those resulting from the extrapolation based on the probability model are not. This poses a well-known problem in the PSHA: whether or not bounded upper limits for the intensity ground motion measures should be placed in the PSHA (e.g., Abrahamson 2000; Bommer et al. 2004; McGarr and Fletcher 2007). Especially for the rare events implied by long return periods, integration along an unbounded range of variability of the ground motion parameters may provide in some cases very (and possibly unrealistically) conservative results. 3DPBNS may be helpful to provide site-specific limits for such upper bounds, if a sufficiently large set of realizations of the scenario is considered.

These arguments support consideration of a different approach for *PSHAe*, which takes full advantage of the 3DPBNS results, without postulating a probability model for ground motion. In this *footprint-based PSHAe* approach (Stupazzini et al. 2015),

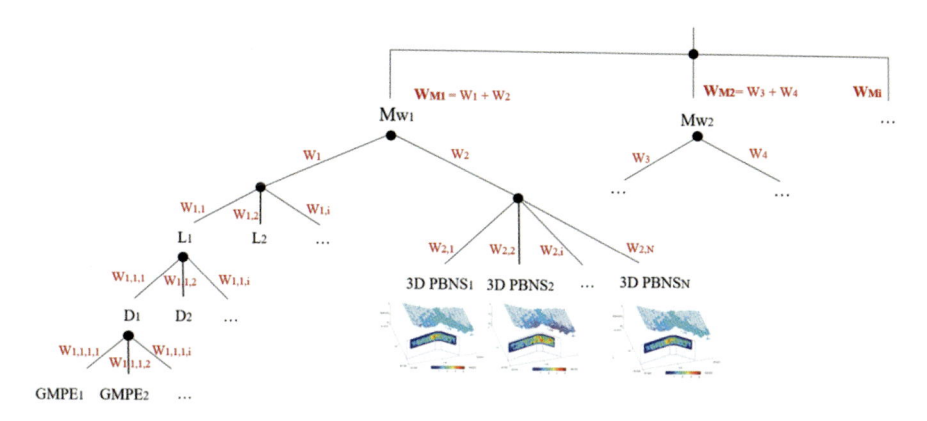

Fig. 8.12 Integration of the 3DPBNSs into a logic tree approach according to Stupazzini et al. 2015 (M_w: magnitude, L_j: location, D_j: depth, $GMPE_j$: ground motion prediction equation, $3DPBNS_j$: 3D physics-based scenario, w_j: weight)

all realizations of the scenario earthquake are considered within a logic-tree framework, each with the same weight (Fig. 8.12). This means that the probability distribution will be the same as provided by the frequency histogram, avoiding the extrapolation to assign non-zero probabilities to values of *IM* not resulting from the 3DPBNS. Furthermore, another potential added value of such an approach is that the spatial correlation structure of the results of PSHA is better maintained than with other approaches, since it does not suffer of the smoothing effects of the assumed probability model.

Figure 8.13 illustrates the results of the *footprint-based PSHAe* based only on 3DPBNSs (i.e. assigning zero weight to the GMPE branch, see Fig. 8.12) in terms of horizontal (geometric mean) PGV hazard maps, for the return periods $T_R = 475$ and 975 years, compared with the corresponding hazard maps obtained using the GMPEs by Chiou and Youngs (2008), Cauzzi et al. (2015) and Bray and Rodriguez-Marek (2004). The latter was calibrated only adopting nearfield accelerograms presenting clear forward-directivity effects. Therefore, the predicted PGV tends to differ substantially from the prediction of Chiou and Youngs (2008) and Cauzzi et al. (2015). Another peculiar feature of the Bray and Rodriguez-Marek (2004) work, is that it distinguishes only between rock and soil and therefore it takes into account the variation of $V_{S,30}$ only up to a limited extent. This explains the more homogeneous maps obtained with this GMPE.

It is worth noting that, for $T_R = 475$ years, the hazard map based on 3DPBNSs is lower than that obtained adopting Bray and Rodriguez-Marek (2004), while it agrees relatively well with the hazard map obtained using the other two GMPEs. On the contrary, for $T_R = 975$ years, the hazard map obtained adopting Bray and Rodriguez-Marek (2004) shows amplitudes of ground shaking similar to the hazard map produced by 3DPBNSs, while the two hazard maps based on the other GMPEs are systematically lower. As already mentioned and discussed in Infantino (2016), these results are due to the lack of back-directivity events of large magnitude (i.e. M_w 7.2 and 7.4), associated with the convex shape of the fault trace and its position with

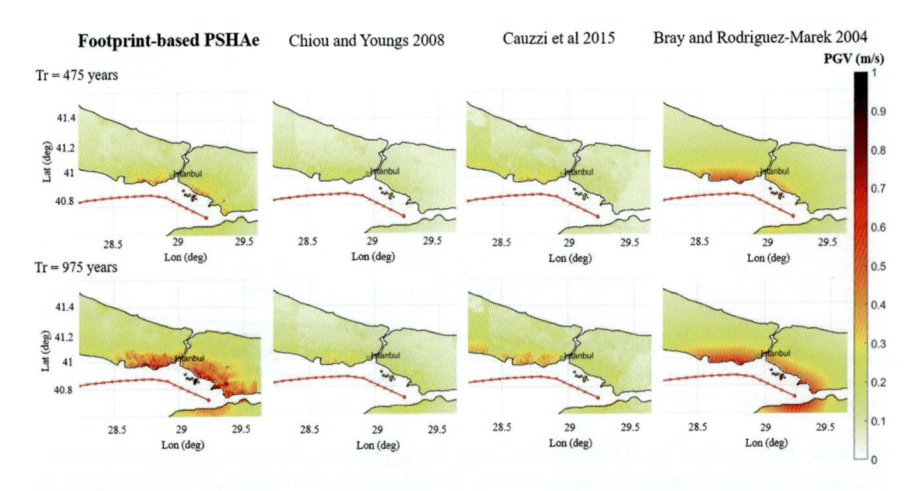

Fig. 8.13 Seismic Hazard Analysis based on 3DPBNSs, Chiou and Youngs (2008), Cauzzi et al. (2015), Bray and Rodriguez-Marek (2004) (from left to right). The hazard maps are presented in terms of PGV (m/s), geometric mean of the horizontal components, for $T_R = 475$ years (top) and 975 years (bottom)

respect to the urban area of Istanbul, that controls the ground motion intensities for the longer return periods.

As already mentioned the footprint-based *PSHAe* is an extremely CPU intensive methodology and therefore the number of the 3DPBNSs to be simulated plays a crucial role in terms of its applicability. In the recent time we are exploring different strategies that seems to be quite promising in order to reduce the computational effort by wisely selecting "a priori" the 3DPBNSs and successively weighting them differently within our logic three.

8.5 Conclusions

"For a large earthquake, the epicenter is not as helpful for engineers as is the footprint" (Housner 1999). With these words, one of the fathers of earthquake engineering commented the evidence from the August 17 1999 Turkish earthquake that the damage distribution could only be understood based on information of the dimensions of the faulted area and its position with respect to the site.

As a matter of fact, creating realistic realizations of future strong earthquake ground motions from known seismic faults and thus providing credible input motions for applications has been for long time a dream of engineering seismology. Within an application to future earthquake ground motions in Istanbul from the North Anatolian Fault branch crossing the Marmara Sea, we have shown in this paper the different and complicated steps involved, on the one side, in the creation of a fully 3D physics-based seismic model, and, on the other side, in the processing of numerical results and in their use for advanced applications of seismic hazard analyses.

Many of such steps are not fully resolved yet, but we have given in this paper some hints on how to cope effectively with them, within the intrinsic limitations of the 3DPBNS approach, but preserving its undeniable advantage over standard empirical approaches based on GMPEs.

More specifically, among the most important indications of this work, we list the following:

- even in the presence of an unlimited computer power and of a vast amount of knowledge, both on the seismotectonic, geological and geophysical context, the level of detail of available input data for 3DPBNS will be very hardly sufficient to solve frequencies larger than about 1.5 Hz: beyond this threshold, taking advantage of stochastic or hybrid methods is almost unavoidable;
- the proposed creation of broad-band synthetics based on an artificial neural network trained on strong motion records to provide a correlation between long and short period spectral ordinates seems to be an effective approach to create earthquake ground motion scenarios presenting a realistic spatial correlation structure and incorporating near-source and 3D site effects in a broad frequency range;
- to use results of 3DPBNS for seismic hazard analyses, a sufficient number of realizations is required to provide reliable distribution models to compute the probability of exceedance of given ground motion intensity measures (*IMs*): for this purpose, realistic random models for the kinematic co-seismic slip distribution along the fault should be provided;
- with respect to the GMPE-based approaches, the probability distribution models from 3DPBNS are expected to be more reliable, because they comply with the specific characteristics of the area.

Besides, it has been shown that the 3DPBNS approach allows one to avoid introducing in the PSHA a more or less arbitrary probabilistic model of the *IM*, with the well-known problem of assigning finite probabilities to extreme and unrealistic values of ground motion.

Acknowledgements This work has been carried out in the framework of the 2014–2016 RELUIS RS2 Project funded by the Italian Department of Civil Protection as well as of research projects sponsored by industrial partners, namely, MunichRe (Germany) and swissnuclear (Switzerland). The third author of the paper, Ilario Mazzieri, has been partially supported by the Project "Modellazione numerica di fenomeni idro/geomeccanici per la simulazione di eventi sismici" funded by GNCS, Italy. The authors are grateful to Filippo Gatti for his contributions for the development of the ANN2BB tool.

References

Abrahamson NA (2000) State of practice of seismic hazard evaluation. In: Proceedings of the GeoEng2000 Melbourne (Australia)

Akinci A, Aochi H, Herrero A et al (2017) Physics-based broadband ground-motion simulations for probable $M_w \geq 7.0$ earthquakes in the Marmara Sea region (Turkey). Bull Seismol Soc Am 107:1307–1323

Ancheta TD, Darragh RB, Stewart JP et al (2014) PEER NGA-West2 database. Earthq. Spectra 30 (3):989–1005
Aochi H, Douglas J, Ulrich T (2017) Stress accumulation in the Marmara Sea estimated through ground-motion simulations from dynamic rupture scenarios. J Geophys Res 122(3):2219–2235
Bohnhoff M, Bulit F, Dresen G et al (2013) An earthquake gap south of Istanbul. Nat Commun 4 (1999)
Bommer JJ, Abrahamson NA, Strasser FO et al (2004) The challenge of defining upper bounds on earthquake ground motions. Seismol Res Lett 75(1):82–95
Boore DM (2003) Simulation of ground motion using the stochastic method. Pure Appl Geophys 160(3–4):635–676
Bray JD, Rodriguez-Marek A (2004) Characterization of forward-directivity ground motion in the near-fault region. Soil Dyn Earthq Eng 24(11):815–828
Brune JN (1970) Tectonic stress and the spectra of seismic shear waves from earthquakes. J Geophys Res 75(26):4997–5009
Causse M, Chaljub E, Cotton F et al (2009) New approach for coupling k-2 and empirical Green's functions: application to the blind prediction of broad-band ground motion in the Grenoble basin. Geophys J Int 179(3):1627–1644
Cauzzi C, Faccioli E, Vanini M, Bianchini A (2015) Updated predictive equations for broadband (0.01-10 s) horizontal response spectra and peak ground motions, based on a global dataset of digital acceleration records. Bull Earthq Eng 13(6):1587–1612
Chaljub E, Moczo P, Tsuno S et al (2010) Quantitative comparison of four numerical predictions of 3D ground motion in the Grenoble valley, France. Bull Seismol Soc Am 100:1427–1455
Chiou BS-J, Youngs RR (2008) An NGA model for the average horizontal component of peak ground motion and response spectra. Earthq Spectr 24(1):173–215
Convertito V, Emolo A, Zollo A (2006) Seismic-hazard assessment for a characteristic earthquake scenario: an integrated probabilistic-deterministic method. Bull Seismol Soc Am 96(2):377–391
Crempien JGF, Archuleta RJ (2015) UCSB method for simulation of broadband ground motion from kinematic earthquake sources. Seismol Res Lett 86(1):61–67
Evangelista L, del Gaudio S, Smerzini C et al (2017) Physics-based seismic input for engineering applications: a case study in the Aterno river valley, Central Italy. Bull Earthq Eng 15 (7):2645–2671
Faccioli E, Maggio F, Paolucci R, Quarteroni A (1997) 2D and 3D elastic wave propagation by a pseudo-spectral domain decomposition method. J Seismol 1(3):237–251
Gallovič F, Brokešová J (2004) On strong ground motion synthesis with k-2 slip distributions. J Seismol 8(2):211–224
Goulet CA, Abrahamson NA, Somerville PG, Wooddell KE (2015) The SCEC broadband platform validation exercise: methodology for code validation in the context of seismic-hazard analyses. Seismol Res Lett 86(1):17–26
Graves R, Jordan TH, Callaghan S et al (2011) CyberShake: a physics-based seismic hazard model for Southern California. Pure Appl Geophys 168(3–4):367–381
Guidotti R, Stupazzini M, Smerzini C et al (2011) Numerical study on the role of basin geometry and kinematic seismic source in 3D ground motion simulation of the 22 February 2011 M_w 6.2 Christchurch earthquake. Seismol Res Lett 82(6):767–782
Herrero A, Bernard P (1994) A kinematic self-similar rupture process for earthquakes. Bull Seismol Soc Am 84(4):1216–1228
Housner GW (1999) The footprint of an earthquake: opinion paper. Earthq Spectr 15(4):825
Infantino M (2016) From 3D physics-based scenarios to advanced methods for seismic hazard assessment: the case of Istanbul. Master thesis in environmental engineering, Politecnico di Milano
Kamae K, Irikura K, Pitarka A (1998) A technique for simulating strong ground motion using hybrid Green's function. Bull Seismol Soc Am 88(2):357–367
Madariaga R, Olsen KB (2002) Earthquake dynamics. In: Lee W, Jennings P, Kisslinger C, Kanamori H (eds) International handbook of earthquake & engineering seismology, part A, vol 81A, 1st edn. Academic, San Diego

Mai PM, Beroza GC (2003) A hybrid method for calculating near-source, broadband seismograms: application to strong ground motion prediction. Phys Earth Planet Inter 137(1–4):183–199

Mai PM, Imperatori W, Olsen KB (2010) Hybrid broadband ground-motion simulations: combining long-period deterministic synthetics with high-frequency multiple S-to-S backscattering. Bull Seismol Soc Am 100(5A):2124–2142

Mazzieri I, Stupazzini M, Guidotti R, Smerzini C (2013) SPEED: SPectral elements in elastodynamics with discontinuous Galerkin: a non-conforming approach for 3D multi-scale problems. Int J Numer Methods Eng 95(12):991–1010

McGarr A, Fletcher JB (2007) Near-fault peak ground velocity from earthquake and laboratory data. Bull Seismol Soc Am 97(5):1502–1510

Motazedian D, Atkinson GM (2005) Stochastic finite-fault modeling based on dynamic corner frequency. Bull Seismol Soc Am 95(3):995–1010

Ordaz M, Martinelli F, D'Amico V, Meletti C (2013) CRISIS2008: a flexible tool to perform seismic hazard assessment. Seismol Res Lett 84(3):495–504

Özgül N (2011) İstanbul il alanının jeolojisi. Yönetici özeti. Directorate of earthquake and geotechnical Investigation, Istanbul Metropolitan Municipality. http://ibb.gov.tr/tr-TR/SubSites/DepremSite/Pages/KentJeolojisiCalismalari.aspx/ İstanbul İl Alanının Jeolojisi Yönetici Özeti. Accessed 25 Feb 2015

Paolucci R, Mazzieri I, Smerzini C, Stupazzini M (2014) Physics-based earthquake ground shaking scenarios in large urban areas. Perspectives on European earthquake engineering and seismology. Chapter 10. Springer International Publishing, pp 331–359

Paolucci R, Mazzieri I, Smerzini C (2015) Anatomy of strong ground motion: near-source records and 3D physics-based numerical simulations of the M_w 6.0 May 29 2012 Po plain earthquake, Italy. Geophys J Int 203(3):2001–2020

Paolucci R, Evangelista L, Mazzieri I, Schiappapietra E (2016) The 3D numerical simulation of near-source ground motion during the Marsica earthquake, central Italy, 100 years later. Soil Dyn Earthq Eng 91:39–52

Paolucci R, Gatti F, Infantino M et al (2018) Broad-band ground motions from 3D physics-based numerical simulations using artificial neural networks. Bulletin of the Seismological Society of America (submitted for review)

Pilz M, Parolai S, Stupazzini M et al (2011) Modelling basin effects on earthquake ground motion in the Santiago de Chile basin by a spectral element code. Geophys J Int 187(2):929–945

Smerzini C, Villani M (2012) Broadband numerical simulations in complex near-field geological configurations: the case of the 2009 M_w 6.3 L'Aquila earthquake. Bull Seismol Soc Am 102 (6):2436–2451

Smerzini C, Galasso C, Iervolino I, Paolucci R (2014) Ground motion record selection based on broadband spectral compatibility. Earthquake Spectra 30(4):1427–1448

Stupazzini M, Paolucci R, Igel H (2009) Near-fault earthquake ground-motion simulation in the Grenoble Valley by a high-performance spectral element code. Bull Seismol Soc Am 99 (1):286–301

Stupazzini M, Allmann A, Käser M et al (2015) PSHAe (probabilistic seismic hazard analysis enhanced): the case of Istanbul. In Proceedings of the 10th Pacific conference on earthquake engineering, Sydney, Australia

Villani M, Faccioli E, Ordaz M, Stupazzini M (2014) High-resolution seismic hazard analysis in a complex geological configuration: the case of Sulmona Basin in Central Italy. Earthq Spectr 30 (4):1801–1824

Chapter 9
Issues with the Use of Spatially Variable Seismic Ground Motions in Engineering Applications

Aspasia Zerva, Mohammad Reza Falamarz-Sheikhabadi, and Masoud Khazaei Poul

Abstract Even though the significance of the spatial variability of seismic ground motions for the response of lifelines and its modeling from array data have been addressed for more than half a century, there are still issues associated with its use in engineering applications, which are the focus of the present paper. Common approaches for the simulation of spatially variable seismic ground motions are reviewed, and their corresponding uncertainties are discussed in detail. The importance of the consideration of rotational ground motions in the seismic excitation of structures, and the significance of the kinematic soil-structure interaction in the modification of the foundation input motions are addressed. In addition, difficulties with absorbing boundary conditions and one-dimensional deconvolution methods, when the spatial variability of the ground motions is considered in the seismic analysis of structures, are elaborated upon, and the necessity of developing three-dimensional coherency models is noted. This critical investigation provides insight into and facilitates the appropriate simulation of spatially variable seismic ground motions in engineering applications.

9.1 Introduction

It is well known that seismic ground motions (SGMs) during earthquakes are generated by the propagation of waves from the source through the earth strata to the ground surface. The seismic excitation of structures is a displacement-based phenomenon depending on the reflection and refraction of the seismic waves transmitted into the structure (Safak 1999). Therefore, the seismic excitation of structures depends both on their geometrical and mechanical characteristics (or kinematic and dynamic properties). The effect of various characteristics of

---

A. Zerva (✉) · M. R. Falamarz-Sheikhabadi · M. K. Poul
Civil, Environmental and Architectural Engineering Department, Drexel University, Philadelphia, PA, USA
e-mail: zervaa@drexel.edu

© Springer International Publishing AG, part of Springer Nature 2018
K. Pitilakis (ed.), *Recent Advances in Earthquake Engineering in Europe*,
Geotechnical, Geological and Earthquake Engineering 46,
https://doi.org/10.1007/978-3-319-75741-4_9

seismic waves, such as phase velocity, incident angle, wave type, and frequency content, on the structural response is, generally, uncertain and not readily quantified. Ignoring the complex characteristics of seismic actions, simplified approaches are proposed by design codes to define the seismic loading pattern of structures. Selecting an appropriate simplified loading pattern for the response evaluation of structures is an influential consideration in their seismic assessment (Zerva 2009). As seismic waves travel from the source to the ground surface, their characteristics, i.e. amplitude and phase, change depending on their path (Zerva and Zervas 2002). The propagation of the seismic waves along different wave paths results in spatially variable seismic ground motions (SVSGMs). The main causes of the SVSGMs, in the near-, middle- and far-field zones, are the wave passage, wave scattering, surface and subsurface topography, soil properties and extended seismic source effects. These differential motions are not influential in the seismic excitation of structures with relatively short footprints, such as ordinary buildings and short-span bridges, if the soil properties are approximately uniform along the structure (Eurocode 8-Part 2 2005), but may significantly affect the seismic behavior of long and extensive structures, such as dams and long-span bridges (Der Kiureghian and Neuenhofer 1992; Falamarz-Sheikhabadi et al. 2016; Harichandran et al. 1996; Todorovska and Trifunac 1990; Vanmarcke et al. 1993; Zerva 1990). There is no consensus whether the effect of the SVSGMs on the seismic response of structures is beneficial or detrimental compared to that of uniform excitations. The only common agreement in the literature is that the effect of the SVSGMs on the seismic behavior of structures is very complicated, and depends on the characteristics of both the structure and the ground motions.

Performance-based earthquake engineering requires the accurate consideration of two main sources of uncertainty in the structural response evaluation, namely modeling strategy and excitation scenario (Falamarz-Sheikhabadi and Zerva 2017). In earthquake engineering, either the effect of both modeling and excitation uncertainties is taken into account in the structural response evaluation (Falamarz-Sheikhabadi and Zerva 2017; Kwon and Elnashai 2006), or the excitation-induced uncertainty is considered to be the only dominant contributor to the uncertainty and the effect of modeling-induced uncertainty is ignored (Kim and Rosowsky 2005; Wong and Harris 2012). Even though the former approach results in a better estimation of the lower and upper bounds of the structural response, the reported numerical results indicate that the uncertainty in the definition of the seismic excitation remarkably surpasses that of the structural modeling strategy (Falamarz-Sheikhabadi and Zerva 2017; Kwon and Elnashai 2006). For example, Falamarz-Sheikhabadi and Zerva (2017) reported in their seismic assessment of a tall, curved, long-span, reinforced-concrete bridge, the Mogollon Rim Viaduct, that the uncertainty in the finite-element modeling may modify the bridge response in the order of 100% (on average), but the excitation-induced uncertainty may lead to a response variability in the order of 1000% (worst-case scenario), if an appropriate method is not adopted for the scaling and the loading pattern of the selected earthquake records. This example clearly illustrates the importance of the appropriate consideration of uncertainty in the definition of any simplified seismic loading pattern for a reliable structural assessment.

Many factors may lead to uncertainties in the selection/simulation of SGMs, which are associated with the earthquake magnitude, seismotectonic environment, source-to-site distance and site response, and the representation of their characteristics, e.g. the selection of appropriate seismic intensity parameter. These uncertainties are not discussed herein, because they have been comprehensively investigated in research efforts related to seismology, seismic risk and reliability analyses (e.g. Baker and Cornell 2006; Bradley 2012). The main focus of the preset study is on uncertainties that are particularly associated with the simulation/consideration of the spatial variability of the seismic ground motions in engineering applications. Even though these uncertainties may lead to significant variability in the structural response, their importance has not been thoroughly discussed. Section 9.2 in the following discusses the simulation of spatially variable ground motions, incorporating the effect of the selection of the coherency model and its parameters, as well as differentiating between the characteristics of spatially variable ground motions at uniform sites and sites with irregular subsurface topography. Section 9.3 describes the significance of rotational seismic ground motions and a new approach for evaluating them, including limitations in the parameter evaluations. Section 9.4 presents soil-structure interaction effects pertinent to extended structures, and includes a discussion on absorbing boundary conditions and deconvolution approaches.

9.2 Simulation of Spatially Variable Seismic Ground Motions

The basic descriptor of the spatial variability is the coherency. Coherency is a statistical estimate derived in the frequency domain from the power and cross spectral densities of the shear-wave window of seismic data recorded at pairs of stations. (For the mathematical derivation of coherency, the reader is referred to, e.g., Zerva (2009)). Coherency is a complex function consisting of two terms:

$$\gamma(\Delta x, \omega) = |\gamma(\Delta x, \omega)| \exp\left[-i\omega \frac{\Delta x}{V_x}\right] \quad (9.1)$$

where $|\gamma(\Delta x,\omega)|$ is the lagged coherency (loss) function; $i = \sqrt{-1}$; ω is the circular frequency; Δx the separation distance between the two stations, and V_x is the apparent propagation velocity of the waves along the x-axis, presumably the one along which coherency is to be estimated. The complex exponential term in Eq. 9.1 reflects the propagation of the waves along the prescribed direction invoking the assumption of plane-wave propagation in a homogeneous and isotropic medium. The lagged coherency in the equation is purely statistical and its values range between 0 and 1. It represents the correlation of the motions at each frequency: At low frequencies ($\omega \to 0$), motions are fully correlated and $|\gamma(\Delta x,\omega)| = 1$, as the wavelength of the seismic waves is very long; at large frequencies and long station separation distances, the lagged coherency tends to 0, because the wavelength of the motions becomes short, and, at intermediate frequencies, it declines from 1 to 0. It is worth noting that the physical representation of the second term in the above equation represents the deterministic phase delay due to wave propagation, and the first, lagged coherency, term reflects the random phase variability due, mostly but not entirely to, scattering effects.

Even though it was well understood that seismic ground motions vary in both time and space, the investigation of their spatial variability started, basically, after the SMART-1 (Strong Motion Array in Taiwan – Phase 1) in Lotung, Taiwan, started recording seismic ground motions in the early 1980's. The array consisted of a center station (C00), and 36 additional ones arranged in three concentric circles around C00, each with 12 equispaced stations, the inner at a 200 m radius, the middle at a 1 km radius and the outer at a 2 km radius. Later on (1985), a smaller-scale array, the Lotung Large Scale Seismic Test (LSST) array, was constructed within the SMART-1 array. This three-dimensional array consisted of 15 ground-surface and 8 downhole instruments, and two small-scale (1/4 and 1/22, also instrumented) models of a reactor containment vessel. On the ground surface, the array was composed of three radially extended arms at 120° intervals, with the smallest distance being, approximately, 3 m and the largest 50 m. Four stations were placed at depth with minimum and maximum distances of 6 m and 47 m, respectively. The unique set of the two arrays provided a plethora of data for the investigation of coherency. Some characteristics of the lagged coherency at this uniform soil site became apparent. These include the behavior of coherency at shorter and longer separation distances and the effect of source-finiteness: By comparing the lagged coherency of data at the shorter and longer separation distances of the LSST and SMART-1 arrays, Abrahamson et al. (1991) noted that coherency is affected by separation distance in the sense that the lagged coherency extrapolated from the SMART-1 array to shorter distances tended to overestimate those evaluated from the LSST array. Regarding the effect of the earthquake magnitude on the spatial variability of the seismic ground motions, even though, theoretically, the effect of source finiteness on the ground surface coherency is expected, it does not appear to be significant (Abrahamson 1993). Spudich (1994) provided a possible explanation for this observation in the case of large earthquakes with unilateral rupture propagation, which constitutes the majority of earthquakes, based on the fact that the waves radiated from the source originate from a spatially compact region that travels with the rupture front, and, thus, at any time instant, a relatively small fraction of the total rupture area radiates. However, the effect of the bilateral rupture on the spatial variability of ground motions is still unmeasured. It should be noted that coherency models have not been derived for near-fault ground motions, which may be significantly affected by the rupture directivity, fling-step and hanging wall effects. Furthermore, it is anticipated that, in this case, the effect of the source finiteness would be significant.

9.2.1 Selection of Lagged Coherency Model

Numerous lagged coherency models have appeared in the literature, most of them based on the SMART-1 records. These empirical coherency models are highly dependent on the researchers' identification of the dominant shear-wave window in the motions, the processing approach used to estimate the lagged coherency, and, mostly, the selected smoothing-window length, as well as the functional form adopted. These issues will not be further elaborated upon, as they have been described in detail by Zerva and Zervas (2002) and Zerva (2009). It should be emphasized, however, that most empirical lagged coherency models produce values lower than unity at low frequencies and tend to a constant value, namely the coherency of noise smoothed with the window length used (Abrahamson 1993), at high frequencies. An example of such an empirical coherency model, one of the first developed from the SMART-1 array data, is that of Harichandran and Vanmarcke (1986), presented in Fig. 9.1. It can be clearly seen in the figure that the lagged coherency is less than unity at zero frequencies, decreases as the separation distance increases, and tends to finite values as frequency and separation distance increase. This behavior is not physically justified but an artifact of the signal processing of the data. Hence, it should be cautioned that the use of such coherency models may result in an erroneous evaluation of the structural response of extended structures. The main reason for this is that spatially variable excitations, in addition to modifying the dynamic response of the structures as compared to the one induced by the uniform ones, also excite the pseudo-static response, which the uniform ones do not. Partial

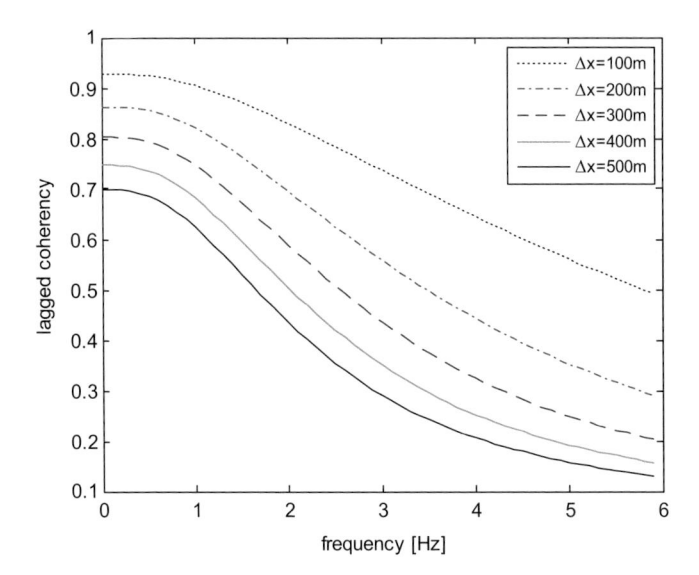

Fig. 9.1 Harichandran and Vanmarcke's (1986) lagged coherency model

correlation at low frequencies will yield displacement waveforms with significant variability, thus increasing the pseudo-static response of the structures.

Perhaps the most frequently used coherency model in the literature is the semi-empirical one of Luco and Wong (1986). Furthermore, this model is the one recommended by Eurocode 8-Part 2 (2005) in its informative annex for use in the simulation of SVSGMs. The model is based on shear-wave propagation through random media developed by Uscinsky (1977), but requires recorded data for the estimation of its decay. The functional form of the lagged coherency of the model is as follows:

$$|\gamma(\Delta x, \omega)| = \exp\left[-\left(\frac{\lambda_x \omega \Delta x}{V_S}\right)^2\right] \quad (9.2)$$

where λ_x is a dimensionless constant termed herein as the incoherency coefficient, and V_S is the shear wave velocity in the random medium. Luco and Wong (1986) used this coherency model for both horizontal and vertical input motion components in the evaluation of the kinematic soil-foundation interaction of rigid, rectangular, shallow foundations. Based on comparisons with recorded data, Luco and Wong (1986) suggested an average value for λ_x/V_S equal to 2.5×10^{-4} (s/m). Kim and Stewart (2003) also recommended a relation for the determination of the incoherency coefficient as $-0.037 + 7.4 \times 10^{-4}\ V_S$ (m/s), which results in higher values for the ratio λ_x/V_S in comparison to the average value recommended by Luco and Wong (1986). For example, shear wave velocities in the range of 100 m/s to 600 m/s, which are close to the values of the 29 sites examined by Kim and Stewart (2003), result in λ_x/V_S in the range of $(3.70–6.78) \times 10^{-4}$ (s/m). Figures 9.2a, b present the exponential decay of the lagged coherency of the model with frequency and station separation distance for the average value of λ_x/V_S suggested by Luco and Wong (1986), and the highest value of the parameter reported by Kim and Stewart

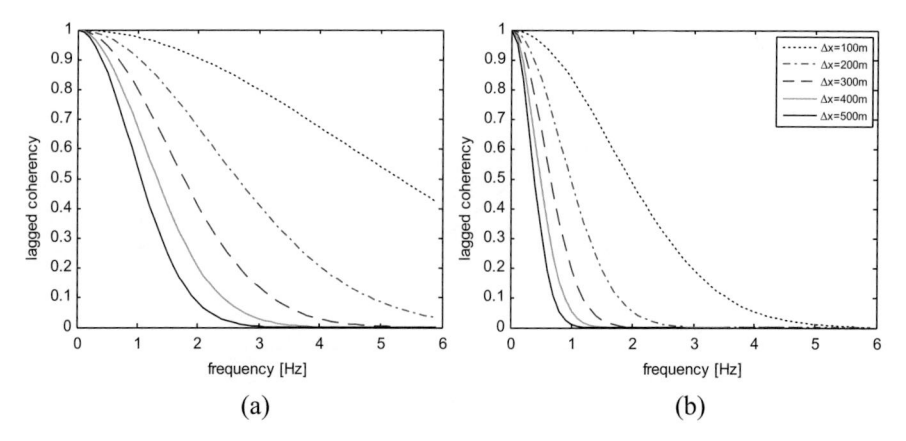

Fig. 9.2 Luco and Wong's (1986) lagged coherency model. (**a**) $\lambda_x/V_S = 2.5 \times 10^{-4}$ (s/m). (**b**) $\lambda_x/V_S = 6.78 \times 10^{-4}$ (s/m)

(2003). Clearly, the lagged coherency satisfies the physical criteria at low and high frequencies, but the model in Fig. 9.2b will induce a more significant pseudo-static response, and the difference in the degree of exponential decay of the models will affect the dynamic response of extended structures.

It should be noted that the average value recommended by Luco and Wong (1986) is based on limited data, and the relation recommended by Kim and Stewart (2003) may be only applicable for the evaluation of the kinematic soil-structure interaction and not the simulation of SVSGMs, because the effects of the foundation flexibility and wave inclination have been also implicitly considered in its derivation. Clearly, with the increase of the soil stiffness (or increase of the shear wave velocity), the effect of the foundation flexibility on its response increases, and this makes the applicability of the proposed relation for stiff soils questionable. Hence, the main difficulty with the model is that empirical data are necessary to determine the value of its incoherency coefficient. In practice, this issue makes the application of the Luco and Wong's (1986) coherency model difficult for different site conditions.

9.2.2 Selection of Apparent Propagation Velocity

In developing coherency models, it is commonly assumed that shear waves transfer the main energy of seismic waves, and, therefore, the shear-wave window is only considered in their derivation. In this approach, the dispersion of scismic waves is ignored, and a constant apparent propagation velocity corresponding to shear waves is used to describe the time delay in the arrival of seismic waves. Because body waves are non-dispersive, except in highly attenuated media, the assumption that they have the same velocity over a wide range of frequencies may be valid. This approach leads to a constant time delay in the multi-support seismic excitation of structures, and ignores the random time delay fluctuations around the wave passage delay due, mainly, to the upward traveling of the seismic waves through horizontal variations of the geologic structure underneath the site (Spudich 1994), and the deviations of the propagation pattern of the waves from that of plane wave propagation (Boissieres and Vanmarcke 1995).

It should be noted that, for the estimation of the lagged coherency (Eq. 9.1), the dominant shear-wave window of the spatial array data is, generally, aligned, i.e., shifted in time, with respect to that of a reference station, so that the apparent propagation effects are removed. The resulting time delays exhibit the variability due to the two aforementioned causes. However, the assumption of a constant apparent propagation velocity has been widely adopted in the simulation of SVSGMs. Various approaches have been proposed for the evaluation of the average apparent propagation velocity from array data (e.g. Loh and Penzien 1984; Goldstein and Archuleta 1991). Frequency-wavenumber (F-K) spectra or

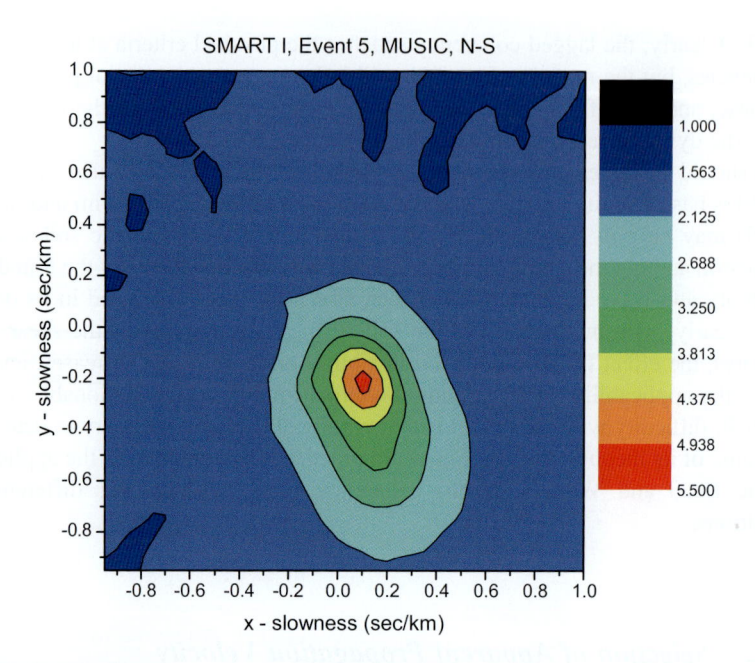

Fig. 9.3 Slowness spectra

stacked slowness spectra, the latter for non-dispersive waves, appear to be the most common approach in the estimation. The techniques are based on the triple (two space and a time) Fourier transform of the window analyzed. An illustration of the slowness spectrum of the shear-wave window of the N-S component of the data recorded at the SMART-1 array during Event 5 (a 6.3 magnitude earthquake at an epicentral distance of 30 km from Lotung and at a focal depth of 25 km) is presented in Fig. 9.3. The direction of the line connecting the peak of the spectrum (red area) with the origin of the slowness plane reflects the direction of propagation of the waves, and its length is equal to the inverse of the propagation velocity (in this case 4.5 km/s).

In seismic assessments, the selection of the value of the apparent propagation velocity is a controversial issue, because it can theoretically range between the values of the shear wave velocity of the surface soil layer (horizontal propagation) to infinity (vertical propagation). Clearly, this may lead to a significant variability in the response of structures, particularly the ones located at sites with soft soil conditions. In a probabilistic analysis, a uniform distribution in the interval [1 km/s, 4 km/s] may be used to consider the effect of the uncertainty in the apparent propagation velocity for the structural response evaluation due to the significant uncertainty in the evaluation of this quantity. The higher values of the apparent propagation velocity can be practically ignored in the seismic assessment of most structures, even long-span bridges.

9.2.3 Lagged Coherency at Uniform Sites

Coherency models have been evaluated mostly from data recorded at soil sites and, particularly, the SMART-1 array site. As already indicated, empirical lagged coherency models carry a significant amount of uncertainty based on the approach by which they were developed. Furthermore, most empirical coherency models were evaluated at a single site for a single event. Even though this approach has advantages, the derived models cannot be readily extrapolated to different sites.

Taking the perspective that lagged coherency, or, more rigorously, the plane-wave coherency, a term coined by Abrahamson (1993) and describing exactly what the word reflects, is independent of the seismic intensity, distance from source as well as site conditions, Abrahamson (1993) developed perhaps the most reliable coherency model, both for the horizontal and the vertical components of the SGMs, in the literature. The model is based on an extensive set of data at various sites, corrected to reproduce unity at zero frequencies, tending, as plane-wave coherency, to zero values at high frequencies and accommodating the fact that coherency depends on station-separation distance. The variation of this model with frequency and station-separation distance is presented in Figs. 9.4a, b. In the figure, the model is cited as plane-wave coherency at soil sites, due to the fact the most of the data used in its development were recorded at soil sites. However, considering that local scattering cannot be the same at "soil" and "rock" sites, grouping all data together cannot be validated. Indeed, Zerva and Zhang (1996), analyzing data at the SMART-1 array, and later on, Liao (2006), estimating lagged coherencies at spatial array data at a variety of site conditions, noted a correlation between amplitude and phase variability at soil and rock sites: Because rock sites produce their peak response at higher frequencies than soil sites, coherency at rock sites should be higher than that at soil sites. Abrahamson (2007) developed another model, based, mainly, on data at the rock site of the Pinyon Flat array, and derived the plane-wave coherencies presented in Figs. 9.4c, d. The comparison of the plots in Fig. 9.4a and c and b and d clearly reflects the fact that scattering at soil sites tends to reduce the values of the plane-wave coherency faster than that at rock sites. It should be noted that spatial variability is also of importance in the seismic safety assessment of nuclear power plants: Their large, mat, rigid foundations tend to average, and, consequently, reduce the translational motions induced by spatially variable seismic excitations, even though they excite some rotational response (Luco and Wong 1986). Considering the higher frequency content of rock sites, the more significantly correlated motions diminish the effect of the translational reduction.

The above being said, there are significant issues concerning the models in Fig. 9.4. One basic issue is the broad classification of coherency at "soil" and "rock" sites, as there is significant variability in the properties at these uniform sites. However, given that sites instrumented with arrays are limited, a designer needs to rely on available information, and the models in Fig. 9.4 are, to the authors' opinion, the most valid ones.

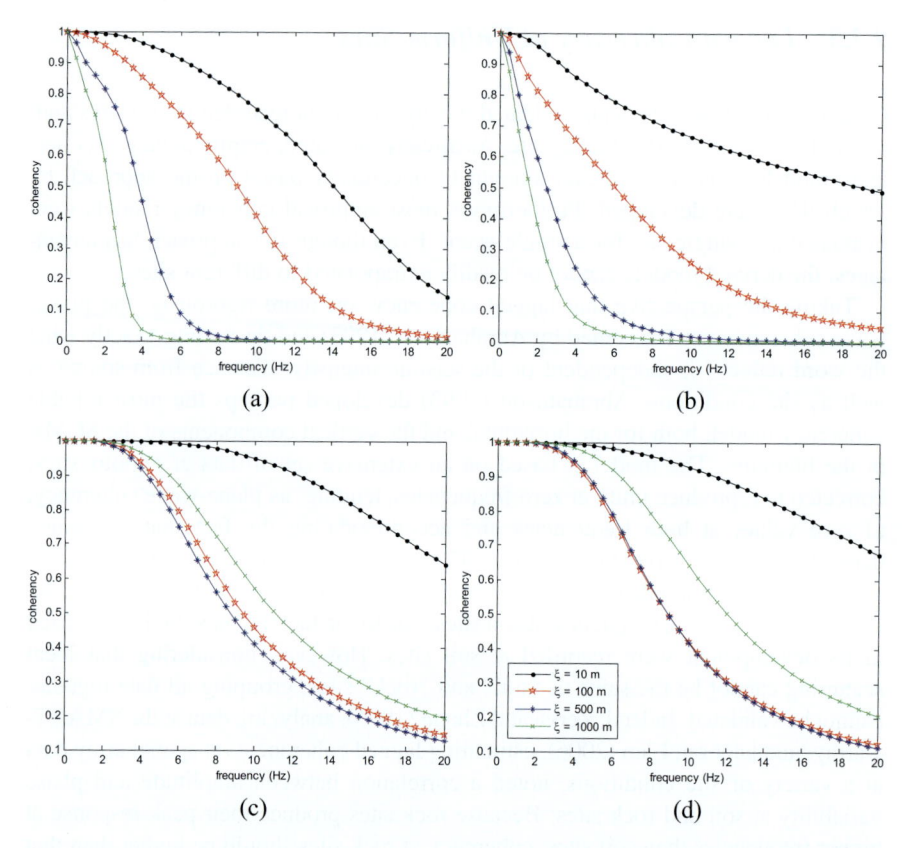

Fig. 9.4 Abrahamson (1993, 2007) plane-wave lagged coherency models (**a**) horizontal coherency at soil sites (**b**) vertical coherency at soil sites (**c**) horizontal coherency at rock sites (**d**) vertical coherency at rock sites. (ξ in the figure legend indicates station separation distance)

Additional sources of uncertainty are caused, especially, by the soil properties, e.g. (JCSS 2001): spatial variability of soil properties (patterns of variability may be either continuous or discrete); limited soil survey and laboratory or in-situ testing; inaccuracy of soil investigation methods or erroneous interpretation of investigation results, and nonlinear response due to ground shaking. The last source not only causes spatial and temporal variations in the shear wave velocity, but also affects the values of the apparent propagation velocity and incoherency coefficient, and results in fluctuation of their values. Soil becomes nonlinear and irreversible at very small strains, and its nonlinear nature highly depends on the characteristics of the seismic excitation (Kolymbas 2003). This feature of soil can significantly affect the incoherency of ground motions due to the non-uniform soil nonlinearity at the site and indicates that coherency models should be functions of both soil properties and seismic intensity. Currently, none of these two factors have been considered in proposing coherency functions or their derivation. Therefore, the use of existing coherency functions for simulating severe earthquakes in the near field is questionable due to the significant effect of the soil nonlinearity on the spatial variability of seismic ground motions. Furthermore, it should be noted that soil becomes more nonlinear with the increase of the seismic intensity, and, as a consequence, the value of the shear wave velocity is reduced. FEMA 356 (2000) proposes relations for modifying the shear wave velocity based on the soil type and the peak ground acceleration (PGA). It is also worth noting that coherency models are usually derived assuming that the site of interest is a homogeneous and isotropic medium. Therefore, the effect of non-uniform site conditions on the SVSGMs is commonly ignored in simulating asynchronous seismic excitations, and no recommendation can be made to even partly mitigate the effect of this uncertainty in the seismic analysis of structures. This lack of availability of coherency models at non-uniform sites is discussed in the following.

9.2.4 Irregular Subsurface Topography

As illustrated in the previous sections, many coherency models have been developed for uniform sites. Such coherency models cannot be utilized for the simulation of SVSGMs at sites with variable subsurface topography, as, it is well known, that the pattern of wave propagation at such sites differs fundamentally from that at uniform ones (e.g. Bard and Bouchon 1985). Furthermore, due to the uniqueness of the geometric characteristics and material contrasts in these regions, even if coherency models can be estimated at a single site for one or multiple events, their extrapolation to different sites is meaningless. This section presents, first, some attempts to remedy the lack of sufficient data for coherency estimation via analytical approaches, and, then, highlights physical insights in coherency estimation at narrow valleys from spatially recorded data.

9.2.4.1 Analytical Evaluations

For variable site conditions, researchers usually adopt two approaches for simulating the SVSGMs: (1) using simplified one-dimensional wave propagation models (e.g. Der Kiureghian 1996; Zembaty and Rutenberg 2002; Bi and Hao 2011), and (2) using complicated two- or three-dimensional numerical simulation methods, such as finite-element, finite-difference and boundary-element methods (e.g. Assimaki et al. 2005; Kamalian et al. 2008). The former approach partly considers dynamic effects of the site response and is more common in seismic analysis of structures because of its simplicity, but does not have a solid theoretical justification, and simulated ground motions cannot be readily compared with reality. The latter approach is more accurate because it can incorporate dynamic and kinematic characteristics of the site and wave propagation into the seismic analysis, but is also more complicated and computationally expensive. Figures 9.5a, b show two typical site conditions for bridges.

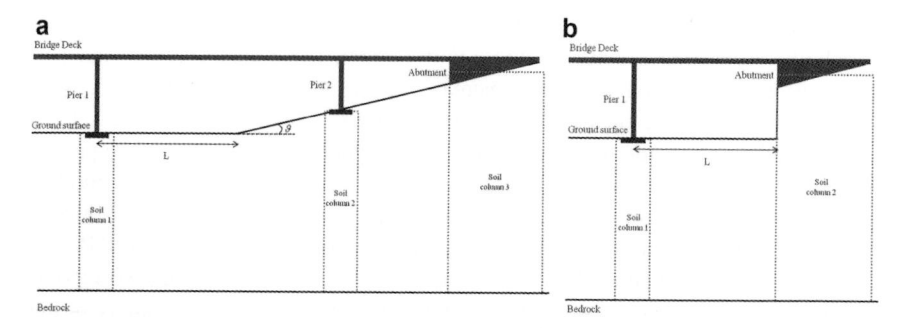

Fig. 9.5 Configuration of site topography: (**a**) slope, and (**b**) vertical slope

In Fig. 9.5a for $\vartheta < 15°$, one may ignore the kinematic effects of the topography on the SVSGMs based on the provisions of AFSP (1995) and Eurocode 8-Part 5 (2004). In such a case, depending on the variations in the fundamental period of the soil columns shown in Fig. 9.5a, one may ignore or consider the dynamic characteristics of the site in the coherency model using simplified one-dimensional wave propagation models (e.g. Der Kiureghian 1996). However, in Fig. 9.5a for $\vartheta > 15°$, the application of the complicated two- or three-dimensional numerical simulation methods is necessary to determine the SVSGMs. In such cases, the effect of the topography may be only ignored on the input seismic motions of Pier 1, if it is far enough from the bottom corner of the slope (Bouckovalas and Papadimitriou 2005; Gatmiri et al. 2009). Figure 9.5b shows a typical configuration of the abutment and pier in highway bridges, for which the seismic input of the pier and abutment cannot be considered the same, but such a consideration is usually ignored in seismic analyses. For ordinary standard bridges, this approach may be adequate due to the level of accepted uncertainty in their design if at least a simplified one-dimensional wave propagation model for simulating ground motions is taken into account. However, the use of the complicated two- or three-dimensional numerical simulation methods is essential for important and irregular bridges located in regions with high seismicity due to the fact that their nonlinear response highly depends on the characteristics of the input excitation.

9.2.4.2 Empirical Coherency at Narrow Valleys

In 1995, a temporary dense array of digital seismographs was deployed in the Parkway Valley, Wainuiomata, New Zealand (Stephenson 2000). This flat-floored valley is approximately 400 m wide and surrounded by greywacke outcrops. The configuration of the array and the Parkway Valley are shown in Fig. 9.6a. Station 01 (not shown in the figure) was installed approximately 2 km NE of the basin on firm rock. Four stations (stations 22–25) were deployed on the soft rock (weathered greywacke) surrounding the valley. Station 13 was never installed; 19 stations (stations 02–21 except 13) were installed on the soft sediments of the valley. The

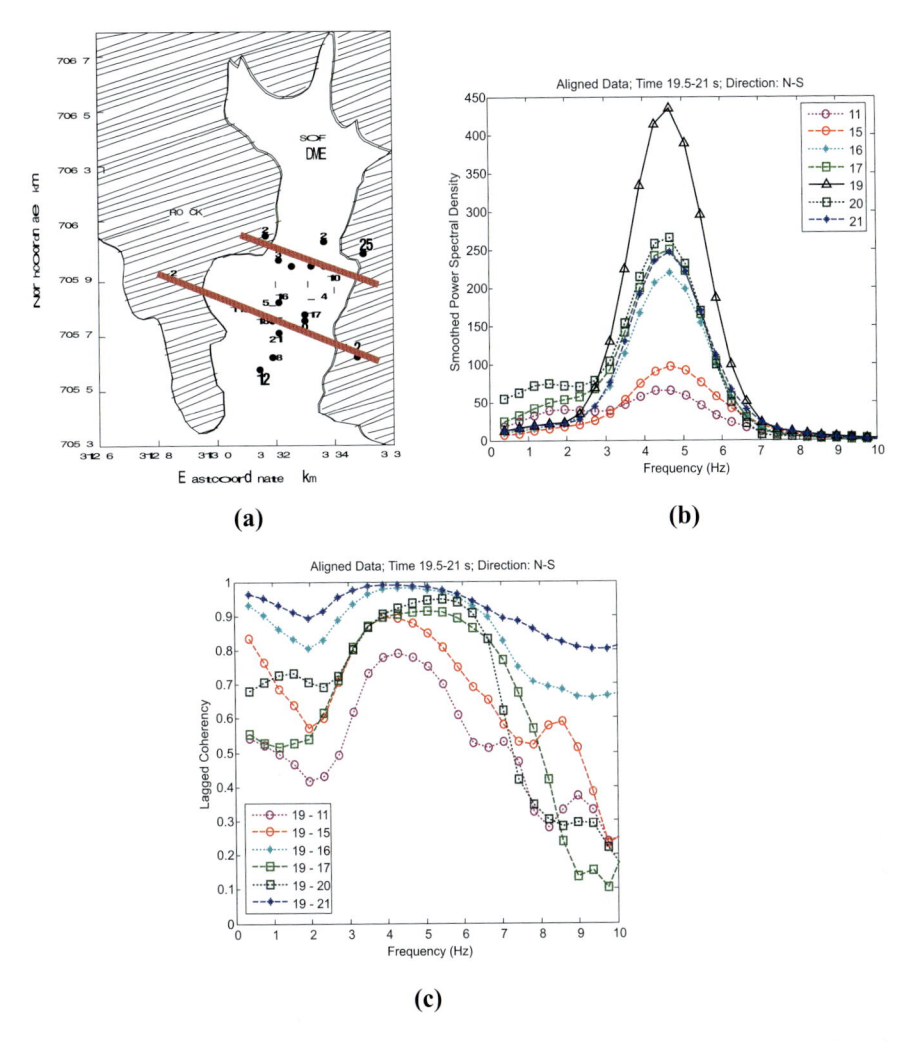

Fig. 9.6 (a) The Parkway array, and (b) representative power spectral densities and (c) lagged coherencies (Adopted from Zerva and Stephenson 2011)

minimum and maximum separation distances between the stations within the valley and surrounding the valley were 22.8 m and 665.7 m, respectively, with an average distance between stations being less than 40 m. Hence, the station-separation distances are pertinent for engineering applications, and the two orange lines in the figure were constructed to mimic footprints of bridges with their abutments located at soft rock and their piers in the soft sediments of the valley.

A very complex pattern of wave propagation was observed during the analyzed event (magnitude 4.9, depth of 28 km, epicentral distance of 80 km, and a geometrical backazimuth of 59° clockwise from north) by Zerva and Stephenson (2011): The E-W component of the data during the onset of the excitation in the valley stations appears to be composed of Love-type waves, in addition to some shear-wave amplification, whereas the N-S component is composed of Rayleigh waves at the higher frequencies and, in part, shear-wave amplification at the lower frequencies. The vertical motions during this window also contain Rayleigh waves at the higher frequencies, but appear to be more erratic than the horizontal motions. During the later windows of the ground motions, high energy waves at low frequencies control the motions in both horizontal directions. These appear to be Love-type waves arriving from the north for the E-W component of the motions and from the east for the N-S component. The vertical motions during these latter windows exhibit low amplitudes. The coherency pattern of the motions also appear significantly different that those at uniform sites. As an illustration, Figs. 9.6b, c present power spectral densities and coherencies at the onset of the excitation of the N-S component of the valley stations along the lower bridge footprint of Fig. 9.6a. It can be clearly seen from the figures that the amplitude variability of the motions (Fig. 9.6b) is dramatic, and a single power or response spectrum cannot describe the data at such irregular topographies. Furthermore, consistently throughout the study, as can also be seen by comparing Figs. 9.6b, c, the correlation in the motions appeared as "hills" in the lagged coherency estimates when the energy in the motions at the participating station pairs peaked at similar frequency ranges. These "hills" appeared irrespective of whether the energy in the motions was caused by shear or surface waves. Another very important observation of the study, corroborating previous results (Bard and Bouchon 1985; Frankel et al. 1991), is that the duration of the records in sedimentary basins is longer than those at rock sites. The study suggested that, whereas the motions at the rock stations 22 and 23 (reflecting the position of the abutments of the bridge) have ceased, the stations in the valley underwent the most severe part of the seismic excitation. The effect of the significant differences in the duration of the motions at sites where surface waves may form is not currently addressed in the modeling of spatially variable excitations at irregular subsurface site conditions.

Recently, Imtiaz et al. (2017) analyzed data recorded at the dense array at the Koutavos-Argostoli area, Cephalonia Island, Greece, which was operational from September 2011 to April 2012. Koutavos-Argostoli is a relatively small alluvial valley approximately 3 km long and 1.5 km wide, surrounded by hills of limestone and marl. The array stations were positioned on four concentric circles, with radii of 5, 15, 40 and 80 m, around the central station A00. Five stations, branching off from A00 in five directions, N 39°, N 112°, N 183°, N 255° and N 328°, were placed on each concentric circle. The station-separation distances ranged between 5 and 160 m, making the results of the evaluation pertinent for engineering applications. Imtiaz et al. (2017) conducted an elaborate sensitivity analysis from a set of 46 local and regional earthquakes (magnitudes ranging between 1.9 and 5.2, over an epicentral distance of a 200 km radius from the center station). Their analysis indicated that the motions in the valley are a mixture of different types of waves. The results from all data sets suggested that lagged coherency is independent of the geometrical (E-W and N-S) or rotated (about the source or valley main axis) directions, slight or no systematic dependence of coherency on magnitude, hypocentral distance or backazimuth, but large coherency resulted from data recorded at station pairs oriented in the valley-parallel direction and low coherency for those recorded at station pairs in the valley-normal direction.

The results of both studies clearly indicate that, at sites with irregular subsurface topography, the amplitude, coherency and duration of the motions can vary significantly, and the common approaches to remedy this problem may not provide satisfactory results for the seismic evaluation of extended structures crossing such regions.

9.2.5 Simulation Approach

Because records of closely spaced strong ground motions are rare, seismic analysis of bridges is usually conducted using simulated ground motions consistent with a prescribed spatial variability model for the region of interest (Hao et al. 1989). In this case, the SVSGMs are artificially generated by means of unconditional or conditional simulation methods (e.g. Deodatis and Shinozuka 1989; Liao and Zerva 2006; Vanmarcke et al. 1993). In most research studies, as well as the recommended analytical approach of Eurocode 8-Part 2 (2005), the SVSGMs are unconditionally simulated as stationary, obeying a coherency model and propagating with a prescribed apparent propagation velocity on the ground surface. They are then multiplied by an envelope function, which gives them a beginning and an end, and iteratively modified to become compatible with the target response spectra. Such SVSGMs depict some of the dominant physical characteristics of ground motions, but do not fully comply with physical considerations. On the other hand, conditional simulations are based on a reference (seed) time series, which can be an actual recorded accelerogram or a synthetic time series. The conditional simulations are generated such that they conform to the reference time series, obey the selected coherency model and propagate with the prescribed propagation velocity on the ground surface. Even though conditionally simulated waveforms represent more realistic ground motions than simulated ones (Zerva 2009), one should be cautious regarding the segmentation error, which is usually considered using a trial and error method (Vanmarcke and Fenton 1991). It is also worth noting that there are conditional simulation methods (e.g. Abrahamson 1988) that, first, decompose the recorded wavefield to its signal and noise components, using F-K analysis, and then estimate the interpolated time series at any location by recombining the signal and noise components.

The seismic response of structures highly depends on the characteristics of the input excitation, and, hence, the processing of earthquake time series with the use of a consistent scheme is an influential factor in the reduction of the variability of the structural output, and, consequently, a reliable structural analysis (Akkar and Bommer 2006; Zerva et al. 2012). The magnitude of the induced pseudo-static forces in the bridge structure highly depends on the shape of the displacement time series and its peak value. The most common techniques for the processing of earthquake records, which may significantly modify the waveform of the conditionally simulated ground motions and cause a manipulation error, are baseline correction and high-pass filtering. Uncertainties in the processing of earthquake time series, and the selection of the high-pass corner frequency have been discussed in detail by Falamarz-Sheikhabadi (2017). It should be emphasized that the processing scheme should be the same for all spatially variable seismic excitations (seed and generated ones), as differences in processing would affect the displacement time series. Notably, most finite element codes require displacement time series as input excitations for the spatially variable ground motion case for exactly this reason.

9.3 Rotational Seismic Ground Motions

The SVSGMs not only cause differential excitations, but also induce rotational seismic ground motions (RSGMs), which may affect the seismic structural response (Fig. 9.7).

In spite of the fact that all six components, three translational and three rotational, are needed to describe the SGMs, very few research studies have investigated the effects of the rotational (rocking or torsional) components on the seismic behavior of

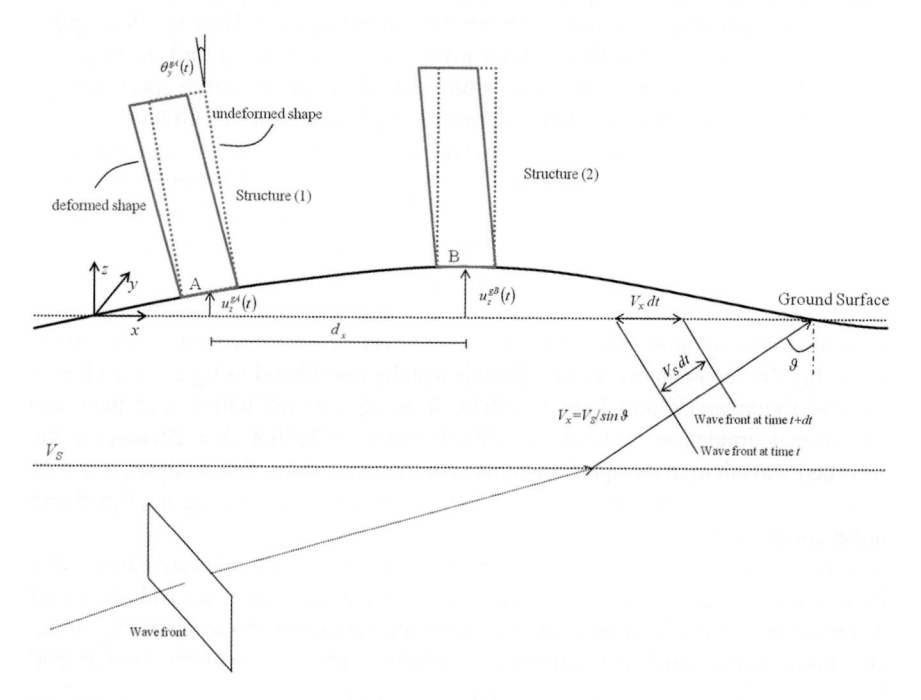

Fig. 9.7 Geometric layout for the seismic wave propagation and the spatial variability of ground motions

structures as compared to those of the translational ones. However, it has been shown that the RSGMs are influential on the seismic response of nuclear reactors (Rutenberg and Heidebrecht 1985), tall asymmetric buildings and irregular frames (Ghafory-Ashtiany and Singh 1986), slender tower-shape structures (Zembaty and Boffi 1994), multiple-support structures (Falamarz-Sheikhabadi et al. 2016), both vertically irregular and regular structures (Falamarz-Sheikhabadi 2014), particularly in the near field (Trifunac 2009). In spite of these studies, most codes do not appropriately address the effect of the seismic rotational loading in the design of structures (Falamarz-Sheikhabadi 2014). This may be justified because of the shortage of earthquake records of RSGMs. Ring laser gyroscopes (RLGs) are one of the most essential instruments for the direct measurement of the rotational components (Nigbor 1994; Takeo 1998). The RSGMs have also been estimated utilizing seismic data from dense arrays of accelerometers (Niazi 1986; Oliviera and Bolt 1989). For short separation distances of accelerometers, the overall characteristics of array-derived rotational time series are in fairly good agreement with the rotational components measured by RLGs (Suryanto et al. 2006). The method of tilt evaluation may be also used to estimate the rocking component using one seismic station by filtering the low-frequency components of its uncorrected strong-motion accelerogram starting from a characteristic frequency (Graizer 2006).

In engineering practice, the RSGMs are mostly simulated, using the classical elasticity theory (Basu et al. 2013; Falamarz-Sheikhabadi and Ghafory-Ashtinay 2012; Trifunac 1982; Zembaty and Boffi 1994), in terms of spatial derivatives of their corresponding translational ones as follows:

$$\vec{\theta}(t) = \frac{\partial u_z(t)}{\partial y}\vec{i} - \frac{\partial u_z(t)}{\partial x}\vec{j} - \frac{1}{2}\left(\frac{\partial u_x(t)}{\partial y} - \frac{\partial u_y(t)}{\partial x}\right)\vec{k} \quad (9.3)$$

where $\vec{i}, \vec{j}$, and $\vec{k}$ are unit vectors in the direction of the x-, y-, and z-axes, respectively (z is the vertical axis); t is time, and u and θ represent the translational and rotational ground motions, respectively. The first two terms on the right-hand side of Equation (9.3) are the rocking components related to the vertical ground motion, and the third term, the torsional component, is related to the horizontal motions. In the frequency domain, the rotational ground displacements may be obtained as:

$$\vec{\Theta}(\omega) = 2\pi i \frac{U_z(\omega)}{t_y(\omega)}\vec{i} - 2\pi i \frac{U_z(\omega)}{t_x(\omega)}\vec{j} - \pi i \left(\frac{U_y(\omega)}{t_x(\omega)} - \frac{U_x(\omega)}{t_y(\omega)}\right)\vec{k} \quad (9.4)$$

where t_j is the wavelength of the seismic waves along the jth-direction on the horizontal surface ($j = x$ or y). Considering the fact that the contribution of body waves to ground accelerations are dominant at close distances to the seismic source, the dispersion effect of surface waves may be ignored (surface waves mostly contribute to the rotational ground displacements in the far field). In this case, by introducing apparent propagation velocities along the x- and y-directions, as the velocity at which a plane wave appears to travel on the horizontal surface, the rotational acceleration components may be approximated by:

$$\vec{\Theta}(\omega) = i\omega \frac{\ddot{U}_z(\omega)}{V_y} \vec{i} - i\omega \frac{\ddot{U}_z(\omega)}{V_x} \vec{j} + i\frac{\omega}{2}\left(\frac{\ddot{U}_x(\omega)}{V_y} - \frac{\ddot{U}_y(\omega)}{V_x}\right)\vec{k} \quad (9.5)$$

with V_j being the constant apparent propagation velocity along the jth-direction. If the coordinate system is oriented along the principal axes, Eq. 9.5 may be simplified as:

$$\vec{\theta}(t) = 0\vec{i} - \frac{\ddot{u}_z(t)}{V_x}\vec{j} + \frac{\ddot{u}_y(t)}{V_x}\vec{k} \quad (9.6)$$

with the x- and y-axes now being defined along the radial and transverse axes, respectively. This simple relation (Eq. 9.6) is usually considered as a first-order approximation for the simulation of the rotational acceleration components. In Eq. 9.6, only the effect of the time delay in the propagation of seismic waves is considered to simulate the RSGMs, even though the loss of coherency may also contribute to the RSGMs. Falamarz-Sheikhabadi et al. (2016), for the first time, considered the combined action of both time delay and lagged coherency, modeled after the Luco and Wong (1986) lagged coherency model (Eq. 9.2), in the simulation of the RSGMs as follows:

$$\{\vec{\theta}(t)\} = \frac{0\vec{i} - R_\theta \ddot{u}_z(t)\vec{j} + R_\theta \ddot{u}_y(t)}{2}\vec{k}; \quad R_\theta = \sqrt{\frac{2\lambda_x^2}{V_S^2} + \frac{1}{V_x^2}} \quad (9.7)$$

In rotational seismology, $1/R_\theta$ is commonly referred to as the equivalent phase velocity. It can be deduced from Eq. 9.7, that the value of the apparent propagation velocity is always greater than the equivalent phase velocity. The ambiguity in the definition of these two characteristics of the seismic wave propagation led some studies to consider very small values for the apparent propagation velocity in the simulation of the SVSGMs.

Herein, it should be noted that the accuracy of the simulation of the RSGMs for practical applications depends heavily on engineering judgment. As indicated earlier in Sects. 9.2.1 and 9.2.2, empirical data are required for the selection of the values of V_x and λ_x, and such information is generally not provided in the seismological or geotechnical literature. In addition, there is no method for the evaluation of the value of the apparent propagation velocity in the near field. This difficulty usually results in the assumption of using the shear wave velocity of the surface layer instead of the apparent propagation velocity, which has been also adopted by design codes, such as Eurocode8 (2005). Certainly, such an assumption leads to large, unrealistic amplitudes for the RSGMs. In a seismic assessment, if the interaction of the rotational and translational components is ignored, this assumption results in an anti-optimal design of structures, particularly those located in high-seismicity zones. However, if they are simultaneously applied to the structure, the aforementioned assumption may result in the under- or over-estimation of the seismic forces applied to the structure due the inherent phase shift between the rotational and translational components. This issue may be even amplified if the effects of the soil-structure interaction, possible pounding of adjacent structures, and differential ground motions are also considered in the seismic assessment. Hence, further research on this topic is necessary for the safe design/retrofit of structures subjected to six earthquake components.

9.4 Soil-Structure Interaction

The deviation of the foundation input motion from the free-field ground motion is attributed to kinematic and dynamic soil-structure interaction effects, which are discussed in the following with emphasis on spatially variable input excitations.

9.4.1 Kinematic Soil-Structure Interaction

The kinematic constraint of the foundation movement, the foundation stiffness relative to that of its surrounding soil, and the SVSGMs result in the kinematic soil-structure interaction (KSSI). For "point" foundations, because their dimensions are relatively small compared to the wavelength of the seismic motions, it is reasonable to assume that the ground motions over the entire foundation area are the same. However, for shallow foundations with large-dimensions, the SVSGMs cause different movement at the soil-foundation interface locations. In this case, the effect of the KSSI on the input motion is usually evaluated via averaging the free-field motions within the footprint area of the foundation (Falamarz-Sheikhabadi and Ghafory-Ashtiany 2015; Veletsos et al. 1997). For deep foundations, the effects of the scattering and embedment should be also taken into account (Luco 1986). It has been shown that the effect of the KSSI is most noticeable at the higher frequencies of the excitation. This observation may be attributed to the fact that higher-frequency waves have shorter wavelengths in comparison to the lower-frequency ones, and, hence, the averaging effect is more pronounced for the former (Todorovska and Trifunac 1992). Herein, it should be noted that there is no unique approach for the consideration of the KSSI in seismic analysis, and most design codes or guidelines do not provide any recommendation for the incorporation of this phenomenon in the definition of the seismic excitation. One exception may be NEHRP (2015) that addresses the problem of the KSSI for shallow, embedded shallow and pile foundations in detail.

For multiple-support structures with extensive foundations, the effect of both the KSSI (multi-point effect) and asynchronous seismic excitation (multi-support effect) should be considered in the evaluation of the input motion at each structure's support. Hence, the SVSGMs need to be described at very short as well as long separation distances on the ground surface level and various soil depths. For this reason, the model of Luco and Wong (1986), described in Eq. (9.2), or the models developed by Abrahamson (1993, 2007), presented in Fig. 9.4, are the most appropriate. The difficulty in selecting appropriate values for the Luco and Wong (1986) model parameters has been discussed in Sect. 9.2.1. It should be noted, however, that essentially all coherency models have been developed from ground-surface records, and the variability of seismic ground motions at depth has not attracted attention.

For deep foundations, such as piles, the effect of the time delay, the incoherence variation with depth and the soil response should be considered to simulate the SVSGMs. In this case, care should be taken in appropriately estimating the time delay in the propagation of shear waves along the vertical direction, which is not the same as their apparent propagation in the horizontal direction. The apparent propagation velocity along the vertical direction (V_z) should be evaluated as follows:

$$V_z = V_x \tan(\beta) \quad (9.8)$$

where β is the incident angle of the shear waves at ground surface layer. This relationship between the apparent propagation velocities along the horizontal and vertical axes is, generally, ignored, and V_z is simply considered equal to the average shear wave velocity of the top 30 m underneath the ground surface (V_{S30}). In extreme loading scenarios, such an approach may result in contradictory values for the velocities, particularly if it is simultaneously assumed $V_x = V_z = V_{S30}$. This is due to the fact that when $V_x = V_{S30}$, then $V_z \to \infty$, and when $V_z = V_{S30}$, then $V_x \to \infty$. However, for reasonable values of the apparent propagation velocity (along the horizontal direction), the assumption that $V_z \cong V_{S30}$ may be acceptable. For example, for a site with $V_{S30} = 300$ m/s, if it is assumed that $V_x = 1000$ m/s and $\beta = 18°$, one may evaluate V_z using Eq. 9.8 as 320 m/s. The incoherence with depth, as noted earlier, is yet to be determined.

9.4.2 Dynamic Soil-Structure Interaction

The foundation vibration due to the soil flexibility affects the foundation input motion, leading to dynamic soil-structure interaction (DSSI). For structures supported on hard soil or rock it is common to ignore the DSSI. However, the effect of the soil/rock domain in the seismic response of important structures, e.g. bridges, tall buildings, dams, and nuclear power plants, can be significant. The DSSI may influence beneficially the seismic behavior of structures by increasing the structural damping and dissipating the input energy (Wolf 1985). On the other hand, it may be have a detrimental effect due to resonance and amplified P-Δ effects, which may result in structural instability (Wolf 1985). To consider the DSSI effect properly, the input motion should be applied at depth in the soil/rock domain, when a numerical (finite element) model of the structure is constructed. Additionally, to absorb the outgoing waves, appropriate conditions should be placed at the boundary between the finite computational domain and the infinite medium. In the following, two important issues regarding the consideration of the spatial variability of seismic ground motions in the numerical modeling of DSSI problems are briefly addressed.

9.4.2.1 Modeling of the Unbounded Soil/Rock Domain

Since the numerical modeling of the infinite soil/rock domain in DSSI problems is not feasible, the most common approach is to model only the domain near the structure (near-field domain) and apply appropriate boundary conditions at the outer end of the truncated model (Fig. 9.8). These conditions, commonly termed absorbing boundary conditions (ABCs), should satisfy the radiation condition at infinity, so as to ensure that no energy is reflected back into the system. Viscous dashpots are the easiest ABCs to implement, because they are independent of the finite domain, and work well for both time and frequency domain analyses (Kellezi 2000). Furthermore, they can perfectly absorb normally incident waves at the boundary. However, their ability to absorb inclined waves with a large angle of incidence ($\theta > 30°$), Rayleigh waves, and evanescent waves is very limited. Hence, for problems considering spatially variable ground motions, for which the apparent propagation of the motions is caused by the inclined wave incidence, their use is not recommended.

The use of infinite elements is an alternative approximate method to model the far-field unbounded domain. The mapped infinite elements work well for static

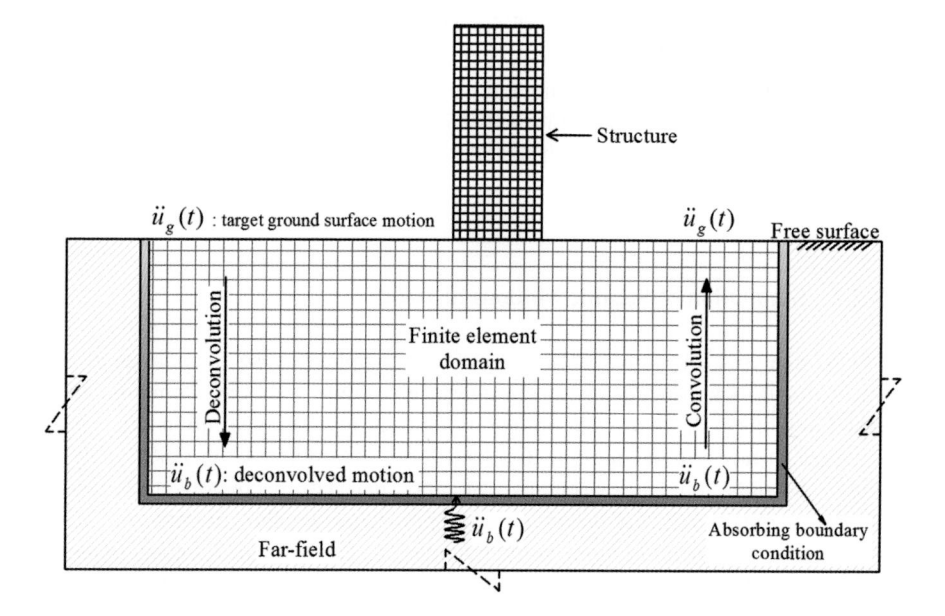

Fig. 9.8 Typical model of soil-structure-interaction (Adopted from Poul and Zerva 2018a)

problems in elastic media (Zienkiewicz et al. 1983). For wave propagation problems, this method cannot be used directly and should be combined with additional techniques (as, e.g. incorporating outwardly propagating wave-like factors in the formulation (Nenning 2010)). This common ABC cannot, also, appropriately model the radiation damping due to inclined seismic wave propagation.

A fairly recent approach, the Perfectly Matched Layer (PML) method, is one of the most powerful tools to fully absorb all outgoing waves at any angle of incidence and any frequency (Chew and Weedon 1994; Zheng and Huang 2002). PML was first proposed by Berenger (1994) to absorb electromagnetic waves in the finite-difference time domain. PML is a layer of an artificial material that is placed around the computational domain. Conditions that there is no reflection at the interface between the finite and PML domains and that the characteristics of the PML elements are such that they fully dissipate the outgoing waves are the key characteristics of the method. Furthermore, ABCs based on the PML approach are more adaptable because they can be expanded to heterogeneities and/or anisotropy in the material, as well as geometrical complications like corners and conformal boundaries (Zheng and Huang 2002). Theoretically, this method is exact and the outgoing waves are fully absorbed. However, when the problem is discretized (either with the finite-element or the finite-difference approaches), the wave equation is solved approximately and some small reflections may occur at the interface (Johnson 2010). In spite of this issue, PML is still the most recommended ABC when SVSGMs are considered in DSSI analyses.

To illustrate the performance of PML to absorb outgoing waves, the 2D model of the soil domain in Fig. 9.8 is subjected to a vertical impulse load at its free surface. The depth and length of the soil profile are 150 m and 300 m, respectively. The shear wave velocity and Poisson's ratio of the soil are considered as 276 m/s and 0.3, respectively. To absorb the outgoing waves, a PML layer with a width of 75 m is placed at the boundaries of the computational domain. The propagation of the waves at different times is presented in Fig. 9.9. The numerical results in the figure clearly demonstrate that the PML approach can effectively absorb the outgoing waves at any angle of incident.

9.4.2.2 Simulation of Seismic Motions at Depth

In DSSI analyses, the input motions need to be applied at the boundaries of the truncated soil/rock domain, i.e. at the interface of finite and ABC domains. Such ground motions at depth are determined by performing a deconvolution analysis of the target surface ground motion. Evidently, the reliability of the DSSI analysis highly depends on the accuracy of the derived deconvolved motions. One-dimensional analytical deconvolution approaches are the common tool used to determine the ground motions at the boundaries of the bounded domain (Kramer 1996). The frequency-domain solution, obtained with the SHAKE program (Schnabel 1972), is usually utilized to deconvolve the surface ground motions for the dynamic

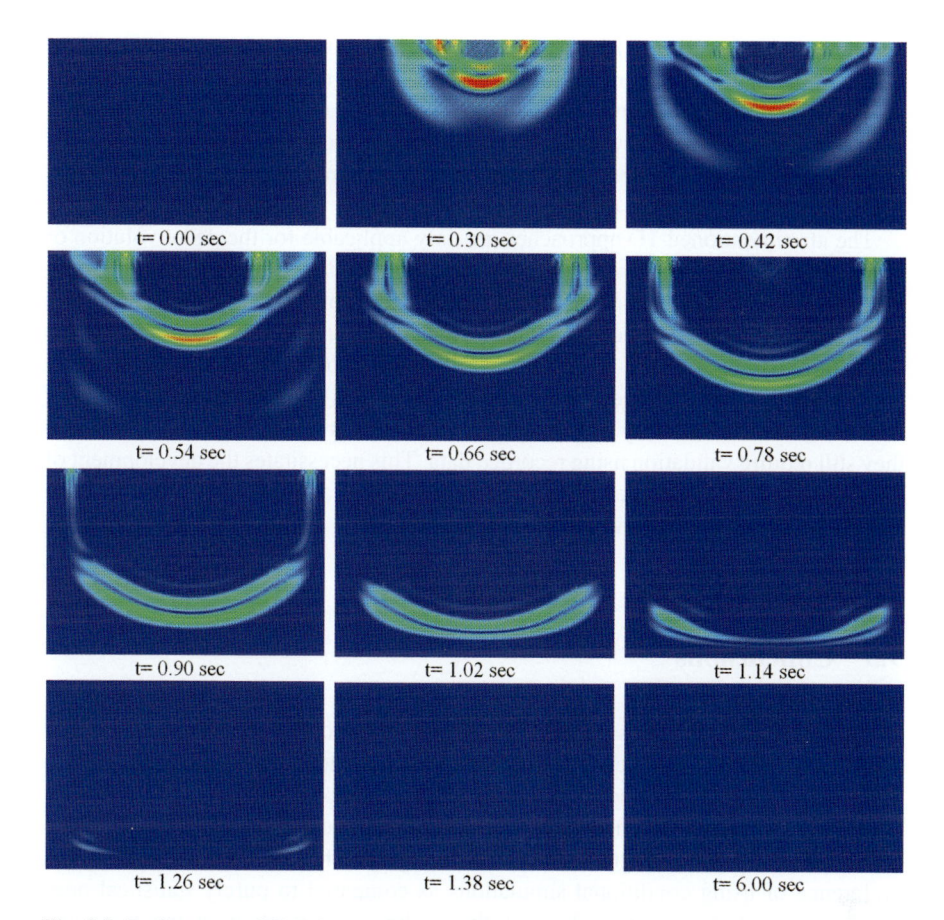

9 Issues with the Use of Spatially Variable Seismic Ground Motions in... 247

Fig. 9.9 Performance of PML in absorbing outgoing waves at any angle of incidence

analysis of structures. The basic idea is to generate motions at a given depth in the soil profile by utilizing the concept of inverse transfer functions. It should be noted that, if the deconvolved motion resulting from SHAKE is applied at the base of the discrete finite-element model of the soil profile, discrepancies between the convolved motion from the finite-element method (FEM) analysis and the target one should be expected (Poul and Zerva 2018b). The main reason is that the damping formulation and the solution approach in the FEM and SHAKE codes are quite different. Furthermore, the boundary conditions in the FEM and SHAKE are also different. Another issue regarding SHAKE is that it cannot be used for the deconvolution of the vertical component (compressional P-waves) of the seismic motion (Schnabel 1972). However, the vertical component should be also considered in the response evaluation of structures, and, thus, also needs to be deconvolved.

The deconvolution process can also be performed directly in finite-element, time-domain analyses in linear (Remier 1973) and equivalent linear-viscoelastic media (Poul and Zerva 2018a). The aim here is to determine the seismic motions at the appropriate depth of the soil profile by modifying the target surface ground motions based on the mathematical model of the system, which is assessed by analyzing the input-output data and determining the finite domain's gain function and phase shift. Poul and Zerva (2018a) showed that the time-domain deconvolution approach yields better estimates for the convolved motion as compared to those resulting from SHAKE over a wide range of frequencies.

The aforementioned 1D approaches are only applicable for the deconvolution of vertically propagating, fully coherent seismic waves. To partially overcome these difficulties, two- and three-dimensional numerical deconvolution approaches, such as the domain-reduction method (Bielak et al. 2003), have been proposed for the evaluation of the seismic excitation at the model boundaries. However, such approaches still have three significant drawbacks: (i) they neglect the loss of coherency effects; (ii) they are complex and computationally expensive, and (iii) they still require validation using recorded data. This necessitates the development of three-dimensional coherency models to appropriately address the problem of the simulation of seismic ground motions at depth.

9.5 Conclusions

This paper addresses issues pertinent to the use of spatially variable excitations for the seismic response of extended structures. Recommendations on the use of appropriate lagged coherency models and apparent propagation velocities at uniform sites have been proposed. The inadequacy of current analytical approaches to reproduce SVSGMs at sites with irregular subsurface topography have been documented. The preference of using conditional simulations as compared to purely statistical ones has been noted. New approaches for the analytical evaluation of the rotational components of the seismic excitation and the need to incorporate them in the seismic assessment of structures have been highlighted. The most appropriate ABC for use in cases when the SVSGMs are of importance was provided and the lack of validity of the 1D wave propagation in such problems was illustrated. The need of three-dimensional coherency models was also emphasized; research towards this goal is currently underway.

Acknowledgments This investigation was supported in part by the U.S. National Science Foundation under Grants No. CMMI-0600262, CMMI-0900179 and CMMI-1129396. Any opinions, findings, and conclusions or recommendations expressed herein are those of the authors and do not reflect the views of the U.S. National Science Foundation.

References

Abrahamson NA (1988) Spatial interpolation of array ground motions for engineering analysis. In: Proceedings of the ninth world conference on earthquake engineering, Tokyo

Abrahamson NA, (1993) Spatial variation of multiple support inputs. In: Proceedings of the first U.S. seminar on seismic evaluation and Retrofit of Steel Bridges. A Caltrans and University of California at Berkeley Seminar, San Francisco

Abrahamson NA (2007) Hard-rock coherency functions based on the Pinyon Flat array data, Draft report to EPRI. Electric Power Research Institute, Palo Alto

Abrahamson NA, Schneider JF, Stepp JC (1991) Spatial coherency of shear waves from the Lotung, Taiwan large-scale seismic test. Struct Saf 10:145–162

Akkar S, Bommer JJ (2006) Influence of long-period filter cut-off on elastic spectral displacements. Earthq Eng Struct Dyn 35:1145–1165

Assimaki D, Kausel E, Gazetas G (2005) Soil-dependent topographic effects: a case study from the 1999 Athens earthquake. Earthq Spectra 21:929–966

Baker JW, Cornell CA (2006) Spectral shape, epsilon and record selection. Earthq Eng Struct Dyn 35:1077–1095

Bard P-Y, Bouchon M (1985) The two-dimensional resonance of sediment-filled valleys. Bull Seismol Soc Am 75:519–541

Basu D, Constantinou MC, Whittaker AS (2013) Extracting rotational components of earthquake ground motion using data recorded at multiple stations. Earthq Eng Struct Dyn 42:451–463

Bérenger P (1994) A perfectly matched layer for the absorption of electromagnetic-waves. J Comput Phys 114(2):185–200

Bi K, Hao H (2011) Influence of irregular topography and random soil properties on coherency loss of spatial seismic ground motions. Earthq Eng Struct Dyn 40:1045–1061

Bielak J, Loukakis K, Hisada Y, Yoshimura C (2003) Domain reduction method for three-dimensional earthquake modeling in localized regions, part I: theory. Bull Seismol Soc Am 93(2):817–824

Boissieres HP, Vanmarcke EH (1995) Estimation of lags for a seismograph array: wave propagation and composite correlation. Soil Dyn Earthq Eng 14:5–22

Bouckovalas GD, Papadimitriou AG (2005) Numerical evaluation of slope topography effects on seismic ground motion. Soil Dyn Earthq Eng 25:547–558

Bradley BA (2012) A ground motion selection algorithm based on the generalized conditional intensity measure approach. Soil Dyn Earthq Eng 40:48–61

Chew WC, Weedon WH (1994) A 3d perfectly matched medium from modified maxwells equations with stretched coordinates. Microw Opt Technol Lett 7(13):599–604

Deodatis G, Shinozuka M (1989) Simulation of seismic ground motion using stochastic waves. J Eng Mech 115:2723

Der Kiureghian A (1996) A coherency model for spatially varying ground motions. Earthq Eng Struct Dyn 25:99–111

Der Kiureghian A, Neuenhofer A (1992) Response spectrum method for multi-support seismic excitations. Earthq Eng Struct Dyn 21:715–740

Eurocode 8: Design of Structures for Earthquake Resistance – Part 2: Bridges, BS EN 1998–2:2005, Brussels, Belgium (2005)

Eurocode 8: Design of Structures for Earthquake Resistance – Part 5: Foundations, retaining structures and geotechnical aspects, BS EN 1998–5:2004, Brussels, Belgium (2004)

Falamarz-Sheikhabadi MR, Ghafory-Ashtiany M (2012) Approximate formulas for rotational effects in earthquake engineering. J Seismol 16:815–827

Falamarz-Sheikhabadi MR (2014) Simplified relations for the application of rotational components to seismic design codes. Eng Struct 69:141–152

Falamarz-Sheikhabadi MR (2017) Excitation, model and analysis uncertainties in seismic assessment of bridges. Ph.D. dissertation, Department of Civil, Architectural and Environmental Engineering, Drexel University, Philadelphia, PA

Falamarz-Sheikhabadi MR, Ghafory-Ashtiany M (2015) Rotational components in structural loading. Soil Dyn Earthq Eng 75:220–233

Falamarz-Sheikhabadi MR, Zerva A (2017) Analytical seismic assessment of a tall long-span curved reinforced-concrete bridge. Part II: structural response. J Earthq Eng 21:1335. https://doi.org/10.1080/13632469.2016.1211566

Falamarz-Sheikhabadi MR, Zerva A, Ghafory-Ashtiany M (2016) Mean absolute input energy for in-plane vibrations of multiple-support structures subjected to horizontal and rocking components. J Probab Eng Mech 45:87–101

Frankel A, Hough S, Friberg P, Busby R (1991) Observations of Loma Prieta aftershocks from a dense array in Sunnyvale, California. Bull Seismol Soc Am 00:1900–1922

French Association for Earthquake Engineering (1995) Recommendations AFPS95 for the reduction of rules relative to the structures and installations built in regions prone to earthquakes, Saint-Remy-les-Chevreuse, France

Gatmiri B, Pouneh M, Arson C (2009) Site-specific spectral response of seismic movement due to geometrical and geotechnical characteristics of sites. Soil Dyn Earthq Eng 29:51–70

Ghafory-Ashtiany M, Singh MP (1986) Structural response for six correlated earthquake components. Earthq Eng Struct Dyn 14:103–119

Goldstein P, Archuleta RJ (1991) Deterministic frequency wave number methods and direct measurements of rupture propagation during earthquakes using a dense array: data analysis. J Geophys Res 96:6187–6198

Graizer V (2006) Tilts in strong ground motion. Bull Seismol Soc Am 96:2090–2102

Hao H, Oliveira CS, Penzien J (1989) Multiple-station ground motion processing and simulation based on SMART-1 array data. Nucl Eng Des 111:293–310

Harichandran RS, Vanmarcke EH (1986) Stochastic variation of earthquake ground motion in space and time. J Eng Mech 112:154–174

Harichandran RS, Hawwari A, Sweidan BN (1996) Response of long-span bridges to spatially varying ground motion. J Struct Eng 122:476–484

Imtiaz A, Cornou C, Bard P-Y, Zerva A (2017) Effects of site geometry on short-distance spatial coherency in Argostoli, Greece. Bull Earthq Eng. https://doi.org/10.1007/s10518-017-0270-z

Johnson S (2010) Notes on perfectly matched layers. Technical report, Massachusetts Institute of Technology, Cambridge, MA

Kamalian M, Jafari MK, Sohrabi-Bidar A, Razmkhah A (2008) Seismic response of 2-D semi-sine shaped hills to vertically propagating incident waves: amplification patterns and engineering applications. Earthq Spectra 24:405–430

Khazaei Poul M, Zerva A (2018a) Efficient time-domain deconvolution of seismic ground motions using the equivalent-linear method, Soil Dyn Earthq Eng (under review)

Khazaei Poul M, Zerva A (2018b) Nonlinear dynamic response of concrete gravity dams considering the deconvolution process. Soil Dyn Earthq Eng (under review)

Kellezi L (2000) Local transmitting boundaries for transient elastic analysis. Soil Dyn Earthq Eng 19(7):533–547

Kim S, Stewart JP (2003) Kinematic soil-structure interaction from strong motion recordings. J Geotech Geoenviron Eng, ASCE 129:323–335

Kim JH, Rosowsky DV (2005) Fragility analysis for performance-based seismic design of engineered wood shearwalls. J Struct Eng ASCE 131:1764–1773

Kolymbas D (2003) Advanced mathematical and computational geomechanics. Springer, New York city

Kramer SL (1996) Geotechnical earthquake engineering. Prentice Hall, Upper Saddle River

Kwon O-S, Elnashai A (2006) The effect of material and ground motion uncertainty on the seismic vulnerability curves of RC structure. Eng Struct 28:289–303

Luco JE (1986) On the relation between radiation and scattering problems for foundations embedded in an elastic half-space. Soil Dyn Earthq Eng 5:97–101

Liao S (2006) Physical characterization of ground motion spatial variation and conditional simulation for performance-based design. Ph.D. dissertation, Department of Civil, Architectural and Environmental Engineering, Drexel University, Philadelphia, PA

Liao S, Zerva A (2006) Physically-compliant, conditionally simulated spatially variable ground motions for performance-based design. Earthq Eng Struct Dyn 35:891–919

Loh C H Penzien J (1984) Identification of wave types, directions, and velocities using SMART-1 strong motion array data. In: 8th world conference on earthquake engineering, San Francisco, pp 191–198

Luco J, Wong H (1986) Response of a rigid foundation to a spatially random ground motion. Earthq Eng Struct Dyn 14:891–908

Nenning M (2010) Infinite elements for elasto- and poroelastodynamicsMonographic Series TU GRAZ – Computation in Engineering and Science, Vol. 8, pp. 1-171, Verlag der Technischen Universität Graz.

Niazi M (1986) Inferred displacements, velocities and rotations of a long rigid foundation located at el centro differential array site during the 1979 imperial valley, California, earthquake. Earthq Eng Struct Dyn 14:531–542

Nigbor RL (1994) Six-degree-of-freedom ground-motion measurement. Bull Seismol Soc Am 84:1665–1669

NEHRP (2015) Recommended seismic provisions for new buildings and other structures (FEMA P-1050-1). Building Seismic Safety Council, A council of the National Institute of Building Sciences, Washington, DC

Oliveira CS, Bolt BA (1989) Rotational components of surface strong ground motion. Earthq Eng Struct Dyn 18:517–528

Prestandard and commentary for the seismic rehabilitation of buildings, Federal Emergency Management Agency (FEMA 356), Washington, DC (2000)

Probabilistic Model Code, Joint Committee on Structural Safety, JCSS-OSTL/DIA/VROU-10-11-2000, (2001)

Reimer RB (1973) Deconvolution of seismic response for linear systems. Ph.D. thesis, Earthquake Engineering Research Center, University of California, Berkeley

Rutenberg A, Heidebrecht AC (1985) Response spectra for torsion, rocking and rigid foundations. Earthq Eng Struct Dyn 13:543–557

Safak E (1999) Wave propagation formulation of seismic response of multi-story buildings. J Struct Eng 125:426–437

Schnabel PB. SHAKE a computer program for earthquake response analysis of horizontally layered sites. EERC report. 1972:72–12

Spudich P (1994) Recent seismological insights into the spatial variation of earthquake ground motions, in new developments in earthquake ground motion estimation and implications for engineering design practice, ATC 35–1, Redwood City, CA

Stephenson WR (2000) The dominant resonance response of parkway basin. In: Proceedings of 12thworld conference on earthquake engineering, Auckland, New Zealand

Suryanto W, Igel H, Wassermann J, Cochard A, Schuberth B, Vollmer D, Scherbaum F, Schreiber U, Velikoseltsev A (2006) First comparison of array-derived rotational ground motions with direct ring laser measurements. Bull Seismol Soc Am 96:2059–2071

Takeo M (1998) Ground rotational motions recorded in near-source region of earth-quakes. Geophys Res Lett 25:789–792

Todorovska MI, Trifunac MD (1990) A note on excitation of long structures by ground waves. J Eng Mech Div ASCE 116:952–964

Todorovska MI, Trifunac MD (1992) The system damping, the system frequency and the system response peak amplitudes during in-plane building-soil interaction. Earthq Eng Struct Dyn 21:127–144

Trifunac MD (1982) A note on rotational components of earthquake motions on ground surface for incident body waves. Soil Dyn Earthq Eng 1:11–19

Trifunac MD (2009) The role of strong motion rotations in the response of structures near earthquake faults. Soil Dyn Earthq Eng 29:382–393

Uscinski BJ (1977) The elements of wave propagation in random media. McGraw-Hill International Book Co., New York

Vanmarcke EH, Fenton GA (1991) Conditioned simulation of local fields of earthquake ground motion. Struct Saf 10:247–264

Vanmarcke E, Heredia-Zavoni E, Fenton G (1993) Conditional simulation of spatially correlated earthquake ground motion. J Eng Mech 119:2333–2352

Veletsos AS, Prasad AM, Wu WH (1997) Transfer functions for rigid rectangular foundations. Earthq Eng Struct Dyn 26:5–17

Wolf JP (1985) Dynamic soil structure interaction. Prentice-Hall, Englewood Cliffs

Wong KKF, Harris JL (2012) Seismic damage and fragility analysis of structures with tuned mass dampers based on plastic energy. Struct Des Tall Special Build 21:296–310

Zembaty Z, Boffi G (1994) Effect of rotational seismic ground motion on dynamic response of slender towers. Eur Earthq Eng 1:3–11

Zembaty Z, Rutenberg A (2002) Spatial response spectra and site amplification effects. Eng Struct 24:1485–1496

Zerva A (1990) Response of multi-span beams to spatially incoherent seismic ground motions. Earthq Eng Struct Dyn 19:819–832

Zerva A (2009) Spatial variation of seismic ground motions: modeling and engineering applications, advances in engineering. CRC Press/Taylor & Francis Group, Boca Raton

Zerva A, Stephenson WR (2011) Stochastic characteristics of seismic excitations at a non-uniform (rock and soil) site. Soil Dyn Earthq Eng 31:1261–1284

Zerva A, Zervas V (2002) Spatial variation of seismic ground motions: an overview. J Appl Mech Rev, ASME 55:271–297

Zerva A, Zhang O (1996) Estimation of signal characteristics in seismic ground motions. Probab Eng Mech 11:229–242

Zerva A, Morikawa H, Sawada S (2012) Criteria for processing response-spectrum-compatible seismic accelerations simulated via spectral representation. Earthq Struct 3:341–363

Zheng Y, Huang X (2002) Anisotropic perfectly matched layers for elastic waves in cartesian and curvilinear coordinates. MIT Earth Resources Laboratory, Consortium Report. http://hdl.handle.net/1721.1/67857

Zienkiewicz OC, Emson C, Bettess P (1983) A novel boundary infinite element. Int J Numer Methods Eng 19:393–404

Chapter 10
Bridging the Gap Between Seismology and Engineering: Towards Real-Time Damage Assessment

Stefano Parolai, Michael Haas, Massimiliano Pittore, and Kevin Fleming

Abstract The development of earthquake early warning systems over the last decade has seen a number of studies that have focused either on improving the real-time estimation of seismological parameters, or on the rapid characterization of the possible damage suffered by a structure. However, the rapid increase in real-time seismic networks with stations installed in both the free field and inside buildings now offers the opportunity to combine the experience gained from these activities to develop a comprehensive real-time damage assessment scheme that, depending upon the time frame and spatial scale of interest, can provide useful information for a risk-based early warning system or for rapid loss assessment. Furthermore, newly developed instruments, with their enhanced computing capabilities, also offer the chance to combine early-warning procedures with the monitoring (during seismic crises) of a structure's behavior. In this paper, an overview of the state of the art in this multidisciplinary field will be given, and an outlook provided as to possible future developments.

10.1 Introduction

Since the beginning of industrialization and even more so in recent decades, the exponential increase in the human population and globalization has resulted in an accelerating trend for urbanization. The associated rapid growth of the extent of urban areas has led to a densification and concentration of people, infrastructure, and wealth within a relatively small extent. Often, these areas are located within regions where seismic hazard is significant, leading to increased levels of seismic risk, often without adequate preparedness or appropriate urban planning. This problem is

S. Parolai (✉)
Istituto Nazionale di Oceanografia e di Geofisica Sperimentale – OGS, Sgonico, Italy
e-mail: sparolai@inogs.it

M. Haas · M. Pittore · K. Fleming
GFZ German Research Centre for Geosciences, Potsdam, Germany

especially acute in developing countries, where there is a strong discrepancy between actual risk and risk awareness. Thus, this global problem calls for improved methodologies for seismic risk assessment (also in real or near-real time), and measures and tools to appropriately and efficiently improve preparedness and mitigate the risks.

One of the pillars of disaster risk reduction is the implementation of early warning systems. For this reason, the development of Earthquake Early Warning Systems (EEWS) has been a focus of seismologists and engineers who have shown that the effectiveness of these systems could significantly contribute to seismic risk mitigation (e.g., Wenzel and Zschau 2014; Wu et al. 2016; Strauss and Allen 2016; Clinton et al. 2016). In step with scientific advances, the interest of decision makers in this topic has grown, and the United Nations stress the relevance of early warning systems in general, as recently documented in the Sendai Framework that recommended the support of the development of multi-hazard early warning, disaster risk information and loss assessment systems by 2030.

The identification of early warning and/or rapid loss assessment tools as important components for tackling natural disasters, and therefore earthquake loss mitigation, calls for a new generation of procedures and methodologies to be developed. Given the modern computing capabilities and scientific knowledgebase, these methodologies should not only be able to identify the expected local intensity of the hazard, i.e., ground motion (for on-site systems) or its spatial distribution (for regional systems), but should be able to forecast the impact of the hazard on the built environment itself. This will allow decision makers to base their decisions on the anticipated losses themselves, rather than only using the individual components of the risk equation as an indicator, for example, only considering the expected ground motion. Such a relatively new and more focused way of thinking calls for closer cooperation between the seismological and engineering communities. Furthermore, more effective down-stream risk communication will be required, therefore requiring the stronger involvement of the social scientist community.

In the following, we illustrate how the movement in the direction of holistic earthquake early warning and rapid response systems could allow for a better and a more flexible way of tailoring such systems to the needs of different potential stakeholders and end users within the public or private sectors. In order to provide the right context, we first provide a short overview of the general state of the art regarding earthquake early warning and rapid response systems, and, afterwards describe some recent advancements that we, to the best of our knowledge, consider significant for future developments in the field. Finally, we will indicate which directions should be considered to improve the performance of EEWS and their suitability for helping to limit human and economic loses after an earthquake, including areas where only sparse strong motion networks are available.

10.2 Earthquake Early Warning Systems

As far back as 1868, the American physician J. D. Cooper (1868) proposed the idea of an early warning system based on the use of the electric telegraph to transmit information. Since information transmitted by telegraph is much faster than seismic waves (having a velocity of a few km/s), a warning can be issued at a target site. In practice, once an earthquake is detected by a nearby seismic network, an early warning system can alert residents in the vicinity before the destructive seismic waves reach them. In addition to the warning time gained this way, these systems can exploit the fact that most of the destructive energy is carried by lower-velocity S-waves and surface waves, while the higher-velocity P-waves, which generate lower levels of ground motion, can provide advance warning (Bormann et al. 2012: Chapter 2: 1–105). The possible amount of advance warning from this effect thus depends on the difference in the time it takes for P-waves and S-waves to arrive at a particular location.

Nowadays, we differentiate between regional network-based early warning systems and those installed as on-site standalone stations. The regional type of early warning system relies on the rapid data transmission capabilities provided by modern communication technologies. When *network stations* register seismic vibrations near the source of an earthquake, they use the measured arrival times of the seismic waves and the amplitude of ground motion at the stations to estimate in real time the location and strength of the earthquake. The accuracy of such estimates increases with the number of stations that record the seismic waves. This, however, also implies that a larger area will be affected by the seismic waves as time passes (e.g., Parolai et al. 2015). Using the estimated earthquake magnitude and location, it is then possible, through the application of so-called attenuation functions or ground motion prediction equations, to estimate the expected ground motion in a region and, depending on the estimated value, issue an alert when necessary. If the stations are located close enough to the earthquake source and population centers are far enough, these systems may allow sufficient time (*lead time*, meaning the time between the issuing of an alert and the arrival of the S-waves, although considering the arrival of the shaking phases that exceed a predefined level of shaking that can be dangerous for a structure might be more appropriate) for safety measures to be undertaken (including automatic actions such as closing gas lines). Such systems are therefore of primary value for alerting a potentially endangered population.

On-site early warning systems (OSEW) on the other hand are based on analyzing in real time the information collected by a ground motion sensor installed directly at the location where the warning is to be triggered. In some cases, the analysis of the data and the decision as to whether to issue an alarm or not is taken directly by the sensor itself (see, e.g., Parolai et al. 2015, 2017), leading to what is referred to as a Decentralized On Site Early warning system (DOSEW). The general process followed involves the sensor at the target site first registering the arriving P-waves, the amplitude of which is then used to estimate, through empirical relationships, the expected amplitude of the approaching S-waves. The lead time for both types of systems depends on the distance to the epicenter. In regional systems, the lead time also depends on how early the earthquake is detected and hence on the proximity of a monitoring station to the earthquake's focus. In general, OSEW systems provide short lead times (a few seconds) but are important for cases of a sparse seismic network for which detection times for, e.g., moderate crustal events from less well monitored seismogenic sources can be relatively long and can find important applications, in particular for critical facilities and industrial infrastructure where automatic procedures can be activated within a very short time. The area around a certain target, within which if an earthquake occur the lead time will be less than or equal to zero, is termed the *blind zone*. The size of this area can be strongly reduced, and the lead time increased for more distant events, if a combination of regional and onsite early warning is used. This is even the case when the network is sparse and not optimized for alarming.

Figure 10.1 compares for the case of the city of Almaty in Kazakhstan as to how the combination of a regional system, using recordings collected by a sparse strong motion network installed in nearby Kyrgyzstan (Parolai et al. 2017), with an on-site one based on recordings made by a single station within the city, yields considerable lead times.

The lead-times here are estimated by simulating the performance of a regional approach (left panel) by calculating the P-wave and the S-wave travel times between the hypocenter (after having discretized the area into a 0.1° grid where each point represents a potential epicenter) and the target location, and between the hypocenter and the stations' locations (assuming a depth of 10 km, Parolai et al. 2017). Under the assumption that a reasonable location and magnitude estimate of the earthquake

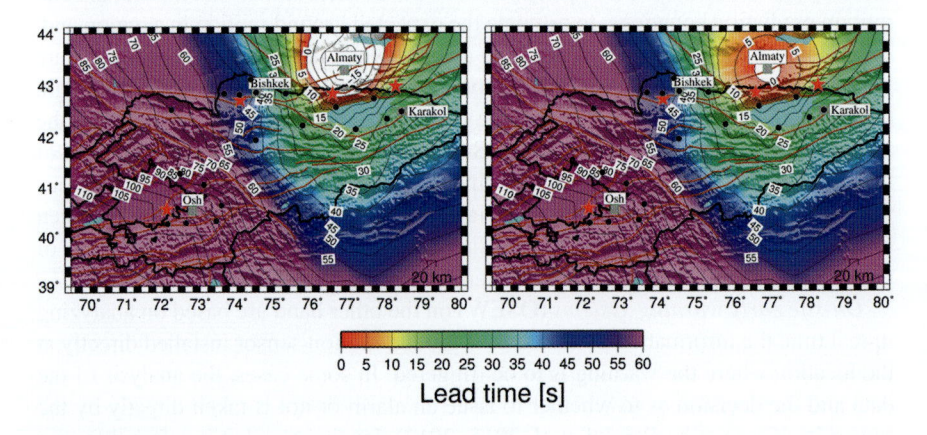

Fig. 10.1 *Left*: The lead-times for Almaty (Kazakhstan) when using a regional approach as a function of the location of possible seismic events (i.e., a 0.1 by 0.1° grid). *Right*: The lead-times for Almaty using a combined regional and on-site approach, as a function of the location of possible seismic events (i.e., a 0.1 by 0.1° grid). Also shown are the main cities (gray squares) and a selection of historical or hypothetical earthquakes (red stars). The brown lines indicate the main tectonic alignments and faults existing in the area

can be obtained when at least three stations are reached by the P-waves, the lead-time is then calculated by subtracting from the estimated S-wave travel time (Parolai et al. 2017) at the target site the P-waves travel time at the last triggered station, after taking into account another 4 s for data analysis and transmission. However, as can be seen by the white area surrounding Almaty representing the blind zone, there are limitations to this approach, These, in turn, can be partially rectified by incorporating on-site stations in Almaty (right panel), leading to some warning time.

Recently, Hoshiba (2013) presented a method for the correction of site amplification factors by applying an infinite impulse response filter to correct the site factor in real time and therefore to improve the forecasting of shaking at target localities. Pilz and Parolai (2016) further developed this approach by including the real-time correction of frequency-dependent site-response factors. This approach therefore accounts not only for the modulus, but also for the changes in the signal phase related to local site conditions and is particularly suitable to account for basin effects.

Stankiewicz et al. (2015) investigated the possibility of using the spectral content of the first few seconds after the P wave's arrival in a regional earthquake early warning framework to forecast the shaking at target localities independent of knowledge of the magnitude and location of the event.

The ongoing development of computationally powerful and energy-efficient wireless sensing units (e.g., Fleming et al. 2009; Boxberger et al. 2017) allows for the implementation of processing tools for specific real-time risk assessment tasks directly in the computational layer of the sensors. This is expected to be more informative than seismological earthquake early warning alone because it makes use of vulnerability and loss models specific for the structures being served by the system. In fact, earthquake early warning is of increasing interest to engineers owing to its potential to trigger real-time actions to reduce seismic risk. The basic concepts of this Performance-Based Earthquake Early Warning have been developed, for example, in Iervolino (2011). Such an approach can obviously combine seismological and engineering expertise and therefore can find applications in both regional and on-site systems (Bindi et al. 2015).

10.3 Loss-Driven Earthquake Early Warning and Rapid Risk Assessment

An efficient cooperation between the seismological and the engineering community has led to efforts to develop early warning/rapid response systems that allow the triggering of real-time actions (either at local or regional scales) based on rapid loss and risk assessments. Pittore et al. (2014), for example, proposed embedding a risk estimation module that extracts from a portfolio of precomputed scenarios those matching the characterization of the event detected by a real time seismic network (Picozzi et al. 2013, 2014).

Such a system is composed of offline and online components. The former includes information regarding the hazard and risk modules (input data) and the ground motion and loss scenarios results (output data). In particular, the risk module includes all available information about the exposure and vulnerability of the target area. In the same block are included all the (pre-calculated off-line) simulations and optimization processing stages, which are used to produce the scenarios, starting from a library of events' parameters input. The real-time block includes both an on-site and regional monitoring network. The above mentioned scenario can also be refined dynamically by taking into account the observations from the on-site components of the network.

The main task of the system is the identification of the most suitable loss scenario from the portfolio, dependent upon the characteristic of the identified seismic event. Note that the risk scenario is able to provide information about the possible spatial distribution of loss, providing the opportunity for decision makers either to spatially differentiate the level of alarm to be issued, or, in the post-event phase, to optimize the response and rescue team intervention.

In summary the system can work in the following manner:

- Automatic identification of the occurrence of an earthquake when the minimum number of stations required is triggered.
- Using evolutionary approaches (e.g., Satriano et al. 2008) for rapid location and magnitude estimation (e.g., Allen and Kanamori 2003). Alternatively, the system can directly estimate the expected shaking scenario at the target site using the approach of Stankiewicz et al. (2015).
- Using the estimated magnitude and location or the ground motion scenario as in Stankiewicz et al. (2015) to restrict the suitable candidates in the scenario portfolio.

The decision on the level of alarm to be issued is then made based on the expected damage and loss distribution.

In contrast to precomputed scenarios, the combination of modern computing power, efficient algorithms, and sufficiently resolved exposure and vulnerability models will allow for ad-hoc computations of impact scenarios in near real-time. Fig. 10.2 presents an example of an impact assessment for a possible scenario (calculated using the software tool CARAVAN, Parolai et al. (2016) in the event of the repetition of the $M = 7.8$ 1911 Kemin earthquake (the easternmost event indicated by a star symbol in Fig. 10.1). Parolai et al. (2017) showed that for the area with estimated macroseismic intensities less than VII (and for some with.

Intensities greater than VII located in the western part of the Issy kul lake, south of Almaty and west of Karakol, see Fig. 10.1) several seconds (including up to more than 20 s for the city of Karakol) of lead time would be possible with the existing configuration of the regional strong motion network ACROSS (the name is derived by the initiative supporting it namely within the framework of the Advanced Remote Sensing—Ground Truth Demo and Test Facilities project). This means that it would be possible to mitigate the forecasted loss indicated for the western part of the area (Fig. 10.2). Note, that the more focused information provided by the loss map (compared to that shown by the

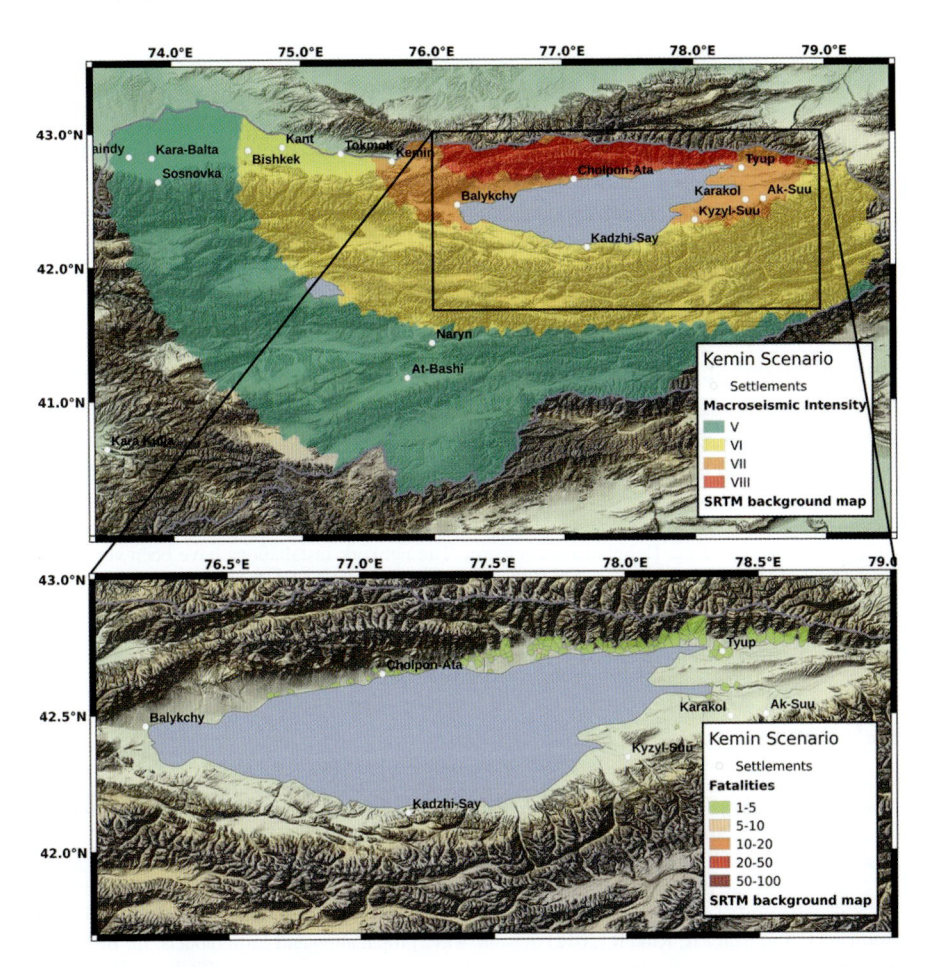

Fig. 10.2 Top: Macroseismic intensity scenario for the case of a repetition of the M = 7.8 1911 Kemin earthquake. Bottom: The corresponding loss scenario (smaller extent, marked by rectangle in the Top)

ground motion intensity map, is certainly of great benefit for decision makers and authorities also in terms of improving post-disaster response actions.

10.4 Conclusions

The combination of the seismological and engineering expertise is becoming a necessity for earthquake early warning systems so that they can provide the necessary tools to decision makers to allow them to select the appropriate alarms criteria by considering the consequences of the hazard resulting an event. Moreover, recent developments in technology allow sophisticated calculations in real time to be run, and the capability of rapidly manipulating large data bases and portfolios has made it possible to not only use such results for early warning at regional and onsite scales, but also for developing holistic systems where the early warning function is coupled with the rapid response information one.

To this regards, recent attempts at the on-site level of combining the traditionally separate tasks of earthquake early warning and health status monitoring of structures (e.g., Bindi et al. 2015) are another example of efforts to bridge the seismological and engineering disciplines. These attempts are promising, especially when considering the necessity of extending the performance of an early warning/rapid risk assessment system over the whole duration of a seismic sequence (and therefore considering the possible cumulative damage to a structure to which aftershocks may contribute) or for systems developed ad-hoc for the case of human induced seismicity (Megalooikonomou et al. 2018).

Acknowledgements The installation of the strong motion network has been financed by the ACROSS initiative of the Helmholtz association. The network installations have been supported by the Global Change Observatory Central Asia of GFZ. Figure 10.1 has been draw using the GMT software (Wessel and Smith 1991).

References

Allen RM, Kanamori H (2003) The potential for earthquake early warning in Southern California. Science 300(5620):786–789. https://doi.org/10.1126/science.1080912

Bindi D, Boxberger T, Orunbaev S, Pilz M, Stankiewicz J, Pittore M, Iervolino I, Ellguth E, Parolai S (2015) On-site early-warning system for Bishkek (Kyrgyzstan). Ann Geophys 58(1):S0112. https://doi.org/10.4401/ag-6664

Bormann P, Engdahl ER, Kind R (2012) Seismic wave propagation and earth models. In: Bormann P (ed) New manual of seismological observatory practice 2 (NMSOP2). Deutsches GeoForschungsZentrum (GFZ), Potsdam. https://doi.org/10.2312/GFZ.NMSOP-2_CH2

Boxberger T, Fleming K, Pittore M, Parolai S, Pilz M, Mikulla S (2017) The Multi-Parameter Wireless Sensing System (MPwise): its description and application to earthquake risk mitigation. Sensors 17(10):2400. https://doi.org/10.3390/s17102400

Clinton J, Zollo A, Marmureanu A, Zulfikar C, Parolai S (2016) State-of-the art and future of earthquake early warning in the European region. Bull Earthq Eng 14:2441. https://doi.org/10.1007/s10518-016-9922-7

Cooper JD (1868) Letter to editor. Daily Evening Bulletin, San Franciso

Fleming K, Picozzi M, Milkereit C, Kuhnlenz F, Lichtblau B, Fischer J, Zulfikar C, Ozel O, The SAFER and EDIM Working Groups (2009) The self-organizing seismic early warning information network (SOSEWIN). Seismol Res Lett 80(5):755–771. https://doi.org/10.1785/gssrl.80.5.755

Hoshiba M (2013) Real-time correction of frequency-dependent site amplification factors for application to earthquake early warning. Bull Seismol Soc Am 103(6):3179–3188. https://doi.org/10.1785/0120130060

Iervolino I (2011) Performance-based earthquake early warning. Soil Dyn Earthq Eng 31(2):209–222. https://doi.org/10.1016/j.soildyn.2010.07.010

Megalooikonomou KG, ParolaiS Pittore M (2018) Analytical fragility embodied in on-site early warning system for induced seismicity. In 16th European Conference on Earthquake Engineering (16 ECEE) (submitted, this issue). Thessaloniki

Parolai S, Bindi D, Boxberger T, Milkereit C, Fleming K, Pittore M (2015) On-site early warning and rapid damage forecasting using single stations: outcomes from the REAKT project. Seismol Res Lett 86(5):1393–1404. https://doi.org/10.1785/0220140205

Parolai S, Boxberger T, Pilz M, Bindi D, Pittore M, Wieland M, Haas M, Oth A, Milkereit C, Dahm T, Lauterjung J (2016) Auf dem Weg zur Schadensabschätzung in Echtzeit. System Erde (GFZ Journal) 6(1):32–37

Parolai S, Boxberger T, Pilz M, Fleming K, Haas M, Pittore M, Petrovic B, Moldobekov B, Zubovich A, Lauterjung J (2017) Assessing earthquake early warning using sparse networks in developing countries: case study of the Kyrgyz Republic. Front Earth Sci 5. https://doi.org/10.3389/feart.2017.00074

Picozzi M, Bindi D, Pittore M, Kieling K, Parolai S (2013) Real-time risk assessment in seismic early warning and rapid response: a feasibility study in Bishkek (Kyrgyzstan). J Seismol 17(2):485–505

Pilz M, Parolai S (2016) Ground-motion forecasting using a Reference Station and complex site-response functions accounting for the shallow geology. Bull Seismol Soc Am 106(4):1570–1583. https://doi.org/10.1785/0120150281

Pittore M, Bindi D, Stankiewicz J, Oth A, Wieland M, Boxberger T, Parolai S (2014) Toward a loss-driven earthquake early warning and rapid response system for Kyrgyzstan (Central Asia). Seismol Res Lett 85(6):1328–1340. https://doi.org/10.1785/0220140106

Satriano C, Lomax A, Zollo A (2008) Real-time evolutionary earthquake location for seismic early warning. Bull Seismol Soc Am 98(3):1482–1494. https://doi.org/10.1785/0120060159

Stankiewicz J, Bindi D, Oth A, Parolai S (2015) Toward a cross-border early-warning system for Central Asia. Ann Geophys 58(1.) Retrieved from http://www.annalsofgeophysics.eu/index.php/annals/article/view/6667

Strauss JA, Allen RM (2016) Benefits and costs of earthquake early warning. Seismol Res Lett 87(3):765–772. https://doi.org/10.1785/0220150149

Wenzel F, Zschau J (eds) (2014) Early warning for geological disasters: scientific methods and current practice. Springer, Berlin

Wessel P, Smith WHF (1991) Free software helps map and display data. EOS Trans Am Geophys Union 72(41):441–448

Wu Y, Liang W, Mittal H, Chao W, Lin C, Huang B, Lin C (2016) Performance of a low-cost earthquake early warning system (*P*-Alert) during the 2016 *M* L 6.4 Meinong (Taiwan) earthquake. Seismol Res Lett 87(5):1050–1059. https://doi.org/10.1785/0220160058

Chapter 11
Earthquake Geotechnics in Offshore Engineering

Amir M. Kaynia

Abstract This paper presents a number of geotechnical issues encountered in earthquake design of offshore structures and subsea facilities. Parallel with construction of traditional structures such as jackets and gravity-based structures, a considerable effort has recently been put to field developments in deep water. This has brought about other challenges that are largely dependent on geotechnical knowledge. This paper addresses some of the more recent approaches and solutions in geotechnical earthquake design of both shallow water and deep-water structures and facilities such as platforms with large bases, pipelines traversing slopes and seabed installations. It is demonstrated how incorporation of radiation damping and nonlinear soil-structure interaction in offshore installations could optimize the design. Considering the importance of earthquake stability of slopes in deep water development, special attention is given to highlighting several key issues in the earthquake response of submarine slopes including strain softening and three-dimensional shaking.

11.1 Introduction

This paper addresses a number of key geotechnical issues encountered in earthquake design of offshore structures and subsea facilities. The topics include fixed offshore platforms, including jacketed and gravity-based structures, earthquake analysis and stability of submarine slopes, response of pipelines, and subsea facilities. For offshore wind turbines, the reader is referred to other publications (e.g. Kaynia 2018). The main geotechnical issue in the earthquake analysis of jacketed offshore platforms is design of piles. Use of analytical solution and empirical methods based on p-y concept are discussed, and the effect of liquefaction on pile analysis is

---

A. M. Kaynia (✉)
Norwegian Geotechnical Institute (NGI), Oslo, Norway

Norwegian University of Science and Technology (NTNU), Trondheim, Norway
e-mail: amir.m.kaynia@ngi.no

© Springer International Publishing AG, part of Springer Nature 2018
K. Pitilakis (ed.), *Recent Advances in Earthquake Engineering in Europe*,
Geotechnical, Geological and Earthquake Engineering 46,
https://doi.org/10.1007/978-3-319-75741-4_11

reviewed. In gravity-based structures, the main issue is analysis of soil-structure interaction (SSI). To this end, the existing solutions for SSI analyses are reviewed and extension of the conventional three-step method to cases with flexible bases and nonlinear response are discussed.

With recent trends in deep water oil and gas development, the earthquake response of submarine slope has become a major issue. In this connection, the role of several factors such as strain softening, multi-direction earthquake shaking and 3-D slope geometry on earthquake response of slopes are evaluated by numerical solutions. Moreover, the response of pipelines traversing slopes are investigated using a numerical model developed for this purpose. Finally, the earthquake response of seabed facilities, such as manifolds are presented and the critical role of radiation damping in reducing the earthquake loading is demonstrated.

11.2 Seismic Design Philosophy

The seismic design philosophy in offshore design generally follows ISO standards for offshore structures. They require achieving acceptable low risks with respect to Health, Safety, and the Environment (HSE), economic loss, and interruption to normal operations (Younan et al. 2015). This design philosophy is reflected in the following two performance expectations:

1. Little or no damage or interruptions to normal operations during frequent earthquakes referred to as Extreme Level Earthquake (ELE) with return periods typically in the range 300–600 years.
2. No serious HSE consequences in rare earthquakes referred to as Abnormal Level Earthquake (ALE) with typical return periods in the range 2500–4000 years although the facility could be irreparable and result in an economic loss.

ISO 19901-2 (2004) has established a procedure for determination of the design return periods based on the seismic hazard condition at the site and ductility performance of the structure. In addition to these events, some operators demand demonstration of sufficient residual capacity to survive nominal post-ALE events, including aftershocks, to allow safe shutdown of facilities and rescue of personnel (Younan et al. 2015).

The input to earthquake analyses consists of time-history records, often representing bedrock or stiff soil outcrop, and matched to a uniform hazard spectrum (UHS) established by a probabilistic seismic hazard analysis (PSHA) covering both ELE and ALE events.

11.3 Earthquake Response of Submarine Slopes

Slopes are often encountered in the development of offshore fields in deep water. Even in cases where the seabed is flat at the location of the platforms and wells, there is often a need to connect a series of subsea clusters, or transport the oil and gas by pipeline to a near offshore platform or an onshore processing and storage facility. In such cases, the pipelines often have to cross or traverse slopes. Large downslope movement in slopes under earthquake can potentially damage the pipeline, and in the event of landslide, the moving failed mass can impact and destroy the subsea facilities located downstream from the escarpment. Figure 11.1 shows an example of a deep-water site in which pipelines are considered to run upslope in order to connect to the platform.

A variety of codes are available for numerical simulation of slope response under earthquake loading. The codes include 1-D solutions, 2-D models (e.g. PLAXIS and FLAC) and 3-D tools (e.g. PLAXIS 3D 2015; FLAC3D 2006). Analyses of slopes are therefore standardized in practice. This section addresses issues either not covered by these codes or ignored in practice often due to their complexity. They include: (a) strain softening, (b) multi-directional shaking, and (c) three-dimensional geometry.

11.3.1 Strain Softening

Geophysical surveys and geotechnical investigations at several deep offshore sites have indicated landslides in slopes with small angles. Some of the compelling

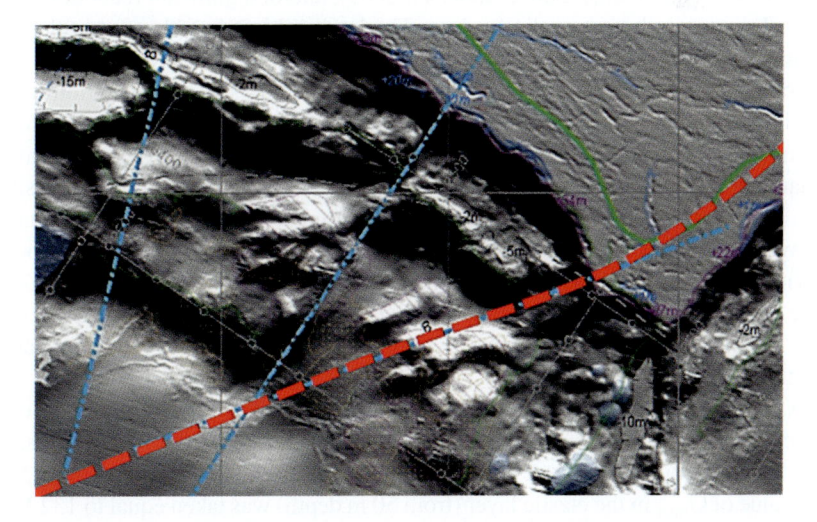

Fig. 11.1 Pipeline (dashed line) traversing submarine slope displaying earlier landslides

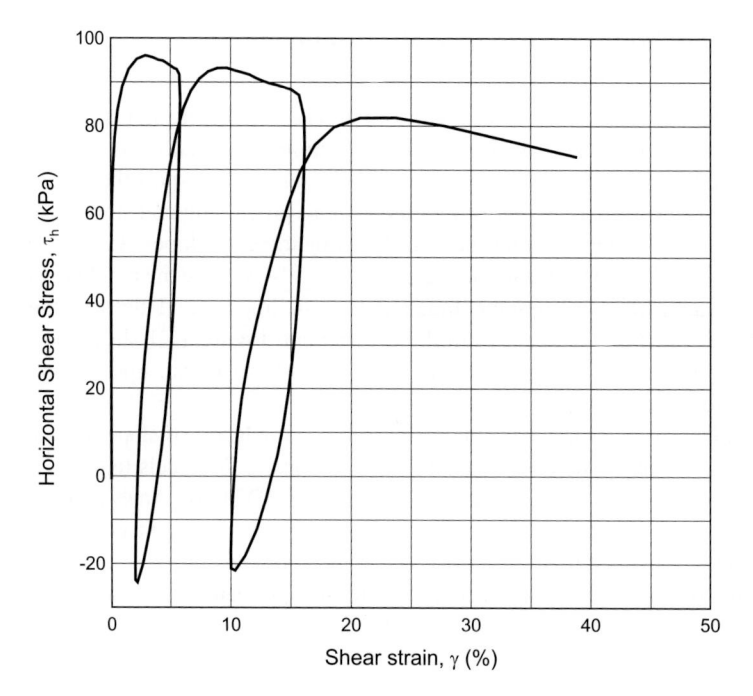

Fig. 11.2 Stress-strain curve for direct simple shear, DSS, test on sensitive clay

arguments for triggering of these landslides are reduction of the soil's shear strength due to strain softening under cyclic earthquake loading, post-earthquake failure due to creep (e.g. Nadim et al. 2007; Andersen 2009), and/or significant reduction of the static shear strength post cyclic loading. The latter subject is addressed in the next section. The strain softening behavior is illustrated in Fig. 11.2 by the stress-strain diagram for a DSS test on a clay sample from an offshore site. The intention was to run the test at constant stress, but the early failure of the sample helped capture the strain-softening response of the soil during subsequent cycles. A series of centrifuge dynamic tests have recently been carried out by Park and Kutter (2015) on sensitive clay. The results will provide a valuable opportunity to understand the dynamic behavior of sensitive clays and to calibrate or verify the existing numerical models.

To assess the effect of strain softening numerically, a number of 1-D analyses were performed using the numerical code *QUIVER_slope* (Kaynia 2012a). To this end, the software was first validated against a commercial 2D code for elastic-perfectly plastic (Mohr-Coulomb) soil behavior. An earthquake excitation with PGA = 0.3 g was applied at bedrock at 100 m depth. The top 50 m of the soil is NC clay with the normalized static shear strength $s_u^{DSS} = 0.20\sigma_v'$ and $G_{max}/s_u^{DSS} = 1100$. To account for the rate effect, the peak shear strength was increased by 30%. The value of G_{max} in the elastic layer (from 50 m depth) was taken equal to 132 MPa, which is 50% larger than G_{max} at the bottom of the NC clay layer.

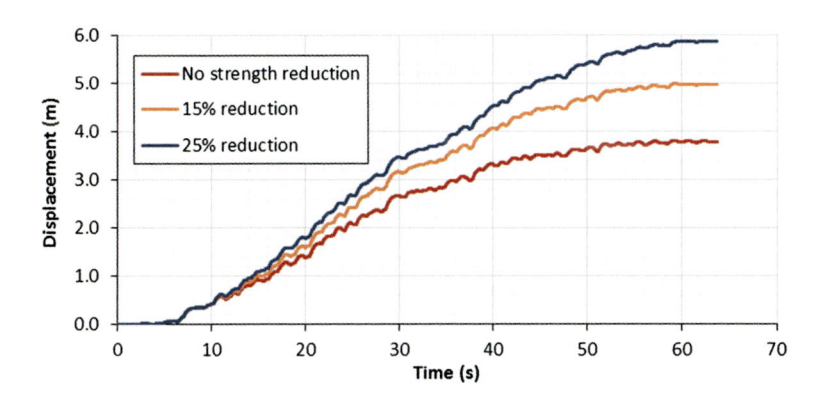

Fig. 11.3 Response of slope for different levels of strain softening

To highlight the role of strain softening, Fig. 11.3 presents the results of two analyses with strain softening behavior in which the peak shear strength was kept unchanged up to shear strain of 5% and reduced linearly to 85% and 75% of the peak strength at shear strain of 15%. The larger displacements compared to the elastic-perfectly plastic results (also shown in the figure), clearly show the importance of strain softening on the earthquake response of slopes.

Taiebat and Kaynia (2010) have developed a simple and practical version of the plasticity model SANICLAY (Dafalias et al. 2006) that accounts for de-structuration (strain softening) and anisotropy, and have implemented it as a user-defined model in FLAC3D (2006). In this model, the plastic potential surface in the triaxial p-q stress space is a rotated and distorted ellipse. The amount of rotation and distortion reflects the extent of anisotropy, and is controlled by an evolving variable α, which is scalar-valued in triaxial, and tensor-valued in multiaxial stress space. The model uses a non-associated flow rule that allows simulation of softening response under undrained compression following oedometric consolidation. Taiebat et al. (2011) used this model to compute 3-D earthquake response of a generic soft clay slope for different earthquake loading and material parameters. The results again highlight the important role of strain softening on the earthquake response of slopes.

11.3.2 Multi-directional Shaking

Early research into multi-directional shaking was primarily concerned with liquefaction analyses for level ground conditions or stability of earth dams. Pyke et al. (1975) performed a number of multi-directional shaking table tests on clayey sand. They found that the total settlement caused by the two (simultaneous) horizontal ground motion components was the same as the sum of the settlements caused by the ground motion components applied separately. Stewart and Yee (2012) found similar results for a series of 1-D and 2-D simple shear tests on dune sand from a nuclear power plant in Japan. Seed et al. (1978) found the same was true for excess pore pressure generation in sands. In addition, Pyke et al. (1975) noted that applying one horizontal and one vertical component increased the total settlement by 20–50% over the settlement caused by applying the horizontal component alone. Kammerer et al. (2003) performed extensive laboratory stress-controlled cyclic tests on granular soil and found that the soil response under multi-directional shearing tended to generate pore pressure faster than that of unidirectional shearing. Su and Li (2003) applied both unidirectional and multi-directional shaking to level saturated sand deposits in a centrifuge and found that the maximum pore pressure at great depths for multi-directional shaking was about 20% larger than in one-directional shaking and the difference reduced to about 10% near the surface.

Multi-directional shaking of slopes has more recently gained interest. Anantanavanich et al. (2012) performed seismic slope stability analyses for two generic offshore soft clay sites with depths of 20 m and 100 m and a slope angle of 10 degrees. They compared the estimated permanent displacements and excess pore pressures generated from applying one or both horizontal components of a ground motion at the same time. They found that multi-directional shaking predicted 20–40% increase in permanent displacements over unidirectional shaking. For the 100 m deep soil profile, multi-directional shaking predicted 30% increase in excess pore pressure at large depths, which reduced to 10% at the soil surface, whereas the 20 m deep soil profile predicted an increase in excess pore pressure between 20% and 40%.

Carlton and Kaynia (2016) conducted a number of numerical simulations in which 3-D slopes of NC clay with simple Mohr-Coulomb failure criterion were subjected to one-component and 3-component earthquake excitations. Through a number of case studies, they observed that inclusion of the earthquake component perpendicular to the slope direction increases the permanent downslope displacements and shear strains in the slope by 25–50% and by 10–50%, respectively. Figure 11.4 shows an example of the results for 3-component (above) and one-component (below) shaking. The shear strength increases linearly with depth with strengths of 5 kPa on surface and 300 kPa at depth 300 m. The slope angle is 15 degrees, and the earthquake record is the magnitude 6.5 California earthquake of 1954 at Ferndale City Hall scaled to 0.6 g on bedrock.

The results of the analyses were also used to assess the accuracy of the displacement predictive equations by Bray and Travasarou (2007) and Kaynia and Saygili (2014). The former equation uses PGA on bedrock and the latter uses the peak acceleration on ground surface. The results showed that while the equation by Bray and Travasarou (2007) provided a good estimate of the displacements, the one by Kaynia and Saygili (2014) predicted lower values. The corresponding equations for permanent strains in Kaynia and Saygili (2014) gave generally larger values than those simulated. The permanent shear strains are used to assess the static stability of slopes post earthquake loading. The mechanism is illustrated in Fig. 11.5 for cyclic shear tests on a sensitive clay in DSS (Andersen 2009). The figure shows both the stress-strain curve under a monotonic static loading (black curve) and two static tests right after a number of cyclic loads (red and blue curves). The results show a

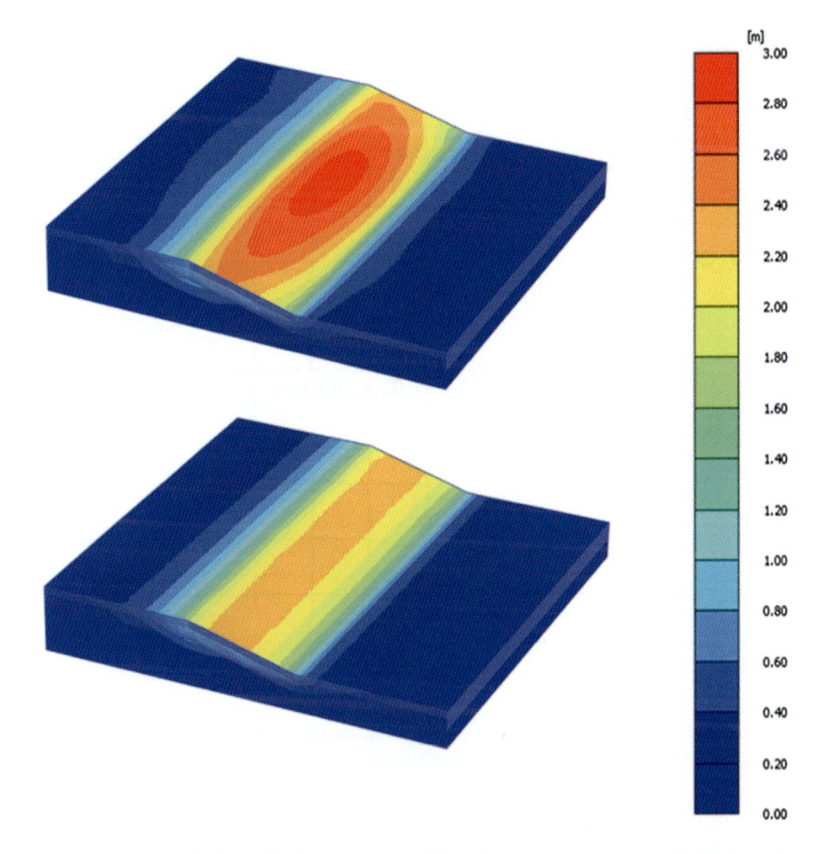

Fig. 11.4 Response of slope for 3-component (above) vs one-component (below) earthquake (Adapted from Carlton and Kaynia 2016)

significant reduction of the static shear strength after the cyclic loading. The reduction of the strength depends on the sensitivity of the clay and the accumulated shear strains during the cyclic loading, and varies with the type of clay, plasticity, consolidations stress and over-consolidation ratio, OCR.

11.3.3 Three-Dimensional Geometry

Although three-dimensional numerical codes have been available for some time, very few analyses of earthquake slope response with three-dimensional geometries have been reported in the literature. This could be attributed in part to the complexity of generating 3-D element meshes and demanding computational power, both of which have been extensively improved in recent years.

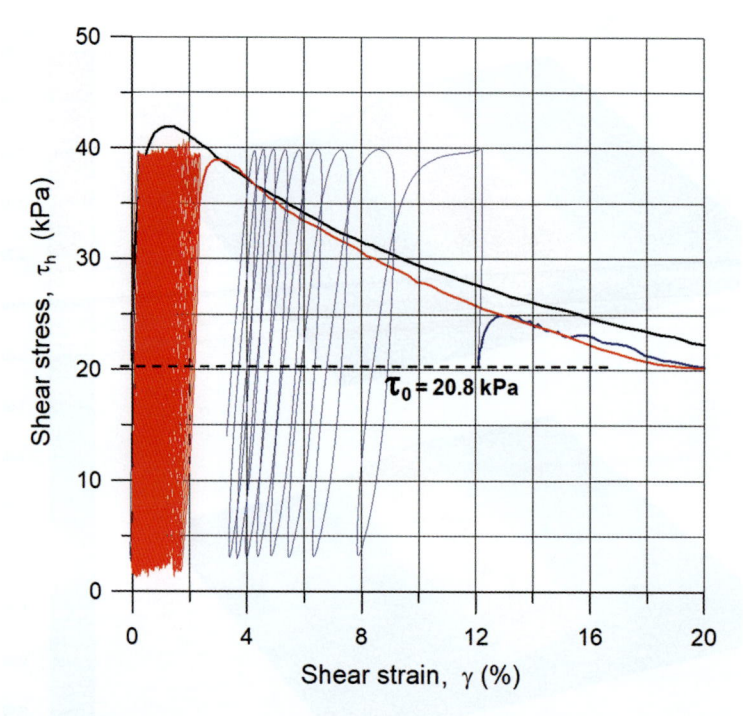

Fig. 11.5 Direct simple shear test on normally consolidated sensitive clay showing effect of cyclic strain accumulation on post cyclic shear strength (After Andersen 2009)

Another reason for ignoring 3-D effects is the findings from a number of studies on the static stability of slopes (e.g. Duncan 1996) that have concluded that consideration of 3-D response improves the static safety factor. The studies reported in Azizian and Popescu (2006) for 3-D seismic analyses of submarine slopes have also concluded that the results of 2-D and 3-D analyses are generally close. The above studies have focused on the effect of 3-D model extension on the critical failure surface.

A different approach has been used by Ferrari (2012) to address the issue of 3-D slope response. Figure 11.6 shows a 3-D slope configuration with a homogeneous soil with $s_u = 25$ kPa and $G_{max} = 25$ MPa. In addition to the usual slope angle used in 2-D models (long side in the figure), there is also a sloping face normal to the main direction (short side in the figure). The two slope angles are denoted as α and β in the following.

This model was excited in the long direction by ten cycles of harmonic (sine) wave with different frequencies. Figure 11.7 displays contours of the computed permanent slope displacements for the case $\alpha = \beta = 1:4$, frequency of 2 Hz and peak acceleration equal to 0.15 g. Due to slope angle β, the static shear stresses in the soil elements in this model are larger than those in the corresponding 2-D models; this is expected to increase the nonlinear response of the soil and permanent displacements.

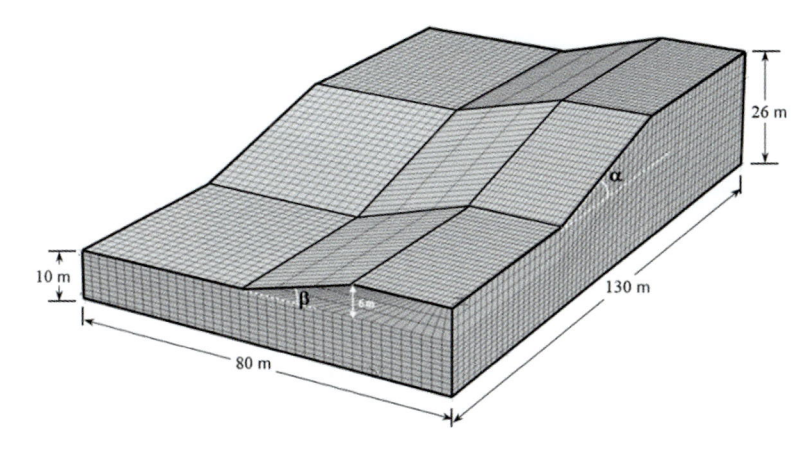

11 Earthquake Geotechnics in Offshore Engineering

Fig. 11.6 Geometry and parameters of 3-D slope model

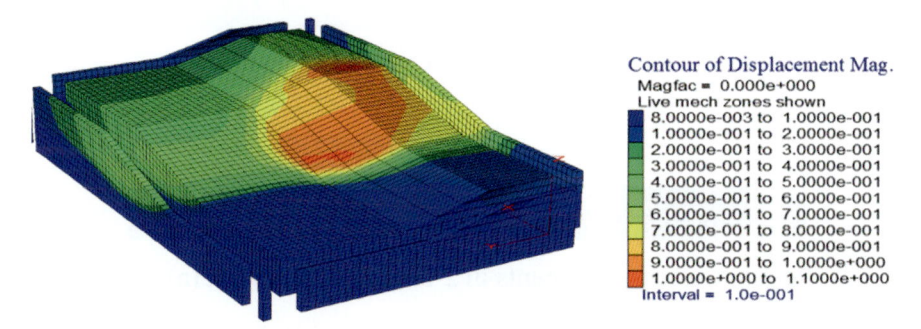

Fig. 11.7 Contours of permanent slope displacements for slope angles 1:4 and input peak acceleration 0.15 g

Figure 11.8 shows two cross-sections through the slope and normal to the direction of excitation. In order to assess the effect of 3-D geometry, the responses of these sections were evaluated separately as 2-D models under the same earthquake excitation. Figure 11.9 displays the contours of earthquake induced displacements in section (a) together with the corresponding results in the same sections of the original 3-D model. Figure 11.10 shows the corresponding comparison for responses in section (b). Comparing the maximum displacements for these sections, one can observe that the additional slope normal to the main slope amplifies the displacements by a factor ranging from about 20% to 50%. These results point out the importance of considering 3-D slope geometry in earthquake analyses.

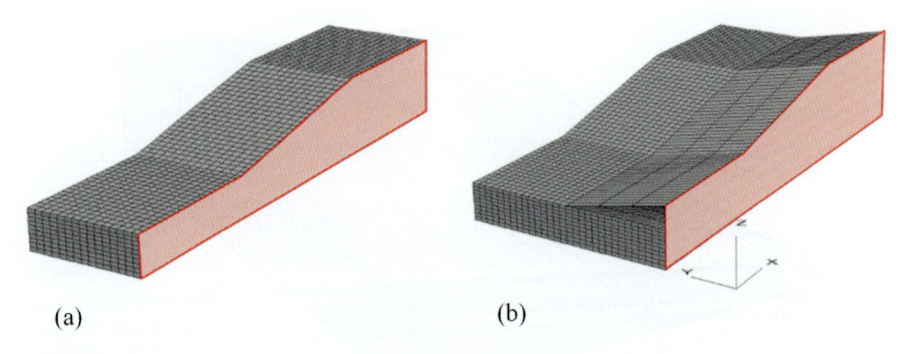

Fig. 11.8 Sections through 3-D slope and equivalent 2-D models

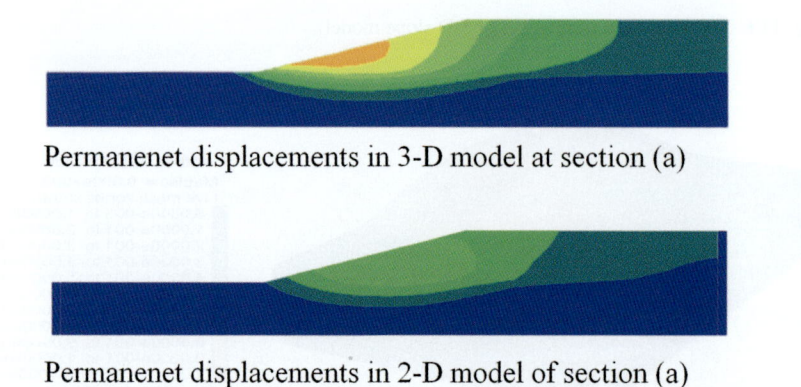

Permanenet displacements in 3-D model at section (a)

Permanenet displacements in 2-D model of section (a)

Fig. 11.9 Comparison between response of 3-D model at section (a) in Fig. 11.8 with response of 2-D model (for color scale, see legend in Fig. 11.7)

Permanenet displacements in 3-D model at section (b)

Permanenet displacements in 2-D model of section (b)

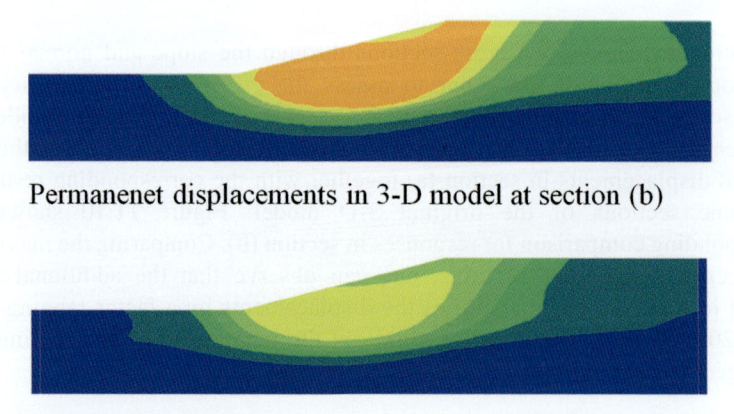

Fig. 11.10 Comparison between response of 3-D model at section (b) in Fig. 11.8 with response of 2-D model (for color scale, see legend in Fig. 11.7)

11.4 Earthquake Response of Pipelines

Extensive research has been conducted on the static response of pipelines on soft seabed in the last decade (e.g. SAFEBUCK JIP). This is primarily due to increased oil and gas development in deep water where typically soft seabed is found. Observations and monitoring of the behavior of pipelines in these environments have unveiled challenging subjects related to the interaction between pipeline and seabed, such as buckling and the so-called pipeline walking. The SAFEBUCK JIP (e.g. White et al. 2011) has supported a program of research for tackling the uncertainties associated with the design of pipelines against lateral buckling and axial walking. During the pipe laying process, pipelines are subject to small amplitude vertical and horizontal oscillations, driven by the sea state and lay vessel motions. In the soft soils found in deep water, pipe embedment can exceed a pipe diameter, and this embedment has a significant effect on the lateral pipe–soil interaction and axial resistance (Westgate et al. 2013).

Research has also been carried out on the earthquake response of pipelines. Observed damages to pipelines in seismic events (e.g. O'Rourke and Liu 1999) have generally been attributed to two hazards: (a) permanent ground deformation (PGD); and (b) soil strain due to seismic wave propagation. Permanent ground deformation can be either localized and abrupt, as in fault rupture, or spatially distributed, such as in landslides and liquefaction-induced lateral spreading. Soil strains induced by seismic wave propagation are generally small (Younan 2012). Depending on the pipe-soil coefficient of friction, the resulting pipeline strains can be equal to or less than ground strains in the case of a straight pipeline segment. However, it is possible that strain localizations may be induced by geometric discontinuities such as pipeline bends, tees and/or valves. One of the approaches for estimating the strains induced by seismic wave propagation is the ASCE approach (ASCE 1984, 2001) that is based on the assumption that a buried pipeline follows the ground motion. The maximum axial strain in the pipe can be approximated by the maximum ground strain, estimated as $\varepsilon = PGV/C$ where PGV is the peak ground velocity and C is the apparent wave propagation velocity.

Kaynia et al. (2014) have investigated two key elements not covered by the above studies: (a) earthquake response of pipelines on sloping seabed excited by asynchronous motions, and (b) strain-softening behavior of soil along the pipeline. The following describes implementation of these features in a computational method.

Figure 11.1 shows an example of seabed topographic features that can be due to geological processes or earlier landslides. Due to the seabed topography and the long extension of the pipeline, points along the pipeline route experience different motions during an earthquake. These motions can be computed by 3-D models of the ground by a suitable software. Alternatively, if the pipeline does not have extensive bends out of plane, one can model only 2-D section of the ground along the pipeline using a 2-D software. The numerical model, *QUIVER-pipe* (Kaynia 2012b) addresses both the asynchronous earthquake motions and the strain-softening soil behavior at the soil-pipe interface (Kaynia et al. 2014).

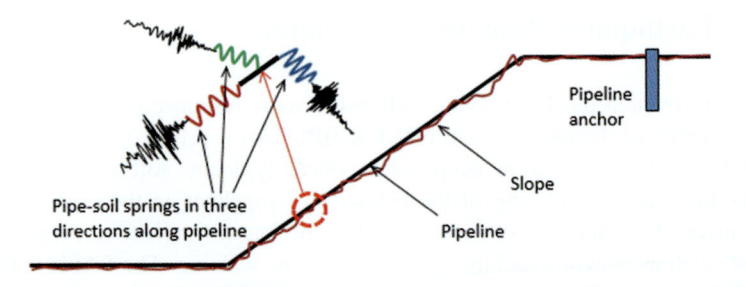

Fig. 11.11 Pipe-soil interaction model for analysis of response due to asynchronous earthquake motions in three directions

Figure 11.11 illustrates an idealization of pipe-soil interaction in a pipeline under earthquake loading. The pipeline has an arbitrary geometry (for simplicity only a 2-D geometry is shown in Fig. 11.11) and is placed on a number of springs representing pipe-soil interaction. The springs are excited at their bases by different acceleration time histories computed by appropriate 2-D/3-D numerical tools. A common practice in pipeline analysis is to ignore the pipeline mass and apply the earthquake motion as static displacements under the springs. While the submerged weight of pipes is often small, the total weight, which steers the dynamic response through inertia forces, could be quite large compared to the soil resistance against pipe movement. Ignoring the mass of the pipe might thus result in un-conservative conditions. This has been illustrated by an analysis in Kaynia et al. (2014). Moreover, by treating the earthquake excitation statically, one cannot capture the out-of-phase movements, which could lead to larger differential soil movements along the pipeline. This effect could also lead to un-conservative results.

The soil springs in *QUIVER_pipe* are specified along the pipeline (axial springs) and perpendicular to its axis (lateral springs). The pipeline can be laid freely on the ground surface and might additionally be anchored at a point on the ground surface on top of the slope, as shown in Fig. 11.11. The anchor has the function of resisting the axial load resulting from the tendency of the pipe to 'walk' down the slope. In practice, the anchor is installed such that the connection has a slack length (typically in the range 200–300 mm).

The pipe-soil interaction in *QUIVER_pipe* is represented by strain-softening springs with either concave or convex forms exemplified in Fig. 11.12. These springs are distributed in the three directions along the pipeline. The developed model is based on the finite element method consisting of 3-D beam elements and distributed soil springs along the pipe in the three directions. The pipe masses and the distributed springs are lumped at the nodes. The main source of damping after reaching the peak strength is energy dissipation through the hysteretic nonlinear response in the springs that follows Masing's rule (Masing 1926). The model is excited by three independent acceleration time histories under the soil springs at each node. A model with N nodes contains 6 N degrees of freedom corresponding to the six degrees of freedom at each node.

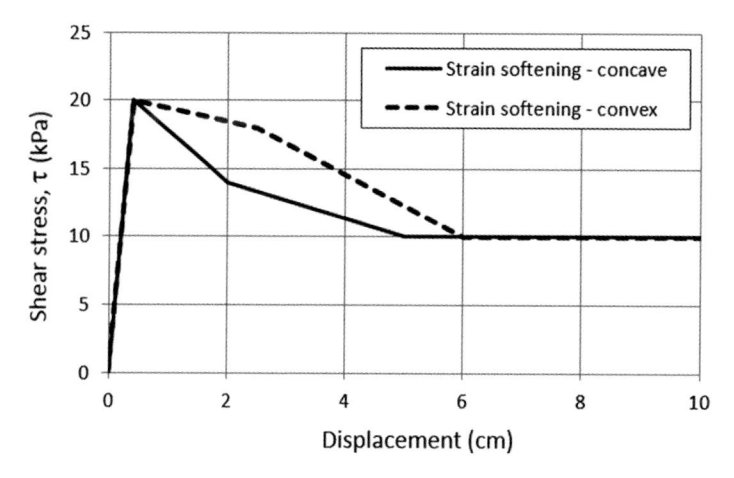

Fig. 11.12 Strain softening springs for pipe-soil interaction

Figure 11.13 presents an example of results from an earthquake pipe-soil interaction analysis. See Kaynia et al. (2014) for details of the soil and pipe model and earthquake excitation. Figure 11.13a shows the FE mesh, and Fig. 11.13b displays the maximum earthquake-induced axial forces computed along the pipeline. It should be noted that the forces are plotted in absolute value; moreover, they are not simultaneous. The maximum permanent displacement on the slope surface is about 2 m. As expected, the axial force reaches a peak value close to the top of the slope.

Figure 11.13c presents the variation of the corresponding maximum bending moments along the pipeline. As opposed to the axial force, the bending moment displays a very sharp increase at top of the slope. These results are physically justifiable in view of the remarkable gradient of the displacements around this location.

11.5 Response of Seabed Facilities

Due to recent trends in the offshore industry to move to deeper water, most of the operations related to oil and gas production and processing are being carried out close to the wells. This requires installation of different facilities and structures directly on the seabed and connecting them by pipelines, jumpers and spools. These installations vary in size from wellhead trees (albeit relatively heavy) to heavy manifolds and templates. Templates are often large steel structures used to support or protect manifolds. Manifolds vastly vary in shape, size and function, and could reach as high as 30 m in height and even larger in plan dimensions.

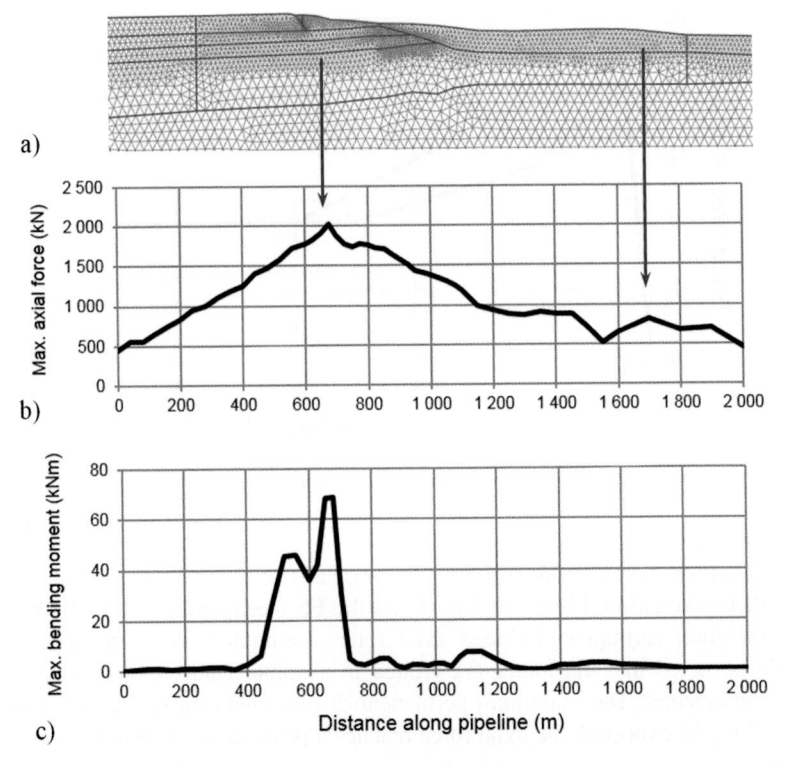

Fig. 11.13 Absolute values of maximum pipe forces along pipeline: (**a**) FEM mesh of slope – pipeline lies on top surface, (**b**) axial force, (**c**) bending moment

Depending on soil conditions and available installation technique and schedule, manifolds could be founded on several (typically 4) piles, on large single bucket foundations, or on steel mudmats. The piles/buckets are often installed by base suction, while mudmats are placed directly on the seabed with skirts penetrating into the seabed to provide additional lateral resistance. The design of these foundations are often driven by size requirements and for loads rather than by weight. Therefore, they end up being quite stiff for the mass they carry which results in relatively high natural frequencies, typically in the range 3–5 Hz.

The large natural frequencies represent some challenges in the dynamic and earthquake analyses of the foundations, including handling of added soil mass and damping. Designers often ignore these parameters or assign arbitrary values without any rigorous analyses. For large foundations, high natural frequency corresponds to large radiation damping that theoretically could be translated to damping ratios exceeding 50%. The radiation damping, which is often ignored in design, would considerably reduce the earthquake-induced accelerations (and loads on the manifold and foundation).

11 Earthquake Geotechnics in Offshore Engineering

In order to highlight the importance of rigorous modelling, three realistic foundations were considered in a realistic soft soil profile representative of deep water locations. The three designs and their key parameters are (Kaynia and Wang 2017):

(a) 4-pile foundation: Diameter, length and wall thickness, D = 3 m, L = 12 m, t = 15 mm, center spacing of piles 9 m
(b) Bucket foundation: D = 8 m, L = 9 m, t = 25 mm
(c) Mudmat: Plan dimensions and length, 17 m × 17 m × 1.0 m.

The soil profile considered in the analyses represents a realistic soft soil site with approximately a linear variation of shear strength with depth. The strain-compatible shear modulus, that is, after consideration of the reduction due to shear strains induced by a strong earthquake, varies from approximately 2 MPa on the seabed to 75 MPa at 100 m depth. The mass density of the soil is relatively constant with depth, except for the 2 m soil below the seabed, and is equal to 1580 kg/m^3. The template mass is 500 tons.

Computation of the pile-group impedances was carried out by PILES (Kaynia 1982). The impedances of the bucket foundation were computed using the FE solution by Tassoulas (1981), and the impedances of the mudmat were compted by the Green's function-based solution of Kaynia et al. (1998). All these methods are rigorous tools based on analytical solutions of wave propagation in layered media and perfectly handling of infinite boundaries. The methods work in the frequeny domain and result in impedances as complex quantities with the real parts reflecting the combined effect of foundation stiffness and added soil mass, and the imaginary part reflecting the hysteretic and radtiation damping. The imaginary part was used in this paper to compute the equivalent damping ratio of the foundations.

Figure 11.14 plots the real and imaginary parts of the computed vertical impedances of the three foundations as functions of frequency. The three foundations were slightly

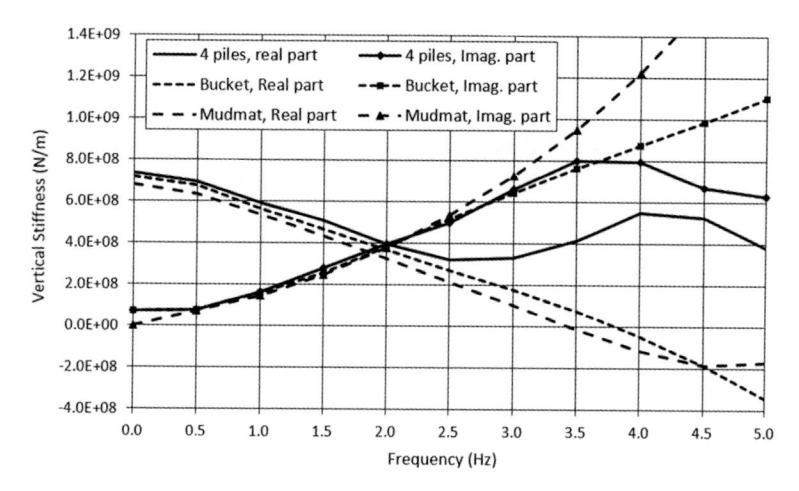

Fig. 11.14 Dynamic impedances of three foundation types studies in this paper: (i) 4-pile group, (ii) Bucket foundation, (iii) Mudma

Fig. 11.15 Damping ratio as function of frequency for three foundation types studied above: (i) 4-pile group, (ii) Bucket foundation, (iii) Mudmat

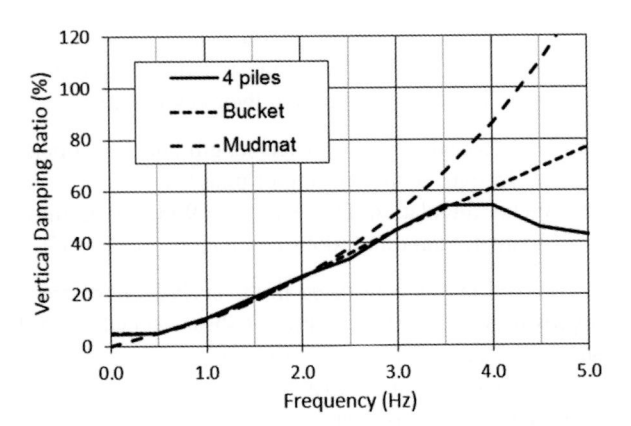

modified to give about the same vertical static stiffness (at f = 0.0 Hz) equal to about 7 MN/m.

It is interesting to note that, although the three foundations have the same static stiffness, their dynamic stiffnesses (real part) are quite different. While the dynamic stiffness of the pile group shows relatively moderate variation with frequency, indicating a small added soil mass, the dynamic stiffnesses of the bucket foundation and mudmat are strongly frequency-dependent, representing a large added soil mass. For example, using the parabolic form of the stiffness of the bucket foundation, one could compute an equivalent added mass of about 1200 tons. This is more than double the manifold mass and 1.7 times the mass of the soil plug (700 tons). Despite such analyses and observations, most engineers assume the added soil mass equal to the soil plug mass. Using the added soil mass and the mass of the manifold (500 tons), one could compute a natural frequency of vertical vibration equal to 3.3 Hz.

Figure 11.15 plots the variation of the computed foundation damping for the above foundation designs. The figure shows that the three designs have fairly similar damping values. For the computed natural frequenncy of 3.3 Hz, one gets a damping ratio of about 50%. As stated earlier, this damping is so large that it prohibits the foundation from oscillation and consequently reduces the vertical earthquake load dramatically.

While damping can dramatically reduce the earthquake loads on subsea structures, its impact on a system of assembled subsea facilities could be even more significant. Manifolds and other installations, such as PLEMs and PLETs, are often connected by elements such as spools that lie on the seabed between the facilities. These elements are relatively light and flexible and follow the motions of the facilities at the two ends. Large earthquake displacements in the structures could result in overstress in these elements at the contact points and the point of touching

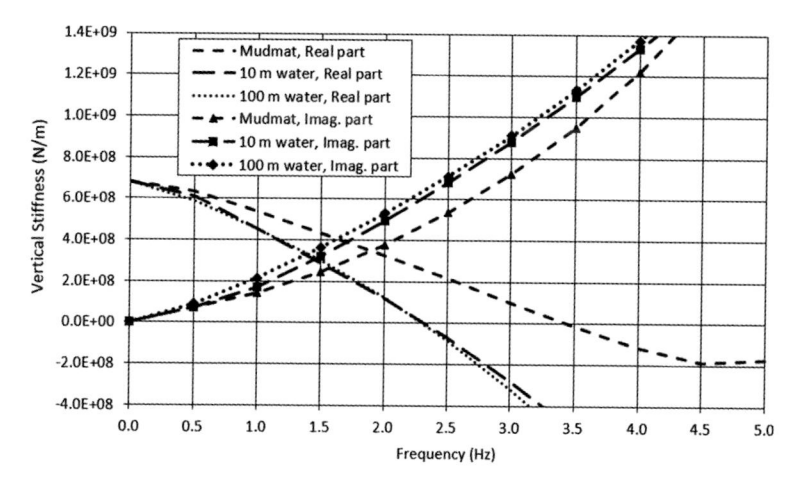

Fig. 11.16 Dynamic impedances of mudmat foundation for different water depth including results for case of no water for comparison

the ground. A realistic, lower displacement of the seabed installations could thus have a major design implication for these sensitive elements.

Another topic in this discussion is the effect of water on the earthquake response of seabed foundations. The study by Kaynia et al. (1998) has shown that water can practically be ignored in the horizontal and rocking responses of seabed foundations. However, the effect is relatively large in the vertical direction at higher frequencies, especially in the form of added mass. Figure 11.16 displays the effect of different water depth on the vertical impedance of the mudmat studies above. The results are in agreement with those in Kaynia et al. (1998) which show that water has an effect only up to a depth of the order of the foundation dimensions.

11.6 Response of Platforms

Earthquake analysis of fixed platforms follow the standard models used in traditional soil-structure interaction (SSI) analyses. Therefore, two approaches are utilized: (a) one-step solutions based on integrated models of soil and structure, and (b) - sub-structuring solutions where the soil and structure domains are modelled and analyzed as separated domains (Kausel 2010). One of the commonly used sub-structuring methods is the well-known three–step solution (Kausel et al. 1978) which was developed for cases with rigid foundations. In this method, the analysis is divided into the three steps, namely, *kinematic interaction*, *impedance calculation* and *inertial interaction*. While it is ideal to perform earthquake analyses by the one-step method, there are several obstacles for its use in practice. The first obstacle is that structural designers rarely use the required state-of-the-art SSI softwares, for example SASSI (Lysmer et al. 1981). The second issue is that engineering usually involves several EPC contractors, each with their own sets of tools and procedures that make compatibility between the analyses a challenge for the project. Therefore, despite its limitations, the three-step methods as described in the following, is the preferred approach in most SSI analyses.

11.6.1 Jacket Structures on Piles

Offshore jackets are made of tubular steel members, and are usually piled into the seabed. The earthquake analysis of these types of platforms are commonly based on a special version of the three-step method in which the piles are included in the structural model and the interaction with the soil is accounted for by the use of nonlinear pile-soil springs commonly known as p-y, t-z and Q-z springs. The p-y curves, in which p denotes the lateral soil reaction per unit length of the pile and y denotes the lateral deflection, have evolved from research in the oil and gas industry during the 1970s. The research has been based on tests using 324 mm diameter steel-pipe piles in soft clay by Matlock (1970), 610 mm diameter steel-pipe piles in stiff clays by Reese et al. (1975), 914 mm diameter RC drilled piles in stiff clays by Reese and Welch (1975), and 619 mm diameter steel-pipe piles in sands by Cox et al. (1974). The p-y curves established from these studies led to recommendations in the American Petroleum Institute standards for oil and gas installations (API 1993). These curves were established with special focus on storm loading which is often the dominating environmental load on jacketed structures. Because the dominant period of storm load is typically 10 s, the p-y curves have been defined with low initial stiffness to ensure higher SSI natural periods that would represent more conservative conditions. Recognizing the importance of capturing the stiffer response for other loads, especially earthquake, the latest version of API (2011) has increased the initial stiffness of the p-y curves. Moreover, considering that the initial pile load tests were performed on smaller diameter piles (for example 0.61 m in Reese et al. 1974), attempts have been made to modify the p-y curves for the effect of pile diameter and stiffer initial stiffness of the piles (e.g., Stevens and Audibert 1979; Jeanjean 2009). Other issues in connection with earthquake loading relate to (1) modelling of cycling response using p-y curves, (2) incorporation of radiation damping at higher frequencies, and (3) consideration of liquefaction. These are briefly discussed in the following.

The centrifuge experiments by Boulanger et al. (1999) were used to verify a model based on cyclic p-y curves. To this end, a nonlinear p-y element was developed based on the results of Matlcock (1970) in which the nonlinear p-y spring is replaced by three springs representing elastic, plastic and gap components in series. The radiation damping was then added in parallel to the elastic spring representing the far-field response. The p-y spring set at each depth is excited by the earthquake motion in the free field computed from site response analyses at that depth. The radiation damping could be estimated by simple models (e.g. Gazetas 1991). In cases that the earthquake loading is not very strong, hence the response can be captured by an equivalent linear method, one can resort to the standard three-step method in which the piles are represented by their dynamic impedances at their pile head. Rigorous numerical tools (e.g. Kaynia 1982) or simple solutions (e.g. Dobry and Gazetas 1988) could be used for the computation of pile impedances that can be converted to equivalent stiffness-mass-damping elements.

Incorporation of liquefaction in analysis of piles is a complex and uncertain issue that is still under research. The uncertainty is primarily due to the complexity of liquefaction and its quantification. Dash et al. (2008) presented a summary of practical solutions used for capturing behavior of piles in liquefiable soil. The solutions include those that completely ignore soil resistance during liquefaction (as proposed in some design codes) and those methods that reduce the strength of p-y curves for (non-liquefied) sands. Among the latter approach, one could mention the well-known p-multiplier method based on SPT data and the Cu-factor method proposed by Liu and Dobry (1995) based on centrifuge test data that assumes a strength degradation factor, $Cu = 1-r_u$, where r_u is the excess pore pressure ratio. Alternatively, one could compute friction angle corresponding to the residual shear strength (e.g. Boulanger and Idriss 2016) and use it to define p-y curves. Some experimental results have shown different forms in the shape of p-y curves. For example, the full-scale tests by Rollins et al. (2005) on single piles and pile groups subjected to blast induced liquefaction and back-calculation of the data have shown that the p-y curves from the test results display a concave pattern at full liquefaction. It should be noted that while the condition of liquefaction often represents a more critical (conservative) case for the pile design, it might create a more favorable case for design of the platform. Therefore, in such cases, one should also analyze the platform assuming that the soil would not liquefy.

11.6.2 Gravity Based Structures

Earthquake analysis of gravity-based structures (GBS) is commonly performed by the three-step method. The conventional three-step method is based on the assumption of rigid foundation and linear soil response. Younan et al. (2015) presented a benchmark three-step earthquake analysis of a concrete GBS with an apparently stiff foundation caisson. The objective of the benchmark study was to assess the accuracy of the three-step method by comparing the results of a one-step integrated analysis with those of the three-step method in which the frequency-dependent complex impedances of the foundation are converted into real-valued sets of spring-mass-dashpot elements, so-called lumped parameter foundation model, LPFM. The procedure often used in practice for calculating the parameters of LPFM is as follows: (1) compute the static stiffness, K_{st}, and added soil mass, M, by fitting a parabola in the form $K_{st} - M \cdot \omega^2$ to the real part of the foundation impedance (ω is frequency in rad/s), (2) compute the damping constant, C, from the imaginary part of the foundation impedance by fitting a line $C \cdot \omega$ to the imaginary part of the impedance.

Fig. 11.17 Finite element model of example concrete GBS

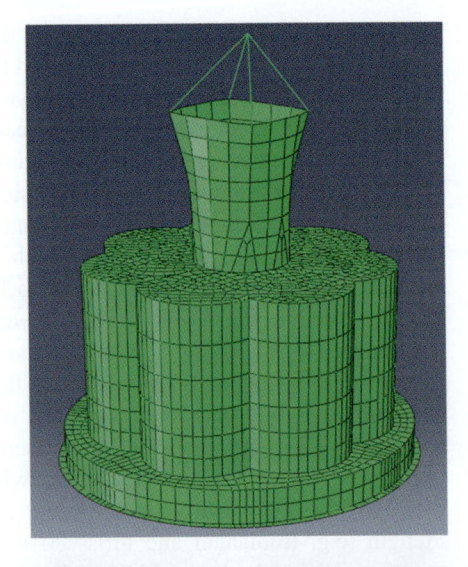

Fig. 11.18 Finite element model of base and shafts in a large concrete GBS with approximate base dimensions 100 m by 130 m

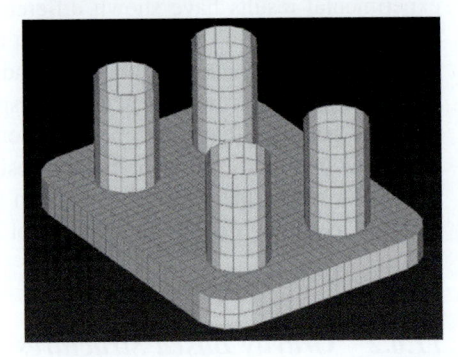

Figure 11.17 illustrates the finite element model of the GBS caisson used in SASSI (Lysmer et al. 1981). A number of points were selected on the model for computing the response spectra by the two methods. The comparison was successful indicating that the assumption of rigid base was satisfactory for the earthquake analysis of this structure (Younan et al. 2015).

In cases of flexible bases, one needs to develop distributed LPFM that would give foundation impedance parameters at predefined nodes or per square meter of the base. As an example, Fig. 11.18 shows the FE model of a concrete GBS with approximate base dimensions 110 m by 130 m. In view of the relatively low height of the caisson, this GBS cannot accurately be modelled as rigid. The distribution of the impedance parameters, namely stiffness, added soil mass and damping, is not unique and depends on the details of the structure and the mode of response, that is, horizontal or vertical. For this purpose, one should establish the parameters by accounting for the loading together with the foundation and structural details.

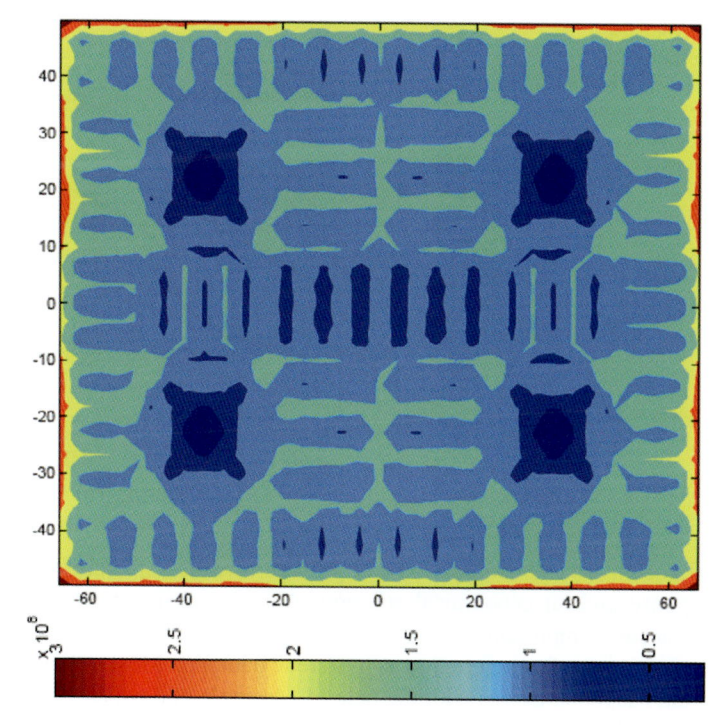

11 Earthquake Geotechnics in Offshore Engineering 283

Fig. 11.19 Distribution of vertical springs over base of GBS shown in Fig. 11.16 for vertical earthquake loading

One of the practical solutions developed for this purpose is due to Tabatabaie and Ballard (2006). The solution consists of the following steps: (a) perform a one-step SSI analysis of soil-structure in the frequency domain for a given earthquake excitation (for example horizontal or vertical) using a coarse model of the foundation and structure, (b) compute the complex-valued forces and corresponding displacements at the nodes of the base, (c) divide the forces and displacements and compute the complex impedances at the nodes, and derive the parameters of LPFM at the nodes following the procedure described above for rigid foundations. The distributed LPFM computed by this procedure can then be used in an SSI model of the structure with refined mesh as required for the detailed design. Figure 11.19 display an example of this type of computation for the distributed vertical spring values of the GBS platform shown in Fig. 11.18 for vertical earthquake loading. The results in Fig. 11.19 show that the stiffness is lowest where the foundation is stiffest (for example under the shafts and internal stiffeners in the GBS caisson) and are largest outside, and close to the edges.

Medium to large earthquake shaking induce larger inertial loads in the platform causing nonlinear soil response and permanent lateral displacement of the platform. In such cases, a performance-based design in which the soil-structure interaction is handled by using nonlinear force-displacement relationships at soil-foundation

Fig. 11.20 Nonlinear hysteretic response following Masing's rule (dashed line) and modified Iwan model (solid line) for given backbone curve (dotted line)

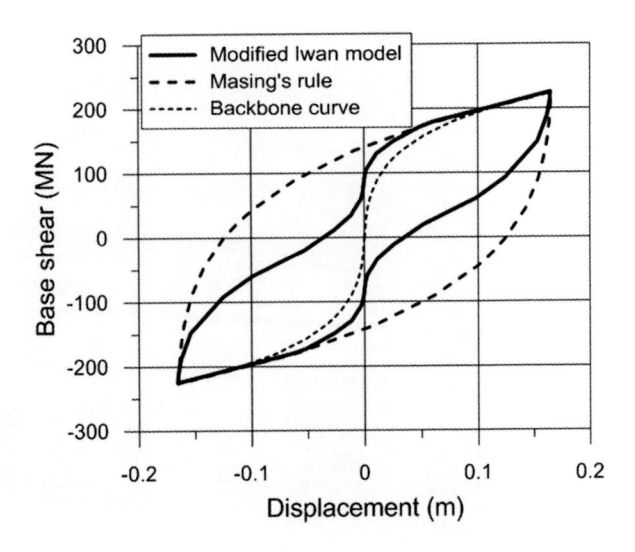

interface (so-called backbone curves) can provide a realistic picture of the earthquake response and an economical solution. A major challenge in this type of analyses is accurate representation of damping in the nonlinear cyclic response by the backbone curves, often referred to as hysteretic damping. Most available models represent the backbone curves with the help of Masing's rule which is a kinematic hardening model easily represented by a series of parallel elasto-perfectly plastic spring first proposed by Iwan (1967). Figure 11.20 shows an example of the nonlinear hysteretic response (dashed line) in a horizontal foundation spring of a platform following Iwan's model. The amount of hysteretic damping, which is directly related to the area circumscribed in a closed response loop, is about 33% which is too large for this displacement.

Different solutions have been proposed for limiting the foundation hysteretic damping. One of these solutions, which has been implemented and verified against actual measurements of Troll Platform (Kaynia et al. 2015), is based on modifying the curvature of the backbone curve (Kaynia and Andersen 2015). Another solution that has been tried by Younan et al. (2015) in the nonlinear SSI analysis of Hebron GBS is to deviate from the Masing's rule by defining different unloading rules that would result in slimmer hysteresis loops. A simple way to achieve this is through a modified Iwan's mechanical model in which a selected number of the elastic-plastic springs are replaced by corresponding nonlinear elastic springs. The resulting model will follow same backbone curve but with a pinched hysteretic response. This is shown in Fig. 11.20 (solid line). For the curve shown in the figure, the damping ratio is reduced to 19% by this modification. The figure also plots the backbone curve (dotted line) for reference.

11.7 Summary and Conclusions

This paper presented a number of geotechnical issues encountered in earthquake design of offshore structures and subsea facilities. The paper addressed some of the more recent approaches and solutions in geotechnical earthquake design of both shallow water and deep-water structures and facilities such as platforms with large bases, pipelines traversing slopes and seabed installations. It was demonstrated how incorporation of radiation damping and nonlinear soil-structure interaction in offshore installations could optimize the design. It was also highlighted how consideration of several factors such as strain softening and three-dimensional shaking that are often ignored in design, could affect the response of submarine slopes and pipelines. Finally, various solutions for earthquake SSI analyses of platforms were reviewed, and solutions were proposed for realistic representation of the foundation nonlinear response including hysteretic damping.

Acknowledgment The author would like to acknowledge partial support from the project "Reducing cost of offshore wind by integrated structural and geotechnical design (REDWIN)" funded by the Norwegian Research Council, grant number 243984.

References

American Petroleum Institute (API) (1993) Recommended practice for planning, designing, and constructing fixed offshore platforms – working stress design, 20th edn. API RP2A-WSD, Washington, DC

American Petroleum Institute (API) and International Organization for Standardization (ISO) (2011) ANSI/API specification RP 2GEO. Geotechnical and foundation design considerations for offshore structures. API, Washington, DC

Anantanavanich T, Pestana J, Carlton B (2012) Multidirectional site response analysis of submarine slopes using the constitutive model MSimpleDSS. In: Proceedings of 15th world conference on earthquake engineering, Lisbon, Portugal, 24–28 September

Andersen KH (2009) Bearing capacity of structures under cyclic loading; offshore, along the coast and on land. Can Geotech J 46:513–535

ASCE (1984) Guidelines for the seismic design of oil and gas pipeline systems. Committee on Gas and Liquid Fuel Lifelines, New York

ASCE (2001) Guidelines for the design of buried steel pipe, American lifelines Alliance

Azizian A, Popescu R (2006) Three-dimensional seismic analysis of submarine slopes. Soil Dyn Earthq Eng 26:870–887

Boulanger RW, Idriss IM (2016) CPT-based liquefaction triggering procedure. J Geotech Geoenviron Eng ASCE 142(2):04015065

Boulanger R, Curras C, Kutter B, Wilson D, Abghari A (1999) Seismic soil-pile-structure interaction experiments and analyses. J Geotech Geoenviron Eng 125(9):750–759

Bray JD, Travasarou T (2007) Simplified procedure for estimating earthquake-induced deviatoric slope displacements. J Geotech Geoenviron Eng ASCE 133(4):381–392

Carlton BD, Kaynia AM (2016) Comparison of the seismic response of offshore slopes using 1, 2, or 3 ground motion components. Paper OTC-26961. In: Proceedings offshore technology conference, Houston, Texas, 2–5 May

Cox WR, Reese LC, Grubbs BR (1974) Field testing of laterally loaded piles in sand. In: Proceedings of 6th offshore technology conference, Houston, pp 459–472

Dafalias YF, Manzari MT, Papadimitriou AG (2006) SANICLAY: simple anisotropic clay plasticity model. Numer Anal Methods Geomech 30(12):1231–1257

Dash SR, Bhattacharya S, Blakeborough A, Hyodo M (2008) P-y curve to model lateral response of pile foundations in liquefied soils. In: Proceedings of 14th world conference on earthquake engineering. Oct 12–17, 2008, Beijing, China

Dobry R, Gazetas G (1988) Simple method for dynamic stiffness and damping of floating pile groups. Geotechnique 38(4):557–574

Duncan JM (1996) State-of-the-art: limit equilibrium and finite element analysis of slopes. ASCE J Geotech Geoenviron Eng 122(7):577–596

Ferrari G (2012) Three-dimensional earthquake response of slopes. MSc thesis, Department of Civil Engineering, University of Bologna and Norwegian Geotechnical Institute

FLAC3D (2006) Fast lagrangian analysis of continua in 3 dimensions. Itasca Consulting Group, Inc., Minneapolis

Gazetas G (1991) Foundation vibrations. In: Fang HY (ed) Foundation engineering handbook. Van Nostrand Reinhold, New York, pp 553–593

ISO 19901-2:2004 (2004) Petroleum and natural gas industries – specific requirements for offshore structures – part 2: seismic design procedures and criteria

Iwan WD (1967) On a class of models for the yielding behavior of continuous and composite systems. J Appl Mech 34(3):612–617

Jeanjean P (2009) Re-assessment of p-y curves for soft clays from centrifuge testing and finite element modelling. In: Proceedings of offshore technology conference, paper OTC 20158

Kammerer A, Pestana J, Seed R (2003) Behavior of monterey 0/30 sand under multidirectional loading conditions. Geomechanics: testing, modeling, and simulation, geotechnical special publication 143

Kausel E (2010) Early history of soil–structure interaction. Soil Dyn Earthq Eng 30:822–832

Kausel E, Whitman RV, Morray JP, Elsabee F (1978) The spring method for embedded foundations. Nucl Eng Des 48:377–392

Kaynia AM (1982) Dynamic stiffness and seismic response of pile groups. Research report R82–03, Department of Civil Engineering, MIT, Cambridge, USA

Kaynia AM (2012a) QUIVER_slope – numerical code for one-dimensional seismic response of slopes with strain softening behavior. NGI report 20071851–00-79-R, 8 June 2012

Kaynia AM (2012b) QUIVER_pipe: numerical code for nonlinear seismic analysis of pipeline-soil interaction. NGI report 20071851–00-83-R, 8 June 2012

Kaynia AM (2018) Seismic considerations in design of offshore wind turbines. Soil Dyn Earthq Eng. In press.

Kaynia AM, Andersen KH (2015) Development of nonlinear foundation springs for dynamic analysis of platforms. In: Meyer V (ed) Proceedings of frontiers in offshore geotechnics III, ISFOG. Taylor & Francis Group, London, ISBN: 978–1–138-02848-7, pp 1067–1072

Kaynia AM, Saygili G (2014) Predictive models for earthquake response of clay and sensitive clay slopes (Chapter 18). In: Ansal A (ed) Perspectives on European earthquake engineering and seismology. Springer, New York, pp 557–584

Kaynia AM, Wang LZ (2017) Seismic response of subsea facilities. In: Proceedings of internation symposium coastal and offshore geotechnics, ISCOG 2017, Hangzhou, China, 5–7 July, pp 25–27

Kaynia AM, Kausel E, Madshus CM (1998) Impedances of underwater rigid square foundations. In: Proceedings of ASCE Spec. Conference geotechnical earthquake engineering, Seattle, WA, USA, pp 1283–1293

Kaynia AM, Dimmock P, Senders M (2014) Earthquake response of pipelines on submarine slopes. In: Proceedings of offshore technology conference, Paper OTC-25186, Houston, Texas, 5–8 May

Kaynia AM, Norén-Cosgriff K, Andersen KH, Tuen KA (2015) Nonlinear foundation spring and calibration using measured dynamic response of structure. In: Proceedings of ASME 2015 34th internation conference on ocean, offshore and arctic engineering, OMAE2015, paper OMAE2015–41236. May 31–June 5, St. John's, Newfoundland, Canada

Liu L, Dobry R (1995) Effect of liquefaction on lateral response of piles by centrifuge model tests. NCEER Bull 9(1):7–11

Lysmer J, Tabatabaie-Raissi M, Tajirian F, Vahdani S, Ostadan F (1981) SASSI – a system for analysis of soil-structure interaction, Report UCB/GT 81-02. University of California, Berkeley

Masing G (1926) Eigenspannungen und Verfestigung beim Messing. In: Proceedings of 2nd international. Congress on applied mechanics, Zurich, pp 332–335

Matlock H (1970) Correlations for design of laterally loaded piles in soft clay. In: Proceedings of 2nd annual offshore technology conference, pp 577–594

Nadim F, Biscontin G, Kaynia AM (2007) Seismic triggering of submarine slides. In: Proceedings of offshore technology conference, paper 18911, Houston, Texas, April 30 – May 3, 2007

O'Rourke MJ, Liu X (1999) Response of buried pipelines subject to earthquake effects. MCEER monograph # 3, University at Buffalo, New York

Park DS, Kutter BL (2015) Static and seismic stability of sensitive clay slopes. Soil Dyn Earthq Eng 79:118–129

PLAXIS 3D (2015) PLAXIS B.V., Delft, The Netherlands

Pyke RM, Seed HB, Chan CK (1975) Settlement of sands under multidirectional shaking. J Geotech Eng Div ASCE 101(GT4):379–398

Reese LC, Cox WR, Koop FD (1974) Field testing and analysis of laterally loaded piles in sand. In: Proceedings of Offshore Technology Conference, Houston, Texas, paper OTC 2080, pp 473–485.

Reese LC, Welch R (1975) Lateral loading of deep foundations in stiff clay. J Geotech Engrg Div 101(7):633–649

Reese L, Cox W, Koop F (1975) Field testing and analysis of laterally loaded piles in stiff clay. In: Proceedings of 7th offshore technology conference, Houston, pp 671–690

Rollins KM, Gerber TM, Lane JD, Ashford S (2005) Lateral resistance of a full-scale pile group in liquefied sand. J Geotech Geoenviron Eng ASCE 131(1):115–125

Seed HB, Pyke RM, Martin GR (1978) Effect of multi-directional shaking on pore-pressure development in sands. J Geotech Geoenviron Eng ASCE 104(GT1):27–44

Stevens JB, Audibert JME (1979) Re-examination of the p-y curve formulations. In: Proceedings of 11th offshore technology conference, Houston, pp 397–403

Stewart JP, Yee E (2012) Nonlinear site response and seismic compression at vertical array strongly shaken by 2007 Niigata-ken Chuetsu-oki earthquake. Report, National earthquake hazards reduction program, Reston, VA

Su D and Li X (2003) Centrifuge tests on earthquake response of sand deposit subjected to multi-directional shaking. In: Proceedings of 16th ASCE engineering mechanics conference, University of Washington, Seattle

Tabatabaie M, Ballard T (2006) Distributed parameter foundation impedance model for time domain SSI analysis. In: Proceedings of 8th US national conference on earthquake engineering commemorating 100th anniversary of the 1906 San Francisco earthquake, San Francisco

Taiebat M, Kaynia AM (2010) A practical model for advanced nonlinear analysis of earthquake effects in clay slopes. Paper no. 1.09a. In: Proceedings of recent advances in geotechnical earthquake engineering and soil dynamics, San Diego, California, 24–29 May

Taiebat M, Kaynia AM, Dafalias YF (2011) Application of an anisotropic constitutive model for structured clay to seismic slope stability. J Geotech Geo-environ Eng ASCE 137(5):492–504

Tassoulas JL (1981) Elements for numerical analysis of wave motion in layered media. Research report R81-12, Department of Civil Engineering, MIT, Cambridge, USA

Westgate ZJ, White DJ, Randolph MF (2013) Modelling the embedment process during offshore pipe-laying on fine-grained soils. Can Geotech J 50:15–27

White DJ, Ganesan SA, Bolton MD, Bruton DAS, Ballard JC, Langford TE (2011) SAFEBUCK JIP – observations of axial pipe-soil interaction from testing on soft natural clays. In: Proceedings of offshore technology conference, paper OTC 21249. 2–5 May 2011, Houston, Texas, USA

Younan AH (2012) Simulating seismic wave motions for pipeline design. In: Proceedings of twenty-second Internation offshore and polar engineering conference, ISOPE. Rhodes, Greece, June 17–22, 2012

Younan AH, Kaynia AM, Loo MM, Widianto and Khalifa J (2015) Seismic design of Hebron platform – an integrated soil-structure-interaction approach. In: Proceedings of ASME 2015 34th international conference on ocean, offshore and arctic engineering, OMAE2015, paper OMAE2015-42134. May 31–June 5, St. John's, Newfoundland, Canada

Chapter 12
The Dynamics of Rocking Isolation

Nicos Makris

Abstract The uplifting and rocking of slender, free-standing structures when subjected to ground shaking may limit appreciably the seismic moments and shears that develop at their base. This high-performance seismic behavior is inherent in the design of ancient temples that consist of slender, free-standing columns which support freely heavy epistyles together with the even heavier frieze atop. While the ample seismic performance of rocking isolation has been documented with the through-the-centuries survival of several free-standing ancient temples; and careful post-earthquake observations in Japan during the 1940's suggested that the increasing size of slender free-standing tombstones enhances their seismic stability; it was Housner (Bull Seismol Soc Am 53 (2):404–417, 1963) who more than half century ago elucidated a size-frequency scale effect and explained that there is a safety margin between uplifting and overturning and as the size of the column or the frequency of the excitation increases, this safety margin increases appreciably to the extent that large free-standing columns enjoy ample seismic stability. This article revisits the important implications of this post-uplift dynamic stability and explains that the enhanced seismic stability originates from the difficulty of mobilizing the rotational inertia of the free-standing column. As the size of the column increases the seismic resistance (rotational inertia) increases with the square of the column size; whereas, the seismic demand (overturning moment) increases linearly with size. The same result applies to the articulated rocking frame given that its dynamic rocking response is identical to the rocking response of a solitary free-standing column with the same slenderness; yet larger size. The article concludes that the concept of rocking isolation by intentionally designing a hinging mechanism that its seismic resistance originates primarily from the mobilization of the rotational inertia of its members is a unique seismic protection strategy for large, slender structures not just at the limit-state but also at the operational state.

N. Makris (✉)
University of Central Florida, Orlando, FL, USA

University of Patras, Patras, Greece

Office of Theoretical and Applied Mechanics, Academy of Athens, Athens, Greece
e-mail: Nicos.Makris@ucf.edu

12.1 Introduction

The design of most structural framing systems is based on three basic concepts which are deeply rooted in modern structural engineering. The first concept is that of creating statically indeterminate (redundant) framing systems. When a "statically indeterminate" structure is loaded by strong lateral loads and some joints develop plastic hinges, there is enough redundancy in the system so that other joints maintain their integrity. In this way, recentering of the structures is achieved to some extent and stability is ensured. The second concept, known as ductility, is the ability of the structure to maintain sufficient strength at large deformations. In this way, even in the event of excessive lateral loads that may convert all joints to plastic hinges, all modern seismic codes demand that these hinges shall develop sufficient ductility so that collapse is prevented; however, in this case the structure may experience appreciable permanent displacements. The third concept that dominates modern structural engineering is that of positive stiffnesses. When a structure behaves elastically, forces and deformations are proportional. When yielding is reached the forces are no longer proportional to the deformations; however, in most cases the stiffnesses at any instant of the deformation history of the structure remain positive—that is if some force is needed to keep the structure away from equilibrium at some displacement; then, a larger force is needed to keep the structure away from equilibrium at a larger displacement. Figure 12.1 (left) illustrates the deformation pattern of a moment-resisting, fixed-base frame when subjected to a lateral load capable to induce yielding at the joints. The force-deformation curve (P-u) is nonlinear; nevertheless, the lateral stiffness of the system remains positive at all times.

Figure 12.1 (right) illustrates the deformation pattern of a free-standing rocking frame (two free-standing rigid columns capped with a freely supported rigid beam) when subjected to a lateral load capable to induce uplifting of the columns. The force-displacement relationship (P-u) of the rocking frame shown at the bottom of Fig. 12.1 (right) indicates that the articulated system has infinite stiffness until uplift is induced and once the four-hinge frame is set into rocking motion, its restoring force decreases monotonically, reaching zero when the rotation of the column $\theta = a = \arctan(b/h)$. Accordingly, the free-standing rocking frame shown in Fig. 12.1 (right) is a four-hinge mechanism that exhibits negative lateral stiffness. Figure 12.1 indicates that while most modern structural engineers are trained to design statically indeterminate structures that exhibit positive stiffnesses and hopefully sufficient ductility (Fig. 12.1 left); ancient builders were designing entirely different structural systems—that is articulated mechanisms that exhibit negative stiffnesses and low damping (Fig. 12.1 right). What is remarkable about these "unconventional" articulated structures is that they have endured the test of time by surviving several strong seismic motions during their 2.5 millennia life. For instance, Fig. 12.2 shows a view of the late archaic Temple of Apollo in Corinth, Greece (Powell 1905).

12 The Dynamics of Rocking Isolation

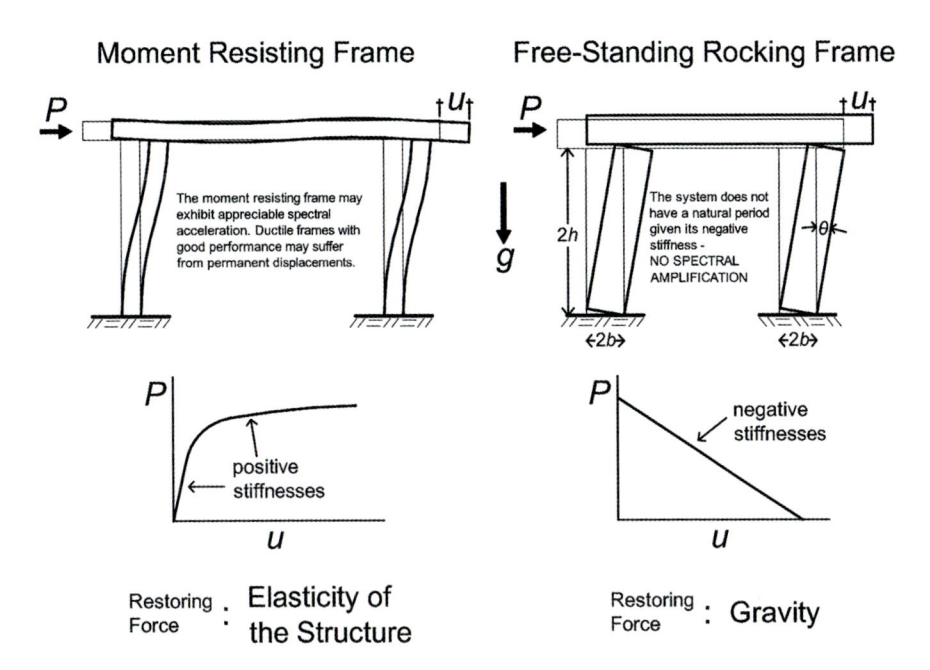

Fig. 12.1 The fundamental difference in the behavior of a traditional moment-resisting frame (left) and a rocking frame with free-standing columns which are allowed to rock (right)

Fig. 12.2 View of the Temple of Apollo in Corinth, Greece. Its monolithic, free-standing columns support massive epistyles and the frieze atop, and the entire rocking frame remains standing for more than 2500 years in a region with high seismicity

The unparallel seismic performance of the free-standing rocking frames shown in Fig. 12.2 is due to the very reason that they are articulated mechanisms. In this way: (a) given their negative stiffnesses they are not subject to any resonance, (b) recentering (elimination of any permanent displacement) is achieved unconditionally with gravity; and (c) the rocking frames, while slender and emblematic, they are large in size to the extent that their rotational inertia, when mobilized, is sufficient to resist the 2500 years seismic hazard.

Analytical studies on the seismic response of slender, free-standing columns have been presented as early as in 1885 by Milne (1885) in an effort to estimate levels of ground shaking. His reasoning is entirely within the context of an equivalent static analysis and by taking moment equilibrium about the imminent pivoting point, he concludes that when the ground acceleration, $\ddot{u}_g$, exceeds the value of g·(width/ height), the column overturns. Four decades after Milne's work, Kirkpatrick (1927) published a remarkable paper on the seismic stability of rocking columns. His work brings forward the two key quantities other than the peak ground acceleration that are responsible for the stability of a slender, free-standing column: (a) the size of the column which enters the equations via the moment of inertia; and (b) the duration of the period of the excitation. Kirkpatrick (1927) after correctly deriving the minimum acceleration amplitude of a harmonic excitation that is needed to overturn a free-standing column with a given size and slenderness, proceeds by presenting the first minimum-acceleration overturning spectrum (Fig. 6 of Kirkpatrick 1927 paper) and shows that as the period of the excitation decreases, a larger acceleration is needed to overturn a free-standing column. While P. Kirkpatrick worked in Hawaii, it appears that his contributions were not known in Japan. Nevertheless, in the late 1940's Ikegami and Kishinouye published two important papers, one following the December 21, 1946 Nankai Earthquake (Ikegami and Kishinouye 1947) and the other following the December 26, 1949 Imaichi Earthquake (Ikegami and Kishinouye 1950). These two papers come to confirm Kirkpatrick's theoretical findings on the rocking response of free-standing columns; since they indicate that the static threshold, g·(width/height), is too low and is not able to explain the observed stable response of more slender; yet, larger tombstones. In their own words Ikegami and Kishinouye (1950) write "In our field investigations, we often met with cases where gravestones had not overturned because of their large dimensions in spite of the small value of the ratio between width and height".

About a decade later Muto et al. (1960) build upon the work of Ikegami and Kishinouye (1947, 1950) and show explicitly that the dynamic response of a rocking column is governed by a negative stiffness; therefore, its free-vibration response is not harmonic; rather it is described by hyperbolic sines and cosines.

The pioneering work of Kirkpatrick (1927) in association with the systematic work conducted in Japan on rocking and overturning during the first-half of the twentieth century matured the knowledge on this subject to the extent that Housner (1963) after introducing the concept of pulse-excitations elucidated a size-frequency scale effect that explained why (a) the larger of two geometrically similar columns can survive the excitation that will topple the smaller column and (b) out of two same acceleration amplitude pulses, the one with longer duration is more capable to induce overturning. While the exact dynamic rocking response of the free-standing slender column turns out to be rather complex, the following section offers a qualitative explanation of the size-frequency scale effect initially identified by Kirkpatrick (1927) and made popular to the earthquake engineering community by Housner (1963).

12.2 A Notable Limitation of the Equivalent Static Lateral Force Analysis

12.2.1 Seismic Resistance of Free-Standing Columns Under "Equivalent Static" Lateral Loads

Consider a free-standing rigid column with size $R = \sqrt{b^2 + h^2}$ and slenderness $b/h = \tan\alpha$ as shown in Fig. 12.3 (left). Let us first assume that the base of the column is moving (say to the left) with a "slowly" increasing acceleration, $\ddot{u}_g$ (say a very long-duration acceleration pulse which allows for an equivalent static analysis). Uplift of the column (hinge formation) happens when the seismic demand (overturning moment) $= m\ddot{u}_g h$ reaches the seismic resistance (recentering

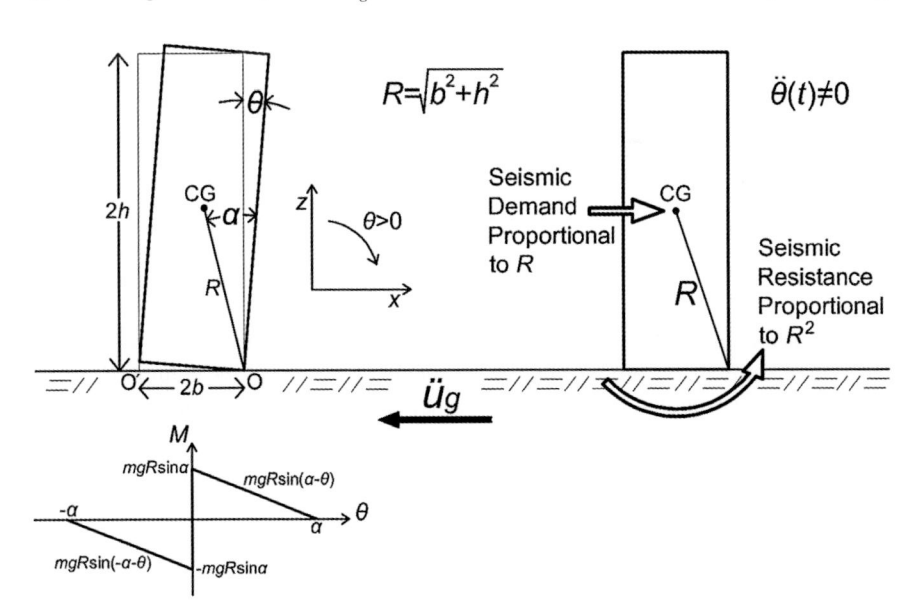

Fig. 12.3 Left: Geometric characteristics of a free-standing rocking column together with its moment rotation diagram. Right: During earthquake shaking which sets the column in rocking motion ($\ddot{\theta}(t) \neq 0$) the seismic resistance is proportional to R^2; while, the seismic demand is proportional to R. Consequently, when a free-standing column is sufficiently large it can survive large horizontal accelerations even if it is very slender

moment) $= mgb$. When uplifting is imminent, "static" moment equilibrium of the column about the pivoting point O gives

$$\underbrace{m\ddot{u}_g h}_{\text{demand}} = \underbrace{mgb}_{\text{resistance}} \quad \text{or} \quad \underbrace{\ddot{u}_g}_{\text{demand}} = g\frac{b}{h} = \underbrace{g\tan\alpha}_{\text{resistance}} \quad (12.1)$$

Equation (12.1), also known as West's formula (Milne 1885; Kirkpatrick 1927), shows that the column <b, h> will uplift when $\ddot{u}_g \geq g\tan\alpha$. Now, given that this is a "quasistatic" lateral inertial loading, the inertia moment due to the nearly zero rotational accelerations of the columns is negligible ($\ddot{\theta} = 0$). Upon uplift has occurred, the rocking column experiences a positive rotation, $\theta(t)$; therefore, the seismic demand is $m\ddot{u}_g R\cos(\alpha-\theta(t))$; while the seismic resistance is merely $mgR\sin(\alpha-\theta(t))$ since $\ddot{\theta} = 0$. For $\theta > 0$, the resistance of the rocking column upon uplifting under quasistatic lateral loading is $\tan(\alpha-\theta(t))$ which is smaller than $\tan\alpha$. Accordingly; once the column uplifts, it will also overturn. From this analysis one concludes that under quasistatic lateral loading the stability of a free-standing column depends solely on its slenderness ($g\tan\alpha$) and is independent to the size ($R = \sqrt{b^2 + h^2}$).

12.2.2 Seismic Resistance of Free-Standing Columns Subjected to Dynamic Loads

In reality, earthquake shaking, $\ddot{u}_g$, is not a quasistatic loading and upon uplifting has occurred the column will experience a finite rotational acceleration ($\ddot{\theta}(t) \neq 0$). In this case, dynamic moment equilibrium gives

$$\underbrace{-m\ddot{u}_g(t)R\cos[\alpha - \theta(t)]}_{\text{seismic demand}} = \underbrace{I_o\ddot{\theta}(t) + mgR\sin[\alpha - \theta(t)]}_{\text{seismic resistance}} \quad \theta > 0 \quad (12.2)$$

where I_o is the rotational moment of inertia of the column about the pivot point at the base—a quantity that is proportional to the square of the size of the column R. As an example, for rectangular columns, $I_o = \frac{4}{3}mR^2$, and Eq. (12.2) simplifies to

$$\underbrace{-\ddot{u}_g(t)R\cos[\alpha - \theta(t)]}_{\text{seismic demand}} = \underbrace{\frac{4}{3}R^2\ddot{\theta}(t) + gR\sin[\alpha - \theta(t)]}_{\text{seismic resistance}} \quad \theta > 0 \quad (12.3)$$

Equation (12.3) indicates that when a slender free-standing column is set into rocking motion the seismic demand (overturning seismic moment) is proportional to R (first power of the size); whereas, the seismic resistance (opposition to rocking) is proportional to R^2 (second power of the size) as shown in Fig. 12.4. Consequently, Eq. (12.3) dictates that regardless how slender a column is (small α) and how intense the ground shaking, $\ddot{u}_g$, is (seismic demand), when a rotating column ($\ddot{\theta}(t)$=finite) is large enough, the second power of R in the right-hand-side (seismic resistance) can always ensures stability. Simply stated, Housner's (1963) size effect is merely a reminder that a quadratic term eventually dominates over a linear term regardless the values of their individual coefficients (see Fig. 12.4).

From its very conception the "equivalent static lateral force analysis" is not meant to deal with any rotational acceleration term; therefore, its notable failure to capture the seismic stability (resistance) of tall free-standing structures. Simply stated, ancient builders were designing structures that their seismic resistance originates primarily from the mobilization of their rotational inertia—a truly dynamic design. It is worth emphasizing that slender rocking structures have moderate strength (uplifting initiates when $\ddot{u}_g > g(b/h) = g\tan\alpha$), negative stiffness; whereas, damping during rocking happens at the instant of impact; therefore, the ductility of these systems is zero. Table 12.1 compares the basic design concepts together with the main response-controlling quantities that are associated with: (a) the traditional earthquake resistant (capacity) design; (b) seismic isolation; and (c) rocking isolation.

It is worth noting that during the last decade there has been a series of publications which aim to direct the attention of engineers to the unique advantages associated with allowing structures to uplift. The underlying concept in this class of publications is the intentional generation of uplifting mechanisms in traditional moment resisting frames (Ajrab et al. 2004; Harden et al. 2006; Kawashima et al. 2007; Gajan and Kutter 2008; Anastasopoulos et al. 2010; Hung et al. 2011; Deng et al. 2012;

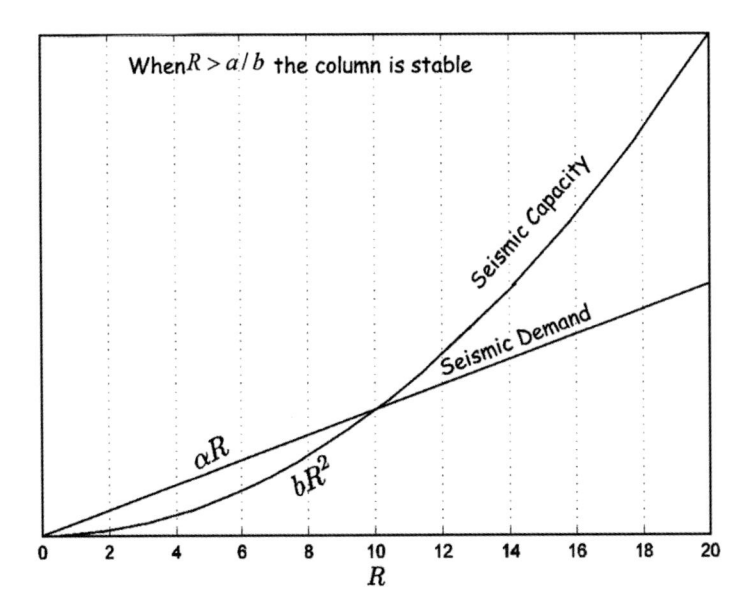

Fig. 12.4 Housner's (1963) "counter intuitive" size effect on the rocking stability of free-standing slender columns is merely a reminder that a quadratic term eventually dominates over a linear term regardless the values of their individual coefficients

Table 12.1 Basic design concepts and response-controlling quantities associated with: (a) the traditional earthquake resistant (capacity) design; (b) seismic isolation; and (c) rocking isolation

	Traditional Earthquake Resistance Design	Seismic isolation	Rocking isolation
	Moment resisting frames		
	Braced frames		
Strength	Moderate to appreciable $$\ddot{u}_g^y = \frac{Q}{m} = 0.1g - 0.25g$$	Low $$\ddot{u}_g^y = \frac{Q}{m} = 0.03g - 0.09g$$	Moderate to appreciable $$\ddot{u}_g^{up} = g\frac{b}{h} = g\tan a$$
Stiffness	Positive and variable due to yielding	Positive, low and constant	Negative, constant
Ductility	Appreciable $\mu = 3-6$	Very large/immaterial[a] LRB[b]: $\mu = 10-30$ CSB[c] $\mu = 1000-3000$	Zero
Damping	Moderate	Moderate to high	Low (only during impact)
Seismic resistance originates from:	Appreciable strength and ductility	Low strength and low stiffness in association with the capability to accommodate large displacements	Low to moderate strength and appreciable **rotational inertia**
Equivalent static lateral force analysis is applicable?	Yes	Yes	No
Design philosophy	Equivalent static	Equivalent static	Dynamic

[a]Makris and Vassiliou (2011)
[b]LRB = Lead Rubber Bearings
[c]CSB = Concave Sliding Bearings

Gelagoti et al. 2012, among others) either at the bottom of shear walls or even at the foundation level by allowing appreciable rotations of the footings due to eccentric loading. In this way, the seismic resistance of these "hybrid" structural systems originate primarily from the intentional creation of a lower failure mechanism which once mobilized it reduces the seismic demand on other critical locations of the structure; while, rocking motion in the way that is illustrated in Fig. 12.3 happens only to some individual members of the overall moment-resisting yielding frame. Consequently, in this class of hybrid systems the development of rotational accelerations of the individual rocking members is somehow suppressed since their motion needs to be compatible with the lateral motion of the overall yielding frame (Makris and Aghagholizadeh 2016; Aghagholizadeh and Makris 2017).

Accordingly, the seismic resistance of these yielding frames is "in-between" that of a traditional moment-resisting yielding frame and that of a rocking frame. In most cases the ductile behavior of the overall moment-resisting yielding frame dominates the system behavior and in this case a "capacity" design approach may be applicable (Gajan et al. 2008).

12.3 Equation of Motion of the Free-Standing Rocking Column

For negative rotations ($\theta(t) < 0$), the equation of motion of a rocking column is

$$-m\ddot{u}_g(t)R\cos\left[-\alpha - \theta(t)\right] = I_o\ddot{\theta}(t) + mgR\sin\left[-\alpha - \theta(t)\right] \quad \theta < 0 \quad (12.4)$$

Equations (12.2) and (12.4) are well known in the literature (Yim et al. 1980; Makris and Roussos 1998, 2000; Zhang and Makris 2001 and references reported therein) and are valid for arbitrary values of the slenderness angle $\alpha = arctan(b/h)$. Equations (12.2) and (12.4) can be expressed in the compact form

$$\ddot{\theta}(t) = -p^2\left\{\sin\left[\alpha\text{sgn}[\theta(t)] - \theta(t)\right] + \frac{\ddot{u}_g}{g}\cos\left[\alpha\text{sgn}[\theta(t)] - \theta(t)\right]\right\} \quad (12.5)$$

In Eq. (12.5), the quantity $p = \sqrt{mRg/I_o}$ is the frequency parameter of the column and is an expression of its size. For rectangular columns $p = \sqrt{3g/(4R)}$.

Figure 12.3 (bottom) shows the moment–rotation relationship during the rocking motion of a free-standing column. The system has infinite stiffness until the magnitude of the applied moment reaches the value $mgR\sin\alpha$, and once the column is rocking, its restoring force decreases monotonically, reaching zero when $\theta = \alpha$. This negative stiffness, which is inherent in rocking systems, is most attractive in earthquake engineering in terms of keeping base shears and moments low (Makris and Konstantinidis 2003), provided that the rocking column remains stable, thus the need for a formula that will offer a safe design value for its slenderness (Makris and Vassiliou 2012).

During the oscillatory rocking motion, the moment–rotation curve follows the curve shown in Fig. 12.3 without enclosing any area. Energy is lost only during impact, when the angle of rotation reverses (Housner 1963).

Following Housner's seminal paper, a number of studies have been presented to address the complex dynamics of one of the simplest man-made structures—the free-standing rigid column. Yim et al. (1980) conducted numerical studies by adopting a probabilistic approach; Aslam et al. (1980) confirmed with experimental studies that the rocking response of rigid columns is sensitive to system parameters, whereas Psycharis and Jennings (1983) examined the uplift of rigid bodies supported on viscoelastic foundation. Subsequent studies by Spanos and Koh (1984) investigated the rocking response due to harmonic steady-state loading and identified 'safe' and 'unsafe' regions together with the fundamental and subharmonic modes of the system. Their study was extended by Hogan (1989, 1990) who further elucidated the mathematical structure of the problem by introducing the concepts of orbital stability and Poincare sections. The steady-state rocking response of rigid columns was also studied analytically and experimentally by Tso and Wong (1989a, b). Their experimental work provided valuable support to theoretical findings.

Depending on the level and form of the ground acceleration, in association with the interface conditions at the base, a free-standing rigid column may translate with the ground, slide, rock, or slide-rock. Analytical and numerical studies on the possible motions of a rigid body were presented by Ishiyama (1982) and Sinopoli (1989). These studies were followed by Scalia and Sumbatyan (1996) and Shenton (1996), who independently indicated that, in addition to pure sliding and pure rocking, there is a slide-rock mode and its manifestation depends not only on the width-to-height ratio and the static friction coefficient but also on the magnitude of the base acceleration.

12.4 The Dynamics of the Rocking Frame

While the ample dynamic stability of a solitary free-standing column as its size increases (large values of ω_p/p) has been documented in a series of publications (Housner 1963; Yim et al. 1980; Makris and Roussos 2000; Zhang and Makris 2001; Makris and Black 2002; Acikgoz and DeJong 2012 among others and references report therein), the concept of rocking isolation becomes attractive and implementable once the dynamics of the rocking frame like the one shown in Fig. 12.1 (right) or Fig. 12.2 is delineated and explained to the extent that it can be easily used by the design engineers.

In an effort to explain the seismic stability of ancient free-standing columns that support heavy epistyles together with the even heavier frieze atop, Makris and Vassiliou (2013) studied the planar rocking response of an array of free-standing columns capped with a freely supported rigid beam as shown in Fig. 12.5.

The free-standing rocking frame shown in Fig. 12.5 is a single DOF structure with size $R = \sqrt{b^2 + h^2}$ and slenderness $\alpha = \mathrm{atan}(b/h)$. The only additional parameter that influences the dynamics of the rocking frame is the ratio of the mass of the cap beam, m_b, to the mass of all the N rocking columns, m_c, $\gamma = m_b/Nm_c$. For the Temple of Apollo in Corinth where the frieze is missing, γ is as low as 0.3, whereas in prefabricated bridges, $\gamma > 4$. As in the case of the single rocking column, the coefficient of friction is large enough so that sliding does not occur at the pivot point at the base and at the cap beam. Accordingly, the horizontal translation

12 The Dynamics of Rocking Isolation

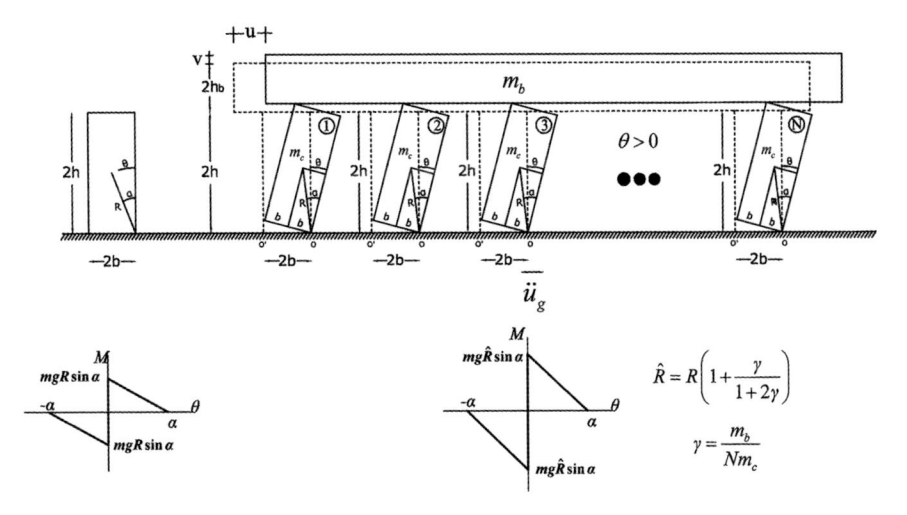

Fig. 12.5 The free-standing rocking frame with columns having size R and slenderness α is more stable than the solitary free-standing column shown on the left having the same size and slenderness

displacement $u(t)$ and the vertical lift $v(t)$ of the cap beam are functions of the single DOF $\theta(t)$. Following a variational formulation Makris and Vassiliou (2013) showed that the equation of motion of the rocking frame shown in Fig. 12.5 is

$$\ddot{\theta}(t) = -\frac{1+2\gamma}{1+3\gamma}p^2\left(\sin[\alpha\operatorname{sgn}[\theta(t)] - \theta(t)] + \frac{\ddot{u}_g(t)}{g}\cos[\alpha\operatorname{sgn}[\theta(t)] - \theta(t)]\right) \quad (12.6)$$

Equation (12.6), which describes the planar motion of the free-standing rocking frame, is precisely the same as Eq. (12.5), which describes the planar rocking motion of a single free-standing rigid column with the same slenderness α, except that in the rocking frame, the term p^2 is multiplied with the factor $(1 + 2\gamma)/(1 + 3\gamma)$. Accordingly, the frequency parameter of the rocking frame, $\hat{p}$, is

$$\hat{p} = \sqrt{\frac{1+2\gamma}{1+3\gamma}}p \quad (12.7)$$

where $p = \sqrt{3g/4R}$ is the frequency parameter of the solitary rocking column and $\gamma = m_b/Nm_c$ is the mass of the cap beam to the mass of all N columns.

According to Eq. (12.7), the rocking response and stability analysis of the free-standing rocking frame with columns having slenderness, α, and size, R, is described by all the past published work on the rocking response of the free-standing single column (Housner 1963; Yim et al. 1980; Aslam et al. 1980; Ishiyama 1982; Spanos and Koh 1984; Zhang and Makris 2001; Makris and Konstantinidis 2003; Vassiliou and Makris 2012, among others), where the column has the same slenderness, α, and a larger size $\widehat{R}$ given by

$$\widehat{R} = \frac{1+3\gamma}{1+2\gamma} R = \left(1 + \frac{\gamma}{1+2\gamma}\right) R \qquad (12.8)$$

The remarkable result offered by Eq. (12.6) – that the heavier the cap beam is, the more stable is the free-standing rocking frame despite the rise of the center of gravity of the cap beam – has been also confirmed by the author after obtaining Eq. (12.7) for a pair of columns with the algebraically intense direct formulation after deriving the equations of motion of the two-column frame through dynamic equilibrium (Makris and Vassiliou 2014). Furthermore, numerical studies with the discrete element method by Papaloizou and Komodromos (2009) concluded to the same result – that the planar response of free-standing columns supporting epistyles is more stable than the response of the solitary, free-standing column. This finding has also been confirmed in the experimental studies of Mouzakis et al. (2002), and Drosos and Anastasopoulos (2014). Figure 12.6 summarizes the increasing seismic stability as we go from the solitary free-standing column to the free-standing rocking frame.

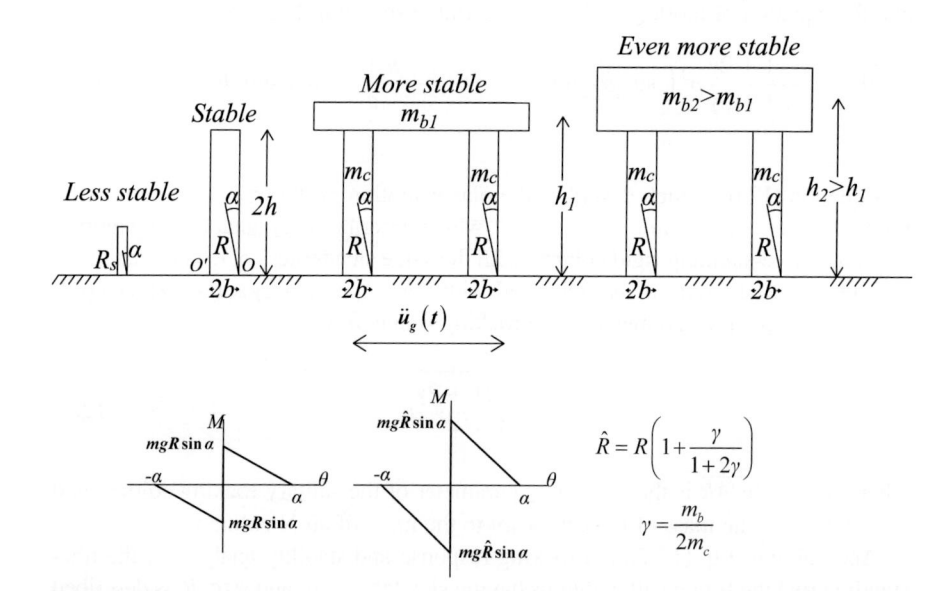

Fig. 12.6 The large free-standing column with size R and slenderness α is more stable than the geometrically similar smaller column shown at the far left of the figure. The free-standing rocking frame with columns having the same size R and same slenderness α is more stable than the solitary rocking column. A heavier freely supported cap-beam renders the rocking frame even more stable regardless of the rise of the center of gravity of the system

12.5 The Emerging Concept of Rocking Isolation for Bridges

The concept of allowing the piers of tall bridges to rock is not new. For instance, the beneficial effects that derive from uplifting and rocking have been implemented since the early 1970s in the South Rangitikei Bridge in New Zealand (Beck and Skinner 1974). Nevertheless, despite the successful design of the South Rangitikei bridge and the ample dynamic stability of the rocking frame as documented by Eq. (12.6) and further confirmed by numerical and experimental studies (Ishiyama 1982; Psycharis et al. 2003; Papaloizou and Komodromos 2009; Mouzakis et al. 2002; Drosos and Anastasopoulos 2014; Makris 2014a, b) most modern tall bridges (with tall slender piers) are protected from seismic action via base (shear) isolation of the deck, rather than via (the most natural) rocking isolation. Part of the motivation of this work is to show in the simplest possible way that in the event that a rocking system is selected, the heavy deck atop the tall slender columns not only does not harm the stability of the columns but in contrast enhances the stability of the entire rocking system as shown by Eq. (12.6) and Fig. 12.6.

This work comes to support the emerging design concept (mainly advanced by the prefabricated bridge technology) of concentrating the inelastic deformations of bridge frame at the locations where the bridge-piers meet the foundation and the deck (Mander and Cheng 1997; Sakai and Mahin 2004; Wacker et al. 2005; Mahin et al. 2006; Cheng 2008; Cohagen et al. 2008; Yamashita and Sanders 2009, among others). It shall however be stressed that in the prefabricated bridge technology, the bridge piers and the deck are not free standing, therefore, the structural system is essentially a hybrid system in-between the rocking frame examined in this work and a traditional ductile moment-resisting frame.

At present, the equivalent static lateral force procedure is deeply rooted in the design philosophy of the structural engineering community which is primarily preoccupied on how to improve the ductility and performance of the seismic connections; while the ample dynamic rocking stability that derives from the beneficial coexistence of large rotational inertia, negative stiffness and gravity as described by Eq. (12.6) is ignored. At the same time, it shall be recognized that during the last decade there have been several publications which have voiced the need to go beyond the elastic response spectrum and the associated equivalent static lateral force procedure (Makris and Konstantinidis 2003; Lagomarsino et al. 2004; Apostolou et al. 2007; Resemini et al. 2008; Anastasopoulos et al. 2010; Makris 2014b among others). In addition to these studies, Acikgoz and DeJong (2012) and Vassiliou et al. (2013) have examined in depth the rocking response of flexible, slender structures and the main conclusion is that the flexure of a tall rocking structure further increases its seismic stability. To this end, it is worth mentioning the recent theoretical work on the three-dimensional rocking response of free-standing columns (Konstantinidis and Makris 2007; Zulli et al. 2012; Chatzis and Smyth 2012a, b; Vassiliou et al. 2017) which confirms the seismic stability of free-standing columns in three dimensions. The time is therefore ripe for the development of new, physically motivated alternative seismic protection technology for the design of large, slender structures. Part of the motivation for this work is to bring forward the ample seismic stability associated with the free rocking of large, slender structures and the corresponding rocking frame.

12.6 Seismic Protection of Structures with Supplemental Rotational Inertia

For a rectangular column the frequency parameter is $p = \sqrt{3g/(4R)}$, therefore Eq. (12.5) assumes the form

$$\ddot{\theta}(t) + \frac{3g}{4R} \sin\left(\alpha \operatorname{sgn}[\theta(t)] - \theta(t)\right) = -\frac{3}{4}\frac{\ddot{u}_g(t)}{R} \cos\left(\alpha \operatorname{sgn}[\theta(t)] - \theta(t)\right) \quad (12.9)$$

By expressing Eq. (12.5) in the form of Eqs. (12.9), Housner's (1963) scale effect becomes most apparent since among two geometrically similar free-standing columns (same slenderness α), the one with the larger size R, experiences an apparent suppressed shaking since the input acceleration on the right hand-side of Eq. (12.9) is multiplied with $1/R$. The suppression of the input ground acceleration with the inverse of the size of the column $= R$ is a direct result of the rocking (rotational) motion of the rigid column and originates from the difficulty to mobilize the rotational inertia of the column which is proportional to R^2 (see Eq. 12.3 and Fig. 12.4).

This unique feature of the dynamics of rigid bodies has been the motivation for proposing supplemental rotating devices to suppress the seismic response of traditional framing systems where the dominant motion of their masses (floors) is translational—not rotational (Hwang et al. 2007; Ikago et al. 2012; Takewaki et al. 2012; Marian and Giaralis 2014; Lazar et al. 2014; Makris and Kampas 2016; Makris 2017 and references reported therein).

As an example, Fig. 12.7a shows a single-degree-of-freedom structure with stiffness k and mass m. A stiff chevron frame supports a flywheel with radius R, and mass m_w that can rotate about an axis O. Concentric to the flywheel there is an attached pinion with radius, ρ_1, engaged to a linear rack connected to the bottom of the vibrating mass, m, of the SDOF. With this arrangement when the mass m undergoes a positive displacement, $u(t)$, the flywheel is subjected to a clockwise rotation, $\theta_1(t)$. Given that there is no slipping between the rack and the pinion,

$$\theta_1(t) = \frac{u(t)}{\rho_1} \quad (12.10)$$

Figure 12.7b shows the free-body diagrams of the vibrating mass, m, and the rotating pinion-flywheel system. For a positive displacement, $u(t)$, to the right, the internal force, F_1, at the rack-pinion interface opposes the motion (to the left). Accordingly, dynamic equilibrium of the vibrating mass when subjected to a ground acceleration, $\ddot{u}_g(t)$, gives

12 The Dynamics of Rocking Isolation

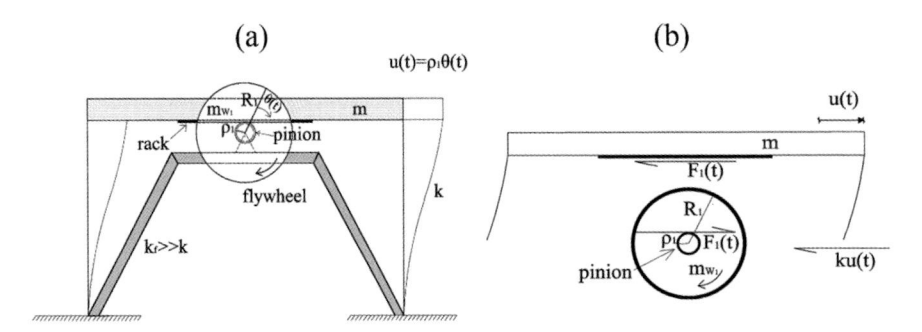

Fig. 12.7 (a) Single-degree-of-freedom structure with mass m and stiffness k with supplemental rotational inertia from flywheel with radius R supported on a chevron frame with stiffness k_f that is much larger than k; (b) free-body diagram of the vibrating mass m when engaged to the pinion of the flywheel pictured

$$m[\ddot{u}(t) + \ddot{u}_g(t)] = -ku(t) - F_1(t) \qquad (12.11)$$

where the internal force F_1, needs to satisfy the moment equilibrium of the flywheel about point O.

$$I_{W1}\ddot{\theta}_1(t) = F_1(t)\rho_1 \qquad (12.12)$$

In Eq. (12.12), $I_{W1} = (1/2)m_{W1}R_1^2$, is the moment of inertia of the flywheel about point O. Substitution of Eq. (12.12) into Eq. (12.11) in association with Eq. (12.10) gives

$$\left(1 + \frac{1}{2}\frac{m_{W1}}{m}\frac{R_1^2}{\rho_1^2}\right)\ddot{u}(t) + \omega_o^2 u(t) = -\ddot{u}_g(t) \qquad (12.13)$$

where, $\omega_o = \sqrt{k/m}$, is the natural frequency of the structure when the pinion-flywheel is disengaged. Upon dividing with the acceleration coefficient, Eq. (12.13) gives,

$$\ddot{u}(t) + \underbrace{\frac{\omega_o^2}{1 + \dfrac{1}{2}\dfrac{m_{W1}}{m}\dfrac{R_1^2}{\rho_1^2}}}_{\text{Lengthening of the period}} u(t) = -\underbrace{\frac{1}{1 + \dfrac{1}{2}\dfrac{m_{W1}}{m}\dfrac{R_1^2}{\rho_1^2}}}_{\substack{\text{Suppression of the input}\\\text{ground motion}}}\ddot{u}_g(t)$$

$$(12.14)$$

Equation (12.14) indicates that the engagement of the flywheel in a rotational motion lengthens the vibration period of the structure; and most importantly it suppresses the level of ground shaking in a way that resembles the dynamics of the free-standing rocking columns expressed by Eq. (12.9).

More on the amplification of the rotational inertia effect and the efficiency of supplemental rotational inertia to suppress the earthquake response of long-period structures can be found in Makris and Kampas 2016.

12.7 Conclusions

Half a century ago George Housner's 1963 seminal paper marked the beginning of a series of systematic studies on the dynamic response and stability of rocking structures which gradually led to the development of rocking isolation—an attractive practical and economical alternative for the seismic protection of tall, slender structures which originates from the mobilization of their large rotational inertia.

After revisiting Housner's size-frequency scale effect for the solitary column which merely explains that when a free-standing column is sufficiently large it can survive any strong shaking, the article builds upon a recent remarkable result—that the dynamic rocking response of an array of free-standing columns capped with a rigid beam is identical to the rocking response of a solitary column with the same slenderness; yet, with larger size, which is a more stable configuration (Eq. 12.6). Most importantly, the dynamics of the rocking frame reveals that the heavier the freely supported beam is, the more stable is the rocking frame regardless of the rise of the center of gravity of the cap beam, concluding that top-heavy rocking frames are more stable than when they are top-light.

This "counterintuitive" behavior renders rocking isolation a most attractive alternative for the seismic protection of bridges given that the heavier is the deck, the more stable is the rocking bridge. The realization of a truly rocking frame which can fully mobilize its rotational inertia with neither post-tensioning nor continuation of the longitudinal reinforcement through the rocking interfaces shall remove several of the concepts associated with the seismic connections of prefabricated bridges such as buckling and fracture of the longitudinal reinforcing bars or spallings of the concrete corners.

The unique feature of the dynamics of free-standing rocking bodies, where the input ground acceleration is suppressed with the inverse of the size of the body has been the motivation for proposing supplemental rotational inertia systems to suppress the seismic response of traditional frames where the dominant motion of their floors is translational.

Acknowledgments Partial financial support for this work has been provided by the research project "SeismoRockBridge" which is implemented under the "ARISTEIA" Action of the "OPERATIONAL PROGRAMME EDUCATION AND LIFELONG LEARNING" and is co-funded by the European Social Fund (ESF) and Greek National Resources.

References

Acikgoz S, DeJong MJ (2012) The interaction of elasticity and rocking in flexible structures allowed to uplift. Earthq Eng Struct Dyn 41(15):2177–2194

Aghagholizadeh M, Makris N (2017) Seismic response of a yielding structure coupled with a Rocking Wall. J Struct Eng ASCE (in press)

Ajrab J, Pekcan G, Mander J (2004) Rocking Wall–frame structures with supplemental tendon systems. J Struct Eng 130(6):895–903

Anastasopoulos I, Gazetas G, Loli M, Apostolou M, Gerolymos N (2010) Soil failure can be used for seismic protection of structures. Bull Earthq Eng 8(2):309–326

Apostolou M, Gazetas G, Garini E (2007) Seismic response of slender rigid structures with foundation uplift. Soil Dyn Earthq Eng 27(7):642–654

Aslam M, Scalise DT, Godden WG (1980) Earthquake rocking response of rigid blocks. J Struct Eng Div (ASCE) 106:377–392

Beck JL, Skinner RI (1974) The seismic response of a reinforced concrete bridge pier designed to step. Earthq Eng Struct Dyn 2(4):343–358

Chatzis MN, Smyth AW (2012a) Modeling of the 3D rocking problem. Int J Non Linear Mech 47:85–98

Chatzis MN, Smyth AW (2012b) Robust modeling of the rocking problem. J Eng Mech 138:247–262

Cheng CT (2008) Shaking table tests a self-centering designed bridge substructure. Eng Struct 30 (12):3426–3433

Cohagen L, Pang JBK, Stanton JF, Eberhard MO (2008) A precast concrete bridge bent designed to recenter after an earthquake, Research report, Federal Highway Administration

Deng L, Kutter B, Kunnath S (2012) Centrifuge modeling of bridge systems designed for rocking foundations. J Geotech Geoenviron 138(3):335–344

Drosos V, Anastasopoulos I (2014) Shaking table testing of multidrum columns and portals. Earthq Eng Struct Dyn 43(11):1703–1723

Gajan S, Kutter B (2008) Capacity, settlement, and energy dissipation of shallow footings subjected to rocking. J Geotech Geoenviron 134(8):1129–1141

Gajan S, Hutchinson TC, Kutter BL, Raychowdhury P, Ugalde JA, Stewart JP (2008) Numerical models for analysis and performance-based design of shallow foundations subject to seismic loading. Report no. PEER-2007/04, Pacific Earthquake Engineering Research Center, University of California, Berkeley

Gelagoti F, Kourkoulis R, Anastasopoulos I, Gazetas G (2012) Rocking isolation of low-rise frame structures founded on isolated footings. Earthq Eng Struct Dyn 41(7):1177–1197

Harden C, Hutchinson T, Moore M (2006) Investigation into the effects of foundation uplift on simplified seismic design procedures. Earthquake Spectra 22(3):663–692

Hogan SJ (1989) On the dynamics of rigid-block motion under harmonic forcing. Proc R Soc A 425 (1869):441–476

Hogan SJ (1990) The many steady state responses of a rigid block under harmonic forcing. Earthq Eng Struct Dyn 19(7):1057–1071

Housner GW (1963) The behaviour of inverted pendulum structures during earthquakes. Bull Seismol Soc Am 53(2):404–417

Hung HH, Liu KY, Ho TH, Chang KC (2011) An experimental study on the rocking response of bridge piers with spread footing foundations. Earthq Eng Struct Dyn 40(7):749–769

Hwang JS, Kim J, Kim YM (2007) Rotational inertia dampers with toggle bracing for vibration control of a building structure. Eng Struct 29(6):1201–1208

Ikago K, Saito K, Inoue N (2012) Seismic control of single-degree-of-freedom structure using tuned viscous mass damper. Earthq Eng Struct Dyn 41(3):453–474

Ikegami R, Kishinouye F (1947) A study on the overturning of rectangular columns in the case of the Nankai earthquake on December 21, 1946. Bull Earthquake Res Inst Tokyo Univ 25:49–55

Ikegami R, Kishinouye F (1950) The acceleration of earthquake motion deduced from overturning of the gravestones in case of the Imaichi earthquake on Dec. 26, 1949. Bull Earthquake Res Inst Tokyo Univ 28:121–128

Ishiyama Y (1982) Motions of rigid bodies and criteria for overturning by earthquake excitations. Earthq Eng Struct Dyn 10(5):635–650

Kawashima K, Nagai Y, Sakellaraki D (2007) Rocking seismic isolation of bridges supported by spread foundations. In: Proceedings of the 2nd Japan-Greece workshop on seismic design, observation and retrofit of foundations, Japan Society of Civil Engineers, vol 2. Tokyo, pp 254–265

Kirkpatrick P (1927) Seismic measurements by the overthrow of columns. Bull Seismol Soc Am 17(2):95–109

Konstantinidis D, Makris N (2007) The dynamics of a rocking block in three dimensions. In: Proceedings of the 8th HSTAM international congress on mechanics, Patras, Greece

Lagomarsino S, Podestà S, Resemini S, Curti E, Parodi S (2004) Mechanical models for the seismic vulnerability assessment of churches. In: Proceedings of IV SAHC, vol 2. A. A. Balkema, Padova, Italy, London (UK), pp 1091–1101

Lazar IF, Neild SA, Wagg DJ (2014) Using an inerter-based device for structural vibration suppression. Earthq Eng Struct Dyn 43(8):1129–1147

Mahin S, Sakai J, Jeong H (2006) Use of partially prestressed reinforced concrete columns to reduce post-earthquake residual displacements of bridges. In: Proceedings of the 5th national seismic conference on bridges and highways, San Francisco

Makris N (2014a) The role of the rotational inertia on the seismic resistance of free-standing rocking columns and articulated frames. Bull Seismol Soc Am 104(5):2226–2239

Makris N (2014b) A half century of rocking isolation. Earthq Struct 7(6):1187–1221

Makris N (2017) Basic response functions of simple inertoelastic and inertoviscous models. J Eng Mech ASCE 143(11):04017123

Makris N, Aghagholizadeh M (2016) The dynamics of an elastic structure coupled with a rocking wall. Earthq Eng Struct Dyn 46(6):945–962

Makris N, Black CJ (2002) Uplifting and overturning of equipment anchored to a base foundation. Earthq Spectra 18(4):631–661

Makris N, Kampas G (2016) Seismic protection of structures with supplemental rotational inertia. J Eng Mech ASCE 142(11):04016089

Makris N, Konstantinidis D (2003) The rocking spectrum and the limitations of practical design methodologies. Earthq Eng Struct Dyn 32(2):265–289

Makris N, Roussos Y (1998) Rocking response and overturning of equipment under horizontal pulse-type motions. Rep. No. PEER-98/05, Pacific Earthquake Engineering Research Center, University of California, Berkeley, CA

Makris N, Roussos Y (2000) Rocking response of rigid blocks under near-source ground motions. Geotechnique 50(3):243–262

Makris N, Vassiliou MF (2011) The existence of "complete similarities" in the response of seismic isolated structures and their implication in design. Earthq Eng Struct Dyn 40:1103–1121

Makris N, Vassiliou MF (2012) Sizing the slenderness of free-standing rocking columns to withstand earthquake shaking. Arch Appl Mech 82(10–11):1497–1511

Makris N, Vassiliou MF (2014) Are some top-heavy structures more stable? J Struct Eng ASCE 140(5):06014001

Mander JB, Cheng CT (1997) Seismic resistance of bridge piers based on damage avoidance design, Tech. Rep. no. NCEER-97-0014, National Center for Earthquake Engineering Research, Department of Civil and Environmental Engineering, State University of New York, Buffalo

Marian L, Giaralis A (2014) Optimal design of a novel tuned mass-damper-inerter (TMDI) passive vibration control configuration for stochastically support-excited structural systems. Probab Eng Mech 38:156–164

Milne J (1885) Seismic experiments. Trans Seismol Soc Jpn 8:1–82

Mouzakis HP, Psycharis IN, Papastamatiou DY, Carydis PG, Papantonopoulos C, Zambas C (2002) Experimental investigation of the earthquake response of a model of a marble classical column. Earthq Eng Struct Dyn 31(9):1681–1698

Muto K, Umemure H, Sonobe Y (1960) Study of the overturning vibration of slender structures. In: Proceedings of the second world conference on Earthquake Engineering, Japan, pp 1239–1261

Papaloizou L, Komodromos K (2009) Planar investigation of the seismic response of ancient columns and colonnades with epistyles using a custom-made software. Soil Dyn Earthq Eng 29:1437–1454

Powell B (1905) The temple of Apollo Corinth. Am J Archaeol 9(1):44–63

Psycharis IN, Jennings PC (1983) Rocking of slender rigid bodies allowed to uplift. Earthq Eng Struct Dyn 11(1):57–76

Psycharis IN, Lemos JV, Papastamatiou DY, Zambas C (2003) Numerical study of the seismic behaviour of a part of the Parthenon Pronaos. Earthq Eng Struct Dyn 32(13):2063–2084

Resemini S, Lagomarsino S, Cauzzi C (2008) Dynamic response of rocking masonry elements to long period strong ground motion. In: Proceedings of the 14th world conference on Earthquake Engineering, Beijing, China

Sakai J, Mahin S (2004) Analytical investigations of new methods for reducing residual displacements of reinforced concrete bridge columns. PEER report 2004/02, Pacific Earthquake Engineering Research Center, University of California, Berkeley, CA

Scalia A, Sumbatyan MA (1996) Slide rotation of rigid bodies subjected to a horizontal ground motion. Earthq Eng Struct Dyn 25:1139–1149

Shenton HW III (1996) Criteria for initiation of slide, rock, and slide-rock rigid-body modes. J Eng Mech 122(7):690–693

Sinopoli A (1989) Kinematic approach in the impact problem of rigid bodies. Appl Mech Rev (ASME) 42(11):233–S244

Spanos PD, Koh AS (1984) Rocking of rigid blocks due to harmonic shaking. J Eng Mech 110(11):1627–1642

Takewaki I, Murakami S, Yoshitomi S, Tsuji M (2012) Fundamental mechanism of earthquake response reduction in building structures with inertial dampers. Struct Control Health Monit 19(6):590–608

Tso WK, Wong CM (1989a) Steady state rocking response of rigid blocks part 1: analysis. Earthq Eng Struct Dyn 18(1):89–106

Tso WK, Wong CM (1989b) Steady state rocking response of rigid blocks part 2: experiment. Earthq Eng Struct Dyn 18(1):107–120

Vassiliou MF, Makris N (2012) Analysis of the rocking response of rigid blocks standing free on a seismically isolated base. Earthq Eng Struct Dyn 41(2):177–196

Vassiliou MF, Mackie KR, Stojadinović B (2013) Rocking response of slender, flexible columns under pulse excitation. In: Proceedings of the 4th ECCOMAS thematic conference on computational methods in structural dynamics and earthquake engineering, Kos Island, Greece, paper no. C 1084

Vassiliou MF, Burger S, Egger M, Bachmann JA, Broccardo M, Stojadinovic B (2017) The three-dimensional behavior of inverted pendulum cylindrical structures during earthquakes. Earthq Eng Struct Dyn 46:2261–2280

Wacker JM, Hieber DG, Stanton JF, Eberhard MO (2005) Design of precast concrete piers for rapid bridge construction in seismic regions. Research report, Federal Highway Administration

Yamashita R, Sanders D (2009) Seismic performance of precast unbonded prestressed concrete columns. ACI Struct J 106(6):821–830

Yim CS, Chopra AK, Penzien J (1980) Rocking response of rigid blocks to earthquakes. Earthq Eng Struct Dyn 8(6):565–587

Zhang J, Makris N (2001) Rocking response of free-standing blocks under cycloidal pulses. J Eng Mech (ASCE) 127(5):473–483

Zulli D, Contento A, Di Egidio A (2012) 3D model of rigid block with a rectangular base subject to pulse-type excitation. Int J Non Linear Mech 47(6):679–687

Chapter 13
Multistory Building Frames and Shear Walls Founded on "Rocking" Spread Footings

G. Gazetas, D. Dais, F. Gelagoti, and R. Kourkoulis

Abstract The seismic performance of a two-story 2D frame and a five-story 3D frame–shear-wall structure founded on spread (isolated) footings is investigated. In addition to footings conventionally designed in accordance with "capacity-design" principles, substantially under-designed footings are also used. Such unconventional ("rocking") footings may undergo severe cyclic uplifting while inducing large plastic deformations in the supporting soil during seismic shaking. It is shown that thanks to precisely such behaviour they help the structure survive with little damage, while experiencing controllable foundation deformations in the event of a really catastrophic seismic excitation. Potential exceptions are also mentioned along with methods of improvement.

13.1 Introduction: Isolation Via Rocking Foundation

In the last 15 years, "Rocking Foundations" have been found to be not only an economic but, in many cases, a technically superior seismic solution to conventionally designed foundations. Their superiority stems from the fact that they constrain the transmitted to the superstructure accelerations (thanks to the cutoff provided by their reduced moment capacity), and that they lead to increased natural period and hysteretic damping (Pecker 1998; Gajan et al. 2005; Kawashima et al. 2007; Anastasopoulos et al. 2010; Deng et al. 2012; Makris 2014; Gazetas 2015; Kutter et al. 2016). A nearly full-scale bridge pier-foundation seismic experiment, conducted by the teams of Professors Panagiotou and Kutter (Antonellis et al. 2016) on the UC San Diego shaking table, demonstrated the outstanding performance of highly under-designed foundations against very strong seismic shaking. Equally-supportive conclusions have been drawn by numerous experimental campaigns with small-scale shaking-table tests under both 1-g and centrifuge conditions.

G. Gazetas (✉) · D. Dais · F. Gelagoti · R. Kourkoulis
National Technical University of Athens, Athens, Greece
e-mail: gazetas@central.ntua.gr

Most of the theoretical and experimental studies on rocking foundations refer to a single footing supporting a simple inverted-pendulum type structure like a single-column bridge pier. A more limited number of studies have dealt with simple frame structures, as well as with frame–with–shear-wall structures (Gelagoti et al. 2012; Kourkoulis et al. 2012; Anastasopoulos et al. 2014, 2015; Antonaki 2013; Dais 2015), with also encouraging results for the beneficial role of under-designed "rocking" foundations in protecting the superstructure. The only drawback is the possible remaining settlement and rotation of the foundation. And whereas footing settlement of a statically determinate structure may not be a major problem, for the highly indeterminate multi-story and multi-spam frames the consequences may be difficult to absorb in design. Hence the need to investigate the feasibility and usefulness of "rocking" spread foundations in such structures.

Two such systems are examined here: (a) a plane two-story 2-span moment-resisting frame; and (b) a three-dimensional five-story building frame with no and with four shear walls.

13.2 Two-Story Frame on two Types of Footings

The frame of Fig. 13.1 was structurally designed according to EC8 for an effective ground acceleration A = 0.36 g and ductility-dependent "behavior" factor q = 3.9. The soil is stiff clay with S_u = 150 kPa and G_o = 105 MPa. Two alternative foundation schemes are examined (Anastasopoulos et al. 2014):

(a) Conventionally over-designed footings that can mobilize a maximum moment resistance M_u from the underlying soil *larger* than the bending moment capacity of the corresponding column M_{RD}. For static vertical loads a factor of safety $F_S \geq 3$ is required against bearing capacity failure. For seismic load combinations a factor of safety $F_E = 1$ is acceptable. The maximum allowable seismic eccentricity criterion is also enforced: e = M/N ≤ B/3. For the investigated soil–structure system this eccentricity criterion was found to be the controlling one, leading to minimum required footing widths B = 2.7 m, 2.5 m and 2.4 m for the left, middle, and right footing, respectively. Notice that the left corner footing is required to be the largest because of its smallest axial load, an hence a tendency for larger eccentricity. Bearing capacities and safety factors are computed according to the provisions of EC8, which are basically similar to those typically used in foundation design practice around the world.

(b) Under-sized footings of the rocking isolation scheme whose geotechnical capacity is smaller than the structural capacity of the columns, "guiding" the plastic hinge at or below the soil–footing interface, instead of at the base of the columns. The small width of the footings promotes full mobilization of foundation moment capacity with substantial uplifting. The eccentricity criterion is completely relaxed, while $F_E < 1$ is allowed. The static $F_S \geq 3$ remains a requirement as a measure against uncertainties regarding soil strength.

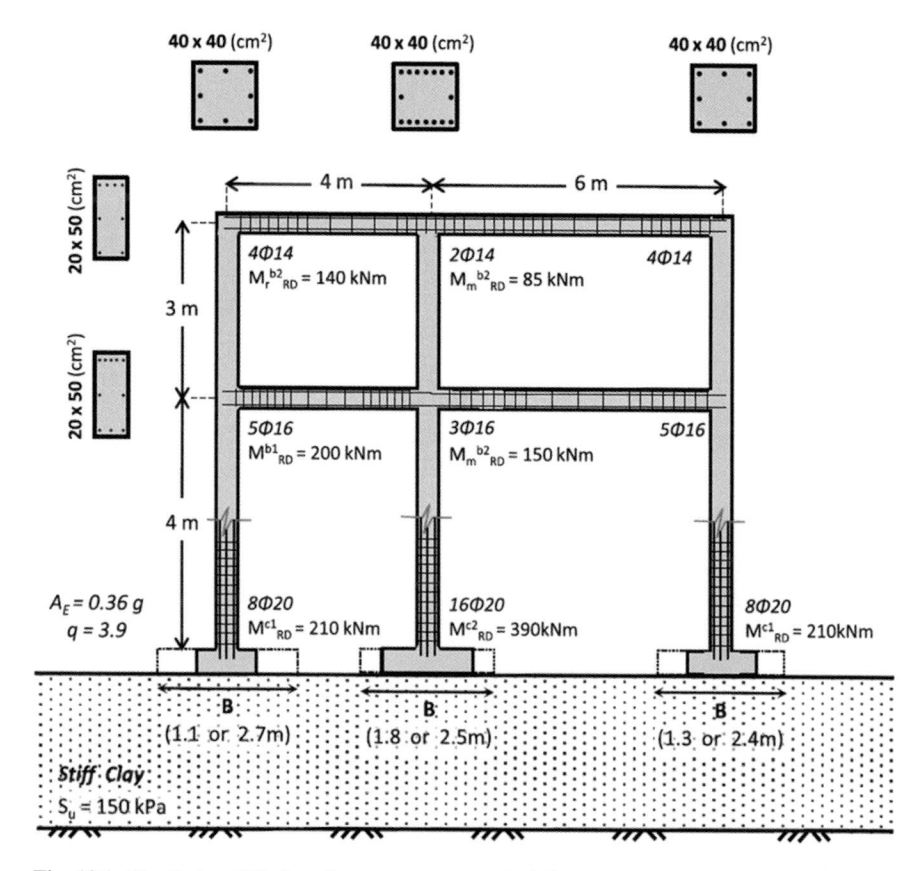

13 Multistory Building Frames and Shear Walls Founded on "Rocking" Spread Footings

Fig. 13.1 The 2-story RC plane frame: geometry and reinforcement

Moreover, it turns out that $F_S \geq 4$ might be desirable in order to promote uplifting–dominated response, and thereby limit seismic settlements and increase re-centering. Applying the methodology which has been outlined in Gelagoti et al. (2012), the footings were designed to be adequately small to promote uplifting, but large enough to limit the settlements. Aiming to minimize differential settlements stemming from asymmetry, the three footings were dimensioned in such a manner so as to have the same F_S. Based on the above criteria, the resulting footing widths for the rocking–isolation design alternative are $B = 1.1$ m, 1.8 m, and 1.3 m, for the left, middle, and right footing, respectively: indeed, substantially smaller than those of the code-based design. Footing dimensions and static factors of safety against vertical loading of the two designs are summarized in Table 13.1.

The performance of the two design alternatives is compared in Fig. 13.2. The deformed mesh with superimposed plastic strain contours of the two alternatives is portrayed on the figure. With the relentless seismic shaking of the Takatori motion, the conventionally designed frame collapses under its gravity load (due to excessive

Table 13.1 Footing dimensions and corresponding factors of safety (computed with the provisions of EC8) against vertical loading, for the two design alternatives of Fig. 13.1

Conventional design			Rocking isolation		
Footing	B (m)	F_S	Footing	B (m)	F_S
Left	2.7	32.0	Left	1.1	5.5
Middle	2.5	11.0	Middle	1.8	5.5
Right	2.4	18.0	Right	1.3	5.5

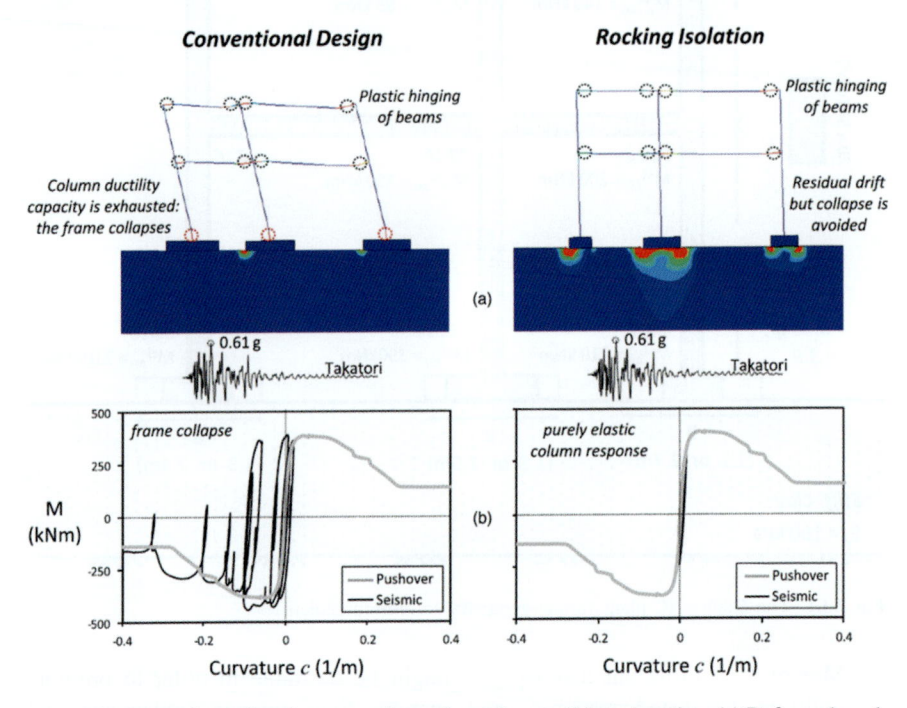

Fig. 13.2 Comparison of performance of two alternatives to Takatori motion. (**a**) Deformed mesh with plastic strain contours; (**b**) column moment-curvature response

drift of the structure, the moments produced by P–δ effects cannot be sustained by the columns, leading to loss of stability and total collapse). As expected, plastic hinges firstly develop in the beams and subsequently at the base of the three columns, while soil under the footings remains practically elastic. The collapse is also evidenced by the substantial exceedance of the available curvature ductility of the columns (Fig. 13.2b). Conversely, the rocking–isolated frame withstands the shaking, with plastic hinging taking place only in the beams, leaving the columns almost unscathed (moment-curvature response: elastic). Instead, plastic hinging now develops within the underlying soil in the form of extended soil plasticization (indicated by the red regions under the foundation.

Thanks to the larger bending moment capacity of the column than of the footing, damage is guided "below ground" and at the soil–foundation interface in the form of detachment and uplifting – evidenced in Fig. 13.2b by the zero residual rotation, unveiling the re-centering capability of the under-designed foundation scheme.

The price to pay: large accumulated settlements. Moreover, despite the fact that the three footings have been dimensioned to have the same static factor of safety F_S (in an attempt to minimize differential settlements exacerbated from asymmetry), the central footing settles more than the two side footings, leading to a differential settlement of the order of 3 cm. The difference in the settlement stems of course from their differences in width. As previously discussed, the central footing was made larger (B = 1.8 m, compared to 1.1 m and 1.3 m of the two side footings) in order to maintain the same FS. Since the latter is common for the three footings, if the loading is more-or-less the same, their response should be similar. However, such equivalence refers to dimensionless quantities, not absolute values. In other words, while the three footings sustain almost the same dimensionless settlement w/B, which is roughly equal to 0.025 ($\approx$ 3 cm/1.2 m) for the two side footings and 0.033 ($\approx$ 6 cm/1.8 m) for the central one, the latter is substantially larger in width and hence its settlement is larger in absolute terms. Naturally, the three footings are not subjected to exactly the same loading, something which further complicates the response. Such differential settlements may inflict additional distress in the superstructure, and are therefore worthy of further investigation. Pertinent amelioration measures are discussed later.

13.3 Five-Story Existing Frame: Seismic Petrofit with Shear Walls

13.3.1 Existing Building

A five-story reinforced-concrete frame (shown in Fig. 13.3) consists of 12 columns connected with beams and carrying 5 slabs. It has been designed according to the norms and practices of the 1970s, with a base-shear coefficient of 0.06. The soil is a stiff clay with uniform with depth $S_u \approx 150$ kPa and the footings (squares or rectangles) were designed to have a static factor of safety against bearing-capacity, $F_s > 3$ and a seismic one $F_E > 2$; The resulting dimensions are given in Tables 13.1 and 13.2.

The frame–foundation–soil system is modelled with 3D finite elements using ABAQUS (Fig. 13.4a). We subjected this frame to the motion recorded in Lefkada 2003 earthquake, which according to our current understanding roughly corresponds to the assumed design base-shear coefficient. The structure fails. Fig. 13.4b shows the computed time histories at the top and the first floor. The two curves are almost identical and reveal failure of the "soft first story" type, with the upper structure moving as a block. Therefore, there is a need for seismic retrofit to upgrade the structure.

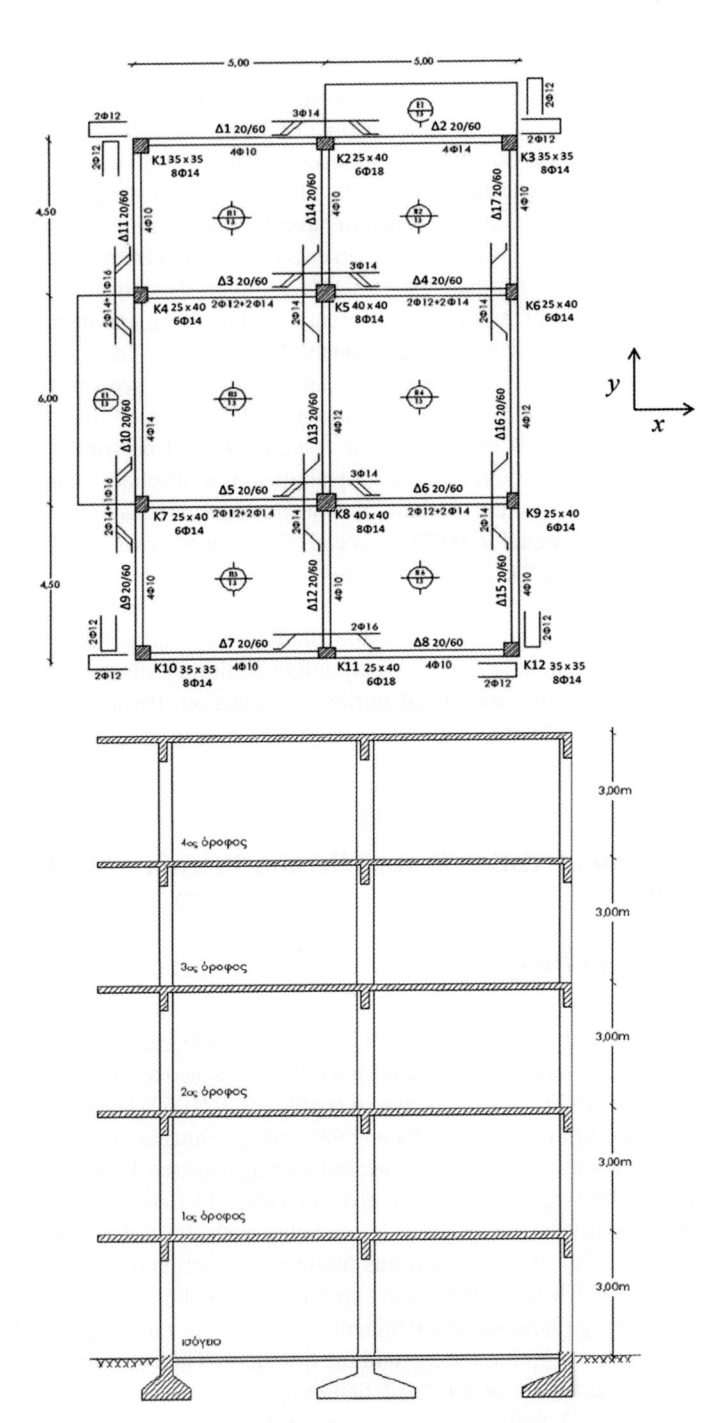

Fig. 13.3 The 5-story RC framed building: plan view and elevation

Table 13.2 Dimensions of half the footings (advantage of symmetry)

Direction	K1	K2	K3	K4	K5	K6
X	1.8	2.0	1.8	1.4	2.4	1.4
Y	1.8	1.4	1.8	2.0	2.4	2.0

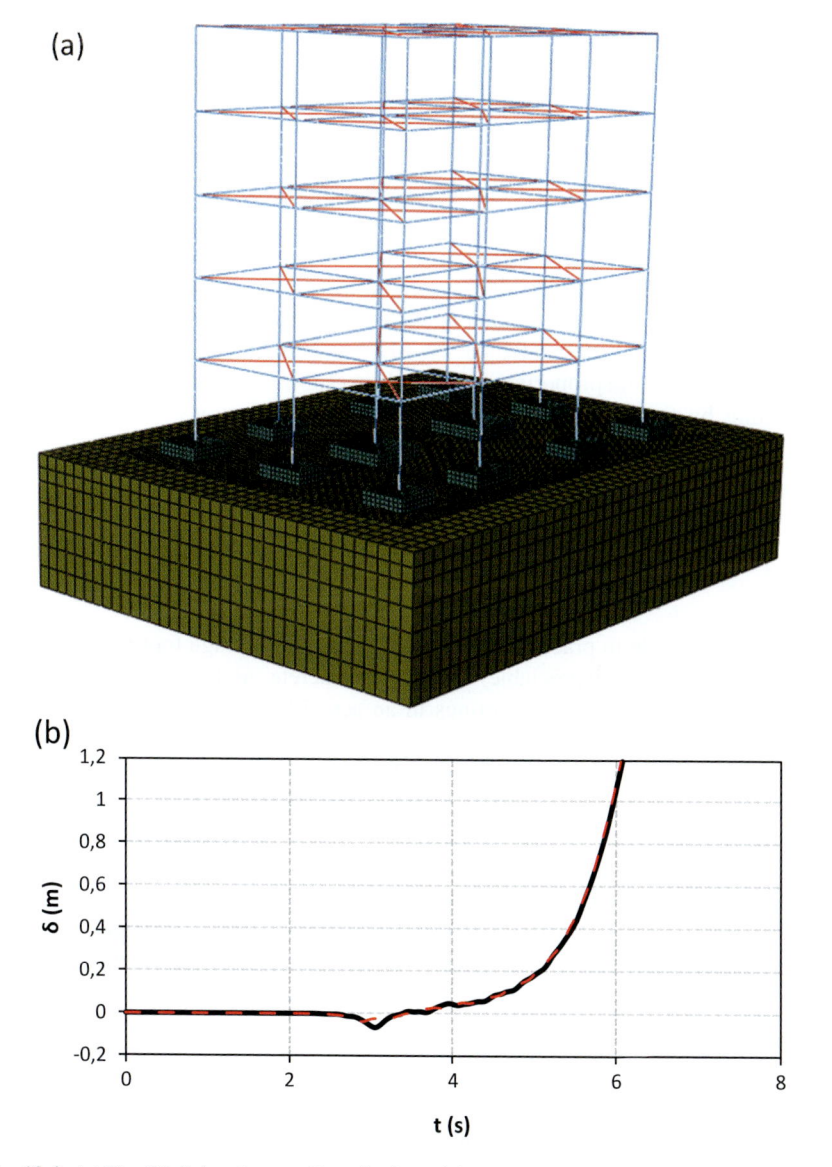

Fig. 13.4 (a) The 3D finite element discretisation of the system; (b) displacement time history of first floor (red dashed line) and roof (black bold line) during the Lefkada 2003 earthquake

13.3.2 Upgrading with Shear Walls on Conventional Foundations

For a simple retrofitting scheme, we construct four shear (structural) walls, two in each direction, in place of the columns K2, K4, K9 and K11 (as shown in Fig. 13.5). The structural design of each wall was based on the current greek codes (EAK and EKΩΣ) and resulted in a wall with $l_w = 1.7$ m, $b_w = 0.4$ m, and $h_w = 9$ m. The latter is the height of the building. The (ultimate) moment capacity of each wall, M_{RD} was computed equal to 2 MNm (in its long direction). The design effective ground acceleration is $A = 0.24$ g and the ductility-depended factor $q = 3$. Details can be found in the thesis od Dais (2015).

The footing of each shear wall obeys the standard capacity-design rules: $F_s \geq 3$, $F_E \geq 1$, $e = M/N \leq L/3$, and loading increased by an over-strength factor $a_{CD} \approx 1.3$. The latter ensures that the foundation system's maximum moment resistance M_u exceeds the structural moment capacity, $M_{RD} \approx 2$ MNm. As a result of the small axial load and the (disproportionately) high overturning moment transmitted onto the footing by the wall, the required footing plan dimensions are $L = 6$ m and $B = 2.2$ m. This is the "conventional" foundation.

13.3.3 Unconventional (Rocking) Foundation

It is highly desirable in practice to be able to reduce these huge footing dimensions. Not so much for the (appreciable) savings in concrete, as for the frequent lack of space between closely-spaced footings in an actual building. All kinds of utilities may exist passing through this space. Hence, it is interesting to investigate the feasibility of solution with a rocking foundation.

To this end, we decided to reduce only the dimensions of the new structural wall while leaving the spread footings of the columns intact. At first, one might expect that such an action would *"shed load"* from the walls to the columns, as their overall stiffness has increased relative to the stiffness of the walls. And, hence, the columns may suffer from disproportionally high moments. Yet, when retrofit is attempted, this is by far the most desirable and easy solution, even though not technically optimal.

So after some trials we select for the structural walls footings:

$$L = 3.2 \, \text{m and} \, B = 1.8 \, \text{m}$$

which are indeed much smaller than those imposed by capacity design, $L = 6$ m and $B = 2.2$ m. The small footing nevertheless complies with the static requirement since $F_s = 6$. It violates only the aforementioned criterion of (seismic) eccentricity:

$$e = M_{RD}/N \approx 2000/820 \approx 2.4 \, \text{m}$$

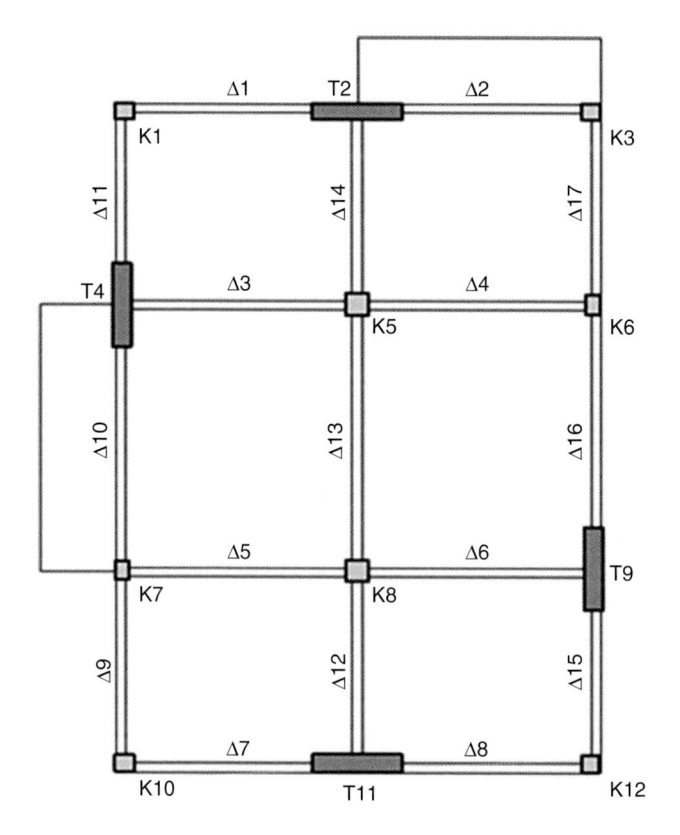

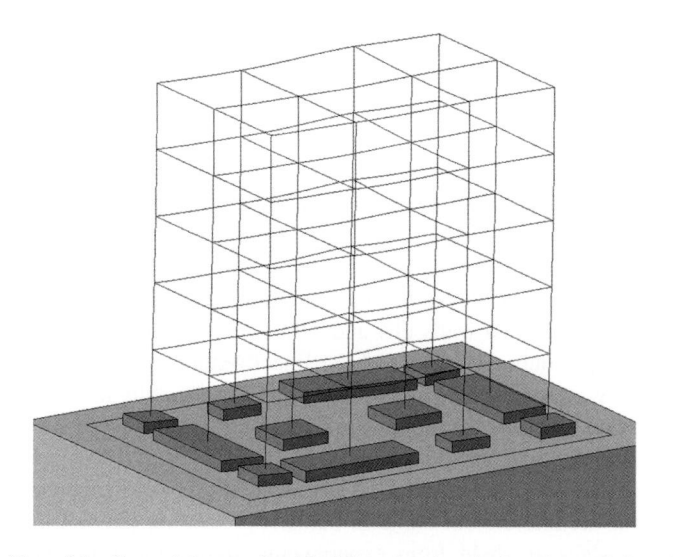

Fig. 13.5 Plan of 1st floor of the retrofitted building showing the location and size of the shear walls (top); first mode deformation of the structure showing the conventional footings (bottom)

which by far exceeds not only $L/3 \approx 1.1$ m but also $L/2 = 1.6$ m! Clearly, a pseudo-static way of thinking would not have allowed the resultant force to fall outside the foundation as this comparison implies.

13.3.4 Comparison of the Seismic Performance of the Two Alternatives

The retrofitted five-story building whose structural walls are supported with (a) the conventional ($L = 6$ m, $B = 2.2$ m) and (b) the unconventional ($L = 3.2$ m, $B = 1.8$ m) foundation is subjected to two ground motions:
- the Lefkada 2003 earthquake record in Lefkada.
- the San Salvador 1986 earthquake record at CIG.

The first is a moderately strong motion with PGA $= 0.42$ g; its response spectral value at $T \approx 0.75$ s (the natural period of the retrofitted structure) only slightly exceeds the (EAK) design spectral value, while its spectral values at larger periods ($T > 0.8$ s) drop below those of the (EAK) design spectrum. It is therefore a design-level excitation.

The San Salvador motion is fairly strong, exceeding the (EAK) design spectrum for all periods, and being some 50% to 100% larger than the Lefkada spectrum at the periods of interest $T > 0.75$ s. Hence it is a higher than design excitation.

The comparison for the Lefkada excitation is given in Figs. 13.6 and 13.7, referring to the response: of the whole structure, of one of the shear structural walls, and of one representative column. Specifically, Fig. 13.6a, b compares the moment-curvature relations and the shear force time histories at the base of wall T11, from which it is evident that the wall of the "rocking isolation" design responds more favorably. Indeed Fig. 13.6c, d shows that this unconventional solution results in smaller roof displacement and small drift ratio of the first floor.

On the other hand, the columns pay a very small penalty despite their increased share of the load. Indeed as seen in Fig. 13.7, column K12, the most severely stressed, experiences an increased ductility demand that is easily within the acceptable range. The axial load carried by the column also slightly increases.

With the stronger San Salvador CIG excitation, Fig. 13.8 shows the moment-curvature relations for two walls (T4 and T9), as well as the time history of roof displacement. No doubt the walls of the unconventional system respond more favorably. They remain in the linear range with maximum moment limited to about 1 MNm. In addition, the top displacement time-history of the unconventional system is consistedly smaller. Evidently, the differences between the two systems are larger with the stronger excitation. And in spite of the rocking of the foundation the shear walls act like the *backbone* in humans preventing the ribs from slipping out: the floor slabs are held from experiencing significant drift, and hence the un-improved columns do not suffer much.

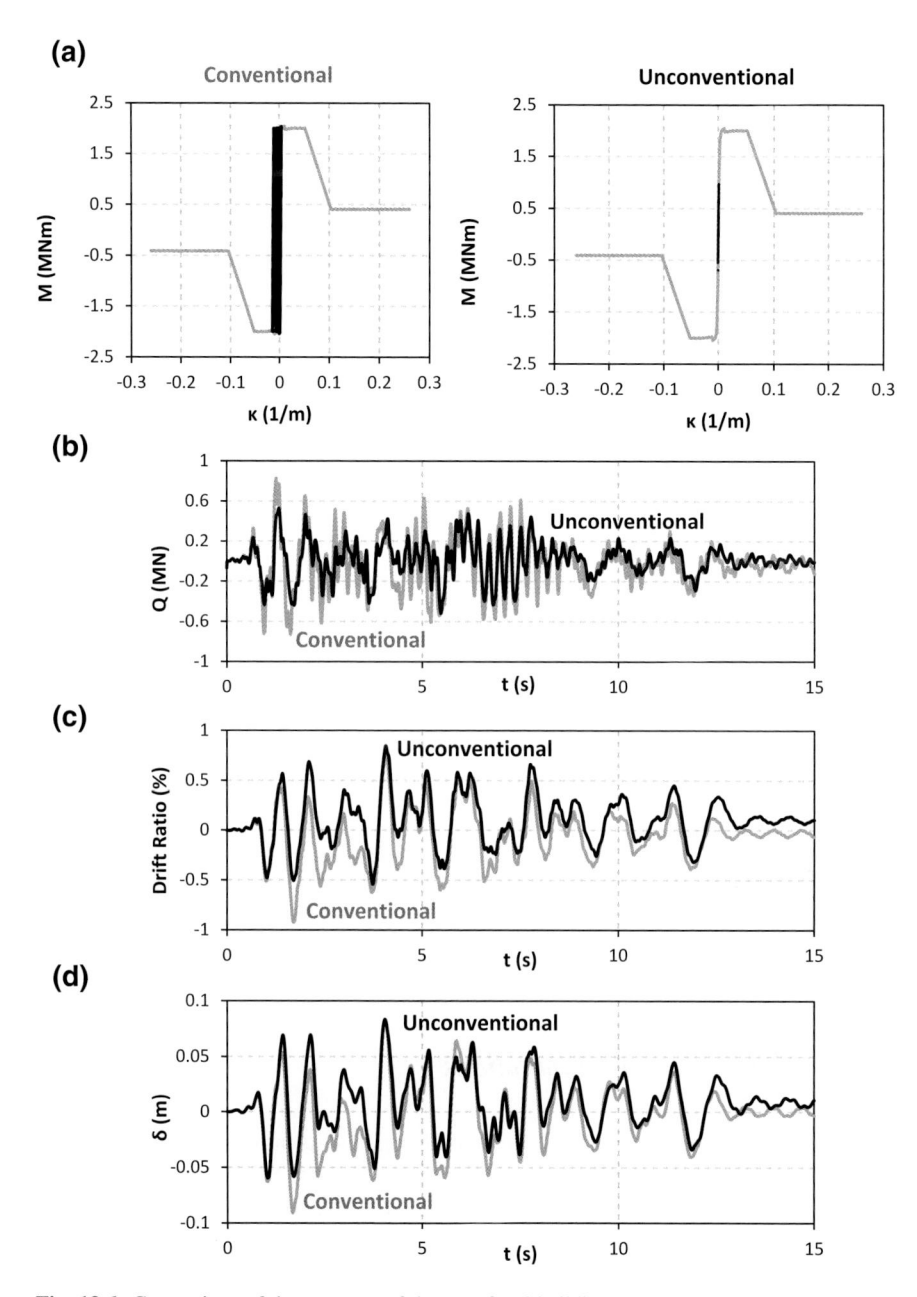

13 Multistory Building Frames and Shear Walls Founded on "Rocking" Spread Footings 319

Fig. 13.6 Comparison of the response of the retrofitted building on conventional and unconventional foundations induced by the 2003 Lefkada motion: (**a**) moment-curvature relation of shear wall T11; (**b**) shear force time history at the base of shear wall T11; (**c**) drift ratio time history of the 1st floor; and (**d**) roof displacement time history

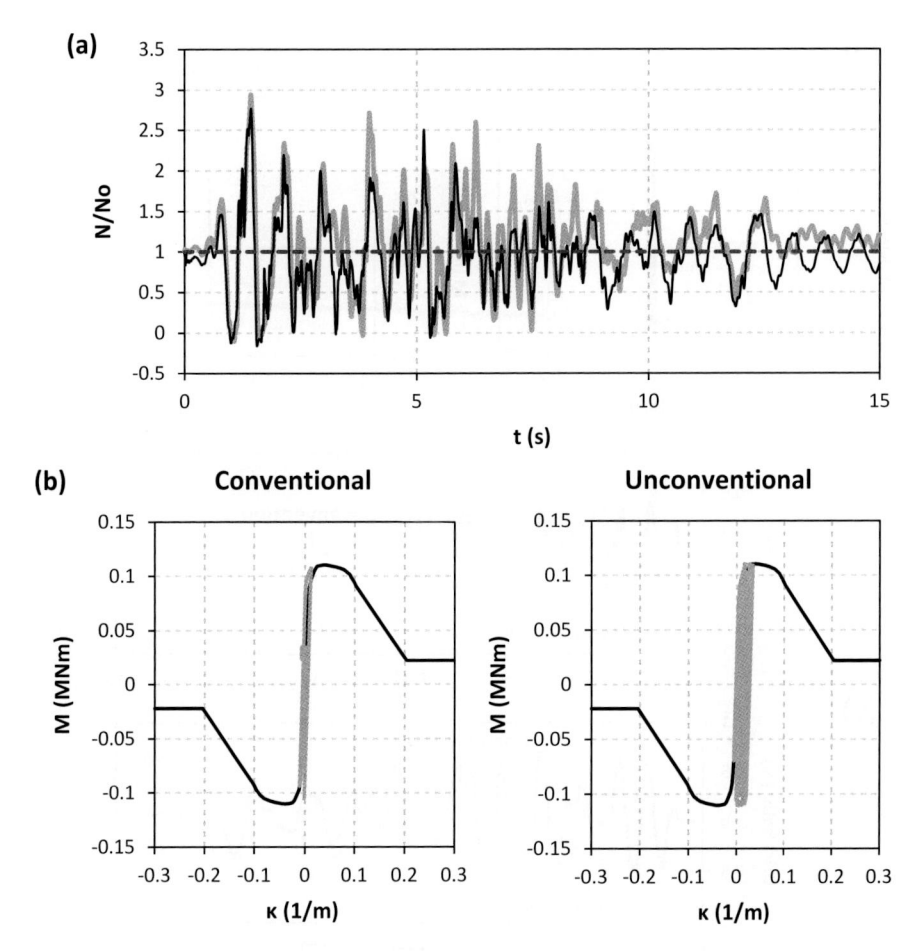

Fig. 13.7 Response of column K12 to 2003 Lefkada: (**a**) time history of normalised axial force; (**b**) moment-curvature relations at the base of the column

13.3.5 Comments and Limitations

The analysis presented above, the specific building, and its proposed retrofit are only an example aimed to show the potential benefits of rocking foundations, even when upgrading existing buildings. The solution investigated is by no means optimal. But it does reinforce the conclusion reached in many studies, experimental and theoretical, that being overly conservative in foundation design does not lead to increased seismic safety of the structure they support. Recall the wisdom of the seminal 1977 article by the late Professor Ralf Peck on "The Pitfalls of Over-Conservatism in Foundation Design."

One of the limitations of the "rocking isolation" for multicolumn buildings on spread footings is that the settlements and rotations of the individual footings will

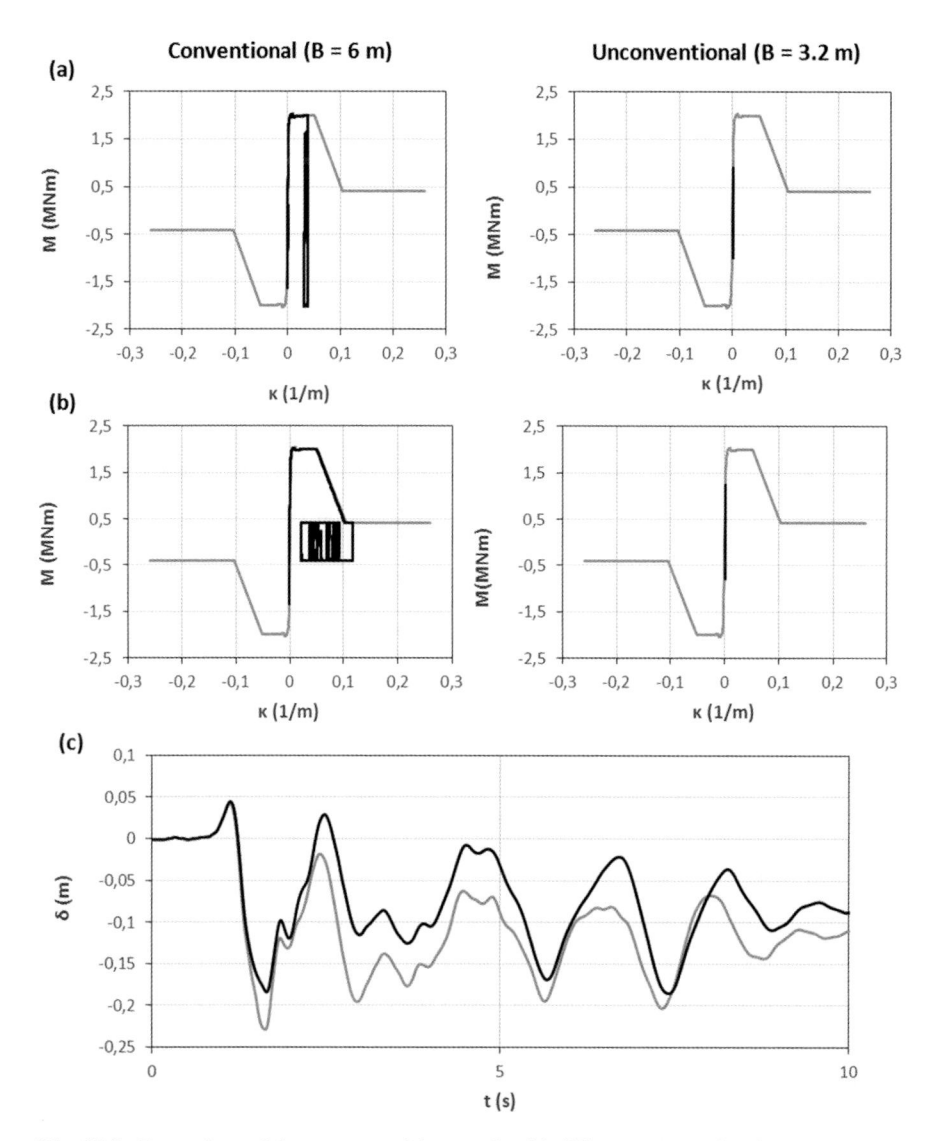

Fig. 13.8 Comparison of the response of the retrofitted building on conventional and unconventional foundations induced by the 1986 San Salvador motion: (**a**) moment-curvature relation of shear wall T9; (**b**) moment-curvature relation of shear wall T4; (**c**) time history of roof displacement

induce differential displacements between the columns of the structural system, and thereby cause damage. This is indeed a potential that must be investigated during analysis and its consequences must be accounted for in the design of the framing system. One solution may be the use of *tie beams*. In many cases their use is compulsory. But if their construction, as usual, fixes them on the top of the footing, rocking will be severely hindered and the "isolation" it provides will practically vanish. Continuous tie beams hinged at the base of the columns have been proposed by Anastasopoulos et al. (2014) which allow the beneficial rotation while they minimize differential settlement and permanent rotation of the footings. However, implementing such ideas in practice requires detailed thorough analysis with realistic modelling of the hinged connections – not a trivial task for engineering practice.

References

Anastasopoulos I, Gazetas G, Loli M, Apostolou M, Gerolymos N (2010) Soil failure can be used for seismic protection of structures. Bull Earthq Eng 8(2):309–325

Anastasopoulos I, Gelagoti F, Spyridaki A, Sideri T, Gazetas G (2014) Seismic rocking isolation of asymmetric frame on spread footings. J Geotech Geoenviron 140(1):133–151

Anastasopoulos I, Drosos V, Antonaki N (2015) Three-storey building retrofit: rocking isolation versus conventional design. Earthq Eng Struct Dyn 44:1235–1254

Antonaki N (2013) Experimental study of rocking isolation: application to retrofitting an existing building. Master's thesis, NTUA

Antonellis G, Gavras AG, Panagiotou M, Kutter BL, Guerini G, Sander AC, Fox PJ (2016) Shake table test of large-scale bridge columns supported on nrocking shallow foundations. J Geotech Geoenviron. https://doi.org/10.1061/(ASCE)GT.1943-5606.0001284, 04015009

Dais D (2015) Seismic retrofit of a 5-storey rxisting building with shear walls: conventional and rocking foundations. Diploma Thesis, School of Civil Engineering, National Technical University of Athens, Greece

Deng L, Kutter BL, Kunnath SK (2012) Centrifuge modeling of bridge systems designed for rocking foundations. Soils Dyn Earthq Eng 138:335–344

Gajan S, Kutter BL, Phalen JD, Hutchinson TC, Martin GR (2005) Centrifuge modeling of load-deformation behavior of rocking shallow foundations. Soils Dyn Earthq Eng 25:773–783

Gazetas G (2015) 4th Ishihara lecture: soil-foundation-structure systems beyond conventional seismic failure thresholds. Soils Dyn Earthq Eng 68:23–39

Gelagoti F, Kourkoulis R, Anastasopoulos I, Gazetas G (2012) Rocking–isolated frame structures: margins of safety against toppling collapse and simplified design approach. Soil Dyn Earthq Eng 32(1):87–102

Kawashima K, Nagai T, Sakellaraki D (2007) Rocking seismic isolation of bridges supported by spread foundations. In: Proceeding, 2nd Japan-Greece workshop on seismic design, observation, and retrofit of foundations, Japan Society of Civil Engineers, Tokyo, pp 254–265

Kourkoulis R, Gelagoti F, Anastasopoulos I (2012) Rocking isolation of frames. J Earthq Eng 138(1):1–14

Kutter BL, Moore M, Hakhamaneschi M, Champion C (2016) Rational for shallow foundation rocking provisions in ASCE 41-13. Earthq Spectra 32(2):1097–1119

Makris N (2014) A half-century of rocking isolation. Earthq Struct 7:1187–1221

Pecker A (1998) Capacity design principles for shallow foundations in seismic areas. 11th European conference on earthquake engineering

Chapter 14
Seismic Design of Foundations in Difficult Soil Conditions: Examples of Solutions

Alain Pecker

Abstract The development of large civil engineering projects in active seismic areas often face the challenge of designing foundations that must sustain large seismic forces while preserving the functionality of the superstructure. The natural solution for such foundations seems to lie in the adoption of piles. However, end bearing piles are not always feasible and piled foundations are also subject to adverse effects which may not make them so attractive. Recent projects have shown that alternative, often innovative solutions, may lie in a combination of solutions coupling at least two of the following elements: shallow foundation, soil improvement, caissons, piles, etc...

The lecture details the pros and cons of the "classical" foundation solutions and illustrate on actual projects how combination of solutions may advantageously get rid of adverse effects while still providing a safe design and preserving constructability of the foundations.

14.1 Introduction

The development of large civil engineering projects in active seismic areas often face the challenge of designing foundations in difficult soil conditions which must sustain large seismic forces while preserving the functionality of the superstructure. The natural solution for such foundations seems to lie in the adoption of piles. However, end bearing piles are not always feasible when the bearing layer lies at considerable depths, and piled foundations are also subject to adverse effects which may not make them so attractive: floating piles may experience a degradation of the skin friction

---

A. Pecker (✉)
Ecole des Ponts ParisTech, Sceaux, France

and therefore a reduction of their bearing capacity. Recent projects have shown that alternative, often innovative solutions, may lie in a combination of solutions combining at least two of the following elements: shallow foundation, soil improvement, caissons, piles, etc.

The lecture details the pros and cons of the "classical" foundation solutions and illustrate on actual projects how combination of solutions may advantageously get rid of adverse effects such as excessive settlements, large bending moments induced by kinematic or inertial effects in piles, lateral spreading due to liquefaction, while still providing a safe design and preserving constructability of the foundations.

14.2 Classical Foundations Types

By classical foundations types, one must understand the most commonly used schemes like shallow foundation, slightly embedded foundations or deep foundations. These have been used since the advent of geotechnical engineering and their design and behavior under permanent loading is well understood although sometimes complex to analyze. On the one hand, when the structure to design is located on soft soil deposits the natural tendency is to rely on deep foundations; as explained below, this may not be the optimal solution in seismic areas. On the other hand, when the structure is located on medium dense deposits, shallow foundations may be appropriate for permanent loading but hazardous in seismic areas.

14.2.1 Shallow or Slightly Embedded Foundations

The main advantages of shallow foundations lie in their moderate cost, ease to construct and simplicity to design. They are commonly used on medium dense, preferably cohesionless, deposits for which differed settlements are negligible. Usually, their design is governed by settlements control.

When placed in a highly seismic environment, they may be subjected to large foundations forces (shear force, overturning moment), soil settlement or even liquefaction. Large foundation forces may induce sliding, uplifting which, despite the convincing technical literature advocating the benefits of such phenomena on the overall structure behavior (e.g. for instance Pecker et al. 2014; Gazetas 2015; Deng et al. 2012), are hardly accepted in the design of conventional structures which is essentially based on the applications of standards. The example of the Rion Antirion Bridge or of the Panama Atlantic Bridge described later in the paper will show that combined with additional features shallow foundations might be a very efficient foundation scheme, even in highly seismic areas.

In addition to developing large forces on the ground, shallow foundations may be sensitive to the ground response which may manifest itself in terms of soil settlement

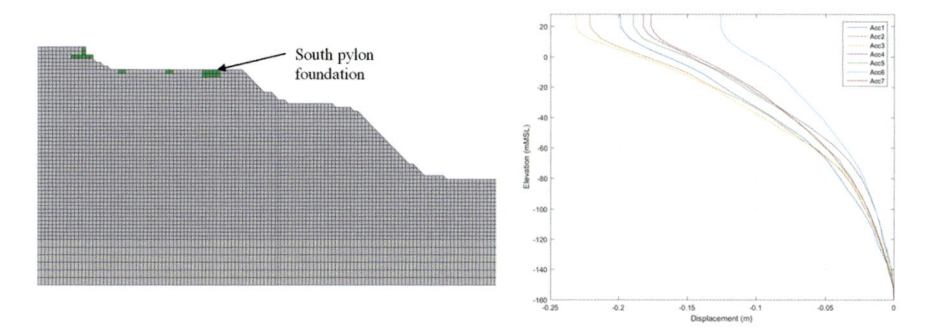

Fig. 14.1 Settlement of a shallow foundation – Chacao bridge (Chile) project

or liquefaction, with or without lateral spreading. To illustrate the previous statement, let us consider the example of the Chacao bridge in Chile (Pich et al. 2017). The bridge is a two–span suspended bridge, 2.2 km in total, located in probably the most seismic area of the world, near Puerto Mont at the place of the 1960 Valdivia earthquake ($M_w = 9.5$). The first design of the foundation for the south pylon was a shallow foundation resting on top of cohesionless medium dense layers: Fig. 14.1 shows the location of the foundation and the calculated vertical displacement profiles at the end of shaking for the seven time histories representing the design motion (return period 1300y, horizontal peak ground acceleration at rock outcrop 0.57 g). In addition to the large scatter in the surface displacements (0.13 m–0.23 m), the values have been considered too large for a reliable design and the foundations replaced by piles. The discussion on piled foundations (Sect. 14.2.2) will nevertheless point out that design of piles foundation for that project is also problematic.

Finally, needless to point out that shallow foundations by themselves are totally inadequate for foundations on liquefiable deposits, especially in the advent of lateral spreading; the example of the Fort de France prefecture presented below shows however that shallow foundation can be used even if surrounded by a laterally spreading layer.

14.2.2 Pile Foundations

It has been shown in the previous sections that, in many occasions, foundations cannot be designed as shallow footings and must be replaced by piles. However, piles may be subjected to large internal forces (shear forces and bending moments) which may lead to prohibitive steel reinforcement for concrete piles. The example of the Chacao Bridge presented in the previous section is used to illustrate this issue: if the original shallow footings are replaced by piles (18 concrete piles, 2.5 m in diameter) the same dynamic analyses yield the distribution, depicted in Fig. 14.2,

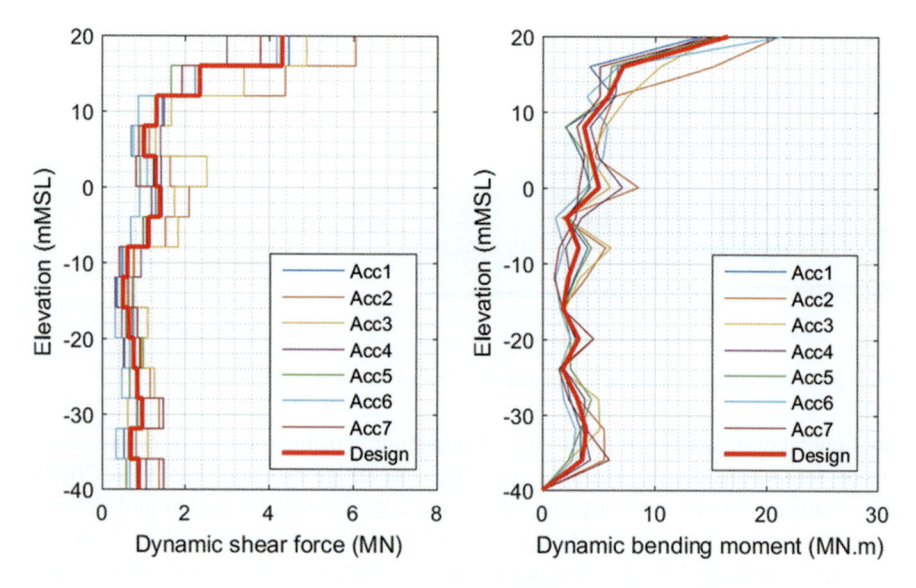

Fig. 14.2 Maximum dynamic forces in one pile – Chacao bridge (Chile) project

of maximum bending moments and shear forces in one of the piles. Very large values, are attained at the pile cap with average bending moments of 16 MN.m and shear force of 4 MN. It must be noticed that these values arise only from kinematic interaction as the superstructure is not modelled; obviously, the location of the foundation close to the slope is responsible of these large forces as permanent downslope displacements take place during the earthquake. However, even in the absence of significant permanent soil displacements, kinematically induced interaction forces may be very large due to a peculiar soil stratigraphy (very soft layer interbedded in–between stiffer layers) and, when combined to inertial interaction forces, prohibit the use of piles as the steel reinforcement ratio exceeds the upper limit imposed by standards (for instance 4% as per AASHTO (2016)). The example of the Panama Atlantic Bridge presented later illustrates this aspect and presents a possible alternative solution.

Large internal forces can also de developed in piles as quasi–static loading when lateral spreading due to liquefaction of some layers takes place. Although methodologies exist to design piles against lateral spreading, they usually are very uncertain, because of the required input data, and lead to large construction costs for the piles. Alternative solutions, as shown below with the foundation of an LNG tank in Australia, may be advantageous.

Development of large internal forces is not the only phenomenon that may adversely affect the behavior of piles during an earthquake. It is well known that floating piles may be very sensitive to degradation of their skin friction under cyclic loading (Puech and Garnier 2017) and this loss of shear resistance may eventually create a bearing capacity failure. The tilt of the building shown in Fig. 14.3 was caused by failure of floating piles in Mexico City during the 1985 Michoacán–Guerrero earthquake.

Fig. 14.3 Loss of skin friction in a piled foundation (Davidovici et al. 1985)

Finally, construction and quality control of piles are always more difficult than for shallow foundations.

14.3 Mitigation Against Lateral Spreading

This is one of the most challenging situation to face when designing foundations: not because of the lack of possible solutions to prevent liquefaction, but because the volume of soil to be treated is very large and leads to high construction costs. As noted previously it is always possible to design piles to withstand displacements imposed by lateral spreading, but, based on our own experience, more economical and less uncertain techniques may be contemplated. The two examples described below do not pretend to be the only possible ones, but they have successfully been implemented on two recent projects.

14.3.1 Piles and Inclusions

The example chosen to illustrate the combination of piles and inclusions consists of large capacity LNG tanks (storage capacity of 180 000m^3, diameter 90 m) constructed by Vinci – Entrepose Contracting at Wheatstone in Australia. The tanks are located on top of thick layers of more or less cohesionless soils, with a 3.3 m thick liquefiable layer located at 3.5 m below the top grade. For other than seismic issues, the tank is founded on 948 steel driven piles with a diameter 0.6 m and length 25 m. In order to limit the dynamic forces induced in the piles by the loss of stiffness and resistance of the liquefiable layer during the earthquake, and by the imposed quasi–static horizontal displacements due to lateral spreading, while maintaining the volume of improved soil to a minimum, the scheme depicted in Fig. 14.4 has been adopted: soil improvement is implemented below the tank but limited to the vicinity of the tank footprint; improvement consists in densification of the layer by means of closed-ended steel inclusions driven through the layer in–between the piles (Fig. 14.4). To limit the length of the inclusions to 5 m, they have been driven through the liquefiable layer from its top before a handmade backfill is placed on top of the layer; then the piles are driven from the top grade. The right part of Fig. 14.4 shows a view of the piles; inclusions cannot be seen because they are covered by the backfill. Analyses have shown that the densification achieved by driving the inclusions, plus the small contribution to the layer stiffness of the inclusions, were adequate to prevent liquefaction within the foot print of the tank. Lateral spreading cannot affect the tank because the large diameter caisson (90 m) constituted of the piles plus the inclusions is self-stable. Therefore, at the expense of inclusions over a limited cross area, the tanks foundations were feasible and able to withstand the seismic forces (kinematic + inertial) induced in the piles and to mitigate the effect of lateral spreading, which can nevertheless takes place outside the tanks footprints.

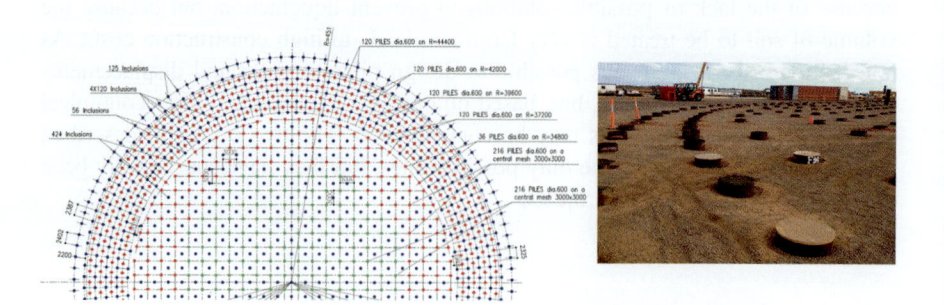

Fig. 14.4 Layout of the foundations of the LNG tank – Wheatstone (Australia)

14.3.2 Shallow Foundations and Geomix Caissons

The new Prefecture in Martinique (French Caribbean islands) is located on top of loose cohesionless deposits, 9 to 17 m thick, resting on top of a soft rock layer exhibiting a marked slope towards the sea. According to the French seismic regulations, the project has to be designed for a peak ground surface acceleration of 0.36 g associated with a magnitude 7.5 earthquake. The site being close to the sea shore, the water table is at 1 m depth, the soft soil deposits are prone to liquefaction and lateral spreading would certainly take place. The objective of the foundation design was two–fold: keep if possible a surface foundation to minimize the cost and duration of construction, and prevent the adverse effects of lateral spreading. The contractor, Solétanche–Bachy, proposed a deep soil mixing improvement (Benhamou and Mathieu 2012) that proved successful during the 1995 Yoko–ken–Nanbu earthquake: the Oriental Hotel in Kobe was founded on piles surrounded by a grid-type structure made of deep mixing walls; during the earthquake, the Oriental Hotel structure did not suffer from any damage despite important displacements on surrounding structures.

The Geomix® technique is a Deep Soil Mixing technique based on the Hydrofraise technology combined with Cutter Soil Mixing (CSM) principles. The process is illustrated in Fig. 14.5 (right). When drilling under the action of two counter rotative cutting wheels, the natural soil is bulked and mixed with a specific injected fluid and then displaced above the cutting head. During withdrawal, rotation of the wheels is reversed to displace the mix below the cutting head and final homogenization is achieved while completing the slurry incorporation. Figure 14.5 (left) shows the plane layout of the Geomix grid: each panel is 6 m long, 19 m deep and 0.5 m thick. A total area of 1 440m^2 has been treated. Figure 14.6 presents a view of the site during construction. On top of the grid a gravel layer is placed on which the building is founded with a direct foundation. The role of the gravel layer is to prevent "hard points" that may damage the building raft due to differential settlements.

With the strength properties of the panel (unconfined compressive strength of 1 to 2 MPa), the improved soil was able to carry the building weight; static settlements were less than 2 cm. With a dynamic shear modulus of the order of 500 MPa and the square arrangement of the panels, each caisson is stiff enough to limit the earthquake

Fig. 14.5 The Geomix technique

Fig. 14.6 Foundation of Fort de France Prefecture

induced shear strain within a cell and therefore to prevent significant pore pressure buildup. Even in case of localized liquefaction within a cell, the raft was designed to bridge the soil between two adjacent panels. Lateral spreading is prevented below the building, like for the LNG tanks presented in the previous section, by making the overall caisson (36 m by 40 m) self–stable under the pressure applied on its periphery by the liquefied laterally spreading soil. This technical solution is definitely competitive and can be applied to other projects in seismic areas.

14.4 Shallow Foundations on Improved Soil

It has been pointed out in several instances in this paper that shallow foundations are very attractive solutions in terms of simplicity for analyses, ease of construction, quality control and cost effectiveness. However, in high seismic areas, settlements, differential settlements creating a tilt of the foundation and even bearing capacity failures may overshadow these advantages. One possibility to improve the solution is to reinforce the soil to minimize settlements and to increase the bearing capacity. However, it is suggested to go a step further and to accept some amount of settlement, uplift, sliding of the foundation provided these quantities remain acceptable for the functionality of the superstructure. Presently, on the one hand, most of the seismic design building codes do not accept any foundation yielding (e.g. Eurocode 8 – Part 5 2004) and, on the other hand, several analytical and experimental studies suggest that foundation yielding may be beneficial to the structure (Gazetas 2015; Deng et al. 2012). One of the most emblematic civil engineering project, and the first one in the world, for which these concepts were implemented is the Rion – Antirion Bridge in Greece (Fig. 14.7, Pecker 2006); this bridge was designed and constructed by a joint venture lead by Vinci. For this project, not only the soil was reinforced with steel inclusions, but the reliability of the foundation behavior was enhanced by implementing the concept of capacity design borrowed from structural engineering (Pecker 1998).

14 Seismic Design of Foundations in Difficult Soil Conditions: Examples... 331

Fig. 14.7 The Rion Antirion bridge

The foundation concept combines soil improvement and shallow foundation, which in the particular geotechnical and environmental conditions of the bridge turned out to be the almost unique possible solution.

Since the project has been described in many publications just a brief description of the concept is given below (Fig. 14.8, left): under each of the foundation, except one, steel hollow inclusions, 2 m in diameter and 30 m long, are driven in the existing soils at a spacing of 7 m by 7 m; given the size of a footing (90 m) this leads to approximately 200 inclusions under each footing.

Theoretical, analytical and experimental studies have shown that, with a proper design of the inclusions, the combination of the inclusions plus the gravel layer increases the bearing capacity of the foundation, favor a "failure" mode corresponding to sliding on the gravel layer and bounds the forces transmitted to the rest of the superstructure: the inertia forces in the superstructure cannot exceed the friction force at the base, which is equal to the weight times the friction coefficient. In addition, since uplift of the footing is possible and allowed in design, the overturning moment at the foundation is also limited. The important aspect of this concept is that permanent tilt of the structure is very limited due to the horizontal sliding. This is exactly the concept of capacity design where the role of the plastic hinge is played by the gravel layer and the overstrength is provided by the inclusions. Therefore, the adopted concept allows for an efficient, cost effective and reliable design of the foundations provided foundation yielding (sliding) and uplift are

Fig. 14.8 The Rion Antirion bridge foundations

Fig. 14.9 Iceda (Le Bugey) site during construction

allowed, which eventually lead to permanent displacements. For the design earthquake (pga at sea bed level equal to 0.48 g) the calculated sliding amounts to 0.25 m.

This project is exemplary because it was the first one in the world and it helped to convey the ideas of accepting some yielding in the foundation. Since the completion of the Rion – Antirion Bridge, at least two projects used exactly the same design concept: the Izmit Bridge in Turkey (Steenfelt et al. 2015), a suspended bridge with a main span of 1550 m and a nuclear building, Iceda, at Le Bugey in France (Okyay et al. 2012). The Iceda building, designed and operated by EDF, is founded on a thick (50 m–70 m) layer of clay soil before reaching a more competent layer; for the reasons explained previously, floating piles were not contemplated and end bearing piles would have been too expensive given the large internal forces that would develop due to both kinematic and inertial interaction forces. The adopted solution consists of long concrete inclusions drilled through the clay layer (Fig. 14.9), topped with a gravel layer on which the mat of the building is founded.

14 Seismic Design of Foundations in Difficult Soil Conditions: Examples... 333

14.5 Caisson and Shallow Foundation

The design concept of the example presented in this section is intended to combine the advantages of shallow foundations while reducing the kinematic effects associated with piles foundations. The Panama Atlantic Bridge, presently under construction by Vinci, is the third bridge across the Panama Canal on the Atlantic side, north of the new locks of Gatun (Joly et al. 2017). It is located in a highly seismic area for the which the design ground motion is defined by a return period event of 2500 years with a pga on rock equal to 0.57 g and a 5% damped spectral acceleration at the plateau of 1.3 g. The site conditions at 2 piers of the access viaducts were particularly difficult with a 10 m thick hydraulic fill layer over a layer of very soft clay, 12 m thick called Atlantic Muck, resting on top of stiffer materials grading from a residual soil to a weathered and soft rock (Fig. 14.10). The Atlantic Muck exhibits a shear wave velocity of 80 m/s and the

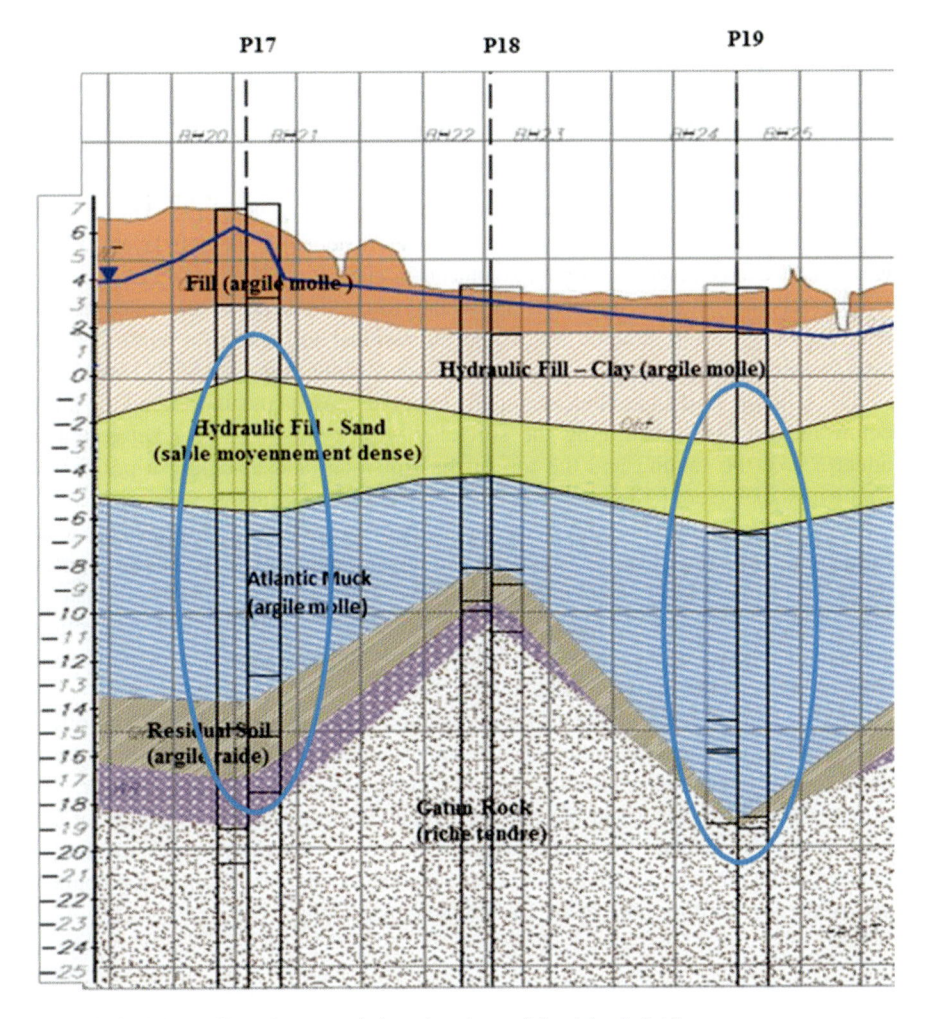

Fig. 14.10 Soil profile at 2 access viaducts locations of the Atlantic bridge

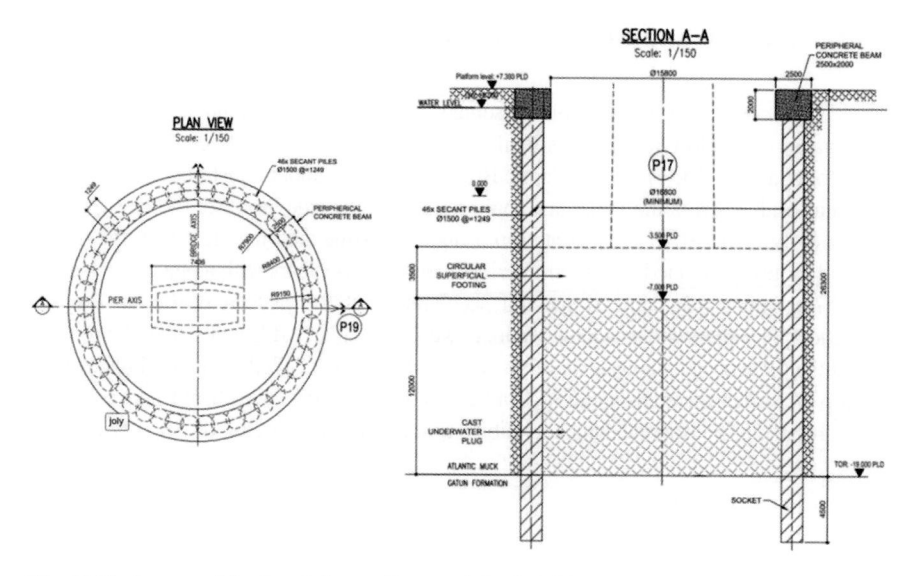

Fig. 14.11 Layout of the foundations – Panama Atlantic bridge

underlying layers a shear wave velocity of 500 m/s up to 800 m/s. This special configuration creates large displacements and deformations in the soft layers which would induce large kinematic bending moments in piles.

The seismic bending moments and shear forces induced in the piles at the interface between the Atlantic Muck and the residual soil lead to a reinforcement ratio exceeding 4%, an upper acceptable limit according to the AASHTO standard.

The solution which has been eventually adopted consists in suppressing the piles below the footing, substituting the soft clay with a mass concrete between the underface of the footing and the top of the soft rock, and installing a peripheral wall with a diameter of 18 m to protect the foundation and limit the forces created by kinematic interaction. The wall is constructed with secant piles anchored in the rock formation, working like individual cantilever beams without hoop forces (Figs. 14.11 and 14.12). A peripheral annular beam at the top of the wall ensures the continuity between the piles and stiffens the wall. No mechanical connection exists either between the footing and the underlying mass concrete or peripheral wall, or between the mass concrete and the rock.

The sequential construction of the foundation involves realization of the piles, liaison of the piles with the annular top beam, excavation underwater, pouring of the mass concrete underwater, dewatering and construction of the footing and pier.

Very detailed sophisticated 3D nonlinear finite element analyses were run for the final design to calculate the foundations displacements (rocking, uplift), the forces induced in the protecting wall, annular beam and the footing. For instance, the uplifted area of the footing does not exceed 15% of the cross section.

Fig. 14.12 View of the foundations during construction – Panama Atlantic bridge

This solution is a real innovation because the caisson does not act to carry the seismic loads from the superstructure but is only used for protecting the foundation from the soil movements around the pier. A similar concept has been used in the Monestier bridge in France but not for earthquake issues: the caisson around the foundation, realized with a cast–in situ diaphragm wall, was designed to protect the foundation from soil creep (5 mm/year). In addition, the concept allows for a foundation on a surface footing with all the merits that have been described and taken advantage of for the Rion–Antirion Bridge: allowance for rocking and uplift of the foundation that limits the forces transmitted to the soil and to the superstructure; however, the beneficial effect of sliding cannot be relied on because the footing is in contact with the peripheral wall.

In conclusion, this concept combines two existing, already innovative, concepts and this combination makes it attractive and potentially useful in other seismic areas.

14.6 Conclusions

In this paper several foundations solutions adopted in actual projects have been presented. The choice of these solutions was driven by the poor quality of the foundation soils which could not allow for the use of classical foundation types like piles. It is shown that combination of different solutions (piles + inclusions, shallow foundation + soil improvement, shallow foundation + caisson), which are used per themselves on other projects, may be very efficient. Efficiency should be evaluated not only in terms of feasibility but also considering reliability, ease for construction, quality control and costs. These examples which have been presented do not pretend to constitute an exhaustive review of existing solutions but were taken from the author's experience to illustrate their potential.

Acknowledgments The author takes the opportunity of this paper to thank all the contractors, Vinci, Entrepose Contracting, Solétanche–Bachy, EDF for involving him in these fascinating and innovative projects.

References

AASHTO (2016) LRFD bridge – design specifications. American Association of State Highway and Transportation Ofiicials, Washington, DC
Benhamou L, Mathieu F, (2012) Geomix Caissons against Liquefaction. In: Proceedings of the international symposium on ground improvement ISSMGE–TC 211, 31 May–1 June, Brussels, Belgium
CEN. European Standard EN 1998-5 (2004) Eurocode 8: design of structures for earthquake resistance. Part 5: foundations, retaining structures, geotechnical aspects. Comité Européen de Normalisation, Brussels
Davidovici V, Despeyroux J, Duretz C, Etchepare B, Pecker A, Sollogoub P (1985) Le séisme du Mexique du 19 Spetembre 1985, Compte Rendu de Mission, AFPS
Deng L, Kutter BL, Kunnath S (2012) Centrifuge modeling of bridge systems designed for rocking foundations. J Geotech Geoenviron Eng ASCE 138:335–344
Gazetas G (2015) 4th Ishihara lecture: soil-foundation-structure systems beyond conventional seismic failure thresholds. Soils Dyn Earthq Eng 68:23–39
Joly E, Agostini L, Lhuissier S, Decoux G (2017) Troisième Traversée sur le Canal de Panama: des viaducts sur tous les fronts. Revue Travaux 936:36–43
Okyay US, Dias D, Billion P, Vandeputte D, Courtois A (2012) Impedance functions of slab foundations with rigid piles. Geotech Geol Eng 30(4):1013–1024
Pecker A (1998) Capacity design principles for shallow foundations in seismic areas. Keynote Lecture. In: Bisch P, Labbé P, Pecker A (eds) Proceedings 11th european conference on earthquake engineering, Balkema, Rotterdam, pp 303–315
Pecker A (2006) Design and construction of the Rion Antirion bridge foundations. International symposium on sea crossing long span bridges, 15–17 February, Mokpo, Korea

Pecker A, Paolucci R, Chatzigogos CT, Correia AA, Figini R (2014) The role of non-linear dynamic soil-foundation interaction on the seismic response of structures. Bull Earthq Eng 12:1157–1176

Pich B, Mauris G, Gogny E, Cheikh–Mhamed A, Jakobsen SE (2017) Pont Suspendu de Chacao au Chili. Revue Travaux 936:76–81

Puech A, Garnier J (2017) Design of piles under cyclic loading – Solcyp recommendations, Civil engineering and geomechanics series. ISTE – Wiley, London

Steenfelt JS, Foged B, Augustesen AH (2015) Izmit Bay bridge: geotechnical challenges and innovative solutions. Int Bridg Eng 3(3):53–68

Chapter 15
Structural Health Monitoring for Seismic Protection of Structure and Infrastructure Systems

Oreste S. Bursi, Daniele Zonta, Emiliano Debiasi, and Davide Trapani

Abstract Structural Health Monitoring (SHM) of civil-engineering structures is becoming more and more popular both in Europe and worldwide mainly because of the opportunities that it offers in the fields of construction management and maintenance. More precisely, SHM offers several advantages in terms of reduction of inspection costs, because of a better understanding of the behavior of both structures and infrastructures under dynamic loads, seismic protection, observation in real or near real-time, of the structural response and of evolution of damage. Therefore, it is possible to produce post-earthquake scenarios and support rescue operations. In this context, this paper provides a review of different technical aspects of SHM summarizing some sensor validation methodologies for SHM. Following that, recent progresses on SHM of buildings subjected to seismic actions and relevant ways to detect damage are recalled. Moreover, some aspects of SHM of tunnels and bridges are covered. Some related applications that use sensor networks designed by the University of Trento and a startup are described, pointing out the solutions adopted to build reliable SHM systems. Finally, concluding remarks and promising research efforts are underlined.

15.1 Introduction

SHM and damage identification nowadays represent important tools in structural engineering. In particular, SHM can be defined as a process of implementing in situ, non-destructive sensing and performing analysis of structural characteristics in order to identify if damage has occurred, define its location, estimate its severity and

---

O. S. Bursi (✉) · D. Zonta
Department of Civil, Environmental and Mechanical Engineering, University of Trento, Trento, Italy
e-mail: oreste.bursi@unitn.it; info@i-kubed.com

E. Debiasi · D. Trapani
Intelligent Infrastructure Innovation srl, Trento, Italy

evaluate its consequences on a structure residual life (Huston 2011). Even if SHM is a relatively new paradigm in civil engineering, the assessment of the health state of a structure by means of tests and relevant measurements is a common practice, so that evaluation and inspection guidelines are available since several years (Rücker et al. 2006). SHM objectives are consistent with this practice but it takes advantage of the new technologies in sensing, instrumentation, communication and modeling in order to integrate them into a decision support system (DSS).

Data obtained from these systems are very useful in several aspects: (1) structural safety evaluation of existing structures/infrastructures and maintenance; (2) rapid evaluation of the state conditions of damaged structures after an earthquake; (3) estimation of residual life of structures; (4) repair and retrofitting. As a result, reduction of downtime and improvement in reliability are expected and can enhance the performance of a structure; moreover, results of monitoring can be used to obtain a deeper insight into the structural behavior, which is useful for design improvements of future structures. In order to get all these objectives, an effective SHM system has to be based on integration of several types of sensors in a modular architecture. Moreover, the advances in the field of information technology and communications assure data transmission also in critical conditions.

In this contribution, some aspects related to the research carried out by the authors based on SHM and implementation of integrated SHM systems covering different structure and infrastructure systems are analyzed in detail, and some results are discussed. Moreover, techniques that rely on early-warning (EW) systems and related aspects, like false alarm rate and real-time probabilistic seismic hazard assessment are under investigation (Brown et al. 2011) but out of the scope of this contribution. Finally, some suggestions on ways to address future research are offered.

15.2 Sensor Fault Type and Validation Methodologies

In this section, we introduce typical sensor types and actions that allow for sensor validation methodologies based on fault detection, isolation and reconstruction as suggested by Yi et al. (2017).

15.2.1 *Sensor Fault Types*

A sensor is considered to be faulty when its measurements display unacceptable deviations from the true values of a measured variable (Yi et al. 2017). Various sensor fault modes can occur due to different factors, such as a malfunction or failure of sensor components and the effects of electromagnetic interference. According to Kullaa (2013), there are seven typical sensor fault types: bias, drift, gain, precision degradation, complete failure 1 (constant), complete failure 2 (constant with noise)

15 Structural Health Monitoring for Seismic Protection of Structure and... 341

and complete failure 3 (bottom noise). Bias, drift, gain and precision degradation are usually called soft sensor faults, in which the sensor has partially failed; whereas, the three types of complete failure are usually called hard sensor faults.

Figure 15.1 schematically shows the aforementioned sensor fault types.

A careful reader can observe that a sensor exhibits bias if its outputs differ from the normal values by a constant. Sensor drift refers to the case where the differences between sensor outputs and normal values change linearly with time. Gain occurs when the normal values of a sensor are multiplied by a constant. Moreover, the case

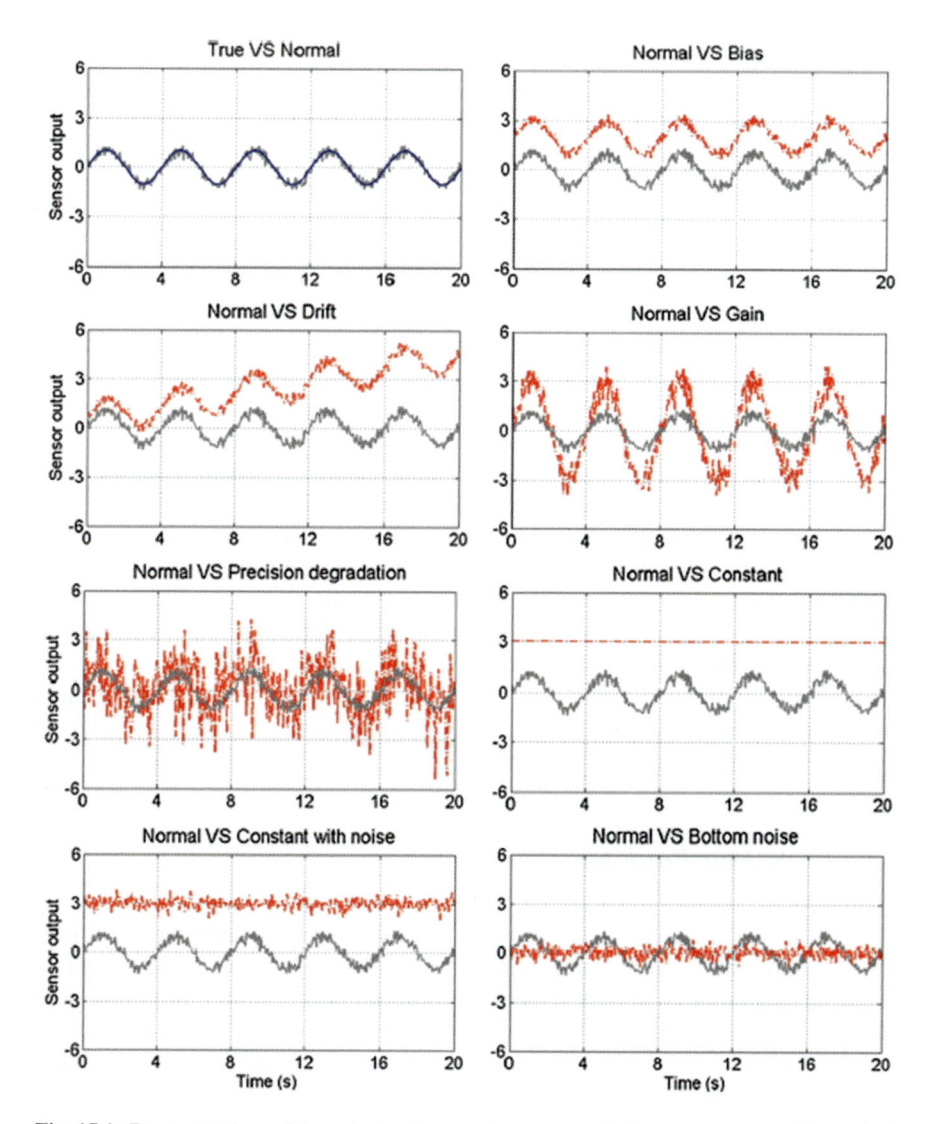

Fig. 15.1 Representations of the outputs of a normal sensor or a faulty sensor corrupted by typical fault types (After Yi et al. 2017)

in which sensor outputs are added to an excessive random noise is referred to as precision degradation. A fault is classified as complete failure 1 when sensor outputs remain constant with time and as complete failure 2 or 3 if sensor outputs are constant with noise or only noise regardless of the change of normal values.

15.2.2 Sensor Fault Detection

The purpose of sensor fault detection is to determine whether a potential fault has occurred in the sensor network. The general approach used for sensor detection consists of two phases. The first phase is the definition of a normal-work condition by building a detection model through the fault-free health-monitoring data; while, the second phase is the evaluation of a sensor fault detection index. The most widely used methods for building a detection model are (Yi et al. 2017): (1) univariate control chart-based; (2) multivariate statistical analysis-based; (3) model residual-based. More precisely, the control chart method, developed from statistical theories, is commonly employed to detect monitor data quality, e.g. shifts in sensor data. The second method, instead, is based on the modelling of correlations among the sensor network. As regards to the last method, a novel technique was developed by Li et al. (2007), that divided the sensor network into two groups: (1) the reference sensor group, which correctly measures structural responses; (2) the uncertain sensor group, which may fail to correctly measure structural responses. Then, a sensor error function can be derived for each corresponding uncertain sensor.

Once the normal-work condition is defined by means of one of the aforementioned methods, a fault detection index can be computed for the currently measured sensor data and compared with a decision threshold. The potential sensor fault is determined to occur after the fault detection index exceeds its corresponding threshold. However, the key issue for sensor fault detection is how to define a proper fault detection index and, on this matter, the interested readers may refer to Yi et al. 2017).

15.2.3 Sensor Fault Isolation

A faulty sensor is relatively easy to isolate when sensor fault detection is performed through univariate control chart or model residual-based approach. However, more involved methods need to be developed to isolate the faulty sensor if the sensor fault detector is derived from multivariate statistical analysis. Several fault isolation methods, such as contribution analysis, missing variable approach and probability quantification, have been developed in recent years.

Contribution analysis represents the most commonly used approach (Alcala and Qin 2011) and, in particular, it examines the contributions of an observed sensor output to a monitoring statistic. Then, the sensor which corresponds to a very large contribution value, generally the largest one, is considered to be the faulty sensor.

The missing variable approach divides the sensor network into two groups: the normal sensor group and the faulty sensor group. It can be used to isolate single or multiple faulty sensors. In the case where a single sensor is faulty, it firstly assumes that each sensor is the missing variable in turn. The corresponding fault detection index is then calculated by removing the missing variable from both the reference and current health-monitoring datasets.

Differently from the contribution analysis and the missing variable approach, a novel probability quantification-based sensor fault isolation method was proposed by Sharifi et al. (2010). A sensor failure index was firstly defined to investigate the isolability of each faulty sensor in the sensor network. Then, the effect of measurement noise on the isolability was analyzed in detail. It should be noted that the measurement noise considered here must be Gaussian distributed. To isolate a specific faulty sensor, a probabilistic decision process based on Bayesian theory was used to quantify the fault probability for each sensor.

15.2.4 *Sensor Fault Reconstruction*

After a faulty sensor has been detected and isolated, sensor fault reconstruction follows as the structural response measured by this sensor contains important information. There are several methods for reconstructing or correcting the faulty sensor outputs, e.g. optimization-based, regression-based and robust denoising-based ones (Yi et al. 2017).

A basic assumption of optimization-based methods is that the reference health-monitoring data contain sufficient information to cover the normal operating process. Then, the reconstructed outputs of the faulty sensor are obtained by solving an optimization problem based on the evaluation of the most likely value of the faulty sensor, i.e. the value that minimizes the magnitude of the deviation between the sensor measurement and its estimation.

As regards to regression-based methods, they can be performed using the Minimum Mean Square Error (MMSE) model constructed from the training dataset. In fact, the MMSE estimation is derived as a class of regression methods, which employs the observed variable to estimate the missing variable.

Finally, robust denoising-based methods are introduced to deal with both small dense noise and remarkable sparse outliers, which may be caused by sensor imperfections, instrumentation errors or sensor failures. An effective method was derived by Yang and Nagarajaiah (2013), which is based on principal component pursuit also called robust principal component analysis (Candès et al. 2011).

15.3 Health Monitoring of Buildings Subjected to Seismic Excitation

After-earthquake damage assessment represents a critical aspect in civil engineering, being crucial the quick identification of safe and unsafe structures for occupancy. In particular, there is interest with regard to possible aftershocks for the protection of public safety and for the estimation of economic losses (Çelebi et al. 2004). More precisely, damage identification in the emergency phase following an earthquake is not an easy task due to unnecessary evacuation and downtime, especially for critical facilities such as schools, hospitals or industrial facilities (Günay and Mosalam 2013). The relevant reason is mainly ascribed to the fact that damage assessment is typically performed by visual inspection procedures; and despite guidelines and usability forms (Baggio et al. 2007), subjectivity is always introduced in the evaluation. Furthermore, in many cases visual inspections are not able to detect damages, especially if they are not severe or when damaged elements are hidden by non-structural elements. These reasons have led in recent years to a growing interest in seismic monitoring systems, which provide objective real-time useful information for reliable damage evaluation.

In this respect, the most common strategies for the estimation of structural damages caused by earthquakes are: (i) monitoring systems based on changes of structural parameters; (ii) monitoring systems based on response monitoring during an earthquake. In the following subsections a brief overview is reported. Successively, a new method developed by the authors that evaluates the interstory drift (ID) ratio based on acceleration and tilt measurements is shortly discussed.

15.3.1 Monitoring Systems Based on Changes of Structural Parameters

Damage detection methods based on changes of modal parameters of a structure require vibration-based techniques for the identification of relevant modal parameters, i.e. natural frequencies, mode shapes, and damping ratios, before and after a seismic event. It is well known that the dynamic behavior of a structure can be expressed as a combination of modes characterized by a set of parameters depending on structure's physical parameters (Rainieri and Fabbrocino 2014). The basic idea of seismic monitoring based on changes of modal parameters is that the presence of damage, included the damage induced by an earthquake, can be assessed by means of a comparison between the dynamic characteristics of the structure before and after a seismic excitation; and the subsequent extraction of a state variable or damage feature sensitive to damage extension and possibly its location. The approach was studied in depth by many authors both from a theoretical viewpoint and practical implementations on realistic structures, particularly on bridges and viaducts. A comprehensive state of the art about these methods can be found in Sohn et al. (2004).

15.3.2 Monitoring Systems Based on Response Monitoring During an Earthquake

The second family of seismic monitoring systems includes systems able to monitor in real-time or in quasi real-time, being different the time needed to compute the dynamic response of the system, the dynamic response of a structure to an earthquake; and, in particular, displacement and deformation demands. In fact, it is well known from displacement-based design theory, that structural damage can be related to seismic displacement demands and in particular to interstory drift ratios (Calvi 1999; Çelebi et al. 2004; Priestley et al. 2007; Sullivan et al. 2012).

In absence of data related to an earthquake, the system can be used to record ambient vibration data used to monitor modal parameters of the building (Celebi 2007). Porter et al. (2006) propose an integrated system based on acceleration measurements at the building's base. Relevant results indicate that the method is suitable to quantify damage but it is not able to localize damage. The only relevant accelerometer devoted to loss estimation is the one placed at the base level. (Ponzo et al. 2010) presents a simplified method based on a statistical approach that uses the data recorded at the top of the building to extract the maximum interstory drift, used as damage indicator.

Currently, the methods based on displacement estimation from acceleration data only are the most popular and appear to be the only valid effective for low-rise buildings. However, displacement calculations from acceleration data only present two fundamental issues: i) total loss of information about structural residual displacements, thus about residual interstory drift (RID); ii) underestimation of peak deformation, i.e. of peak interstory drift (PID).

15.3.3 A Monitoring System Based on Acceleration and Tilt Measurements

Based on the aforementioned considerations, the authors developed a sensing bar which is suitable for the seismic SHM of framed RC buildings. In these buildings, damage assessment can be performed by monitoring in real time the value of the ID ratio, which is defined as the relative displacement between two consecutive floors over the interstory height. Moreover, also the RID, when present, can be considered as a clear indicator of structural damage.

More precisely, the sensing bar consists of a hinged bar -for example an L or C steel section- instrumented with 2 biaxial accelerometers that measure accelerations, one at each end of the beam and remaining parallel to the floors; and one biaxial inclinometer or accelerometer measuring the tilt of the beam as indicated in Fig. 15.2. The accelerometers are fastened to steel cubic supports, in turns rigidly fixed to the floors by means of dowels. The beam is linked to the supports by means of mechanical hinges, i.e. one spherical hinge and one Cardan joint, or two Cardan

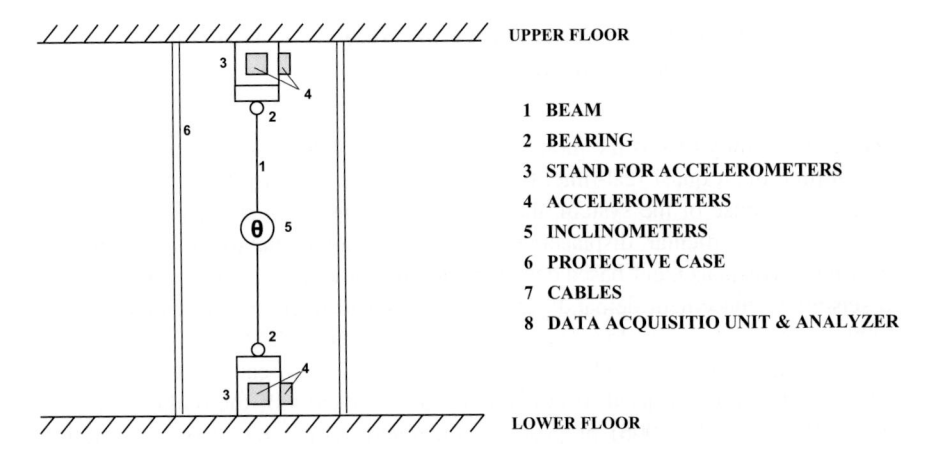

Fig. 15.2 Sketch of the sensing beam in the undeformed configuration

1 BEAM
2 BEARING
3 STAND FOR ACCELEROMETERS
4 ACCELEROMETERS
5 INCLINOMETERS
6 PROTECTIVE CASE
7 CABLES
8 DATA ACQUISITIO UNIT & ANALYZER

joints. An inclinometer or capacitive accelerometer is fastened to a horizontal plate welded to any point of the beam, e.g. at the mid span.

The rigid constraint between supports and floors ensures that accelerometers measure the horizontal accelerations of the floors. The mechanical hinges ensure that the beam is free to rotate during the seismic event and tilts of an angle equal to the ratio of the residual differential displacements between the floors and the distance between beam's ends.

The described sensing bar is suitable for the application of different algorithms which permit to overcome limitations of the PID estimation method based on double integration of acceleration measurements. If the assumption of a hinge behavior between the sensing bar and the floor to which the bar is fixed is valid, the inclination of the bar at a given instant with respect to the initial configuration is equal to the relative displacement between the two floors over by the height of the bar.

The inclinometer installed on the sensing bar allows to directly measure this inclination only in the static configuration, that is, at the end of the seismic excitation. During the seismic motion, the inclinometer measures an apparent angle, function of instantaneous inclination and instantaneous acceleration to which the instrument is subject to. It is worth noting that MEMS inclinometers are practically accelerometers estimating pitch and roll angles from acceleration measurements. For this reason, when an inclinometer is acquired dynamically, acceleration experienced by the inclinometer must be compensated.

15.3.3.1 Prototype System and Applications

The proposed sensing bar was applied to a prototype monitoring system named SafeQuake that is installed in several buildings. One example of installation is the Elementary School in Stenico, see in this respect, Fig. 15.3, in the Autonomous Province of Trento (Italy). The building was completed in the 80s. The main part, in

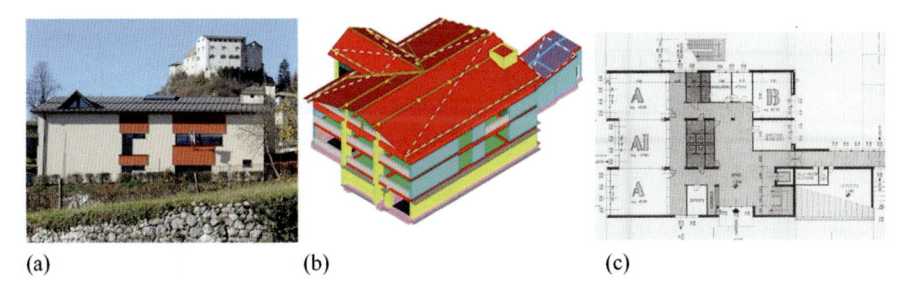

Fig. 15.3 (**a**) Main building; (**b**) FE of the elementary school of Stenico – south-east view-; (**c**) plan of the ground floor

which the monitoring system is installed, contains the classrooms, the laboratories, offices and – in the basement – a fire station. This part occupies 480 m^2, divided in two floors, an attic and a basement with a maximum height of 10.6 m. From a structural viewpoint, the main building is made of composite masonry walls and concrete columns, concrete slabs and a timber roof.

The seismic SHM records in real-time the accelerations at the ground and at each of the two floors, detects seismic events and estimates floor displacements; these last quantities can be compared to thresholds for a preliminary seismic vulnerability assessment. Therefore, in order to map sensor measures to the damage state, before the installation of the seismic SHM, a finite element (FE) model of the main building was carried out. This analysis is required also to identify the key parts of the structure for which significant measurements must be acquired. The FE was also used to perform a non-linear static analysis, in order to determine for each structural element the interstory drift that leads to serviceability (operational) and ultimate (safe life) limit state. A different 3D FE model was also used to perform a pushover analysis, in which the masonry walls were modeled as 1D macroelements characterized by elastic-perfectly plastic behavior. In this model, the resistance of the masonry walls was calculated based on the elastic limit of the material, while the ultimate displacement was calculated based on its flexural and shear response.

A total of 4 sensing bars was installed, two at the first floor and two at the second. Based on the assumption of a rigid floor behavior, only the sensing bar of the lower floor was instrumented with 4 monoaxial accelerometers and 1 biaxial inclinometer, while the sensing bar of the upper floor was instrumented with 2 monoaxial accelerometers and 1 biaxial inclinometer as depicted in Fig. 15.4.

15.3.4 Use of MEMS for Building Monitoring

A cost-efficient and effective monitoring system based on microelectromechanical systems (MEMS) for strain and acceleration sensors, named MEMSCON was proposed by Trapani et al. (2012) for reinforced concrete buildings. The

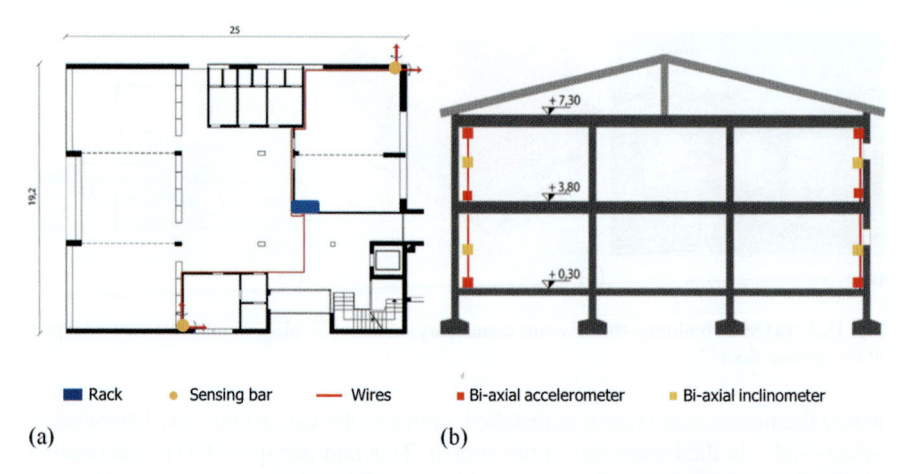

Fig. 15.4 Layout of the monitoring system

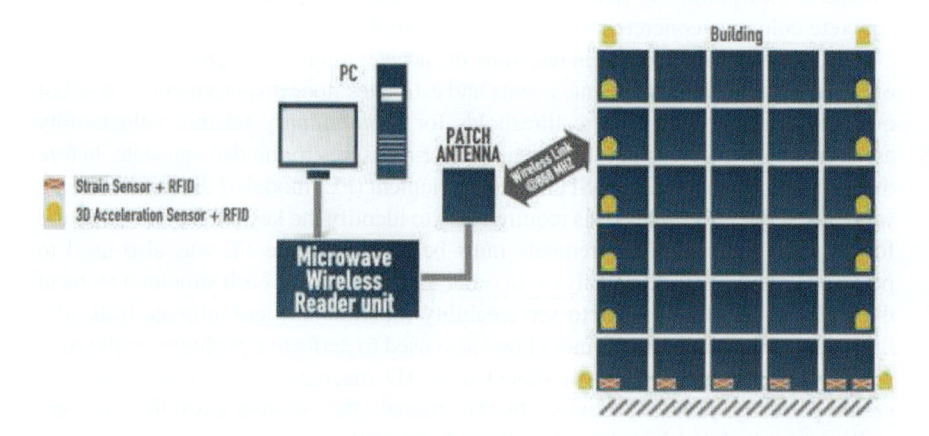

Fig. 15.5 The MEMSCON layout after Trapani et al. (2012)

MEMSCON system illustrated in Fig. 15.5, consists of a network of strain and acceleration sensors installed inside a building for recording data when a severe seismic event occurs or when it is scheduled by the user. In particular, the sensors wirelessly transmit data to a remote acquisition unit where data are stored, processed and interpreted using a purposely-developed software (Santana et al. 2011; Torfs et al. 2011). This software consists of a damage assessment module and a decision support module, which provide an estimation of the damage level inside the building and an insight on the rehabilitation methodologies and costs, respectively.

The designed sensors are 3D capacitive MEMS accelerometers based on different technologies: for in plane acceleration, the sensor has a interdigitated-comb structure while for the out of plane acceleration a pendulum system is used. These MEMS devices are packaged inside a plastic housing together with a lithium battery, an

ASIC, an ADC, a Zigbit module and an antenna. Features of the device are 200 Hz sampling rate, a 54 seconds maximum acquisition period, a 16 bit resolution, and 2.5 V range for the ADC. Moreover, the acceleration sensor in characterized by a reduced energy consumption, due to the fact that it remains in the idle state until a seismic event particularly relevant occurs. As regards to the strain sensors, they are designed to be directly bonded to the reinforcing bars embedded in the RC columns at the ground level. Moreover, they provide both the deformed shape of the structure, from which the structure undergoes damage due to an earthquake and the strain field after the seismic event.

More precisely, these sensors consist of two parts: the front-end sensor embedded into the RC element and the mote to be externally attached. The front-end sensor is a capacitive MEMS strain sensor connected to an ASIC, both embedded in a PDMS substrate 4 cm long and 5 mm thick. The whole package is bonded to a polyimide carrier 8 cm long and 1 cm wide, to be bonded in turn to the reinforcing bars, by using a cyanoacrylate glue.

The Memscon technology was tested by Trapani et al. (2012) under laboratory conditions on a 3D reinforced concrete frame through dynamics tests, that reproduced an actual earthquake of increasing amplitude. More precisely, the installed accelerometers are validated comparing their response to the response of a set of reference piezoelectric-based accelerometers adopted to measure in plane accelerations along both directions. The observed discrepancy between MEMSCON device responses and reference devices is in the order of 20 mg, evaluated in terms of root mean square, which is of the order of the observed background noise amplitude. Based on a preliminary filtering of signals, in order to remove low frequencies, MEMSCON accelerometers allow also for the estimation of displacement time histories starting from recorded acceleration measures. The comparison between these estimations and the time histories of horizontal displacements actually induced by the horizontal actuator indicates that the estimation is reliable, being the discrepancy less than a few tenths of a millimeter.

15.4 Problems of Health Monitoring of Infrastructure Systems

Nowadays, owners and operators of modern civil engineering infrastructures need information about integrity and reliability of the network, infrastructural system, and/or components, possibly in real time. As a result, they can evaluate the state of the structure and also assess when preventive actions need to be taken. This information allows for taking timely decisions on whether functionality has been impaired or whether it can remain in operation with a pre-specified level of reliability (Karbhari and Ansari 2009). The monitoring systems presented herein are installed with this perspective and can be used to track seismic structural responses too.

Fig. 15.6 Locations of the monitored points in a generic tunnel cross-section. The origin of θ-coordinate (rotation) and the direction of monitored deformations are depicted too

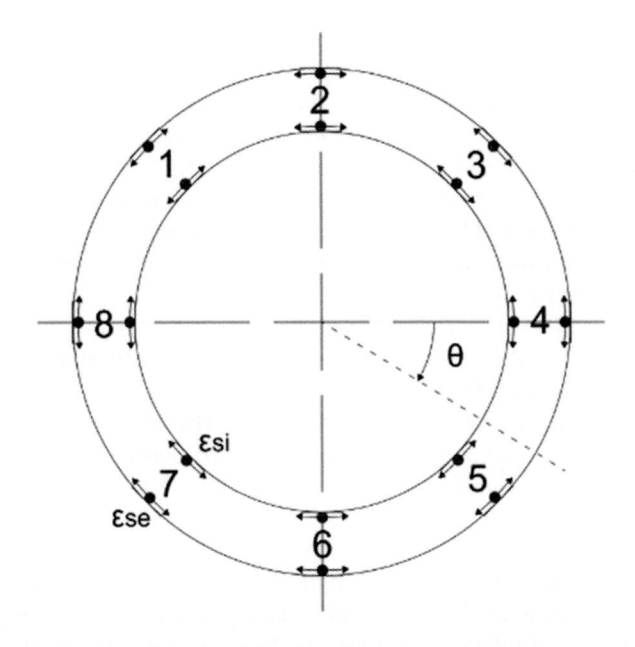

15.4.1 A FBG Sensor System for Monitoring the Earthquake Response of Tunnel Linings

In order to investigate the capabilities of FBG sensors in monitoring the structural inelastic response under seismic events, Bursi et al. (2015) studied in depth both with numerical and experimental tools a circular concrete tunnel lining. The tunnel is located in Rome, Italy. The outside diameter is 4.8 m, lining thickness 0.2 m and the tunnel axis is at 20 m below ground level. The whole tunnel is subdivided into 21 rings; and the relevant deformations are recorded in 8 sections, as depicted in Fig. 15.6, at the centroidal cross-section of each ring.

A specimen of a tunnel ring, characterized by a width of 1 m, was reproduced in the laboratory. More precisely, the experimental campaign was split into three parts: (i) tests on materials, (ii) tests on substructures; (iii) one test on the full-scale specimen of the tunnel lining. Tests on materials were carried out in order to characterize the material mechanical properties; while, test on substructures were performed in order to calibrate the cyclic test to be applied to the tunnel-ring specimen. Moreover, through tests on substructures the best FBG configuration was selected among three package solutions: FBG sensors attached to reinforcement bars, either (1) bonded to or (2) unbonded in concrete; and (3) FBG sensors externally mounted on metal holders welded to reinforcement bars through access holes left during casting. It was found that the unbonded solution provided the best estimation of rebar strains.

For the full-scale test FBG sensors were positioned symmetrically at the inner and outsider part of each section indicated in Fig. 15.6. Further sensors were added in

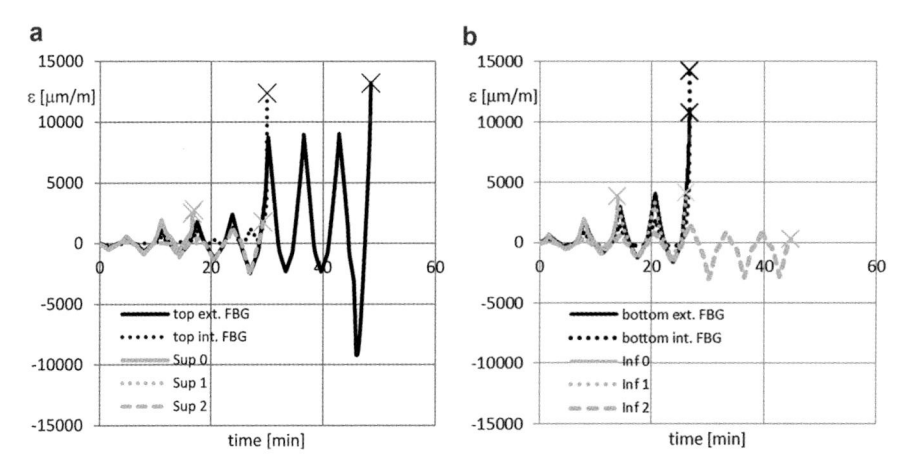

Fig. 15.7 SSC4 test: (**a**) top side, and (**b**) bottom side FBG measurements vs. conventional strain gauges -Sup 0–2 and Inf 0–2

two sections where plastic hinges were expected, i.e. in correspondence with the inner actuator. Further, temperature sensors were used to compensate for temperature-induced wavelength shifts of all fibre sensors. Hence, the FBG sensor system finally consisted of a total number of 40 sensors. The cyclic test performed on the tunnel specimen showed that, fibres measured greater deformation in Sections #2, 4, 6 and #8, where plastic hinges appeared. More precisely, external FBG fibres approached a maximum value of about 0.6% in Sect. #8; while, internal FBG fibres reached a maximum value of 1.2% at Sect.#2. Some responses can be observed in Fig. 15.7. The satisfactory results obtained through both substructure tests and the full-scale test proved the effectiveness of a FBG sensor system to monitor the inelastic response of structures like tunnels during seismic events.

15.4.2 SHM of the Colle Isarco Viaduct

The Colle Isarco viaduct is a segmental prestressed concrete box girder located in northern Italy (Bolzano). Overall, the viaduct comprises two structurally independent decks, the so-called North and South carriageways, with 13 spans, for a total length of 1028.2 m. The main span of the viaduct, 163 m long, consists of two symmetric reinforced concrete Niagara box girders, which support a suspended beam of 45 m, as depicted in Fig. 15.8. Each box girder ends with a 59 m–long cantilever, counterbalanced by a back arm with a length of 91 m.

The SHM system recently installed on the viaduct consists of three different sets of instruments, each based on a different technology. The first set, installed and

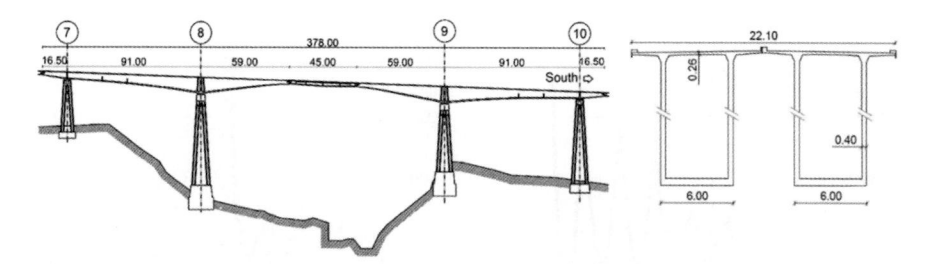

Fig. 15.8 Elevation of the three main spans of the Colle Isarco viaduct and generic cross section of the box-girder. Dimensions in m

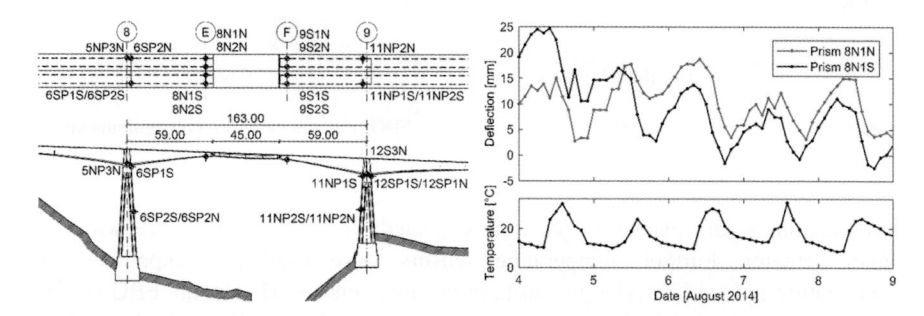

Fig. 15.9 (a) Configuration of prisms between Piers #8 and #9. Dimensions in m; (b) time histories of deflection and temperature field data

activated in early 2014, is made of two Leica TM50 topographic total stations and 72 GPR112 prisms. The second and third sets, installed on June 2016 are made of 56 fiber optic sensors (FOSs) implementing fiber Bragg gratings (FBGs) and 74 PT100 platinum resistance thermometers connected to their respective reading units. The topographic network was designed to monitor the deflection of the decks between Pier #7 and #10 during the last structural intervention of 2014 and afterwards. The systems based on FOSs and PT100 sensors were instead designed to monitor the long-term effects of the recent post-tensioning intervention. More precisely, these last systems record the strain of both the top and bottom slabs of the box girders and the temperature pattern between Piers #7 and #10, respectively.

The total stations started acquiring data on June 9, 2014. Figure 15.9 shows the vertical displacement of prisms 8N1N and 8N1S, along with the air temperature, recorded from August 4 to 9, 2014. These prisms are placed at the edge of the North girders, i.e. a location that is sensitive to variations in loads, temperature and mechanical properties. By observing these measurements, we can conclude that the behavior of the two decks before post-tensioning was similar, and mostly affected by temperature rather than live loads. Based on Fig. 15.9, we can also argue that when the air temperature increases in the morning, the edge of each deck moves down, with a short time delay. This occurs because the source of heat, i.e. the sun, increases the temperature of the top

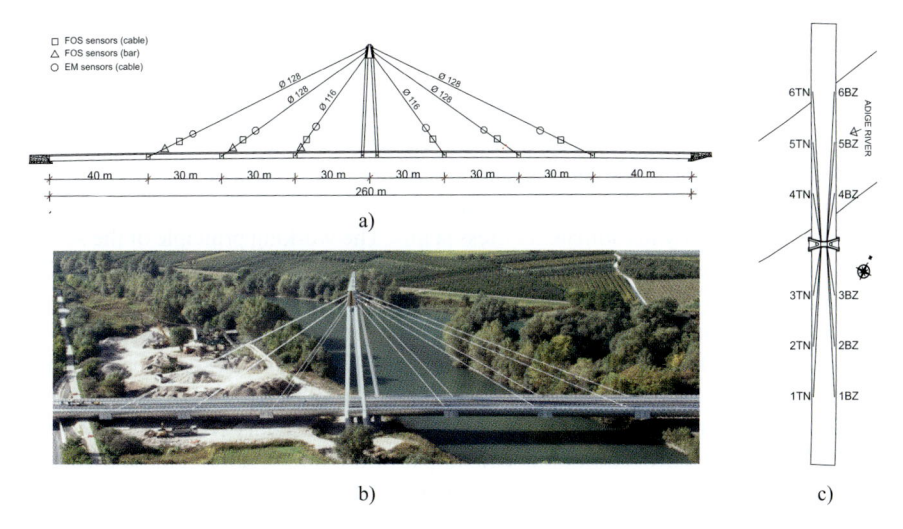

Fig. 15.10 (**a**) Longitudinal section of the bridge and sensors layout; (**b**) overview of the bridge; (**c**) plan view of the bridge and names of the cables

slab more than that of the bottom slab, and so leads the top slab to elongate more than the bottom one. Other details can be found in Beltempo et al. 2017).

15.4.3 SHM of the Ponte Adige Cable-Stayed Bridge

The Adige Bridge depicted in Fig. 15.10, is a cable-stayed bridge spanning the Adige River 10 km north of the town of Trento, Italy. It is a statically indeterminate structure, having a composite steel-concrete deck of length 260 m overall, supported by 12 stay cables, 6 per deck side. The cross section of the deck consists of 4 "I" section steel beams 2 m high with variable flange dimensions along the span and a 25 cm thick concrete slab. The deck bears on the abutments and is anchored every 30 m to the cable stays. The bridge antenna consists of 4 pylons 45 m tall, located in the middle of the bridge. The stays are full locked steel cables of diameters 116 mm and 128 mm, designed for operational loads varying from 5000 to 8000 kN.

On completion of the construction work, in 2008, the owner of the bridge, the Autonomous Province of Trento, decided to install a monitoring system, to continuously record both the tension and elongation of the cable-stays. Load measurements are performed using a relatively new technique based on elasto-magnetic principles, developed by Intelligent Instrument Systems Inc. Strain measurements are collected by a fiber optic sensor network, provided by Smartec SA. The system was installed and has been operating since March 2011. The sensors described herein are also used to acquire measurements during a seismic event.

15.4.4 Elasto-Magnetic Sensors

When a magnetic field H is applied to a ferromagnetic medium, the resulting magnetic flux density B is proportional to a constant μ which is a feature of both the medium and the stress applied, and is referred to as magnetic permeability. An elasto-magnetic (EM) sensor measures the magnetic permeability of the steel cable and uses this quantity to estimate its stress status. The working principle of the sensor was first proposed by Jarosevic (1998), and later developed by Sumitro et al. (2002) into an industrial prototype.

The EM sensor consists of two coils wound on-site around the cable, separated by plastic shells, and protected by an epoxy and metal cover. To make a measurement, the interrogation unit charges the primary coil, and applies a magnetic field H to the cable. The resulting B-field produces a current on the secondary coil, which in turn is recorded by the interrogation unit. The output of the interrogation unit is a voltage proportional to the magnetic flux B and, H being constant, to the magnetic permeability of the cable. Based on the nominal value provided by the producer, the estimated life of this model of sensor is 50 years and the operating temperature range − 20 °C to +80 °C; the accuracy is estimated at 3–10% (depending on the size of the EM sensors). Because of the interrogation procedure, the sampling period is normally higher than 10 sec. The response of an EM sensor is based on the elasto-magnetic properties of the specific steel of the cables, and is also sensitive to the cable cross-section and size, to the temperature and to the manufacturing process of the sensor. Therefore, these sensors require calibration before use. Site-fabricated sensors, as in this case, should ideally be calibrated onsite, for example loading and unloading the cable, while simultaneously comparing the response of the EM sensor with a reference gauge, e.g. a load cell. Because the installation was carried out two years after the opening of the bridge, and the owner did not allow cables to be unloaded in order to perform tests, calibration was carried out in the laboratory; more precisely, both a prototype sensor and a cable segment identical to that of the bridge were reproduced.

15.4.5 Fiber optic sensors

The fiber optic sensors installed on the Adige Bridge were supplied by Smartec SA, specifically the "MuST" model based on the Fiber Bragg Grating (FBG) technology. The system installed on the Adige bridge includes both strain, i.e. the MuST 12.1010 model and temperature sensors, i.e. the MuST 12.1081 model. Each MuST sensor is a long gauge-length device, which includes two independent FBGs, packed into a polycarbonate tube: the first one is embedded in a segment of fiber, pretensioned between the two ends of the sensor, elongating with the structure being monitored; the second one is kept loose and serves to record temperature for thermal compensation. Fig. 15.10a shows an overview of the optical sensor network deployed on the

Adige Bridge. Each of the 12 cables is directly fitted with a 1 m length strain gauge sensor. In addition, the lower anchorages of three cables are fitted each with a set of six strain sensors: the sensors are arranged in triplets around the two threaded bars of the anchorage in such a way as to record their axial deformation and bending. Besides the thermometers embedded in the strain sensor, the temperature field of the bridge is recorded through five temperature sensors: four of these are embedded in the concrete slab, two at each side, symmetrically with respect to the antenna, and one is mounted outside the antenna. The read-out unit is a dynamic instrument with a sweep laser source and a photodiode measuring the wavelengths reflected from the instruments; it can acquire at 500 Hz. The monitoring system can operate in two regimes: static and dynamic. The expression "static data" means data acquired at very low frequency, e.g. 1 sample every 15 min, continuously in time. In the Adige Bridge configuration, every static measurement is an average of 4000 samples at about 500 Hz; this acquisition strategy reduces the impact of bridge traffic over data collection.

Conversely, dynamic data acquisition take place whenever required, at any frequency up to 500 Hz. Therefore, the system can also be used in the case of an earthquake. In this case, the fiber optic system records the cable elongations during a seismic event. These measures can track potential bridge damage. The accuracy of the aforementioned instruments is $5\mu\varepsilon$ in terms of strain and $0.5\,°C$ for temperature, while the range of measurement is $\pm 5\permil$ strain and $-20\,°C$ to $+60\,°C$ temperature. Fiber optic strain sensors, as all strain sensors, measure only changes of strain between two states: as a result, every measurement defines the difference with respect to the value at the time of installation. Because the fiber optic system was installed only 2 years after bridge construction, absolute values of strain with respect to the entire bridge life are unavailable. In any case, the whole sensing system is quite useful in the case of seismic events.

15.5 Conclusions

Structural health monitoring (SHM) is becoming more and more important. We can say that its ultimate target is the ability to monitor a structure throughout its working life in order to reduce maintenance requirements and subsequent downtime, also caused by seismic loadings. Currently, visual inspection represents the standard method used for structural assessment, along with non-destructive evaluation techniques. Nonetheless, most of these techniques require a lot of manual work and a significant downtime. Thus, an increasing interest in SHM is rising, because of cost savings due to the reduced number of manual inspections. In this contest, MEMS and optical fibers are becoming desirable features in SHM systems and there has been a large development of new sensors during the last years. Along these lines, we have shown some applications both to buildings and infrastructures like tunnels and bridges.

As regards sensors, we have seen that sensor faults may occur in SHM systems due to various factors, e.g., malfunctions or failures of sensor components, the effects of electromagnetic interference, etc. Therefore, we have emphasized the importance of sensor validation whose purpose is to detect, isolate and reconstruct potential sensor faults for practical SHM applications. In this respect, we have to underline that complex systems or systems subjected to earthquake excitations tend to exhibit strong nonlinearities. Therefore, relevant health-monitoring data always exhibit non-Gaussian behavior. Thus, validation techniques for data treatment have to take into account this non-Gaussianity.

Then, we have reviewed some sample cases taken for buildings and infrastructures, and we have analysed some aspects related to the implementation of sensors and SHM systems.

Generally, typical SHM systems do not use an integrated approach to the design, implementation, and operation of the SHM system, resulting in partial benefits from the system. Too often, a disproportionate emphasis is placed on the collection of data rather than on the management of these data and the use of decision-making tools, i.e. decision support systems (DSS) that assist the ultimate aim of using the collected data to entail a better management of the infrastructure system. Typically, systems collect data on a continuous or periodic basis and transmit data to a common point. The data are then compared with results from a numerical model that simulates the original structure. The weak point is that most systems do not attempt to update the model to reflect ageing and deterioration, or changes made through routine maintenance or even rehabilitation. Thus, the chosen approach is not able to predict future responses nor to identify critical areas that may need further monitoring. All in all, we need SHM systems endowed with DSS based on systems identification and/or non-destructive damage evaluation algorithms to rapidly process data and capable of fulfilling the ultimate goals of estimating capacity and service life of structures and infrastructure systems.

Acknowledgments The work presented herein was carried out with different financial grants from the European Union through the projects (MEMSCON), FP7-NMP, (MONICO) FP7-SME-2007-1, under the research agreement between Autostrada del Brennero SpA/Brennerautobahn AG and the University of Trento; and several agreements between the Autonomous Province of Trento, Italy and the University of Trento.

References

Alcala CF, Qin SJ (2011) Analysis and generalization of fault diagnosis methods for process monitoring. J Process Control 21(3):322–330
Baggio C et al (2007) Field manual for post-earthquake damage and safety assessment and short terms countermeasures. Scientific and technical reports, EUR 22868 EN
Beltempo A, Bursi OS, Cappello C, Zonta D, Zingales MA (2017) Viscoelastic model for the long-term deflection of segmental prestressed box girders. Computer-aided civil and infrastructure engineering, page doi:https://doi.org/10.1111/mice.12311, ISSN 1467-8667

Brown HM, Allen RM, Hellweg M, Khainovski O, Neuhauser D, Souf A (2011) Development of the ElarmS methodology for earthquake early warning: real time application in California and offline testing in Japan. Soil Dyn Earthq Eng 31:188–200

Bursi OS, Bonelli A, Fassin M (2015) Capabilities of a fiber bragg grating sensor system to monitor the inelastic response of concrete sections in new tunnel linings subjected to earthquake loading. Comput Aided Civ Inf Eng 30(8):636–653

Calvi GM (1999) A displacement-based approach for vulnerability evaluation of classes of buildings. J Earthq Eng 3(3):411–438

Candès EJ, Li X, Ma Y, Wright J (2011) Robust principal component analysis. J ACM 58(3):1–37

Celebi M (2007) Developments in seismic monitoring for risk reduction. J Risk Res 10(5):715–727

Çelebi M, Sanli A, Sinclair M, Gallant S, Radulescu D (2004) Real time seismic monitoring needs of a building owner and the solution – a cooperative effort. Earthq Spectra EERI 20(2):333–346

Günay S, Mosalam KM (2013) PEER performance-based earthquake engineering methodology, revisited. J Earthq Eng 17(6):829–858

Huston D (2011) Structural sensing, health monitoring, and performance evaluation. CRC Press/Taylor & Francis, Boca Raton

Jarosevic A (1998) Magnetoelastic method of stress measurement in steel. Proceedings of the NATO

Karbhari VM, Ansari F, (2009) Structural health monitoring of civil infrastructure systems. Woodhead Publishing Series in Civil Engineering, Amsterdam. ISBN: 978-1-84569-392-3

Kullaa J (2013) Detection, identification, and quantification of sensor fault in a sensor network. Mech Syst Signal Process 40(1):208–221

Li Z, Koh BH, Nagarajaiah S (2007) Detecting sensor failure via decoupled error function and inverse input-output model. J Eng Mech 133(11):1222–1228

Ponzo FC, Ditommaso R, Auletta G, Mossucca a (2010) A fast method for structural health monitoring of Italian reinforced concrete strategic buildings. Bull Earthq Eng 8(6):1421–1434

Porter K, Mitrani-Reiser J, Beck JL (2006) Near-real-time loss estimation for instrumented buildings. Struct Design Tall Spec Build 15(1):3–20

Priestley N, Calvi GM, Kowalsky M (2007) Displacement-based seismic damage of structures, 1st edn. Edited by IUSS press, Pavia

Rainieri C, Fabbrocino G (2014) Operational modal analysis of civil engineering structures. Springer, New York

Rücker W, Hille F, Rohrmann R (2006) F08b guideline for structural health monitoring, SAMCO Final Report

Santana J, Van den Hoven R, Van Liemp C, Colin M, Saillen N, Trapani D, Zonta D, Torfs T, Van Hoof C (2011) A 3-axis accelerometer and strain sensor system for building integrity monitoring. Sensors Actuators A Phys 188:141–147

Sharifi R, Kim Y, Langari R (2010) Sensor fault isolation and detection of smart structures. Smart Mater Struct 19(10):105001

Sohn H, CR Farrar, FM Hemez (2004) A Review of structural health monitoring literature: 1996–2001. Los Alamos National Laboratory report, LA-13976-MS

Sullivan TJ, Priestley MJN, Calvi GM (2012) A model code for the displacement-based seismic design of structures. IUSS Press, Pavia

Sumitro S, Kurokawa M, Wang ML (2002) Elasto-magnetic sensor utilization on steel cable stress measurement. Proceedings of the first fib congress, Concrete structures in the 21th Century, Osaka, 13–19 Oct 2002

Torfs T, Sterken T, Brebels S, Santana J, van den Hoven R, Saillen N, Trapani D, Zonta D, Marmaras P, Bimpas M, Van Hoof C (2011) Low power wireless sensor network for building monitoring. IEEE Sensors, Limerick, Ireland, 28–31 Oct 2011

Trapani D, Zonta D, Molinari M, Amditis A, Bimpas M, Bertsch N, Spiering V, Santana J, Sterken T, Torfs T, Bairaktaris D, Bairaktarisg M, Camarinopulos S, Frondistou-Yannas M, Ulieru D (2012) Full-scale laboratory validation of a MEMS-based technology for post-earthquake damage assessment. Proceedings of the 15 world conference earth engineering, Lisbon

Yang Y, Nagarajaiah S (2013) Time-frequency blind source separation using independent component analysis for output-only modal identification of highly damped structures. ASCE J Struct EngJ Struct Eng 139(10):1780–1793

Yi T-H, Huang H-B, Li HN (2017) Development of sensor validation methodologies for structural health monitoring: a comprehensive review. Measurement 109:200–214

Chapter 16
Large Scale Testing Facilities – Use of High Gravity Centrifuge Tests to Investigate Soil Liquefaction Phenomena

Gopal S. P. Madabhushi

Abstract Soil liquefaction following earthquake events causes severe damage to Civil Engineering Infrastructure as witnessed in many of the recent earthquake events. High gravity centrifuge tests are able to simulate earthquake induced liquefaction in saturated soils and allow us to study the physics behind liquefaction phenomena and the behaviour of structures that are located on such sites. In this paper, the use of large scale testing facilities in studying the problems in geotechnical earthquake engineering will be highlighted. Soil liquefaction problems are used as a vehicle to illustrate the use of these large scale testing facilities. Some of the recent investigations that were carried out at University of Cambridge will be presented. These include the novel testing that was carried out which involved creation of triaxial chambers within centrifuge models to delineate drainage effects on liquefiable soils. Direct comparisons are made between free-field soil and the soil enclosed within the triaxial chamber. Similarly the reduction in settlement of foundations on liquefiable soils due to air injection *a priori* to earthquake loading will be presented. The differences in the failure mechanisms of shallow foundations caused by the injected air are presented.

16.1 Large Scale Testing Facilities

In geotechnical earthquake engineering, it is very attractive to conduct large scale testing of physical models to understand the failure mechanisms created by earthquake loading in a specific boundary value problem such as retaining walls, pile foundations or embankment dam failures. As the soil exhibits highly non-linear behaviour under the action of earthquake loading which begets large stresses and large strains, it is imperative that the physical models are tested at prototype stresses and strains. A convenient way to generate full scale, prototype stresses and strains in

G. S. P. Madabhushi (✉)
Professor of Civil Engineering, University of Cambridge, Cambridge, UK
e-mail: mspg1@cam.ac.uk

Fig. 16.1 A view of the 10 m diameter Turner Beam Centrifuge at University of Cambridge

small scale physical models is by the use a high gravity centrifuge, such as the one at the Schofield Centre, University of Cambridge, shown in Fig. 16.1. This is a balanced beam geotechnical centrifuge that is classified as 150 g-ton machine with a payload capacity of about 1 ton. For earthquake simulation tests the maximum centrifugal acceleration is restricted to about 100 g's. The principles of centrifuge modelling are described later in Sec. 2.

16.1.1 Earthquake Actuators

In order to model earthquake loading on centrifuge models in-flight, powerful earthquake actuators are required. These actuators need to deliver large forces (of the order of several kN) in a very short time scale (of the order of fractions of seconds) due to the scaling laws presented later in Table 16.1. In Cambridge there are two types of earthquake actuators that are available to the modellers. These are described next.

16.1.1.1 Stored Angular Momentum (SAM) Earthquake Actuator

Much of the research in the last decade in Cambridge in the area of earthquake geotechnical engineering has been carried out using the Stored Angular Momentum (SAM) actuator. This device was developed by Madabhushi et al. (1998) and shown in Fig. 16.2. SAM actuator has been prolific over the last decade and produced more than 10 PhD theses. The SAM actuator operates by storing all the required energy for firing a model earthquake in a set of fly wheels, which are spun up to the required RPM using a 3 phase electric motor. The fly wheels have the stored angular momentum at the frequency of the required earthquake. The fly wheels are enclosed in a crank case and drive a reciprocating rod. A fast acting clutch was developed at Cambridge that can engage the reciprocating rod in under 20 ms. When an earthquake is desired, the fast acting clutch is activated by using high pressure nitrogen. This engages the shaking table on which the centrifuge model package is mounted. The magnitude of the earthquake can be adjusted by moving the pivot point on the

Table 16.1 Scaling laws

	Parameter	Scaling law model/prototype
General scaling laws (slow events)	Length	1/N
	Area	$1/N^2$
	Volume	$1/N^3$
	Mass	$1/N^3$
	Stress	1
	Strain	1
	Force	$1/N^2$
	Work	$1/N^3$
	Energy	$1/N^3$
	Seepage velocity	N
	Time (consolidation)	$1/N^2$
Dynamic events	Time (dynamic)	1/N
	Frequency	N
	Displacement	1/N
	Velocity	1
	Acceleration / Acceleration due} To gravity (g's)	N

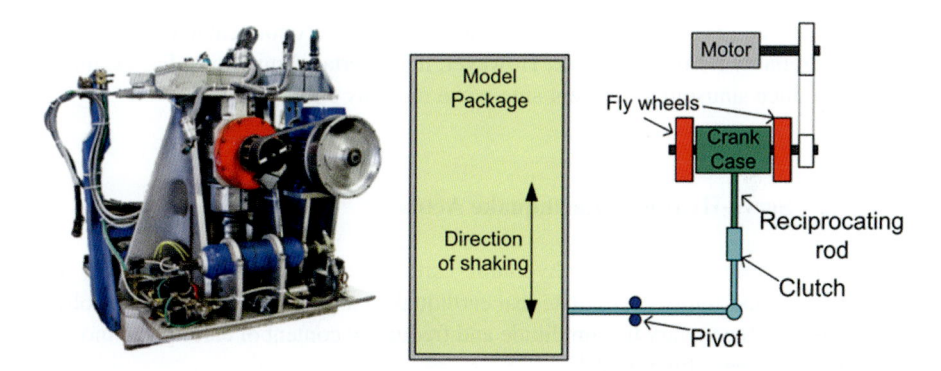

Fig. 16.2 A view of the Stored Angular Momentum (SAM) earthquake actuator

cross rod. Thus the frequency, duration and intensity of the model earthquakes can be chosen by the centrifuge modeller. An example of a typical earthquake that has been produced by the SAM actuator is shown in Fig. 16.3. In this example, the centrifuge modeller fired the earthquake in a 50 g test at a frequency of 40 Hz. The duration of the earthquake was chosen as 500 ms. In Fig. 16.3 the equivalent prototype earthquake is shown which has a magnitude of about 0.2 g applied at the base of the model (i.e. bedrock motion). The FFT of the motion in Fig. 16.3 shows that most of the energy of the earthquake is concentrated at 0.8 Hz although a higher harmonic is present at 2.1 Hz. The duration of the prototype earthquake is

Fig. 16.3 An example of the sinusoidal input motion from SAM actuator

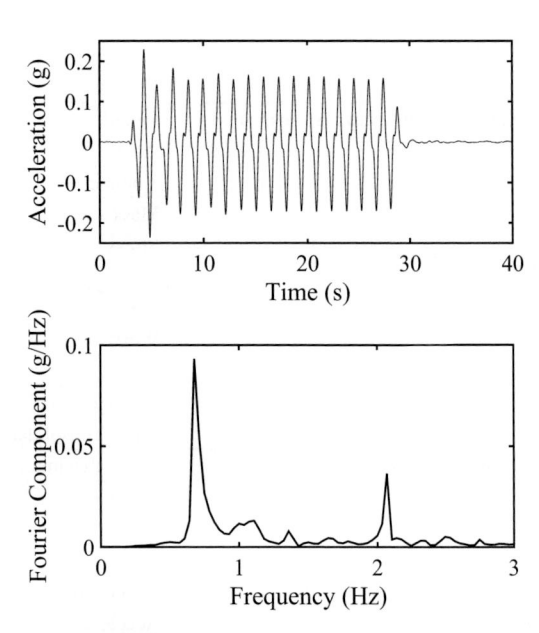

about 25 s. The dynamic scaling laws presented in Table 16.1 were used to convert the model earthquake characteristics into an equivalent prototype event.

SAM actuator is a simple, mechanical actuator that is economical to build and has operated reliably for over a decade. However, it has certain limitations. For example, it can produce sinusoidal or swept sine wave motions only.

16.1.1.2 Servo-Hydraulic Earthquake Actuator

The main advantage of a servo-hydraulic earthquake actuator is that it offers the researchers opportunity to simulate real earthquake motions. These servo-controlled actuators are able to vary the amplitude and frequency content of earthquake motion applied to the centrifuge models.

In Cambridge a new servo-hydraulic shaker has been developed that was commissioned in late 2011. Madabhushi et al. (2012) describe the construction and performance of this actuator. A view of this earthquake actuator is shown in Fig. 16.4. The main operating principle of this earthquake actuator is that the energy required to fire the model earthquakes is stored in highly pressurised hydraulic oil. The hydraulic oil is pressurised to about 260 bar and pumped into four main accumulators. The pressurised oil is then directed through a double acting actuator directly attached to the shake table. The spent oil is then collected in a low pressure (7 bar) accumulator. The movement of the double acting actuator is servo-controlled through a close loop by the servo-amplifier.

The servo-hydraulic earthquake actuator built at Cambridge uses many of the features of the Turner beam centrifuge shown in Fig. 16.1. For example, the main

Fig. 16.4 A view of the Servo-Hydraulic Earthquake Actuator

reaction to the earthquake shaking force imparted to the centrifuge model will be provided by the main body of the beam centrifuge. The entire shaker assembly is mounted on a self-contained swing which can be loaded and unloaded like any other centrifuge package tested on the centrifuge. The hydraulic power pack that supplies the high pressure fluid is outside the centrifuge and is supplied to the earthquake actuator through high pressure fluid slip rings.

One of the main advantages of using a servo-hydraulic earthquake actuator is that we can simulate more realistic motions as mentioned earlier. In Fig. 16.5a an example input motion generated by the servo-hydraulic earthquake actuator in a 50 g centrifuge test to simulate a scaled Kobe earthquake motion of 1995 is presented. The peak amplitude of the input motion in this case was about 0.2 g and the duration of the earthquake was about 12 s. In this figure the FFT of the input motion is also presented which shows the presence of multiple peaks corresponding to the frequency components in the earthquake motion. Similarly in Fig. 16.5b an example of the input motion of the Imperial Valley motion in a 50 g centrifuge test is presented. This motion is much longer i.e. nearly 80 s in prototype scale. The FFT of this input motion also shows the high frequency content of this motion captured by the servo-hydraulic earthquake actuator.

16.1.2 Model Containers

Dynamic centrifuge modelling requires the use of specialist model containers. As the centrifuge models are subjected to earthquake motions at the base, the ends of the container, if rigid, can impose additional, spurious P waves in the soil body. Several

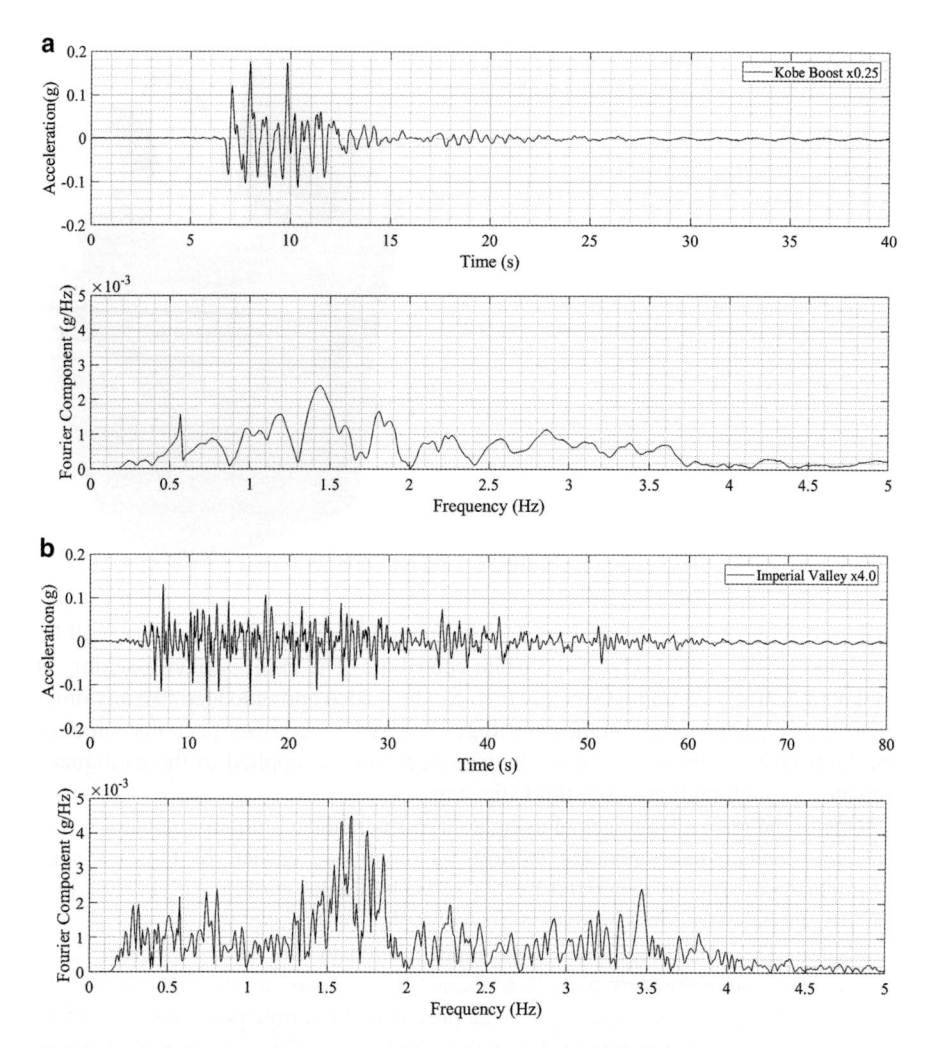

Fig. 16.5 Example input motions generated by the Servo-Hydraulic earthquake actuator (**a**) Kobe Earthquake motion (**b**) Imperial Valley motion

researchers have focused on developing specialised model containers that reduce the impact of these P waves from the end walls of the container.

16.1.2.1 Laminar Model Container

One concept is to use laminae that are separated by cylindrical bearings that allow free displacement of each lamina relative to the next. Such containers undergo free,

Fig. 16.6 Laminar model container

Fig. 16.7 Equivalent Shear Beam Model container

lateral displacement allowing the soil lateral movements of the soil. A view of the Cambridge laminar box is shown in Fig. 16.6. This container is able to model a depth of about 25 m of soil depth in a 80 g centrifuge test. More details of the laminar box and its performance is described by Brennan et al. (2006). The most effective use of laminar model container is in modelling of lateral spreading problems following soil liquefaction. As the container has effectively no stiffness laterally, it is able to deform following the spreading of the liquefied soil on a slope.

16.1.2.2 Equivalent Shear Beam (ESB) Model Container

Another concept of model container is to mimic the lateral deformations observed in a free, vertical column of soil modelled as a shear beam. The lateral deflections in the soil are calculated for a given magnitude earthquake. These are matched by constructing the equivalent shear beam (ESB) container with alternating rings of aluminium and rubber. The thickness of the rubber can be changed with the depth of the model container. A view of one the ESB model containers at Cambridge is shown in Fig. 16.7. This containers construction and performance are described in detail by Zeng and Schofield (1996). This container is able to model a depth of about 16 m of soil in a 80 g test. Larger ESB model containers were also constructed at Cambridge, which follow the same design principle, but can model depths of about 34 m in a 80 g centrifuge test. The ESB model containers are useful in modelling level ground problems to study soil structure interaction, for example.

Fig. 16.8 Transparent sided model container with a 45° mirror

16.1.2.3 Transparent Sided Model Container

While the above model container simulate the free-field conditions well in specific centrifuge tests, they have the disadvantage of being opaque when viewed from the side. With the recent advances in high speed imaging and the development of the geo-PIV software (White et al. 2003), it is now possible to obtain high resolution images using high speed cameras that can acquire images at 1000 frames per second. However, this requires the side of the model containers to be transparent. Initial attempts at using high speed cameras in dynamic centrifuge testing were carried out by Cilingir and Madabhushi (2011) using a transparent sided model container shown in Fig. 16.8. This set up used a vertically mounted Phantom Camera and a 45° mirror and was quite successful in obtaining soil displacements and soil strains during earthquake loading next to model tunnels.

More recently, a more compact Motion Blitz cameras became available, that can be mounted directly in front of the transparent window and removing the need for the 45° mirror.

16.2 Principle of Centrifuge Modelling

The basic premise in centrifuge modelling is that we test $1/N$ scale model of a prototype in the enhanced gravity field of a geotechnical centrifuge (Madabhushi 2014). The gravity is increased by the same geometric factor N relative to the normal earth's gravity field.

This can be illustrated using a simple example. Let us consider a block structure of mass M and with dimensions $L \times B \times H$ sited on a horizontal soil bed as shown in Fig. 16.9. The average vertical stress exerted by this block on the soil can be easily calculated as;

$$\sigma_v = \frac{M\,g}{L \times B} \quad (16.1)$$

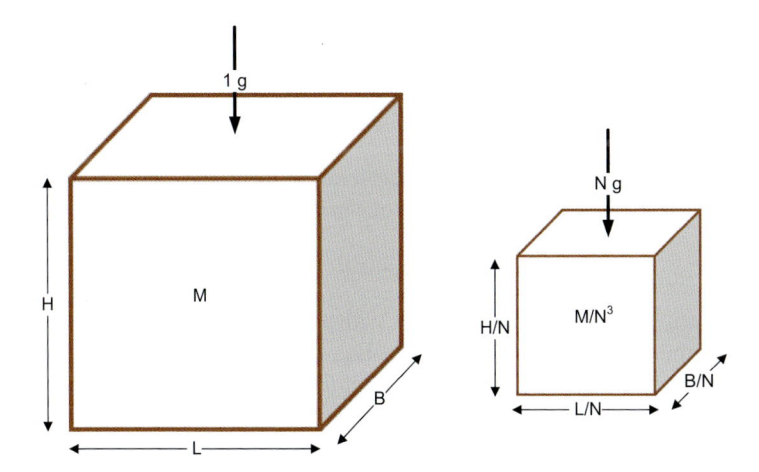

Fig. 16.9 Principle of centrifuge modelling

Similarly the vertical strain induced in the soil at any given point can be calculated as;

$$\varepsilon = \frac{\delta \alpha}{\alpha} \quad (16.2)$$

where α is a characteristic length in the soil body. Now let us consider a scale model of this block in which all the dimensions are scaled down by a factor N as shown in Fig. 16.1. As all the dimensions are scaled down by a factor N, the mass of this scaled down block will be M/N^3. Let us now imagine that this scale model of the block is placed in the increased gravity field of $N \times$ earth's gravity. If we now recalculate the vertical stress underneath this scale model of the block, we can see that;

$$\sigma_v = \frac{\frac{M}{N^3} \times Ng}{\frac{L}{N} \times \frac{B}{N}} = \frac{M\,g}{L \times B} \quad (16.3)$$

Thus the vertical stress below this scale model of the block is same as that below the larger block obtained in Eq. 16.1.

Similarly, if we consider strains in the soil;

$$\varepsilon = \delta \alpha \Big/ \frac{N}{\alpha/N = \frac{\delta \alpha}{\alpha}} \quad (16.4)$$

we can see that the prototype strain in Eq. 16.2 is recovered, as the changes in displacements and the original length are both scaled by the same factor N.

We increase the 'gravity' acting on our scaled model by placing it in a geotechnical centrifuge. The centrifugal acceleration will give us the 'N g' environment in which the scaled model will behave in an identical fashion to the prototype in the field. We can relate the angular velocity of the centrifuge to the required 'g' level.

When the centrifuge is rotating with an angular velocity of $\dot{\theta}$, the centrifugal acceleration at any radius 'r' is given by;

$$\bar{a} = r\dot{\theta}^2 \qquad (16.5)$$

We wish to match this centrifugal acceleration to be the same geometric scale factor as the one we used to scale down our prototype by i.e. N.

$$Ng = r\dot{\theta}^2 \qquad (16.6)$$

The centrifugal acceleration changes with the radial distance from the axis of rotation of the centrifuge as indicated in Eq. 16.6. We will normally arrange the speed of the centrifuge such that the model at the desired radius (say a typical point in the model like its centroid) will experience the desired centrifugal acceleration 'Ng'. This will give us the angular velocity $\dot{\theta}$ with which we have to rotate our centrifuge. For example, for the Turner beam centrifuge at Cambridge the nominal working radius is 4.125 m. If we need to create a centrifugal acceleration of '100 g' on a centrifuge model, then using Eq. 16.6, we can calculate the angular velocity as;

$$\dot{\theta} = \sqrt{\frac{100 \times 9.81}{4.125}} = 15.42 \ rad/s$$

$$\dot{\theta} = 147.3 \ RPM \qquad (16.7)$$

So by spinning the centrifuge at 147.3 RPM, we can create the required 100 g of centrifugal acceleration.

In centrifuge modelling the model behaviour observed during testing must be related to the behaviour of the equivalent field sized structure. This is achieved by a set of scaling laws that link the model and prototype parameters. The scaling laws were originally proposed by Schofield (1980, 81) and are easy to derive as shown by Madabhushi (2014). A set of important scaling laws are reproduced in Table 16.1.

In Table 16.1 it is seen that the scaling law for dynamic events are somewhat different from those for slower general events. For example, the scaling for dynamic time and consolidation time differ by a factor of $1/N$. This conflict in model scales is normally avoided by scaling the viscosity of pore fluid (Adamidis and Madabhushi 2015). Depending on the type of problem being modelled, centrifuge modellers are able to make suitable adjustments to the models to capture the most representative prototype behaviour.

16.3 Soil Liqueafaction

In this paper soil liquefaction is used as an example of a complex problem that can be modelled effectively using centrifuge modelling. Many of the recent earthquakes have caused extensive damage to infrastructure due to soil liquefaction. Some examples of these are presented below.

16.3.1 Examples of Soil Liquefaction

There are many examples of damage to civil engineering structures due to soil liquefaction. The settlement and rotation of the Harbour Master's tower at Kandla port due to soil liquefaction is shown in Fig. 16.10 following the Bhuj earthquake of 2001. This building was supported on pile foundations that pass through a soil profile that is susceptible to liquefaction. Further details of this case history can be found in Madabhushi et al. (2005). A similar failure mechanism was observed in New Zealand earthquake of 2011 in the case of a low-rise residential building supported on shallow foundations shown in Fig. 16.10. Here the super-structure showed very little damage but the building was a write-off due to foundation failure. Madabhushi and Haigh (2009) argued that the super-structure stiffness plays an important role in determining the failure mechanism of the structure located on liquefiable soil. In Fig. 16.11 the lateral spreading caused by soil liquefaction in Portoveijo in the recent Ecuador earthquake is presented. Soil liquefaction and subsequent lateral spreading causes a different type of failure in pile foundations. During the Haiti earthquake, the pile foundations of a wharf structure in Port au Prince suffered formation of plastic hinges at pile heads as shown in Fig. 16.11. A similar mechanism was proposed *a priori* based on dynamic centrifuge model tests by Knappett and Madabhushi (2009) also shown as an inset in Fig. 16.11.

On the other hand, underground structures like tunnels, pipe lines or fluid storage tanks (when partially empty) are naturally buoyant structures. Soil liquefaction can cause floatation of such structures. During the Tohuku earthquake of Japan, an underground tank has suffered floatation as shown in Fig. 16.12. Chian et al.

Fig. 16.10 Settlement and rotation of a tall structure and a residential structure due to soil liquefaction

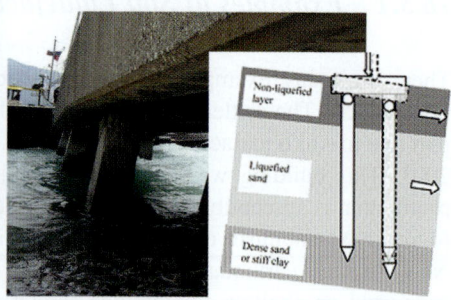

Fig. 16.11 Lateral spreading due to soil liquefaction and damage to pile foundations

Fig. 16.12 Flotation of an underground tank following soil liquefaction

(2014) investigated the effects of floatation of underground structures due to soil liquefaction in detail.

Although all these failures are different, the underlying cause of their predicament is soil liquefaction. The last two decades have seen a great advancement of both the scientific understanding of liquefaction phenomena and of modelling liquefaction using numerical and centrifuge modelling, particularly with the establishment of the George E Brown Network of Earthquake Engineering Simulation (NEES) in the USA and similarly the UK-NEES network. The failure mechanisms of buildings, piles, retaining walls and bridge foundations have been widely investigated. Despite these advancements, there are several aspects of liquefaction that remain unclear. The definition of liquefaction may be considered as a specific example. Soil liquefaction may be defined, using Terazaghi's effective stress principle, as the state of saturated soils when the pore pressure matches the total stress, thereby reducing the effective stress to zero.

$$\sigma'_v = \sigma_v - (u_{hyd} + u_{excess}) \tag{16.8}$$

where u_{hyd} and u_{excess} are hydrostatic and excess pore pressures respectively. The total stress σ_v, is usually considered to be the geostatic vertical stress. This definition is appropriate for level ground with no buildings or other structures. When considering a soil element below a building, the total stress in the soil is affected by the bearing pressure exerted by the building and therefore a higher excess pore pressure may be required to liquefy the soil. This is, however, difficult to determine, as the stress distribution due to the structure changes with the onset of liquefaction. Coelho et al. (2007) show that the stress distribution below a shallow foundation narrows down with liquefaction, forming a column of highly stressed soil underneath the foundation that remains non-liquefied while the free field soil fully liquefies. Similar observations were also made by Ghosh and Madabhushi (2007), who investigated excess pore pressure generation underneath a heavy foundation for a nuclear reactor building. Underneath a building, the vertical effective stress therefore changes with the evolution of excess pore pressures generated by earthquake loading from two viewpoints. Firstly, using Eq. 16.8, the effective stress decreases as excess pore pressures increase. Secondly, the change in stress distribution below the building causes the total and hence effective stresses to change. Thus the definition of liquefaction, given earlier, needs to be updated. It must be understood that the value of effective stress in Eq. 16.8 is not the free field effective stress or the initial effective stress. It must be the effective stress at any given point and at any given time, where the excess pore pressure is known. It must also be pointed out that in this paper the subtle differences between 'initial liquefaction' and 'flow liquefaction' (Kramer 1996) have not been considered.

16.3.2 Theoretical Framework of Liquefaction

Casagrande (1936) proposed the existence of a 'critical void ratio' for sands, based on his load-controlled drained shear box tests. He envisaged that when a natural soil deposit has a void ratio equals to or greater than this 'critical void ratio', it is susceptible to liquefaction failure. Casagrande (1971) described the observation of liquefaction in undrained cyclic loading of saturated sands in triaxial tests as the point at which there is a substantial loss of shear strength when the sand is subjected to continuous shear strains. Further, he described the point at which the pore pressure in the sample equals the cell pressure in a cyclic triaxial test on a dense sand sample as 'cyclic mobility'. Castro (1969) associated liquefaction with a sudden loss of shear strength resulting in a catastrophic failure. In laboratory tests, he observed that a sample of sand subjected to cyclic or monotonic loading exhibited liquefaction failure only if the driving stresses were larger than the undrained shear strength of the sample. Following earthquake loading and the subsequent generation of excess pore pressures in saturated sands, the driving shear stresses below a building can be greater than the undrained shear strength. Castro considered the steady state deformations that occur in the presence of elevated pore pressures following earthquake loading as liquefaction failure. This was thought to be justified, as many dams such as the upper and lower San Fernando dams were known to have failed after the end of earthquake loading, (Dixon and Burke 1973). Dixon and Burke observed that there was a possibility of liquefaction occurring at great depths below these dams contrary to the opinion of Casagrande (1971).

Roscoe et al. (1958) and Schofield and Wroth (1968) established the Critical State soil mechanics framework based on the postulation that a soil element that has reached a Critical State will continuously deform without further changes in stress or volume. This state can be depicted as a single line in q-p'-v space. Schofield (1980, 1981) and later Muhunthan and Schofield (2000) applied the Critical State framework to soil liquefaction. Consider the stress state of a soil element on the loose or 'wet' side of the Critical State. When this soil element is subjected to cyclic shear stresses under undrained conditions, the propensity to suffer volumetric contraction is manifested as an increase in excess pore water pressures. This causes the effective confining stress to reduce, as shown in Fig. 16.13. Eventually, the stress path will cross the tensile rupture or fracture surface resulting in a disaggregation of the continuum into a clastic body with unstressed grains free to slide apart. This results in the massive loss of strength seen during liquefaction.

One of the manifestations of the soil stress path reaching the fracture surface shown in Fig. 16.13 is that the soil permeability increases at these very low effective stresses. Haigh et al. (2012) demonstrated that this is possible based on simple soil column experiments with upward hydraulic gradients causing fluidization of sand layer and resulting in near 'liquefaction' state with very low vertical effective stresses. It must be pointed out that in these experiments there was no earthquake induced shear stress that causes excess pore water pressures. However the loss of effective stress due to upward hydraulic gradients is considered to take the soil's stress path into the same low effective stress regime as that of an earthquake induced soil liquefaction state. In Fig. 16.14 the changes in permeability with effective stresses are presented for three different types of sands. The increased permeability for effective stress values of <0.2 kPa can be clearly seen in this figure. In addition it can be seen in Fig. 16.14 that the finer soils such as Fraction E sand show a larger reduction in the permeability in this low effective stress range compared to relatively coarser grained sands such as Fraction D or Hostun sand. In addition to the changes

Fig. 16.13 Critical State framework for soil liquefaction. (After Muhunthan and Schofield 2000)

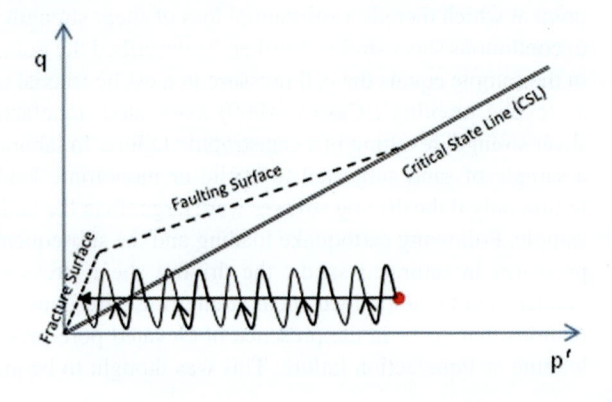

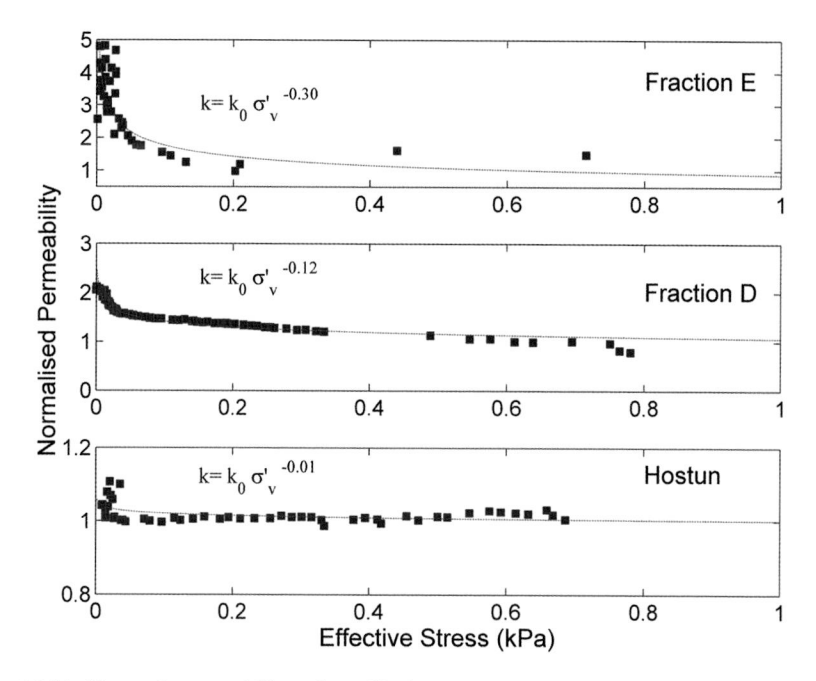

Fig. 16.14 Change in permeability at low effective stresses

in permeability of sands, at very low effective stresses their compressibility also changes. More details on this aspect are discussed by Haigh et al. (2012) and by Adamidis and Madabhushi (2016).

16.4 Dynamic Centrifuge Testing of Soil Liquefaction Problems

16.4.1 Simple, Level Sand Beds

Dynamic centrifuge tests were carried out on loose and dense sand layers by Coelho et al. (2007). These were horizontal, fully saturated sand beds tested at 50 g with prototype dimensions of 33.6 m long and 18.2 m deep. The soil used was uniformly graded Fraction E sand (Leighton Buzzard 100/170). This silica sand was extensively used in many research projects at Cambridge and its properties are well established. While the models were heavily instrumented as reported by Coelho et al., in this paper only three instruments will be considered as shown in Fig. 16.15. These will be the base accelerometer (ACC) that records the input acceleration, a pore pressure transducer (PPT) at a depth of 14.6 m (292 mm at model scale) that

Fig. 16.15 Cross-sectional view of the centrifuge model in an ESB model container

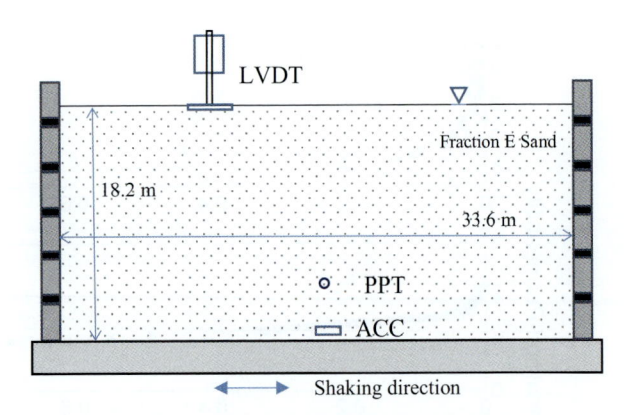

records excess pore pressures and a surface LVDT that measures soil settlement. Again only two tests with relative density of soil model of 50% (loose) and 80% (dense) will be considered here, although more tests were carried out at intermediate relative densities, Coelho et al. (2007) (Fig. 16.15).

In Figs. 16.16 and 16.17, the results from dynamic centrifuge tests on soil deposits with relative densities of 50% and 80% are presented. Both models were subjected to very similar earthquake loading with a peak horizontal acceleration of 5 g with nominally 10 cycles. This peak acceleration of 5 g is equivalent to 0.2 g of peak acceleration applied at the bedrock level at prototype scale.

16.4.1.1 Excess Pore Pressures

In Fig. 16.16, it can be seen that both soil models experience excess pore water pressures of about 140 kPa, equivalent to the total vertical stress at the depth of the instrument. This $\sigma_v = u_{excess}$ line is plotted in these figures as a dashed line to indicate soil liquefaction following the definition given by Eq. 16.1. The main difference in the excess pore pressure traces is that for the case of dense sand shown in Fig. 16.16b, the dilation is stronger, manifested as larger amplitude suction cycles being superposed on the excess pore pressure generated. It may also be noted that during these large suction cycles, the excess pore pressure temporarily exceeds the dashed line suggesting that the excess pore pressures are greater than the total stress for those brief moments. This is only possible if vertical equilibrium is not maintained at those moments, i.e. the soil body has to accelerate vertically upwards. Further it can be seen in Fig. 16.16 that the soil starts to reconsolidate after the end of earthquake as the excess pore pressures slowly start to dissipate. The rates of excess pore pressure dissipation are very similar for both loose and dense sands. Brennan and Madabhushi (2011) showed that the co-efficient of consolidation can be calculated for the liquefied soil in this period.

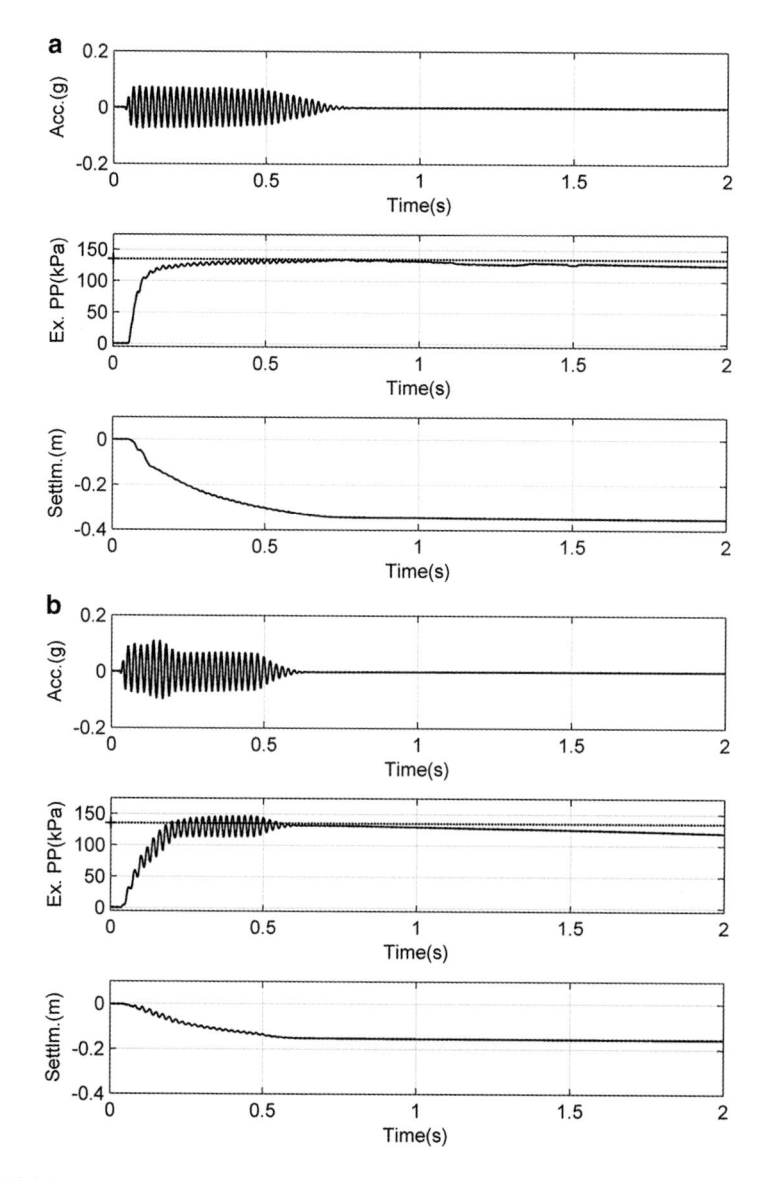

Fig. 16.16 Results from the centrifuge test on a soil model with a RD of 50% & 80% (**a**) RD = 50% (**b**) RD = 80%

16.4.1.2 Soil Settlements

In Fig. 16.16, the settlements suffered by loose and dense sands are presented. It can be seen that the loose sand suffers a total settlement of about 7 m while that for dense sand is less than half of this value being about 3 m. This is to be expected, as the dense sand suffers much smaller volumetric strains compared to loose sands even in the triaxial tests. In Fig. 16.16 it can also be seen that the rate of settlement is steepest in the co-seismic period, reducing to a much smaller value in the post-seismic period. This is true for both loose and dense sands.

This observation is important, as settlement during the co-seismic period is only possible if the liquefied soil is not behaving in an undrained fashion. As these are level sand beds with no driving shear stresses induced by foundations etc., the rapid co-seismic settlements imply that some drainage of pore fluid is occurring to allow for the soil settlements. Thus the hypothesis of liquefaction being a partially drained event based on the soil stress state reaching the fracture line, as discussed in the previous section, is at least a plausible explanation for these rapid rates of settlement. A corollary to this observation is that thorough introspection is needed in using undrained cyclic triaxial tests to investigate the liquefaction behaviour of saturated sands.

Further, if one considers the soil stress state immediately after the end of the earthquake loading in Fig. 16.16, the excess pore pressures in the soil at this stage are still high and closely match the total stress. However, the rate of settlement changes abruptly after the end of the earthquake loading. Applying the definition given in Eq. 16.8, both soils are 'liquefied' at this stage. There must be a change in the behaviour of the soil to cause a change in the settlement rate. This aspect is further considered in developing a micro-mechanical model for soil liquefaction.

16.4.2 Shallow Foundations on Liquefiable Sand Beds

The problem of shallow foundations on liquefiable soil layers has been addressed by many researchers using dynamic centrifuge modelling before e.g. Mitrani and Madabhushi (2011), Marques et al. (2013). More recently Adamidis and Madabhushi (2017a) have investigated the effect of the thickness of liquefiable layers on shallow foundation behaviour (see Fig. 16.17). This is a relevant issue as many practical applications for shallow foundations encounter liquefiable layers where the thickness of these layers can be the order of the width of the foundation.

In Fig. 16.17 the cross-sectional view of two centrifuge models OA4 and OA6 with a shallow foundation supporting a single degree freedom structure is shown.

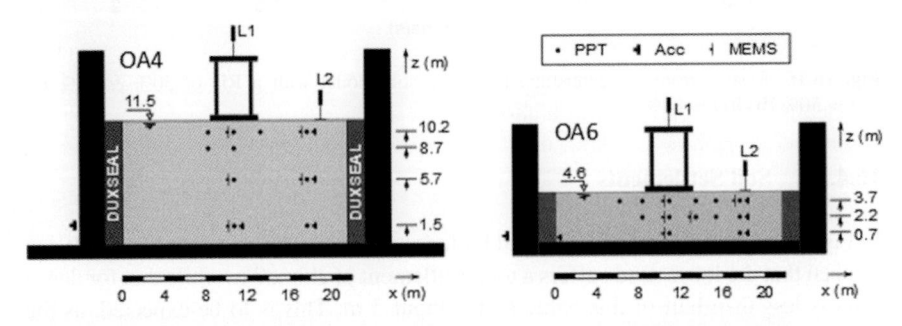

Fig. 16.17 Cross-section of centrifuge models with varying depths of liquefiable layers

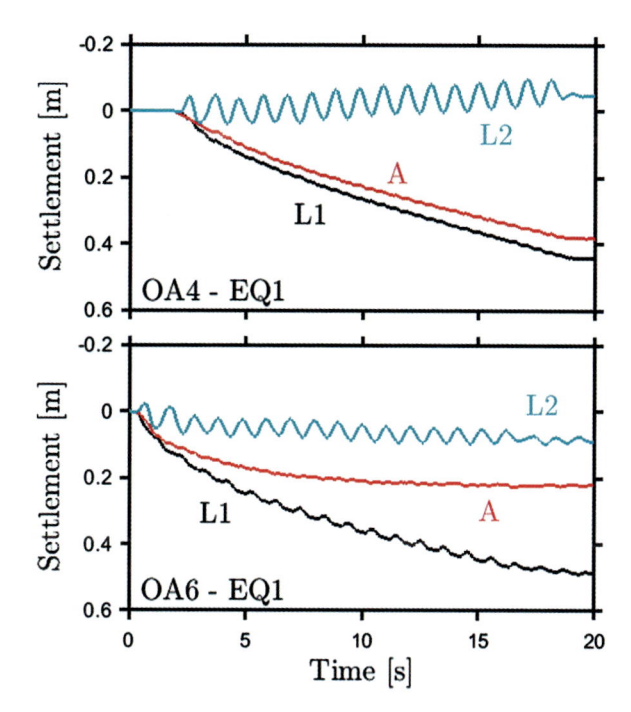

Fig. 16.18 Settlement of foundations on deep and shallow liquefiable layers

The bearing pressure exerted by the foundation on the soil is ~50 kPa. The only difference between these centrifuge models was the thickness of the liquefiable layers. In Fig. 16.18 the settlement of the structure recorded in each centrifuge test is presented during a sinusoidal earthquake with a peak bedrock acceleration of 0.23 g.

The settlement of the structure is given by L1 and that of the free field is given by L2. 'A' gives the settlement of soil next to the foundation (obtained from PIV analysis of images). In both centrifuge tests the foudation settles much more than the free field during the co-seismic period. Surprisingly the magnitude of settlement of structure given by L1 in both tests during this earthquake were quite comparable (about 0.5 m) despite the differences in the thickness of the liquefiable layers. Free-field settlement L2 shows some heave in test OA4 and some settlement in test OA6. This suggests that the failure mechanism is wider in test OA4 and is much more narrower and focused below the foundation in test OA6.

The differences in the deformation mechanisms that drove structural settlement can be achieved by examining the total volumetric and shear strains at the end of each event. Strains for tests OA4 and OA6 are depicted in Fig. 16.19. These were calculated using the displacement fields computed through PIV (White et al. (2003)). In these figures it can be seen that the volumetric strains, in general, are distributed over the entire soil model. It can also be seen that the volumetric strains are larger at the surface and reduce with depth. This is to be expected as the volumetric strains at the surface accumulate as one moves towards the surface. In contrast, the shear

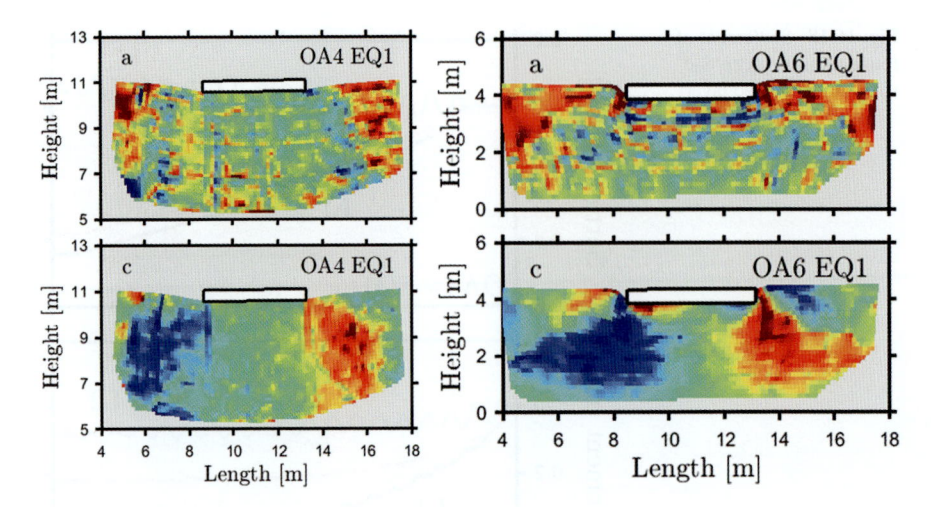

Fig. 16.19 Volumetric and shear strains below foundation on deep liquefiable layer

strains are very sharply focused and emanate from the edges of the shallow foundation, where one would anticipate the largest stress concentration. Further, the shear strains are much larger and focused in test OA6 with a shallow depth of liquefiable layer compared to the test OA4 that had a deeper soil layer. This also confirms the earlier observation that the deformation mechanism is much more focused and narrow in the case of a thin, liquefiable layer. More details of this set of tests can be found in Adamidis and Madabhushi (2017a).

The main observation from these tests was that deformation of the liquefied soil and the shallow foundation are governed by both volumetric strains and shear strains. Presence of volumetric strains suggests that the liquefaction cannot be treated as an 'undrained' event even in the co-seismic period.

16.4.3 Drainage During Liquefaction Events

Recently Adamidis and Madabhushi (2017b) have carried out novel centrifuge tests in which an attempt was made to create 'triaxial' chambers within a centrifuge model. This was done by having a latex membrane isolated zone of saturated soil as shown in Fig. 16.20. Instrumentation such as accelerometers and pore pressure transducers were used both within the triaxial chamber as well as the free-field. It must be pointed out that this triaxial cell represents quite a large sample at prototype scale. The soil was deposited at a relative density of ~40% and was saturated with 50 cS methylcellulose. The volume flow in and out of the triaxial chamber was controlled through a valve system as shown in Fig. 16.20. Valves 1 and 2 are left open during saturation of the model after which they were shut. During centrifuge flight valve 1 was kept shut, but valve 2 was open after the end of the earthquake and

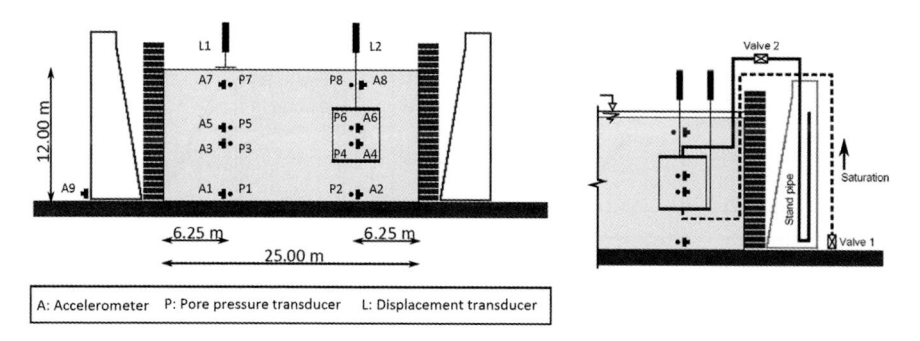

Fig. 16.20 Cross-section of a centrifuge model with enclosed 'triaxial cell'

any fluid outflow was monitored along with the settlement of the triaxial chamber. Two types of centrifuge tests were conducted. In the first case (test OA2) the triaxial cell had simple, latex boundary and therefore was able to expand laterally into the liquefied soil. In the second case (test OA3) the latex boundary was surrounded by very fine steel wire to prevent any lateral expansion of the triaxial cell. However, the triaxial cell is still able to suffer lateral contraction, if the soil behaviour dictated it. More details of the model preparation and testing can be found in Adamidis and Madabhushi (2017b). It must be pointed out that the data for these tests are shown at model scale.

16.4.3.1 Response of the Soil in the Triaxial Chamber & in the Free-Field

The excess pore pressure traces at different locations within the chamber and in the free-field are presented in Fig. 16.21 for the two tests OA2 and OA3 respectively along with the settlement data and the input accelerations applied at the bedrock level. The time scales on x-axis are partitioned to show initial cycles, co-seismic and post-seismic periods. For the centrifuge test OA2 shown in Fig. 16.21 the excess pore pressure build-up in the initial cycles is comparable between free-field and within the triaxial chamber. In the co-seismic period the free-field excess pore pressures are quite different at P3 & P5 locations, while those within the triaxial chamber quickly equalize. This is even clearer in the start of the post-seismic period. This is attributed to the drainage of pore fluid from the base of the model to the soil surface in the free-field which is maintained throughout the co-seismic period and also a few seconds into the post-seismic period. This is explained by using liquefaction and solidification front concepts by Adamidis and Madabhushi (2017a). Similar behaviour is also observed in test OA3 where the chamber is not allowed to bulge out into the liquefied soil. This slightly delayed the excess pore pressure equilibrating within the chamber as seen as a slow drop in excess pore pressure at P4 in Fig. 16.21.

In Fig. 16.21 the settlement of the triaxial chamber is also compared to the free-field settlements. In case of test OA2 we can see that the chamber settles a lot less

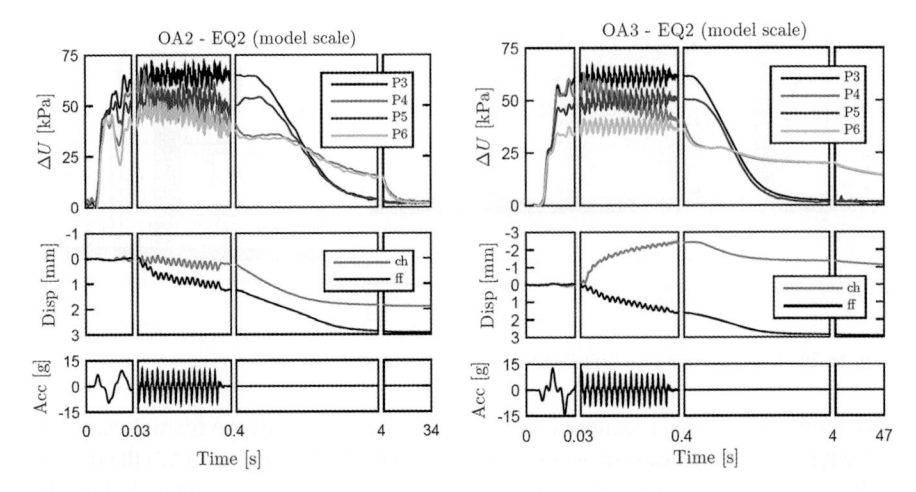

Fig. 16.21 Response in the free-field and in the triaxial chamber in test OA2 & OA3

than the free-field even in the co-seismic period. This again emphasizes the importance of drainage during earthquake loading, which is prevented artificially in a triaxial chamber. For test OA3, which is constrained from expanding laterally outwards, the free-field settles as before but the triaxial chamber actually heaves up. This is due to the radial compression of the chamber by the liquefied soil outside the chamber causing a rise in the top cap due to 'constant volume' condition that was imposed artificially.

16.4.3.2 Stress-Strain Behaviour of Liquefied Soil in the Chamber & in the Free-Field

The stress-strain behaviour of liquefying sands can also be obtained in these centrifuge tests from the acceleration-time histories measured in the free-field and in the chamber. The shear stress and shear strain plots for tests OA2 and OA3 are shown in Fig. 16.22. In Fig. 16.22c the τ–σ'_v plot is shown for the whole earthquake loading period from starting circles to finishing squares, for both free-field and in the chamber. Similarly Fig. 16.22d and e show the τ–γ plot for the initial cycles (left) and post liquefaction (right). Equivalent plots for test OA3 with constrained lateral boundary for the triaxial chamber are also shown in Fig. 16.22. In this test the drop in shear strains in the free-field relative to the chamber (see Fig. 16.22) is even more significant clearly suggesting the drop in shear stiffness in the free field is far more than in the chamber. Overall the stress paths in Fig. 16.22 for OA2 and OA3 tests are very different as are the stress-strain plots.

These comparisons shows quite clearly that by creating and using triaxial samples to study liquefaction, different behaviour can be elicited by imposing different drainage boundary conditions and by imposing radial constraints against lateral

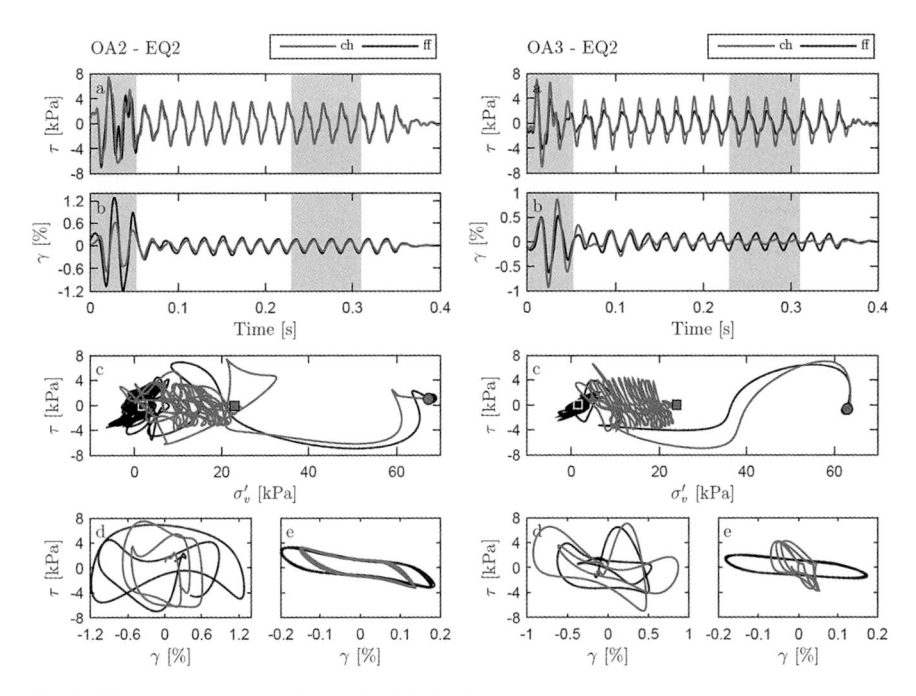

Fig. 16.22 Stress-strain plots in test OA2 & OA3

expansion of the triaxial chamber. The soil in the free field seems to suffer a greater degradation in its stiffness compared to the soil enclosed in the triaxial chamber.

16.4.4 Novel Liquefaction Mitigation Methods

Liquefaction mitigation methods of various kinds have been investigated previously in Cambridge. The efficacy of drains in relieving the excess pore pressures was investigated by Brennan and Madabhushi (2005). Similarly use of impermeable barriers or solidification of liquefiable sands using cementation was investigated by Mitrani and Madabhushi (2013). In a very recent study at Cambridge, liquefaction mitigation was attempted by partially saturating a sand bed, Zeybek and Madabhushi (2016, 2017). This was achieved by using in-flight air injection at the base of the soil model as shown in Fig. 16.23. A shallow foundation was placed at the soil surface that applied a bearing pressure of ~ 50 kPa. The soil was saturated using 50 cS methylcellulose as usual as the centrifuge testing was carried out at 50 g's. A bench mark test FS-1 was first conducted with no air injection. This was followed by two other tests PS-1 and PS-2 in which high pressure air was injected into the soil to cause partial saturation. This was carried out over a period of about 180 seconds. Different types of injection devices were used in tests PS-1 and PS-2. Using digital imaging obtained for PIV analysis, it was possible to perform further digital image

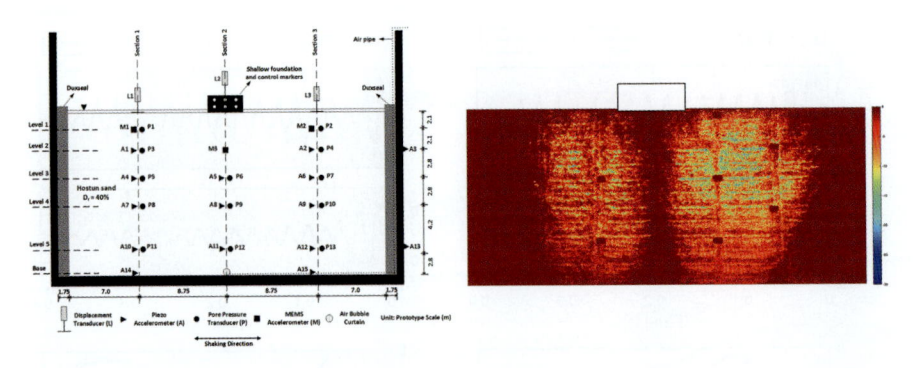

Fig. 16.23 Cross-section of the centrifuge model with air injection & image analysis showing air-injected region below the foundation

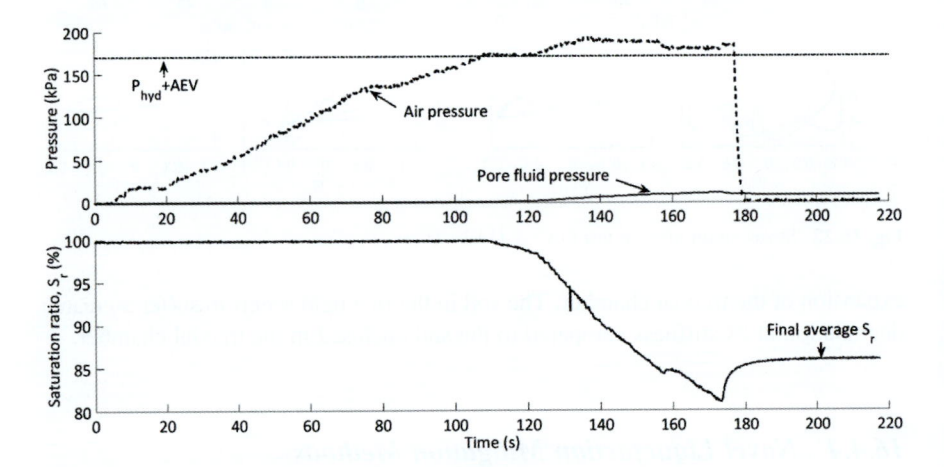

Fig. 16.24 Changes in degree of saturation during air injection

analysis to observe the areas of the foundation soil affected by air-injection. In Fig. 16.23, an example showing the region into which air-injection took place is shown (lighter yellow indicates more air presence.

In this testing program, the first step was the air injection prior to application of any earthquake loading. The centrifugal acceleration was increased to reach 50 g's. At this stage high pressure air was injected at the base of the soil model. Details of the air injection system are described by Zeybek and Madabhushi (2016). Also the results from these tests are shown at prototype scale. In Fig. 16.24 the increase in air pressure is plotted along with the decrease in the degree of saturation. In this figure it can be seen that the degree of saturation drops from 100% to about 81% during air injection, before recovering slightly to a value of 85%.

Soon after the air injection process was completed earthquake loading was applied. In Fig. 16.25 the response recorded by the pore pressure transducers is plotted both below the structure and in the free-field. In this figure it can be seen that while the bench mark test FS-1 shows full levels of liquefaction both PS-1 and PS-2 show much lower levels of excess pore pressure generation both at shallow level and deep level. The settlement suffered by the shallow foundation in each test are also plotted in Fig. 16.25 (note +displacement is taken as settlement in this plot). The settlements observed in this 0.2 g earthquake are seen to be much smaller in the air injected models PS-1 and PS-2 (about 40 mm settlement) compared to the fully saturated case of FS-1 (about 750 mm settlement). Also the air injection device in PS-2 worked better than in the case of PS-1.

It can therefore be concluded that the air-injection and subsequent drop of degree of saturation by about 15%, has successfully reduced the settlement suffered by the shallow foundation. In these tests high speed imaging was carried out during earthquake loading and resulting images were analysed using the geo-PIV software. This produces the displacement vectors below the shallow foundation as shown in Fig. 16.26. In this figure a direct comparison of the deformation suffered in each of the centrifuge tests are presented. It can be seen in this figure that the benchmark case

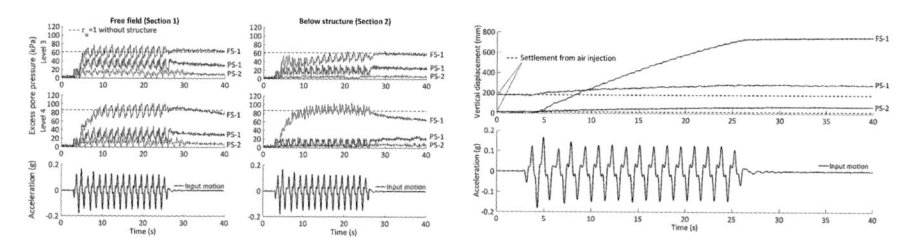

Fig. 16.25 Excess pore pressures & settlements in bench mark test (FS-1) and the air-injection tests (PS-1 and PS-2)

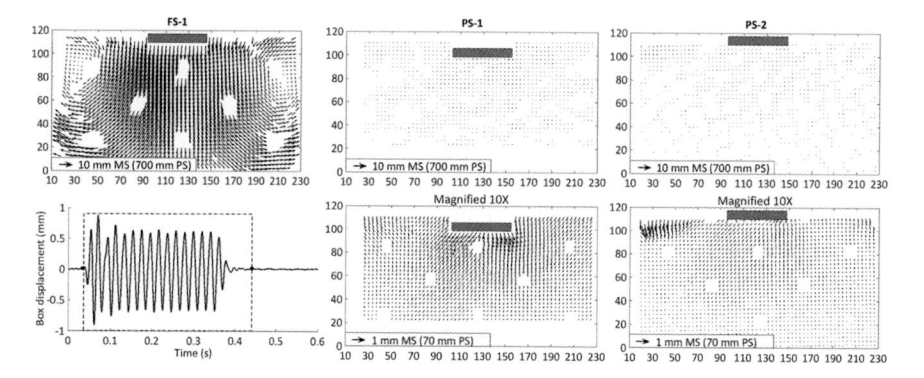

Fig. 16.26 Soil deformations below the foundation in the bench mark test (FS-1) and the air-injection tests (PS-1 & PS-2). Note the deformations in second row are magnified by ×10

of FS-1 shows enormous deformations as the foundation settles by 750 mm following soil liquefaction. In comparison for the centrifuge tests PS-1 and PS-2 the deformations are barely noticeable. In fact in the second row of this figure the deformations are magnified by a factor of $\times 10$ and replotted to reveal the deformations. In these second row figures the deformations become more visible. For the case of PS-1, one can decipher a bearing capacity failure type mechanism evolving. For the case of PS-2 the deformations are still small and this figure is dominated by the free-field settlements on the left hand corner. It must be pointed that the PIV field shown here is much smaller than the actual soil sample shown in Fig. 16.26. This is due to the limited visual field of the high speed camera.

16.5 Conclusions

Liquefaction of soil following earthquake events continue to have serious consequences to civil engineering infrastructure. Research into soil liquefaction and development of theoretical frameworks is predominantly driven by observations from cyclic triaxial testing while dynamic centrifuge tests continue to provide new and more detailed information on liquefaction phenomena. In this paper the basic assumption that liquefaction events are largely undrained is questioned. Earlier work at Cambridge has shown that the permeability and compressibility of sands under low effective stresses can increase significantly. This leads to drainage of liquefied sands during the earthquake loading events. Dynamic centrifuge test data from tests on level beds of sand confirm that the excess pore pressure generation is similar for loose and dense sands but the settlements are much lower for dense sands. This can be attributed to the fact that loose sands when liquefied are much more compressible than dense sands. Also the majority of settlements occur during the co-seismic period confirming increased permeability and a consequent drainage occurring even in the co-seismic period. Recent centrifuge testing conducted at Cambridge on shallow foundations on liquefiable layers of different thicknesses show that both volumetric and shear strains occur below shallow foundations once the ground has liquefied.

Novel experiments were also conducted on level beds of sand in which a 'triaxial chamber' was created within the centrifuge model. It was shown that the behaviour of liquefied sand within the chamber is very different to that in the free-field. This illustrates the importance of drainage in the co-seismic period. Further the centrifuge data was also used to show the post-liquefaction reconsolidation process and the changes in permeability and compressibility of sands during this period. Finally, the dynamic centrifuge testing was used to investigate the efficacy of air injection to reduce the liquefaction potential of soils. The data from this series of tests show that injection of air reduces both the excess pore pressure generation and consequently the settlement of shallow foundations on such partially saturated sands. The deformation mechanisms obtained through PIV analysis also confirm the reduced settlement of such foundations.

References

Adamidis O, Madabhushi SPG (2015) Use of viscous pore fluids in dynamic centrifuge modeling. Int J Phys Modell Geotech 15(3):141–149

Adamidis O, Madabhushi GSP (2016) Post-liquefaction reconsolidation of sand. Proc R Soc A Math Phys Eng Sci 472:20150745. https://doi.org/10.1098/rspa.2015.0745

Adamidis O, Madabhushi GSP (2017a) Deformation mechanisms under shallow foundations resting on liquefiable layers of varying thickness, Geotechnique, Thomas Telford, https://doi.org/10.1680/jgeot.17.P.067

Adamidis O, Madabhushi GSP (2017b) Drainage during earthquake-induced liquefaction, Geotechnique, Thomas Telford, https://doi.org/10.1680/jgeot.16.p.090

Brennan AJ, Madabhushi SPG (2005) Liquefaction and drainage in stratified soil. ASCE J Geotech Geo Environ Eng 131(7):876–885

Brennan AJ, Madabhushi SPG (2011) Measurement of coefficient of consolidation during reconsolidation of liquefied sand. ASTM Geotech Test J 34(2):64–72

Brennan AJ, Madabhushi SPG, Houghton NE (2006) Comparing laminar and ESB Containers for Dynamic Centrifuge Modelling. Proceedings of Centrifuge'06, International Conference on physical modelling, Hong Kong

Casagrande A (1936) Characteristics of cohesionless soils affecting the stability of slopes and earth fills. J Boston Soc Civil Eng. pp 257–294

Casagrande A (1971) On liquefaction phenomena. Reprinted in Geotechnique 21(3)

Castro G (1969) Liquefaction of sands. PhD Thesis, Harvard University. Reprinted as Hazard Soil Mechanics Series (8):1–112

Chian S, Tokimatsu K, Madabhushi S (2014) Soil liquefaction–induced uplift of underground structures: physical and numerical modeling. ASCE J Geotech Geoenviron Eng 140(10):04014057

Cilingir U, Madabhushi SPG (2011) A model study on the effects of input motion on the seismic behaviour of tunnels. J Soil Dyn Earthq Eng 31:452–462

Coelho PALF, Haigh SK, Madabhushi SPG, O'Brien AS (2007) Post-earthquake behaviour of footings when using densification as a liquefaction resistance measure. Ground Improv J 11(1):45–53

Dixon SJ, Burke JW (1973) Liquefaction case history. ASCE J Soil Mech Found Eng SM10:823-840

Ghosh B, Madabhushi SPG (2007) Centrifuge modelling of seismic soil-structure interaction effects. J Nucl Eng Des 237:887–896

Haigh SK, Eadington J, Madabhushi SPG (2012) Permeability and stiffness of sands at very low effective stresses. Geotechnique 62(1):69–75

Knappett JA, Madabhushi SPG (2009) Seismic bearing capacity of piles in liquefiable soils. Soils Found 49(4):525–536

Kramer SL (1996) Geotechnical earthquake engineering. Prentice Hall Inc., Upper Saddle River. ISBN 013374943-6

Madabhushi SPG (2014) Centrifuge modelling for civil engineers. CRC Press, London. 9780415668248

Madabhushi SPG, Haigh SK (2009) Effect of superstructure stiffness on liquefaction-induced failure mechanisms. Int J Geotech Earthq Eng 1(1):72–88

Madabhushi SPG, Schofield AN, Lesley S (1998) A new stored angular momentum (SAM) based earthquake actuator. Proceedings of Cenrifuge'98, International conference on Centrifuge Modelling, Tokyo, Japan

Madabhushi SPG, Patel D, Haigh SK (2005) Geotechnical aspects of the Bhuj earthquake, Chapter 3, EEFIT Report, Institution of Structural Engineers, London. ISBN 0901297 372

Madabhushi SPG, Haigh SK, Houghton NE, Gould E (2012) Development of a servo-hydraulic earthquake actuator for the Cambridge turner beam centrifuge. Int J Phys Modell Geotech 12(2):77–88

Marques ASP, Coelho PALF, Haigh SK, Madabhushi SPG (2013) Centrifuge modelling of liquefaction effects on shallow foundations. Special edition on seismic evaluation and rehabilitation of structures. In: Ilki A, Fardis MN (eds) Geotechnical, geological and earthquake engineering, vol 26. Springer, Dordrecht, pp 425–441

Mitrani H, Madabhushi SPG (2011) Rigid containment walls for liquefaction remediation. J Earthq Tsunami 6(3). https://doi.org/10.1142/S1793431112001309

Mitrani H, Madabhushi SPG (2013) Geomembrane containment walls for liquefaction remediation. Proc ICE Ground Improv J 166(GI1):9–20

Muhunthan B, Schofield AN (2000) Liquefaction and dam failures. Proceedings of *GeoDenver*, Denver, Colorado, USA

Roscoe KH, Schofield AN, Wroth CP (1958) On the yielding of soils. Geotechnique 8:22–53

Schofield AN (1980) Cambridge geotechnical centrifuge operations. Geotechnique 25(4):743–761

Schofield AN (1981) Dynamic and earthquake geotechnical centrifuge modelling. Proceedings International conference on recent advances in geotechnical earthquake engineering and soil dynamics, St Louis, vol III, pp 1081–1100

Schofield AN, Wroth CP (1968) Critical state soil mechanics. McGraw-Hill, London

White DJ, Take WA, Bolton MD (2003) Soil deformation measurement using particle image velocimetry (PIV) and photogrammetry. Géotechnique 53(7):619–663

Zeng X, Schofield AN (1996) Design and performance of an equivalent-shear-beam container for earthquake centrifuge modelling. Géotechnique 46(1):83–102

Zeybek A, Madabhushi SPG (2016) Effect of bearing pressure and degree of saturation on the seismic liquefaction behaviour of air-induced partially saturated air-sparged soils below shallow foundations. Bull Earthq Eng 15:339. https://doi.org/10.1007/s10518-016-9968-6

Zeybek A, Madabhushi SPG (2017) Influence of air injection on the liquefaction-induced deformation mechanisms beneath shallow foundations. J Soil Dyn Earthq Eng 97:266–276

Chapter 17
Seismic Analysis and Design of Composite Steel/Concrete Building Structures Involving Concrete-Filled Steel Tubular Columns

Konstantinos A. Skalomenos, George D. Hatzigeorgiou, and Dimitri E. Beskos

Abstract Composite construction in steel and concrete offers significant advantages over the conventional one based exclusively on either steel or concrete. This paper provides a comprehensive overview of the state of research in analysis and design of composite steel/concrete building structures involving concrete-filled steel tubular (CFT) columns and steel beams. Experimental and analytical/numerical research on the seismic behavior and simulation of CFT columns and composite framed structures under strong ground motions are all considered with emphasis on recent works of the authors. The paper also discusses seismic analysis/assessment methodologies and performance-based seismic design (PBSD) methods that enable engineers to produce composite structures with deformation and damage control.

17.1 Introduction

Composite construction in steel and concrete of building structures subjected to seismic loading offers significant advantages over construction either in steel or concrete separately (Shanmugan and Lakshmi 2001; Deierlein 2000; Leon et al. 2011; Zhao et al. 2010; Triantafillou 2013). Structural steel has high strength, ductility and reduces erection time. Reinforced concrete provides high rigidity and

---

K. A. Skalomenos
Disaster Prevention Research Institute, Kyoto University, Kyoto, Japan

G. D. Hatzigeorgiou
School of Science and Technology, Hellenic Open University, Patras, Greece

D. E. Beskos (✉)
Institute of Structural Engineering and Disaster Reduction, Tongji University, Shanghai, People's Republic of China

Department of Civil Engineering, University of Patras, Patras, Greece
e-mail: d.e.beskos@upatras.gr

© Springer International Publishing AG, part of Springer Nature 2018
K. Pitilakis (ed.), *Recent Advances in Earthquake Engineering in Europe*,
Geotechnical, Geological and Earthquake Engineering 46,
https://doi.org/10.1007/978-3-319-75741-4_17

compressive strength and is economical, fire resistant and durable. Composite members combining steel and concrete enjoy the advantages of both materials. Composite construction usually appears in a wide variety of applications, including composite bridges, composite slabs, shear connectors, composite columns, innovative composite structural systems and fire and seismic resistant composite systems (Tomii et al. 1977; Sheikh et al. 1989; Zhou and Zhu 1997; Tawil and Deierlein 2001; Green et al. 2004). Extensive experimental, computational and analytical research on these topics has been conducted worldwide over the last 30 years or so.

The present paper is concerned with composite steel/concrete columns and building frames consisting of composite columns and steel beams. There are basically two kinds of composite columns: concrete encased composite columns and steel tubes filled with concrete (Fig. 17.1). In the case of concrete-encased composite columns, one or more standard shape steel sections are encased in concrete and the concrete encasement stiffens the steel member, thereby restraining local and overall buckling of the steel member, while fireproofing it at the same time. This type of members requires formwork and additional transverse and longitudinal reinforcement to provide confinement to the concrete encasement. In the case of concrete filled tubes (CFT) (Zhao et al. 2010), the steel tube replaces the formwork during construction and confines the concrete infill during service, while the concrete infill restrains the local buckling of the steel tube and reduces the construction costs as well as the required amount of transverse and longitudinal reinforcement. Moreover, the tube improves effectively the stiffness, strength, ductility and energy absorption capacity of the column and increases the stability of the column as a system. In addition, the concrete infill enhances the fire resistance of the CFT column by acting as a heat sink (Shakir-Khalil 1988). For these reasons, structural systems with CFT components have been the subject of extensive research over the last decades for understanding their behavior under seismic loading conditions and have been applied to bridges and to low-rise as well as high-rise or super-high-rise structures, such as industrial buildings, structural frames and supports, electricity transmitting poles, decks of railways and spatial construction.

The present paper provides a comprehensive overview of the state of research in CFT columns including experimental and analytical studies with emphasis on recent works of the authors. More specifically, the present paper covers: (a) experimental

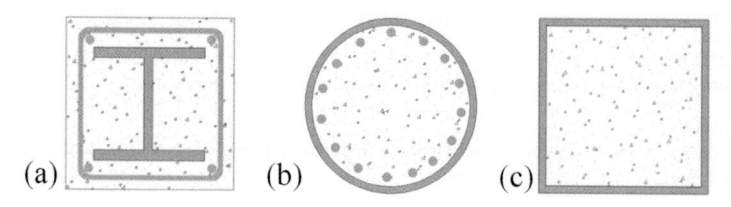

Fig. 17.1 Configuration of composite columns: (**a**) Structural steel shape fully encased in reinforced concrete; (**b**) Concrete-filled tube with reinforcement bars; and (**c**) Concrete-filled tube

research on CFT columns using conventional and high-performance steels; (b) analytical/numerical methods for simulating the cyclic behavior of CFT columns and model development methodology; (c) modelling, seismic analysis and design of composite concrete-filled steel tube/moment resisting frame (CFT-MRFs) structures; and (d) new performance-based seismic design methods and seismic damage assessment methodologies for CFT-MRFs.

17.2 Concrete-Filled Steel Tube Columns

17.2.1 *Experimental Research on CFT Columns*

The study of Furlong (1967) was one of the first studies on CFT columns carried out in the U.S., providing experimental data on CFT beam-column specimens and numerical estimations of their ultimate capacities by applying a constant concentric axial load and increasing the moment to failure. Their study showed that the increased capacity of CFT columns is based on the delay of local buckling of the steel tube arising from the concrete infill and not from the confinement of the concrete. A few years later, an extensive experimental investigation into CFT columns was carried out by Tomii et al. (1977) in Japan, which included over 270 tests on CFT stub column specimens with various cross-sections, width-to-thickness ratios, axial load levels and material strengths. Their test results and subsequent studies on cyclic behavior of CFT columns (Sugano et al. 1992; Gourley and Hajjar 1993; Aho and Leon 1997; Elremaily and Azizinamini 2002; Varma et al. 2002; Han and Yang 2005; Zhang et al. 2009; Nie et al. 2012; Perea et al. 2014; Skalomenos et al. 2016) demonstrated the significant advantages of CFT columns over steel or reinforced concrete conventional columns. Based on the existing experimental database in this research area, current design codes have included relevant provisions for the design of CFT columns, such those of AIJ (1997), EC4 (2004), ACI (2011) and AISC (2010). These code provisions specify width-to-thickness ratio limits and procedures for estimating the elastic stiffness and the axial load and moment capacities of CFTs.

In recent years, considerable effort has been put into improving the seismic behavior of CFT columns, such as increasing the compressive strength of the concrete (Portolés et al. 2011) or inserting an inner steel tube into the concrete core (Han et al. 2006). Both of these strategies increase the stiffness and strength capacities of CFT columns. However, a CFT column would rapidly lose its stiffness and strength capacity after its steel tube begins to fail, e.g., by local buckling, followed rapidly by fracture through the cross-section. The use of high-performance steel, such as high-strength steels, in a CFT column, can further improve its seismic behavior. Compared with conventional steel, high-strength steel provides a larger elastic deformation capacity for the column and local buckling may be delayed until the occurrence of a large rotation.

17.2.1.1 Use of High-Strength Steel in CFT Columns

Few experimental studies have been carried out on CFT columns made of high-strength steel subjected to cyclic loading. The most recent are those of Varma et al. (2002), who examined the behavior of square CFTs made of 317 and 552 MPa steel tube filled with 110 MPa concrete and Inai et al. (2004), who examined the behavior of circular and square CFT beam columns with 400, 590 and 780 MPa steel tubes and 40–90 MPa concrete. These studies resulted in different conclusions regarding the influence of high-strength steel. Varma et al. (2002) did not observe a significant influence of the yield stress of steel on curvature ductility of CFT columns, while Inai et al. (2004) concluded that the ductility of CFTs increases as the steel strength increases.

In order to improve further the seismic performance of CFT columns, Skalomenos et al. (2016) recently conducted an experimental investigation into CFT columns made of the Japanese ultra-high strength (HS) steel, namely the H-SA700. This HS steel has a specified yield stress and tensile strength range from 700 to 900 MPa and 780 to 1000 MPa, respectively. HS steel offers approximately two times higher yield stress than that of conventional steel and therefore, a higher elastic region. However, HS steel results in an increase of the yield to ultimate tensile strength ratio and a reduction in rupture elongation.

Seven square and circular specimens made of HS and conventional steel were subjected to constant compressive axial load and cyclic flexural load protocols. The lateral loading history consisted of several story drifts, namely 0.25, 0.5, 1, 2, 3, 4, 6, 8 and 10%, with two cycles imposed at each drift level. The high value for the last drift level was assumed to consider the case of a maximum earthquake and to evaluate the low ductility capacity of HS steel. The cross section of the specimens was 150 × 150 × 6. The applied axial loads were equal to 0.25 of the axial yield strength capacity of the column.

17.2.1.2 Behavior of CFT Columns Made of Ultra-High Strength Steel

Response curves of the normalized moment (by ultimate moment M_u) versus story drift and buckling failure mode in 10% story drift of two representative square CFT specimens are shown in Fig. 17.2. The symbol Δ denotes occurrence of local buckling in the steel tube. Both specimens exhibited cyclic strength deterioration after local buckling occurred (Fig. 17.2). The story drifts at first yielding of steel tubes and at maximum flexural strength were, respectively, 1.0 and 3.0% for conventional CFT and 1.9 and 6.0% for HS CFT. The maximum flexural strengths achieved by these specimens were 98 and 173 kNm, respectively. Local buckling occurred during the second cycle at 4.0% story drift in conventional CFT and during the second cycle at 6.0% story drift in HS CFT. Their residual strengths were 0.86 and 0.76 times their maximum strength, while their total axial shortenings (δ_{max})

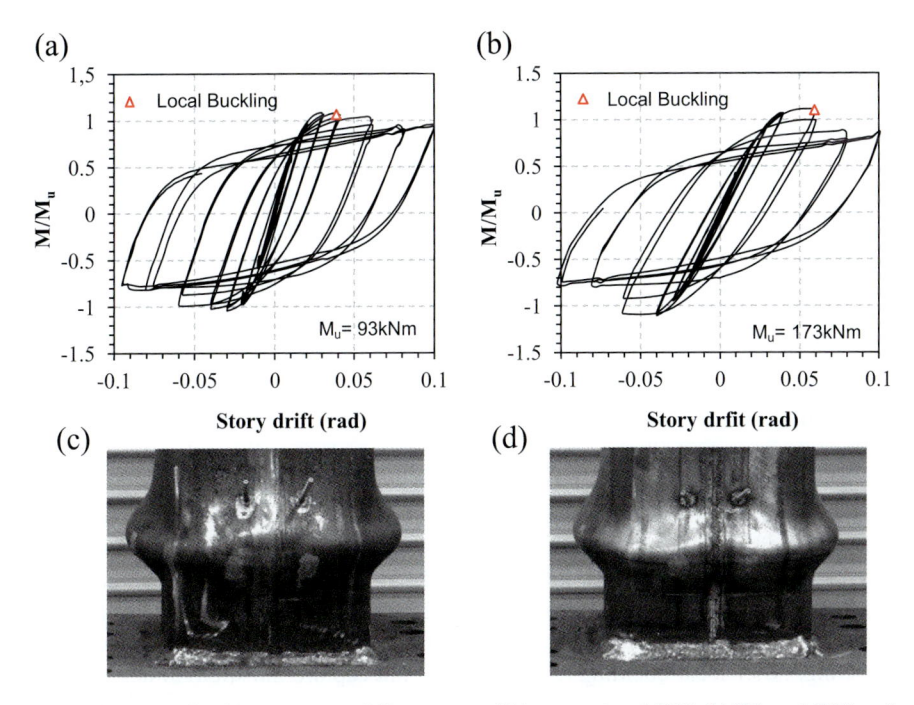

Fig. 17.2 Normalized moment-story drift response of (**a**) conventional CFT; (**b**) HS steel CFT; and buckling failure mode of (**c**) conventional CFT and (**d**) HS CFT (Skalomenos et al. 2016)

were 12.15 and 8.94 mm, respectively. In these two specimens, although significant local buckling occurred, cracks or fracture of steel tubes were not observed.

Based on the test results, the HS steel provided a larger elastic deformation capacity to the CFTs, keeping the steel tube elastic up to 1.5 times the larger story drift (nearly at 2.0%) than the conventional steel. Furthermore, the HS CFT did not suffer strength deterioration until a 6.0% story drift due to local buckling delay. For this reason, the amount of axial shortening in the HS steel specimen was lower. Taking also into account the fact that according to the current seismic design provisions, the limits for collapse prevention (CP) in the preliminary design stage are drift ratios close to 4 and 5%, the use of HS steel in CFT columns appears to be an option, considering that up to a 10% story drift no fracture occurred. The results demonstrated that the inherent low ductility of HS steel did not reduce the deformation capacity of HS CFTs until very high story drifts.

17.2.2 Numerical Simulation of CFT Columns

Many analytical models have been presented so far to predict the cyclic behavior of CFT columns, which have been mostly based on beam theory in conjunction with concentrated or distributed plasticity modelling, e.g. (Hajjar and Gourley 1997; Varma et al. 2002; Inai et al. 2004; Skalomenos et al. 2014a; Serras et al. 2017). The distributed plasticity model also includes the fiber or layered models. Hajjar et al. (1998) and Aval et al. (2002) developed a fiber-based distributed plasticity element formulation for CFT beam-column members, which accounts for the effects of geometric and material nonlinearities, including slip between the steel and concrete surfaces (ranging from perfect bond to immediate slip), steel yielding, concrete strength and stiffness degradation and the effect of confinement and cyclic loading on the concrete core. Furthermore, Tort and Hajjar (2010) presented a beam finite-element formulation to analyze composite square CFT beam-column members with interlayer slip subjected to monotonic, cyclic and transient dynamic loads. This was derived by the mixed finite-element principles within a distributed plasticity approach.

In a study on partially concrete-filled steel tube columns, Susantha et al. (2002) proposed a stress-strain curve for the infill concrete based on a model developed by Mander et al. (1988), with a modified post-peak descending branch and a uniaxial cyclic rule using a focal point for the unloading. The steel was modeled using a modified uniaxial two-surface plasticity model (Shen et al. 1995), which captures hardening, but not local buckling of the steel tube. The model was compared against several experimental results, showing good agreement up to the onset of local buckling. After local buckling, the model did not capture the strength degradation properly.

Varma et al. (2002) developed a fiber-based model for modeling the plastic hinge region of a CFT member in combined bending and axial load, where finite element-based effective stress-strain curves were used for the concrete and steel fibers. Simple finite element models based on beam theory cannot directly take into account the cyclic load buckling behavior of a steel tube, the interface action between steel tube and in-filled concrete and the confinement of concrete infill. These features can be taken into account in a direct manner only by using a refined finite-element model. However, refined models, such as the fiber or layered models increase greatly the computational time of the seismic analysis of a structure even if its inelastic behavior is limited at the plastic hinge region only.

Han et al. (2003) and Inai et al. (2004), by using their experimental results on circular and rectangular CFT columns under constant axial load and increasing flexural loading suggested simplified models for the moment versus curvature response and the lateral load versus lateral displacement relationship. Simplified relationships between force and deformation are accurate enough to capture the overall behavior with respect to the design quantities, but they are inadequate on estimating the deteriorating behavior of a CFT column caused by its failure modes, e.g., cracking of concrete or local buckling of steel tubes.

The recent studies of Skalomenos et al. (2013, 2014a) and Serras et al. (2016) developed concentrated plasticity models for square and circular CFT columns, respectively, for a wide variety of steel and concrete grades taking into account in a direct manner strength and stiffness degradation.

17.2.2.1 Simplified Hysteretic Nonlinear Models for CFTs

A missing aspect in the research field of the CFTs is the unavailability of simplified hysteretic models for simulating the inelastic response of CFTs including strength and stiffness degradation. Recently, Skalomenos et al. (2014a) introduced a methodology based on efficient plastic hinge models. Three hysteretic models for square CFT columns which consider all the aspects of their seismic behavior were developed. The proposed hysteretic models are simple, yet accurate and take into account strength and stiffness deterioration. For a given geometry (width and thickness) and material strengths (steel and concrete) of a square CFT, one can estimate the nonlinear cyclic response of the column in a direct manner without having to utilize more complicated distributed plasticity or fiber type of models. The proposed models can easily be implemented in nonlinear dynamic analysis programs (RUAUMOKO, OpenSees etc.). The computational time of the nonlinear analysis is greatly reduced due to the plastic hinge modeling. Currently, the proposed methodology was extended to the construction of hysteretic models to simulate circular CFTs (Serras et al. 2016). Circular CFTs made of conventional steels seem to exhibit more stable cyclic behavior than the corresponding square CFTs due to their higher confinement of the concrete core and delay of local buckling of the circular steel tube (Skalomenos et al. 2016).

17.2.2.2 Methodology for Model Development

Existing simple hysteretic models that include strength and stiffness degradation are chosen at the beginning. Then, calibration of the hysteretic model parameters is made on the basis of an extensive response databank obtained by determining the cyclic response of a large number of square and circular CFT columns using a refined finite element (FE) analysis (ATENA 2012). Thus, advanced three-dimensional (3-D) FE models are established for analyzing the cyclic behavior of CFTs that involve both steel and concrete hysteretic behavior under a cyclic load-protocol of variable flexural intensity (ATC 1992) and constant axial loads. The accuracy of those advanced FE models is first confirmed by comparing their results with those of experiments. Then, a response databank in terms of various geometrical and strength parameters is produced and empirical expressions providing in functional form all the required parameters of the considered hysteretic models are derived with the aid of regression analysis. Using the calibrated hysteretic models, comparisons with experimental and numerical results are made for further adjustments. These modified hysteretic models are: (a) the Bouc-Wen model, (b) the Ramberg-Osgood model and (c) the Al-Bermani model and take into account in a direct manner all the important quantities which affect the degrading cyclic behavior of CFT columns such as, the nonlinear kinematic hardening and the cyclic local buckling of steel tube, the nonlinear behavior of confined concrete in cyclic loading and the interface action between steel tube and in-filled concrete. One can directly use these models for the simulation of CFT columns alone or as members of composite MRFs frames to determine their response to cyclic loading.

17.2.2.3 Validation of the Proposed Hysteretic Models

The accuracy of the proposed calibrated hysteretic models of Bouc-Wen, Ramberg-Osgood and Al-Bermani is confirmed by comparing their results with those of experiments on square and circular CFT columns under cyclic load protocols with variable intensity and several levels of constant axial load. Figure 17.3 compares the behavior of the calibrated proposed models with that of CFT column experiments (Varma et al. 2002; Inai et al. 2004). The comparisons demonstrate the accuracy of the proposed hysteretic models to simulate the behavior of CFT columns. The stiffness, the ultimate strength and the phenomena of strength and stiffness deterioration are simulated effectively.

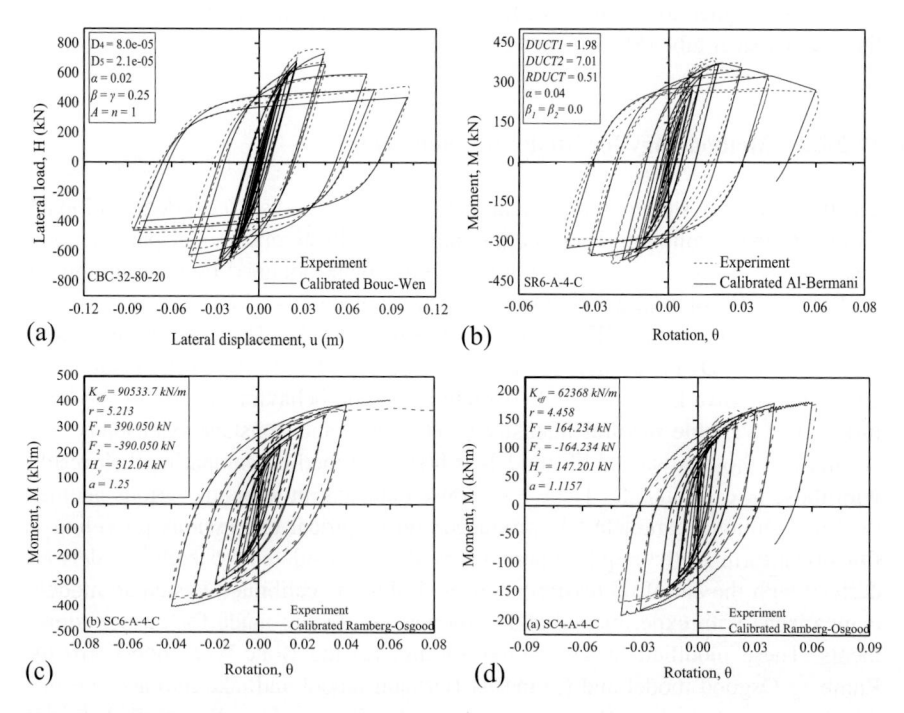

Fig. 17.3 Proposed hysteretic models (Bouc-Wen, Ramberg-Osgood and Al-Bermani models) against experimental data: (**a**) square; (**b**) square; (**c**) circular; and (**d**) circular (Skalomenos et al. 2014a; Serras et al. 2017)

17.3 Concrete-Filled Steel Tube/Moment Resisting Frames

Composite moment resisting frames (MRFs) having concrete-filled steel tube (CFT) columns and steel girders (CFT-MRFs) are modern structures that offer significant advantages in building construction as primary resistance systems, such as large strength and stiffness to control story drifts and great dissipation energy capacity. Those systems have been popular for constructing mid-rise and high-rise buildings in Japan, China and the U.S (Hajjar 2002; Zhao et al. 2010). This increasing trend in using CFT columns in MRF buildings led to the intensification of experimental and analytical research on CFT-MRF systems or their individual components (Kawaguchi et al. 1996; Schneider and Alostaz 1998; Ricles et al. 2004; Herrera et al. 2008; Skalomenos et al. 2014a, 2015a) to comprehend the seismic performance of this novel and innovative structural system.

17.3.1 Experimental Investigation on CFT-MRFs

Kawaguchi et al. (2000) tested ten reduced-scale one story, one-bay CFT-MRFs, under constant vertical loads on columns and alternately repeated horizontal load. Each frame was designed as a strong beam-weak column system, with the beams designed to remain elastic. Local buckling at the base of the columns occurred, leading to strength degradation. They concluded that the above CFT-MRFs showed better hysteretic characteristics than weak beam-strong column MRFs with wide flange steel columns. Chen et al. (2004) performed pseudodynamic tests on a full-scale three-bay composite frame, which included a braced frame in the central bay and moment connections between steel beams and CFT columns in the exterior bays. The study reported that the presence of the braces limited the inelastic demand imposed on the moment connections and that most of the energy was dissipated by the buckling restrained braces, with only minor yielding observed in all the moment connections. Herrera et al. (2008) conducted pseudodynamic tests on an approximately half scale two bay-four story composite MRF with a basement level, using performance-based design concepts. A weak-beam strong column concept in the design and a split tee beam-to-column moment connection were used to enable plastic hinges to form in the beams. A series of tests performed under different seismic hazard levels revealed the efficient seismic performance of CFT-MRFs. Finally, Tsai et al. (2008) conducted tests on a full scale three story, three bay CFT buckling restrained braced (BRB) frame. Square CFTs were used for the two exterior columns and circular CFTs for the two center columns. A series of six pseudodynamic tests were conducted subjecting the frame to ground motions of various intensities and then a quasi-static cyclic load protocol was applied until fracture of the braces.

17.3.2 Modelling of CFT-MRFs for Seismic Analysis

17.3.2.1 Square CFT Columns

The modified hysteretic models developed by Skalomenos et al. (2014a) are examined in this Section. These models appear to have two advantages: (1) the models were calibrated for a wide range of values of geometrical and strength parameters and (2) they are simple, yet effective concentrated plasticity models that can provide very good predictions of the force-deformation responses of CFT columns to cyclic loading. The accuracy of these predictions is shown in a wide range of comparisons between numerical and experimental results concerning beam-columns in a previous section.

17.3.2.2 Steel Beams and Beam-Column Connections

In order to analyze entire composite frames, especially seismically designed moment resisting frames where inelasticity is expected to be primarily limited to the steel beams, accurate models for these beams are necessary. In steel beams the onset of local buckling of the web or flange will cause deterioration in strength and stiffness. The Ramberg-Osgood hysteresis model can be selected for simulating the seismic behavior of a steel beam. Degradation effects can be included in the Ramberg-Osgood model with the aid of the Ruaumoko's strength degradation model, with a backbone curve based on ductility demands. The parameters which define the strength reduction variation of Ruaumoko's strength degradation model can be determined using the proposed relationships by Lignos and Krawinkler (2011) in conjunction with the PEER/ATC 72–1 (2010) guidelines. Details can be found in Skalomenos et al. (2015a).

Regarding the modeling of connections, a simplified model for all types of connections recommended by EC3 (2005), can be adopted. The connection is modeled by using only two springs, one vertical (K_V) and one rotational (K_θ). EC3 (2005) recommends values for the elastic stiffness K_θ and K_V and the moment M_{conn} and shear V_{conn} of the springs in that model. Both the rotational and the vertical springs are modeled as elastic-perfectly plastic.

17.3.2.3 Panel Zone

The panel zone model is used to account for the inelastic shear deformation in the beam-CFT column joint panel. An inelastic-trilinear shear-deformation (V-γ) model for the panel zone in the steel beam to the CFT column connection is adopted here, which is based on the superposition of a bilinear shear-deformation relationship for the steel tube to a trilinear for the concrete core where the total shear strength is equal to the sum of the two component strengths (steel and concrete) at identical

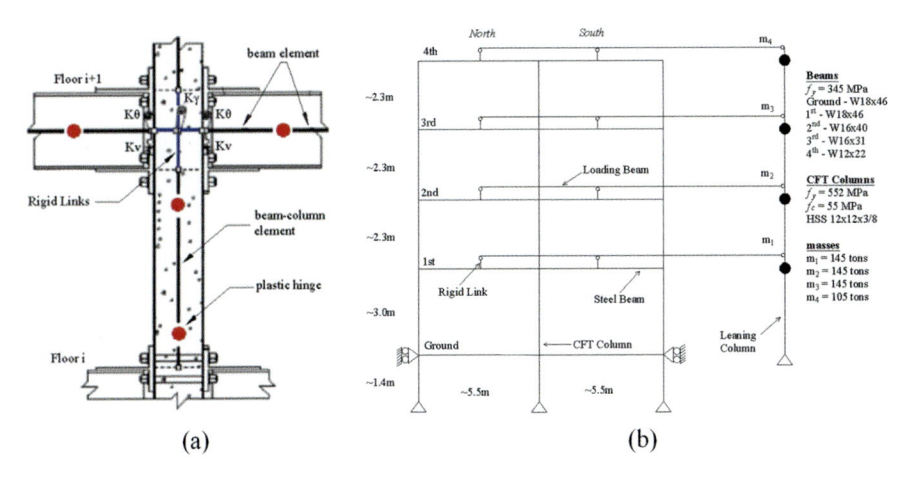

Fig. 17.4 (a) Modelling of CFT-MTFs; (b) Test CFT-MRF by Herrera et al. (2008)

deformation. The shear yield force and the post yielding stiffness of this superposed model are determined on the basis of the studies of Sheet et al. (2013) and Muhummud (2004). The CFT panel zone model, shown in Fig. 17.4a, consists of a rotational spring (K_γ) between the beam and the column, representing the relative rotation between them and two rigid links in the vicinity of the panel zone in order to model the rigid extension of the beam and the column. This model is commonly referred to as the *scissors* model.

17.3.2.4 Model Validation

The accuracy of the proposed hysteretic model of Ramberg-Osgood for simulating the seismic behavior of steel beams and that of trilinear model for simulating the panel zone behavior, are confirmed by comparing their results against those of experiments on steel beams and panel zones under a cyclic load protocol with variable intensity (Skalomenos et al. 2015a). The composite moment resisting frame investigated experimentally by a pseudodynamic testing method by Herrera et al. (2008), is used here to verify the validity of the aforementioned models for individual components (columns, beams, connections and panel zone) of a CFT-MRF by comparing its response to acceleration records at different seismic hazard as obtained experimentally and numerically with the aid of Ruaumoko (Carr 2006). This frame is a 0.6-scale model of a four stories and two bays frame being one of the perimeter CFT-MRFs of a three dimensional prototype building. In addition to the four stories above the ground, the frame also has a basement level and consists of rectangular CFT columns, wide flange beams and split tee moment connections, as shown in Fig. 17.4b. The experiment was conducted under three different hazard levels, the frequently occurring earthquake (FOE) level, the design basis earthquake (DBE) level and the maximum considered earthquake (MCE) level. Figure 17.5a, b shows the computational and experimental results for the 4th floor

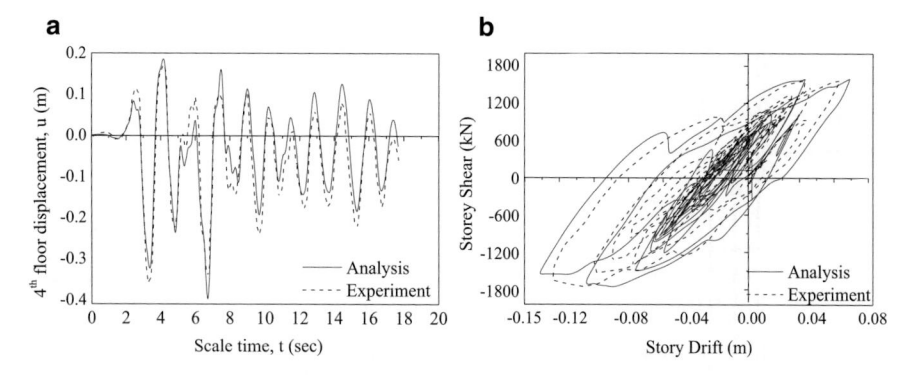

Fig. 17.5 Comparisons of experimental and computational results: (**a**) 4th floor displacement history; (**b**) first story shear-story drift response for MCE test (Skalomenos et al. 2015a)

displacement-time history and for the 1st storey shear-drift response corresponding to the MCE, respectively. As it can been seen from these figures the analysis results closely follow the experimental ones both for displacements and shears forces.

17.4 Seismic Design

From analyses of the effects of significant earthquakes (since the early 1980s), engineers have concluded that seismic risks in urban areas are increasing and are far from socio-economically acceptable levels. There is an urgent need to reverse this situation and it is believed that one of the most effective ways of doing this is through: (a) the development of more reliable seismic standards and code provisions than those currently available and (b) their stringent implementation for the design and construction of new engineering facilities and also for the seismic vulnerability assessment of the existing facilities and the upgrading of those considered hazardous. Thus, a design approach should not only consider design aspects but also deal with the damage control both in structural and non-structural elements.

A promising approach towards the above development has been proposed by the SEAOC Vision 2000 Committee in 1995 in its "Performance-Based Seismic Engineering of Buildings" report, which will be denominated as the "performance-based seismic engineering" (PBSE). This report presented a conceptual framework for PBSE, as well as the different methodologies that have been proposed for the application of such framework to the design, construction, occupancy and maintenance with particular emphasis on the design named "performance-based design" (PBD) or "performance-based seismic design" (PBSD) of new buildings. The emerging need to consider different criteria associated with various levels of performance has led to a recent emphasis on, and important developments in, PBSD. In concept, PBSD provides the opportunity for society and owners to choose performance goals and obliges the designer to formulate and implement a design process

Table 17.1 Recommended drift and ductility limits by performance levels (SEAOC 1999)

	Fully operational PL1	Operational PL2	Life safety PL3	Near collapse PL4
IDR	0.005	0.018	0.032	0.040
μ_θ	1.0	3.6	6.2	8.0

that fulfills the stated goals. In nutshell, the main objective of this design philosophy is to achieve the desired behavior of the structure for different levels of seismic action (SEAOC 1999; Bozorgnia and Bertero 2004). This section provides a comprehensive description of basic definitions and the mathematical description of the damage state. Then, on the basis of an extensive response database the section suggests performance-based design methods and controlled-damage design approaches for composite steel/concrete framed structures with CFT columns that are missing aspects in the modern seismic design of building structures.

17.4.1 Mathematical Description of the Damage State

Existing seismic codes in an implicit way and more recent performance-based seismic design methods in an explicit and more systematic way employ the concept of damage to establish structural performance levels corresponding to increasing levels of earthquake actions. These performance levels mainly describe the damage of a structure through damage indices, such as the inter-story drift ratio (*IDR*) or the member plastic rotations (μ_θ). The *IDR* is the relative displacement between two successive floors divided by the floor height and corresponds mainly to damage of non-structural elements. In contrast, the local ductility refers to the structural elements and is given by the equation

$$\mu_\theta = \frac{\theta_{max}}{\theta_y} \quad (17.1)$$

where θ_{max} is the maximum member rotation due to seismic excitation and θ_y is the yield member rotation. Values of *IDR* and local ductility are given in FEMA356 (2000) and SEAOC (1999). Table 17.1 gives indicative values of those quantities based on SEAOC (1999) for four performance levels.

Another damage index (D_{PA}) has been developed by Park and Ang (1985). The D_{PA} is expressed as a linear combination of the damage caused by excessive deformation and that contributed by repeated cyclic loading effects, as shown in the following equation

$$D_{PA} = \frac{\delta_m}{\delta_u} + \frac{\beta}{Q_y \delta_u} \int dE \quad (17.2)$$

In the above equation, the first part of the index is expressed as the ratio of the maximum experienced deformation δ_m to the ultimate deformation δ_u under monotonic loading. The second part is defined as the ratio of the dissipated energy $\int dE$ to the term $(Q_y\delta_u)/\beta$, where Q_y is the yield strength and the coefficient β is a non-negative parameter determined from experimental calibration. It should be pointed out that several improved damage indices have been proposed by Bozorgnia and Bertero (2004). However, there are still many uncertainties in the calculation of the damage index both at the level of ground motion that will be taken into account, as well as of the real material strength used in the structure.

17.4.2 Creation of an Extensive Response Database for CFT-MRFs

A family of 96 plane regular CFT-MRFs was examined using the aforementioned modeling techniques for the parametric seismic analyses conducted for the development of rational and efficient performance-based seismic design methods (PBSD). The examined framed structures are regular, have story heights and bay widths equal to 3 m and 5 m, respectively and square CFT columns. The frames have the following structural characteristics: number of stories, n_s, with values 1, 3, 6, 9, 12, 15, 18 and 20 and three bays, n_b, steel yielding strength ratio $e_s = 235/f_s$ with the yielding stress f_s taking the values of 275 and 355 MPa, concrete strength ratio $e_c = 20/f_c$ with the compressive strength f_c taking the values 20 and 40 MPa, the beam-to-column stiffness ratio, ρ (calculated for the story closest to the mid-height of the frame) and column to beam strength ratio, α (taking various values within practical limits) defined as

$$\rho = \frac{\sum (I/l)_b}{\sum (I/l)_c}, \quad \alpha = \frac{M_{RC,1,av}}{M_{RB,av}} \tag{17.3}$$

In these relations, I and l are the second moment of inertia and the length of the steel member (column c or beam b), respectively, $M_{RC,1,av}$ is the average of the plastic moments of the columns of the first storey and $M_{RB,av}$ the average of the plastic moments of the beams in all the stories. Every frame was designed for seismic loads according to European codes by spectrum analysis. The Spectrum Type 1 with PGA = 0.36 g, soil type B and behavior factor $q = 4$ was selected. Then, an ensemble of 100 ordinary (far-field type) ground motions of soil type B and with an average spectrum as close as possible to the Eurocode 8 (EC8 2004) elastic spectrum are selected (without any scaling) and are used for the nonlinear time history analyses of this study. A full list of all the ground motions and the frames with their modeling characteristics can be found in detail in Skalomenos (2014).

The nonlinear time histories are based on incremental dynamic analysis (IDA) which drives the structures to five performance levels. These five performance levels

(SEAOC 1999; FEMA356 2000) are: (a) $IDR_{max} = 0.5\%$, (b) occurrence of the first yielding, (c) $IDR_{max} = 1.8\%$, (d) $IDR_{max} = 2.5\%$, (e) $IDR_{max} = 3.2\%$ and (f) $IDR_{max} = 4.0\%$. The response quantities of interest are as follows: (1) the IDR_{max} along the height of the frame, (2) the maximum roof displacement $u_{r,max}$, (3) the maximum rotation ductility μ_θ along the height of the frame, (4) the displacement profile (storey-i) at first yielding $u_{i,y}$, (5) the maximum roof displacement ductility, $\mu = u_{r,max}/u_{r,y}$ and (6) the behavior factor q calculated as the ratio of the scale factor of the accelerogram which drives the structure to the specific performance level over the scale factor of the accelerogram which corresponds to first yielding (Elnashai and Broderick 1996).

17.4.3 Performance-Based Seismic Design and Assessment Methods for CFT-MRFs

Using the abovementioned response database, efficient performance-based seismic design methods were developed (Skalomenos et al. 2014b, 2015c) on the basis of a displacement-based design method, which determines the seismic design forces setting specific values for displacements and that of a force-based design method, which uses behavior factors q for spectrum analysis. The influence of specific parameters on the maximum structural response, such as the number of stories, the beam-to-column stiffness ratio, the column-to-beam strength ratio, the level of inelastic deformation induced by the seismic excitation and the material strengths, are taken into account in the preliminary stage of the design.

17.4.3.1 Direct Displacement-Based Design Method for CFT-MRFs

A relatively new performance-based seismic design procedure called the direct displacement-based design (DDBD) has been adopted by modern seismic codes. The DDBD procedure developed over the past 20 years with the aim of mitigating the deficiencies in current force-based design. The difference from force-based design is that DDBD method employs the displacement as the fundamental parameter and models the structure to be designed by an equivalent single-degree-of-freedom (SDOF) system, rather than by its initial characteristics. This is based on the *Substitute Structure* approach pioneered by Shibata and Sozen (1976).

The DDBD method was proposed by Pristley (1993), while its further development and improvement has been the subject of numerous research efforts (Sullivan et al. 2006; Pristley et al. 2007). The philosophy behind this design approach is to design a structure which will achieve, rather than be bounded by, a given performance limit state under a given seismic intensity. The method defines the design performance level of the structure in terms of displacement limits. Therefore, displacement is the key parameter of the design method. Since damage is directly related to displacements, seismic design methods based on displacements in a direct or indirect manner have an advantage over force-based methods, namely an easy and direct damage control.

17.4.4 Basic Steps of DDBD

This section briefly describes the basic steps of the DDBD procedure for a multi-degree-of-freedom (MDOF) framed building. The n-degree-of-freedom system with one concentrated mass m_i per every floor i and its associated lateral displacement Δ_i is replaced by an equivalent single-degree-of-freedom (SDOF) system with mass m_e, stiffness K_e, viscous damping ξ_e and design ductility μ_e. To determine ductility μ_e the equivalent yield displacement $\Delta_{e,y}$ is required. $\Delta_{e,y}$ is obtained by expressions like those proposed in the next section. When these equivalent system properties have been determined, the design base shear V_b for the substitute structure can be estimated. The base shear is then distributed between the mass elements of the real structure as inertia forces and the structure is analyzed under these forces to determine the design moments at locations of potential plastic hinges (Priestley et al. 2007).

17.4.5 Expression for Estimating the Yield Displacement

The response of each frame at first yielding to each accelerogram is obtained. The occurrence of the first plastic hinge in a CFT-MRF, which always occurs in beams because of the capacity design, is defined as the state of first yielding. By analyzing the response databank for the CFT-MRFs, the effect of the structural characteristics of the frames on their floor yield displacements $\Delta_{y,i}$ is identified (Skalomenos et al. 2015a). The expression

$$\Delta_{y,i} = h_i^{b_1} \cdot \left(\frac{h_i}{H}\right)^{b_2} \cdot n_s^{b_3} \cdot e_s^{b_4} \cdot e_c^{b_6} \tag{17.4}$$

is selected as a good candidate for approximating the response databank with i being the ith floor. The grade of steel and concrete strength have been included in Eq. (17.4) in the parameters $e_s = 235/f_s$ and $e_c = 20/f_c$ together with the h_i that denotes the height of floor and the H which is the total height of the frame. A nonlinear regression analysis leads to explicit values of the constants b_1–b_5 of Eq. (17.4) as given in Table 17.2. With respect to the databank, Eq. (17.4) offers a ratio $\Delta_{y,ap}/\Delta_{y,ex}$ (ap means "approximate" and ex means "exact") with a mean value of 0.97, a median value of 0.97 and dispersion equal to 0.28. The approximation of the median values $\Delta_{y,ex}$ of the databank compared to those resulting from the proposed Eq. (17.4) gives a correlation factor R^2 equal to 0.97.

Table 17.2 Indices of Eq. (17.4)

n_s	b_1	b_2	b_3	b_4	b_5
≤ 4	−3.474	4.504	4.528	−1.150	0.022
> 4 & ≤ 12	−2.557	3.461	2.912	−1.040	0.023
> 12 & ≤ 20	−3.374	4.268	4.007	−1.117	0.025

17.4.5.1 Behavior Factor q (or Strength Reduction R) for CFT-MRFs

A nonlinear regression analysis of the response dataset gave the following formula for the behavior factor q (Skalomenos et al. 2015b):

$$q = 1 + 1.90 \cdot \left(\mu_r^{0.76} - 1\right) \quad (17.5)$$

The above equation is relatively simple and satisfies the fundamental principle $q = 1$ for $\mu_r = 1$. The roof displacement ductility μ_r, is taken as $\min(\mu_{r,IDR}, \mu_{r,\theta})$, where $\mu_{r,IDR}$ is the roof displacement ductility at a target *IDR*, while $\mu_{r,\theta}$ is the roof displacement ductility when a member of the structure experiences local ductility μ_θ. The $\mu_{r,IDR}$ is defined as the ratio $u_{r,max(IDR)}/u_{r,y}$, where $u_{r,max}$ is the maximum roof displacement for the target IDR_{max} of a predefined performance level. The $u_{r,y}$ is estimated by using Eq. (17.4). Then, by analyzing the response database, the ratio $\beta = u_{r,max(IDR)}/(H \cdot IDR_{max})$ was found to depend on the number of stories n_s, the parameters ρ, α and the grade of steel f_s. The following approximation for the ratio β is proposed:

$$\beta = 1 - 0.547 \cdot \left(n_s^{0.259} - 1\right) \cdot \rho^{-0.154} \cdot \alpha^{-0.334} \cdot e_s^{-0.130} \quad (17.6)$$

The relation between $\mu_{r,\theta}$ and μ_θ is given by

$$\mu_{r,\theta} = 1 + 1.078 \cdot \left(\mu_\theta^{0.936} - 1\right) \cdot n_s^{-0.044} \cdot \rho^{0.144} \cdot \alpha^{0.375} \cdot e_s^{0.211} \quad (17.7)$$

The generation of the proposed Eq. (17.5) to calculate the behavior factor q is based on a large amount of data and randomness introduced by the seismic records. Vamvatsikos and Cornell (2002) employed appropriate summarization techniques that reduce this response data into the distribution of damage measures and the probability of exceeding any specific limit-state for a given intensity measure level. Following their techniques here, the whole response databank is summarized to produce the 16% and 84% $q-\mu_r$ curves where the 16%, 84% fractiles are defined considering the responses of the databank to be normally distributed. The values of predicted factors, q_{app}, are taken from Eq. (17.5) and shown in Fig. 17.6 through comparisons with the 'exact' values, q_{exact}, as obtained from the response databank, while the outliers exceeding the limits of 16% and 84% have been removed. This figure shows that the 16% and 84% fractiles are limited by constant multipliers of the factor q equal to 0.77 and 1.23, which satisfies the lower and upper limit of the design value, respectively (Skalomenos et al. 2015b).

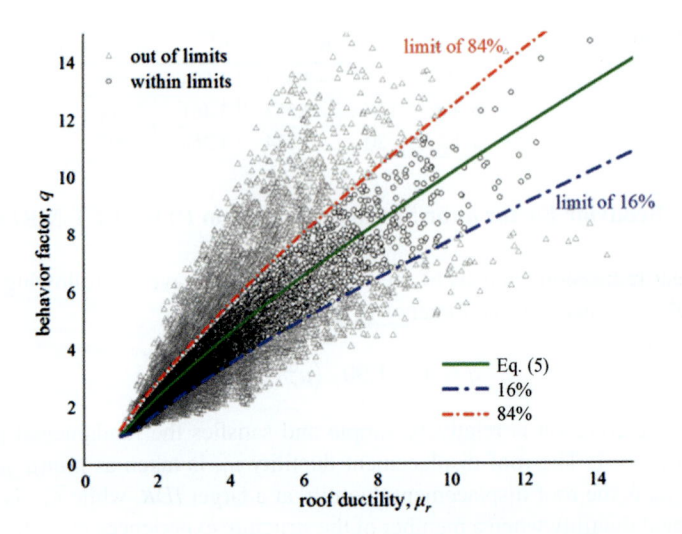

Fig. 17.6 Graphical approximation of the response databank with the proposed Eq. (17.5), 16% and 84% fractile values (Skalomenos et al. 2015b)

17.4.5.2 Hybrid Force-Displacement (HFD) Seismic Design Method

This section presents the PBSD methodology for CFT-MRFs which combines the advantages of the well-known force-based (FBD) and displacement-based seismic design methods in a hybrid force/displacement (HFD) design scheme. The method has been proposed in the works of Karavasilis et al. (2006) and Tzimas et al. (2013, 2017) for steel frames and its full description for CFT-MRFs can be found in Skalomenos et al. 2015c.

The proposed hybrid method uses the equations shown in the previous section. It starts by transforming the target values of *IDR* and local ductility μ_θ to a target roof displacement and then calculates the behavior factor q associated with the roof displacement ductility μ_r. The proposed method: (a) uses both drift and ductility demands as input variables for the initiation of the design process; (b) avoids the use of a substitute single degree of freedom system done by the DBD; (c) uses conventional elastic response spectrum analysis and design as the FBD does; (d) includes the influence of the number of stories; (e) includes the influence of the structural parameters and (f) the material strengths.

The HFD seismic design procedure can be summarized in the following steps:

1. Definition of the basic building attributes

With reference to the type of frame examined here, definition of the number of stories, n_s, number of bays, n_b, bay widths and storey heights and limits on the depth of beams and columns due to architectural requirements, are needed.

2. Definition of the performance level

The performance levels considered here are the immediate occupancy (IO) under FOE, the life safety (LS) under DBE and the collapse prevention (CP) under MCE. The earthquake intensity level is represented by the appropriate for each performance level elastic acceleration response spectrum. For the three performance levels, spectra are defined based on the design response spectrum of EC8 (2004).

3. Definition of input parameters (performance metrics)

Performance Metrics Definition of limit values for the maximum inter-storey drift ratio (IDR_{max}) and maximum local ductility (rotation ductility, μ_θ, for beams/columns) along the height of the frame. These limit values can be selected based on the performance level defined by FEMA356 (2000) or SEAOC (1999).

4. Estimation of the input variables (yield roof displacement and mechanical characteristics)

Yield Roof Displacement, u_{ry} An initial estimation of the yield roof displacement can be obtained by using Eq. (17.4). *Mechanical characteristics.* For a CFT-MRF, an initial estimation for the parameters column-to-beam strength ratio, α, beam-to-column stiffness ratio, ρ has to be made (Eq. (17.3)). These parameters control the period T of the structure. The characteristics a and ρ vary along the height of a steel frame and therefore, their nominal values are taken equal to those of the storey closest to the mid-height of the building (Chopra 2007).

5. Determination of the behavior (or strength reduction) factor q

The behavior factor q is determined by Eq. (17.5).

6. Design of the structure

The ordinates of the elastic design spectrum are divided by the q factor and design the building on the basis of an elastic response spectrum analysis along with the capacity and ductile design rules of seismic codes is performed (EC8 2004).

7. Iterative process

Iterations with respect to the input variables u_{ry}, ρ, a and period T follow. The sufficient number of iterations for achieving convergence depends on the initial estimations of the input variables (see step 4).

17.4.6 Seismic Damage Assessment of CFT-MRFs

This section quantifies the damage on beams and columns of CFT-MRFs through simple expressions that relate the damage index D_{PA} introduced in a previous section (Eq. (17.2)) with the characteristics of the frames and the ground motions. Similar expressions for CFT-MRFs have been proposed by the authors for the most commonly used damage indices of the literature, such as those proposed by Roufaiel and Meyer (1987) and Bracci et al. (1989). All expressions with a detailed description of the procedure can be found in Kamaris et al. (2016) together with application examples that verify their efficiency and rationality.

By analysing the response database of CFT-MRFs, the following relationship was identified as a good candidate to correlate the maximum damage, D, of column bases or beams in terms of characteristics of the structure (n_s, a, f_s) and the ground motions that excite them (spectral acceleration S_a):

$$D = b_1 \cdot n_s^{b_2} \cdot a^{b_3} \cdot \left(\frac{S_a}{g}\right)^{b_4} \left(\frac{235}{f_s}\right)^{b_5} \tag{17.8}$$

with b_1, b_2, b_3, b_4 and b_5 constants to be determined. The aforementioned relation is relatively simple and satisfies the physical constraint $D = 0$ for $S_a = 0$. However, for nonzero values of S_a for which structural members behave elastically, Eq. (17.8) gives non-zero values of damage either at columns or beams, while in reality damage there is zero. Thus, before using this equation, one should compute the internal forces (bending moments M and axial forces P) of the columns or beams by performing a linear elastic analysis and check if their values are in the elastic range. If the members are in the elastic range, the value of damage is equal to zero by default, otherwise the proposed relationship can be used. The nonlinear regression analysis of the results of the parametric study led to the following expressions for the damage index D_{PA}:

(a) for CFT column bases:

$$D_{PA}^c = 0.24 \cdot n_s^{0.16} \cdot a^{-0.30} \cdot \left(\frac{S_a}{g}\right)^{0.39} \left(\frac{235}{f_s}\right)^{0.05} \tag{17.9}$$

(b) for beams:

$$D_{PA}^b = 0.31 \cdot n_s^{0.33} \cdot a^{-0.12} \cdot \left(\frac{S_a}{g}\right)^{0.43} \left(\frac{235}{f_s}\right)^{-0.08} \tag{17.10}$$

The mean, median and standard deviation of the ratio of the "exact" value of D obtained from inelastic dynamic analyses over the approximate one calculated from Eqs (17.9) and (17.10), respectively., i.e., D_{exact}/D_{ap}, are 1.00, 0.96 and 0.29 for CFT column bases, respectively and 0.99, 0.96 and 0.29 for beams, respectively. The values of those metrics show that the proposed formulae are fairly accurate. Thus, one can use the proposed expressions to have a rapid damage assessment of CFT-MRF structures without using the more sophisticated and time consuming non-linear dynamic analysis.

17.5 Conclusions and Future Developments

The present paper provides a comprehensive overview of the state of research in composite steel/concrete structural fames consisting of concrete-filled steel tubular (CFT) columns including experimental and analytical/numerical studies. Main conclusions are as follows:

(a) Recent experimental research on CFT columns made of ultra-high strength (HS) steel revealed that HS steel effectively improves the seismic performance of columns by increasing their elastic deformation capacity and delaying the failure of local buckling. HS CFTs need further experimental investigation to find efficient member configurations that mitigate the rupture failure caused by the low HS steel ductility capacity.
(b) Efficient analytical/numerical methods for simulating the cyclic behavior of CFT columns and a model development methodology were presented. Simple, yet accurate, plastic-hinge models were developed for square and cyclic CFT columns including strength and stiffness deterioration. Development of simplified hysteretic models for other types of composite members, such as beam/columns with fully or partially concrete encased steel sections is a subject of future studies.
(c) Modelling techniques for seismic analysis and design of composite concrete-filled steel tube/moment resisting frame (CFT-MRFs) structures are introduced based on concentrated plasticity theory for conducting less time-consuming non-linear time history analysis. The seismic performance of a wide range of CFT-MRF structures under several levels of seismic hazard was investigated.
(d) On the basis of the displacement-based design method and that of a force-based design method, new performance-based seismic design and damage assessment methods were developed. The methods are fairly accurate in predicting actual building seismic response and in designing rationally CFT-MRFs.
(e) The behavior factors q proposed through this study for CFT-MRF structures, are valuable and important. According to Eurocode 8, the selection of behavior factors for the preliminary design of those structures is based on the factors proposed for all steel MRFs. This study recommends larger behavior factors compared with those proposed for steel structures, allowing therefore some advantages to be gained using composite columns.

Acknowledgments The first author would like to express his gratitude to the Japan Society for the Promotion of Science (JSPS) for awarding him the "FY2015 JSPS Postdoctoral Fellowship for Overseas Researchers" and to his host researcher, Professor Masayoshi Nakashima, for his encouragement and support during the past 2 years at Kyoto University. All the authors would like to thank Dr. G. Kamaris and Mr. D. Serras for their valuable assistance and helpful discussions.

References

ACI (2011) Building code requirements for structural concrete and commentary. American Concrete Institute, Farmington Hills

Aho M, Leon RT (1997) A database for encased and concrete-filled columns. Report no. 97-01, School of Civil and Environmental Engineering, Georgia Institute of Technology, Atlanta, Georgia

AIJ (1997) Architectural Institute of Japan. Recommendations for design and construction of concrete filled steel tubular structures

AISC (2010) American institute of steel construction Inc., Seismic Provisions for Structural Steel Buildings. AISC, Chicago

ATC (1992) Guidelines for cyclic seismic testing of components of steel structures. Report no. ATC-24, Applied Technology Council, Redwood City, California

ATENA (2012) Advanced tool for engineering nonlinear analysis, version 4.3.1g. Červenka Consulting Ltd., Prague, Czech Republic

Aval SBB, Saadeghvaziri MA, Golafshani AA (2002) Comprehensive composite inelastic fiber element for cyclic analysis of concrete-filled steel tube columns. J EngMech ASCE 128(4):428–437

Bozorgnia Y, Bertero VV (2004) Earthquake engineering: from engineering seismology to performance-based engineering. CRC Press, Boca Raton

Bracci JM, Reinhorn AM, Mander JB (1989) Deterministic model for seismic damage evaluation of reinforced concrete structures. Technical report NCEER 89–0033, State University of New York at Buffalo

Carr AJ (2006) HYSTERES and RUAUMOKO-2D inelastic time-history analysis of two-dimensional framed structures. Department of Civil Engineering, University of Canterbury, New Zealand

Chen CH, Hsiao BC, Lai JW, Lin ML, Weng YT, Tsai KC (2004) Pseudo-dynamic test of a full-scale CFT/BRB frame: part 2 – construction and testing. In: Proceedings of 13th WCEE, Vancouver, Canada

Chopra AK (2007) Dynamics of structures. Pearson Prentice Hall, Berkeley

Deierlein G (2000) New provisions for the seismic design of composite and hybrid structures. Earthquake Spectra 16(1):163–178

EC3 (2005) Eurocode 3 - Design of steel structures, part 1.1: general rules and rules for buildings, European Prestandard ENV 1993-1-1. European Committee for Standardization (CEN), Brussels

EC4 (2004) Eurocode 4 - Design of composite steel and concrete structures, part 1.1: general rules and rules for buildings, European Prestandard ENV 1994-1-1. European Committee for Standardization (CEN), Brussels

EC8 (2004) Eurocode 8: Design of structures for earthquake resistance - part 1: general rules, seismic actions and rules for buildings. European Committee for Standardization, Brussels, Belgium

Elnashai AS, Broderick BM (1996) Seismic response of composite frames-II, calculation of behaviour factors. Eng Struct 18(9):707–723

Elremaily A, Azizinamini A (2002) Behavior and strength of circular concrete-filled steel tube columns. J Constr Steel Res 58:1567–1591

FEMA-356 (2000), Federal Emergency Management Agency, "Prestandard and commentary for the seismic rehabilitation of buildings". Report no. FEMA-356, prepared by the SAC Joint Venture for the FEMA, Washington, DC

Furlong RW (1967) Strength of steel-encased concrete beam-columns. J Struct Eng ASCE 93(5):113–124

Gourley BC, Hajjar JF (1993) "A synopsis of studies of the monotonic and cyclic behaviour of concrete-filled steel tube beam columns". Department of Civil Engineering. University of Minnesota, Minneapolis, USA

Green TP, Leon RT, Rassati GA (2004) Bidirectional tests on partially restrained, composite beam-to-column connections. J Struct EngASCE 130(2):320–327

Hajjar JF (2002) Composite steel and concrete structural systems for seismic engineering. J Constr Steel Res 58:703–728

Hajjar JF, Gourley BC (1997) A cyclic nonlinear model for concrete-filled tubes cross-section strength. J Struct Eng ASCE 11:1327–1336

Hajjar JF, Schiller PH, Molodan A (1998) A distributed plasticity model for concrete-filled steel tube beam-columns with interlayer slip. Eng Struct 20(8):663–676

Han L-H, Yang Y-F (2005) Cyclic performance of concrete-filled steel CHS columns under flexural loading. J Constr Steel Res 61(4):423–452

Han LH, Zhao XL, Yang YF, Feng JB (2003) Experimental study and calculation of fire resistance of concrete-filled hollow steel columns. J Struct Eng ASCE 129(3):346–356

Han LH, Huang H, Tao Z, Zhao XL (2006) Concrete-filled double skin steel tubular (CFDST) beam–columns subjected to cyclic bending. Eng Struct 28(12):1698–1714

Herrera RA, Ricles JM, Sause R (2008) Seismic performance evaluation of a large-scale composite MRF using Pseudodynamic testing. J Struct Eng ASCE 134(2):279–288

Inai E, Mukai A, Kai M, Tokinoya H, Fukumoto T, Mori K (2004) Behavior of concrete-filled steel tube beam columns. J Struct Eng ASCE 130(2):189–202

Kamaris G, Skalomenos KA, Hatzigeorgiou GD, Beskos DE (2016) Seismic damage estimation of in-plane regular steel/concrete composite moment resisting frames. Eng Struct 115:67–77

Karavasilis TL, Bazeos N, Beskos DE (2006) A hybrid force/displacement seismic design method for plane steel frames, behavior of steel structures in seismic area. In: Mazzolani FM, Wada A (eds) Proceedings of STESSA conference, Yokohama, Japan, 14–17 august, Taylor & Fransis, pp 39–44

Kawaguchi J, Morino S, Sugimoto T (1996) Elasto-plastic behavior of concrete-filled steel tubular frames, composite construction in steel and concrete: proceedings of an Engineering Foundation conference, Irsee, Germany

Kawaguchi J, Morino S, Sugimoto T, Shirai J (2000), Experimental study on structural characteristics of portal frames consisting of square CFT columns. In: Proceedings of the conference composite construction in steel and concrete IV, Banff, Canada, May 28–June 2

Leon RT, Perea T, Rassati GA, Lange J (2011) Composite construction in steel and concrete VI. In: Proceedings of the 2008 composite construction in steel and concrete conference VI. ASCE, Reston, Virginia, USA

Lignos DG, Krawinkler HK (2011) Deterioration modeling of steel components in support of collapse prediction of steel moment frames under earthquake loading. J Struct Eng ASCE 137 (11):1291–1302

Mander JB, Priestley M, Park R (1988) Theoretical stress-strain model for confined concrete. J Struct Eng ASCE 117(8):1804–1826

Muhummud T (2004) Seismic design and behavior of composite moment resisting frames constructed of CFT columns and WF beams. Ph.D. thesis, Lehigh University, Bethlehem, Pa

Nie JG, Wang YH, Fan JS (2012) Experimental study on seismic behavior of concrete filled steel tube columns under pure torsion and compression-torsion cyclic load. J Constr Steel Res 79:115–126

Park YJ, Ang HS (1985) Mechanistic seismic damage model for reinforced concrete. J Struct Eng ASCE 111(4):722–739

PEER/ATC (2010) Modeling and acceptance criteria for seismic analysis and design of tall buildings. PEER/ATC-72-1 report, Applied Technology Council, Redwood City, California

Perea T, Leon RT, Hajjar JF, Denavit MD (2014) Full-scale tests of slender concrete-filled tubes: interaction behavior. J Struct Eng ASCE 140(9):04014054

Portolés JM, Romero ML, Bonet JL, Filippou FC (2011) Experimental study of high strength concrete-filled circular tubular columns under eccentric loading. J Constr Steel Res 67 (4):623–633

Priestley MJN (1993) Myths and fallacies in earthquake engineering - conflicts between design and reality, bulletin of the New Zealand National Society for earthquake. Engineering 26 (3):329–341

Priestley MJN, Calvi GM, Kowalsky MJ (2007) Displacement-based seismic design of structures. IUSS Press, Pavia

Ricles JM, Peng SW, Lu LW (2004) Seismic behavior of composite concrete filled steel tube column-wide flange beam moment connections. J Struct Eng ASCE 130(2):223–232

Roufaiel MSL, Meyer C (1987) Analytical modeling of hysteretic behaviour of R/C frames. J Struct Eng ASCE 113(3):429–457

Schneider SP, Alostaz YM (1998) Experimental behavior of connections to concrete-filled steel tubes. J Constr Steel Res 45(3):321–352

SEAOC (1999) Recommended lateral force requirements and commentary, 7th edn. Structural Engineers Association of California, Sacramento

Serras DN, Skalomenos KA, Hatzigeorgiou GD, Beskos DE (2016) Modeling of circular concrete-filled steel tubes subjected to cyclic lateral loading. Struct ICE 8:75–93

Serras DN, Skalomenos KA, Hatzigeorgiou GD, Beskos DE (2017) Inelastic behavior of circular concrete-filled steel tubes: monotonic versus cyclic response. Bull Earthq Eng:1–22 (in press)

Shakir-Khalil H (1988) Steel-concrete composite columns-I. In: Narayanan R (ed) Steel-concrete composite structures: stability and strength. Elsevier Applied Science, London

Shanmugan NE, Lakshmi B (2001) State of the art report on steel-concrete composite columns. J Constr Steel Res 57:1041–1080

Sheet IS, Gunasekaran U, MacRae GA (2013) Experimental investigation of CFT column to steel beam connections under cyclic loading. J Constr Steel Res 86:167–782

Sheikh TM, Deierlein GG, Yura JA, Jirsa JO (1989) Beam-column moment connections for composite frames: part 1. J Struct Eng ASCE 115(11):2858–2876

Shen C, Mamaghani I, Mizuno E, Usami T (1995) Cyclic behaviour of structural steels, II: theory. J Eng Mech ASCE 121(11):1165–1172

Shibata A, Sozen M (1976) Substitute structure method for seismic design in R/C. J Struct Div ASCE 102(1):1–18

Skalomenos KA (2014) "Seismic performance of plane moment resisting frames with concrete filled steel tube columns and steel I beams". Ph.D. thesis, Department of Civil Engineering, University of Patras, Greece

Skalomenos KA, Hatzigeorgiou GD, Beskos DE (2013) Determination of Bouc-Wen hysteretic model parameters for simulating the seismic behavior of square CFT columns: in proceedings of 10th HSTAM international congress on mechanics, Chania, Crete, 23–27 May

Skalomenos KA, Hatzigeorgiou GD, Beskos DE (2014a) Parameter identification of three hysteretic models for the simulation of the response of CFT columns to cyclic loading. Eng Struct 61:44–60

Skalomenos KA, Hatzigeorgiou GD, Beskos DE (2014b), Seismic yield displacement of composite steel/concrete plane frames: in proceedings of 8th Hellenic National Conference of steel structures, 2–4 October, 2014, Tripoli, Greece

Skalomenos KA, Hatzigeorgiou GD, Beskos DE (2015a) Modelling level selection for seismic analysis of CFT/MRFs by using fragility curves. Earthq Eng Struct Dyn 44(2):199–220

Skalomenos KA, Hatzigeorgiou GD, Beskos DE (2015b) Seismic behavior of composite steel/concrete MRFs: deformation assessment and behavior factors. Bull Earthq Eng 13 (12):3871–3896

Skalomenos KA, Hatzigeorgiou GD, Beskos DE (2015c) Application of the hybrid force/displacement (HFD) seismic design method to composite steel/concrete plane frames. J Constr Steel Res 115:179–190

Skalomenos KA, Hayashi K, Nishi R, Inamasu H, Nakashima M (2016) Experimental behavior of concrete-filled steel tube columns using ultrahigh-strength steel. J Struct Eng ASCE 142 (9):04016057

Sugano S, Nagashima T, Kei T (1992) Seismic behaviour of concrete-filled tubular steel columns, 10th structures congress, Morgan J ed, ASCE, San Antonio, Texas, USA, 914–917

Sullivan TJ, Priestley MJN, Calvi GM (2006) Direct displacement-based design of frame-wall structures. J Earthq Eng 1:91–124

Susantha KAS, Ge H, Usami T (2002) Cyclic analysis and capacity prediction of concrete-filled steel box columns. Earthq Eng Struct Dyn 31(2):195–216

Tawil E-T, Deierlein GG (2001) Nonlinear analysis of mixed steel-concrete frames. I: element formulation. J Struct Eng ASCE 127(6):647–655

Tomii M, Yoshimura K, Morishita Y (1977), Experimental studies on concrete filled steel tubular stub columns under concentric loading. In: Proceeding of the international colloquium on stability of structures under static and dynamic loads, SSRC/ASCE, Washington, DC, March, 718–741

Tort C, Hajjar JF (2010) Mixed finite element for three-dimensional nonlinear dynamic analysis of rectangular concrete-filled steel tube beam-columns. J Struct Eng ASCE 136(11):1329–1339

Triantafillou TC (2013) Steel-concrete composite structures, Department of Civil Engineering, University of Patras Press, Patras, Greece (in Greek)

Tsai KC, Hsiao PC Wang KJ, Weng YT, Lin ML, Lin KC, Chen CH, Lai JW, Lin SL (2008) Pseudo-dynamic tests of a full-scale CFT/BRB frame-part I: specimen design, experiment and analysis. Earthq Eng Struct Dyn 37(7):1081–1098

Tzimas AS, Karavasilis TL, Baseos N, Beskos DE (2013) A hybrid force/displacement seismic design method for steel building frames. Eng Struct 56:1452–1463

Tzimas AS, Karavasilis TL, Baseos N, Beskos DE (2017) Extension of the hybrid force/displacement (HFD) seismic design method to 3D steel moment-resisting frame buildings. Eng Struct 147:486–504

Vamvatsikos D, Cornell CA (2002) Incremental dynamic analysis. Earthq Eng Struct Dyn 31(3):491–514

Varma AH, Ricles JM, Sause R, Lu LW (2002) Experimental behavior of high strength square concrete-filled steel tube columns. J Struct Eng ASCE 128(3):309–318

Zhang GW, Xiao Y, Kunnath S (2009) Low-cycle fatigue damage of circular concrete-filled-tube columns. ACI Struct J 106(2):151

Zhao XL, Han LH, Lu H (2010) Concrete-filled tubular members and connections. Spon Press, New York

Zhou P, Zhu ZQ (1997) Concrete-filled tubular arch bridge in China. Structural Engineering International, International Association for Bridge and Structural Engineering, 7(3):161–166

Chapter 18
Seismic Design of Steel Structures: New Trends of Research and Updates of Eurocode 8

Raffaele Landolfo

Abstract The European standards for the design and verification of structures are currently under revision. Indeed, after 10 years from their final issue, some criticisms arose as a consequence of both the scientific findings and the design experience gained in Europe. For these reasons, an European program for the revision and harmonization of Eurocodes (mandate M/15 "Evolution of the Structural Eurocodes") is currently ongoing within the framework of CEN/TC 250, which shall be completed by 2020. The revision process of the rules for the seismic design of steel and steel-concrete structures (currently given by Chaps. 6 and 7 of Eurocode 8 part 1) is supported by a specific working group (WG2) of the commission TC250/SC8 that, in cooperation with the ECCS Technical Committee 13 (TC13), will provide the relevant preliminary documents to the Project Team in charge of drafting the new version of the code. In this perspective, this paper summarizes the critical issues and main aspects that require further insights into the field of steel structures in seismic areas, as well as the recent and currently ongoing relevant research activities that justify the relevant normative updates.

18.1 Introduction

Eurocodes are the European standards for structural and geotechnical design, which have been largely adopted by the national regulations of each European country. The Eurocodes were published about 10 years ago, after more than 20 years of studies and preparatory work, and represented the excellence of knowledge by the time of their publication. As a matter of fact, currently some parts of the codes are obsolete if compared to the latest scientific findings and technological advancements achieved over the last 10 years. In addition, during this period, the European structural

R. Landolfo (✉)
Department of Structures for Engineering and Architecture, University of Naples Federico II, Naples, Italy
e-mail: landolfo@unina.it

designers highlighted some critical issues in the interpretation and application of some rules and recommendations. All these considerations led to a process of amendment of all Eurocodes, which shall be completed within the next 2020.

Nowadays, the main objective of the European Commission is to develop a new generation of harmonized structural and geotechnical design provisions that will replace the different rules that are mandatory in the various EU nations, in order to ensure both uniformity in the level of safety and efficiency of building heritage and to promote and facilitate the free movement and professional integration of technicians (i.e. engineers and architects) into the European Community.

The revision program is coordinated by CEN/TC 250, which is the European Technical Committee for the Standardization (CEN) of "Structural Eurocodes". Given the wide scope and complexity of the work program, the TC 250 is organized in Sub-Committees (SC), one per Eurocode. Within each SC, there are usually several working groups (WGs) dealing with specific topics, whose members are appointed by the EU countries belonging to CEN. WGs generally carry out preliminary activities in their field of expertise and represent the reference environment for Project Teams (PT). The latter will officially draft the next version of Eurocodes, under the Mandate M/515 "Evolution of Eurocodes", that will officially become the new standards once approved by all competent bodies and Member States. In this dynamic and multicultural context, the revision of the rules and provisions for the seismic design of steel structures is a very important part of the overall update of EN1998-1 (hereinafter also indicated as Eurocode 8 or EC8).

In this regard, the research activity in the field of seismic design and seismic analysis of steel structures has been very active during the last years and highly development-oriented to support designers.

Since the 1980s, many researchers devoted their efforts to these goals and, in this framework, the Technical Committee (TC) 13 - Seismic Design of Steel Structures of European Convention for Constructional Steelwork (ECCS) played a key role. In particular, under the former Chairmanship of Federico M. Mazzolani, TC13 published the "European Recommendations for steel structures in seismic zones" in 1988 (see Fig. 18.1a). These recommendations were incorporated as the "Steel section" of the first edition of Eurocode 8 and constituted the framework of the Chap. 6 of current version of the code.

In 2007 all ECCS Technical Committees were restructured and the organization and the membership of TC13 were renovated. Following this process, the Chairmanship changed, and the Author was appointed as the new Chair of the Committee. Nowadays, the TC13 brings together most of the European experts in the field of seismic design and assessment of steel and steel-concrete composites and is representative of almost all European countries.

After 5 years since its restructuring, one of most interesting outcome is the publication of the book "Assessment of EC8 Provisions for Seismic Design of Steel Structures", see Fig. 18.1b. This book can be considered as the first preparatory document oriented towards the next version of Eurocode 8, since it summarizes, from the TC13 perspective, all the critical issue of EN 1998-1 needing clarification

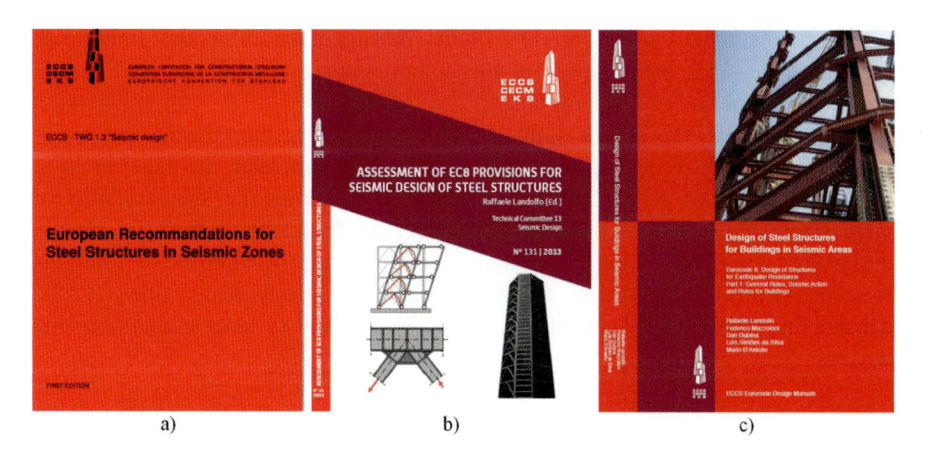

Fig. 18.1 First ECCS "European recommendations for steel structures in seismic zones" (**a**); ECCS/TC13 No. 131/2013: Assessment of EC8 provisions for seismic design of steel structures (**b**); ECCS Manual on "Design of Steel Structures for Buildings in Seismic Areas" (**c**)

and/or further development (Landolfo 2013, 2015). TC13 members actively contributed to the book providing not only their comments but also valuable contributions from their personal researches carried out both within the TC13 action and their research programs. With this regard, it is worth to highlight that all TC13 members voluntarily offered their contributions, which constitutes the actual added value of the Committee and testifies the crucial and strategic importance of the addressed topics, as well.

In line with the mission of TC13, a new ECCS Manual on "Design of Steel Structures for Buildings in Seismic Areas" (see Fig. 18.1c) was recently published with the aim to address practitioners with the basic principles of the codes and summarized the developments and the most updated research results on the seismic design of steel structures in the framework of Eurocodes.

In 2015, the sub-commission SC8 of CEN/TC250, which is responsible for EN1998, established the WG2 – Steel and Composite Structures – with the aim of carrying out a preliminary investigation concerning the criticisms and the open issues related to Chaps. 6 and 7 of the EC8. At the same time, the Writer was appointed as WG2 coordinator and the ECCS TC13 officially became a technical committee supporting WG2.

More recently, in the Autumn 2017, the SC8 procedure to select the experts of the Phase 2 Project Team (PT) was over. This PT, whose official denomination is "SC8.T2 Material-dependent sections of EN 1998-1", is composed by six members, among which the Writer, and it will be in charge to revise the chapters of EC8 concerning the various building materials (i.e. reinforced concrete, steel, steel-concrete, timber, masonry and aluminium) within the next 3 years.

This brief summary of the current evolutive process of the Eurocodes' framework highlights the tight commitment of the commissions currently involved in this project over a period of only 6 years.

In this context, this article summarizes the criticisms, the fallacies and weak points of current EC8 dealing with steel structures. In particular, the paper looks at the following topics: (i) the influence of material properties (i.e. the variability of yield stress and toughness), the local ductility of dissipative zones, (ii) seismic design and pre-qualification of beam-to-column joints, (iii) innovative solutions for eccentrically braced frames, (iv) the revision of capacity design rules, (v) the damage limit state and P-Delta effects, (vi) new structural systems and anti-seismic devices.

18.2 Criticisms of Ec8 and Potential Upgrading

18.2.1 Material Overstrength

A well-established general concept in seismic design of structures is that non-dissipative members must be designed on the basis of the expected material strength of the dissipative zones. The ratio between the average yield stress and the characteristic yield value for a given steel class is called γ_{ov} by EC8. There is no specific information about the values to be attributed to γ_{ov}, but National Authorities have the possibility to select the most appropriate ones. However, a constant value of $\gamma_{ov} = 1.25$ is currently suggested, which is contradictory with the available experimental evidence of the dependence on the yield strength of the steel, as also recently confirmed by the results of OPUS project. In this research project statistical data on characteristics of steel products in Europe have been collected and processed in order to define normative values of material overstrength factors, which would account for such aspects as steel grade, type of steel product, etc. The analysis of tensile yield strength of European steel by different producers clearly showed that the material overstrength factors specified in EN 1998-1 seems to be conservative only for higher-grade steels. Moreover, the results of tensile coupon tests sampled from hot-rolled sections showed that lower-grade steels are characterised by the larger values of the actual yield strength, and consequently larger material overstrength factors. On the contrary plated products have smaller overstrength in comparison with long hot rolled products.

Another criticism of Eurocode 8 about material overstrength is the unclear identification of the sources of overstrength, so that the reader is not aware about the actual meaning of the coefficients suggested to cover the undesired effects due to overstrength. The code should provide the values of the overstrength coefficient with reference to each source of overstrength, that are: (1) overstrength of delivered material; (2) code required overstrength; (3) strain-hardening overstrength; (4) random material variability overstrength; (5) overstrength due to strain rate.

Lastly, considering the recent trends of the steel market and the significant development in steel processing, the use of high strength steel (HSS) is very attractive and detailing rules on HSS shall be included in the next version of the code. Indeed, seismic applications potentially represent the rational field to exploit the high performance of HSS, because the combined use of HSS for non-dissipative members and of mild carbon steel (MCS) for dissipative members may allow an easier application of capacity design criteria (Dubina et al. 2008; Longo et al. 2014; Tenchini et al. 2014, 2016). The expected design improvement would be obtained in terms of smaller member sizes than those obtained when using MCS only. Structures designed using the combination of HSS and MCS are called "dual-steel" structures. The results of the international research project HSS-SERF highlighted the advantages of dual-steel concept, especially for what concerns the control of seismic response of multi-storey buildings to achieve overall ductile mechanism (Dubina et al. 2014).

18.2.2 Selection of Steel Toughness

Steel toughness characterizes the safety of steel structures to brittle failure and it is quantified by the fracture energy determined by means of Charpy tests on standardised V-notched specimens. According to EN 1993-1-10:2005 "The toughness of the steels and the welds should satisfy the requirements for the seismic action at the quasi-permanent value of the service temperature". In the current Eurocodes (i.e. EN 1998-1 and EN 1993-1-10) it is unclear how to determine material toughness requirements in the seismic design situation. Indeed, no load combination to be used for determination of the stress level σ_{Ed} is provided in the seismic design situation. However, it can be considered that in the seismic design situations the stress level in dissipative zones will reach the yield strength ($\sigma_{Ed} = f_y(t)$). In such conditions the prescription given by EN 1993-1-10 cannot be applied, as this code provides data only up to stress levels of $\sigma_{Ed} = 0.75 f_y(t)$, and no extrapolation is allowed. In addition, one of the parameters contributing to the determination of the reference temperature T_{Ed} is the strain rate. No guidance is given in EN 1993-1-10 or EN 1998-1 on strain rate values to be considered in the seismic design situation. Anyway, some rules on the strain rate values can be found in the EN15129, which states that "... The differences due to strain rate shall be evaluated with reference to a frequency variation of ±50 %".

Background information on toughness oriented rules in EN 1993-1-10 is available in a joint JRC-ECCS report (Sedlacek et al. 2008). According to this document, the $\sigma_{Ed} = 0.75 f_y(t)$ value corresponds to the maximum possible "frequent stress", where for the ultimate limit state verification yielding of the extreme fibre of the elastic cross-section has been assumed ($\sigma_{Ed} = f_y(t)/1.35 = 0.75 f_y(t)$). Consequently, a tentative conclusion can be drawn, stating that the value $\sigma_{Ed} = 0.75 f_y(t)$ given by EN1993-1-10 would correspond to the possibly yielded cross-section, and presumably could be used for selection of material toughness and thickness in the seismic design situation.

However, other open issues remain, like the reference temperature in the seismic design situation and strain rate to be used for determination of the reference temperature, for which no guidance exists either in EN 1993-1-10 or in EN 1998-1.

18.2.3 Local Ductility

In order to guarantee adequate ductility of the dissipative members, EN 1998-1 provides restrictions of width-thickness ratio b/t of steel profiles according to the cross-sectional classes specified in EN 1993-1-1. However, the use of EC3 classification arises some criticisms, namely (1) the small number of parameters considered to characterize the beam performance and (2) the Eurocode classification is based on monotonic loading while it must be different for seismic loading, because of strength deterioration induced by the repetition of inelastic deformations.

In the recent past, different classification criteria of bare steel profiles accounting for both cross section slenderness and member slenderness were early proposed (e.g. Mazzolani and Piluso 1992; D'Aniello et al. 2012) and it has been pointed out that shifting from a section-based to a member-based classification would represent a significant advancement of the code. Indeed, the member-based classification is based on a relationship between the member ductility and the stress ratio $s = f_c/f_y$ the ratio between the peak collapse stress f_c and the yield stress f_y). A number of studies proposed different formulations for the stress ratio s (e.g. D'Aniello et al. 2012, 2014a, b, 2015a, b; Güneyisi et al. 2013, 2014), but in all cases the main parameters influencing s are the flange and web slenderness (λ_f and λ_w), as well as the shear length L^* (namely the distance of the cross section subjected to the peak bending moment from the contra-flexure cross section). Among the others, the formulation provided in D'Aniello et al. (2012) provides accurate empirical equations to predict the rotation capacity of steel profiles accounting for the influence of cyclic loading on the basis of experimental results. The tests showed that a cross section classification based on monotonic rotation capacity (as the case of EN 1993:1-1) could be non-conservative for seismic applications. Figure 18.2 clarifies this issue showing the reduction of rotation capacity due to cyclic loading as respect to monotonic loading for steel beams with wide flange and hollow section profiles, respectively.

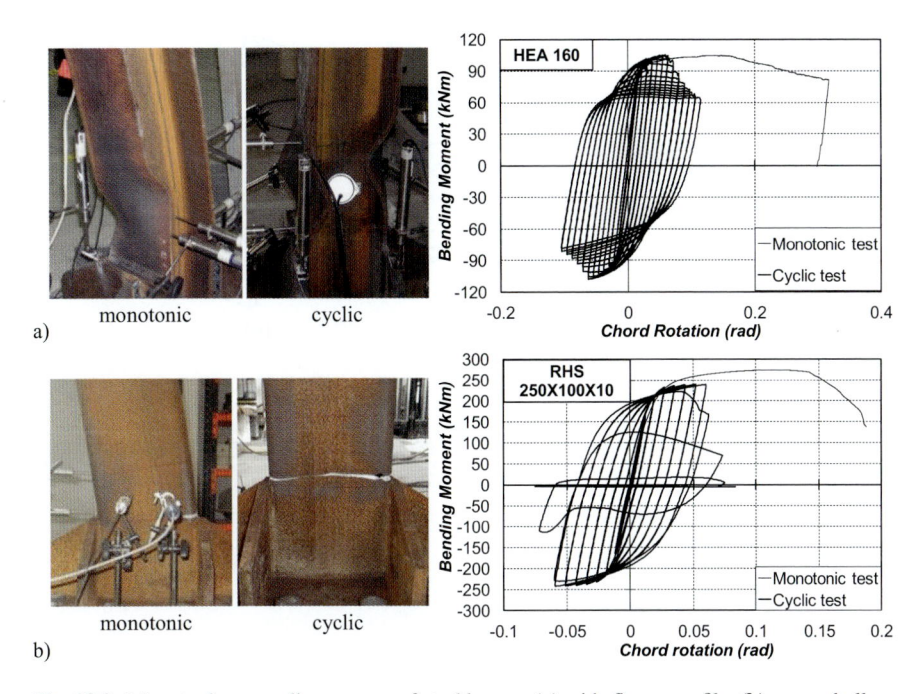

Fig. 18.2 Monotonic vs. cyclic response of steel beams: (**a**) wide flange profile; (**b**) square hollow section (D'Aniello et al. 2012)

18.2.4 Design Rules and Pre-qualification of Beam-to-Column Joints

The design of seismic resisting beam-to-column joints is one of the most troublesome issue encountered by designers of steel structures, especially if partial strength joints are considered. Indeed, no easy-to-use design tools able to predict the seismic performance of dissipative beam-to-column joints and to meet code requirements are currently available. With this regard, EN 1998-1 prescribes design supported by experimental tests, resulting in impractical solutions within the time and budget constraints of real-life projects (Landolfo 2016). Nowadays, this is one of the main reason that limits the use of steel in seismic resistant structures. On the other hand, also for full-strength joints reliable design tools are necessary. Owing to the variability of steel strength, these connections could not have enough overstrength (e.g. min $1.1 \times 1.25\ M_{b.rd}$, being $M_{b.rd}$ the beam bending strength), and in such cases their plastic rotation capacity must be prequalified by relevant tests and numerical procedures.

A recently concluded European research project "EQUALJOINTS (European pre-QUALified steel JOINTS, RFSR-CT-2013-00021) developed the first European prequalification procedure for the moment resisting connections in seismic resistant steel frames, in compliance with EN1998-1 requirements. Further aims of the project

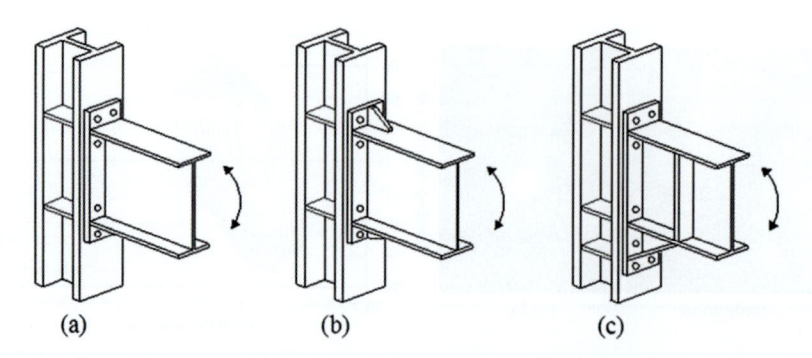

Fig. 18.3 Beam-to-column joints qualified in the framework of EQUALJOINTS project

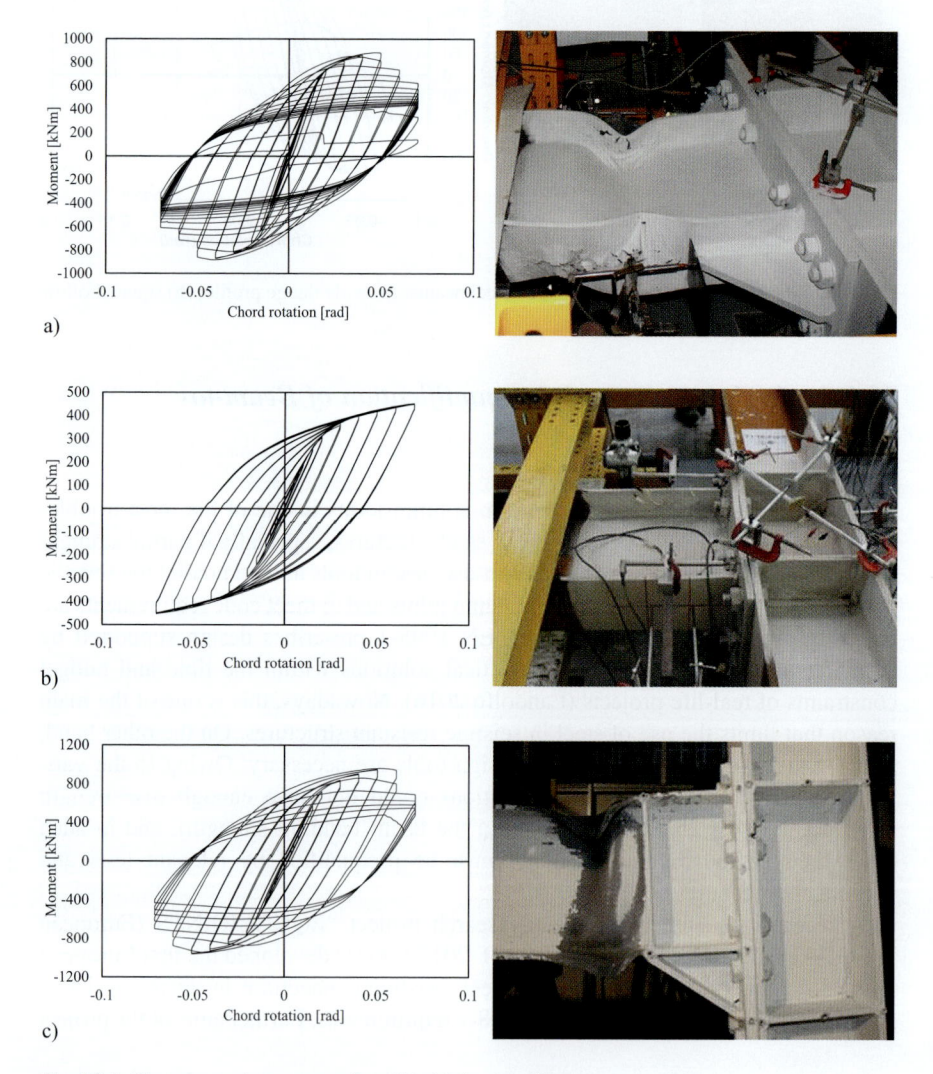

Fig. 18.4 Experimental response of qualified joints qualified

were to qualify a set of standards for all-steel beam-to-column joints (see Figs. 18.3 and 18.4), and to develop prequalification charts and design tools that can be easily used by designers. The project was also intended as a pre-normative research aimed at proposing relevant design criteria to be included in the next version of EN 1998-1.

The developed design criteria extend the philosophy of hierarchy of resistances to the joint by establishing a hierarchy among the strength of its macro-components (e.g. the web panel, the connection, the beam), and the corresponding sub-components (e.g. end-plate, bolts, welds, etc.), as well. The capacity design is applied between each macro-component in order to obtain three different design objectives defined comparing the joint (i.e. web panel and connection) strength to the beam flexural resistance, namely: full, equal and partial strength (D'Aniello et al. 2017a, b, c, d; Tartaglia et al. 2018). In the first case, the joint is designed to be more resistant respect to the beam, in order to concentrate all the plastic deformation in the connected element. Equal strength joints are designed to have a balanced strength between the connection and the beam bending capacity. Partial strength joints are designed to have the strength of the connection and the column web panel lower than 0.8 times the strength of the connected beam.

The capacity design requirements to obtain the required joint behaviour can be guaranteed by satisfying the following inequality:

$$M_{wp,Rd} \geq M_{con,Rd} \geq M_{con,Ed} = \alpha \cdot (M_{B,Rd} + V_{B,Ed} \cdot s_h) \quad (18.1)$$

Where $M_{wp,Rd}$ is the flexural resistance corresponding to the strength of column web panel; $M_{con,Rd}$ is the flexural strength of the connection; $M_{con,Ed}$ is the design bending moment at the column face; α depends on the design performance level. It is equal to $\gamma_{sh} \times \gamma_{ov}$ for the full-strength joints, while equal to 1 for equal strength joints. $M_{B,Rd}$ is the plastic flexural strength of the connected beam; s_h is the distance between the column face and the tip of the rib stiffener; $V_{B,Ed}$ is the shear force corresponding to the occurring of the plastic hinge in the connected beam; it is given by:

$$V_{B,Ed} = V_{B,Ed,M} + V_{B,Ed,G} \quad (18.2)$$

where $V_{B,Ed,G}$ is the contribution due to the gravity loads that is calculated in the section where theoretically the plastic hinge is expected to form. $V_{B,Ed,M}$ is the shear force due to the formation of plastic hinges at both beam ends, spaced by the length L_h (the approximate distance between plastic hinges) and calculated as:

$$V_{B,Ed,M} = \frac{2 \cdot M_{B,rd}}{L_h} \quad (18.3)$$

γ_{ov} is overstrength factor due to the material randomness, depending on the steel grade (assumed equal to 1.25, as recommended by EC8); γ_{sh} is the strain hardening factor corresponding to the ratio between the ultimate over the plastic moment of the beam. Based on the characteristic yield and ultimate strength of European mild

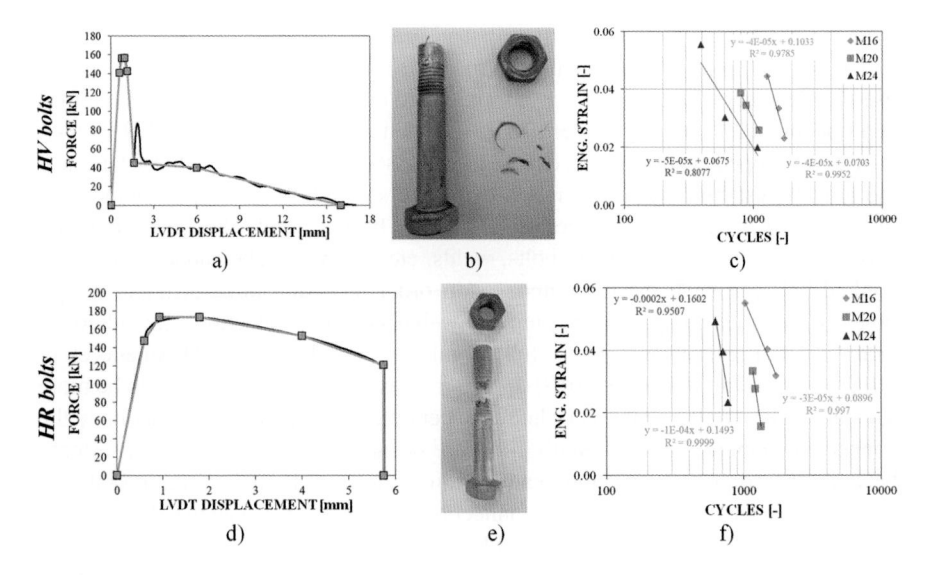

Fig. 18.5 HV vs. HR bolts: force – displacement, failure mode and low-cycle fatigue resistance

carbon steel grades and in line with AISC358-16 prescriptions, in the present study γ_{sh} is conservatively assumed equal to 1.20.

For equal and partial strength joints the following ductile criterion was also introduced in order to avoid failure mode 3 per bolt row (D'Aniello et al. 2017a, b, c, d; Cassiano et al. 2017, 2018), namely:

$$F_{t,Rd} \geq \gamma_{ov} \cdot \gamma_{sh} \cdot F_{p,Rd} \qquad (18.4)$$

Where $F_{t,Rd}$ is the resistance of the bolts in tension and $F_{p,Rd}$ is the resistance of the entire line.

The EQUALJOINTS project also investigated the type of pre-loadable bolts to be used for the prequalified joints. Indeed, in Europe both HV and HR assemblies can be used for high strength pre-loadable bolts, but their tensile failure mode is significantly different (D'Aniello et al. 2016, 2017a, b, c, d), namely nut stripping in the first case (see Fig. 18.5a, b) and tensile tearing in the second (see Fig. 18.5d, e). Considering that the shear resistance is kept once the nut stripping is activated, HV bolts were used to qualify the joints. In addition, low-cycle fatigue resistance of bolt assemblies was also investigated by imposing constant amplitude displacement cycles and measuring the number of cycles N until failure was achieved, hence allowing determining the ε-N curves (see Fig. 18.5c, f). It is interesting to observe that in both types of bolts the fatigue resistance inversely increases with the diameter, namely being larger for the smaller diameters. This feature can be explained considering that larger size of the crests (which increase with the diameter of the shank) of the threaded zone corresponds to an increase of the stress concentration.

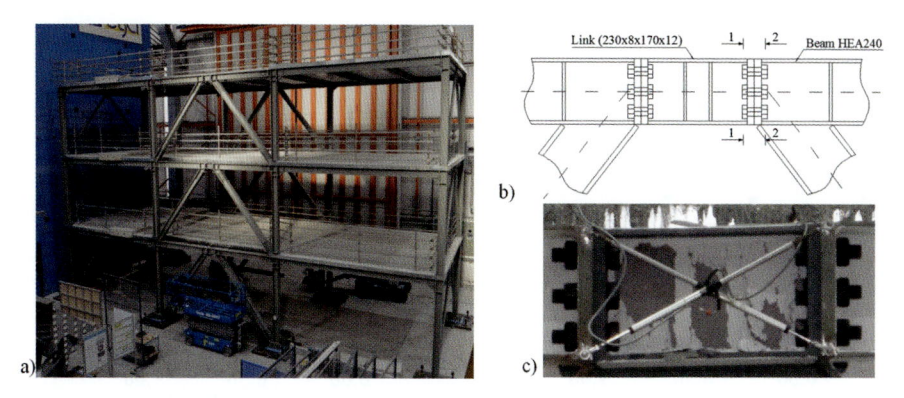

Fig. 18.6 DUAREM (FP7/2007–2013 SERIES n° 227887) project: (**a**) full scale mock-up; (**b**) removable link details and (**c**) response

18.2.5 *Innovative Solutions for Eccentrically Braced Frames*

Provisions for design of eccentrically braced frames (EBFs) in EC8 refer to links made of I sections only. Other cross sections, such as tubular ones can also be used (Berman and Bruneau 2008). Tubular links are an attractive solution because they can provide a stable response without the need of bracings to restrain lateral-torsional buckling, as the case of traditional I-shape links. However, requirements for this type of link are currently missing in the EC8.

Conventional links are continuous with the containing beam. However, replaceable links can be used providing effective performance (Dubina et al. 2008; Mazzolani et al. 2009). Bolted flush end-plate connections at both link ends and the rest of the structure can be used, thus allowing the replacement of the dissipative elements damaged in the aftermath of moderate to strong earthquakes, with corresponding reduction of the repair costs.

Detachable links, very effective in dual frames, are obtained by combination of moment resisting frames (MRFs) and eccentrically braced frames (EBFs).

Within the recent European project "DUAREM" (which was part of the FP7 Project – Seismic Engineering Research Infrastructures for European Synergies – SERIES) the performance of Dual-EBFs designed to provide re-centering capability to the structure has been experimentally investigated (Ioan et al. 2016). The objectives of this project were (i) to validate the re-centring capability of dual structures with removable shear links; (ii) to evaluate the overall seismic performance of dual EBFs; (iii) to investigate the interaction between the link and the composite slab and (iv) to validate the link removal technology after seismic damage. To achieve these aims, full-scale pseudo-dynamic tests were carried out on a three-story building (see Fig. 18.6a) with one bay equipped with EBF and two spans with MRF per side of the test direction. Figure 18.6b shows the details of the replaceable links, while

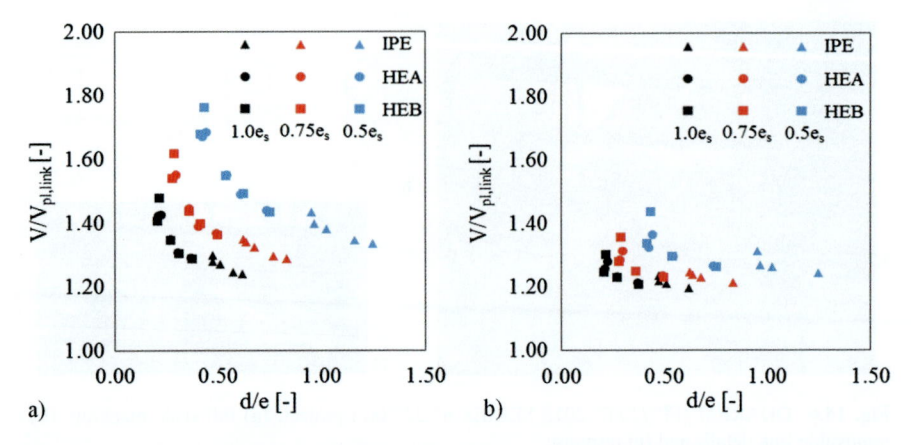

Fig. 18.7 Link shear overstrength with (**a**) and without (**b**) axial restraints (Zimbru et al. 2017)

Fig. 18.6c depicts the damaged condition of a link after testing. The experimental tests showed that the effectiveness of dual systems and the need to improve the EC8 provisions for these systems. However, a crucial aspect influencing the effectiveness of the system is the resistance of the link end-connections, which can be prone to premature failure.

Indeed, considering solely the first order theory, the flush end-plate connections are theoretically subjected to link ultimate shear force (e.g. $V_u = \gamma_{ov} 1.5 V_p$) and the corresponding bending moment (e.g. $M_u = 0.5 e V_u$, where e is the link length). However, as shown by Della Corte et al. (2013), significant tensile axial forces may develop due to restraints to axial deformations and nonlinear geometric effects, similarly to catenary actions developing in the case of joints under column loss (Cassiano et al. 2017; Tartaglia and D'Aniello 2017). These tensile forces may significantly modify the link shear overstrength, which can be significantly larger than the value 1.5 currently recommended by EC8.

Figure 18.7 clarifies this concept. In fact, it can be seen the variation of the "d/e" ratio (with "d" depth of the link section and "e" its length) the "V/V_p" link overstrength can increase up to 1.8 in the case of compact links and in the presence of axial restraints (see Fig. 18.7a), while the codified value equal to 1.5 is conservative without axial restraints (see Fig. 18.7b).

Based on the results of comprehensive parametric analyses, Della Corte et al. (2013) proposed the following equation to estimate the link shear overstrength for plastic rotation $\gamma_p = 0.08$ rad:

$$\frac{V_{0.08}}{V_y} = 1 + \beta \left(\frac{A_f}{A_v}\right)\left(\frac{d}{e}\right) \tag{18.5}$$

Where A_f is the area of both flanges of the link, A_v the shear area and β is a coefficient depending on (i) the stiffness of the axial restraints, (ii) the shape of cross section, (iii) the material strain-hardening properties. For all European HE shapes it can be

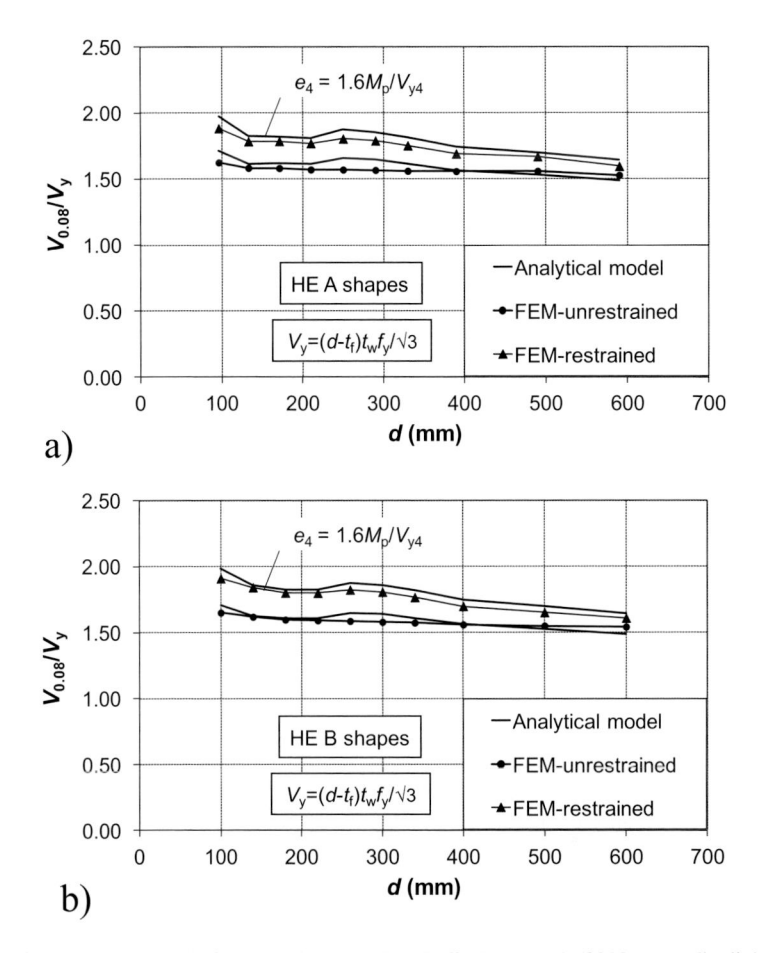

Fig. 18.8 Accuracy of analytical model proposed by Della Corte et al. (2013) to predict link shear overstrength

assumed values of $\beta = 1.70$ for links with axial restraint and $\beta = 1.35$ for links without axial restraint. In the case of IPE shapes, it can be used $\beta = 1.70$ for links with axial restraint and $\beta = 1.60$ for links without axial restraint. Figure 18.8 shows the satisfactory agreement between the link shear overstrength ratios predicted by Eq. (18.5) and those obtained from finite element simulations.

Therefore, extension of code provisions and development of design guidelines for links with axial restraints as well as detachable links are necessary and should be provided for the next version of EC8.

18.2.6 Detail Rules for Traditional Structural Systems

The generic design force given by EC8 for non-dissipative members is equal to $R_{Ed,i} = R_{Ed,G,i} + 1.1\, \gamma_{ov}\Omega R_{Ed,E,i}$ where subscripts "G" and "E" indicates the effect of gravity and earthquake loads, respectively; γ_{ov} is the coefficient accounting for yield stress randomness; $\Omega = \min(R_{pl,Rd,i}/R_{Ed,i})$, where $R_{pl,Rd,i}$ is the design strength of the i-th plastic zone and $R_{Ed,i}$ is the required strength. However, the research of last 10 years highlighted some controversial aspects in this approach. For Moment resisting frames (MRF) in case of small gravity load effects, amplifying the earthquake-induced counterpart of the required strength by the factor $\gamma_{ov}\Omega$ means that the internal actions corresponding to the first real plastic hinge formation are being calculated. However, in case of large gravity load effects, the proposed Ω factor markedly underestimates the real overstrength (Tenchini et al. 2014; Cassiano et al. 2016). One proposal of correction has been reported by Elghazouli (2008). The meaning of the multiplicative coefficient 1.1 is not clearly stated in the code. According to Elghazouli (2008) it was introduced to take into account strain hardening of steel and strain rate effects. However, the meaning of this coefficient 1.1 and its rational background behind are unknown.

For braced structures the general EC8 capacity design procedure is not effective to avoid soft storey mechanism and to guarantee adequate energy dissipation into the ductile elements. One of the main criticism about the design rules of dissipative members is the requirement on the variation of capacity-to-demand ratio of bracing members or overstrength factor $\Omega_i = N_{pl,br,Rd,i}/N_{Ed,br,i}$ that should vary within the range Ω to 1.25 Ω, where $\Omega = \min(\Omega_i)$, $N_{pl,br,Rd,i}$ is the plastic axial strength of bracing members at the i-th storey and $N_{Ed,br,i}$ is the relevant seismic demand. The recent studies carried out by (Costanzo et al. 2017a, b, 2018; Costanzo and Landolfo 2017) clearly show that this rule (which is not included in North American seismic codes) does not ensure uniform distribution of plastic deformations along the building height. With reference to non-dissipative members, EC8 allows using simplified procedure for calculating the forces acting in the columns belonging to the braced bays. Indeed, plastic mechanism analysis is not required for columns and it is sufficient to perform only an elastic analysis without specifically accounting for the distribution of internal actions occurring in the post-buckling range. This calculation method gives less conservative estimation of internal forces than those calculated according to both North American codes, thus largely underestimating the earthquake-induced effects into the columns in the most of cases. Another criticism of EC8 concerns the design of the brace-intercepted beam of inverted-V or chevron CBFs. Indeed, the rules prescribed by EN 1998 lead to design weaker and more deformable beams as respect to AISC 341 and CSA-S16. This aspect significantly influences the poor dissipative behaviour of chevron CBFs. As shown by D'Aniello et al. 2015a, b, the flexural stiffness of the brace-intercepted beam plays a key role in the performance of chevron CBFs, being the flexural response of the beam and the brace deformation in compression correlated phenomena (D'Aniello et al. 2015a, b; Costanzo et al. 2016). Indeed, the elastic deflection caused by the unbalanced force can be large enough to prevent yielding of the brace in tension and to concentrate the damage in the compression diagonal, thus leading to very poor overall performance due to the brace deterioration. The governing parameter is the beam-to-brace stiffness ratio (K_F) defined as the ratio k_b/k_{br}, where k_b is the beam flexural stiffness at the intersection with braces and k_{br} is the vertical stiffness of the bracing members (D'Aniello et al. 2015a, b). On the basis of interpolating curve fitting of numerical data, the relationship between the interstorey drift ratio corresponding to the yielding of the brace in tension (θ_y) and K_F was derived as follows:

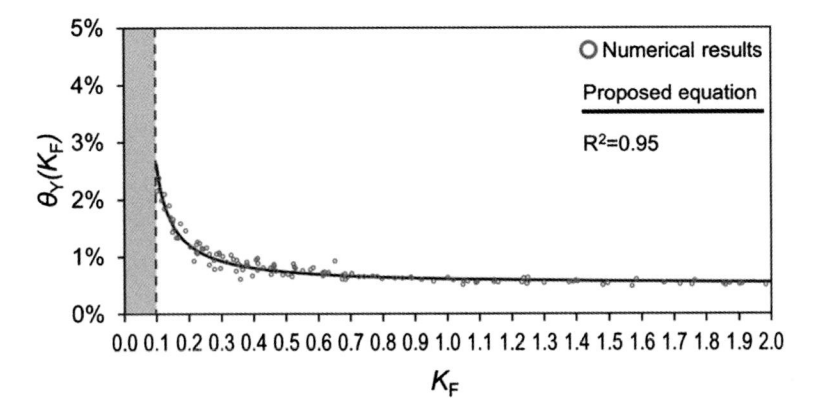

Fig. 18.9 Brace yielding drift ratio vs. K_F: numerical results and proposed equation (D'Aniello et al. 2015a, b)

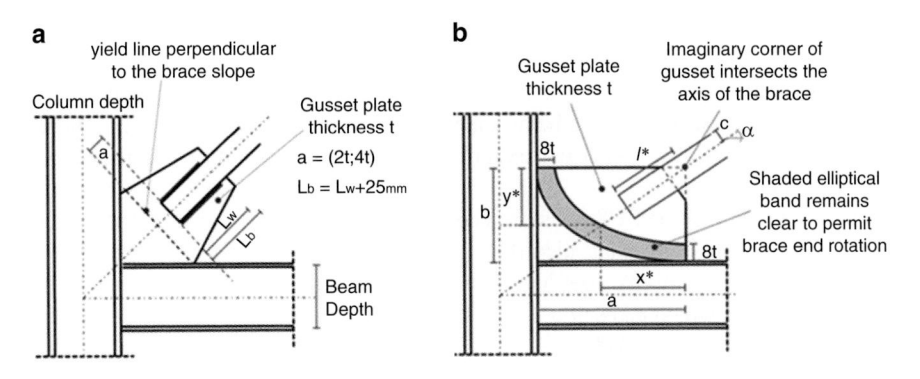

Fig. 18.10 Gusset plate connections for CBFs: (**a**) linear and (**b**) elliptical yield line (Landolfo et al. 2017)

$$\theta_y(K_F) = \frac{0.008 \cdot K_F + 0.0013}{1.6 \cdot K_F - 0.08} \quad \forall K_F \geq 0.1 \tag{18.6}$$

As it can be observed in Fig. 18.9, Eq. (18.6) is limited to $K_F > 0.1$, because bracing member cannot yield in tension for smaller values. Based on this relation, $K_F = 0.2$ can be assumed as rational value to allow brace yielding at 1% of interstorey drift ratio.

Another issue needing further improvement is the design of gusset plates of ductile concentrically braced frames, which should be detailed to favour the out-of-plane buckling of brace as well as adequate ductility and low-cycle fatigue strength. With this regard, US guidelines (e.g. AISC341-16) provides criteria and examples of appropriate details and corresponding rules for gusset plate connections that can be designed to develop ductile yield line (either linear, as in Fig. 18.10a, or elliptical, as in Fig. 18.10b) where the brace rotates following its buckling.

18.2.7 Drift Limitations and Second-Order Effects

A very general problem with the current codified EN 1998-1 rules for the design of MRFs is the compatibility of maximum drifts imposed at the damage limitation limit state and the large behaviour factors for the ultimate limit state. It has long been recognized that the design of MRFs is dictated by drift limitations of EN 1998-1, which produces strong overstrength and, consequently, reduced ductility demand as well as increase of costs (Elghazouli 2008; Aribert and Vu 2009). The imposed limits for P-Delta effects could also be a source of significant frame overstrength (Elghazouli 2008; Aribert and Vu 2009). Indeed, the EC8 stability coefficient severely penalizes sway frames as MRFs. It is interesting to note that EC8 and North American rules (e.g. NEHRP) adopts different stability coefficients. In particular, in the EC8 the stability coefficient is a function of the secant stiffness at each storey, namely it depends on the behaviour factor assumed in the design. On the contrary, North American provisions considers that this coefficient is dependent on the elastic storey stiffness, thus being independent of the structural ductility. Therefore, this substantial difference can explain the larger sensitivity of EC8 frames to the stability checks.

In order to revise the EC8 rules on second-order effects Vigh et al. (2016) carried out a numerical study that identified the main criticisms related to the EC8 interstorey drift sensitivity coefficient and proposed simplified formulation for the stability index θ, which takes into account for both the design overstrength factor and the redundancy factor as follows:

$$\theta = \frac{P_{tot} \cdot q_d \cdot d_e}{V_{tot} \cdot h \cdot \Omega_S} \tag{18.7}$$

Where P_{tot} is the total gravity load above the floor under investigation, V_{tot} is the total global shear load above the floor, d_e is the elastic displacement (determined by a linear analysis based on the design response spectrum; i.e. not including q_d), h is the storey height. The factor Ω_S is equal to $\Omega \times (\alpha_u/\alpha_1)$, being Ω the design overstrength factor and α_u/α_1 the ratio between the collapse multiplier and that corresponding to the first non-linear event.

On the basis of the comprehensive numerical investigation, this formulation of θ appears appropriate when the moment resisting frames are designed for ductility class DCH using a behaviour factor of about 6.

18.2.8 Rules for Low-Dissipative Structures

The analysis of the current EC8 shows that it is necessary to develop simplified rules and design aids for the design of structures in low-seismicity regions. Besides, the development of intermediate rules, less severe than DCM and DCH rules, seems desirable thus allowing the justification of a low-to-moderate ductility of the structure. Moreover, the work undertaken by the Committee showed that background studies are still necessary to detect weaknesses of DCL design in moderate to high seismicity regions in case of an earthquake with PGA higher than expected and definition of required elastic overstrength providing the same safety level as DCM or DCH design.

To cover this lack, the recent RFCS MEAKADO research program (Degee et al. 2015) systematically investigated the design options for steel and steel-concrete structures located in low and moderate seismic areas (e.g. Germany, France, Belgium, etc.) with the aim to propose design rules that are optimised for those levels of seismic action, trying to relax some of the most constraining rules of the DCM and DCH requirements of EC8. In order to develop less stringent capacity design rules for moderate ductile structures, this project has been organized in two main actions. The first investigates the main phenomena that contribute to energy dissipation in steel structures subjected to earthquake, but whose knowledge is not yet sufficient to quantify their contribution such as slip in bolted connections; plastic ovalization of bolt holes; post-buckling strength of diagonals in compression; energy dissipation capacity in beams with class 3 and 4 cross-sections. To achieve this purpose both experimental tests and numerical simulations have been carried out. The second action investigates the possibilities of tuning EC8 design rules given for DCM to the actual seismicity zone and to the targeted behaviour factor. The research focuses on moment resisting frames and concentrically braced frames, as being the most common configurations in practice. The following DCM-EC8 design rules are being re-considered, mainly on the base of numerical simulations calibrated and validated by the results of the experimental campaign. The experimental and numerical evidence confirmed that for structures that should withstand accelerations at Significant Damage (SD) limit state not larger than 0.5 g it is feasible to adopt simplified rules and relaxed requirements of global hierarchy of resistances.

18.2.9 New Structural Types

The rules for non-conventional structural typologies are missing in the current version of EC8. Therefore, there is a need to incorporate design provisions for concentrically braced frames with buckling restrained braces (BRBs), MRFs with free from damage joints, concentrically braced frames with dissipative connections in EN 1998-1, and cold formed structures (CFS).

BRBs are widely used for the seismic retrofit of existing reinforced concrete structures (e.g. D'Aniello et al. 2014a, b; Della Corte et al. 2015). Therefore, this type of bracing is often considered as a hysteretic anti-seismic device. However, BRBs can be also considered as special type of dissipative concentrically braced system. In line with that, design provisions for BRBs already exist in other seismic design codes, such as AISC 341-16. Availability of design provisions for these structural systems in EN 1998-1 would enlarge the opportunities of European designers. Recently, Bosco et al. (2015) proposed a design procedure for steel frames equipped with BRBs consistent with the framework of EC8. Indeed, their design procedure is obtained by modifying and updating the rules stipulated in EC8 for steel chevron braced frames. On the basis of pre-defined allowable ductility demand Significant Damage and Near Collapse limit states, Bosco et al. (2015) calibrated the behaviour factor for BRBs by nonlinear dynamic analysis for both moderate and high seismic excitation levels. More recently, Vigh et al. (2017) refined this design procedure and demonstrated the feasibility of the proposed design procedure in a reliable manner with a methodology based on FEMA P-695.

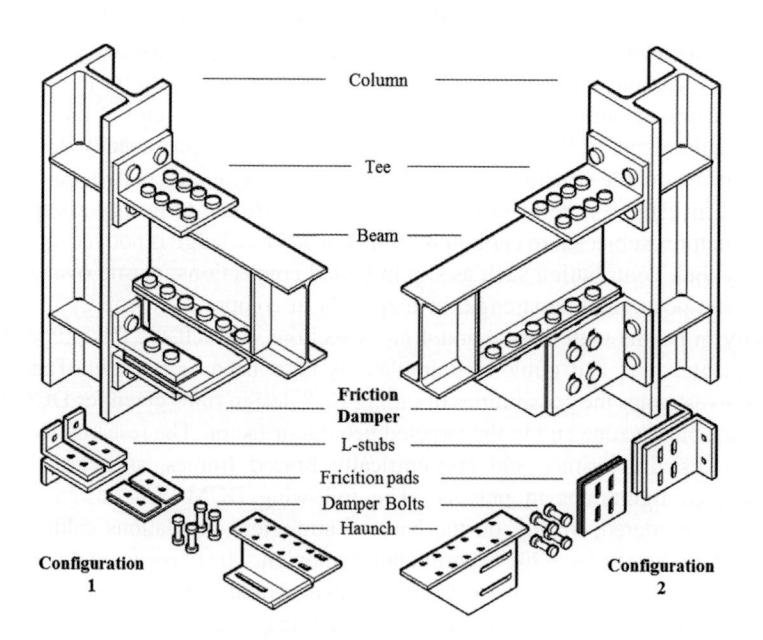

Fig. 18.11 The two types of FREEDAM joints

18 Seismic Design of Steel Structures: New Trends of Research and...

The use of free from damage joints is very promising. This new system allows replacing the plastic damage into plastic hinges at the end of the beams with the friction dissipation developed into special types of friction dampers located at the bottom flange of the beam (Piluso et al. 2014; Latour et al. 2015). Specific design rules and prequalification criteria for this type of joints are currently under development within the ongoing RFCS FREEDAM project. Figure 18.11 depicts the two types of the FREEDAM joints (D'Aniello et al. 2017a, b). As it can be noted, the main difference is the shape of the friction damper (horizontal and vertical, respectively). In particular, the sliding part of the friction damper is detailed with slotted holes that allow the slippage with respect to the fixed parts (friction pads and L-stubs). Under bending actions, both joint configurations rotate around the centre of compression, which is located in the mid thickness of the Tee web under hogging and in the mid distance between the L-stubs/bolts of the damper under sagging. Therefore, in order to avoid the damage of steel components, those elements should be designed to resist the maximum forces transferred by the friction damper under both static and dynamic slippage. The required design strength of FREEDAM joints can be easily obtained by calibrating the slippage force of the friction damper. The slippage force is computed as the product of the friction coefficient (which is a mechanical feature of the interface between the friction pads and the haunch), the number of friction interfaces (e.g. two in case of a symmetrical damper) and the sum of the pre-tightening forces applied by means of preloaded bolts.

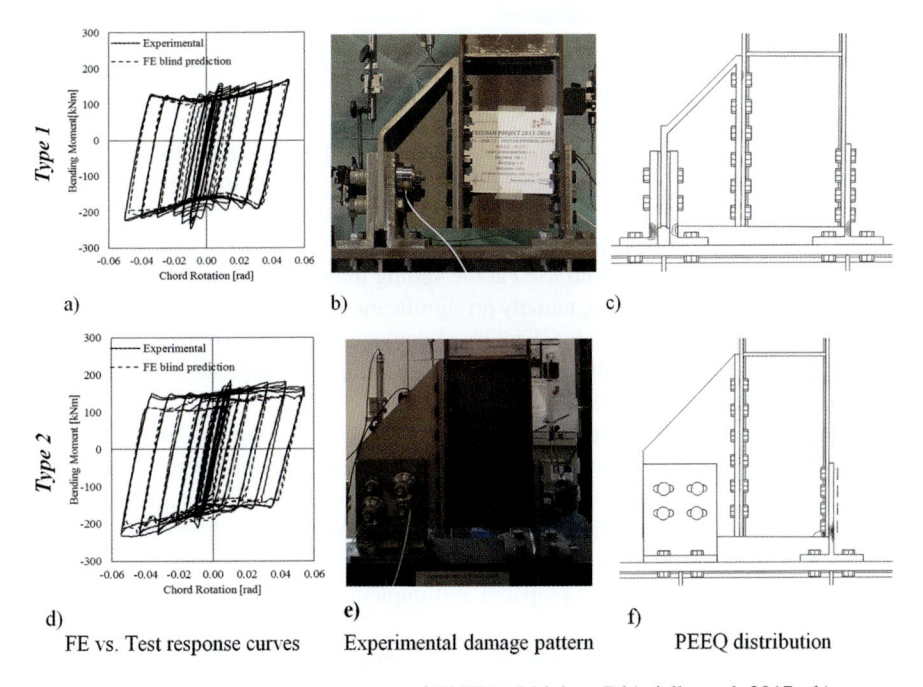

Fig. 18.12 Experimental vs. FE response of FREEDAM joints (D'Aniello et al. 2017a, b)

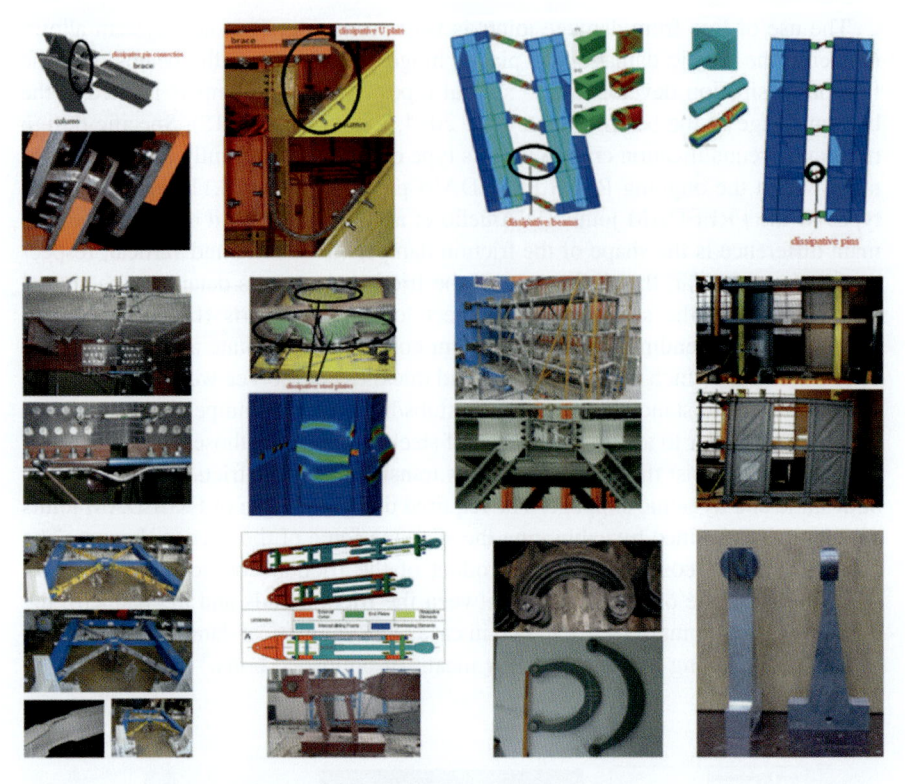

Fig. 18.13 RFCS INNOSEIS project: anti-seismic devices and dissipative systems (Vayas 2017)

Figure 18.12 depicts the comparison between the FE analyses and the cyclic tests in terms of response curves (i.e. bending moment at the column axis and chord rotation) and failure mode, respectively. The experimental tests showed that the cyclic behaviour of these joints is stable with low degradation, even though it depends on the friction material used at the sliding interface. FE analyses confirmed the experimental observations, namely no significant damage occurs even though the numerical models show some localization of damage in the base of the webs of both Tee and L-stubs. However, the equivalent plastic strain (PEEQ) are very small and fully acceptable for large rotational demands.

Another structural system that currently lacks code support is concentrically braced frames with dissipative connections. Experimental and theoretical investigations on concentrically braced frames with dissipative connections exist (Plumier et al. 2006) that indicate an excellent seismic performance in terms of strength, stiffness and ductility. Design guidelines for concentrically braced frames with dissipative connections can be prepared and implemented in future versions of EN 1998-1.

The use of anti-seismic devices is another important topic that is not clearly covered by the current codes. Indeed, there is a lack of uniformity and rules of applications using both EC8 and the EN 15129. The recent European project INNOSEIS (Vayas 2017) aims at filling this gap. In this project several innovative anti-seismic devices (see Fig. 18.13) are investigated in order to draft pre-normative design recommendations that will allow them to receive the status of code-approved systems. In addition, a reliability based methodological procedure to define values of behaviour factors (i.e. q-factors) for building structures will be established and applied to determine q-factors for structural systems with the examined anti-seismic devices.

For what concerns the cold formed structures (CFS), the current EC8 does not provide any specific design rules. In addition, considering the current limitations for DCM and DCH, the CFS can be theoretically used only for low seismic areas as DCL structures, since CFS are made of class 4 members. However, this condition is far from being valid. Indeed, if properly designed, CFS can guarantee adequate level of ductility and can be used even for high seismicity areas (Fiorino et al. 2009, 2011, 2012, 2013, 2014, 2016a, b, 2017a, b; Macillo et al. 2017a, b; Iuorio et al. 2014a, b).

CFS has been growing in popularity all over the world, because it represents a suitable solution to the demand for low-cost high-performance houses (Dubina et al. 2013). The members adopted for CFS are obtained by cold rolling, namely produced by pressing or bending steel sheets with thickness ranging between 0.4 and 7 mm. This construction process provides several advantages such as the lightness of systems, the high quality of end products, thanks to the production in controlled environment, and the flexibility due to the wide variety of shapes and section dimensions that can be obtained by the cold rolling process. Moreover, the CFS systems, being dry constructions, ensure short execution time. Besides, economy in transportation and handling, the low maintenance along the life time together with the high strength to weight ratio (which is an essential requirement for a competitive behaviour under seismic actions) represents additional benefits that CFS are able to achieve.

Over the last decades the research team under the leadership of the Author, at the University of Naples Federico II, has been involved in several research programs (e.g. FP7 ELISSA project, KNAUF project, LAMIEREDL project, etc.) focusing on the assessment of the seismic behaviour of low-rise buildings built with CFS members (see Fig. 18.14) by means of comprehensive experimental and numerical campaign carried out in three scale levels: micro-scale, meso-scale and macro-scale.

Micro-scale level consisted of monotonic and cyclic tests on main connecting systems (see Fig. 18.14a–f), namely panel-to-steel connections, clinching steel-to-steel connections, ballistically nailed gypsum panel-to-steel connections, nails, etc.

Meso-scale tests consisted of monotonic and cyclic tests on shear walls (see Fig. 18.14g–i), namely full-scale shear walls, CFS bracings, partition walls, etc.

Finally, in order to evaluate the global building seismic response, shake table (i.e. macro-scale) tests on a full-scale two-storey 2.7 m × 4.7 m (plan dimensions) × 5.3 m (high) building (macro-scale level) were carried out (see Fig. 18.14l–n).

The main findings of all these studies will contribute to the background documents of the relevant new section of the next EC8.

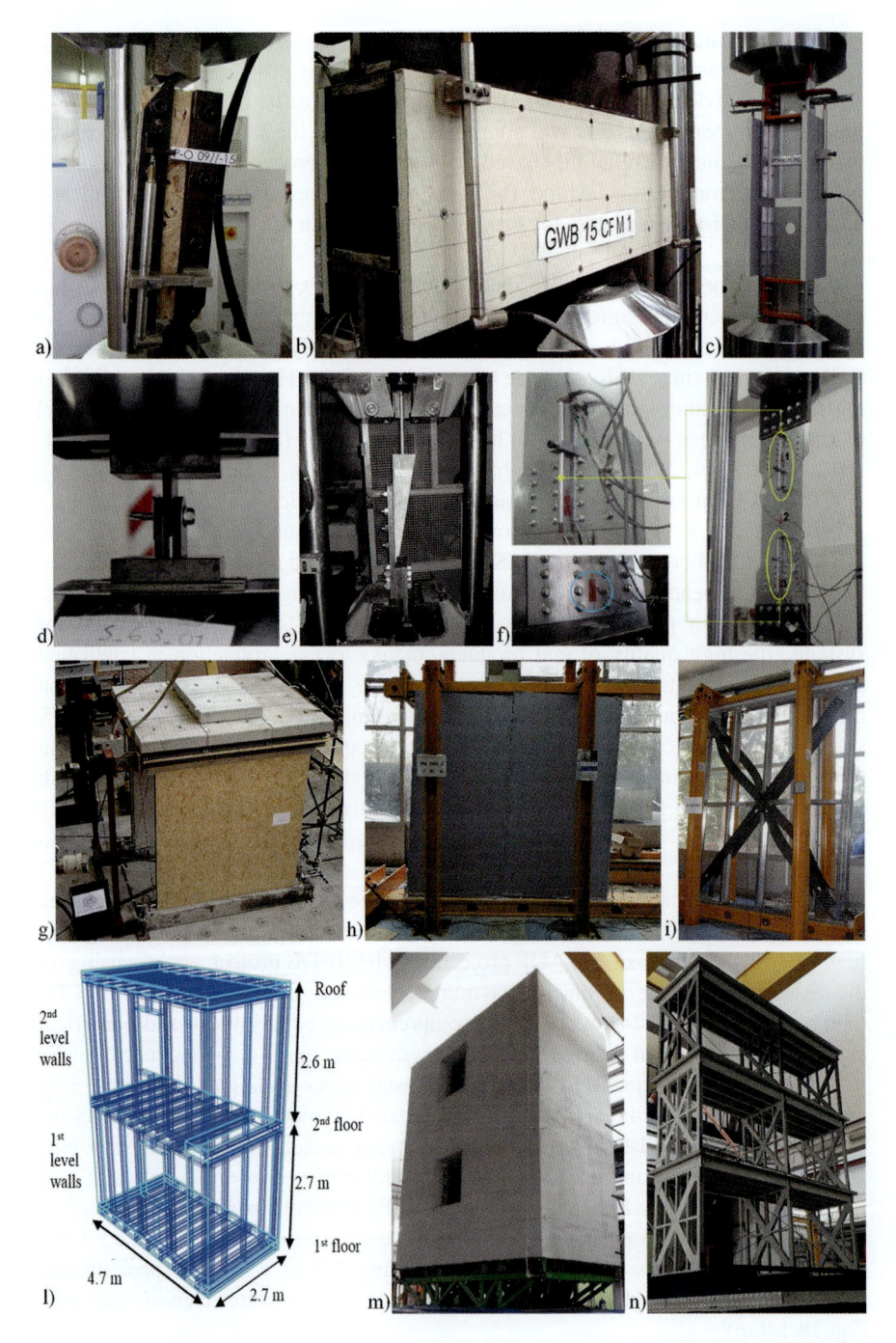

Fig. 18.14 Some examples of the experimental activities on CFS carried out at University of Naples Federico II

18.3 Conclusions

This paper briefly summarizes the main criticisms of the provisions and design rules given by EC8 for steel structures that need to be revised and improved in the next generation of the code.

Within this framework, the WG2 of the TC250/SC8, in cooperation with the ECCS TC13, is carrying out a preliminary work aimed at supporting the revision of the Eurocode 8 chapters on steel and steel-concrete composites structures and at introducing a new chapter on aluminium alloy structures. Currently, the main ongoing tasks deals with the analysis of the current state of international codes and the critical examination of the comments received from different European countries after 10 years of application of Eurocodes. The main goals to achieve in the next generation of codes are to overcome the criticisms found in the current version of the code by the different users, to cope the gap between the latest findings of the scientific research and the normative framework and to take into account the needs of construction industry and professional practice.

This preparatory study will provide a background documents for the experts of steel structures involved in the Project Team of SC8.T2 - Material-dependent sections of EN 1998-1, which is the body that will have to finalize the drafting of all chapters of EC8 concerning each building material. Hence, this phase is extremely important and delicate, because it will lead to the definition of the future structure of Eurocode 8. In this regard, the Author, as coordinator of TC13 and TC250/SC8/WG2, wishes to express his deepest gratitude to all members of these committees, which provided their constantly enthusiastic, valuable and voluntary contributions to ensure seismically safer and more resilient steel constructions.

References

Aribert JM, Vu HT (2009) New criteria for taking into account P-Δ effects in seismic design of steel structures. Proceedings of the 6th STESSA conference, 2009

Berman J, Bruneau M (2008) Tubular links for eccentrically braced frames. II: experimental verification. J Struct Eng ASCE 134(5):702–712

Bosco M, Marino EM, Rossi PP (2015) Design of steel frames equipped with BRBs in the framework of Eurocode 8. J Constr Steel Res 113:43–57

Cassiano D, D'Aniello M, Rebelo C, Landolfo R, Da Silva LS (2016) Influence of seismic design rules on the robustness of steel moment resisting frames. Steel Compos Struct Int J 21 (3):479–500

Cassiano D, D'Aniello M, Rebelo C (2017) Parametric finite element analyses on flush end-plate joints under column removal. J Constr Steel Res 137:77–92

Cassiano D, D'Aniello M, Rebelo C (2018) Seismic behaviour of gravity load designed flush end-plate joints Steel Compos. Struct Int J (in press)

Costanzo S, Landolfo R (2017) Concentrically braced frames: European vs. North American seismic design provisions. Open Civil Eng J 11(Suppl-1, M11):453–463

Costanzo S, D'Aniello M, Landolfo R (2016) Critical review of seismic design criteria for chevron concentrically braced frames: the role of the brace-intercepted beam. Ingegneria Sismica: Int J Earthq Eng 1–2:72–89

Costanzo S, D'Aniello M, Landolfo R (2017a) Seismic design criteria for chevron CBFs: European vs North American codes (PART-1). J Constr Steel Res 135:83–96

Costanzo S, D'Aniello M, Landolfo R (2017b) Seismic design criteria for chevron CBFs: proposals for the next EC8 (PART-2). J Constr Steel Res 138C:17–37

Costanzo S, D'Aniello M, Landolfo R De Martino A (2018) Critical discussion on seismic design criteria for X concentrically braced frames. Ingegneria Sismica: Int J Earthq Eng (in press)

D'Aniello M, Landolfo R, Piluso V, Rizzano G (2012) Ultimate behavior of steel beams under non-uniform bending. J Constr Steel Res 78:144–158

D'Aniello M, Güneyisi EM, Landolfo R, Mermerdaş K (2014a) Analytical prediction of available rotation capacity of cold-formed rectangular and square hollow section beams. Thin-Walled Struct 77:141–152

D'Aniello M, Della CG, Landolfo R (2014b) Finite element modelling and analysis of "all-steel" dismountable buckling restrained braces. Open Constr Build Technol J 8(Suppl 1: M4):216–226

D'Aniello M, Costanzo S, Landolfo R (2015a) The influence of beam stiffness on seismic response of chevron concentric bracings. J Constr Steel Res 112:305–324

D'Aniello M, Güneyisi EM, Landolfo R, Mermerdaş K (2015b) Predictive models of the flexural overstrength factor for steel thin-walled circular hollow section beams. Thin-Walled Struct 94:67–78

D'Aniello M, Cassiano D, Landolfo R (2016) Monotonic and cyclic inelastic tensile response of European preloadable gr10.9 bolt assemblies. J Constr Steel Res 124:77–90

D'Aniello M, Tartaglia R, Costanzo S, Landolfo R (2017a) Seismic design of extended stiffened end-plate joints in the framework of Eurocodes. J Constr Steel Res 128:512–527

D'Aniello M, Cassiano D, Landolfo R (2017b) Simplified criteria for finite element modelling of European preloadable bolts. Steel Compos Struct IntJ 24(6):643–658

D'Aniello M, Zimbru M, Landolfo R, Latour M, Rizzano G, Piluso V (2017c) Finite element analyses on free from damage seismic resisting beam-to-column joints. In Papadrakakis M, Fragiadakis M (eds), Proceedings of COMPDYN 2017-16th ECCOMAS thematic conference on computational methods in structural dynamics and earthquake engineering, Rhodes Island, Greece, 15–17 June 2017

D'Aniello M, Zimbru M, Latour M, Francavilla A, Landolfo R, Piluso V, Rizzano G (2017d) Development and validation of design criteria for free from damage steel joints. CE/papers, special issue.In Proceedings of Eurosteel 2017, vol 1(2–3), September 2017, pp 263–271. doi: https://doi.org/10.1002/cepa.57

Degee H, Kanyilmaz A, Castiglioni C, Couchaux M, Martin P-O, Aramburu A, Calderon I, Wieschollek M, Hoffmeister B (2015) Design of steel and composite structures with limited ductility requirements for optimized performances in moderate earthquake areas – the "MEAKADO" project. SECED 2015 conference: earthquake risk and engineering towards a resilient world, 9–10 July 2015, Cambridge, UK

Della Corte G, D'Aniello M, Landolfo R (2013) Analytical and numerical study of plastic overstrength of shear links. J Constr Steel Res 82:19–32

Della Corte G, D'Aniello M, Landolfo R (2015) Field testing of all-steel buckling restrained braces applied to a damaged reinforced concrete building. J Struct Eng ASCE 141(1)

Dubina D, Stratan A, Dinu F (2008) Dual high-strength steel eccentrically braced frames with removable links. Earthq Eng Struct Dyn 37(15):1703–1172

Dubina D, Ungureanu V, Landolfo R (2013) Design of cold-formed steel structures: eurocode 3: design of steel structures. Part 1–3 design of cold-formed steel structures. ECCS – European Convention for Constructional Steelwork, Associacao Portuguesa de/Wiley, Timisoara (Romania)

Dubina D, Landolfo R, Stratan A, Vulcu C (2014) Application of high strength steels in seismic resistant structures. "OrizonturiUniversitare" Publishing House. ISBN 978-973-638-552-0.

Elghazouli AY, Seismic design of steel framed structures to Eurocode 8. Proceedings of the 14th world conference on earthquake engineering, Beijing, China, 12–17 Oct 2008
EN 1993-1. Eurocode 3: design of steel structures—Part 1–1: general rules and rules for buildings. CEN, 2005
EN 1993-1. Eurocode 3: design of steel structures— Part 1–8: design of joints. CEN, 2005
EN 1993-1-10. 2005 – Eurocode 3: design of steel structures – Part 1–10: Material toughness and through-thickness properties. CEN, 2005
EN 1998-1. Design of structures for earthquake resistance – Part 1: General rules, seismic actions and rules for buildings. CEN, 2005
EN 15129. Anti-seismic devices. European committee for standardization. CEN, 2010
Fiorino L, Iuorio O, Landolfo R (2009) Sheathed cold-formed steel housing: a seismic design procedure. Thin-Walled Struct 47(8–9):919–930
Fiorino L, Iuorio O, Landolfo R (2011) Seismic analysis of sheathing-braced cold-formed steel structures. Eng Struct 34:538–547
Fiorino L, Iuorio O, Macillo V, Landolfo R (2012) Performance-based design of sheathed CFS buildings in seismic area. Thin-Walled Struct 61:248–257
Fiorino L, Iuorio O, Landolfo R (2013) Behaviour factor evaluation of sheathed cold-formed steel structures. Adv Steel Constr Int J 9(1):26–40
Fiorino L, Iuorio O, Landolfo R (2014) Designing CFS structures: the new school BFS in Naples. Thin-Walled Struct 78:37–47
Fiorino L, Terracciano MT, Landolfo R (2016a) Experimental investigation of seismic behaviour of low dissipative CFS strap-braced stud walls. J Constr Steel Res 127:92–107
Fiorino L, Iuorio O, Macillo V, Terracciano MT, Pali T, Landolfo R (2016b) Seismic design method for CFS diagonal strap-braced stud walls: experimental validation. J Struct Eng 142 (3):04015154
Fiorino L, Pali T, Bucciero B, Macillo V, Terracciano MT, Landolfo R (2017a) Experimental study on screwed connections for sheathed CFS structures with gypsum or cement based panels. Thin-Walled Struct 116:234–249
Fiorino L, Macillo V, Landolfo R (2017b) Shake table tests of a full-scale two-story sheathing-braced cold-formed steel building. Eng Struct 151:633–647
Güneyisi EM, D'Aniello M, Landolfo R, Mermerdaş K (2013) A novel formulation of the flexural overstrength factor for steel beams. J Constr Steel Res 90:60–71
Güneyisi EM, D'Aniello M, Landolfo R, Mermerdaş K (2014) Prediction of the flexural overstrength factor for steel beams using artificial neural network. Steel Compos Struct Int J 17(3):215–236
Ioan A, Stratan A, Dubină D, Poljanšek M, Molina FJ, Taucer F, Pegon P, Sabău G (2016) Experimental validation of re-centring capability of eccentrically braced frames with removable links. Eng Struct 113:335–346
Iuorio O, Fiorino L, Landolfo R (2014a) Testing CFS structures: the new school BFS in Naples. Thin-Walled Struct 84:275–288
Iuorio O, Macillo V, Terracciano MT, Pali T, Fiorino L, Landolfo R (2014b) Seismic response of CFS strap-braced stud walls: experimental investigation. Thin-Walled Struct 85:466–480
Landolfo R (ed) (2013) Assessment of EC8 provisions for seismic design of steel structures. technical committee 13 —Seismic design, No 131/2013. ECCS
Landolfo R (2015) The activities of the ECCS-TC13 seismic committee: bridging the gap between research and standards. Proceedings of 8th international conference on behavior of steel structures in Seismic areas, Shanghai, China, 1–3 July 2015
Landolfo R (2016) European qualification of seismic resistant steel bolted beam-to-column joints: the equal joints project. Proceedings of 1st EU-Sino workshop on earthquake-resistance of steel structures, Shanghai, China, 27 Oct 2016
Landolfo R, Mazzolani FM, Dubina D, da Silva L, D'Aniello M (2017) Design of steel structures for buildings in seismic areas. Wilhelm Ernst & Sohn/Verlag fürArchitektur und

technischeWissenschaften GmbH & Co. KG, Berlin, 1st Edition. ISBN (Ernst & Sohn): 978-3-433-03010-3

Latour M, Piluso V, Rizzano G (2015) Free from damage beam-to-column joints: testing and design of DST connections with friction pads. Eng Struct 85:219–233

Longo A, Montuori R, Nastri E, Piluso V (2014) On the use of HSS in seismic-resistant structures. J Constr Steel Res 103:1–12

Macillo V, Fiorino L, Landolfo R (2017a) Seismic response of CFS shear walls sheathed with nailed gypsum panels: Experimental tests. Thin-Walled Struct 120:161–171

Macillo V, Bucciero B, Terracciano MT, Pali T, Fiorino L, Landolfo R (2017b) Shaking table tests on cold-formed steel building sheathed with gypsum panels. Ce/papers 1(2017):2847–2856. https://doi.org/10.1002/cepa.336

Mazzolani FM (ed) (1988) European recommendations for steel structures in seismic zones. Technical committee 13 —seismic design, No. 54/1988. ECCS

Mazzolani FM, Piluso V (1992) Member behavioural classes of steel beams and beam-columns. Proceedings of first state of the art workshop, COSTI, Strasbourg

Mazzolani FM, Della Corte G, D'Aniello M (2009) Experimental analysis of steel dissipative bracing systems for seismic upgrading. J Civil Eng Manag 15(1):7–19

Piluso V, Montuori R, Troisi M (2014) Innovative structural details in MR-frames for free from damage structures. Mech Res Commun 58:146–156

Plumier A, Doneux C, Castiglioni C, Brescianini J, Crespi A, Dell' Anna S, Lazzarotto L, Calado L, Ferreira J, Feligioni S, Bursi OS, Ferrario F, Sommavilla M, Vayas I, Thanopoulos P and Demarco T (2006) Two innovations for earthquake resistant design – the INERD project – final report, research programme of the research fund for coal and steel: steel RTD, report EUR 22044EN.

Sedlacek G, Feldmann M, Kühn B, Tschickardt D, Höhler S, Müller C, Hensen W, Stranghöner N, Dahl W, Langenberg P, Münstermann S, Brozetti J, Raoul J, Pope R, Bijlaard F (2008) Commentary and worked examples to EN 1993-1-10 "Material toughness and through thickness properties" and other toughness oriented rules in EN 1993. JRC 47278 EUR 23510 EN, European Communities. ISSN 1018–5593

Tartaglia R, D'Aniello M (2017) Nonlinear performance of extended stiffened end plate bolted beam-to-column joints subjected to column removal. Open Civil Eng J 11(Suppl-1, M6):369–383

Tartaglia R, D'Aniello M, Rassati GA, Swanson JA, Landolfo R (2018) Full strength extended stiffened end-plate joints: AISC vs recent European design criteria. Eng Struct 159:155–171

Tenchini A, D'Aniello M, Rebelo C, Landolfo R, da Silva LS, Lima L (2014) Seismic performance of dual-steel moment resisting frames. J Constr Steel Res 101:437–454

Tenchini A, D' Aniello M, Rebelo C, Landolfo R, da Silva L, Lima L (2016) High strength steel in chevron concentrically braced frames designed according to Eurocode 8. Eng Struct 124:167–185

Vayas I (2017) Innovative anti-seismic devices and systems European convention for constructional steelwork, 1st edn, 2017. ISBN: 978-92-9147- 136-2.

Vigh LG, Zsarnóczay A, Balogh JM, Castro JM (2016) P-Delta effect and pushover analysis: review of eurocode 8–1. Report N.1 of WG2 CEN/TC 250/SC 8, 2/2/2016

Vigh LG, Zsarnóczay Á, Balogh T (2017) Eurocode conforming design of BRBF – part I: proposal for codification. J Constr Steel Res 135:265–276

Zimbru M, D'Aniello M, Stratan A, Dubina D (2017) Finite element modelling of detachable short links. In Papadrakakis M, Fragiadakis M (eds) Proceedings of COMPDYN 2017-16th ECCOMAS Thematic conference on computational methods in structural dynamics and earthquake engineering, Rhodes Island, Greece, 15–17 June 2017

Chapter 19
Unreinforced Masonry Walls Subjected to In-Plane Shear: From Tests to Codes and Vice Versa

Elizabeth Vintzileou

Abstract According to Eurocode 6, the lateral capacity of unreinforced masonry walls is calculated using a simple equation, accounting exclusively of a shear failure mode and based on the shear strength of masonry (friction analogy). With the purpose of checking the efficiency of the Code formula, experimental results are collected from the Literature and an effort is made to predict the lateral capacity of the tested walls. This comparison between predicted and experimental shear resistances shows that the Code formula systematically overestimates the lateral capacity of shear walls. The interpretation of this inconsistency is attempted, a qualitative re-evaluation of selected test results is offered and, finally, proposals for further actions within relevant Code Committees are submitted.

19.1 Introduction

The seismic resistance of unreinforced masonry buildings strongly depends on the in-plane shear and out-of-plane flexural behaviour of walls. Therefore, the-as accurate as possible-calculation of the respective bearing capacities constitutes a key issue for the aseismic design of unreinforced masonry buildings, as well as for the assessment of existing masonry buildings. Furthermore, displacement-based design requires data on the deformation capacity of vertical masonry elements. Last but not least, the behaviour factors to be taken into account in the design or re-design of masonry structures need to be adequately documented. The International Engineering Community has devoted significant effort to the experimental documentation of the aforementioned structural issues, whereas modeling strategies at micro-, meso- and macro-scale were developed and applied with the purpose of reproducing the experimentally observed behaviour. Finally, current European Codes (EC6-Part 1-1 [2005], EC8-Parts 1 and 3 [2004 and 2005]) provide design rules, meant to ensure

E. Vintzileou (✉)
School of Civil Engineering, National Technical University of Athens, Athens, Greece
e-mail: elvintz@central.ntua.gr

the verification of masonry elements with sufficient margins of safety. Among the open issues and the aspects that are not covered by the current versions of the Eurocodes (e.g. displacement-based design of masonry buildings is not an option, despite the progress in the field), this paper deals with the behaviour of unreinforced masonry walls subjected to in-plane cyclic actions under simultaneous vertical (compressive) stress and, more specifically, with the calculation of the lateral capacity of shear walls according to Eurocode 6. For this purpose, several sets of tests on shear walls are collected from the international Literature and their lateral capacity is calculated using the formula proposed by EC6. It should be noted that the creation of a database including all the experimental results published since mid-twentieth century is beyond the scope of this paper, although such an exhaustive database would serve as a basis for further revision of the current regulatory documents. Nonetheless, it is believed that the number of the tests and the overall quality of the respective investigations ensure the reliability of the observations made in this paper.

19.2 Calculation of the In-Plane Capacity of Unreinforced Masonry Walls According to Eurocode 6

According to the §6.2 of Eurocode 6, Part 1-1 (2005), the design value of the bearing capacity of a wall subjected to in-plane shear is calculated using the following Eq. (19.1):

$$V_{Rd} = f_{vd} t l_c \qquad (19.1)$$

where, t denotes the thickness of the wall, l_c is the length of the compressed part of the wall, ignoring any part of the wall that is in tension and f_{vd} denotes the design value of the shear strength of masonry, calculated according to Eq. (19.2):

$$f_{vd} = f_{vk}/\gamma_M = (f_{vk0} + 0,40\sigma_d)/\gamma_M \qquad (19.2)$$

where,
f_{vk0} denotes the characteristic value of shear strength under zero normal stress,
σ_d denotes the design compressive stress perpendicular to the shear (under the considered load combination). The normal stress is the average value calculated on the compressed portion of the section and
γ_M denotes the partial safety factor for masonry, depending on the quality of both the masonry blocks production and the construction of masonry.

Limiting values are provided for f_{vk}, whereas, in case of unfilled perpendicular joints, the shear strength under zero normal stress, f_{vk0}, is reduced by 50%.

The characteristic shear strength of masonry under zero normal stress can be determined on the basis of tests. However, values of that property are given in the

Code, depending on the type of the block, as well as on the type (general purpose or thin layer) and the (mean) compressive strength of the mortar.

Several comments can be made, both regarding Eq. (19.1) and the values of material properties inserted to Eq. (19.2), namely:

(a) Equation (19.1): This equation is based on the occurrence of a diagonal crack within the wall or, alternatively, on the occurrence of a horizontal crack and sliding along it. However, in case of slender walls, flexural failure may occur, whereas in case of squat walls, a different failure mode occurs (as tests have shown). Equation (19.1) does not cover all possible failure modes and, hence, its efficiency in predicting the real capacity of walls is expected to be limited. It has to be noted that the situation is rectified, partly though, in the (under preparation) revised version of Eurocode 6. Actually, another failure mode is added (flexural failure due to in-plane shear). Thus, an equation is added, allowing for the resisting bending moment to be calculated, as follows:

$$M_{Rd} = \left(l^2 t \frac{\sigma_d}{2}\right)\left(1 - \frac{\sigma_d}{0.85 f_d}\right) \tag{19.3}$$

where,
f_d is the design value of the compressive strength of masonry,
t is the thickness of the wall,
l is the total length of the wall, including any part in tension,
σ_d is the average design compressive stress on the overall cross-section of the wall (σ_d is positive when the section is in compression).

(b) Friction coefficient, μ, and shear strength under zero normal stress, f_{vk0}: A unique value of friction coefficient (=0.40) is adopted by EC6 for walls subjected to in-plane shear. Due to the fact that, normally, low rise buildings are constructed using unreinforced masonry in seismic regions, the vertical stress on the walls due to vertical loads is expected to be rather small. Nevertheless, during an earthquake, the vertical compressive stress on the compressed part of the section of a wall may be increased because of the in-plane bending moments, as well as by the cracking of the wall and, hence, by the reduction of the length of the resisting section. It might, therefore, be of significance to account for friction coefficient values in function of the compressive stress on the wall. The adoption of a unique value for the friction coefficient seems to insinuate that it constitutes a masonry property, independent of any mechanical parameter, such as strength of masonry units and mortar, imposed normal stress, etc. (see also Tomaževič 2009).

The method for the assessment of the shear strength under zero normal stress, f_{vk0}, the values adopted by Eurocode 6, as well as the adequacy of this mechanical property for the calculation of the bearing capacity of walls in shear are discussed in several papers (see i.a. Magenes and Calvi 1997, Tomaževič 2009). Tomaževič (2009) expresses the opinion that the shear strength under zero normal stress is not adequate for the calculation of the shear capacity of walls failing after the occurrence of diagonal cracks, since such cracks are due to the attainment of the tensile strength of masonry at the central region of a wall. Tomaževič did perform a systematic experimental program on triplets, using seven types of perforated clay bricks and two classes of mortar, in order to obtain f_{vk0} values. With some exceptions, the results do not seem to confirm the values proposed by Eurocode 6.

With the purpose of checking the efficiency of EC6 provisions in predicting the bearing capacity of unreinforced masonry walls subjected to in-plane shear under simultaneous vertical compression, experimental results were collected and evaluated. The results of this evaluation are presented herein.

19.3 Tests on Unreinforced Masonry Walls Subjected to In-Plane Shear Under Simultaneous Vertical Compression

19.3.1 Brief Description of Tests

Tables 19.1 and 19.2 present a summary of selected published test results. In the tests reported in Table 19.1, unreinforced masonry walls (made of a variety of masonry units and mortars) were subjected to cyclic in-plane shear under simultaneous vertical compressive stress. The aspect ratio of the tested walls (Height/Length, H/L) varies between 0.40 and 2.50. The vertical compressive stress on the walls, acting simultaneously with the in-plane shear takes values between 0.20 MPa and 2.0 MPa. Two schemes are used by the researchers, namely double fixed or cantilever walls. The failure mode of each test is also reported in the respective publications. It should be noted that, in several publications, typical hysteresis loops are provided, as well as pictures of the failure mode for some selected specimens.

Regarding the characterization of materials, in most of the publications evaluated herein, the mechanical properties of masonry units, mortar and masonry are measured (according to relevant Standards), along with the shear strength of masonry under zero normal stress and the friction coefficient.

19.3.2 Calculated Values of Maximum Shear Resistance

Although, as mentioned in the previous paragraph, in most of the publications both f_{vk}, and friction coefficient values measured by the respective researchers are provided, those values were not used in the calculation of the expected maximum resistance of the walls. This is because the purpose was to apply Eq. 19.1 as a

Table 19.1 In-plane shear tests on unreinforced masonry walls: summary of data

Specimen	Length (m)	Height (m)	Thickness (m)	σ_v (MPa)	Boundary conditions
Churilov and Dumova-Jovanoska (2010, 2011)					
UMW1	2.50	1.82	0.25	1.00	DF
UMW2	1.50	1.82	0.25	1.00	DF
UMW3	2.50	1.82	0.25	0.50	DF
UMW4	1.50	1.82	0.25	0.50	DF
Magenes et al. (2010)					
CS00	1.25	2.50	0.40	0.20	DF
CS01	1.25	2.50	0.40	0.50	DF
CS02	1.25	2.50	0.40	0.20	DF
CT01	2.50	2.50	0.40	0.50	DF
CT02	2.50	2.50	0.40	0.20	DF
Magenes et al. (2008b)					
CS01	1.25	2.50	0.175	1.00	DF
CS02	1.25	2.50	0.175	1.00	DF
CS03	1.25	2.50	0.175	0.50	DF
CS04	1.25	2.50	0.175	2.00	DF
CS05	1.25	2.50	0.175	1.00	DF
CS06	1.25	2.50	0.175	1.00	C
CS07	2.50	2.50	0.175	1.00	DF
CS08	2.50	2.50	0.175	1.00	C
Calvi and Magenes (1994a, b)					
MI1	1.50	2.00	0.38	1.36*	DF
MI2	1.50	2.00	0.38	0.70*	DF
MI3	1.50	3.00	0.38	1.35*	DF
MI4	1.50	3.00	0.38	0.72*	DF
Abrams (1992), Abrams and Shah (1992)					
W1	3.60	1.60	0.198	0.55	C
W2	2.70	1.60	0.198	0.37	C
W3	1.80	1.60	0.198	0.37	C
Gouveia and Lourenço (2007)					
W2.1	1.02	1.15	0.143	0.90	DF
W2.2	1.02	1.15	0.143	0.90	DF
Messali et al. (2017)					
COMP-0a	1.10	2.75	0.102	0.70	DF
COMP-1	1.10	2.75	0.102	0.70	C
COMP-2	1.10	2.75	0.102	0.50	C
COMP-3	1.10	2.75	0.102	0.40	DF
COMP-4	4.00	2.75	0.102	0.50	DF
COMP-5	4.00	2.75	0.102	0.30	DF

(continued)

Table 19.1 (continued)

Specimen	Length (m)	Height (m)	Thickness (m)	σ_v (MPa)	Boundary conditions
COMP-6	4.00	2.75	0.102	0.50	C
Costa (2007)					
W1	1.475	2.75	0.300	0.68	DF
W2	3.125	2.75	0.300	0.32	DF
W3	4.375	2.75	0.300	0.23	DF
W4	1.475	2.75	0.300	0.45	DF
Tomaževič (2009)					
B1/1	1.00	1.43	0.28	1.92	C
B1/2	1.00	1.43	0.28	0.96	C
B2/1	1.02	1.51	0.28	1.71	C
B2/2	1.02	1.51	0.28	0.94	C
B2/3	1.02	1.51	0.28	1.37	C
B3/1	1.01	1.42	0.29	1.67	C
B3/2	1.01	1.42	0.29	0.89	C
B4/1	0.99	1.42	0.29	1.62	C
B4/2	0.99	1.42	0.29	1.00	C
B6/1	1.07	1.47	0.25	1.96	C
B6/2	1.07	1.47	0.25	1.01	C
Magenes et al. (2008a)					
CL04	2.5	2.6	0.300	0.68	DF
CL05	2.5	2.6	0.300	0.68	DF
CL06	1.25	2.6	0.300	0.50	DF
CL07	1.25	2.6	0.300	0.50	DF
CL08	2.5	2.6	0.300	0.68	DF
CL09	1.25	2.6	0.300	0.50	DF
CL10	2.5	2.6	0.300	0.68	DF
Morandi et al. (2013)					
MA1	1.25	2.00	0.35	0.50	DF
MA2	1.25	2.00	0.35	0.70	DF
MA3	1.25	2.00	0.35	1.00	DF
MA4	2.50	2.00	0.35	0.50	DF
MA5	2.50	2.00	0.35	0.70	DF
Morandi et al. (2014)					
MB1	1.35	2.14	0.35	0.15	DF
MB2	1.35	2.14	0.35	0.45	DF
MB3	1.35	2.14	0.35	0.65	DF
MB4	2.70	2.14	0.35	0.45	DF
MB5	2.70	2.14	0.35	0.65	DF
Javed et al. (2013)					
WIc	1.36	1.66	0.236	0.71	DF
WIIa	1.36	1.28	0.236	0.42	DF
WIIb	1.36	1.28	0.236	0.42	DF

(continued)

Table 19.1 (continued)

Specimen	Length (m)	Height (m)	Thickness (m)	σ_v (MPa)	Boundary conditions
WIIc	1.36	1.28	0.236	0.42	DF
WIIIa	1.36	1.28	0.236	0.64	DF
WIIIb	1.36	1.28	0.236	0.64	DF
WIVa	1.36	0.90	0.236	0.64	DF
WIVb	1.36	0.90	0.236	0.64	DF

Note *DF* Double fixed, *C* Cantilever
(*) There was significant variation of the vertical compressive stress during testing. The values listed in the table are the maximum ones (recorded at the attainment of the maximum resistance)

Designer would do, i.e. by adopting the value of f_{vk0} proposed by the Code (for a given type of masonry units and for a given mortar quality), as well as a friction coefficient equal to 0.40.

It should be noted here that, as there are no data for the transformation of the characteristic value of shear strength under zero normal stress (f_{vk0}) to its mean value, the characteristic values given in EC6 are used, when Eq. (19.1) is applied. Therefore, the maximum resistances should be slightly higher than those calculated and reported in Table 19.2. Furthermore, it is not clear whether the value of the friction coefficient is a characteristic value as well. Thus, in absence of relevant information, the value of 0.40 was used to calculate the maximum resistances.

It is evident that Eq. (19.1) was not expected to yield accurate results in case of walls that exhibited flexural failure (as stated by the researchers). However, according to the version of EC6, currently in use, in-plane flexural failure and the respective safety verification is not considered.

However, taking into account the sensible amendments to be made to EC6, the in-plane resistance of the walls for flexural failure was also calculated (on the basis of Eq. (19.3)). The respective calculated values are also reported in Table 19.2. The calculations using Eq. (19.3) were based on the following assumptions: (a) The value of vertical compressive stress (σ_v) applied to each specimen was inserted to the equation. Furthermore, (b) the mean value of the compressive strength of masonry (as measured in the respective publication) was used. Thus, the bending moment at flexural failure was first calculated. Subsequently, taking into account the boundary conditions of each distinct specimen (either double fixed or cantilever), the maximum in-plane shear resistance was calculated.

The results of application of Eq. 19.1, irrespectively of the observed failure mode, are illustrated in Fig. 19.1. It is clear that the Code equation overestimates the shear resistance of most of the walls, whereas the scatter is quite large. There are few exceptions in which Eq. 19.1 underestimates the shear capacity of walls.

In order to account for the observed failure mode of the specimens reported in Tables 19.1 and 19.2, the values of maximum shear resistances calculated using

Table 19.2 In-plane shear tests on unreinforced masonry walls: Summary of experimental results and predicted values of maximum resistance

Specimen	$expV_{max}$ (kN)	Failure mode	$predV_{max}$ (kN) Eq. (19.1)	pred/exp. V_{max}	$predV_{max}$ (kN) Eq. (19.3)	pred/exp. V_{max}
Churilov and Dumova-Jovanoska (2010, 2011)						
UMW1	189.1	S	312.5	1.65	634.1	0.30
UMW2	88.5	S	187.5	2.11	228.3	0.39
UMW3	157.4	S	187.5	1.19	317.0	0.50
UMW4	65.5	S	112.5	1.72	114.1	0.57
Magenes et al. (2010)						
CS00	49.0	F	90	1.83	46.4	0.95
CS01	94.0	S	150	1.60	80.16	1.17
CS02	48.0	S/F	90	1.88	32.1	1.50
CT01	234.0	S	300	1.28	320.7	0.73
CT02	154.0	S	180	1.17	128.3	1.20
Magenes et al. (2008a)						
CS01	85.0	S	187.5	2.20	93.1	0.91
CS02	85.0	S	187.5	2.20	93.1	0.91
CS03	50.0	S	125	2.50	46.6	1.07
CS04	150.0	S	312.5	2.08	186.3	0.80
CS05	100.0	H	187.5	1.88	93.1	1.07
CS06	42.5	F	187.5	4.41	46.6	0.91
CS07	220.0	H	262.5	1.19	93.1	2.36
CS08	160.0	H	262.5	1.64	46.6	3.44
Calvi and Magenes (1994a, b)						
MI1	260.0	S	421.8	1.62	495.0	0.53
MI2	239.3	S	273.6	1.14	254.8	0.94
MI3	178.6	V	421.8	2.36	327.6	0.55
MI4	164.3	S	278.2	1.69	174.7	0.94
Abrams (1992), Abrams and Shah (1992)						
W1	422.6	S	299.4	0.71	358.4	1.18
W2	195.8	S/C	186	0.95	135.6	1.44
W3	89.0	F	124	1.39	60.3	1.47
Gouveia and Lourenço (2007)						
W2.1	80.9	S	81.70	1.00	71.4	1.13
W2.2	88.9	S	81.70	0.92	71.4	1.24
Messali et al. (2017)						
COMP0a	29.2	C	32.9	1.13	25.2	1.16
COMP-1	9.5	C	37.3	3.92	12.6	0.76
COMP-2	9.5	C	27.4	2.90	9.0	1.06
COMP-3	14.6	C	24.5	1.68	14.4	1.02
COMP-4	121.0	S	142.8	1.18	237.6	0.51
COMP-5	102.5	SL	110.2	1.08	142.6	0.72
COMP-6	109.5	SJS2	142.8	1.30	118.8	0.92

(continued)

Table 19.2 (continued)

Specimen	expV$_{max}$ (kN)	Failure mode	predV$_{max}$ (kN) Eq. (19.1)	pred/exp. V$_{max}$	predV$_{max}$ (kN) Eq. (19.3)	pred/exp. V$_{max}$
Costa (2007)						
W1	59.0	–(**)	243.1	4.12	82.0	0.72
W2	137.5	S	260.6	1.90	173.8	0.79
W3	210.0	S	317.1	1.51	243.8	0.86
W4	50.0	M	146.4	2.93	54.7	0.91
Tomaževič (2009)						
B1/1	140.6	S	257.5	1.93	141.7	0.99
B1/2	92.0	S	137.4	1.49	70.9	1.30
B2/1	133.7	S	234.7	1.76	124.4	1.10
B2/2	90.9	S	139.8	1.54	68.4	1.30
B2/3	118.0	S	190.0	1.61	99.6	1.18
B3/1	128.7	S	244.4	1.90	131.1	0.98
B3/2	84.2	S	143.0	1.70	69.9	1.20
B4/1	141.7	S	226.5	1.60	122.2	1.16
B4/2	93.9	S	144.7	1.54	75.4	1.24
B6/1	131.0	S	223.7	1.71	143.9	0.91
B6/2	91.6	S	137.3	1.50	74.1	1.24
Magenes et al. (2008b)						
CL04	310.0	S	354	1.14	449.1	1.45
CL05	345.0	S	354	1.03	449.1	1.30
CL06	82.0	F	180	2.20	84.6	1.03
CL07	72.5	S	180	2.48	106.7	1.47
CL08	270.0	S	354	1.31	431.0	1.60
CL09	72.0	S	187.5	2.60	104.1	1.45
CL10	215.0	S	429	1.99	416.4	1.94
Morandi et al. (2013)						
MA1	130.0	S	175	1.35	128.3	0.99
MA2	166.0	S	210	1.27	174.8	1.05
MA3	198.0	S	262.5	1.33	239.6	1.21
MA4	401.0	SJS2 (*)	350	0.87	513.0	1.28
MA5	500.0	SJS2(*)	420	0.84	699.3	1.40
Morandi et al. (2014)						
MB1	46.0	R	122.9	2.67	43.4	0.94
MB2	120.0	H	179.6	1.50	122.7	1.02
MB3	115.0	S	217.4	1.89	169.9	1.48
MB4	184.0	R	359.1	1.95	490.7	2.67
MB5	115.0	H	434.7	3.78	679.4	5.90
Javed et al. (2013)						
WIc	108.4	SJS2	155.3	1.43	152.3	1.41
WIIa	79.2	SJS2	118.1	1.49	127.6	1.61
WIIb	92.6	SJS2	118.1	1.28	127.6	1.38

(continued)

Table 19.2 (continued)

Specimen	expV$_{max}$ (kN)	Failure mode	predV$_{max}$ (kN) Eq. (19.1)	pred/exp. V$_{max}$	predV$_{max}$ (kN) Eq. (19.3)	pred/exp. V$_{max}$
WIIc	84.7	SJS2	118.1	1.40	127.6	1.51
WIIIa	123.7	SJS2	146.4	1.18	182	1.47
WIIIb	105.5	SJS2	146.4	1.39	182	1.73
WIVa	161.4	SJS2/HJS	146.4	0.91	258.8	1.61
WIVb	151.3	SJS2/HJS	146.4	0.97	258.8	1.71

Note 1 Failure modes: *S* Shear, *F* Flexural, *H* Hybrid (shear + rocking), *S/F* Mixed shear-flexural, *V* Vertical cracks, *SL* sliding along the bottom mortar joint, *SJS1* Shear cracks through mortar joints, *SJS2* SJS1+crushing at the toe, *HJS* Horizontal joint sliding, *C* (combined): Flexure with toe crushing and sliding, *M* Shear + rocking

Note 2 (*) Failure mode based on photos of the specimen, (**) Failure not characterized by the Author. Non-typical failure

Fig. 19.1 Comparison between experimental and calculated maximum resistances (Using Eq. 19.1)

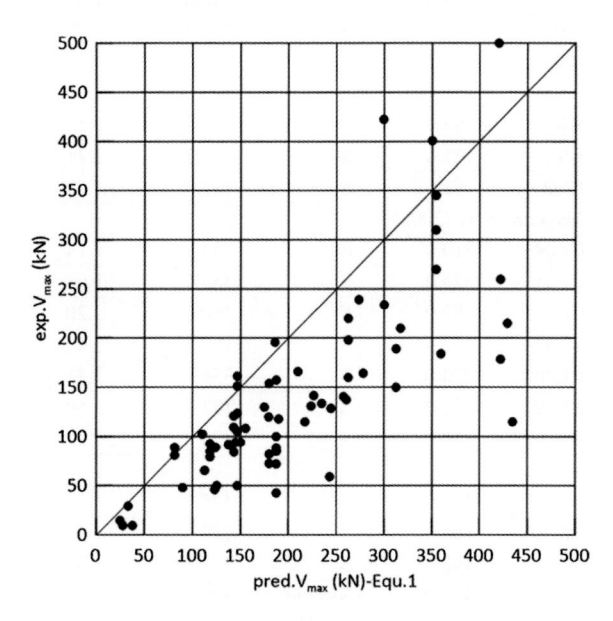

either Eq. (19.1) or Eq. (19.3), depending on the observed failure mode, are plotted against the experimental values (Fig. 19.2).

The picture is expectedly improved. However, the maximum resistance of the walls is still significantly overestimated. It has to be observed though (see Table 19.2) that the maximum resistance in case of flexural failure (Eq. 19.3) is quite accurately predicted. On the contrary, the resistance due to shear failure (Eq. 19.1) is overestimated.

It has to be assumed that the Designer of a plain masonry building does not know whether a wall will fail in shear or in bending. It is therefore expected that the Designer (using the revised form of EC6) will calculate the shear capacity of each wall, using both

Fig. 19.2 Comparison between experimental and calculated maximum resistances (Using Eqs. 19.1 and 19.3)

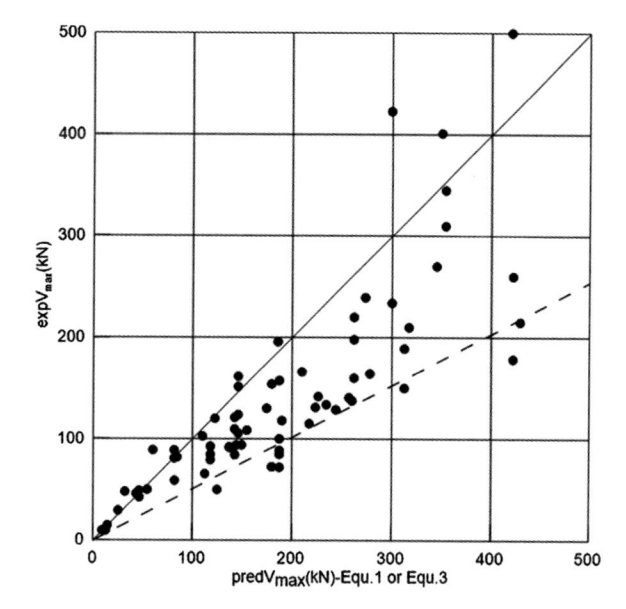

Fig. 19.3 Comparison between experimental values of maximum resistance and minimum values calculated using Eqs. 19.1 and 19.3

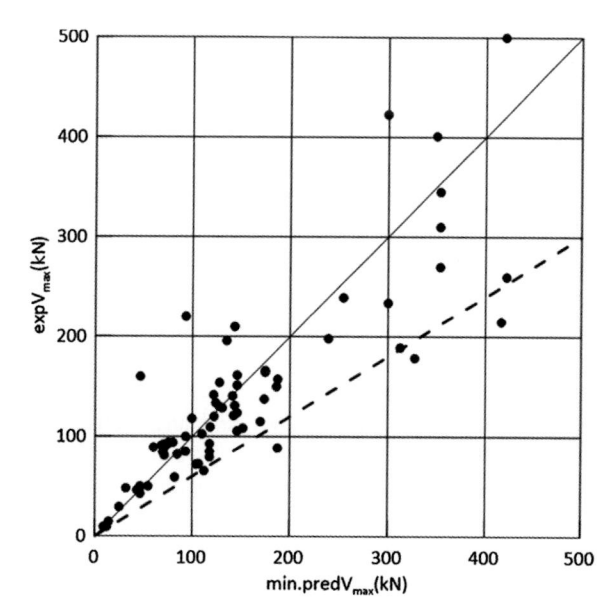

Eqs. (19.1) and (19.3), keeping the smaller of the two calculated values. Thus, in Fig. 19.3, the smaller of the values calculated using Eqs. (19.1) and (19.3) are plotted against the experimental values, irrespectively of the failure mode observed by the researchers for each distinct specimen. A significantly better picture is obtained. Nevertheless, to reach a safe value of the maximum resistance, a γ_{Rd}-value as high as 1.70 should be used.

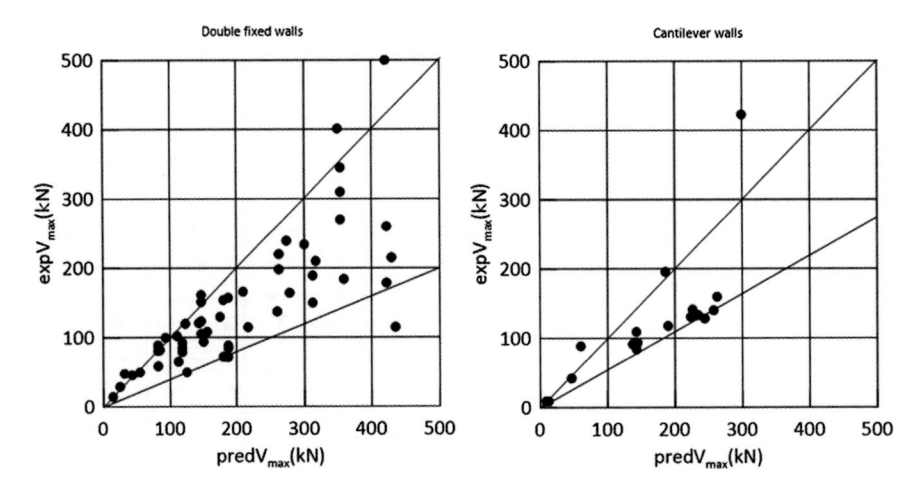

Fig. 19.4 Experimental values of maximum resistance are plotted against the calculated ones, for double fixed and cantilever walls. Equations (19.1) or (19.3) are used, depending on the reported failure mode

The fact that the best approximation of experimental values is reached by ignoring (in a number of cases) the failure mode reported by the researchers is not acceptable. Further calibration of the design models and re-evaluation of the experimental data is needed.

During the evaluation of the experimental data included in this paper, it was observed that the scheme of the tested walls (double fixed or cantilevers) seems to affect the ratio between predicted and experimental values of maximum resistance: In Fig. 19.4, the experimental values of maximum resistance are compared to the calculated ones for double fixed and cantilever walls separately. The lateral resistances were calculated using Eq. 19.1 or Eq. 19.3, depending on the failure mode reported by the respective researchers. It seems that the predicted values of maximum resistance are more scattered in case of double fixed walls than for cantilevers. The author of this paper cannot find any plausible explanation for this finding, which, however, needs to be further investigated.

The effect of the value of the vertical compressive stress, σ_v, on the ratio between the predicted and the experimental value of the maximum shear resistance was evaluated, on the basis of the experimental results (Fig. 19.5). Although the scatter of the ratio of resistances is very large, it seems that there is a clear tendency of the scatter to increase for higher compressive stress values. A plausible explanation for this finding could be that, the higher the vertical stress, the higher the possibility of crushing at the toe of the wall (failure of the inclined masonry strut). As this (local but significant) mode of failure is not considered in Eqs. (19.1) and (19.3), the predicted maximum resistance may be significantly higher than the experimental one. However, the accuracy of this explanation needs to be checked on the basis of relevant information included in the publications. It is believed that, in case the

Fig. 19.5 Ratio of predicted to experimental value of the maximum resistance vs. value of vertical compressive stress on the wall

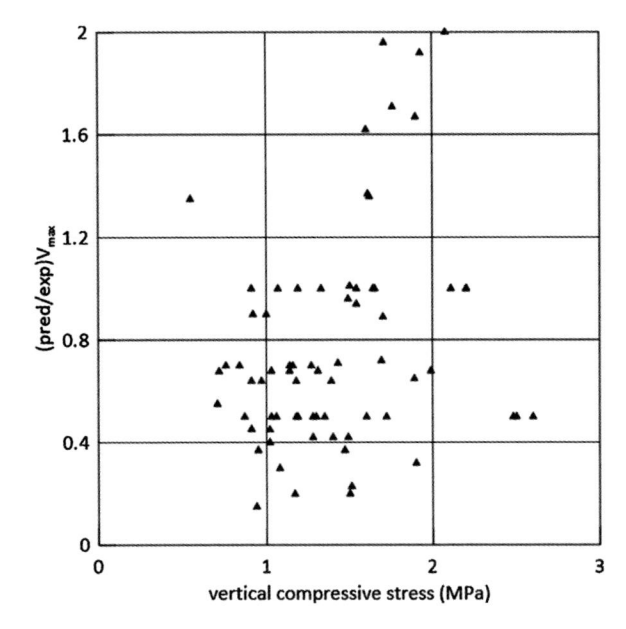

occurrence of bi-diagonal (shear) cracks is followed by crushing of the toe, the maximum resistance of the wall will be closer to its flexural than to its shear resistance. The investigation of this aspect is interesting from the Mechanics of masonry point of view; it is, however, to be checked whether it is of significance from the design point of view. Actually, due to the limited number of storeys of unreinforced masonry buildings, vertical stresses higher than 0.50 MPa are unrealistically high. Unfortunately, a large portion of the available test results were performed under a rather high value of compressive stress acting on the wall.

19.4 Comments on the Failure Modes Observed During Testing

In an effort to interpret the difference observed in several experimental results between the predicted and the measured maximum lateral resistance of walls, the author of this paper examined more closely selected experimental data. More specifically, were examined the detailed description of the failure mode (as well as the pictures showing the wall after completion of the test, wherever available), as well as the compatibility between the reported failure mode and the form of the hysteresis loops.

In Magenes et al. (2008a), the failure mode of specimens (see also Tables 19.1 and 19.2, specimens CS01–08) is described in detail, whereas photographs showing the crack pattern are provided. Hysteresis loops are also included in the paper. As the Authors themselves state in the paper, in most of their specimens, the form of the hysteresis loops is similar to that typical for rocking behaviour. On the other hand, the detailed description of the failure mode shows that immediately after the occurrence of a diagonal crack (extending to the entire length of the diagonal), large horizontal displacements took place, without further increase of the lateral resistance of the piers. Looking at the pattern of the crack, one can observe that the bond length of the units is rather limited. Thus, after the occurrence of the inclined crack, one cannot expect but a very limited contribution of friction along the horizontal unit-masonry joints. Therefore, in this case, the lateral resistance of the piers would be more or less equal to the shear force causing its cracking. Actually, taking into account the value of the tensile strength of masonry measured by the Authors on wallettes subjected to diagonal compression (=0.27 MPa), the application of the well known formula by Turnšek and Čačovič (1971) [for a value equal to 1.50 for the β-factor] yields the following values of lateral resistance of the tested double-fixed walls (that compare quite satisfactorily with the experimental ones):

CS01 and CS02: $predV_{max} = 85.40kN$. Experimental values: 85.00/85.00kN
CS03: $predV_{max} = 66.50kN$. Experimental value: 50.00kN
CS04: $predV_{max} = 114.20kN$. Experimental value: 150.00kN
CS05: $predV_{max} = 85.40kN$. Experimental value: 100.00kN
CS07: $predV_{max} = 170.80kN$. Experimental value: 220.00kN

In Tomaževič (2009), the Author presents the failure mode of the walls (in two photos). In both walls (tested as cantilevers under high compressive stress) shear cracks occurred, along with crushing of the units at the base of the walls. The Author states (for walls B2 and B6 shown in the photos, see also Tables 19.1 and 19.2) that "...In both cases, tensile cracks and crushing of units at support have been observed before the shear failure". Actually, although the Author states that the walls failed in shear, Eq. 19.1 fails to predict their maximum resistance (mean value of the ratio "predicted/experimental maximum resistance" equal to 1.66). If, on the contrary, one takes into account the crushing at the toe, the maximum resistance to flexure can be calculated (see Table 19.2, specimens B1/1 to B6/2). Thus, a mean value of the ratio "predicted/experimental maximum resistance" equal to 1.15 is obtained. Yet, such a calculation may not be accurate. Actually, as a failure mode due to the attainment of the bearing capacity of the oblique masonry strut is not considered in the Code, it is not possible to check whether such a failure mode (based on the strength of masonry under oblique compression and taking into account the fact that the inclined strut is subjected to transverse tension) has contributed to the behaviour of the walls tested by Tomaževič (2009).

A very detailed description of the crack pattern throughout testing is provided in Javed et al. (2013). Thanks to this systematic work, mixed crack patterns and failure modes (involving diagonal or bi-diagonal cracks along with crushing and splitting of masonry units) are identified. As the Authors rightly state, current regulatory documents do not provide design models covering all possible failure modes. The case of failure mode of squat walls, also considered in the paper, will be commented upon in the next section. As shown in Tables 19.1 and 19.2, Eqs. 19.1 and 19.3 fail to predict the bearing capacity of the walls tested by Javed et al. (2013). The form of the hysteresis loops included in the paper seems to confirm the mixed (shear-flexural) failure mode described by the Authors (with some pintching effect-typical for shear-governed behaviour, but with large-area of hysteresis loops-typical for flexure-governed behaviour). However, as there is no design model (included in the regulatory documents) for the verification of the diagonal strut, the Designer cannot detect this type of failure; thus, the lateral resistance of the walls is overestimated.

It should also be noted that there is pronounced inconsistency between predicted and observed failure mode detected by some investigators. For example, Costa (2007) predicts -on the basis of calculated resistances-a flexural failure mode for Walls 1, 2 and 4. However, walls 1 and 4 failed under the occurrence of extensive vertical cracks (throughout the height of the walls), whereas Wall2 failed in shear.

The examples presented in this Section show that (a) the design models included (or to be included) in EC6 should cover all possible failure modes, as well as the interaction between failure modes. Furthermore, (b) the efficiency of those design models (in predicting both the experimental maximum resistance and the observed failure mode) should be checked on the basis of an exhaustive evaluation of the experimental data.

19.4.1 The Case of Squat Walls

The number of squat walls tested in in-plane shear is rather small, although their presence in unreinforced masonry buildings is quite frequent. This lack of sufficient experimental results is coupled with lack of design models for the verification of squat walls in EC6 (2005). The failure mode observed in squat walls (Fig. 19.6) is quite different from that observed in piers with an aspect ratio close to or higher than unity. Actually, squat walls exhibit either two or more sets of diagonal or bi-diagonal cracks (depending on their aspect ratio) or oblique cracks combined with a horizontal crack occurring approximately at mid-height of the wall. Crushing at the toe (under oblique compression) is also observed.

It is also to be noted that in both cases, the hysteresis loops are not typical for shear sensitive elements, although both walls exhibited a failure that would be attributed to shear. For the wall shown in Fig. 19.6 (WIVb-aspect ratio equal to 0.66), Javed et al. (2013) have calculated (on the basis of an idealized hysteresis loops envelope) a displacement ductility factor equal to 18.66 and 32.46 in the positive and the negative loading directions respectively. It has to be admitted that these figures are unexpectedly high, suggesting that the behaviour of the wall was not governed by shear. The identical to the wall shown in Fig. 19.6b (WIVa) exhibited (on the basis of both its failure mode and form of its hysteresis loops) a failure closer to a typical shear failure. Its calculated ductility factor is +8.71/−7.24. Although these values are significantly smaller than the ones calculated for the wall WIVb, they are still non typical for walls failing in shear. Similar observations can be made for the wall shown in Fig. 19.6a (with an aspect ratio H/L = 0.44), as well. A different behaviour was observed by Morandi et al. (2013, 2014), who tested walls

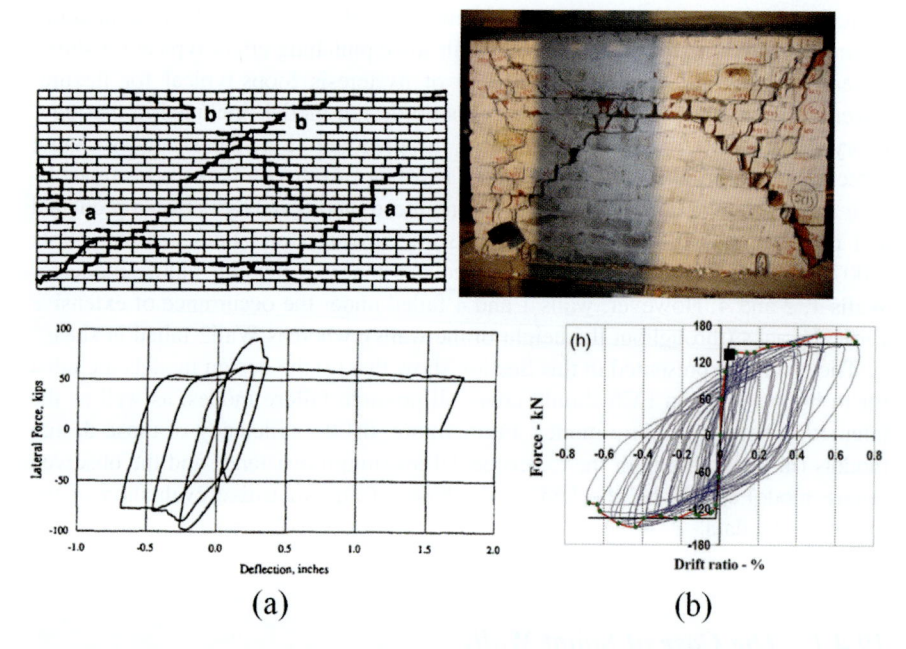

Fig. 19.6 Crack pattern and hysteresis loops for squat walls: (**a**) Wall tested by Abrams and Shah (1992) and (**b**) Wall tested by Javed et al. (2013)

with an aspect ratio equal to 0.80: The tested walls were made of thin shell and web clay bricks with filled or unfilled head joints. The walls with filled head joints exhibited a clear shear behaviour and failure (illustrated by both the crack pattern and the hysteresis loops), whereas those with unfilled head joints exhibited rather a "flexural-rocking" behaviour, as the Authors state.

It seems, therefore, that a closer look to the available experimental results is needed to identify the failure modes occurring in squat walls, to estimate the aspect ratio value adequate to characterize a wall as squat, etc.

19.5 Concluding Remarks

The limited in extent review of relevant experimental data, as well as the results of application of current Codes equations for the calculation of the lateral capacity of unreinforced masonry walls show that

(a) Eurocode 6 and Eurocode 8 do not consider all modes in which walls can fail when subjected to in-plane shear under simultaneous vertical (compressive) stress. Actually, according to the current version of the Eurocode 6, walls are verified exclusively against shear failure, whereas in the (under preparation) version of the Code, flexural failure is also considered. The case of failure of the inclined masonry struts (subjected to simultaneous transverse tension) is not taken into account and, hence, walls are not verified against such a failure that may be critical for buildings in seismic areas.

(b) As a result, Code Equations fail to accurately predict the shear capacity of walls measured during testing. Furthermore, as the capacity corresponding to a pure shear failure is negatively affected by a simultaneous crushing at the toe of the wall, the respective Code equation systematically overestimates the lateral resistance of walls, thus, yielding unsafe results. It has to be admitted though that when a clear flexural failure was observed during testing, the lateral capacity of the wall was quite accurately predicted by the respective design formula.

(c) The inconsistency between experimental and predicted values of the lateral capacity of walls is more pronounced in case of squat walls that seem to exhibit mixed failure modes.

(d) It seems, therefore, that there is an urgent need for the creation of a database including all the available experimental results on walls in shear (both monotonic and cyclic) and for the exhaustive evaluation of the experimental data. The purpose of that work would be to identify the parameters that affect the behaviour of walls (such as monotonic or cyclic actions, boundary conditions, magnitude of the vertical stress, aspect ratio, type of masonry units and mortar, construction aspects [e.g. filled or unfilled perpendicular joints], etc.). The major influencing parameters should appear in the design equations, which should yield not only accurate and safe values of lateral resistance of the walls under verification, but also an accurate prediction of their failure mode. In fact, this is information the Designer should be aware of, as it affects parameters of the overall behaviour of the building (such as its behaviour factor).

(e) A parameter not included to the current version of EC8 is the effect of cycling on the force-response of a wall. This effect might be covered by means of a coefficient (smaller than unity). However, as the degradation due to cycling is expected to depend on the failure mode of the wall (shear, flexural, shear/flexural, etc.), evaluation of the test results regarding this parameter is needed.

(f) Another aspect not considered within EC8 is that shear walls are subjected to simultaneous in-plane and out-of-plane actions. Although this might be a critical safety condition, the available experimental results investigating this realistic case are very limited.

(g) According to the current version of ECs, bearing elements in masonry buildings are verified for action-effects (forces and bending moments). However, the option of displacement-based design should be envisaged as well. Relevant numerical tools have been developed, whereas the available experimental results allow for constitute laws describing the behaviour of masonry to be identified. There are several publications, in which such constitutive laws are derived, whereas some computer codes allow for displacement-based design of masonry buildings. It is also to be mentioned that, although the values of behaviour factors allotted to URM buildings are not discussed in this paper, the experimental results show that isolated walls exhibit significantly higher ductility

(in terms of displacements) than the one corresponding to the low behaviour factor values of EC8. On the other hand,
(h) The collection and the re-evaluation of the available experimental results will allow for identification of parameters that need to be further investigated or are not yet investigated or for information needed for the evaluation but not explicitly provided by the publications, etc. Finally, it is believed that
(i) The fruitful discussion on the calculation of the shear capacity of masonry (at material level) should reach a sensible conclusion regarding the validity of the friction analogy (adopted by EC6) vs. the tensile strength of masonry, the values of shear strength under zero normal stress, as well as the values of friction coefficient to be used in design, taking into account that-although there are standardized tests for the evaluation of those properties-most of the Designers make use of the tabulated values provided by the Code.

References

Abrams DP (1992) Strength and behavior of unreinforced masonry elements. In: Proceedings 10th world conference on earthquake engineering. Madrid, Spain, vol 6, pp 3475–3480

Abrams DP Shah N (1992). Cyclic load testing of unreinforced masonry walls. Technical report, Department of Civil Engineering. University of Illinois at Urbana-Champaign, December, USA, 46pp

Calvi GM Magenes G (1994a) Experimental research on response of URM building systems. In: Abrams DP, Calvi GM (eds) Proceedings US-Italian workshop on guidelines for seismic evaluation and rehabilitation of unreinforced masonry buildings. NCEER-94-0021, Buffalo, New York, pp 41–57

Calvi GM Magenes G (1994b) Experimental results on unreinforced masonry shear walls damaged and repaired. In: Proceedings of the 10th international brick block masonry conference, 5–7 July, Calgary, Canada

Churilov S Dumova-Jovanoska E (2010). In-plane shear behaviour of unreinforced masonry walls. In: Proceedings of the 14th European conference on earthquake engineering, 30 August–3 September, Ohrid, FYROM

Churilov S Dumova-Jovanoska E (2011) Experimental evaluation of in-plane shear behaviour of unreinforced and strengthened brick masonry walls. In: Proceedings of 7th international conference AMCM 2011, 13–15 June, Krakow, Poland ICS, 31(3):491–514

Costa A (2007) Experimental testing of lateral capacity of masonry piers. An application to seismic assessment of AAC masonry buildings. MSc dissertation in earthquake engineering. Rose School, Pavia, Italy, 245pp

EN 1996-1-1 (2005) Eurocode 6: design of masonry structures – Part 1-1: general rules for reinforced and unreinforced masonry structures. CEN/TC250, Brussels

EN 1998-1 (2004) Eurocode 8: design of structures for earthquake resistance-Part 1: general rules, seismic actions and rules for buildings. CEN/TC250. Brussels

EN 1998-3 (2005) Eurocode 8: design of structures for earthquake resistance-Part 4: assessment and retrofitting of buildings, CEN/TC250, Brussels

Gouveia JP Lourenço PB (2007) Masonry shear walls subjected to cyclic loading. Influence of confinement and horizontal reinforcement. In: Proceedings 10th North American masonry conference, 3–5 June, St. Louis, Missouri, USA, pp 838–848

Javed M, Magenes G, Alam B, An K, Ali Q, Syed AM (2013) Experimental seismic performance evaluation of unreinforced brick masonry shear walls. Earthquake Spectra 31(1):215–246

Magenes G, Calvi GM (1997) In-plane seismic response of brick masonry walls. Earthq Eng Struct Dyn 26:1091–1112

Magenes G Morandi P Penna A (2008a) In-plane cyclic tests of calcium silicate masonry walls. In: Proceedings of the 14th international brick block masonry conference, 17–20 February, Sidney, Australia

Magenes G Morandi P Penna A (2008b) Experimental in-plane cyclic response of masonry walls with clay units. In: Proceedings of the 14th world conference on earthquake engineering, 12–17 October, Beijing, China

Magenes G Penna A Galasco A (2010) In-plane cyclic shear tests of undressed double-leaf stone masonry panels. In: Proceedings 8th international masonry conference, 4–7 July, Dresden, Germany

Messali F Ravenshorst G Esposito R Rots J (2017) Large-scale testing program for the seismic characterization of Dutch masonry walls. In: Proceedings 16th world conference on earthquake engineering, 9–13 January, Santiago, Chile, Paper no. 4753

Morandi P Albanesi L Magenes G (2013) In-plane experimental response of masonry walls with thin shell and web clay units. In: Proceedings of the Vienna Congress on recent advances in earthquake engineering and structural dynamics 2013 (VEESD 2013), 28–30 August, Vienna, Austria, Paper no 292

Morandi P Albanesi L Magenes G (2014) URM walls with thin shel/web clay units and unfilled head-joints: Cyclic in-plane tests. In: Proceedings of the Second European conference on earthquake engineering and seismology, 25–29 August, Istanbul, Turkey

Tomaževič M (2009) Shear resistance of masonry walls and Eurocode 6: Shear versus tensile strength of masonry. Mater Struct 42(7). https://doi.org/10.1617/s11527-008-9430-6.889-907

Turnšek V Čačovič F (1971) Some experimental results on the strength of brick masonry walls. In: Proceedings of the 2nd international brick block masonry conference. British Ceramic Society, Stoke-on-Trent, pp 149–156

Chapter 20
Seismic Design of Bridges: Present and Future

Andreas J. Kappos

Abstract A critical overview is provided of current trends in codes for seismic design of bridges, with emphasis on European practice. It is discussed whether the current Eurocode 8-2 provisions are performance-based and what, if anything, is really missing or lagging behind the pertinent state-of-the-art. Two different approaches recently proposed by the author for performance-based design (PBD) of bridges are presented and the feasibility of incorporating them in the next generation of codes, such as the new EC8-2 (currently in the evolution process), is discussed. The first procedure is in line with the exigencies of 'direct DBD' wherein stiffness and subsequently strength of the bridge are determined to satisfy a target displacement profile, with due account of the effect of higher modes. The second procedure is 'deformation-based design' wherein local deformations of dissipating components are an integral part of the design; two versions of this procedure are presented, one for bridges with ductile piers and one for seismically isolated bridges. Both PBD procedures are applied to a code-designed bridge and comparisons are made in terms of feasibility, cost, and performance.

20.1 Introduction

The timing of this contribution seems to be quite appropriate, particularly in a European context, as the revision of Eurocode 8 – Part 2 (CEN [Comité Européen de Normalisation] 2005) has just started (national comments have been submitted to CEN and the Project Team that will draft the first version of the new Code will be appointed in early 2018). So, it is arguably the right time to reflect on whether

A. J. Kappos (✉)
Research Centre for Civil Engineering Structures, Department of Civil Engineering, City, University of London, London, UK

Department of Civil Engineering, Aristotle University of Thessaloniki, Thessaloniki, Greece
e-mail: ajkap@civil.auth.gr; Andreas.Kappos.1@city.ac.uk

the basic approach adopted in this Code is in line with the current international trends in seismic design of bridges or whether something important is missing or lagging behind the pertinent state-of-the-art.

The author and his co-workers have been working over the last decade or so on developing new methods for the design of bridges, which are summarised in Sect. 20.3. A critical review of the most current approaches to seismic design was given by the author in his theme lecture at the 2014 ECEES (Kappos 2015). Since then, new developments took place in deformation-based design for both ductile pier bridges (Gkatzogias and Kappos 2015; Kappos and Gkatzogias 2017) and bridges with seismic isolation (Kappos and Gkatzogias 2017; Gkatzogias and Kappos 2017).

In the remainder of this chapter, a critical overview of PBD aspects in the current Eurocode 8 – 2 (EC8-2) is presented, followed by a brief presentation of the performance-based methods for bridges developed by the author and his co-workers. They are followed by a case study wherein the different methods are applied to the design of an actual, code-designed, bridge and comparisons are made of the ease of use and the cost associated with each design method, as well as of the resulting seismic performance. The final section summarises the key conclusions drawn from the aforementioned comparisons and presents the author's perspective on the feasibility of including the current PBD methods (or elements thereof) in the new EC8-3.

20.2 Critical Overview of Eurocode 8 – 2 and Its Evolution

Eurocode 8 – Part 2 (CEN [Comité Européen de Normalisation] 2005) was released in 2005, and two minor amendments were issued in 2009 and 2011. This means that the basic code will be more than 15 years old when its new edition will come in force around 2021. Over these 15 years the American Code for Bridges (AASHTO) (AASHTO [American Association of State Highway and Transportation Officials] 2017) will have changed five times (a new edition is issued every 3 years) while the AASHTO Guide Specification for Seismic Design (AASHTO 2015), an alternative document based on the displacement-based approach, will have (at least) one new edition and three interim revisions. This is indicative of the difficulty to reach a consensus in Europe, as well as of the organisational issues arising in the revision of European codes of practice. Interestingly, one cannot claim that this is mainly due to the fact that in Europe there are low and high seismicity areas (which is, in fact, a major obstacle to reaching consensus) since significant differences in seismic hazard also exist in the US, but this does not hinder the regular updating of the Code.

In the remainder of Sect. 20.2 selected aspects of EC8-2 are discussed in a critical manner, primarily in the light of PBD. It is not the intention of this section to provide a full presentation of all provisions of this code. At the end of the section a brief reference is made to some key comments made during the national enquiry stage of the evolution of EC8-3.

20.2.1 Performance Requirements and Seismic Actions

Except at the very beginning (clause 1.1.1), the term 'performance requirements' is hardly ever mentioned in EC8-2; in fact one should go to the Isolation clauses, in particular those on bearings (clause 7.5.2) to find again this term (probably because these clauses were drafted based on similar provisions of the American codes). Nevertheless, there are two 'basic' (performance) requirements in clause 2.2:

- *No collapse requirement*: After the occurrence of the design earthquake event, the bridge should retain its structural integrity and sufficient residual resistance (strength), although at some parts of the bridge considerable damage may occur.
- *Minimisation of damage requirement*: A seismic action with a high probability of occurrence may cause only minor damage to secondary components and to those parts of the bridge intended to contribute to energy dissipation. All other parts of the bridge should remain undamaged.

These performance requirements are made more specific in the next clause (2.3), where the two basic approaches to performance are set out, i.e. ductile and 'limited ductile' response under the (single) design action. The concept of bridges with ductile piers is well known and long applied to bridges, and EC8-2 is not really different from other codes in this respect. The 'limited ductile' option, though, is somewhat unique to EC8-2, in the sense that it encompasses all other options (except that of ductile piers), i.e., as explicitly mentioned in clause 2.3.2.3:

- Where detailing of plastic hinges for ductility is not reliable ('convenient' should perhaps be the word here), such as bridges with short piers or high axial forces.
- Where the seismic response may be dominated by higher mode effects, such as in cable-stayed bridges.

    However, through indirect references, EC8-2 seems to include here other cases too, notably:
- All bridges in regions of low seismicity, as the seismic actions are low enough anyway and force reduction factor (q –factor) is not necessary.
- Bridges with seismic isolation; these are basically treated in clause 7 of EC8-2 (and annexes J, JJ and K).

The latter is rather confusing, as designing for limited ductile response is clearly not the same as isolating a bridge, and it would be more transparent to list isolation (or passive systems in general) as a third, distinct, option. Interestingly, active or semi-active control (although already used in a number of bridges all over the world (Gkatzogias and Kappos 2016a)) is not even mentioned in EC8-2.

The difference between the case of $q = 1.0$ and $q = 1.5$ is one worth some discussion. In the note to clause 2.3.2.3 it is recognised that the value of 1.5 is related mainly to the overstrength (rather than the ductility); however, $q = 1.0$ is recommended for bridges with short piers (hence an expected shear-dominated failure mode), as if these bridges did not possess any overstrength, which is clearly not the case.

Regarding the specific requirements for the two performance levels (PLs), they are based on well-established principles, i.e. capacity design and detailing for ductility (when applicable), which are also adopted in most codes. The target plastic mechanism is, of course, that involving yielding in the vertical members only (piers), which is different from that in buildings. Interestingly for a code that is developed by Europeans is that the local ductility criterion is that "the structure should be capable of sustaining at least 5 full cycles of deformation to the ultimate displacement", which is pertinent to ground motions with much longer duration than those used to define design seismic action in Europe.

A well-known weakness of EC8 is that only one level of seismic action is specifically prescribed, corresponding to a 10% in 50 years probability of exceedance. The same reference seismic action, A_{Ek}, is prescribed for both buildings and bridges, which means that an ordinary building and an ordinary bridge should be designed for the same 'design earthquake' (in general, road and railway bridges are considered to belong to importance class II, as do Ordinary buildings). This is not the case in the US and other international bridge codes, where the design lifetime is 100 years for ordinary bridges. Although EC8-2 prescribes that "seismic action with a high probability of occurrence may cause only minor damage to secondary components and to those parts of the bridge intended to contribute to energy dissipation", this lower seismic action is nowhere specified, even as a recommended value.

20.2.2 Control of Displacements

As often done in the current Eurocodes (and other codes), provisions are misplaced in sections that are more or less irrelevant. E.g., in clause 2.3 (Compliance criteria) one also finds all provisions related to modelling the stiffness of reinforced concrete members; nevertheless, the most detailed (and interesting) provisions are given in the corresponding annexes (B and C), both of which are informative (rather than normative). Clause 2.3 also includes an extended sub-clause on 'Control of displacements'. This is an important issue, clearly related to PBD, particularly of bridges. An interesting, albeit confusing, distinction is made there between "clearances provided for protection of critical or major structural members" and clearances related to "detailing of non-critical structural components". Unlike most other codes, EC8-2 provides a specific provision for the size (d_{Ed}) of expansion joints between the deck and the abutments, i.e.

$$d_{Ed} = 0.4 d_E + d_G + 0.5 d_T \tag{20.1}$$

where d_E is the seismic displacement, d_G the displacement from permanent and quasi-permanent actions, including prestressing after losses, shrinkage and creep in concrete bridges, and d_T is the displacement due to thermal movements. The fractions of d_E and d_T are chosen "based on a judgement of the cost-effectiveness of the measures taken to prevent damage". Note that only 40% of the seismic displacement is taken into account, hence closing of the joint is indeed envisaged under the design earthquake; actually the Code (correctly) states that this is the case that abutment backwalls can be considered as 'sacrificial' elements. If the designer decides that the abutments should be treated as 'critical' elements and their backwalls are no more deemed as 'sacrificial' another clearance is in order, namely

$$d_{Ed} = d_E + d_G + 0.5d_T \qquad (20.2)$$

i.e. the full seismic displacement is now taken into account. The same value is also taken to design overlap length between the deck and the abutments (as, obviously, unseating is a 'critical' failure mode).

Except in cases of monolithic deck to pier connections, control of displacements is achieved mainly by providing seismic links (such as shear keys or cables) and holding-down devices, which are described in clause 6.6.3. It is noted that these links must remain inactive (by providing appropriate slack) under the design seismic action when the links are located between adjacent sections of the deck at intermediate separation joints (within the span) and in the longitudinal direction at moveable end-supports between the deck and the abutment or pier of existing bridges being retrofitted, if the requirements for minimum overlap length are not met; the latter are specified in clause 6.6.4. To the author's experience these overlap lengths are more than sufficient in bridges constructed in high seismicity regions of Europe since the 1980s, but can be a problem elsewhere and/or in older bridges.

20.2.3 Some Key Issues Raised in the Evolution Phase

A number of comments on the current EC8-2 have been submitted by 16 national groups during the systematic review phase that ended in February 2017; these comments are included in the 31-page document CEN/TC 250/SC 8 N 607, distributed among CEN members in March 2017. The author had the pleasure to lead the UK/BSI Panel ('Mirror group') for EC8-2 and discuss with the panel members various aspects of this document. Some of the pertinent comments (related directly or indirectly to PBD) are listed and commented upon in the following:

- "Resilient designs should be aimed at, for limiting damage and reducing restoration times after earthquakes":

    Resilience-based design is in a way a broader version of PBD, as the individual structure (e.g. the bridge) is seen as a component of a system (e.g. the road network). Its introduction in EC8-2 would be a major step, but is unlikely to happen in the near future, as such an approach cannot be introduced for bridges only; other structures should also be covered. But it is important to raise the issue, as it will be part of the discussion that will open when the next stage of the evolution of Eurocodes begins.

- "Bridges with abutments rigidly connected to the deck can be designed considering higher values of behaviour factor (up to 2.0) if adequate measures are considered to minimise the adverse bridge-backfill interaction effects".

  This is one of the aspects relating to revisiting a key issue, that of integral bridges. Although perhaps not obvious, this is very much a performance-related issue, as the integral bridge (favoured by several designers on both sides of the Atlantic, arguably more so in the US) is an approach to PBD quite different from that based on displacements. In a nutshell, whereas in DBD the structure is designed to develop substantial drifts while meeting the adopted performance criteria, in integral bridges displacements are drastically reduced (at the expense, of course, of higher forces). Although integral bridges are by no means a panacea, it is also true that they are not properly addressed in the current EC8-2, probably because of the state-of-the-art at the time the Code was drafted.

- "When a short-span bridge with continuous deck has its abutments embedded in stiff natural soil formations over more than 80% of their height, it can be considered as fully locked-in..."

  NOTE: For long-span bridges with abutments rigidly connected to the deck, the backfill soil may be taken into account as additional means of resistance and dissipation, provided that the requirements of 2.2.3 are met.

  This is another aspect of integral bridge design, discussed previously; it relates to the proper modelling of SSI, without which a rational design of integral bridges is not possible.

- "When assessing the irregularity of bridges, the r_i indices (clause 4.1.8) are specified only for the critical section of each ductile member (i), i.e. the location of maximum seismic demand in each principal direction of the bridge".

  The issue of irregularity is critical in seismic design and the current procedure in clause 4.1.8 (Regular and irregular seismic behaviour of ductile bridges) if applied blindly leads to most bridges, even regular ones, being classified as irregular. Limiting the calculation of the 'local force reduction factors' r_i for each pier to the most critical section (usually the pier base) results in a more rational classification. The issue is discussed in more detail in (Gkatzogias and Kappos 2016b), where some other issues, not directly related to PBD are also addressed.

It is important to note that apart from the UK comment for introduction of resilience-based design, there are no other national comments on the introduction of PBD procedures, with explicit verifications of multiple performance objectives.

20.3 Overview of Performance-Based Design Approaches

The methods critically summarised in the following sections are typical examples of PBD procedures that can be applied to a large number of bridge types and hence have the potential to be adopted in future versions of seismic codes for bridges; one should not forget that, as often stressed in meetings of various Eurocode committees, a code should fully cover roughly 80% of the structures to which it refers.

The method presented in Sect. 20.3.1 is based on the Direct Displacement Based Design (DDBD) concept, on which significant contributions have been made in the past by N. Priestley, M. Kowalsky, M. Calvi, and their co-workers (see a critical review in Kappos 2015), while the methods presented in Sect. 20.3.2 are based on the Deformation-based Design (Def-BD) concept, introduced by the author.

20.3.1 Direct Displacement Based Design

The DDBD methods for bridges have been presented in (Kappos 2015) wherein a detailed discussion of their advantages, disadvantages and pitfalls was also given. The 'Janus double face' in these methods is their relative simplicity. On one hand this simplicity (in particular the use of the equivalent SDOF for defining displacement demand and the use of simple analysis methods so long as the base shear has been estimated on the basis of displacement demand) certainly enhances their ease of use (especially for a preliminary design), despite the fact that several of the concepts involved are not familiar to engineers not previously trained in these methods. On the other hand this simplicity is clearly a pitfall when higher modes cannot possibly be ignored, as is typically the case with bridges in their transverse direction. The DDBD method proposed by the author and his associates (Kappos et al. 2012; Kappos et al. 2013) manages to duly account for higher mode effects, at the expense, of course, of loss of simplicity. An overview of this method, that can be considered as the state-of-the-art insofar a broadly applicable DDBD procedure for bridges is concerned, is summarised in the remainder of this section.

Step 0 – Definition of Initial Input Parameters General input parameters are defined including geometry, e.g. column height and cross-section, mass properties, and material properties. As a starting point for the estimation of the column cross-sections, the output of the dimensioning of the deck and the piers for the Ultimate and Serviceability Limit States under the pertinent combinations of permanent and transient actions can be used. Then, single or multiple performance levels are set as design objectives, by designating the targeted damage states ('damage-based' displacements) for selected seismic hazard levels, expressed in terms of elastic displacement response spectra.

Step 1 – Selection of the Displacement Pattern The step prescribed in the 'standard' DDBD procedure (Kowalsky 2002) involves the computation of the relative pier-to-deck stiffness (RS) and the determination of whether the bridge has a rigid or a flexible displacement pattern. Given that the procedure presented here is intended for bridges where higher mode contribution should *not* be ignored, the flexible displacement pattern scenario is adopted, disregarding the relative stiffness parameter. This means that this step is essentially redundant, nevertheless it is deemed advisable to retain it, as it is always useful for the designer to have a proper indication of the relative stiffness of the deck.

Step 2 – Definition of Target-Displacement Profiles The iterative EMS (effective mode shape) method is followed, entailing the following steps:

  (i) *Evaluation of mode shapes (Φ_j)*: Due to the unavailability of the member effective properties at the beginning of the process, a first estimation is required. Based on current seismic design practice for bridges it can be assumed that the superstructure, particularly in the common case that it is prestressed, will respond essentially elastically, regarding its flexural stiffness, while for the torsional stiffness of prestressed concrete box girders 20% of the uncracked value can be assumed. A secant flexural stiffness based on 10% the gross section rigidity (EI_g) should be used for columns expected to deform inelastically, while 60% EI_g is recommended for columns that are expected to remain below yield. The reduction in the effective axial and shear stiffness of the column(s) can be considered proportional to the reduction in the effective flexural stiffness. Once the structural properties have been established, the eigenvalue problem can be solved, hence the mode shapes Φ_j can be obtained.
 (ii) *Evaluation of modal participation factors (Γ_j)*: The modal participation factors are *computed* using standard procedures of modal analysis.
(iii) *Evaluation of peak modal displacements ($u_{i,j}$)*: The peak modal displacements are computed according to Eq. (20.3), where index i represents the DOF associated with a lumped mass, as per the inertial discretization, index j represents the mode number, $\Phi_{i,j}$ is the modal factor of joint i at mode j, and S_{dj} is the spectral displacement for mode j obtained by entering the 5%-damped design spectra with the period obtained from modal analysis.

$$u_{i,j} = \Gamma_j \Phi_{i,j} S_{dj} \tag{20.3}$$

(iv) *Evaluation of expected displacement pattern*: The displacement pattern (δ_i) is obtained by an appropriate combination of the peak modal displacements, such as the SRSS or CQC combinations depending on how closely spaced the natural frequencies of the participating modes are.

It is noted that a displacement pattern derived from the above procedure accounts for the effect of all significant modes; therefore, it does not correspond to an actual inelastic deformed shape of the bridge, particularly so in the case of asymmetric systems. To obtain the target displacement profile (Δ_i), the displacement pattern given from the SRSS combination is scaled in such a way that none of the member (pier or abutment) displacements exceeds the target displacements obtained based on strain or drift criteria:

$$\Delta_i = \delta_i \frac{\Delta_{D,c}}{\delta_c} \qquad (20.4)$$

where $\Delta_{D,c}$ and δ_c are the 'damage-based' displacement and the modal value at the location of the critical member, c, whose displacement governs the design, respectively. Prior to applying (20.4) one iteration might be needed to identify the most critical member, when this is not obvious; alternatively, the ratio could be calculated for all piers and the most critical be used for scaling. Then, peak modal displacements ($u_{i,j}$) are scaled to N modal target-displacement profiles ($U_{i,j}$) using the same scaling coefficient as that used to obtain the target-displacement profile in (20.4), i.e.

$$U_{i,j} = u_{i,j} \frac{\Delta_{D,c}}{\delta_c} \qquad (20.5)$$

An immediate consequence of the aforementioned procedure is that the combination of the N modal target-displacement profiles ($U_{i,j}$) yields the target-displacement profile (Δ_i); hence, when the SRSS combination rule is used:

$$\Delta_i = \sqrt{\sum_j U^2_{i,j}} \qquad (20.6)$$

Step 3 – Definition of N + 1 Equivalent SDOF Structures These idealised structures are established based on equality of the work done by the MDOF bridge and the equivalent SDOF structure. Each of the N SDOF structures is related to the corresponding modal target-displacement profile ($U_{i,j}$), whereas the additional SDOF is related to the (final) target-displacement profile (Δ_i). Utilising Eqs. (20.7) and (20.8), an equivalent system displacement (Δ_{sys}, $U_{sys,j}$), mass (M_{sys}, $M_{sys,j}$), and location (x_{sys}, $x_{sys,j}$) of the SDOF along the MDOF bridge deck is computed for each of the $N + 1$ SDOF structures. The 'location' of the SDOF system (i.e. of the masses M_{sys} or $M_{sys,j}$) coincides with the point at which the resultant of the modal forces is applied, and is one of the criteria used for checking convergence of the procedure. In Eqs. (20.7) and (20.8), m_i is the mass associated with joint i, and n is the number of joints as per the inertial discretization.

$$U_{sys,j} = \frac{\sum_{i=1}^{n} m_i U_{i,j}^2}{\sum_{i=1}^{n} m_i U_{i,j}}, \quad M_{sys(j)} = \frac{\sum_{i=1}^{n} m_i U_{i,j}}{U_{sys,j}}, \quad x_{sys,j} = \frac{\sum_{i=1}^{n} (m_i U_{i,j} x_i)}{\sum_{i=1}^{n} (m_i U_{i,j})} \qquad (20.7)$$

$$\Delta_{sys} = \frac{\sum_{i=1}^{n} m_i \Delta_i^2}{\sum_{i=1}^{n} m_i \Delta_i}, \quad M_{sys} = \frac{\sum_{i=1}^{n} m_i \Delta_i}{\Delta_{sys}}, \quad x_{sys} = \frac{\sum_{i=1}^{n} (m_i \Delta_i x_i)}{\sum_{i=1}^{n} (m_i \Delta_i)} \qquad (20.8)$$

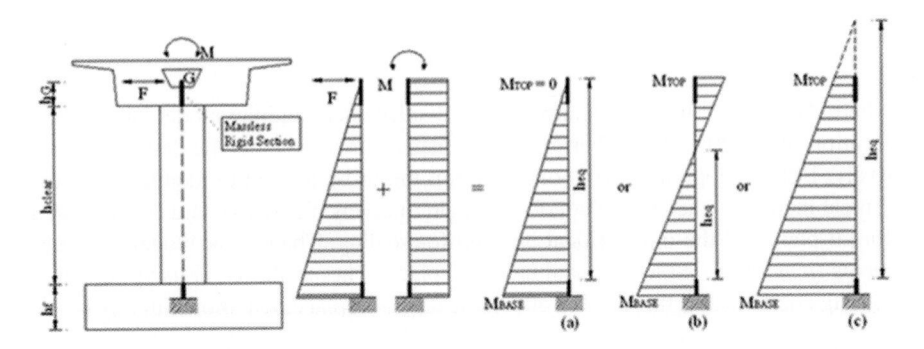

Fig. 20.1 Pier modelling and transverse response accounting for the deck's torsional stiffness

Step 4 – Estimation of Equivalent Viscous Damping Levels Utilising the target displacement (Δ_i) and the modal target-displacement profiles ($U_{i,j}$), the ductility level is calculated for each member (for each of the $N + 1$ profiles), according to Eq. (20.9). Yield curvatures are estimated using Eq. (20.10), where ε_y is the yield strain of reinforcing steel and D is the diameter of a circular section. Similar equations are provided for different section shapes (Priestley et al. 2007).

$$\mu_{\Delta_i} = \Delta_i/\Delta_{yi}, \quad (or\ \mu_{\Delta_i} = U_{i,j}/\Delta_{yi,j}) \tag{20.9}$$
$$\varphi_y = 2.25\varepsilon_y/D \tag{20.10}$$

Figure 20.1 shows the modelling of a pier with a rigid base, whose top is monolithically connected to the deck, whereas possible moment diagrams under transverse loading are also illustrated. A pier moment diagram consists of two different components; the bending moment derived from the inertial horizontal forces F, acting on the mass centroid (G), and the bending moment induced from the eccentricity of the horizontal forces with respect to the shear centre, in the usual case wherein the shear centre does not coincide with the mass centroid. The final moment diagram depends on the cracked torsional stiffness of the bridge deck, the superstructure-abutment connection and the pier-to-superstructure relative stiffness. One should properly account for the degree of fixity at the pier top and hence for the transverse response of the pier regarding its flexural stiffness (k_{pier}) and yield displacement ($\Delta_{y,pier}$), according to Eqs. (20.11) to (20.13), (referring to case (b) in Fig. 20.1; similar relationships apply to the other cases).

$$x_k = \frac{h_{eq}}{h} = \frac{h_{eq}}{h_{clear} + h_G}, \quad k_{eq} = \frac{3EI}{h_{eq}^3}, \quad k_{pier} = x_k k_{eq} \tag{20.11}$$

$$x_{\Delta y} = \frac{L_{eq}}{L_{eff}} = \frac{h_{eq} + 0.022 f_y d_{bl}}{h_{clear} + h_G + 0.022 f_y d_{bl}} \tag{20.12}$$

$$\Delta_{y,eq} = \frac{\varphi_y L_{eq}^2}{3}, \quad \Delta_{y,pier} = \frac{1}{x_{\Delta y}} \Delta_{y,eq} \tag{20.13}$$

In Eqs. (20.11) to (20.13), $\Delta_{y,eq}$ and k_{eq} are the yield displacement and the flexural stiffness of the equivalent cantilever, E is the elastic modulus of the pier material, I is the moment of inertia of the pier cross-section (modified for cracking effects wherever necessary), d_{bl} is the longitudinal reinforcement bar diameter and $0.022 f_y d_{bl}$ is the strain penetration length (where f_y is the yield stress of the longitudinal reinforcement in MPa). The height of the equivalent cantilever (h_{eq}) cannot be determined at the initial stage of design, therefore either preliminary structural analyses should be performed for each of the $N + 1$ equivalent structures under lateral loads compatible with the corresponding profile, or an assumption that the height of the equivalent cantilever equals the height of the pier, be made during the first iteration. The first approach is strongly recommended for the case of significant higher mode effects, since it reduces the number of iterations required to achieve convergence.

A relationship between hysteretic damping and ductility has then to be selected. The one (Dwairi et al. 2007) based on Takeda's hysteretic model is used here (Eq. 20.14), wherein elastic viscous damping (ξ_v) up to 5% is added to the hysteretic damping

$$\xi_i = \xi_v + \frac{50}{\pi}\left(\frac{\mu_\Delta - 1}{\mu_\Delta}\right)\% \tag{20.14}$$

These damping values need to be combined in some form to obtain system damping for each of the $N + 1$ equivalent SDOF structures. A weighted average can be computed, as given by Eq. (20.15), where $W_i/\Sigma W_k$ is a weighting factor, based on the work (W_i) done by each member (Eq. (20.16)), according to (Kowalsky 2002).

$$\xi_{sys} = \sum_{i=1}^{n}\left(\frac{W_i}{\sum_{k=1}^{n} W_k}\xi_i\right), \quad \xi_{sys(j)} = \sum_{i=1}^{n}\left(\frac{W_{i,j}}{\sum_{k=1}^{n} W_{k,j}}\xi_{i,j}\right) \tag{20.15}$$

$$W_i = V_i \Delta_i, \quad W_{i,j} = V_{i,j} U_{i,j} \tag{20.16}$$

Calculation of the weighting factors presupposes knowledge of member forces (V_i), which are not known at the current step. As a starting point, it can be assumed that the seismic force carried by the abutments is equal to 30% of the total seismic force carried by the bridge and column shears are inversely proportional to column heights, as illustrated by Eq. (20.17) (Kowalsky 2002), where μ is less than 1 for elastic columns and equal to 1 for columns that have yielded. In subsequent iterations, system damping is computed using member forces obtained from structural analysis.

$$W_i = \mu_{\Delta_i}\Delta_i/h_{eq,i}, \quad W_{i,j} = \mu_{\Delta_i}U_{i,j}/h_{eq,ij} \tag{20.17}$$

Step 5 – Determination of the Effective Periods of the Equivalent Structures Utilising the $N + 1$ system target displacements (Δ_{sys}, $U_{sys,j}$), levels of system damping (ξ_{sys}, $\xi_{sys,j}$), and elastic response spectra for the chosen seismic demand, the effective periods (T_{eff}, $T_{eff,j}$) of the equivalent structures are determined from the design spectrum. Once the effective periods have been determined, effective stiffnesses (k_{eff}, $k_{eff,j}$) and design base shears (V_B, $V_{B,j}$) are computed by Eqs. (20.18) and (20.19), respectively.

$$k_{eff} = 4\pi^2 M_{sys}/T_{eff}^2, \quad k_{eff,j} = 4\pi^2 M_{sys,j}/T_{eff,j}^2 \tag{20.18}$$
$$V_B = k_{eff}\Delta_{sys}, \quad V_{B,j} = k_{eff,j}U_{sys,j} \tag{20.19}$$

Step 6 – Verification of Design Assumptions Design base shears (V_B, $V_{B,j}$) are distributed in proportion to the inverse of the column height according to Eq. (20.20), which is based on the simplifying assumption that all columns have the same diameter and longitudinal reinforcement ratio, zero post-elastic slope of the force-displacement response, mass small enough for the inertia forces due to self-weight to be neglected, and the same end-fixity conditions. In Eq. (20.20) μ_i and μ_k are less than one for elastic columns and equal to one for columns that have yielded, and F_{Abt} represents the total force carried by the abutments. R/C member cracked section stiffnesses are computed for each of the $N + 1$ profiles, using Eqs. (20.21) and are compared with values assumed at Step 2. If the values related to the target-displacement profile (Δ_i) differ significantly, computed secant stiffnesses ($k_{eff,i}$) are utilised in the EMS to obtain revised target-displacement profiles (Δ_i, $U_{i,j}$). Steps 2 to 6 are repeated by changing column secant stiffnesses until the target profile (Δ_i) stabilises. Although a strict approach requires iteration within Steps 2 to 6 until all profiles (Δ_i and $U_{i,j}$) stabilise, the implementation of the methodology in the next section indicates that whenever Δ_i stabilises, $U_{i,j}$ also practically stabilise, hence Δ_i can be used as the sole convergence criterion.

$$V_{B,k} = (V_B - F_{Abt})\frac{\mu_{\Delta,k}}{h_k}\bigg/\sum_{i=1}^{n}\frac{\mu_{\Delta,i}}{h_i}, \quad V_{B,kj} = (V_{B,j} - F_{Abt,j})\frac{\mu_{\Delta,kj}}{h_k}\bigg/\sum_{i=1}^{n}\frac{\mu_{\Delta,ij}}{h_i} \tag{20.20}$$

$$k_{eff,i} = V_{B,i}/\Delta_i, \quad k_{eff,ij} = V_{B,ij}/U_{i,j} \tag{20.21}$$

Step 7 – Structural Analysis Once the target-displacement profile (Δ_i) stabilises, base shears ($V_{B,j}$) are distributed as inertia forces to the masses of the MDOF structure in accordance with the modal target-displacement profiles ($U_{i,j}$), given by Eq. (20.22). In this equation $F_{i,j}$ are the bent inertia forces, $V_{B,j}$ are the design base shears, indices i and k refer to joint numbers, and n is the number of joints.

$$F_{k,j} = V_{B,j}(m_k U_{i,j}) / \sum_{i=1}^{n}(m_i U_{i,j}) \qquad (20.22)$$

N structural analyses (as many as the significant modes) are performed on the bridge under the inertia loads, to obtain the 'modal' base shear for each column. Secant stiffnesses $k_{eff,ij}$ obtained from the iteration within Step 6, at which stabilisation of Δ_i (hence stabilisation of $U_{i,j}$ as mentioned in Step 6) was observed, should be used in each of the N structural analyses, in order to be consistent with the DDBD philosophy. Afterwards, displacements derived from the N structural analyses are compared with the corresponding profiles $U_{i,j}$. In the case of significantly different displacements, reasonable values for column secant stiffnesses are assumed and analyses are conducted until convergence is achieved. Once the displacement profiles obtained from structural analyses converge to the assumed modal target-displacement profiles, column secant stiffnesses and abutment forces from each analysis are compared with the values assumed at Step 6, at which stabilisation of $U_{i,j}$ was achieved. It is reminded that during the first loop of iterations the seismic force carried by the abutments is assumed equal to 30% of the total seismic force carried by the bridge for all the $N + 1$ cases. In case of significant discrepancy, the target-displacement profile is revised utilising the EMS method and forces from structural analysis. Steps 2 to 7 are repeated, until column secant stiffnesses and abutment forces converge.

To perform the new loop of iterations, the new EMS in particular, previous loop secant stiffnesses ($k_{eff,i}$) (Step 6) can be assumed as the starting point. Furthermore, revised equivalent cantilever heights are computed according to the results of the N structural analyses, which were previously performed, as far as the modal target-displacement profiles ($U_{i,j}$) are concerned, whereas in the case of the (final) target-displacement profile (Δ_i), proper values of the equivalent cantilever heights can be approximately determined by combining the peak 'modal' responses (N structural analyses). Following the same approach, the force carried by the abutments and the base shear distribution for each of the $N + 1$ cases required in the subsequent steps are determined from analysis results, instead of using Eq. (20.20), which, given the diversity of the column end-fixity conditions, is not accurate enough.

Step 8 – Design of the MDOF Structure The MDOF bridge is designed in accordance with capacity design principles (e.g. (Kappos 2015; Priestley et al. 2007)) such that the desired failure mechanism is achieved. The response quantities of design interest (displacements, plastic hinge rotations, action effects for the piers) are determined by combining the peak 'modal' responses (from the N structural analyses), using an appropriate modal combination rule (e.g. SRSS or CQC), and superimposing the pertinent combinations of permanent and transient actions.

20.3.2 Deformation-Based Design

The suggested procedure consists of 5 distinct steps including a preliminary design and subsequent verifications involving nonlinear response-history analysis (NLRHA) and verification of a number of performance objectives (PO) depending on bridge importance (e.g. (CEN [Comité Européen de Normalisation] 2005)), i.e. Non-essential (minor importance), Ordinary (average), Essential (high), and Critical I, II (major). The PO is described with reference to the member (abutment, pier, deck, etc.) structural performance level (SP) (Table 20.1) and the associated seismic action (EQ). Four different SPs and EQs are introduced to formulate the 'performance matrix' depicted in Table 20.2; SPs are quantitatively described in terms of operationality (service), damage, and feasibility of repair, while the considered range of return periods T_R coupled with each SP is in line with the widely varying requirements prescribed in different codes (Gkatzogias and Kappos 2015).

The 'performance matrix' in Table 20.2 assigns a higher PO to ordinary isolated bridges compared to ordinary 'ductile-pier' bridges. This is in line with code specifications (e.g. (CEN [Comité Européen de Normalisation] 2005)) regarding the requirement of limiting the inelastic response of the substructure, aiming at the proper performance of the isolation system, when inelastic action develops and the effectiveness of the isolation system may be reduced, resulting in larger deformation demands in the isolated structure. In the light of the previous consideration, controlled inelastic response of the piers (e.g. associated with spalling of concrete cover) is allowed under the highest SP considered in ordinary bridges (i.e. SP3). On the other hand, the 'critical' PO is not defined in the case of isolated bridges as this would require a practically inactive isolation system under the only considered hazard level, i.e. EQIV. In the remainder of Sect. 20.3.2 the different (albeit based on similar concepts) procedures proposed for bridges where seismic energy dissipation takes place in ductile piers, and bridges where the dissipation takes place in the isolation system, are described.

20.3.2.1 Ductile Pier Bridges

Step 1 – Preliminary Design The purpose of this step is to establish a basic level of strength under the minimum considered hazard level, typically EQI (associated with SP1 in Table 20.2) to ensure that the bridge remains operational during and after an earthquake having a higher probability of exceedance (i.e. EQ(SP2)), taking into consideration the range within which the inelastic deformations should fall. Clearly, this stage involves striking a balance between economy and performance. The required strength of the yielding zones (i.e. normally the pier ends) that will respond inelastically under the least frequent event is calculated through elastic analysis and the use of a selected allowable rotational ductility factor, while inelastic deformations are verified during the next step through NLRHA.

20 Seismic Design of Bridges: Present and Future

Table 20.1 Recommended structural performance criteria for concrete bridge components

Member	Ductile bridges				Isolated bridges			
	SP1	SP2	SP3	SP4	SP1	SP2	SP3	SP4
Pier	Yield	$\varepsilon_c = 3.5$–4‰	$\varepsilon_c = 18$‰	ε_{cu}	–	Yield	$\varepsilon_c = .5$–4‰	ε_c
El. Bearings	–	$\gamma_q = 1 \sim 1.5$	$\gamma_q = 2$	Ultimate	$\gamma_q<1$	$\gamma_q = 1 \sim 1.5$	Ultimate	Link activ.
Deck	–	–	Uncracked	Yield	–	–	Uncracked	Yield
Abutments	–	–	Yield	Ultimate	–	–	Yield	Ultimate
Foundation	–	–	Yield	Ultimate	–	–	Yield	Ultimate

Table 20.2 Performance matrix for the proposed methodology for ductile or isolated bridges

Seismic hazard		Structural performance			
EQ	T_R (yrs)	SP1	SP2	SP3	SP4
EQI	<50	Ductile		–	–
EQII	50–100	Isolated	Ductile		–
EQIII	500–1000		Isolated	Ductile	
EQIV	~2500			Isolated	Ductile
Service		Full	Operational	Limited	Disrupted
Damage		Negligible	Minimal	Significant	Severe
Repair		No/Economic	Economic	Feasible	Non-feasible

Seismic hazard		Ductile bridges: analysis type per PL			
EQI	<50	IMP	L	–	–
EQII	50–100	IMP	L+NL	NL	–
EQIII	500–1000	IMP	L+NL	NL	IMP
EQIV	~2500	L+NL	L+NL	NL	IMP

Seismic hazard		Isolated bridges: analysis type per PL			
EQI	<50	DE	–	–	–
EQII	50–100	NL	DE+NL	–	–
EQIII	500–1000	NL	DE+NL	DE+NL	–
EQIV	~2500	–	DE+NL	DE+NL	NL

Analysis: Design equations (DE), Implicit calulation (IMP), Linear (L), Nonlinear (NL)

Importance: Non-essential, Ordinary, Essential, Critical I, II

To meet the target objective, element forces and chord rotations are obtained from the results of a standard response spectrum (elastic) analysis (RSA) carried out for an appropriate fraction (x) of the earthquake level associated with EQ(SP2); this is done to account (when the verification is carried out at step 2) for issues such as the differentiation of the design values of material strength adopted in commonly available design aids, and the mean values used in NLRHA, member overstrength, and the differences in moments derived from RSA and NLRHA. The pier stiffness considered at this stage is the secant value at yield, accounting for the effects of axial load ratio (e.g. (Priestley et al. 2007)) and either a minimum reinforcement (CEN [Comité Européen de Normalisation] 2005) or that resulting from design for 'non-seismic' loading (if higher than the minimum). Subsequently, elastic chord rotations (θ_{el}) are related to the corresponding inelastic ones (θ_{inel}), using an empirical procedure like the one proposed in (Bardakis and Fardis 2011). The allowable chord rotation ductility factor ($\mu_{\theta,SP}$) for SP2 is estimated based on allowable material strains (Gkatzogias and Kappos 2015), thus providing the yield rotation

Fig. 20.2 Definition of pier yield moments

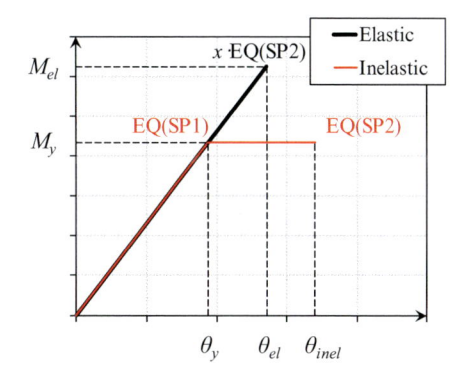

($\theta_y = \theta_{inel}/\mu_{\theta,SP}$) in each pier. Assuming an elastic-perfectly plastic M-θ response having the same initial slope as the 'elastic' M-θ diagram, the yield moment (M_y) used for the (flexural) design of the pier (implicitly related to EQ(SP1)), can be easily computed as the moment corresponding to θ_y (Fig. 20.2). In case pier longitudinal reinforcement demand is found to be less than the minimum, reduction of cross sections is in order (reduction of stiffness), otherwise deformations for the considered SP will be less than the allowable ones.

Deformation control in the piers does not fully guarantee that the bridge will remain operational; it is equally important to check that bearings (which are typically present unless a fully integral solution is adopted) also remain functional. Hence, displacements (or the corresponding strains) of bearings under the full 'operationality' actions EQ(SP2) should conform to the deformation criteria (Table 20.1). An exception to the procedure described above is the case of Critical II bridges wherein member design moments and deformations are retrieved directly from linear analysis under xEQIV (Table 20.2) without reducing M_y (i.e. $\mu_{\theta,SP} = 1$). Verifications aiming to ensure elastic response are performed with the aid of NLRHA in Step 2 under the same seismic intensity (i.e. EQIV).

Step 2 – PL1 Verifications A partially inelastic model (PIM) of the structure is set up, wherein the energy dissipation zones of the piers (e.g. top and bottom of columns if they are monolithically connected to the deck) are modelled as yielding elements, with their strength based on the reinforcement calculated in Step 1. In the same model, the remaining parts of the bridge are modelled as elastic members (with duly reduced stiffness in R/C piers). Since the dissipating zones have been designed for flexure at Step 1, the stiffness of the piers can now be calculated from moment-curvature (M-φ) analysis using the longitudinal reinforcement ratio (ρ_l) of the pier ends and mean values for strength of materials, since deformations are to be checked at this stage. NLRHA of the PIM also requires the definition of a suite of ground motions, which, in a design context, should be compatible with the selected design spectrum. Selection and scaling of input motions can be performed following, as a minimum, relevant code-based specifications (e.g. CEN [Comité Européen de Normalisation] 2005); the selected earthquake motions will be used for both this step and the following ones, and they should be properly scaled to the level associated with the PL considered. Alternatively, different suites of motions can be established for each PL based on different selection criteria and shape of target spectra in a more refined approach expected to be more relevant in the case of bridges of higher importance.

Verifications are associated in this step with 'operationality' (SP2) criteria except for critical II bridges wherein elastic member response is sought (i.e. SP1); performance checks are carried out in terms of specific limits for maximum drifts and/or plastic deformations of critical members, which in turn are derived considering 'acceptable damage' indicators. An appropriate way to define specific limits for R/C piers in line with the refined analysis tools used in Def-BD, is in terms of strains. For instance, the functionality of the bridge will not be impaired if cover concrete does not spall (concrete strain ε_c below 3.5~4‰). Such strain values can then be used to derive deformation limits (e.g. $\mu_{\theta,SP}$), based on the results of the M-φ analysis of piers, and the pier shear span (h_{eq}) estimated from NLRHA. If the adopted performance criteria are not satisfied (e.g. within 5–10%), stiffening or softening of pier columns will be required. Considering the commonly adopted bolted elastomeric bearings, the deformation limit associated with a functionality level could be set in terms of bearing shear strain, e.g. $\gamma_q = 1$~1.5 (Mori et al. 1997). Moreover, the width of joints (located only at the abutments in most modern bridges) should be selected such that they remain open under EQ(SP2), to avoid damage at the backwalls. In the case of Non-essential bridges, SP2 verifications may be performed considering elastic analysis results (Step 1) due to the reduced importance and the low associated seismic actions.

Step 3 – PL2 Verifications Members considered as elastic in setting up the PIM (i.e. deck, abutments, foundation), are designed for flexure in this step using the NLRHA results under EQ(SP3) (i.e. EQIV for Essential and Critical bridges). Design of the superstructure should aim at the non-cracked, rather than yielding, state of deck sections (Gkatzogias and Kappos 2016b) (apart from the case of continuity slabs in beam/girder bridges). In general, it is very likely that the deck and the abutments will have (from 'non-seismic' load combinations) a higher strength than that required on the basis of this analysis. Deformation limit for yielding elements (i.e. the piers) can be computed as in the previous step by adopting relevant material strains (e.g. $\varepsilon_c = 18‰$), noting however, that deformation demand is not expected to be critical at this PL apart from the cases wherein a seismic action higher than the one corresponding to bridges of average importance is adopted. On the contrary, it is essential that bearing deformations be checked at this stage; allowable (seismic) strain values γ_q for typical elastomeric bearings can be set to around 2.0 (CEN [Comité Européen de Normalisation] 2005).

Step 4 – PL3 Verifications To account for the less ductile nature of shear (or - flexure-shear) failure, shear forces should be calculated for seismic actions corresponding to a higher level than EQ(SP3) (except for critical bridges). To simplify the design procedure, design and detailing for shear can be carried out using shear forces calculated from Step 3, implicitly related to EQ(SP4) through appropriately selected magnification factors (γ_v). Recommended γ_v factors, accounting mainly for the strain-hardening effect corresponding to higher plastic rotations at this earthquake level, are between 1.10 and 1.20 (Gkatzogias and Kappos 2015). Detailing of piers for confinement, anchorages and lap splices, is carried out with due consideration of the expected level of inelasticity. Instead of basing the detailing on default curvature ductilities (as in EC8), the actual μ_φ estimated for the earthquake associated with SP4 is used in this method. This results in both more rational and, as a rule, more economical, detailing of the piers. In the case of bearings, it should be verified in terms of both ultimate deformability and stability; toppling considerations are expected to yield the critical (i.e. allowable) γ_q strain limit at this stage. Moreover, non-significant yielding of the deck should be ensured under EQ(SP4).

20.3.2.2 Seismically Isolated Bridges

Step 1 – Preliminary Design In the case of isolated bridges, the first step aims at the identification of the critical (in terms of economy and performance) PL and at a first 'near-optimal' estimation of the basic parameters of the isolation system, namely, its strength ($\bar{v}_0$), post-elastic stiffness (k_p) or isolation period (T_p), and damping ratio (ξ). $\bar{v}_0$ represents the ratio $V_0/(mg)$, where V_0 is the horizontal shear resistance of the isolation system at zero displacement, m the isolated mass and g the acceleration of gravity. The 'near-optimal' isolation solution is defined here as the one that results in 'near-minimum' peak total acceleration ($\ddot{U}_0$) of the deck while keeping within allowable limits the peak relative displacements (u_0) of the isolation system and the deformations of the substructure (piers). Both objectives of the preliminary design are investigated on the basis of 'design equations' (DE) that provide direct estimates of u_0, $\ddot{U}_0$ for an isolated SDOF as a function of ξ, η (i.e. strength normalised to seismic intensity), and T_p, under different PLs associated with (code-based) target spectra with the same frequency content but different amplitude. In general, 'design equations' can be extracted for code-prescribed (or other) target spectra and be provided as ready-to-use tools; see (Gkatzogias and Kappos 2017) for details of this procedure.

$\ddot{U}_0$ and u_0 inelastic spectra for the adopted EQ(SP2) and EQ(SP3) seismic actions are first established by plotting DE in a u_0 and $\ddot{U}_0$ vs $\bar{v}_0$ format (Fig. 20.3), facilitating the identification of isolation schemes (ξ, $\bar{v}_0$, T_p) with a 'near-optimal' performance under different earthquake intensities, and systems consisting of different isolation and energy dissipation devices. Fig. 20.3 indicates that when an isolated structural system designed for optimal performance under a 'rare' event (e.g. optEQIII) is subjected to stronger ground motions (e.g. EQIV), it results in suboptimal u_0 response compared to the displacement response of a system optimised for the higher seismic action (i.e. optEQIV). On the other hand, increased $\ddot{U}_0$ (and hence total shear) are obtained in the case of the optEQIV system when the latter is subjected to shorter return period events (e.g. EQIII). In design terms, the previous observation could be translated into an increased cost of isolators in the first case and an increased cost of reinforcing steel in the concrete piers in the second. Hence, the

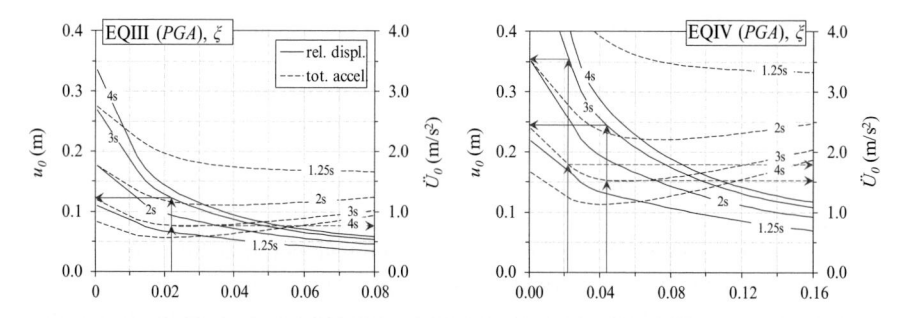

Fig. 20.3 Direct peak response estimation of SDOF systems (m, ξ, $T_p = 3.0$ s,) optimally designed (in blue), and corresponding response of optEQIII under EQIV (in red)

decision on the level of seismic action under which the isolation system is near-optimally selected, i.e. EQ(SP2) or EQ(SP3), should be made cautiously apart from the case of inherently linear systems (i.e. linear viscous dampers and linear isolators) wherein the reference seismic actions have no effect on optimal response, and in Essential bridges wherein the isolation system is activated only under EQ(SP2).

Selecting an isolation and energy dissipation system will result in a first estimation of the geometrical and mechanical properties of devices to be used in subsequent steps, so long as, ξ, η, and $?_p$, of the selected system are properly distributed to a sufficient number of devices located at the piers and abutments, accounting for the deck weight distribution to the substructure, minimisation of torsional effects, system reliability, and cost issues (e.g. cost for testing) that will normally dictate the above decision. It is worth noting that the constraint of maintaining classical normal modes does not apply here due to the use of NLRHA, hence, optimal distributions of dampers may be explored.

Distribution of isolation system properties and determination of DPs of devices will also provide an estimate of the pier strength required to ensure the target performance under the selected reference event, i.e. quasi-elastic response of piers under EQ(SP2) or controlled inelastic response under EQ(SP3). Regarding the second case, the strength at the pier ends should be established to retain the effectiveness of the isolation system through proper consideration of the range within which the inelastic deformations should fall ($\varepsilon_c = 3.5\sim4‰$) following the procedure described in Step 1 for 'ductile-pier' bridges, herein implemented for EQ (SP3) without requiring an elastic analysis. Pier column forces and chord rotations can be estimated from the maximum shear transferred through the isolators to the pier top and a proper estimation of h_{eq}. An issue deserving some further consideration is that convergence of response quantities derived from DE and the MDOF analysis in the following steps, depends on the substructure's stiffness (potentially involving limited inelastic response) and inertial characteristics that are ignored in Step 1. Consideration of the latter parameters is deemed superfluous at this stage since subsequent steps of the method involve NLRHA of a detailed model of the bridge.

Step 2 – PL1 Verifications During this step, the PIM of the structure is set up and SP1 verifications are performed based on NLRHA results under a suite of records selected and scaled using the principles described in Sect. 20.3.2.1. Hysteretic isolators and dampers are modelled as yielding and dashpot elements, respectively, with mechanical properties as defined in Step 1, and the remaining parts of the bridge as elastic members. Pier column stiffness should correspond either to yield (from M-φ analysis) or to the gross section; pier stiffness under EQ(SP1) is expected to have a minor effect on isolator deformations used in the subsequent verifications. Verifications are carried out in terms of both 'operationality' and 'structural performance' to ensure both 'full' service of the bridge (i.e. no closure) and 'negligible' (or preferably zero) damage of the isolation and energy dissipation devices. The 'operationality' requirement can be satisfied by providing an adequate restoring capability assessed on the basis of design charts with a view to limiting residual displacements (u_{res}) to near-zero values (Gkatzogias and Kappos 2017). Considering the requirement for 'negligible' damage of isolators, an upper limit of deformation (combined with an appropriate safety factor S_F) corresponding to the yielding of the steel shims (e.g. $\gamma_q < 1.0/S_F$ Mori et al. 1997) should be applied in the case of elastomer-based isolators. If the adopted performance criteria are not satisfied, mechanical properties of devices should be modified without performing additional NLRHAs; conformity to the requirements of Step 1 can be evaluated using DEs. In any case, operationality verifications at this step are not expected to be critical for piers, as the latter are designed for responding quasi-elastically up to the next PL, whereas for a Non-essential bridge, NLRHA is omitted and required response quantities are derived from DEs.

Step 3 – PL2 Verifications During analysis in this step, the PIM of Step 2 should be used with pier stiffness taken as the secant value at yielding and device properties as modified in Step 2. SP2 verifications should ensure that the extent of damage is such that the bridge can be repaired after the earthquake without significant disruption of service. Regarding the isolation system, the previous requirement can be expressed as an adequate restoring capability allowing for u_{res} that can result in a 'brief' closure of the bridge (e.g. evaluated according to CEN [Comité Européen de Normalisation] 2005), and 'minimal' damage in the isolators (e.g. $\gamma_q = 1\sim1.5$). The performance sought for the substructure at this PL refers to essentially elastic response of piers. When pier design for flexure is carried out in terms of design values of material strength, pier moment derived from analysis (based on mean strength values) should be properly reduced (by the x factor) similarly to 'ductile-pier' bridges. The final reinforcement ratio ρ_l should be selected by adopting the highest demand derived from Steps 1, 3.

Step 4 – PL3 Verifications Verification of the 'near-optimal' performance sought in Step 1 and subsequently modified in Steps 2, 3 constitutes the primary objective in this step. The deformation capacity of isolation and energy dissipation devices should be checked for ultimate deformation, uplift, and stability. With regard to the piers, it should be verified through analysis that the deformation demand is consistent with accurately estimated limit values (i.e. from M-φ analysis using ρ_l ratios determined in Step 3), allowing for controlled inelastic response of piers under EQ(SP3). Members considered elastic in setting up the PIM, are also designed in flexure during this step similarly to SP3 requirements for 'ductile-pier' bridges.

Step 5 – PL4 Verifications The final step involves explicit analysis only in the case of non-essential bridges, wherein the target performance set is properly assessed through NLRHA under EQ(SP4) This involves activation of seismic links (in piers and abutments) and the abutment-backfill system resulting in damage in the foundation system (e.g. in piles), increased inelastic deformations in the piers and reduction of the effectiveness of the isolation system. Analysis complexity along with relevant uncertainties in structural response are disproportionate to the associated low importance, which is one of the main reasons that this SP is excluded from the PO of Ordinary bridges (CEN [Comité Européen de Normalisation] 2005). In all other POs, this step includes detailing of piers for confinement, anchorages and lap splices with due consideration of the expected level of inelasticity, and member shear design under EQ(SP4) which coincides with EQIV in Ordinary and Essential bridges.

20.4 A Comparative Case Study

The efficiency of the proposed Def-BD procedure was evaluated for a 3-span bridge of total length L = 99 m (Fig. 20.4). The 10 m wide prestressed concrete box girder deck has a 7% longitudinal slope and is supported by two single-column piers of cylindrical section and heights of 5.9 and 7.9 m. In the 'ductile-pier' design, the deck is monolithically connected to the piers, and rests on the abutments through elastomeric bearings, while in the isolated bridge it is supported through isolators (described later) on both piers and abutments that allow movement of the deck in

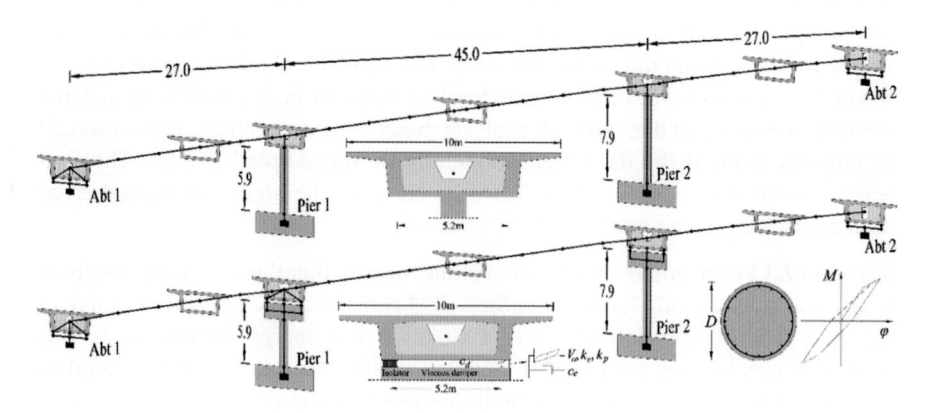

Fig. 20.4 Configuration and modelling of studied bridge; ductile-pier (top) and isolated case

any direction. The bridge lies on firm soil and both piers and abutments have surface foundations. Apart from the modification of the pier-to-deck connection, the clear height of the piers is reduced in the second case to accommodate the (1.5 m deep) pier cap. EQIII was associated with the EC8 'Type 1' elastic spectrum ($T_R = 475$ years) for a *PGA* of $0.21g$, site conditions 'C', and assuming a corner period at the transition to the constant displacement region of the spectrum $T_D = 4.0$ s in line with recent trends (Kappos et al. 2012), whereas EQII and EQIV were selected as half and twice the spectrum of EQIII, respectively. Design was performed adopting the Ordinary bridge PO in all cases, focusing on the transverse response (where higher modes affect the response) and ignoring SSI effects (firm ground conditions, adequate foundation provided). Response-history analysis was carried out using *Ruaumoko 3D* (Carr 2004) and standard modelling techniques (see (Gkatzogias 2017) for details).

20.4.1 Ductile Pier Bridges

The bridge has been designed (Gkatzogias 2017) using all procedures discussed in this chapter, i.e. EC8 (Code-BD), multimodal DDBD and Def-BD. While the Def-BD cases included explicit verifications under multiple performance levels (PLs) and associated seismic actions, Code-BD and MDDBD designs involved response spectrum analysis (RSA) under a single target spectrum associated with $T_R = 475$ years (EQIII in Table 20.2). Code-BD was performed for a seismic hazard zone associated with a *PGA* of $0.14g$ (denoted as Zone I), whereas zones II (*PGA* of $0.21g$) and III (*PGA* of $0.31g$) were considered in MDDBD, similarly to Def-BD. In the assessment of the various designs, the inconsistency in the level of the design seismic actions was treated as described in Sect. 20.4.1.1.

The main features of each design methodology are summarised in Table 20.3. The real bridge forming the starting point for the case study (an overpass in Egnatia Motorway, in N. Greece) had monolithic pier to deck connections and was designed using the National Code in force at the time, which was very similar to EC8-2. However, for a number of reasons, mainly having to do with the design practices used for modern bridges in Greece, the bridge was clearly overdesigned. In fact, while the bridge was designed for a PGA of 0.21g and $q = 3.1$ (hence expected to develop yielding at around 0.07g), first pier yielding occurred at 0.12g indicating the significant overdesign of the piers in the actual bridge. Hence the case that was compared with the various DBD procedures was designated as Code-BDn and is characterised by the fact that the *PGA* that causes the first yielding in the bridges is practically the same in all cases; 'first yielding' means that the pier with the lowest reinforcement reaches its design resistance M_{Rd}.

Table 20.3 Key features of different design methodologies (ductile pier bridges)

Method features	Def-BD	MDDBD	Code-BD
Seismic input	Acceleration spectrum, suite of accelerograms	Displacement spectrum	Acceleration spectrum
Analysis	Linear static/dynamic, nonlinear dynamic	Linear static, modal	Linear static/dynamic
Pier stiffness	Secant stiffness at yield, M-φ analysis	Secant stiffness at maximum response	Secant stiffness at yield
Damping	Fully populated damping matrix	Equivalent viscous damping	Modal damping ratios and behaviour factor
Controlled deformation parameters	No restriction (strains, deformations, displacements)	Strains (implicitly), displacements (explicitly)	Displacements (implicitly)
Explicitly controlled PLs per application	Multiple	Single	Single
Required iterations per application	Limited number	Significant number	Iterative application based on assumed-calculated strength

20.4.1.1 Performance of Alternative Designs

The different designs were assessed in terms of structural performance by modelling each bridge as a structure that can develop inelastic response in every component and evaluating the response of each design to a number of ground motions different from those used at the design stage (see details in (Gkatzogias 2017)). To enable meaningful and consistent comparison between the cases of Def-BD (ZII) and Code-BD (ZI) (i.e. designed for different levels of seismic actions), assessment of Code-BD was performed in terms of a normalised intensity measure A/A_d, where A corresponds to the intensity measure considered (i.e. *PGA*), and A_d is the design seismic action (corresponding to EQI, Table 20.2). The efficiency of each design was examined under three different levels of seismic actions defined with respect to the normalised intensity measure A/A_d and the assumptions adopted in the Def-BD case, i.e. EQII ($A/A_d = 2.3$), EQIII ($A/A_d = 4.5$), and EQIV ($A/A_d = 9.1$). Def-BD (ZII) and Code-BD results derived from the above assessment procedure, are discussed in the following with regard to the 'ordinary bridge' seismic performance objective, i.e. SP2 verifications under EQII, SP3 verifications under EQIII, and SP4 verifications under EQIV. In this context, violated performance requirements reported for the actual bridge refer to the bridge assessed for the normalised seismic actions, referred as Code-BDn to differentiate it from the response of the bridge under non-normalised seismic actions (Code-BD, not presented herein), where all code requirements under the 'design' seismic action were satisfied.

In Fig. 20.5 the displacement envelopes derived from design and assessment of Code-BDn are compared with those computed during the implementation of Def-BD (ZII). Fig. 20.6 shows chord rotation demands in the piers under the normalised levels of seismic actions along with allowable deformation limits (SP requirements);

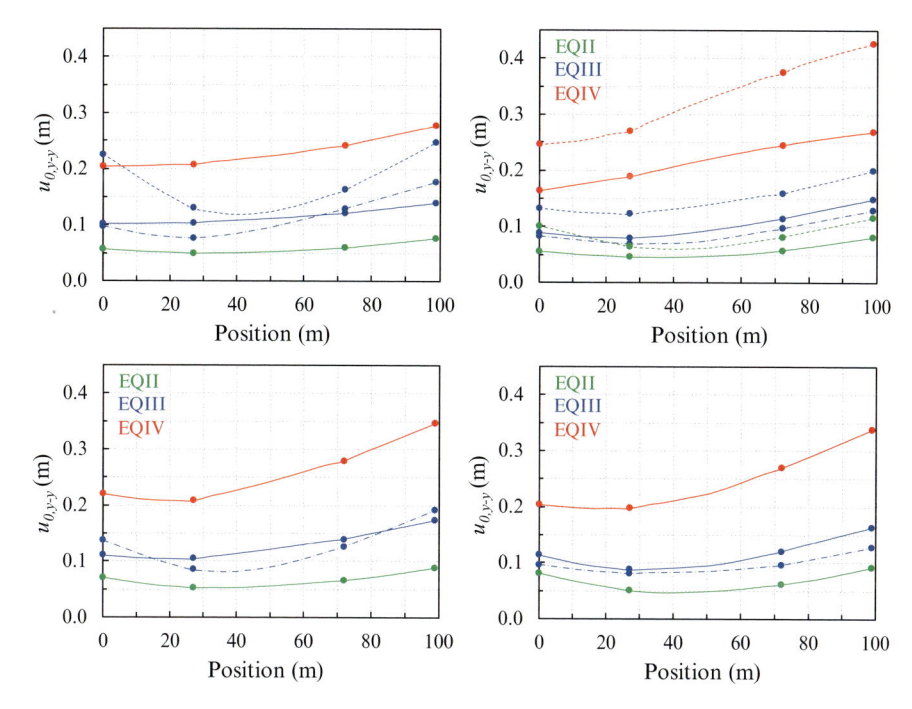

Fig. 20.5 Peak displacement demand derived from design (left) and assessment (right) of Def-BD (ZII), Code-BDn, MDDBD (ZII) (top), and Def-BD (ZIII), MDDBD (ZIII) (bottom)

additional design quantities for the case of Code-BDn are provided in (Gkatzogias 2017). Increased drifts and displacements (Fig. 20.5) were recorded in the case of Code-BDn for all three considered PLs. Pier chord rotation limits were satisfied in the case of Pier 2 under EQII, in both piers under EQIII, and violated in all other cases, while shear resistance was found inadequate under EQIV (Fig. 20.6 bottom). Bearing strain deformation limits were generally violated under the normalised seismic actions (Gkatzogias 2017), indicating that the design of the actual bridge was conservative with regard to the design of piers but not the design of bearings.

Despite the notable differences (Table 20.3) among the design principles adopted by the methodologies discussed herein with regard to the type of analysis, the definition of the seismic input, the type of stiffness and damping used to control design quantities, the range of directly controlled parameters and the number of iterations required (Kappos 2015), both performance-based approaches (i.e. Def-BD, MDDBD) aim at a specific structural performance (defined on the basis of deformations) under single or multiple levels of seismic actions, and not surprisingly yield generally similar drifts and displacements, at least for the PL for which explicit verifications were carried out in both procedures, i.e. SP3 verifications under EQIII. This is evident in Fig. 20.5-right where the deck displacement profiles derived from the assessment of Def-BD and MDDBD designs are compared. MDDBD yields 15 and 20% lower displacements (or drifts) at Pier 2 in the case of

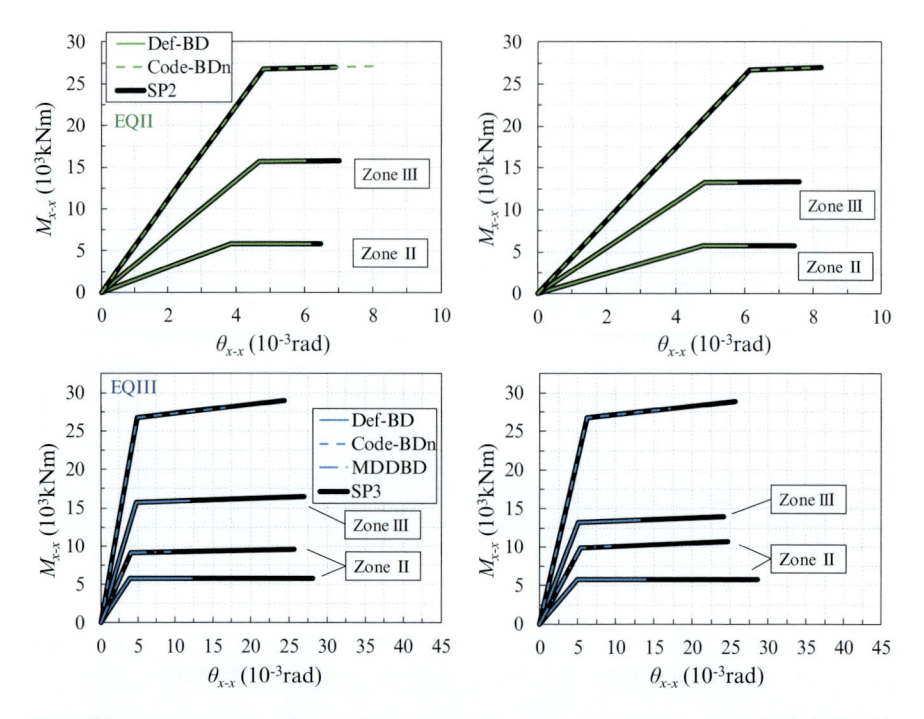

Fig. 20.6 Moment (M) vs. chord rotation (θ) demand curves derived from assessment of Def-BD (ZII, ZIII), Code-BDn, MDDBD (ZII) under EQIII (top) and EQIV (bottom) at the base of Pier 1 (left) and Pier 2 (right), compared with allowable deformation limits

ZII and ZIII, respectively. The relevant reductions in bearing deformations located at Abutment 2 are 13 and 22%, whereas smaller differences are observed for Abutment 1 and Pier 1 (Fig. 20.5-right)). Resulting deviations in the assessed response should be evaluated duly considering the underlying design assumptions; the ability of each methodology to accurately capture the structural response during the design stage, under a specific level of seismic action entailing inelastic response, depends primarily on the type of analysis used (along with the associated seismic input and the definition of stiffness-damping properties) and the complexity (or irregularity) of the studied structure.

Def-BD represents the most refined approach (Table 20.3) resulting in the best match between design and assessment displacement profiles. Deviations are attributed to the sensitivity of analysis results to the seismic input, specifically to ground motion selection and/or scaling procedures, while improved mean predictions (i.e. smaller discrepancies between design and assessment quantities) can be attained at the expense of increased computational effort. On the other hand, modal analysis, forming part of MDDBD, attempts to capture the maximum probable response to a given seismic action based on equivalent properties (i.e. secant pier stiffness at the maximum displacement, equivalent viscous damping) and the statistical combination (e.g. SRSS) of peak 'modal' responses at the instant of maximum response (i.e. after the formation of plastic hinges). In addition to certain concerns regarding the efficiency of equivalent linearisation approaches in predicting inelastic seismic action effects in single-degree-of-freedom systems vibrating mostly at lower than peak displacement amplitudes, the aforementioned type of analysis cannot account for the modification of the dynamic characteristics of the structure during the successive formation of plastic hinges in multi-degree-of freedom systems; hence its efficiency is expected to decrease as the degree of structural irregularity and the level of seismic action increase. It should be stressed that the displacement profiles shown in Fig. 20.5 consist of non-simultaneous peak deformations corresponding to mean values derived from a series of NLRHAs or to statistically combined peak modal values. In either case, the displacement profile curvature in the specific bridge configuration indicates the contribution of higher modes to the seismic response, rather than an 'actual' deformed shape of the bridge deck which mainly exhibits a 'rigid body' translational response due to the unrestrained conditions at the abutments (Kappos et al. 2013). In this context, the increased curvature in the MDDBD displacement profile (i.e. increased contribution of second mode) derived at the design stage, resulted in an overestimation of displacements at the critical elements of the studied bridge, i.e. the elastomeric bearings in Abutment 2 (Fig. 20.5-left). This is similar in nature to the reduced efficiency of the Def-BD method in accurately describing the shape of the displacement profile (or else the dynamic properties) under EQIV when the implicit approach is used, and is even more pronounced in the case of force-based code approaches, like Code-BDn (Fig. 20.5). In Code-BDn modal analysis is performed using the secant stiffness at yield of the piers disregarding altogether the effects of nonlinearity on the dynamic characteristics of the studied system, resulting in significant shape deviations between the 'design' and 'assessment' Code-BDn profiles.

Regarding the efficiency of different design approaches i.e. their ability to satisfy the adopted performance requirements without being over-conservative, Def-BD and MDDBD approaches were found to be safe in the sense of satisfying the relevant performance requirements (i.e. SP2 to SP4 in the case of Def-BD, and SP3 in the case of MDDBD). Nevertheless, anchoring a displacement profile of increased curvature at the target displacement of the critical member (i.e. LDRBs in Abutment 2, Fig. 20.5-left)) and following the iterative equivalent linearisation approach of MDDBD (Kappos et al. 2013), resulted in higher reinforcing steel and concrete area demand in the piers. On the contrary, the rigorous evaluation of inelastic deformations by incorporating refined analysis procedures in the case of Def-BD brought the deformation demand closer to the pertinent deformation limits, leading to cost reduction (see Sect. 20.4.1.2) without jeopardising the desired performance under multiple PLs. Bearing displacements at Abutment 2 are lower compared to the MDDBD case under EQIII (Fig. 20.5-left) because the design of piers and bearings is governed by SP2 and SP4 requirements that are implicitly considered in MDDBD (and Code-BD) but not necessarily satisfied, as demonstrated in the case of Code-BDn (Fig. 20.6) which can be deemed equivalent to a code-type design of the bridge under a higher level of seismic action than that corresponding to Zone I (notwithstanding the previously discussed conservatism). Considering the actual bridge design under the non-normalised seismic action (Code-BD), it is evident that due to the conservatism adopted with regard to the design of the piers and the lower level of seismic action considered, the structural performance will be superior to that of Def-BD in terms of recorded damage in the piers under the 'design' seismic actions (i.e. non-normalised EQIII). Yet, this is achieved at an increased cost, while it does not ensure an overall satisfactory performance of the bridge under a higher level of seismic action unless all bridge members are consistently overdesigned.

In general, the outcomes of Def-BD and MDDBD were affected to a certain degree by the adopted minimum longitudinal reinforcement ratio $\rho_{l,min}$ taken equal to 10‰. Given that the basic requirement to be satisfied by providing a minimum reinforcement ratio is to ensure a minimum local ductility in bridge piers, $\rho_{l,min}$ can be defined as the maximum of the values specified in Eurocode 2 (EN1992-1-1) and the EN1998-2 Handbook (Fardis et al. 2012) which provides a sufficient amount of steel reinforcement to ensure a design value of flexural strength of the pier section not less than the cracking moment.

Overall, what essentially differentiates Def-BD from other methods is its ability to control a broader range of design parameters (i.e. from strains up to flexural deformations and drifts) and PLs (two explicitly and two implicitly considered) within a single application of the method. Clearly, one can run MDDBD and Code-BD for different PLs (i.e. multiple applications of the method) but this would require at least double the computational effort, if at all feasible in the case of MDDBD due to its infeasibility for low and moderate seismic levels of seismic actions (Kappos et al. 2012).

20.4.1.2 Cost Comparisons

Table 20.4 summarises pier geometries and reinforcement for all alternative designs of the bridge shown in Fig. 20.4. For the Code design, data from the real bridge are included as the ones actually representing current practice. It is seen that Def-BD yielded notable reductions in pier longitudinal reinforcing steel (w_{sl}), transverse steel (w_{sw}), and concrete (w_c) weight (or volume) compared to MDDBD, equal to $\Delta w_{sl} = 41\%$, $\Delta w_{sw} = 17\%$, $\Delta w_c = 36\%$ in the case of ZII, and $\Delta w_{sl} = 50\%$, $\Delta w_{sw} = 20\%$, $\Delta w_c = 28\%$ in the case of ZIII, noting that in MDDBD the transverse reinforcement was governed by shear design performed in accordance with capacity design principles (as adopted in current codes like EC8-2 and also suggested by the developers of the DDBD method (Priestley et al. 2007)).

20.4.2 Isolated Bridges

The basic isolated bridge configuration is shown at the bottom of Fig. 20.3 Herein, the case of the bridge isolated with lead rubber bearings (LRBs) is considered, adopting the 'ordinary bridge' performance objective, and accounting for the effect

20 Seismic Design of Bridges: Present and Future

Table 20.4 Design output derived from different methodologies

Method	Zone	Member	Section (m)	h or t_R[1] (m)	$\rho_{l,req}$[2] (‰)	A_l[3] (cm²)	P_i[3] (‰)	$Vol_{l,t}$[4] (m³)	Bars	ρ_w[3] (‰)	Vol_w[5] (m/m)	Hoops (per mm)
Def-BD	ZII	Pier 1	$D_p = 1.2$	5.94	9.7	118	10.4	0.229	24Ø25	12.4	0.021	Ø16/60
		Pier 2	$D_p = 1.2$	7.93	10.2	118	10.4		24Ø25	10.6		Ø16/70
		LDRB[6]	0.35 × 0.45	0.088	–	–	–	–	–	–	–	–
	ZIII	Pier 1	$D_p = 1.7$	5.94	12.5	285	12.5	0.481	58Ø25	13.2	0.047	2Ø16/75
		Pier 2	$D_p = 1.7$	7.93	9.2	216	9.5		44Ø25	10.4		2Ø16/95
		LDRB	0.35 × 0.45	0.088	–	–	–	–	–	–	–	–
Code-BD	ZI	Pier 1	$D_p = 2.0$	5.94	–	471	15.0	0.918	96Ø25	9.8	0.055	(Ø16 + 14)/75
		Pier 2	$D_p = 2.0$	7.93	–	471	15.0		96Ø25	9.8		(Ø16 + 14)/75
		LDRB	0.35 × 0.45	0.044	–	–	–	–	–	–	–	–
MDDBD	ZII	Pier 1	$D_p = 1.5$	5.94	9.8	177	10.0	0.391	36Ø25	9.1	0.026	2014/95
		Pier 2	$D_p = 1.5$	7.93	12.4	221	12.5		45Ø25	7.9		2014/110
		LDRB	0.35 × 0.45	0.088	–	–	–	–	–	–	–	–
	ZIII	Pier 1	$D_p = 2.0$	5.94	11.5	363	11.6	0.960	74Ø25	10.4	0.058	2Ø16/80
		Pier 2	$D_p = 2.0$	7.93	19.0	599	19.1		122Ø25	10.4		2Ø16/80
		LDRB	0.35 × 0.45	0.088	–	–	–	–	–	–	–	–

[1]h : clear height of pier, t_R : total rubber thickness
[2]Required reinforcement at the base of the pier column
[3]Provided reinforcement at the base & top of the pier column
[4]Provided volume of reinforcing steel
[5]Provided volume of reinforcing steel per metre of column length
[6]Low damping rubber bearing

of bidirectional excitation. Design of the bridge according to EN1998-2 (CEN [Comité Européen de Normalisation] 2005) and assessment of the bridge's seismic performance is presented in the following.

For the EC8-2 design, 1D target acceleration spectra associated with $T_R = 475$ years (EQIII in Table 20.2) were used, with the seismic action applied separately in each principal direction of the bridge, followed by directional combination of response quantities (CEN [Comité Européen de Normalisation] 2005). The acceleration spectrum used in RSA was derived by multiplication of the target (i.e. elastic, $\xi = 5\%$) spectrum by the damping modification factor η_{eff} implicitly considering the reduction of seismic accelerations due to the introduction of additional damping at the isolation interface (ξ_{eff}) for periods longer than $0.8T_{eff}$ Fig. 20.7). At the assessment stage, 2D target spectra, artificial records used to represent the seismic action, and scaling of records to different levels of seismic action were defined (Gkatzogias 2017). An iterative analysis scheme was adopted applying sequentially the two spectral approaches prescribed in EC8-2 (i.e. fundamental mode spectrum analysis in Steps i-iii, and multi-mode spectrum analysis in Steps iv-vi) independently along the longitudinal and transverse direction of the bridge); details of the procedure are given in (Gkatzogias 2017).

Application of the aforementioned procedure involves a number of iterations, typically three, doubling if upper bound (UB) and lower bound (LB) properties for the isolators are considered, as per the EC8-2 requirement). This increases significantly when the characteristics of the passive devices required as an input in Step (i) are not known (as is typically the case in the design of a new bridge), and additional design criteria, such as those included in Def-BD, are sought, e.g. 'near-optimal' selection of isolation system properties based on multi-level performance requirements using charts like those of Fig. 20.3. In addition, due to the different analysis principles in EC8-2 and Def-BD, design is expected to yield different properties of isolators even if a common target performance is sought in both methods, thus complicating direct comparisons. In view of the previous consideration, the design outcome of the bridge designed to the Def-BD approach (regarding the isolator

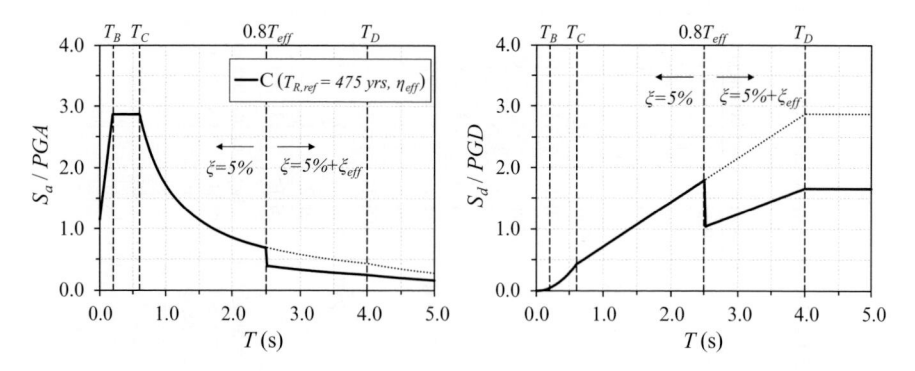

Fig. 20.7 1D design horizontal acceleration (left) and displacement (right) response spectra for site conditions 'C' ($T_{R,EQIII} = 475$ years) used in EN1998-2 design

properties and the diameter of the piers) was used herein, focusing mainly on the predicted response and the design of the piers rather than the appropriate selection of device properties according to the EC8-2 approach. This enables direct comparisons among the two methods while revealing potential pitfalls that would have been encountered if isolators were selected according to the EC8-2 procedure.

Comparing key response quantities (under EQIII) from the EC8-2 procedure and the Def-BD that involved NLRHA for a suite of artificial records, showed that the Code procedure overestimates displacements and forces approximately by 30% (LB properties) and 10% (UB properties), respectively, compared to NLRHA results. Corresponding deviations of responses derived using design equations (DEs) from those from NLRHA results are 4% and 1%. EC8-2 further requires verification of the passive devices for a higher (than EQIII) level of seismic action, considered implicitly through the amplification of relative displacements derived from analysis under EQIII. Herein, an amplification factor of 2.0 was adopted which is higher than the EC8-2 recommended value of 1.5 (a nationally determined parameter) for consistency with the definition of target spectra (i.e. adoption of $S_{F,EQIV} = 2.0$ in Def-BD). Despite the overestimation of displacements under EQIII and the higher amplification factor, the EN1998-2 spectral approach underestimates displacements u_0 by up to 13%. Neither of the above deviations, i.e. overestimation and underestimation of u_0 under EQIII and EQIV, respectively, is on the safety side, since the first results in overestimating the restoring capability of the isolation system and the second in under-designed isolators.

The different features of the EC8-2 and Def-BD methodologies for seismically isolated bridges are summarised in Table 20.5. It is clear from the description of the EC8-2 'spectral' approach (similar to that of other modern codes, e.g. (ASCE

Table 20.5 Key features of Def-BD and EN1998-2 design methodologies (isolated bridges)

Method features	Def-BD	EN1998-2 ('spectral' approach)
Seismic input	Acceleration spectrum, suite of accelerograms	Acceleration/displacement spectrum
Analysis	generalised design eqs., nonlinear dynamic	Linear static/dynamic
Pier stiffness	'Gross' section stiffness/secant stiffness at yield, M-φ analysis	'Gross' section stiffness/secant stiffness at yield
Isolator stiffness	Stiffness degradation models	Secant stiffness at maximum response
Damping	Fully populated damping matrix	Equivalent viscous damping in fundamental mode, behavior factor in substructure design
Controlled deformation parameters	No restriction (strains, deformations, displacements)	Displacements (explicitly)
Explicitly controlled PLs per application	Multiple	Single
Required iterations per application	Limited number	Significant number

[American Society of Civil Engineers] 2016)), that displacements in seismically isolated structures are explicitly considered in the design procedure, as opposed to 'pure' force-based approaches commonly adopted by codes for bridges with energy dissipation in the piers that address displacements at the final stage of design. In this sense, the key features of the EC8-2 approach in Table 20.5 are closer to those of the MDDBD rather than the Code-BD approach in Table 20.2. Although EC8-2 does not include a strict 'design route' aiming at the specification of strength that results in a predefined (target) displacement under a specific level of seismic action (the cornerstone of the direct displacement-based design philosophy), the equivalent linearisation approaches involved in the two procedures share the same basic principles.

20.4.2.1 Performance of Alternative Designs

Figure 20.8-right compares estimates of peak displacements (relative to the ground) of the deck and piers derived from the design and assessment stages according to EC8-2. Different 'design' displacement profiles under the same level of seismic actions and properties (UB, LB) correspond to different cases of combination of actions ('design' case) or to different angle of incidence ('assessment' case). Due to the typically symmetric properties of the isolation interface, directional combination

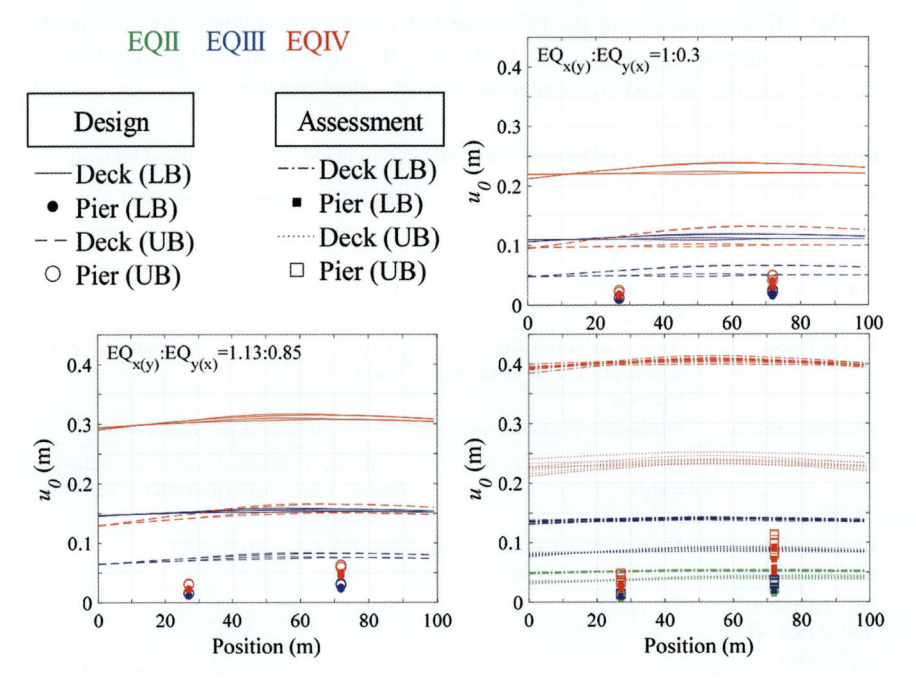

Fig. 20.8 Deck and pier peak displacements u_0 derived from EC8-2 design (top right, bottom left) and assessment for $\theta_{EQ} = 0–180°$ (bottom right)

(100%X + 30%Y) of responses as per EN1998-2 for the case of RSA (design stage), is expected to result in peak response quantities approximately equal to $\sqrt{(1^2 + 0.3^2)} =$ 1.04 times the values derived under unidirectional excitation; i.e., due to the symmetry of the isolated structure, there is no significant differentiation of peak responses along a principal direction of the bridge deriving from the orthogonal component of seismic action, even in the case of non-symmetric substructures. On the other hand, extensive parametric NLRHA of nonlinear isolation systems under bidirectional excitation (Gkatzogias 2017), using records scaled to a target spectrum defined as $\sqrt{2}$ times the target spectra under unidirectional excitation in line with EC8-2 requirements for nonlinear dynamic analysis (i.e. the 2D target spectrum also adopted in the Def-BD case study for consistency), showed that peak displacements resulting from unidirectional excitation are expected to increase approximately by a factor of 1.43 in the biaxial case.

The different approaches in considering the effect of bidirectional excitation during the design and assessment stages resulted in differences of 22% (LB) and 67% (UB) in deck displacements under EQIII (Fig. 20.8-right). Improved convergence can be achieved when the effect of bidirectional excitation is considered at both stages using the same approach. For example, adopting during the design stage the target spectra of horizontal components used at the assessment stage (Fig. 20.8-left) constraints relevant displacement deviations under EQIII within a range of 10–20% attributed to the approximate nature of the equivalent linearisation approach and the assumption that peak responses along the principal axes of the bridge occur simultaneously (as opposed to the case of uncorrelated components of seismic action). This is practically equivalent to the SRSS directional combination of peak responses derived from independent RSA under the 1D target spectrum, an approach permitted by EC8-2 only for bridges with energy dissipation in the piers. The SRSS directional combination will also increase the reinforcing steel demand bringing the 'design' total acceleration (i.e. $\approx \sqrt{2} \cdot 1.34 / 1.5 = 1.26$ m/s^2) closer to the value derived from Def-BD, thus mitigating the effect of the uncontrolled inelastic response in the piers described in the following. In any case, Fig. 20.8 shows significant underestimation of relative displacements under EQIV indicating the risk of under-designing isolators, and the inadequacy of the EC8-2 design approach in estimating the peak displacement response through the implicit amplification of EQIII displacements by a factor $S_{F,EQIV}$, since the increase in peak displacement response is disproportionate to the increase of seismic actions. The fact that the relevant SP requirements of the isolation system are met at the assessment stage (see (Gkatzogias 2017)), is attributed to the adoption of the isolator properties derived from the Def-BD method which is capable of reliably estimating the response under multiple PLs irrespective of the definition of the target spectrum under bidirectional excitation.

Comparison of the peak deck displacements derived during assessment of the EC8-2 and Def-BD designs (Fig. 20.8-bottom right) reveals nearly identical displacement profiles despite the significant reduction in pier strength in the EN1998-2 case, i.e. the substructure response has a negligible effect on the deck relative displacement demand. Nevertheless, the distribution of the deck displacements to the isolators and the piers is different in the considered design cases as implied by the increased pier displacements in Fig. 20.8, more so in the case of Pier 2. Transferring part of the displacement demand from the isolation interface to the substructure reduces the efficiency of the isolation system.

An overview of the isolator deformation demand derived from the entire set of NLRHAs performed at the assessment stage is presented in the polar charts of Fig. 20.9 where mean values of u_0 are plotted against incidence angle θ_{EQ} (for each PL) and compared with relevant SP design criteria. The figure highlights the reduction in the displacement response of isolators located at the top of Pier 2 mainly in the range of $\theta_{EQ} = 0{\sim}135°$, as opposed to the Def-BD case where the relevant demand under EQIV follows the target SP3 requirements (Table 20.2) regardless of the considered incidence angle.

The reduction in the energy dissipated in the isolation system is counterbalanced in the case of EC8-2 by a significant increase in the inelastic deformation of the piers compared to Def-BD as shown in the representative moment vs. curvature plots of Fig. 20.10 showing the case of an artificial record (Art 4), $\theta_{EQ} = 90°$, and UB-DPs; in this case the lower provided pier reinforcing steel ratios reported in Table 20.6 yielded six times larger curvature demand at the base of Pier 2 under EQIV.

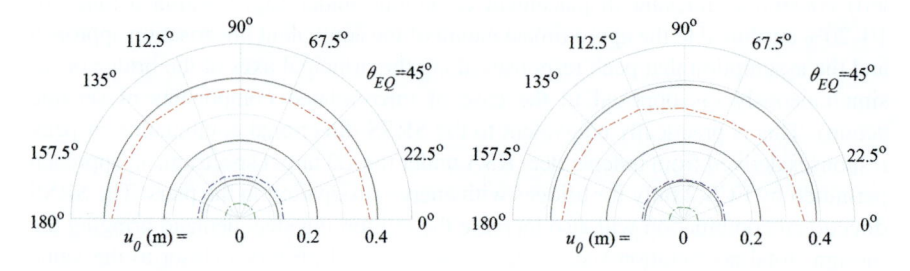

Fig. 20.9 Peak relative displacements u_0 of isolator located on top of Pier 1 (left) and Pier 2 (right) derived from assessment for $\theta_{EQ} = 0{-}180°$, compared with SP requirements per PL

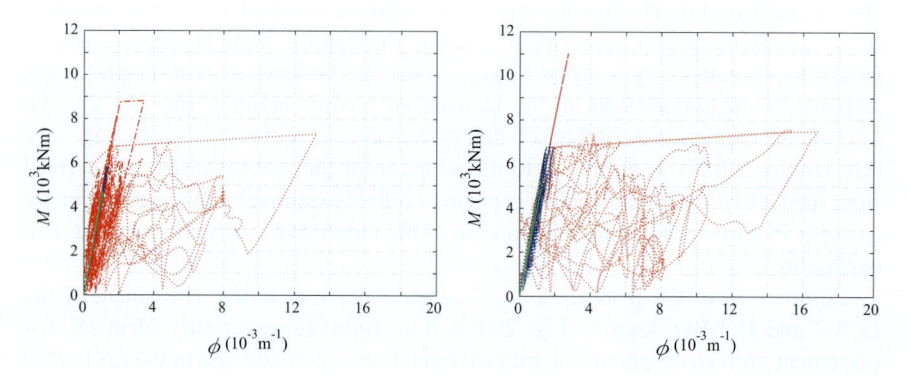

Fig. 20.10 M-φ response histories at the base of Pier 1 (left) and Pier 2 (right) derived from EN1998-2 and Def-BD under record Art 4 (assessment), $\theta_{EQ} = 90°$, and UB-DPs

20 Seismic Design of Bridges: Present and Future

Table 20.6 Outcome of Def-BD and EN1998-2 methodologies

Method	Member		Section (m)	h or t_R^1 (m)	$\rho_{l,req}^2$ (‰)	A_l^3 (cm²)	ρ_l^3 (‰)	$Vol,_l^4$ (m³)	Bars	ρ_w^3 (‰)	$Vol,_w^5$ (m³/m)	Hoops (per mm)
Def-BD	Pier 1	(top)	$D_p = 1.5$	4.44	–	74	4.2	0.304	–	5.9	0.019	⌀14/75
		(base)	$D_p = 1.5$	4.44	8.2	147	8.3		30⌀25			
	Pier 2	(top)	$D_p = 1.5$	6 43	–	133	7.5		27⌀25	6.3		⌀14/70
		(base)	$D_p = 1.5$	6 43	14.7	265	15.0		54⌀25			
	LRB[6]	–	$D_l = 0.75$	0.225	–	–	–	–	–	–	–	–
EN1998-2	Pier 1	(top)	$D_p = 1.5$	4.44	–	72	4.1	0.129	19⌀22	4.4	0.013	⌀14/110
		(base)	$D_p = 1.5$	4.44	–	72	4.1		19⌀22	4.4		
	Pier 2	(top)	$D_p = 1.5$	6 43	–	72	4.1		19⌀22	4.4		⌀14/110
		(base)	$D_p = 1.5$	6 43	–	72	4.1		19⌀22	4.4		
	LRB[6]	–	$D_l = 0.75$	0.225	–	–	–	–	–	–	–	–

[1] h: clear height of pier, t_R: total rubber thickness
[2] Required reinforcement at the base/top of the pier column
[3] Provided reinforcement at the base/top of the pier column
[4] Provided volume of reinforcing steel
[5] Provided volume of reinforcing steel per metre of 1 column length
[6] Lead rubber bearing

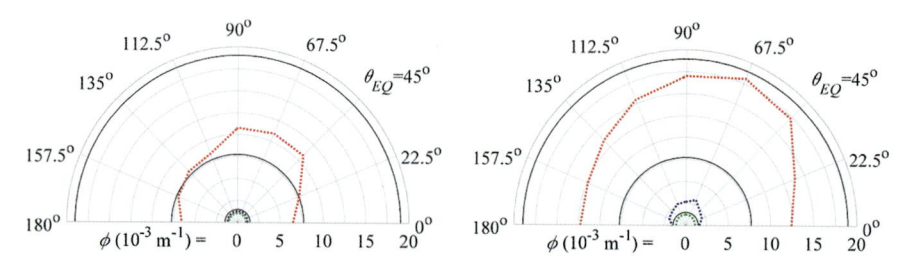

Fig. 20.11 Curvatures φ at the base of Pier 1 (left) and Pier 2 (right) derived from EC8-2 assessment for $\theta_{EQ} = 0$–$180°$ and UB-DPs

Considering the polar diagrams of Fig. 20.11 to summarise analysis results derived from assessment and UB-DPs, it is clear that the SP3 criterion aiming at the controlled inelastic response of the piers is violated in most cases, contrary to the Def-BD method wherein the ductility demand was constrained below the value corresponding to $\varepsilon_c = 3.5‰$. Specifically, the lower pier strength in the case of EN1998-2 design reduces the efficiency of isolators bringing the deformation demand in Pier 2 close to its flexural capacity for $\theta_{EQ} = 45$~$90°$. It is noted that exceedance of SP4 deformation limits (Table 20.2) was avoided only because of the decision to apply the minimum reinforcing steel requirements during the EC8-2 design, thus providing some overstrength, instead of reducing the pier dimensions, in which case the inelastic ductility demand would have further increased. Furthermore, the shear capacity of piers assessed according to EC8-2 was also found inadequate in the case of the EN1998-2 design under EQIV and UB-DPs (Gkatzogias 2017).

The nonlinear response of piers, considered undesirable by modern design codes, further highlights the inconsistent approach of EC8-2 in designing the components of the isolation system under a higher level of seismic action than that corresponding to EQIII, as a means to ensure the 'increased reliability' requirement, without specifically addressing the effect of these actions on pier response. On the grounds that the pier response will control the performance of the isolation system and the bridge overall response, design of the isolation system under higher seismic actions is expected to be meaningful when followed by relevant verifications of the piers, pointing to the necessity in the case of EC8-2 to limit inelastic pier response under actions higher than those associated with the 'design' earthquake.

20.4.2.2 Cost Comparisons

Key results from the application of the EC8-2 and the Def-BD procedures are summarised in Table 20.6. It is seen that significant reductions in longitudinal (i.e. 58%) and transverse (i.e. 28%) reinforcement were observed in the case of EC8-2; ρ_l was governed by minimum requirements and ρ_w by confinement requirements.

Discrepancies in steel demand between EN1998-2 and Def-BD approaches in Table 20.6 are due to the design assumptions made in each method. The shear demand in each pier in Def-BD was calculated considering the deck total acceleration under EQIV and bidirectional excitation, a β-factor of 0.75 (mainly accounting for the difference in mean and design values of material strength), and a target rotational ductility factor under EQIV equal to $\mu_{\theta,SP3} = 1.2$ (the resulting acceleration is about 1.6 m/s^2). On the other hand, the shear demand in the EN1998-2 approach was derived from the deck total acceleration under EQIII and unidirectional excitation, a behaviour factor $q = 1.5$, and consideration of the effect of the orthogonal component of seismic action through the directional combination of response quantities (resulting acceleration is 0.93 m/s^2). These acceleration values suggest that the reason for steel reduction in the case of EC8-2 is the inconsistent safety format adopted in 'response spectrum' and 'response history' analysis methods regarding the effect of bidirectional excitation, the limitation of the inelastic pier response in isolated bridges under the 'design' (EQIII) rather than the 'maximum considered' seismic actions (EQIV), and the differences in response quantities resulting from 'spectral' and rigorous 'direct integration' methods.

20.5 Concluding Remarks and Recommendations

Current trends in the seismic design of bridges were reviewed herein, starting from Eurocode 8 and moving on to new proposals for PBD, such as the direct displacement-based design and deformation-based design, also addressing the issue of whether the latter could be useful within the frame of a 'new generation' of codes. The answer depends on a number of criteria, i.e. whether the new procedures lead to better seismic performance (also for seismic actions higher than those explicitly considered in the design), whether they lead to a more economical design, and whether they contribute to simplicity and ease of use.

To start from the last criterion, the answer is an honest 'no'; there are no magical solutions wherein performance and economy are enhanced, while the procedures applied to achieve them are simpler or easier to use than the existing ones. Having said this, one should also recall that in the last 3–4 decades seismic design of bridges (and other engineering structures) heavily depends on the availability of pertinent software; hence ease of use should be seen from this (pragmatic) perspective. Moreover, as already advocated by this author and others, interesting approaches like DDBD, closely associated with the 'old New Zealand school of thought' (particularly as represented by Paulay and Priestley) that relies on engineering judgement and 'hand solutions', are relevant for preliminary design of bridges (so long as a displacement controlled, rather than an integral bridge, is selected), but cannot address the final design of all bridge types that fall within the scope of modern codes like EC8 or AASHTO. Hence, the more complex, but also covering a broader range of bridges, deformation-based procedures addressed herein emerge as possible candidates for replacing the existing code procedures.

The major criteria of enhanced performance and economy, can only be assessed through a comparative evaluation of designs resulting from the application of alternative procedures to the same bridges, as was done herein (and presented in more detail in Gkatzogias 2017). The comparative study included both bridges with ductile piers and seismically isolated bridges, hence all common types of modern bridges. Of course, only one basic configuration was analysed, hence the following conclusions should be subjected to further scrutiny in the future.

Considering the case of bridges with monolithic pier-to-deck connections, the following conclusions were drawn based on the assessment of designs resulting from the deformation-based design (Def-BD) method, the modal direct displacement-based design (MDDBD) method, and a force-based code-type (Code-BD) method (corresponding to the actual bridge design):

- Adopting refined analysis and modelling approaches, and providing a consistent performance-based design format within the Def-BD framework including explicit consideration of multiple PLs), resulted in superior seismic performance in the case of Def-BD. This was demonstrated by the control of various design parameters and structural performance requirements over multiple performance levels within a non-iterative application of the method.
- Significant cost reductions were achieved compared to the MDDBD procedure, whereas potential cost reductions may also be obtained when compared to force-based code-type procedures due to the adoption of more rational design approaches accounting for the characteristics of the design seismic actions for the specific site, in lieu of 'standard' capacity design considerations.
- The above render Def-BD a rigorous methodology, applicable to most of the common concrete bridge configurations without practical limitations related to the irregularity of the structural system considered. Inevitably, this comes at the cost of additional computational time and effort associated with the use of nonlinear dynamic analysis and the explicit consideration of multiple PLs. Nevertheless, minimum effort for iterations is ensured by estimating pier strength on the basis of allowable deformations, and by providing a logical analysis-design route wherein each step corresponds to a different performance level, which the author deems as one of the key advantages of the method from the clarity perspective.
- MDDBD resulted in similar structural performance as Def-BD for the explicitly verified common PL (i.e. SP3 requirements); nevertheless, the inherent inability of the former method to account for the modification of the dynamic characteristics of the structure during the successive formation of plastic hinges, along with the introduction of an equivalent linearisation approach, resulted in an increase in pier dimensions and reinforcing steel requirements. Disregarding altogether the effect of pier inelastic response on the estimation of the dynamic characteristics of the bridges in the case of Code-BD, resulted in the highest deviation among the displacement response assumed during the design and the response assessed through rigorous nonlinear dynamic analysis.

- Assessment of Code-BD under normalised levels of seismic action (designated as Code-BDn), introduced to enable meaningful comparisons with Def-BD, revealed multiple violations of the adopted performance requirements for the PLs that are implicitly considered in code-type procedures (SP2, SP4). This confirmed that definition of strength on the basis of 'life-safety' design criteria does not necessarily ensure controlled structural performance under different PLs associated with 'operationality'/'functionality' and 'collapse-prevention' requirements. On the other hand, if the performance of the actual bridge under the non-normalised seismic action is considered, all relevant verifications are satisfied due to the adopted over-conservatism in design.

For the case of the seismically isolated bridges, the following conclusions were drawn based on the assessment of designs resulting from the deformation-based design (Def-BD) method, and the EC8-2 method as applied herein:

- Although the EN1998-2 design method for seismically isolated bridges aims at the explicit verification of displacements, the adopted equivalent linearisation approach may involve a significant number of iterative applications when the characteristics of the isolation devices are not known at the beginning of the design procedure, as is the case in the design of a new bridge. Further iterations will be required if additional design criteria, such as those included in the proposed Def-BD, are sought, e.g. by setting a target performance under multiple performance levels with respect to peak deformations, energy dissipation, minimisation of the substructure design cost, etc. On the contrary, preliminary selections in Def-BD are facilitated by using ad-hoc design equations that are easy to derive when the appropriate software is available.
- The introduced iterations in the EC8-2 procedure do not necessarily ensure an accurate estimation of peak response, due to the approximate nature of the equivalent linearisation approach. In the bridge studied herein, i.e. deck isolated through hysteretic isolators (LRBs), the EC8-2 approach, initially applied to an equivalent SDOF system of the bridge and subsequently to the MDOF system accounting also for the substructure, resulted in deviations of displacement response in a range of 10–30% compared to rigorous nonlinear dynamic analysis results. Furthermore, the implicit estimation of deformations under EQIV was found inadequate to capture the peak inelastic response of isolators pointing to the risk of underestimating their required deformation capacity.
- Application of the EC8-2 design procedure resulted in significant reductions in reinforcing steel demands. Nevertheless, subsequent assessment of the design in line with the EC8-2 requirements indicated that the main sources of this reduction, namely (*i*) the inconsistent consideration of the effect of bidirectional excitation in different analysis methods, (*ii*) the limitation of the inelastic pier response in isolated bridges under the 'design' (EQIII) rather than the 'maximum considered' seismic actions (EQIV), and (*iii*) the deviation of response quantities resulting from 'spectral' equivalent linearisation approaches, may compromise the safety of the isolated structure under EQIV by imposing large inelastic

deformations in substructure elements, in contrast to the reliable and stable performance resulting from Def-BD.

The above deficiencies of equivalent linearisation approaches may be more pronounced in isolation systems incorporating velocity dependent energy dissipation devices (e.g. fluid viscous dampers, not addressed in the present case study), since their inherent inability to estimate the peak inertia forces transferred to the substructure will require recourse to more complex approximate procedures, unless nonlinear dynamic analysis is adopted, which is a strategy typically preferred by designers when viscous dampers are involved.

Acknowledgments The Deformation-based Design approach presented in Sect. 20.3.2 of this chapter, as well as the case study summarised in Sect. 20.4, form part of the PhD work of Kostas Gkatzogias, carried out under the supervision of the author at City, University of London. Kostas, and also Yannis Gidaris, have also contributed to the application of the Multi-modal Direct-Displacement Based Design approach before starting their PhD, when they were still MSc students at the Aristotle University of Thessaloniki. It was a pleasure for the author to work with Kostas over these years and his contribution to all these developments is gratefully acknowledged herein. Thanks are also due to the members of the BSI Panel ('Mirror group') for EC8-2 (led by the author) that have contributed to the compilation of UK comments, some of which are included in Sect. 20.2.3, in particular Stergios Mitoulis (Surrey University), John Lane (RSSB), and, again, Kostas Gkatzogias.

References

AASHTO (2015) Guide Specifications for LRFD Seismic Bridge Design, 2nd edition with 2012, 2014, and 2015 Interim Revisions. Washington, DC
AASHTO [American Association of State Highway and Transportation Officials] (2017) LRFD bridge design specifications, 8th edn. AASHTO, Washington, DC
ASCE [American Society of Civil Engineers] (2016) Minimum design loads and associated criteria for buildings and other structures (ASCE/SEI 7-16). ASCE, Reston
Bardakis VG, Fardis MN (2011) A displacement-based seismic design procedure for concrete bridges having deck integral with the piers. Bull Earthq Eng 9(2):537–560
Carr AJ (2004) Ruaumoko 3D: inelastic dynamic analysis program. University of Canterbury, Canterbury
CEN [Comité Européen de Normalisation] (2005) Eurocode 8, design of structures for earthquake resistance — Part 2: bridges. CEN, Brussels
Dwairi HM, Kowalsky MJ, Nau JM (2007) Equivalent damping in support of direct displacement-based design. J Earthq Eng 11:512–530
Fardis MN, Kolias B, Pecker A (2012) Designers' guide to Eurocode 8: design of bridges for earthquake resistance: EN 1998-2. ICE Publishing/Thomas Telford, London
Gkatzogias KI (2017) Performance-based seismic design of concrete bridges for deformation control through advanced analysis tools and control devices. PhD thesis, City, University of London
Gkatzogias KI, Kappos AJ (2015) Deformation-based seismic design of concrete bridges. Earthq Struct 9(5):1045–1067
Gkatzogias KI, Kappos AJ (2016a) Semi-active control systems in bridge engineering: a review. Struct Eng Int (IABSE) 26(4):290–300

Gkatzogias KI, Kappos AJ (2016b) Seismic design of concrete bridges: Some key issues to be addressed during the evolution of Eurocode 8 – Part 2. Concrete structures conference, Hellenic Society of Concrete Research and Technical Chamber of Greece, Thessaloniki

Gkatzogias KI, Kappos AJ (2017) Deformation-based design of seismically isolated con-crete bridges. 16th World Conference on Earthquake Engineering (WCEE), Santiago, Chile, paper no. 0997

Kappos AJ (2015) Performance-based seismic design and assessment of bridges. In: Ansal A (ed) Perspectives on European earthquake engineering & seismology, vol 2. Springer, Heidelberg New York Dordrecht London, pp 163–206

Kappos AJ, Gkatzogias KI (2017) Deformation-based design of concrete bridges. In: Beskos D, Zhou Y, Qian J, Lu X (eds) International workshop on performance-based seismic design of structures. Tongji University, Shanghai

Kappos AJ, Gidaris I, Gkatzogias K (2012) Problems associated with direct displacement-based design of concrete bridges with single-column piers, and some suggested improvements. Bull Earthq Eng 10(4):1237–1266

Kappos AJ, Gkatzogias K, Gidaris I (2013) Extension of direct displacement-based design methodology for bridges to account for higher mode effects. Earthq Eng Struct Dyn 4(4):581–602

Kowalsky MJ (2002) A displacement-based approach for the seismic design of continuous concrete bridges. Earthq Eng Struct Dyn 31(3):719–747

Mori A, Moss PJ, Carr AJ, Cooke N (1997) Behaviour of laminated elastomeric bearings. Struct Eng Mech 5(4):451–469

Priestley MJN, Calvi GM, Kowalsky MJ (2007) Displacement-based seismic design of structures, 1st edn. IUSS Press, Pavia, 720 pp

Chapter 21
Technologies for Seismic Retrofitting and Strengthening of Earthen and Masonry Structures: Assessment and Application

Paulo B. Lourenço

Abstract Earthen and masonry structures are usually heavy and do not possess an integral behavior. A consequence of these characteristics, in combination with the adopted materials featuring low tensile strength and ductility, is that such structures often collapse in a quasi-brittle way, with local failures, usually out-of-plane. This paper first addressed the seismic assessment of these structures, by providing some recent shaking table tests and blind predictions. Obvious limitations were found in providing a good estimate of collapse. Subsequently, techniques for retrofitting and strengthening are addressed, with applications shown in a real case study.

21.1 Introduction

Natural hazards have caused a considerable number of disasters in the last decades. According to the World Bank, from 1975 to 2005 the number of natural disasters increased from approximately 100 to more than 400 (Parker et al. 2007). These events lead to important economic impacts (Noy 2009), deaths and irrecoverable losses due to the collapse of existing masonry buildings. Consequently, earthquakes contribute significantly to these natural hazard disasters. It is predicted that in the current century the total fatalities caused by earthquakes will increase to about 2.57 ± 0.64 million (Holzer and Savage 2013). Recent seismic events caused severe damages to a considerable number of existing masonry constructions, such as the earthquakes in L'Aquila (Italy 2009) (Augenti and Parisi 2010), in Canterbury (New Zealand 2010 and 2011) (Leite et al. 2013) or in Emilia (Italy 2012) (Penna et al. 2014).

Historic earthen and masonry buildings were built for many centuries taking into account mostly vertical static loads according to the experience of the builder, usually, without much seismic concern. The seismic behavior of ancient masonry buildings is

---

P. B. Lourenço (✉)
ISISE, Department of Civil Engineering, University of Minho, Guimarães, Portugal
e-mail: pbl@civil.uminho.pt

© Springer International Publishing AG, part of Springer Nature 2018
K. Pitilakis (ed.), *Recent Advances in Earthquake Engineering in Europe*,
Geotechnical, Geological and Earthquake Engineering 46,
https://doi.org/10.1007/978-3-319-75741-4_21

particularly difficult to characterize and depends on several factors, namely the materials properties, geometry of the structure, connections between structural and non-structural elements, stiffness of the horizontal diaphragms and building condition. However, the different masonry types present common features that lead to high seismic vulnerability of these buildings, such as: (a) low tensile strength and ductility of masonry; (b) weak connections between orthogonal walls and between walls and horizontal diaphragms; (c) high mass of the masonry structural elements; (d) flexible horizontal diaphragms; (e) absence of seismic requirements at the time of their construction (Lagomarsino 2006; Lourenço et al. 2011).

Regarding the out-of-plane behavior of these structures, the low strength/mass ratio of common masonry structures increases their vulnerability in the out-of-plane direction because inertia forces are not restrained due to reduced stiffness and strength of the masonry walls in that direction. Despite the numerous studies carried out so far (a state-of-the-art review is provided by Ferreira et al. 2014 and by Sorrentino et al. 2016), numerous issues are still unresolved and scarce consensus exists amongst experts on the most appropriate methods to use for seismic safety assessment. For this reason, the seismic performance of masonry structures has received great attention in the last decade, mainly for masonry buildings without box-behavior (Lourenço et al. 2011). However, little consensus exists on the most appropriated assumptions and approaches for assessing the seismic safety of unreinforced masonry buildings without box-behavior.

Therefore, the out-of-plane behavior of historic earthen and masonry structures remains, possibly, the most challenging response in case of seismic action. As demonstrated recurrently all over the world by earthquakes, in case of lack of an integral behavior of the building, out-of-plane failure dominates. This is also favored by the fact that many historic buildings possess large spaces inside, with insufficient connections to transverse structural elements. Even if adequate connections between elements can be ensured, e.g. by tying walls, enforcing connections between walls and enforcing wall to floor connections, it is necessary to avoid disintegration of the walls themselves when subjected to out-of-plane actions, especially in case of rubble masonry.

The out-of-plane response of these structures or their local mechanisms, often assumed as macro-blocks with almost rigid behavior that become independent from the global structure is complex. The dynamics of these local mechanisms close to collapse are very sensitive to the seismic input (e.g. frequency contents, duration or directivity of the signal) but also to the structure itself (e.g. boundary conditions). The methods for structural analysis available for this purpose, both for research and engineering applications, are rather different in terms of formulation, input, conceptual complexity and computational efficiency. For these reasons, the results obtained can also vary greatly. This paper considers, first, a blind test exercise on the out-of-plane failure of historic masonry structures, involving prediction, testing and postdiction (Lourenço et al. 2017). This involved about 25 international experts in the field and clearly demonstrates that further developments are needed in the field. Then, one engineering application of seismic safety assessment and strengthening of a historic earthen structure is shown.

21.2 Blind Test: Out-of-Plane Shaking Testing Failure of Masonry

Experts on masonry structures were invited to present their conjectures on the dynamic response of two idealized masonry structures tested on a shaking table and subjected to unidirectional ground motion. One structure was constructed of irregular stone and the other of clay-unit masonry with English bond (Figs. 21.1 and 21.2). The walls of the brick structure were built with perforated brick, and cement-based mortar, whereas the walls of the stone specimen were built with granite stone and lime-based mortar. The configuration of each structure included a single perforated unreinforced wall with a gable, and return walls on both ends. In each structure, an opening was placed in one of the returning walls, resulting in an asymmetry, and consequently, inducing torsional movements. The thickness of the walls was equal to 0.500 m and 0.235 m for the stone and brick structure, respectively. Each structure was tested on the LNEC shaking table in Lisbon (Portugal). For details on the shaking table tests, see Candeias et al. (2017).

21.3 Expert Results

The geometry of the structures, the material properties (specific mass, Young's modulus, tensile and compressive strength), the normalized accelerogram envelopes of the seismic action applied at the base, and the corresponding response spectra were provided to the experts. No specific requirements were given to experts in terms of the types of computed results they needed to provide.

Fig. 21.1 Stone structure: (**a**) general view; (**b**) return wall with opening

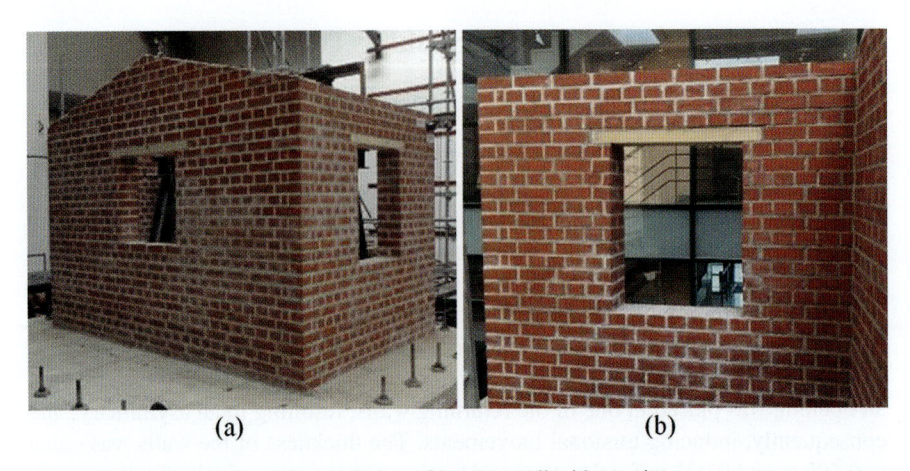

Fig. 21.2 Brick structure: (**a**) general view; (**b**) return wall with opening

The experts presented several modelling approaches, type of structural analysis and assessment criteria for predicting the dynamic behavior of the structures. It is noted that the predictions were made for either or both test structures depending on the expertise of the expert. For details on the predictions of experts, see Mendes et al. (2017) and de Felice et al. (2017).

21.3.1 Predictions

As an example, the different modelling approaches used are given in Table 21.1 for the brick structure. A total of 36 predictions of the seismic capacity were made: 17 for the brick structure and 19 for the stone structure. Most predictions were performed with rigid block analysis. Since no information was available on the experimental failure mechanism, blind predictions were based either on personal judgement or on preliminary finite elements or discrete elements models.

Methods based on rigid block analysis represented the mechanism with a single-degree-of-freedom (SDOF) system subjected to the horizontal static loads in addition to the self-weight. The collapse condition was then calculated by equilibrium equations or with the principle of virtual works, with either a static or a kinematic approach. In the static approach, the capacity was estimated as the PGA that activates the mechanism and the demand was derived from the acceleration response spectrum, taking into account the dynamic amplification through the structure and the reserve of stability from the activation of the mechanism to the out-of-plane overturning. In the kinematic approach, the seismic capacity was identified by means of a non-linear analysis, leading to a capacity curve whose ultimate point identified the maximum attainable displacement at collapse. The demand was derived from the displacement response spectrum, calculating a fundamental period of the equivalent

21 Technologies for Seismic Retrofitting and Strengthening of Earthen... 505

Table 21.1 Example of predictions by experts, for the brick structure

Participant	Mechanism[a]	PGA [g]	Modelling approach	Evaluation of capacity	Assessment method	Notes
Experimental		1.27				
A	2	0.30	RB	S (non-linear)	DB	Failure mechanism identified by personal judgment. Collapse displacement predicted through displacement response spectrum. Collapse PGA estimated through time history analyses. *Postdiction also performed*
	7	0.35	RB	D (non-linear)	FB	
B	4	0.75	RB	S (linear)	FB	Failure mechanism identified by personal judgment.
C	2	1.00	RB	S (linear)	DB	Failure mechanism identified by personal judgment. Collapse PGA predicted with the capacity spectrum method as that corresponding to a spectral displacement equal to 40% of the instable equilibrium displacement.
E	2	0.37	RB	PVW	FB	Failure mechanism identified by personal judgment. Collapse PGA predicted with a force-based approach that makes use of the principle of virtual works, also including dynamic amplification, or with a displacement-based approach.
	2	0.39	RB	PVW	DB	
F	3	0.42	FEM/DEM + RB	IDA	DB	Failure mechanism identified by FE model running non-linear dynamic analyses under artificial accelerogram. Collapse PGA predicted on an equivalent SDOF system with a displacement-based or a force-based approach. *Postdiction also performed*
	3	0.95	FEM/DEM + RB	IDA	FB	
G	1	0.40	FEM-macro	IDA	FB	Failure mechanism identified by FE model. Non-linear dynamic analyses performed. Collapse PGA predicted as that of the accelerogram used in the analysis in which convergence is lost, or with pushover analyses under mass proportional loads.
	1	0.60	FEM-macro	POA	FB	
H	6	0.86	FEM-macro + RB	S (non-linear)	DB	Failure mechanism identified by FE model. Collapse PGA predicted with limit analysis on rigid-block system.

(continued)

Table 21.1 (continued)

Participant	Mechanism[a]	PGA [g]	Modelling approach	Evaluation of capacity	Assessment method	Notes
I	5	0.57	FEM- macro + RB	PVW	FB	Failure mechanism identified by personal judgment. Collapse PGA predicted with a force-based approach that makes use of the principle of virtual works.
	8	0.75	FEM-macro	POA	FB	Failure mechanism identified by FE model. Collapse PGA predicted with pushover analyses under mass proportional loads (two software programs used).
	8	0.76	FEM-macro	POA	FB	
	5	1.00	DEM	IDA	FB	Failure mechanism identified by personal judgement and DE model. Non-linear dynamic analyses performed. Collapse PGA predicted as that of the accelerogram used in the analysis in which convergence is lost.
	8	1.00	FEM-macro	IDA	FB	Failure mechanism identified by personal judgement and FEM model. Non-linear dynamic analyses performed. Collapse PGA predicted as that of the accelerogram used in the analysis in which convergence is lost.
K	2	0.47	FEM-micro	POA	FB	Failure mechanism identified with FE model. Collapse PGA predicted with a force-based approach that makes use of the principle of virtual works, including elastic deformability. *Postdiction also performed*

[a](Mendes et al. 2017) for the predicted failure mechanisms and their numbering

PGA peak ground acceleration at collapse, *RB* limit analysis with rigid-body systems, *FEM-micro/-macro* Finite Element Method with micro-/macro-modelling approaches, *DEM* Distinct Element Method, *S* static analysis, *D* dynamic analysis, *PVW* principle of virtual works, *POA* pushover analysis, *IDA* incremental dynamic analysis (time-step integration under earthquake base motion)

SDOF system. One prediction was performed by integrating the equation of motion of a rigid block under earthquake base motion and the PGA was calculated as that inducing instability.

Numerical models with finite elements (FEM, with either macro- or micro-modelling approaches), distinct elements (DEM), or combined finite-discrete elements (FEM/DEM) were used to predict the mechanism and/or to assess the seismic capacity. In finite element macro-models, the masonry was described as an equivalent homogeneous material, while in micro-modelling, the units were represented explicitly and the joints were described by interfaces, where cracking is allowed. Distinct element models considered some representation of the shape of blocks and joints. Differently from limit analysis based approaches, with FEM and DEM the failure mechanism was identified directly by the model. Analyses were either static (pushover) under horizontal loads, or dynamic with time-step integration under artificial accelerograms compliant with the response spectrum provided. In the former case (pushover), the capacity was assessed as the peak of the response curve, whilst in the latter, simulations were carried out under increasing intensity of the input (incremental dynamic analysis, IDA) up to a given definition of failure, and the PGA of the last run was taken as the seismic capacity.

21.3.2 Failure Mechanisms for Predictions

As an example, eight collapse mechanisms were predicted by the experts for the brick structure, see Fig. 21.3. Predictions varied widely, mainly depending on the assumption by the experts related to (i) the effectiveness of the connections between front and side walls at the corners, (ii) the bending strength of the façade, and (iii) the in-plane strength of side walls. Based on their capability of foreseeing the above features of the experimental failure mechanism, Mendes et al. (2017) assessed as good two (out of 8) failure mechanisms for the Brick House (#2 and #7). Predicting the collapse mechanism of the brick structure resulted more challenging than for the stone structure, due to the higher slenderness and flexibility of the walls.

21.3.3 Prediction of the Seismic Capacity in Terms of Peak Ground Acceleration (PGA)

The predictions of the out-of-plane seismic capacity were scattered, which may be considered by itself a demonstration of the scarce consensus amongst researchers on suitable strategies to handle this problem, as well as of the difficulty of the proposed challenge. For the brick structure, the predicted PGAs ranged between 0.30 g to 1.00 g, with an average of 0.64 g (49% lower than the experimental value) and a Coefficient of Variation (CV) of 39%. If only the good predictions of failure

Experimental (PGA=1.27g)

Mechanism 1 (PGA: 0.40g; 0.60g)

Mechanism 2 (PGA: 0.30g; 0.37g; 0.39g; 0.47g; 1.00g)

Mechanism 3 (PGA: 0.42g; 0.95g)

Mechanism 4 (PGA: 0.75g)

Mechanism 5 (PGA: 0.57g; 1.00g)

Mechanism 6 (PGA: 0.86g)

Mechanism 7 (PGA: 0.35g)

Mechanism 8 (PGA: 0.75g; 0.76g; 1.00g)

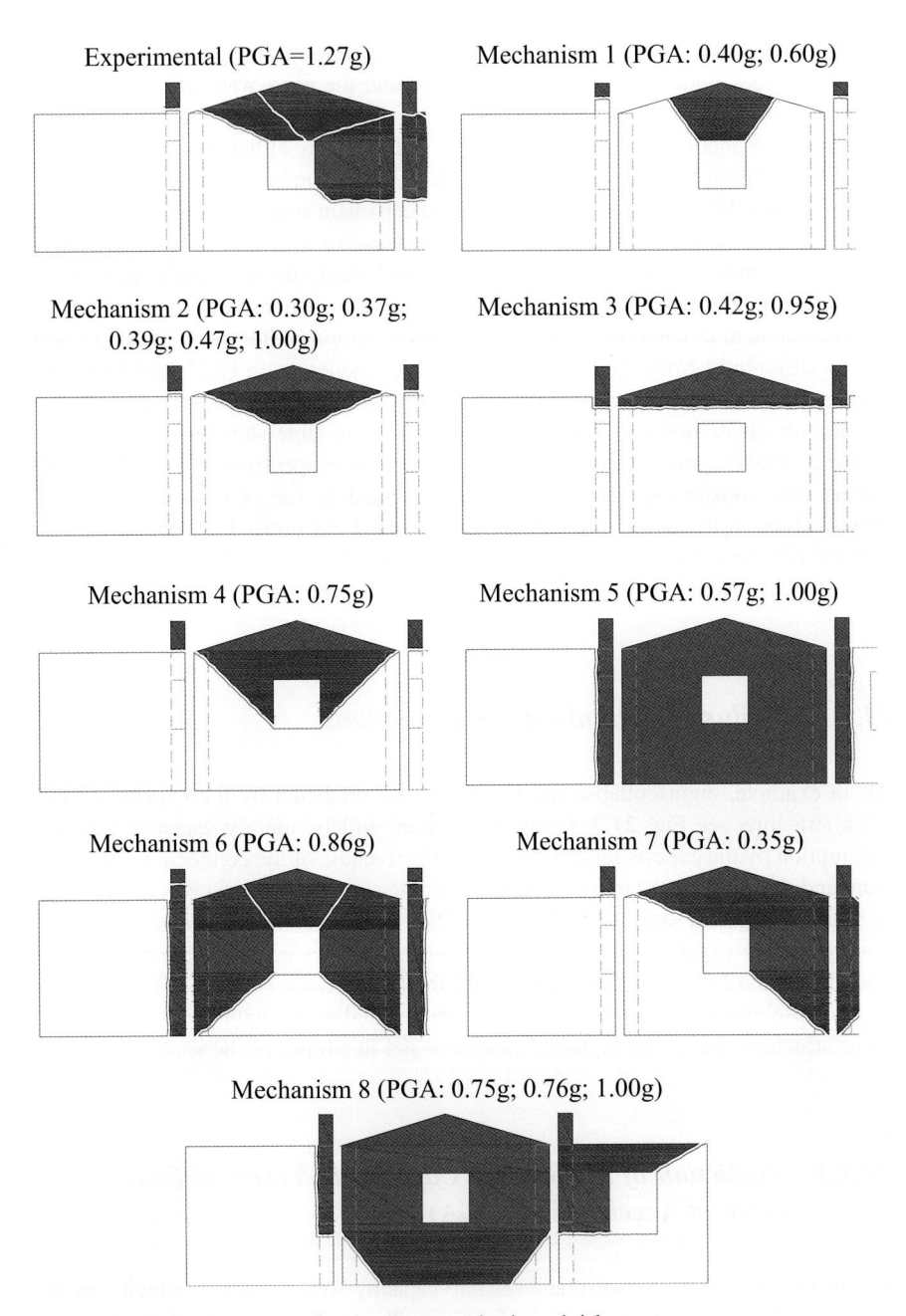

Fig. 21.3 Idealized expert predicted collapse mechanisms, brick structure

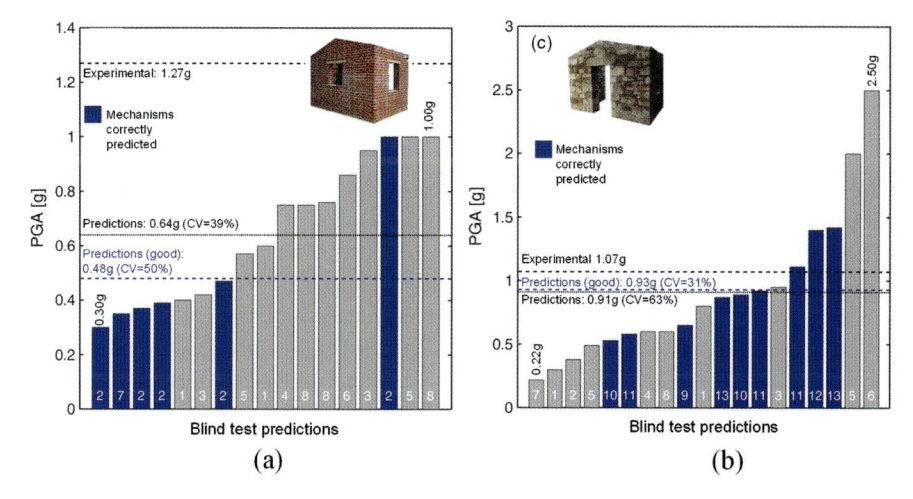

Fig. 21.4 Seismic acceleration (PGA) capacity provided by blind predictions: (**a**) brick structure; (**b**) stone structure

mechanisms are considered, the mean estimate is 0.48 g (CV = 50%), which is worse than the total mean value (Fig. 21.4a). Better predictions were provided for the Stone House on average (PGA = 0.91 g, 15% lower than the experimental value), but with a wider range, from 0.22 g to 2.50 g (CV = 64%). The good predictions of failure mechanism led to a slightly better estimate (0.93 g and CV = 31%, Fig. 21.4b). Note that the graphs indicate the failure mechanism corresponding to each prediction and the bars of those assessed as good are filled in blue.

Even though identifying the correct collapse mechanism has to be considered fundamental for a reliable estimate of the seismic capacity, the results of the blind test predictions indicate that this is not enough, since the estimated collapse PGAs differed largely and were mostly incorrect even when a common mechanism was assumed. There was no clear relationship between accuracy of the prediction and modelling approach.

21.3.4 Postdictions

The shaking table tests were simulated a posteriori by six research groups, making use of various approaches, ranging from analytical methods based on rigid-body mechanisms, to numerical models with finite elements, discrete elements, and combined finite-discrete elements, see de Felice et al. (2017). The previously adopted modelling approaches were not used necessarily again by the same groups, as the time constraints were different. Obviously, at this stage, the experimental results were known, making the task much easier than in blind test predictions. In some cases, e.g., with limit analysis, the failure mechanism was assumed as the starting point of the assessment. Alternatively, the capability of the model to estimate both the failure mechanism and the seismic displacement demand was investigated. The experts that run both predictions and postdictions with the same models had the possibility to update them, e.g., by re-calibrating some parameters, in order to match experimental results.

Nevertheless, the collapse PGA was not provided by many postdictions, which focussed on the assessment of the maximum displacement capacity, the simulation of the displacement response in time or only the identification of the failure mechanism. On the other hand, most postdictions evaluated the seismic displacement demand, which is compared with the displacements recorded in the last tests (at collapse). Taking advantage of the time available for postdictions, the sensitivity of the results to some variables, such as meshing, strength properties, analysis parameters, and input characteristics was also investigated in some cases.

The failure mechanisms provided by postdictions do not differ largely from each other. All of them represented well the torsional response of the structures, whilst the bending strength of the façade resulted underestimated in most cases. Still, this does not allow to ignore the fact that the scatter in the predictions of the experts was too high, independently of the reliability of the single test made in the shaking table.

21.4 Strengthen of Earthen Structures in Peru

After the earthquake of Pisco, Peru in 2006 the conservation of local historic adobe buildings received extensive attention. A collaboration between the Minister of Culture in Peru and the Getty Conservation Institute (GCI) was started in order to assess the post-earthquake damage (Cancino et al. 2009).

Next, the Getty Conservation Institute (GCI) initiated the Seismic Retrofitting Project (SRP), as a collaborative project. The ultimate objective of SRP is to provide low-tech seismic retrofitting techniques and easy-to-implement maintenance programs for historic earthen buildings in order to improve their seismic performance while preserving their historic fabric (Cancino et al. 2012). Using Peruvian building prototypes as case studies, the project aims to design and test these techniques; provide guidance for those responsible for implementation, including architects, engineers, and conservators (Cancino et al. 2012); and, work with authorities to gain acceptance of these methods, with the goal of ultimately including them as part of the Peruvian National Building Code.

Four prototype buildings in Peru have been adopted: Casa Arones is a representative example of a two stories mansion, located in the heart of Cusco, constructed at the end of the sixteenth century; the Church of Kuño Tambo, a religious adobe structure of the seventeenth century, in the Cusco region; Hotel El Comercio, situated in Lima, a typical three-story patio mansion, L-shaped, built around two interior courtyards, dating back to middle nineteenth century, and, Ica Cathedral, a highly complex timber-masonry structure of the eighteenth century, with parts of adobe and brick masonry on the exterior envelope, in the city of Ica. The latter is addressed here and further details are given in Ciocci et al. (2018).

21.5 Safety Assessment Under Current Conditions

The Cathedral of Ica is a highly complex structure with a rectangular plan, with a choir loft, a main nave, transept, altar and two lateral aisles. Two sub-structures are evident; an external masonry envelope and an internal timber frame. Built later in a neoclassical style, the main façade is of fired brick masonry, with a thickness of 2.25 m, flanked on both sides with massive bell towers. The lateral walls are of adobe masonry, over a base course of fired brick and rubble stone masonry. The entire timber frame system is comprised by a series of pillars, pilasters and a complex vaulted roof system, with a traditional rendering technique of cane reeds nailed with leather strips, mud plaster and gypsum, known as *quincha*.

Ica Cathedral has suffered damage due to a series of past earthquakes, namely in 2007 and 2009, with a magnitude 8.0 and 5.8 respectively. The most recent earthquakes led to the collapse of several parts of the vaulted roof and the main dome. The masonry envelope also suffered extensive cracking.

The first experimental vibration mode (2.84 Hz), see Fig. 21.5, corresponds to the dominant mode in the transversal direction of the longitudinal side walls of the Cathedral. Both the longitudinal walls of the nave experience a first order out-of-phase excitation with higher intensity in the northern wall, on the transept area. A model updating process was conducted, regarding the masonry envelope and the updated material properties.

The structural behavior of Ica Cathedral is highly influenced by the interaction between the masonry envelope and the internal timber structure, resulting in a higher lateral capacity, compared to the value obtained from the masonry envelope alone. Out-of-plane stiffness values at early stages are governed by the response of the timber sub-structure. Yet, various stresses in timber connections were found in critical state of failure, justifying the partial collapse of the vaulted roof and the central dome. According to the pushover nonlinear analyses, under a mass proportional lateral load, the out-of-plane capacity of the main façade is 0.45 g. Corresponding cracks in the pediment and at connection areas with the choir loft were well correlated with damage in-situ, as shown in Fig. 21.5. The out-of-plane capacity of the north lateral wall is 0.28 g, much lower than the capacity demand of 0.45 g for the region. Several flexural cracks are formed along the transversal wall, in the interface plane with the rubble stone base course, together with vertical cracks in the intersection with the transept. Those cracks are also present on site.

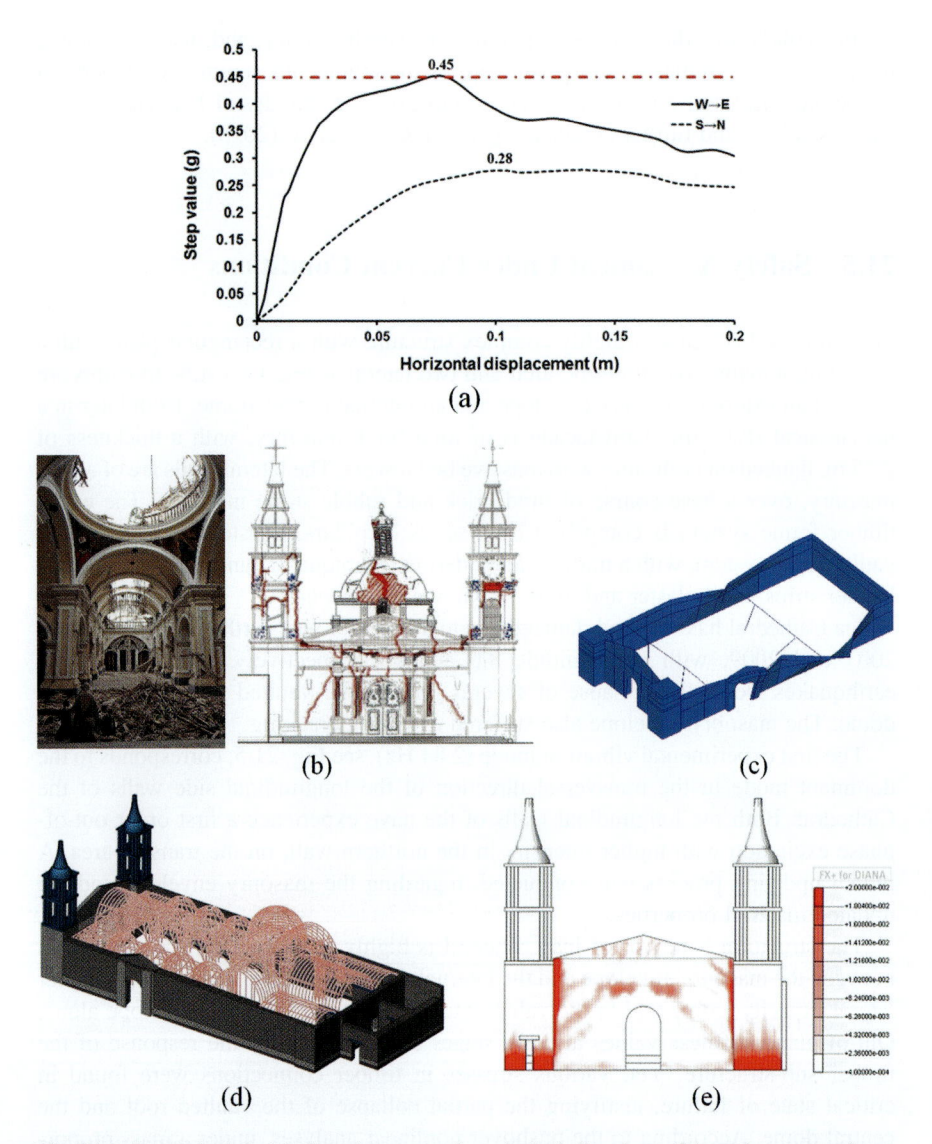

Fig. 21.5 Ica Cathedral: (**a**) Capacity curves for principal directions; (**b**) crack pattern of east facade (right) and collapsed main dome and barrel vault (left); (**c**) 1st mode shape from dynamic in-situ tests (2.84 Hz); (**d**) FE model in 3D view; (**e**) plot of maximum principal strain distribution near collapse

21.6 Traditional Strengthening Techniques

A strengthening plan has to be aligned with conservation principles, namely minimal intervention, authenticity and reversibility. The strengthening philosophy and implementation, through traditional techniques, is long proven and present in many historic earthen buildings, though often disregarded in current strengthening practices. It involves the combination of additional mass and stiffness elements; i.e. buttresses and also bracing elements; i.e. corner keys, horizontal keys, bond beams and anchored tie beams. Material compatibility, consolidation and sufficient care in timber connection details are also of vital importance.

Consolidation measures mostly involve the replacement of highly deteriorated adobe and base course masonry parts. The extent of replacement should be clarified from damage mapping and as limited as possible, so as to preserve as much of the historic fabric. Interlocking between old and new masonry parts should also be established.

Reestablishment or addition of buttresses can efficiently address low or compromised out-of-plane capacity and minimize lateral deflections, especially in walls of large spans. Connectivity between the existing earthen walls and the new buttresses can be ensured e.g. by inserting horizontal timber elements at various heights, as shown in Fig. 21.6a, b.

For ensuring connectivity and substantial stiffness in corners between orthogonal walls and pillars, orthogonal or diagonal timber keys should be used, Ortega et al. (2017). The insertion should be made at horizontal planes of various elevations and involve mostly the upper parts of walls (Fig. 21.6c, f). For existing cracks, several processes that respect the historic fabric can be applicable, such as partial replacement of material, stitching, repointing and grout injections, with mud based grouts, Silva et al. (2012).

Additional lateral restrain, in the case of series of walls in longitudinal alignment, can be offered by means of tie beams. The system is placed at the level of the top eaves, embedded favorably along the entire thickness of the walls. It is subjected to both axial tension and compression, given the dynamic character of seismic loads. Thus, for the system of ties to work, both in tension and compression, an adequate anchoring system is needed. A double system of vertical timber anchors, attached close to the interior and exterior surfaces of each wall is proven effective (Fig. 21.6d, e). Though, the overall activation is expected low in case of walls with similar out-of-plane bending stiffness.

Lastly, the implementation of an internal horizontal timber frame system, at the top level and entire thickness of earthen walls, can enhance substantially the capacity under lateral forces. Namely, a bond-beam or U-beam (Fig. 21.6g, h), which can also serve as support system for roof rafters (wall plates). The whole system can be also connected with a system of tie beams, plates and anchors, forming a combined internal and external timber frame system (Fig. 21.6d). For the tying system to perform, timber elements need to be confined in masonry and subjected to normal vertical stresses, from overlapping masonry parts and roof loads, so that friction or shear action is available.

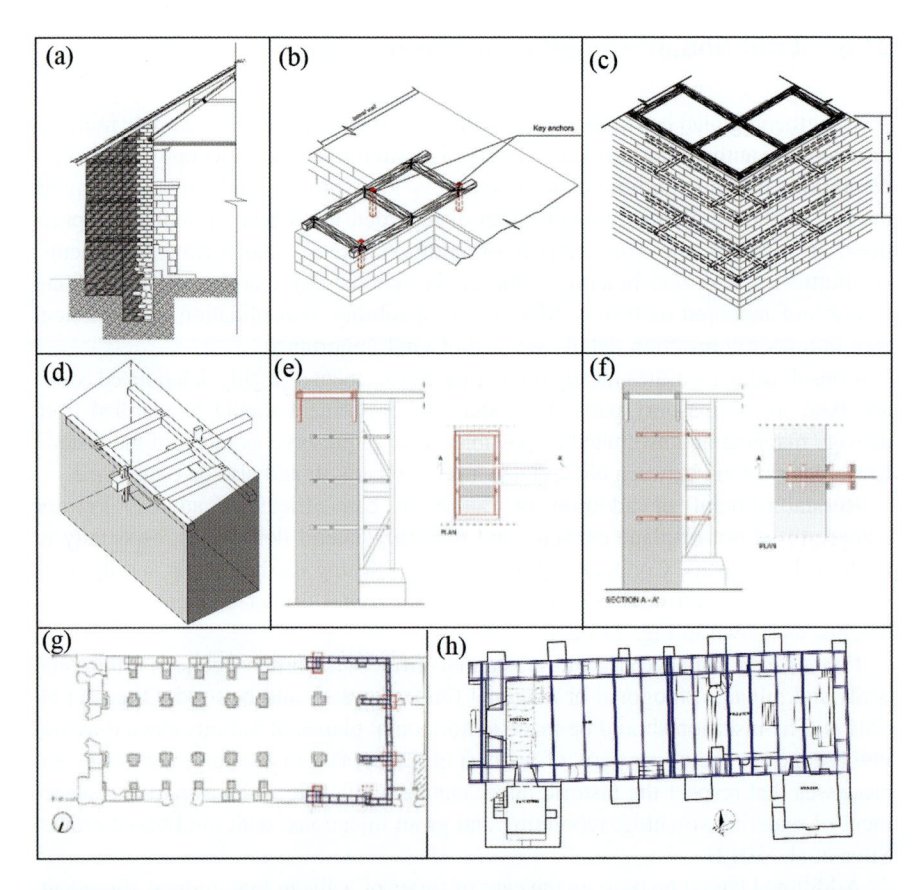

Fig. 21.6 Detailing of traditional strengthening techniques. (**a**) New buttress, with horizontal timber keys and new interlocking, and geo-mesh. (**b**) Horizontal timber key between buttress and adjoining wall. Vertical key anchors to enhance connectivity. (**c**) Timber embedded corner keys in elevation. Continuous bond beam at top eave. (**d**) Timber frame system, with bond beam, tie and vertical timber anchors. (**e**) Connection between masonry wall and timber floor. Tying system with ties and timber anchors. (**f**) Connection between masonry wall and timber pillar. Horizontal timber anchors. (**g**) U-beam at top eaves. (**h**) Bond-beam at top eaves in the periphery of the nave

21.7 Safety Assessment After Strengthening

The implementation of the strengthening provides a minor increase in stiffness and a large increase in capacity, compared to the current state. The two substructures deform more uniformly, under more rigid connections and the seismic demand of 0.45 g in the lateral capacity of the north lateral wall is surpassed (Fig. 21.7). Compared to the current state, the capacity in the north-south direction is increased by 100%. Flexural cracking is observed, less in extent and size, over a distributed area of the north-west corner, compared to extensive diagonal and horizontal fragmentations at current state (Fig. 21.7).

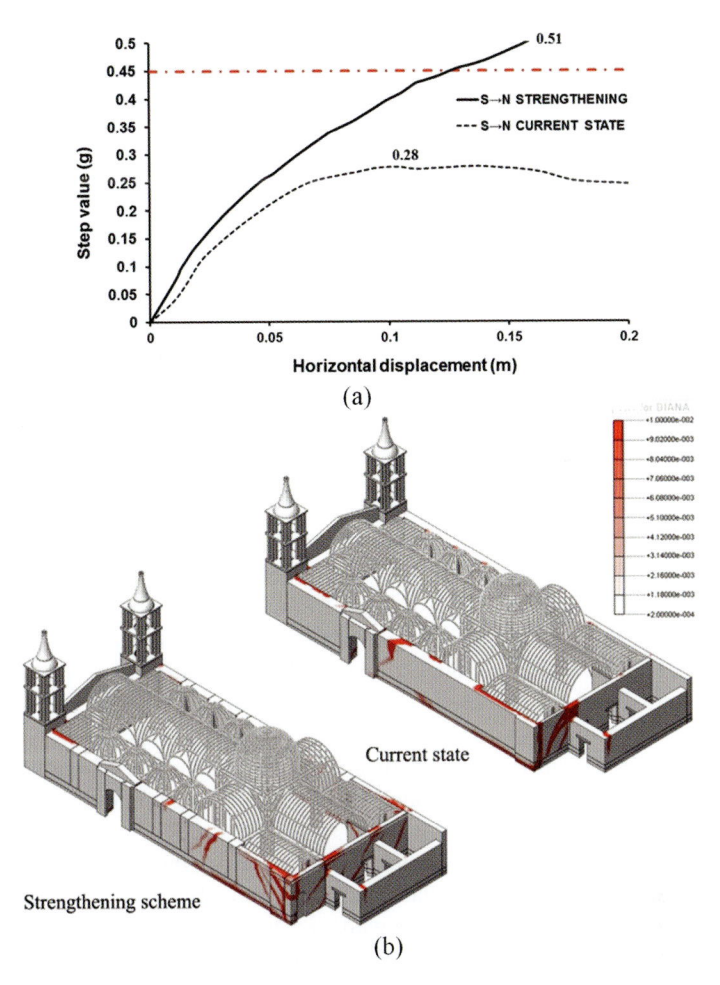

Fig. 21.7 Ica Cathedral: (**a**) load-displacement diagram, at the top of the north wall. Note the seismic demand of 0.45 g (in red); (**b**) distribution of maximum principal tensile strains at ultimate condition

21.8 Conclusions

The out-of-plane seismic response for unreinforced earthen and masonry walls are far from trivial. Capturing all aspects of behavior may easily escape the most proficient of modelers. Considerable variance can occur in assessments done by different modelers due to the complexities of nonlinear dynamic response of these truly three-dimensional structures. Because of this, a user must acknowledge that his or her own model is likely to not represent actual response precisely despite the complexity of the model or the analysis tool.

Despite the modeling challenges, structural earthquake engineers should not forget their primary objective – that being to safely assess the capacity of a given building structure with respect to collapse and thus protect the lives of its occupants and the people outside the building. With regard to this objective, some conservatism is much better than the converse. However, excessive conservatism must be circumvented when economies of retrofit solutions may not be practical or the impact on the cultural heritage may be too severe.

The blind test prediction of the out-of-plane seismic capacity of a masonry structure resulted extremely challenging. Predictions of the failure mechanism displayed a large variability, mainly depending on the assumptions on: (i) the effectiveness of the connections between front and side walls; (ii) the bending strength of the façade; and (iii) the in-plane strength of side walls. A method to achieve good estimates appears unavailable, whilst a combination of numerical discrete models and engineering judgement seems able to provide the best guess of the failure mode. Estimates of capacity also differed largely, with an underestimation, in one case significant, of the actual capacity, even when the correct failure mechanism was assumed.

The tools available for researchers and practitioners may significantly contribute to the estimate of seismic safety of existing earthen and masonry structures, but appear by themselves not yet sufficient for a refined prediction of the failure mechanism and a reliable estimate of the seismic displacement demand. Still, advanced simulations are the current best guess for engineering applications. Investment on dynamic identification and monitoring and well-designed non-destructive field testing in engineering applications are essential, as they allow calibrating the numerical models and increasing the reliability of structural analysis.

Traditional strengthening techniques can improve the integrity of earthen structures, increase the out-of-plane capacity and redistribute seismic loads, between transversal and longitudinal walls, ensuring a so called 'integral behavior'. Commonly used, traditional strengthening techniques involve the use of buttresses, together with systems of timber strengthening elements, such as bond beams, anchors, corner keys and tie beams.

Under the Seismic Retrofitting Project of the Getty Conservation Institute, USA, extensive inspections, surveys, in-situ testing and nonlinear structural analyses of earthen historic structures, assessed the current state and revealed structural deficiencies. For one case studies, Ica Cathedral, a complete design and assessment of the above-mentioned strengthening solutions are incorporated. For the retrofitted structures, performance criteria and seismic local demands were met, with sufficient safety and acceptable levels of repairable damage.

References

Augenti N, Parisi F (2010) Learning from construction failures due to the 2009 L'Aquila, Italy, earthquake. J Perform Constr Facil 24(6):536–555. https://doi.org/10.1061/(ASCE)CF.1943-5509.0000122

Cancino C, Farneth S, Garnier P, Newmann JV, Webster F (2009) Damage assessment of historic earthen buildings after the August 15, 2007, Pisco Earthquake. Research Report, Getty Conservation Institute, Los Angeles

Cancino C, Lardinois S, D'Ayala D, Ferreira CF, Dàvila DT, Meléndez EV, Santamato LV (2012) Seismic retrofitting project: assessment of prototype buildings. Research report, Getty Conservation Institute, Los Angeles

Candeias PX, Campos-Costa A, Mendes N, Costa AA, Lourenço PB (2017) Experimental assessment of the out-of-plane performance of masonry buildings through shaking table tests. Int J Archit Herit 11(1):31–58. https://doi.org/10.1080/15583058.2016.1238975

Ciocci MP, Sharma S, Lourenço PB (2018) Engineering simulations of a super-complex cultural heritage building: Ica Cathedral in Peru, Meccanica (accepted for publication)

de Felice G, de Santis S, Lourenço PB, Mendes N (2017) Methods and challenges for the seismic assessment of historic masonry structures. Int J Archit Herit 11(1):143–160

Ferreira TM, Costa AA, Costa A (2015). Analysis of the out-of-plane seismic behaviour of unreinforced masonry: a literature review. Int J Archit Herit 9(8):949–972. https://doi.org/10.1080/15583058.2014.885996

Holzer TL, Savage JC (2013) Global earthquake fatalities and population. Earthq Spectra 29(1):155–175. https://doi.org/10.1193/1.4000106

Lagomarsino S (2006) On the vulnerability assessment of monumental buildings. Bull Earthq Eng 4(4):445–463. https://doi.org/10.1007/s10518-006-9025-y

Leite J, Lourenço PB, Ingham JM (2013) Statistical assessment of damage to churches affected by the 2010–2011 Canterbury (New Zealand) earthquake sequence. J Earthq Eng 17(1):73–97. https://doi.org/10.1080/13632469.2012.713562

Lourenço PB, Mendes N, Ramos LF, Oliveira DV (2011) Analysis of masonry structures without box behaviour. Int J Archit Herit 5:369–382. https://doi.org/10.1080/15583058.2010.528824

Lourenço PB, Mendes N, Costa AA, Campos-Costa A (2017) Methods and challenges on the out-of-plane assessment of existing masonry buildings. Int J Archit Herit 11(1):1. https://doi.org/10.1080/15583058.2017.1237114

Mendes N, Costa AA, Lourenço PN, Bento R, Beyer K, de Felice G, Gams M, Griffith M, Ingham J, Lagomarsino S, Lemos JV, Liberatore D, Modena C, Oliveira DV, Penna A, Sorrentino L (2017) Methods and approaches for blind test predictions of out-of-plane behavior of masonry walls: a numerical comparative study. Int J Archit Herit 11(1):59–71

Noy I (2009) The macroeconomic consequences of disasters. J Dev Econ 88(2):221–231. https://doi.org/10.1016/j.jdeveco.2008.02.005

Ortega J, Vasconcelos G, Rodrigues H, Correia M, Lourenço PB (2017) Traditional earthquake resistant techniques for vernacular architecture and local seismic cultures: a literature review. J Cult Herit 27:181. https://doi.org/10.1016/j.culher.2017.02.015

Parker R, Little K, Heuser S (2007) Development actions and the rising incidence of disasters. IEG evaluation brief, no. 4. World Bank, Washington, DC. Available at: http://documents.worldbank.org/curated/en/2007/06/11918676/development-actions-rising-incidence-disasters

Penna A, Morandi P, Rota M, Manzini CF, da Porto F, Magenes G (2014) Performance of masonry buildings during the Emilia 2012 earthquake. Bull Earthq Eng 12(5):2255–2273. https://doi.org/10.1007/s10518-013-9496-6

Silva RA, Schueremans L, Oliveira DV, Dekoning K, Gyssels T (2012) On the development of unmodified mud grouts for repairing earth constructions: rheology, strength and adhesion. Mater Struct 45(10):1497–1512

Sorrentino L, D'Ayala D, de Felice G, Griffith M, Lagomarsino S, Magenes G (2016) Review of out-of-plane seismic assessment techniques applied to existing masonry buildings. Int J Archit Herit 11(1):2–21. https://doi.org/10.1080/15583058.2016.1237586

Chapter 22
Seismic Performance of a Full-Scale FRP Retrofitted Sub-standard RC Building

Alper Ilki, Erkan Tore, Cem Demir, and Mustafa Comert

Abstract External jacketing of columns with Fiber Reinforced Polymers (FRPs) is a promising retrofitting technique for improving seismic performance of sub-standard reinforced concrete (RC) buildings. The enhancement in deformation capacity and shear strength of jacketed members helps to prevent the brittle collapse mechanism of buildings with inadequate ductility. This paper provides an overview on the retrofitting of columns with FRP jacketing for particularly ductility enhancement and gives a brief summary of seismic strengthening recommendations of various design documents. In addition, a recent full-scale test conducted simultaneously on an as-built and a FRP retrofitted building, which are identical with design geometry, material quality and seismic deficiencies, is briefly presented and the performance of the retrofitting technique is evaluated. Finally, analytical behavior obtained through nonlinear static analyses executed using FRP-confined concrete models recommended in different technical documents are reviewed in comparison with the experimental behavior.

22.1 Introduction

Fiber reinforced polymer (FRP) composites are becoming popular and widely used in construction industry, thanks to their outstanding features and decreasing costs. High strength to weight ratio, corrosion resistance, ease of application process are some of the advantages of FRP composites besides the drawbacks of low glass transition temperature, high-tech manufacturing requirements, uncertainties on long-

---

A. Ilki (✉) · C. Demir
Faculty of Civil Engineering Istanbul, Istanbul Technical University, Istanbul, Turkey
e-mail: ailki@itu.edu.tr

E. Tore
Department of Civil Engineering, Balikesir University, Balikesir, Turkey

M. Comert
RISE Engineering and Consultancy, Istanbul, Turkey

© Springer International Publishing AG, part of Springer Nature 2018
K. Pitilakis (ed.), *Recent Advances in Earthquake Engineering in Europe*,
Geotechnical, Geological and Earthquake Engineering 46,
https://doi.org/10.1007/978-3-319-75741-4_22

term behavior and durability. Nowadays, several kinds of FRP composite products (sheets, laminates, reinforcements, profiles, tubes etc.) are utilized for retrofitting applications as well as construction of new structures.

Retrofitting of sub-standard structures against potential severe seismic actions is a crucial engineering problem particularly in developing countries for reducing earthquake-induced casualties and economic losses. FRP composites are effective alternatives to conventional materials used for retrofitting applications. In recent years, various retrofitting approaches making use of FRP composites have been developed for eliminating different structural deficiencies and for improving the seismic performance of existing structures. For instance, enhancement in deformation, axial load and shear capacities is obtained when RC (reinforced concrete) columns are externally confined with FRP sheets (Ilki et al. 2009; Ozcan et al. 2010; De Luca et al. 2011; Realfonzo and Napoli 2012; Ghatte et al. 2016). External FRP confinement also restrains buckling of longitudinal reinforcement (Tastani et al. 2006; Bournas and Triantafillou 2011; Giamundo et al. 2014; Bai et al. 2017) and provides clamping effect on insufficient lap-splices to ensure ductile behavior for desired seismic performance (Hamad et al. 2004; Bousias et al. 2007; Harajli and Khalil 2008; ElGawady et al. 2010; Giamundo et al. 2017). Various forms of FRP wrapping leads to significant increase on shear capacity of RC members (Ye et al. 2002; Galal et al. 2005; Barros et al. 2007; Pellegrino and Modena 2007; Chen et al. 2010). Behavior of sub-standard beam-column joints are improved with externally bonded FRP sheets (Antonopoulos and Triantafillou 2003; Ilki et al. 2010; Li and Kai 2011; Akguzel and Pampanin 2012; Cosgun et al. 2012; Sezen 2012). Properly strengthened infill walls with diagonal FRP sheets and anchorage details can be considered in lateral load resisting systems to improve global seismic performance of RC frames (Karadogan et al. 2003; Binici et al. 2007; Yuksel et al. 2010; Ozsayin et al. 2011). Strengthening of structural members through near surface mounting (NSM) technique is an attractive solution towards increasing flexural resistance (De Lorenzis and Teng 2007; Yost et al. 2007; Barros et al. 2008; Bournas and Triantafillou 2009; El-maaddawy and El-dieb 2011; Sharaky et al. 2014; Kaya et al. 2017). Most of these techniques have already been accepted by engineering community, thus the national design guidelines and provisions have been published since the beginning of 2000s to standardize retrofitting design and application (*fib* Bulletin No. 14 2001; ISIS 2001; JSCE 2001; TSDC 2007; GB 50608 2010; S806–12 2012; TR55 2012; CNR-DT 200 R1 2013; ACI 440.2R 2017).

Seismic retrofitting of RC structures with FRP composites were initiated with external jacketing of bridge columns in early 1990s (Mosallam et al. 2013). According to the best knowledge of the authors', first research attempts for investigating the effect of FRP jacketing of RC columns were conducted in the 1980s (Fardis and Khalili 1982; Katsumata et al. 1988). Since then, a vast number of experimental and theoretical studies have been performed to investigate effectiveness of FRP jacketing on seismic performance of RC columns (Saadatmanesh et al. 1996; Ozcan et al. 2008, 2010; Ilki et al. 2009; Napoli et al. 2011; Ghatte et al. 2016). Nowadays, the most well-known seismic retrofitting technique with FRP composites is external FRP confinement of RC columns thanks to the enormous growth in research activities. These research activities include numerous tests conducted at the material, member and system levels, which are generally performed in laboratory conditions with size restrictions. In this study, the performance of external FRP confinement used for retrofitting of columns of a three-story sub-standard building is investigated through open air full-scale tests. One of the test buildings is tested as is and the other after retrofitting with FRP confinement. The evaluation of the observed behavior and the test results point to a very significant enhancement of the seismic performance after retrofitting. Additionally, the analytical predictions made by considering the FRP-confined concrete models given by three current design documents (TSDC 2007; CNR-DT 200 R1 2013; ACI 440.2R 2017) are compared with the test results.

22.2 FRP Jacketing of Columns

22.2.1 External Confinement with FRPs

RC columns are the vitally important primary lateral load resisting members particularly in low-rise and mid-rise RC buildings. Many of the existing buildings have critical defects in columns (e.g. low material quality, large stirrup spacing, and poor reinforcement details) that may lead to low ductility and brittle shear failure. These sub-standard buildings generally do not meet the ductility demands under severe seismic actions and may experience partial or total collapse under such actions. External confinement with FRP composites, which is applied through a rapid and easy installation process, was developed for enhancement of shear, axial load and deformation capacities of such sub-standard RC members. In FRP jacketing, high strength fibers (i.e. carbon, glass, aramid, basalt or hybrid fibers) of FRP composites are oriented along the cross-section perimeter, perpendicular to the longitudinal axis of the retrofitted member, in order to constrain damage of the concrete (i.e. disintegration, inclined shear cracks, delamination of cover) under combined internal forces.

Confinement is a well-known action for substantial enhancement of axial strength and strain capacity of concrete under compression. A passive lateral confinement pressure is provided with external FRP jacketing due to linear elastic tensile behavior of FRP composites. When the FRP confined concrete is subjected to axial compression, encased concrete expands and the FRP jacket is subjected to tension in the hoop direction. Dilation and disintegration of concrete is effectively resisted by the FRP jacket that provides lateral confinement pressure, and significant improvement is achieved on the stress-strain behavior of concrete (Fig. 22.1). Failure of confined concrete abruptly occurs when the FRP jacket ruptures under tensile stress in the hoop direction. General behavior of the FRP confined concrete and confinement mechanism is illustrated at Fig. 22.2 Cross section geometry and aspect ratio, concrete quality, amount of confinement material, loading conditions are some of the critical parameters which can affect the stress-strain relationship of confined

Fig. 22.1 General stress-strain relationship of FRP confined concrete and unconfined concrete

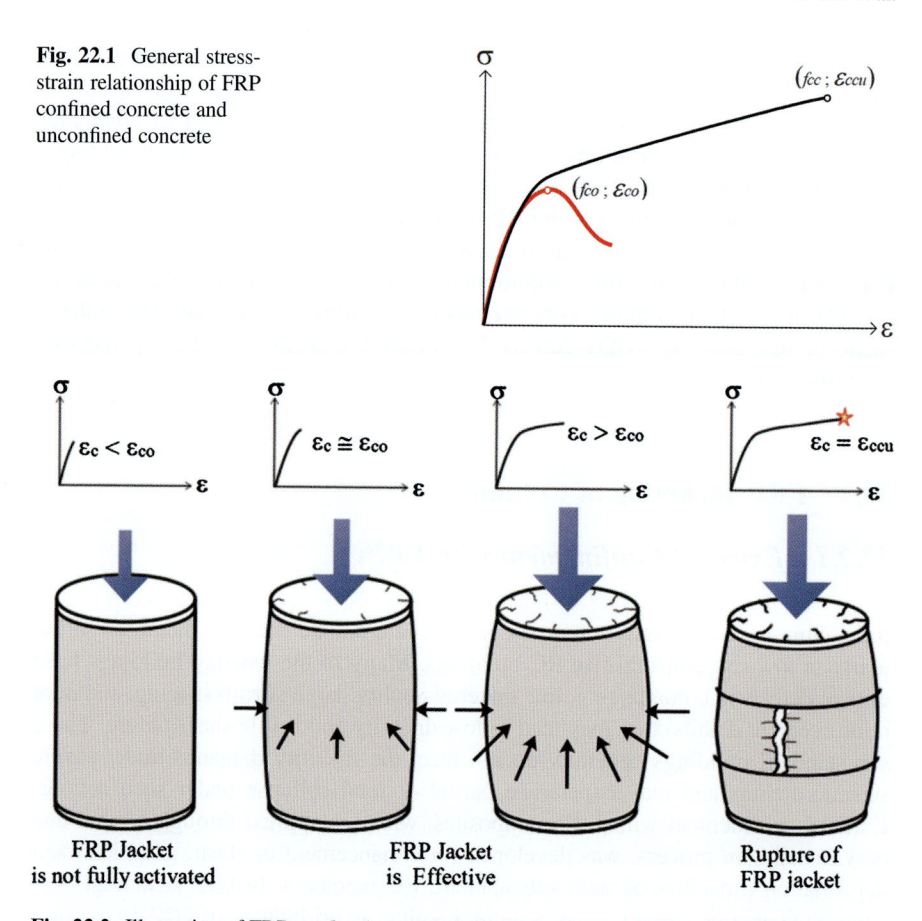

Fig. 22.2 Illustration of FRP confined concrete behavior under concentric axial load

concrete. A significant amount of research has been conducted to investigate the effects of these parameters and mathematical models for confined concrete have been proposed by many researchers (Samaan et al. 1998; Spoelstra and Monti 1999; Toutanji 1999; Shahawy et al. 2000; Lam and Teng 2003a, b; Harajli 2006; Youssef et al. 2007; Eid and Paultre 2008; Ilki et al. 2008; Ozbakkaloglu and Lim 2013).

Geometry of the cross section is among the most effective parameters that can significantly affect the stress-strain relationship characteristics (Mirmiran et al. 1998; Lam and Teng 2003b). FRP confinement is more effective for circular sections since this geometrical configuration allows formation of uniform lateral confining pressure on the entire cross section. In prismatic columns (i.e. columns with square or rectangular cross-sections), confining pressure is concentrated around the corner regions leading to a non-uniform confinement pressure distribution (Campione and Miraglia 2003; Toutanji et al. 2010; Wu and Wei 2010). Thus, corners of square or rectangular cross-sections should be rounded to eliminate premature failure of FRP composites due to stress concentration at corners (Rochette and Labossière 2000;

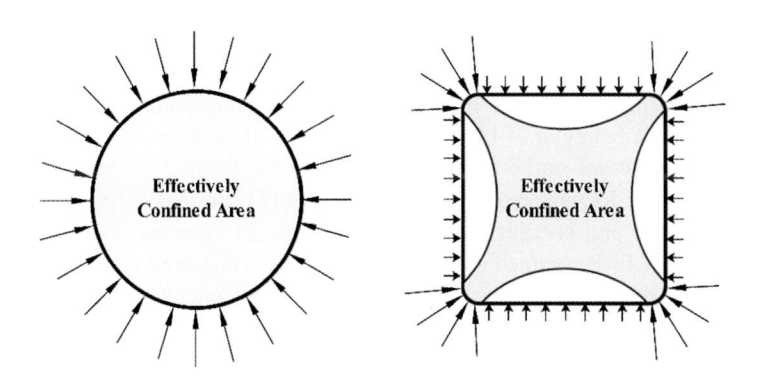

Fig. 22.3 Effectively confined areas for circular and square cross-section shapes

Lam and Teng 2003b; Abbasnia et al. 2012). The effectively confined area and distribution of lateral confining stress for circular and square cross-sections are schematically represented in Fig. 22.3. For non-circular cross-sections, as expected, effectiveness of confinement is higher for square shape, and decreases with the increment of cross-sectional aspect ratio of rectangular members (Harajli 2006; Toutanji et al. 2010; Wu and Wei 2010; De Luca et al. 2011). In addition, the effectiveness is less when the section is larger. Most of the confined concrete models for square and rectangular sections contain a shape efficiency factor for considering the detrimental effect of cross section geometry (Lam and Teng 2003b; Harajli 2006; Youssef et al. 2007; Ilki et al. 2008; Toutanji et al. 2010; Wang et al. 2012a). Shape efficiency factors are generally based on effectively confined area concept proposed by Mander et al. (1988), which has been developed for steel reinforcement confined concrete.

The presence of internal transverse reinforcement can also be effective on the stress-strain relationship of externally FRP confined concrete. Confined concrete models have been proposed to consider combined effect of internal transverse reinforcement and external FRP jacketing (Harajli 2006; Eid and Paultre 2007; Ilki et al. 2008; Chastre and Silva 2010; Lee et al. 2010; Pellegrino and Modena 2010; Wang et al. 2012b). However, sub-standard RC columns of existing buildings mostly have transverse reinforcement with improper details (i.e. 90-degree hooked end) and large spacing, which would provide negligible confinement action (Teng et al. 2016). Therefore, stress-strain models developed for FRP confined plain concrete can be directly used to predict behavior of sub-standard RC columns for seismic strengthening applications. Internal transverse reinforcement contribution to confinement action should be accounted for in analysis of properly designed RC columns (Eid et al. 2009; Lee et al. 2010; Wang et al. 2012b).

For reliable estimation of the seismic performance of FRP jacketed columns, stress-strain relationship of FRP confined concrete under cyclic loading should be well known. Studies on cyclic compression behavior have proven that, cyclic loading did not have an adverse effect on the behavior (Ilki and Kumbasar 2002; Shao et al. 2006; Ilki et al. 2008; Lam and Teng 2009; Wang et al. 2012b; Demir et al. 2015). Therefore, the envelope of cyclic stress-strain relationship can be assumed to be compatible with the monotonic relationships. Utilizing of monotonic FRP confined concrete models in section analysis give reasonable predictions for nonlinear seismic behavior of FRP jacketed columns (Ghatte et al. 2016).

Most of the studies on FRP confined concrete are focused on concentric axial loading. However, RC concrete columns are generally subjected to eccentric loading under axial force and bending moment. In the case of eccentric loading, a strain gradient over the FRP confined column section exists. The lateral confining pressure is not constant over the cross section, owing to uneven concrete dilation caused by the strain gradient (Bisby and Ranger 2010). Results of the experimental and numerical studies indicate that the stress-strain relationship is affected by the strain gradient (Hu et al. 2011; Hadi and Widiarsa 2012; Wu and Jiang 2013).

22.2.2 FRP Confined Concrete Models

Many models for FRP confined concrete have been proposed by different researchers for predicting stress-strain behavior that take into account effects of various parameters. Teng and Lam (2004) classified these models into two categories: Design-oriented models (Samaan et al. 1998; Lam and Teng 2003a, b; Ilki et al. 2004, 2008; Harajli 2006; Youssef et al. 2007; Ozbakkaloglu and Lim 2013) and Analysis-oriented models (Spoelstra and Monti 1999; Yan and Pantelides 2006; Jiang and Teng 2007; Lee et al. 2010; Nisticò and Monti 2013). Design-oriented models comprises a closed-form expression, based on empirical equations for ultimate point of stress-strain curve and are more practical for design practice. Due to mathematical simplicity of models, design documents generally recommended design-oriented models for engineering applications with conservative limitations and safety factors. Design recommendations of ACI 440.2R (2017), CNR-DT 200 R1 (2013) and TSDC (2007) for ductility enhancement are summarized in the following. A common notation is used throughout this study to describe expressions of these design documents.

22.2.3 ACI 440.2R (2017)

American Concrete Institute (ACI) Committee 440 reported the first version of ACI 440.2R in 2002 and noticeable revisions were made in 2008 and 2017. ACI 440.2R (2017) is the most recent design guideline for strengthening system with externally bonded FRP composites. ACI 440.2R (2017) recommends a FRP confined concrete model based on the design-oriented model proposed by Lam and Teng (2003a, b) for ductility enhancement in seismic strengthening applications. The stress strain model consists of parabolic and ascending linear branch which are derived from following equations:

$$f_c = \begin{cases} E_c \varepsilon_c - \left(\dfrac{(E_c - E_2)^2}{4f_{co}}\right)\varepsilon_c^2 & 0 \leq \varepsilon_c \leq \varepsilon_t \\ f_{co} + E_2 \varepsilon_c & \varepsilon_t \leq \varepsilon_c \leq \varepsilon_{cc} \end{cases} \quad (22.1\text{a})$$

$$E_2 = \dfrac{f_{cc} - f_{co}}{\varepsilon_{cc}} \quad (22.1\text{b})$$

$$\varepsilon_t = \dfrac{2f_{co}}{E_c - E_2} \quad (22.1\text{c})$$

The maximum confined concrete strength (f_{cc}) and corresponding strain (ε_{cc}) are obtained from empirical expressions Eqs. 22.2 and 22.3 with an additional reduction factor ($\Psi = 0.95$). In these equations, f_{co} and ε_{co} are compressive strength and corresponding strain of unconfined concrete, κ_a and κ_b are shape factors for square and rectangular sections which are computed by Eqs.22.5 and 22.6, respectively. Ultimate confined concrete strain (ε_{ccu}) is limited to 0.01 for preventing excessive cracking and for ensuring concrete integrity. Additionally, maximum confining pressure provided by FRP jacket (f_l) is limited to at least 0.08 f_{co} for assuring effective confinement action.

$$f_{cc} = f_{co} + \psi 3.3 \kappa_a f_l \quad (22.2)$$

$$\varepsilon_{cc} = \varepsilon_{co}\left(1.5 + 12\kappa_b \dfrac{f_l}{f_{co}} \left(\dfrac{\varepsilon_{fe}}{\varepsilon_{co}}\right)^{0.45}\right) \quad (22.3)$$

The maximum lateral confining pressure (f_l) is calculated using Eq. 22.4. where, E_f, n, t_f, and ε_{fe} are FRP elastic modulus, number of FRP plies, the nominal thickness of one FRP ply and effective FRP strain, respectively; and D is the diameter of the column. Effective strain of FRP (ε_{fe}) represents hoop strain on FRP confinement attained at the instance of assumed failure, which has been observed to be significantly lower than the FRP rupture tensile strain (ε_{fu}) obtained from direct tension tests. In order to account this strain reduction, design documents generally define a strain efficiency factor (κ_e). Although many values and expressions have been proposed for strain efficiency factor depending on FRP materials and experimental conditions, ACI 440.2R (2017) proposes the value of 0.55. In addition to reduction of FRP rupture strain with the efficiency factor, maximum FRP effective strain is also limited to 0.004 for seismic loading cases, to ensure the shear integrity of the concrete. Equation 22.4 is directly implemented for columns with circular sections. For prismatic columns with noncircular cross sections, the maximum confining pressure of equivalent circular cross-section is considered with an equivalent diameter (D) that is equal to diagonal dimension of the cross-section.

$$f_l = \dfrac{2E_f n t_f \varepsilon_{fe}}{D} \quad (22.4)$$

The reduction of confinement efficiency due to non-uniform confining pressure on square and rectangular cross sections is taken account with shape factors. These factors depend on the effectively confined area concept and are calculated by Eqs. 22.5 and 22.6. ACI 440.2R (2017) recommends use of a revised effective confinement area as proposed by Lam and Teng (2003b) which can be computed by Eq. 22.7 In this equation, b and h are width and height of the rectangular cross section, respectively; r_c is the rounding radius of corners, A_g is the gross cross-section area and ρ_s is the longitudinal reinforcement ratio. The sharp corners should be rounded to a minimum 13 mm radius to reduce stress concentration and voids between concrete and FRP jacket. Due to detrimental effect of square and rectangular cross sections on confinement action, unless experimental evidence is available, maximum cross-section aspect ratio (h/b) is limited to 1.5 and maximum side dimension should not exceed 900 mm.

$$\kappa_a = \frac{A_e}{A_c}\left(\frac{b}{h}\right)^2 \tag{22.5}$$

$$\kappa_b = \frac{A_e}{A_c}\left(\frac{h}{b}\right)^{0.5} \tag{22.6}$$

$$\frac{A_e}{A_c} = \frac{1 - \frac{\left[\left(\frac{b}{h}\right)(h-2r_c)^2 + \left(\frac{h}{b}\right)(b-2r_c)^2\right]}{3A_g} - \rho_s}{1 - \rho_s} \tag{22.7}$$

In seismic strengthening applications, complete FRP wrapping should be applied around the perimeter on plastic hinging region of the member with a length not less than the ACI 318 (2014) recommendation for special transverse reinforcement and plastic hinge length (L_p) which is obtained by Eq. 22.8. In this equation, d_{bl} and f_y are the diameter and yield strength of longitudinal reinforcement of column, respectively, and g is the gap at the end of wrapping which is between FRP jacket and adjacent structural member. The maximum length of the gap is limited to 50.8 mm.

$$L_p = g + 0.044 f_y d_{bl} \tag{22.8}$$

22.2.4 CNR-DT 200 R1 (2013)

National Research Council of Italy (CNR) published first version of CNR-DT 200 R1 in 2004 for strengthening design of existing structures with externally bonded FRP and the last version of design document was released in 2013. FRP confined concrete model with parabolic and linear branch is proposed by CNR-DT 200 R1 (2013) for axial load capacity and ductility enhancement. The mathematical expressions of FRP confined concrete is given as follows;

$$\frac{f_c}{f_{co}} = \begin{cases} \dfrac{a\varepsilon_c}{\varepsilon_{co}} - \left(\dfrac{\varepsilon_c}{\varepsilon_{co}}\right)^2 & 0 \leq \varepsilon_c \leq \varepsilon_{co} \\ 1 + b\dfrac{\varepsilon_c}{\varepsilon_{co}} & \varepsilon_{co} \leq \varepsilon_c \leq \varepsilon_{cc} \end{cases} \tag{22.9.a}$$

$$a = 1 + \frac{f_{co} + E_2\varepsilon_{co}}{f_{co}} \tag{22.9.b}$$

$$b = \frac{f_{co} + E_2\varepsilon_{co}}{f_{co}} - 1 \tag{22.9.c}$$

$$E_2 = \frac{f_{cc} - f_{co}}{\varepsilon_{cc}} \tag{22.9.d}$$

The ultimate compressive strength and strain values of confined concrete model are calculated by Eqs. 22.10 and 22.11, respectively.

$$f_{cc} = f_{co} + 2.6 f_{co}\left(\frac{f_l}{f_{co}}\right)^{2/3} \tag{22.10}$$

$$\varepsilon_{cc} = 0.0035 + 0.015\sqrt{\frac{f_l}{f_{co}}} \tag{22.11}$$

The effective lateral confining pressure (f_l) is computed by Eq. 22.12 with the addition of geometric strengthening ratio (ρ_f) which is also known as volumetric ratio of FRP confinement. This ratio can be calculated for circular and rectangular cross sections by Eq. 22.13.a and 22.13.b, respectively. In these equations, b_f is the FRP stripe width and s_f is the center-to-center spacing between stripes for discontinued FRP wrapping, which is considerably less effective than continuous wrapping and not recommended by ACI 440.2R (2017) for seismic strengthening. The confining pressure (f_l) should not be less than $0.05 f_{co}$ to ensure ascending post-peak behavior. The effective FRP strain (ε_{fe}) for ductility enhancement with FRP jacketing can be computed by Eq. 22.14, where η_a is the environmental conversion factor which depends on exposure condition and FRP type, γ_f is the partial safety factor that is 1.10 for confinement action. Additionally, effective strain of FRP should be less than 0.6 times the FRP ultimate tension strain (ε_{fu}).

$$f_l = \frac{1}{2}\rho_f \kappa_e E_f \varepsilon_{fe} \tag{22.12}$$

$$\rho_f = \frac{4t_f b_f}{D s_f} \tag{22.13.a}$$

$$\rho_f = \frac{2t_f(b+h)b_f}{(bh)s_f} \tag{22.13.b}$$

$$\varepsilon_{fe} = \eta_a \frac{\varepsilon_{fu}}{\gamma_f} \tag{22.14}$$

Shape efficiency factors are used in Eq. 22.15 for calculating lateral confining pressure efficiency factor (κ_e). Not only the efficiency of cross-section geometry (κ_H)

is taken into account, but also the effect of vertical discontinuity of confinement with FRP stripes (κ_v) and spirally wrapped FRP confinement (κ_α) is considered. When the plastic hinging region is completely wrapped by FRP, as highly recommended for seismic retrofitting, (κ_v) and (κ_α) factors are equal to 1.0. For square and rectangular sections, the efficiency factor depends on the cross-section shape and (κ_H) is calculated by Eq. 22.16, while it can be taken 1.0 for circular cross-sections. The cross-section aspect ratio (h/b) is limited to 2.0 and maximum side length is considered as 900 mm. The minimum limit of corner rounding radius (r_c) is considered as 20 mm by CNR-DT 200 R1 (2013).

$$\kappa_e = \kappa_H \kappa_v \kappa_\alpha \tag{22.15}$$

$$\kappa_H = 1 - \frac{(b - 2r_c)^2 + (h - 2r_c)^2}{3A_g} \tag{22.16}$$

22.2.5 TSDC (2007)

Turkish Seismic Design Code (TSDC 2007) covers the recommendations for assessment and design rules for retrofitting of existing buildings. Axial strength and ductility enhancement with FRP jacketing is considered by using a bilinear axial stress-strain relationship for FRP confined concrete model. The model can be represented mathematically as given in the following;

$$f_c = \begin{cases} \dfrac{f_{co}}{\varepsilon_{co}} \varepsilon_c & 0 \leq \varepsilon_c \leq \varepsilon_{co} \\ f_{co} + \left(\dfrac{f_{cc} - f_{co}}{\varepsilon_{cc} - \varepsilon_{co}}\right)(\varepsilon_c - \varepsilon_{co}) & \varepsilon_{co} \leq \varepsilon_c \leq \varepsilon_{cc} \end{cases} \tag{22.17}$$

The transition point of bilinear model is the unconfined concrete strength (f_{co}) and corresponding strain (ε_{co}) which can be assumed as 0.002. FRP confined concrete strength (f_{cc}) and ultimate strain (ε_{ccu}) are calculated by using Eqs. 22.18 and 22.19 which is based on the experimental works carried out by Ilki and Kumbasar (2002, 2003) and Ilki et al. (2004) on low and medium strength concrete prisms. The effective lateral confining pressure (f_l) is predicted by Eq. 22.12 similar to CNR-DT 200 R1 (2013). In TSDC (2007), the FRP effective strain (ε_{fe}) is considered as the minimum of 0.004 and $0.5\varepsilon_{fu}$. The effective strain limit generally governs by 0.004 in calculations (ε_{fe}) due to high rupture strains of available FRP materials, and lead to extremely conservative analytical predictions. Only the limit of $0.5\varepsilon_{fu}$ is proposed for the next version of TSDC which is expected to be release soon.

$$f_{cc} = f_{co}\left[1 + 2.4\frac{f_l}{f_{co}}\right] \quad (22.18)$$

$$\varepsilon_{cc} = 0.002\left[1 + 15\left(\frac{f_l}{f_{co}}\right)^{0.75}\right] \quad (22.19)$$

The shape efficiency factor (κ_e) in Eq. 22.12 is computed by Eq. 22.20 for different cross-section shapes. Additionally, upper limit of 2.0 for cross section aspect ratio (h/b) and minimum corner rounding radius (r_c) of 30 mm are also recommended by TSDC (2007). The greater aspect ratio (2.5) limit is also proposed for new version of TSDC based on the recent studies on full-scale RC columns with high aspect ratio (Realfonzo and Napoli 2012; Ghatte et al. 2016).

$$\kappa_e = \begin{cases} 1 & \text{circular sections} \\ \left(\dfrac{b}{h}\right) & \text{square sections} \\ 1 - \dfrac{(b-2r_c)^2 + (h-2r_c)^2}{3bh} & \text{rectangular sections} \end{cases} \quad (22.20)$$

22.3 Experimental Studies and Comparison with Analytical Prediction

22.3.1 Full-Scale Tests

Retrofitting of columns with FRP jacketing for improving seismic performance is a well investigated technique through numerous member level experimental studies. Design recommendations and theoretical approaches for nonlinear modeling are mainly based on these member level laboratory tests which are performed under controlled and idealized conditions. Therefore, full-scale building tests may play a major role as benchmark tests where the effectiveness of the retrofit schemes and design tools is controlled and demonstrated under realistic conditions. Additionally, tracking of the damage propagation in the structural system and observation of deformation development in elements for different loading levels, provides a valuable data for assessment of the seismic behavior of RC buildings. Due to challenges of realizing full-scale experiments, only limited number of studies (Balsamo et al. 2005; Ludovico et al. 2008; Garcia et al. 2014) could have been conducted up to now. In this part of the study, two full-scale tests performed very recently by the authors (Tore et al. 2017) is reported and experimental results are compared to analytical predictions made using the FRP confined concrete models available in design documents.

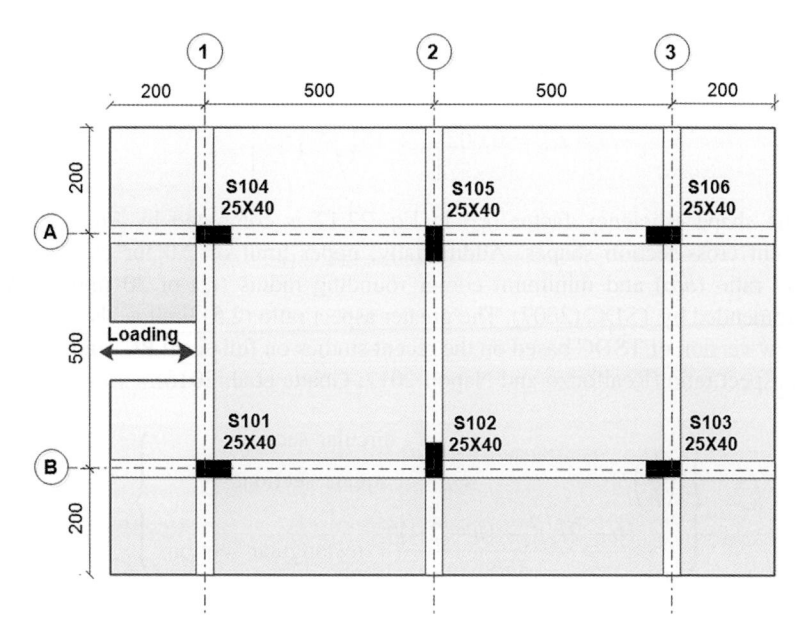

Fig. 22.4 Plan view of the buildings (dimensions are in cm)

22.3.2 Properties of the Test Buildings

Two three-story identical buildings have been designed to represent the sub-standard RC buildings which had been built in 1990's in Turkey. Each building had two bays in the loading direction and one in the perpendicular with a 500 cm span length and 285 cm story height (Fig. 22.4). Lateral load resisting frames consisted of columns with 25 cm x 40 cm and beams with 25 cm × 50 cm cross-section dimensions. All columns were reinforced with eight 14 mm longitudinal bars and 8 mm 90-degree end-hooked stirrups with 32 cm spacing. Deformed longitudinal bars of columns were continued from foundation to first story level. Lap splice length for column longitudinal bars was 60 times the diameter (60Φ). The beams were designed stronger than the columns as generally observed for heavily damaged or collapsed buildings that experienced major earthquakes. The transverse reinforcement of beams were spaced at 15 cm along the beam length for ensuring adequate shear capacity that was required to resist the weight of additional concrete masses. The 25-cm thick slab and supporting beams were elongated as 200 cm long overhangs at each side of the building except the side that actuators were mounted.

Average concrete compression strength was approximately 18 MPa as obtained from uniaxial compression tests on 100 mm × 200 mm cylinders. The average yield strength of the deformed longitudinal and transverse reinforcing bars were 447 and 444 MPa, respectively.

Infill walls were constructed at the end of the overhangs of the second and the third stories with a 5 cm gap between infill walls and RC members, so that the infill

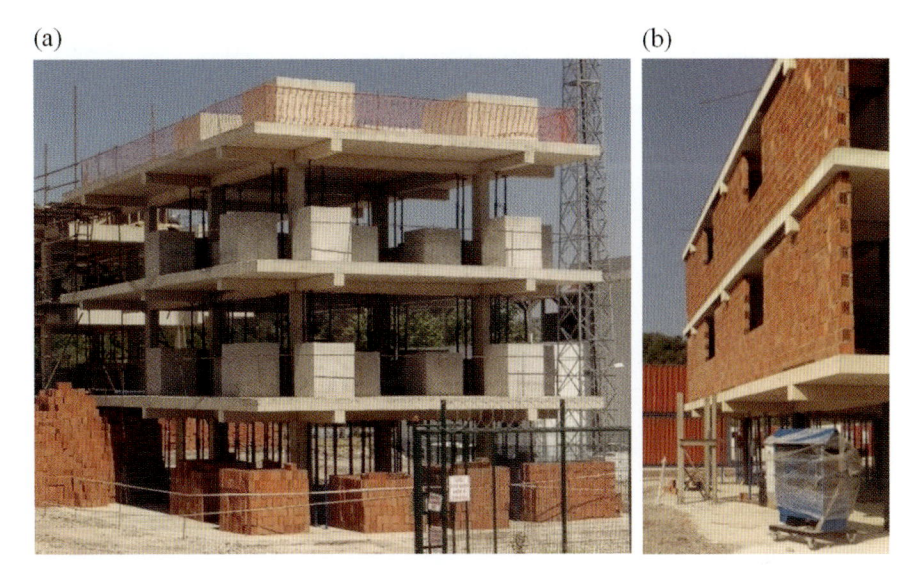

Fig. 22.5 Views of test buildings, concrete weight blocks and infill walls during construction process (Tore et al. 2017)

walls only contributed as additional mass elements (Fig. 22.5). In addition to the thick slab, concrete blocks weighing between 21.6 and 28.8 kN were cast at the upper stories for increasing the axial load levels of the first story columns (Fig. 22.5a). At the lowest story, column axial load ratios without reinforcement contribution to axial capacity of columns were approximately 25% for S101-S104, 45% for S102-S105 and S103-S106 columns.

22.3.3 Retrofitting Strategies

The test buildings had a critical ductility problem in particular for the sub-standard RC columns. The critical plastic deformations were estimated to occur at the first story columns, since the story mechanism was expected to occur under lateral loading due to weak column-strong beam configuration. In addition to the requirement of the ductility, columns were also vulnerable under shear because of insufficient transverse reinforcement. According to Elwood and Moehle (2005) and Yoshimura et al. (2004) models, axial collapse of the flexural-shear critical RC columns follows lateral strength loss caused by shear failure after the flexural yielding.

Carbon Fiber Reinforced Polymer (CFRP) composite sheets were used for external FRP jacketing of the columns. Mechanical properties provided by the manufacturer of CFRP sheets were 4900 MPa for tensile strength, 240 GPa for

Fig. 22.6 Retrofitting scheme of first story columns

elastic modulus and 2.0% for ultimate tensile strain. Unit weight of the CFRP sheets were 1.8 g/cm^3 and thickness of one CFRP ply was 0.165 mm.

The applied retrofitting scheme targeted the enhancement of concrete strain capacity for improving ductility, and increasing shear capacity of columns to ensure flexure-dominated behavior. The first story column upper and lower ends were wrapped with 5 plies (Fig. 22.6) and the second story column end regions were wrapped with 3 plies of CFRP with 60 cm height. Additionally, two plies of CFRP sheets were wrapped to remaining height of columns for increasing shear capacity. Before jacketing with CFRP, firstly column surfaces were grinded and local surface defects were repaired to obtain smooth surfaces and then the corners of the columns were rounded to a radius of 30 mm.

Reference and retrofitted buildings were simultaneously subjected to lateral loading in two consecutive phases: quasi-static reversed cyclic loading and pushover loading. For this purpose, a steel reaction frame was built in between the two buildings. Totally six servo-controlled hydraulic actuators with 300 kN load and 800 mm displacement capacities were mounted on both sides of the reaction frame (Fig. 22.7). For each building, two actuators were attached at the second story and one was at the first story slab levels. During the test, ratio of the lateral loads applied to the first and second stories was kept constant as 0.5. Firstly, static cyclic lateral loading was applied incrementally from 0.125% to 0.9% first story drift ratio levels. Then, the actuators were demounted from the building and both buildings were pushed monotonically.

Fig. 22.7 Reaction frame and loading system (Tore et al. 2017)

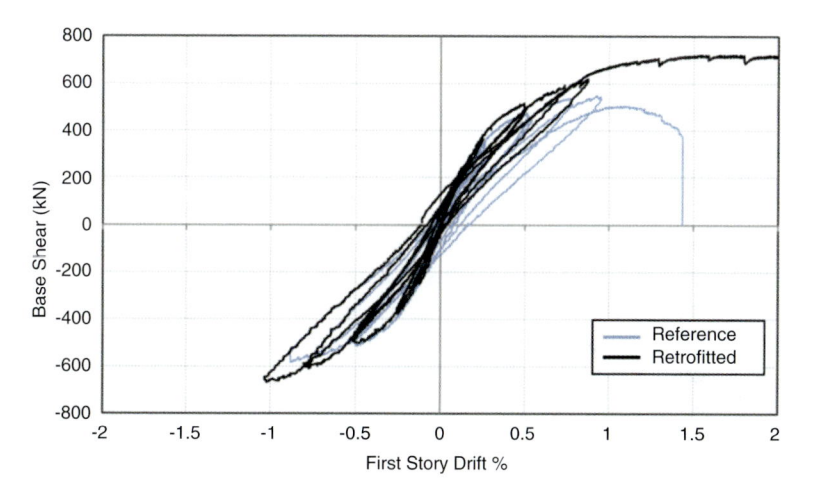

Fig. 22.8 Base shear-1st story drift responses up to collapse of reference building

22.3.4 Experimental Results

The base shear vs. first story drift ratio responses of both test buildings are presented in Fig. 22.8. During the test, damage formation of the reference building started with the flexural cracks at low drift ratio cycles. Then, inclined cracks which were clear signs of shear effects appeared on the columns. Reference building reached its maximum lateral strength at 0.9% drift cycle and quasi-static loading phase was terminated due to the observation of severe damage such as spalling of cover concrete and vertical cracks as an evidence of buckling of longitudinal reinforcements on S103-S106 columns. Most of the cracks on the S101-S104 columns were closed and deformations were concentrated on single shear cracks. Noticeable

Fig. 22.9 Collapsed reference building

widening of the shear crack has been observed just before the collapse of building which is an indication for flexural-shear behavior of columns. The lateral strength loss of the reference building started just after the 0.9% drift ratio in the pushover loading phase and dramatic degradation continue up to 1–1.45% first story drift ratio when the abrupt collapse was observed. The pancake type collapse occurred with the sequential brittle failure of the columns as seen in Fig. 22.9.

In the case of the retrofitted building, no significant damage could be observed during the first phase of the loading. The lateral load capacity of the retrofitted building was still maintained even at 1.45% drift ratio where the reference building collapsed. Moreover, the resisted lateral load continued to increase until 2% drift ratio, where first flexural cracks could be visible at the first story column bases. After that point, lateral strength gradually decreased while the drift ratio continued to increase. The pushover loading phase was resumed until approximately 15% drift ratio (Fig. 22.10) and was terminated due to stroke limitations of the actuators. Plastic deformations of FRP jacketed columns were concentrated on a few flexural cracks at the top and bottom ends of the first story columns (Fig. 22.11). Even at 15% drift, no partial or total collapse occurred. The lateral load capacity loss at high drifts was mainly due to second order effects.

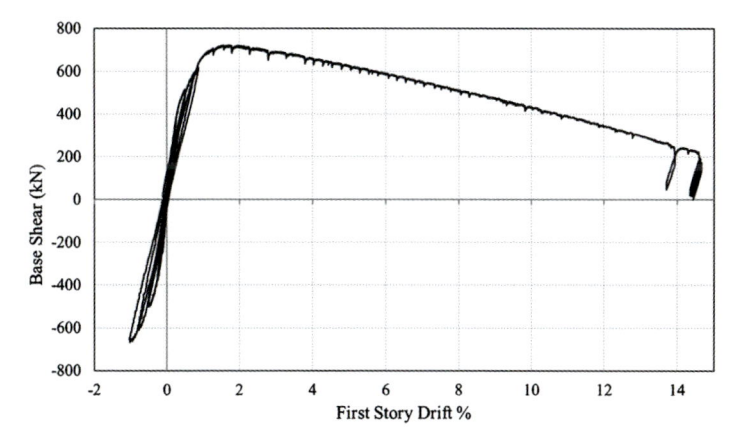

Fig. 22.10 Base shear-1st story drift response of the retrofitted building

Fig. 22.11 Deformed shape of the FRP retrofitted building

22.3.5 Nonlinear Behavior Prediction with Design Document Based FRP Confined Concrete Models

Experimental behavior of the retrofitted building was compared with the analytical predictions obtained through nonlinear pushover analyses which were conducted by using the SAP2000 v19 (2016) structural analysis software. Columns were modelled with nonlinear fiber type hinges located at the top and bottom ends of the elements. The beams were modelled with a conventional plastic hinge assumption. The remaining parts of the columns and beams, in between the assigned plastic hinges,

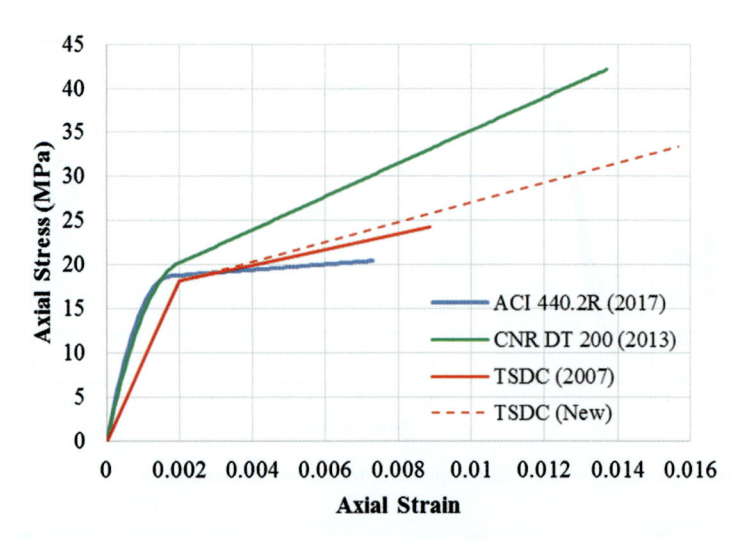

Fig. 22.12 Stress-strain models for FRP confined concrete (Wrapping with 5 plies)

were modelled by using elastic frame elements with cracked section properties. Rigid diaphragm behavior was considered at each story level. In addition to material nonlinearity, geometric nonlinearity was also considered. The distribution of lateral pushover loading in the nonlinear static analysis was assumed to be similar with the tests (P and 2P acting to the first and second stories, respectively).

Material stress-strain relationships were assigned to steel or concrete fibers of the plastic hinge sections. FRP confined concrete models recommended by ACI 440.2R (2017), CNR-DT 200 R1 (2013) and TSCD (2007) were used to obtain the axial stress-strain behavior of FRP confined concrete (Fig. 22.12). The dotted red line in the Fig. 22.12 represents the model of TSDC with a proposal for new version which is previously mentioned. A trilinear steel stress-strain relationship (Fig. 22.13) was assigned to longitudinal reinforcement fibers. Due to the effect of FRP jacketing on prevention of longitudinal bar buckling, stress-strain relationship of steel under compression was assumed to be similar to tension.

One other important parameter that may significantly affect the nonlinear analyses results is the assumption done for the plastic hinge length. Only the ACI 440.2R (2017) provided a particular plastic hinge length expression for FRP jacketed columns. For the analysis with FRP confined concrete model of TSDC (2007), the plastic hinge length was taken as half of the cross-section depth ($h/2$). Additionally, the plastic hinge length equation proposed by the Fédération Internationale du Béton (2012), which is originally given for RC members under cyclic loading, was used for analyses made considering CNR-DT 200 R1 (2013) FRP confinement model. Same plastic hinge assumption was used for FRP jacketed columns before by Biskinis and Fardis (2013). FRP confined concrete parameters of the considered models and considered plastic hinge lengths are given in Table 22.1. In this table, the first value at plastic hinge length column is given for S101-S104 and S103-S106, and

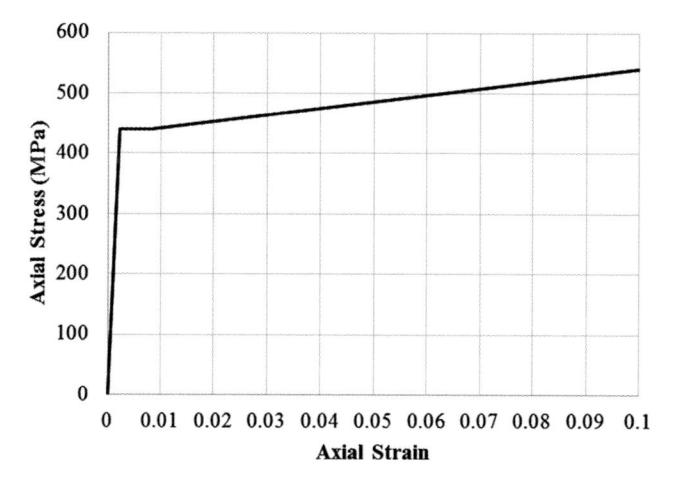

Fig. 22.13 Trilinear stress-strain model for reinforcing steel

Table 22.1 FRP confined concrete model parameters and plastic hinge lengths

Confined concrete model	Confined concrete strength (f_{cc}) (MPa)	Ultimate confined concrete strain (ε_{ccu})	Considered plastic hinge length (L_p) (cm)
ACI 440.2R (2017)	20.45	0.0073	28.6–28.6 cm
CNR-Dt 200 R1 (2013)	42.14	0.0137	15.8–12.8 cm
TSDC (2007)	24.25 (33.41)[a]	0.0089 (0.0156)[a]	20.0–12.5 cm

[a] Strain value obtained model with ($0.5\varepsilon_{fu}$) effective FRP strain limit

the second one is given for S102-S105 due to the different cross section heights at the direction of bending.

Comparisons of the experimental response with the analytical behavior obtained through nonlinear static analyses are given in Fig. 22.14. The curves are ceased when the extreme compression fibers of the columns reach the assigned FRP confined concrete ultimate strain capacities according to the recommendation of the considered document. The initial stiffness and behavior of the retrofitted building at the quasi-static cyclic loading phase was mostly well predicted from analyses with FRP confined concrete models. ACI 440.2R (2017) and TSDC (2007) slightly underestimated the lateral load capacity of the building. The ultimate drift ratios obtained from analysis are approximately 1.12% for analyzed models with ACI 440.2R (2017) and TSDC (2007) approaches, and 1.52% for CNR-DT 200 R1 (2013) approach. The ultimate displacement values of all analysis are conservative when compared with the experimental response due to extraordinary ductile behavior of the test building. The extremely conservative ultimate drift ratio value obtained from the analyses with ACI 440.2R (2017) and TSDC (2007) model is caused by the considerably low ultimate confined concrete strain which is sourced from the limitation of effective FRP strain to 0.004. As noted previously, this limitation is

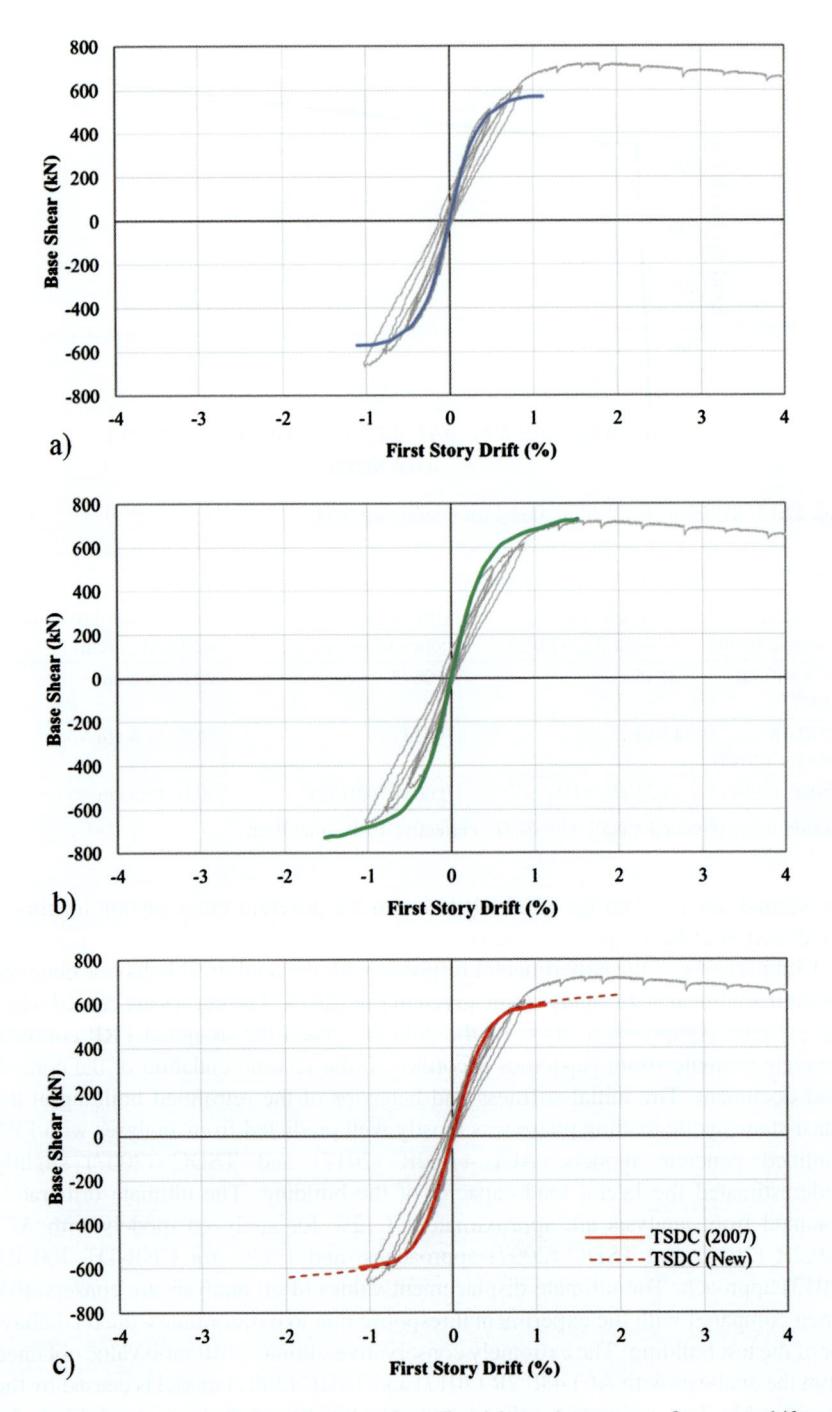

Fig. 22.14 Comparison of experimental and predicted base shear versus first story drift ratio curves. FRP confined concrete model with (**a**) ACI 440.2R (2017), (**b**) CNR-DT 200 R1 (2013) and (**c**) TSDC (2007) and revised version

for ensuring the shear integrity of concrete, however, during the retrofitted building test no shear related deformation could be observed. When the analyses with TSDC (2007) model are revised using ultimate FRP effective strain limit as $0.5\varepsilon_{fu}$, an acceptable and conservative ultimate drift ratio (2.02%) is obtained (dotted line at Fig. 22.14c).

22.4 Conclusions

The efficiency of FRP jacketing for improving ductility and shear capacity of sub-standard columns has been clearly demonstrated through an experimental study executed on two full-scale three story reinforced concrete frame structures. While the as-built reference building suddenly collapsed at just over 1% first story drift ratio, the retrofitted building sustained lateral drifts of approximately 15% without collapsing.

Nonlinear static analyses were conducted to predict the seismic behavior of the tested sub-standard buildings. For the retrofitted building, FRP confined concrete models given by ACI 440.2R (2017), CNR-DT 200 R1 (2013) and TSDC (2007) were used. The envelopes of the cyclic part of the experimental behavior were well predicted through nonlinear analyses. On the other hand, ACI 440.2R (2017) and TSDC (2007) approaches led to quite conservative results for ultimate drift ratio with respect to CNR-DT 200 R1 (2013) model. This is mainly caused by the restriction of effective FRP strain to 0.004. Reasonable and sufficiently conservative results were achieved through revision of analyses, which considered the effective FRP strain as $0.5\varepsilon_{fu}$.

Acknowledgments Authors are thankful to DowAksa Advanced Composites Holding B.V. and Governorship of Yalova for various supports provided for experimental studies. The contributions of Ilgaz Doğan (DowAksa), Çağlar Göksu Akkaya (PhD, ITU), Pınar İnci (PhD Cand., ITU), Ergün Binbir (PhD Cand., ITU), Ali Osman Ateş (PhD Cand., ITU), Ali Naki Şanver (MSc Cand., Rise Eng.), Çağlar Üstün (MSc Cand., RISE Eng.), Duygu Çakır (Civil Eng., Rise Eng.), Gökhan Sarı (Undergrad., Balıkesir University), Emin Amini (Undergrad., Balıkesir University), Tamer Şahna (Undergrad., Balıkesir University), Oğuzhan Sözer (Undergrad., Balıkesir University), Berkay Aldırmaz (MSc Cand., ITU), Ömer Faruk Halıcı (MSc, ITU), Mehmet Aksa (Undergrad., Kultur Uni.) are also gratefully acknowledged.

References

Abbasnia R, Hosseinpour F, Rostamian M, Ziaadiny H (2012) Effect of corner radius on stress-strain behavior of FRP confined prisms under axial cyclic compression. Eng Struct 40:529–535. https://doi.org/10.1016/j.engstruct.2012.03.020

ACI Committee 318 (2014) Building code requirements for structural concrete ACI 318-14. American Concrete Institute, Farmington Hills

ACI Committee 440 (2017) Guide for the design and construction of externally bonded FRP systems for strengthening concrete structures ACI 440.2R-17. American Concrete Institute, Farmington Hills

Akguzel U, Pampanin S (2012) Assessment and design procedure for the seismic retrofit of reinforced concrete beam-column joints using FRP composite materials. J Compos Constr 16:21–34. https://doi.org/10.1061/(ASCE)CC.1943-5614.0000242

Antonopoulos CP, Triantafillou TC (2003) Experimental investigation of FRP-strengthened RC beam-column joints. J Compos Constr 7:39–50

Bai YL, Dai JG, Teng JG (2017) Buckling of steel reinforcing bars in FRP-confined RC columns: an experimental study. Constr Build Mater 140:403–415. https://doi.org/10.1016/j.conbuildmat.2017.02.149

Balsamo A, Colombo A, Manfredi G et al (2005) Seismic behavior of a full-scale RC frame repaired using CFRP laminates. Eng Struct 27:769–780. https://doi.org/10.1016/j.engstruct.2005.01.002

Barros JAO, Dias SJE, Lima JLT (2007) Efficacy of CFRP-based techniques for the flexural and shear strengthening of concrete beams. Cem Concr Compos 29:203–217. https://doi.org/10.1016/j.cemconcomp.2006.09.001

Barros JAO, Varma RK, Sena-Cruz JM, Azevedo AFM (2008) Near surface mounted CFRP strips for the flexural strengthening of RC columns: experimental and numerical research. Eng Struct 30:3412–3425. https://doi.org/10.1016/j.engstruct.2008.05.019

Binici B, Ozcebe G, Ozcelik R (2007) Analysis and design of FRP composites for seismic retrofit of infill walls in reinforced concrete frames. Compos Part B Eng 38:575–583. https://doi.org/10.1016/j.compositesb.2006.08.007

Bisby L, Ranger M (2010) Axial-flexural interaction in circular FRP-confined reinforced concrete columns. Constr Build Mater 24:1672–1681. https://doi.org/10.1016/j.conbuildmat.2010.02.024

Biskinis D, Fardis MN (2013) Models for FRP-wrapped rectangular RC columns with continuous or lap-spliced bars under cyclic lateral loading. Eng Struct 57:199–212. https://doi.org/10.1016/j.engstruct.2013.09.021

Bournas DA, Triantafillou TC (2011) Bar buckling in RC columns confined with composite materials. J Compos Constr 15:393–403. https://doi.org/10.1061/(ASCE)CC.1943-5614.0000180

Bournas DA, Triantafillou TC (2009) Flexural strengthening of reinforced concrete columns with near-surface-mounted FRP or stainless steel. ACI Struct J 106:495–505. https://doi.org/10.14359/56615

Bousias S, Spathis A-L, Fardis MN (2007) Seismic retrofitting of columns with lap spliced smooth bars through FRP or concrete jackets. J Earthq Eng 11(5):653–674

Campione G, Miraglia N (2003) Strength and strain capacities of concrete compression members reinforced with FRP. Cem Concr Compos 25:31–41. https://doi.org/10.1016/S0958-9465(01)00048-8

Chastre C, Silva MAG (2010) Monotonic axial behavior and modelling of RC circular columns confined with CFRP. Eng Struct 32:2268–2277. https://doi.org/10.1016/j.engstruct.2010.04.001

Chen GM, Teng JG, Chen JF, Rosenboom OA (2010) Interaction between steel stirrups andshear-srengthening FRP strips in RC beams. J Compos Constr 14:498–509. https://doi.org/10.1061/(ASCE)CC.1943-5614.0000120

CNR-DT 200 R1 (2013) Guide for the design and construction of externally bonded FRP systems for strengthening existing structures. CNR – Advis Comm Tech Recomm Constr CNR-DT 200 R1/2013 154

Computers and Structures Inc (2016) Integrated software for structural analysis. SAP2000 v19, California

Cosgun C, Comert M, Demir C, Ilki A (2012) FRP retrofit of a full-scale 3D RC frame. In: Proceedings 6th international conference FRP composites civil engineering CICE 2012 1–8

De Lorenzis L, Teng JG (2007) Near-surface mounted FRP reinforcement: an emerging technique for strengthening structures. Compos Part B Eng 38:119–143. https://doi.org/10.1016/j.compositesb.2006.08.003

De Luca A, Nardone F, Matta F et al (2011) Structural evaluation of full-scale FRP-confined reinforced concrete columns. J Compos Constr 15:112–123. https://doi.org/10.1061/(ASCE)CC.1943-5614.0000152

Demir C, Darilmaz K, Ilki A (2015) Cyclic stress-strain relationships of FRP confined concrete members. Arab J Sci Eng 40:363–379. https://doi.org/10.1007/s13369-014-1517-5

Eid R, Paultre P (2007) Plasticity-based model for circular concrete columns confined with fibre-composite sheets. Eng Struct 29:3301–3311. https://doi.org/10.1016/j.engstruct.2007.09.005

Eid R, Paultre P (2008) Analytical model for FRP-confined circular reinforced concrete columns. J Compos Constr 12:541–552. https://doi.org/10.1061/(ASCE)1090-0268(2008)12:5(541)

Eid R, Roy N, Paultre P (2009) Normal- and high-strength concrete circular elements wrapped with FRP composites. J Compos Constr 13:113–124. https://doi.org/10.1061/(ASCE)1090-0268(2009)13:2(113)

El-maaddawy T, El-dieb AS (2011) Near-surface-mounted composite system for repair and strengthening of reinforced concrete columns subjected to axial load and biaxial bending. J Compos Constr 67:602–614. https://doi.org/10.1061/(ASCE)CC.1943-5614.0000181

ElGawady M, Endeshaw M, McLean D, Sack R (2010) Retrofitting of rectangular columns with deficient lap splices. J Compos Constr 14:22–35. https://doi.org/10.1061/(asce)cc.1943-5614.0000047

Elwood KJ, Moehle JP (2005) Axial capacity model for shear-damaged columns. ACI Struct J 102(4):578–587

Fardis MN, Khalili HH (1982) FRP-encased concrete as a structural material. Mag Concr Res 34:191–202. https://doi.org/1982.34.121.191

Fédération Internationale du Béton (FIB) (2012) Model code 2010—final draft, vol 1, Bulletin 65, and vol 2, Bulletin 66. Lausanne, Switzerland

Fédération Internationale du Béton (FIB) Bulletin 14 (2001) Externally bonded FRP reinforcement for RC structures. Lausanne, Switzerland

Galal K, Arafa A, Ghobarah A (2005) Retrofit of RC square short columns. Eng Struct 27:801–813. https://doi.org/10.1016/j.engstruct.2005.01.003

Garcia R, Hajirasouliha I, Guadagnini M, Helal Y, Jemaa Y, Pilakoutas K, Mongabure P, Chrysostomou C, Kyriakides N, Ilki A, Budescu M, Taranu N, Ciupala MA, Torres L, Saiidi M (2014) Full-scale shaking table tests on a substandard RC building repaired and strengthened with post-tensioned metal straps. J Earthq Eng 18(2):187–213. https://doi.org/10.1080/13632469.2013.847874

GB 50608 (2010) Technical code for infrastructure application of FRP composites. China Metallurgical Construction Research Institute Co.

Ghatte HF, Comert M, Demir C, Ilki A (2016) Evaluation of FRP confinement models for substandard rectangular RC columns based on full-scale reversed cyclic lateral loading tests in strong and weak directions. Polymers (Basel) 8:1–24. https://doi.org/10.3390/polym8090323

Giamundo V, Lignola GP, Fabbrocino F et al (2017) Influence of FRP wrapping on reinforcement performances at lap splice regions in RC columns. Compos Part B Eng 116:313–324. https://doi.org/10.1016/j.compositesb.2016.10.069

Giamundo V, Lignola GP, Prota A, Manfredi G (2014) Analytical evaluation of FRP wrapping effectiveness in restraining reinforcement bar buckling. J Struct Eng 140:4014043. https://doi.org/10.1061/(ASCE)ST.1943-541X.0000985

Hadi MNS, Widiarsa IBR (2012) Axial and flexural performance of square RC columns wrapped with CFRP under eccentric loading. J Compos Constr 16:640–649. https://doi.org/10.1061/(ASCE)CC.1943-5614.0000301

Hamad BS, Rteil A, Salwan BR, Soudki K (2004) Behavior of bond-critical regions wrapped with fiber-reinforced polymer sheets in normal and high-strength concrete. J Compos Constr 8:248–257. https://doi.org/10.1061/(ASCE)1090-0268(2004)8:3(248)

Harajli MH (2006) Axial stress-strain relationship for FRP confined circular and rectangular concrete columns. Cem Concr Compos 28:938–948. https://doi.org/10.1016/j.cemconcomp.2006.07.005

Harajli MH, Khalil Z (2008) Seismic FRP retrofit of bond-critical regions in circular RC columns: validation of proposed design methods. ACI Struct J 105:760–769

Hu B, Wang JG, Li GQ (2011) Numerical simulation and strength models of FRP-wrapped reinforced concrete columns under eccentric loading. Constr Build Mater 25:2751–2763. https://doi.org/10.1016/j.conbuildmat.2010.12.036

Ilki A, Kumbasar N (2002) Behavior of damaged and undamaged concrete strengthened by carbon fiber composite sheets. Struct Eng Mech 13:75–90. https://doi.org/10.12989/sem.2002.13.1.075

Ilki A, Kumbasar N (2003) Compressive behaviour of carbon fiber composite jacketed concrete with circular and non-circular cross-sections. J Earthq Eng 7(3):381–406

Ilki A, Kumbasar N, Koc V (2004) Low strength concrete members externally confined with FRP sheets. Struct Eng Mech 18(2):167–194

Ilki A, Peker O, Karamuk E et al (2008) FRP retrofit of low and medium strength circular and rectangular reinforced concrete columns. J Mater Civ Eng 20:169–188. https://doi.org/10.1061/(ASCE)0899-1561(2008)20:2(169)

Ilki A, Demir C, Bedirhanoglu I, Kumbasar N (2009) Seismic retrofit of brittle and low strength RC columns using fiber reinforced polymer and cementitious composites. Adv Struct Eng 12:325–347. https://doi.org/10.1260/136943309788708356

Ilki A, Bedirhanoglu I, Kumbasar N (2010) Behavior of FRP retrofitted joints built with plain bars and low-strength concrete. J Compos Constr:312–326. https://doi.org/10.1061/(ASCE)CC.1943-5614.0000156

ISIS (2001) Design manual no. 3: reinforcing concrete structures with fibre reinforced polymers. ISIS (Canadian network of centres of excellence on intelligent sensing for innovative structures), Winnipeg

Jiang T, Teng JG (2007) Analysis-oriented stress-strain models for FRP-confined concrete. Eng Struct 29:2968–2986. https://doi.org/10.1016/j.engstruct.2007.01.010

JSCE (2001) Recommendations for the upgrading of concrete structures with use of continuous fiber sheets. Concrete Engineering Series 41. Jpn Soc Civ Eng, Tokyo

Karadogan F, Yusel E, Ilki A (2003) Structural behaviour of ordinary RC bare and brittle partitioned frames with and without lap splice deficiency. In: Wasti ST, Ozcebe G (eds) seismic assessment and rehabilitation of existing buildings, NATO science series (Series IV: earth and environmental sciences). Springer, Dordrecht, pp 335–356

Katsumata H, Kobatake Y, Takeda T (1988) A study on strengthening with carbon fiber for earthquake-resistant capacity of existing reinforced concrete columns. In: Proceedings 9th world conference on earthquake

Kaya E, Kütan C, Sheikh S, İlki A (2017) Flexural retrofit of support regions of reinforced concrete beams with anchored FRP ropes using NSM and ETS methods under reversed cyclic loading. J Compos Constr 21(1):04016072

Lam L, Teng JG (2003a) Design-oriented stress–strain model for FRP-confined concrete. Constr Build Mater 17:471–489. https://doi.org/10.1016/S0950-0618(03)00045-X

Lam L, Teng JG (2003b) Design-oriented stress–strain model for FRP-confined concrete in rectangular columns. J Reinf Plast Compos 22:1149–1186pp. https://doi.org/10.1177/073168403035429

Lam L, Teng JG (2009) Stress-strain model for FRP-confined concrete under cyclic axial compression. Eng Struct 31:308–321. https://doi.org/10.1016/j.engstruct.2008.08.014

Lee JY, Yi CK, Jeong HS et al (2010) Compressive response of concrete confined with steel spirals and FRP composites. J Compos Mater 44:481–504. https://doi.org/10.1177/0021998309347568

Li B, Kai Q (2011) Seismic behavior of reinforced concrete interior beam-wide column joints repaired using FRP. J Compos Constr 15:327–338. https://doi.org/10.1061/(ASCE)CC.1943-5614.0000163

Di Ludovico M, Prota A, Manfredi G, Cosenza E (2008) Seismic strengthening of an underdesigned RC structure with FRP. Earthquake Engng Struct Dyn 37:141–162. https://doi.org/10.1002/eqe.749

Mander JB, Priestley MJN, Park R (1988) Theoretical stress–strain model for confined concrete. J Struct Eng 114(8):1804–1826. https://doi.org/10.1061/(ASCE)0733-9445(1988)114:8(1804)

Mirmiran A, Shahawy M, Samaan M et al (1998) Effect of column parameters on FRP-confined concrete. J Compos Constr 2:175–185

Mosallam AS, Bayraktar A, Elmikawi M et al (2013) Polymer composites in construction: an overview. SOJ Mater Sci Eng 2(1):25

Napoli A, Nunziata B, Realfonzo R (2011) Cyclic behaviour of FRP confined RC rectangular columns with high aspect ratio. In: CICE 2010 – the 5th international conference on FRP composites in civil engineering, Beijing, China

Nisticò N, Monti G (2013) RC square sections confined by FRP: analytical prediction of peak strength. Compos Part B Eng 45:127–137. https://doi.org/10.1016/j.compositesb.2012.09.041

Ozbakkaloglu T, Lim JC (2013) Axial compressive behavior of FRP-confined concrete: experimental test database and a new design-oriented model. Compos Part B Eng 55:607–634. https://doi.org/10.1016/j.compositesb.2013.07.025

Ozcan O, Binici B, Ozcebe G (2010) Seismic strengthening of rectangular reinforced concrete columns using fiber reinforced polymers. Eng Struct 32:964–973. https://doi.org/10.1016/j.engstruct.2009.12.021

Ozcan O, Binici B, Ozcebe G (2008) Improving seismic performance of deficient reinforced concrete columns using carbon fiber-reinforced polymers. Eng Struct 30:1632–1646. https://doi.org/10.1016/j.engstruct.2007.10.013

Ozsayin B, Yilmaz E, Ispir M et al (2011) Characteristics of CFRP retrofitted hollow brick infill walls of reinforced concrete frames. Constr Build Mater 25:4017–4024. https://doi.org/10.1016/j.conbuildmat.2011.04.036

Pellegrino C, Modena C (2007) Fiber-reinforced polymer shear strengthening of reinforced concrete beams: experimental study and analytical modeling. ACI Struct J 103(5):720–728

Pellegrino C, Modena C (2010) Analytical model for FRP confinement of concrete columns with and without internal steel reinforcement. J Compos Constr 14:693–705. https://doi.org/10.1061/(Asce)Cc.1943-5614.0000127

Realfonzo R, Napoli A (2012) Results from cyclic tests on high aspect ratio RC columns strengthened with FRP systems. Constr Build Mater 37:606–620. https://doi.org/10.1016/j.conbuildmat.2012.07.065

Rochette P, Labossière P (2000) Axial testing of rectangular column models confined with composites. J Compos Constr 4:129–136

S806–12 (2012) Design and construction of building structures with fibre-reinforced polymers. CSA, Toronto

Saadatmanesh H, Ehsani MR, Jin L (1996) Seismic strengthening of circular bridge pier models with fiber composites. ACI Struct J 93:639–647

Samaan M, Mirmiran A, Shahawy M (1998) Model of concrete confined by fiber composites. J Struct Eng 124:1025–1031. https://doi.org/10.1257/0002828041464551

Sezen H (2012) Repair and strengthening of reinforced concrete beam column joints with fiber reinforced polymer composites. J Compos Constr 16:499–506. https://doi.org/10.1061/(ASCE)CC.1943-5614.0000290

Shahawy M, Mirmiran A, Beitelman T (2000) Tests and modeling of carbon-wrapped concrete columns. Compos Part B Eng 31:471–480. https://doi.org/10.1016/S1359-8368(00)00021-4

Shao Y, Zhu Z, Mirmiran A (2006) Cyclic modeling of FRP-confined concrete with improved ductility. Cem Concr Compos 28:959–968. https://doi.org/10.1016/j.cemconcomp.2006.07.009

Sharaky IA, Torres L, Comas J, Barris C (2014) Flexural response of reinforced concrete (RC) beams strengthened with near surface mounted (NSM) fibre reinforced polymer (FRP) bars. Compos Struct 109:8–22. https://doi.org/10.1016/j.compstruct.2013.10.051

Spoelstra BMR, Monti G (1999) FRP-confined concrete model. J Compos Constr 3:143–150

Tastani SP, Pantazopoulou SJ, Zdoumba D et al (2006) Limitations of FRP jacketing in confining old-type reinforced concrete members in axial compression. J Compos Constr 10:13–25. https://doi.org/10.1061/(Asce)1090-0268(2006)10:1(13)

Teng JG, Lam L (2004) Behavior and modeling of fiber reinforced polymer-confined concrete. J Struct Eng 130(11):1713–1723

Teng JG, Lam L, Lin G et al (2016) Numerical simulation of FRP-jacketed RC columns subjected to cyclic and seismic loading. J Compos Constr 20:4015021. https://doi.org/10.1061/(ASCE)CC.1943-5614.0000584

Tore E, Comert M, Demir C, Ilki A (2017) Collapse testing of full-scale RC buildings with or without seismic retrofit of columns with FRP jackets. In: COST action TU1207 end of action conference. Budapest, Hungary

Toutanji H (1999) Stress-strain characteristic of concrete columns externally confined with advanced fiber composite sheets. ACI Mater J 96:397–404

Toutanji H, Han M, Gilbert J, Matthys S (2010) Behavior of large-scale rectangular columns confined with FRP composites. J Compos Constr 14:62–71. https://doi.org/10.1061/(ASCE)CC.1943-5614.0000051

TR55 (2012) Design guidance for strengthening concrete structures using fibre composite materials, 3rd edn. Concrete Society, London

TSDC (2007) Specification for the buildings to be constructed in disaster areas. Ministry of Public Works and Settlement, Ankara

Wang Z, Wang D, Smith ST, Lu D (2012a) CFRP-confined square RC columns. I: Exp Investig J Compos Constr 16:150–160. https://doi.org/10.1061/(ASCE)CC.1943-5614.0000245

Wang Z, Wang D, Smith ST, Lu D (2012b) Experimental testing and analytical modeling of CFRP-confined large circular RC columns subjected to cyclic axial compression. Eng Struct 40:64–74. https://doi.org/10.1016/j.engstruct.2012.01.004

Wu YF, Jiang C (2013) Effect of load eccentricity on the stress-strain relationship of FRP-confined concrete columns. Compos Struct 98:228–241. https://doi.org/10.1016/j.compstruct.2012.11.023

Wu YF, Wei YY (2010) Effect of cross-sectional aspect ratio on the strength of CFRP-confined rectangular concrete columns. Eng Struct 32:32–45. https://doi.org/10.1016/j.engstruct.2009.08.012

Yan Z, Pantelides CP (2006) Fiber-reinforced polymer jacketed and shape-modified compression members: II-Model. ACI Struct J 103:894–903

Ye L, Yue Q, Zhao S, Li Q (2002) Shear strength of reinforced concrete columns strengthened with carbon-fiber-reinforced plastic sheet. J Struct Eng 128:1527–1534. https://doi.org/10.1061/(ASCE)0733-9445(2002)128:12(1527)

Yost R, Gross SP, Dinehart DW, Mildenberg JJ (2007) Flexural behavior of concrete beams strengthened with near-surface-mounted CFRP strips. ACI Struct J 104:430–437

Youssef MN, Feng MQ, Mosallam AS (2007) Stress-strain model for concrete confined by FRP composites. Compos Part B Eng 38:614–628. https://doi.org/10.1016/j.compositesb.2006.07.020

Yuksel E, Ozkaynak H, Buyukozturk O et al (2010) Performance of alternative CFRP retrofitting schemes used in infilled RC frames. Constr Build Mater 24:596–609. https://doi.org/10.1016/j.conbuildmat.2009.09.005

Yoshimura M, Takaine Y, Nakamura T (2004) Axial collapse of reinforced concrete columns. In: 13th world conference on Earthquake Engineering Vancouver, Canada

Chapter 23
Advances in the Assessment of Buildings Subjected to Earthquakes and Tsunami

Tiziana Rossetto, Crescenzo Petrone, Ian Eames, Camilo De La Barra, Andrew Foster, and Joshua Macabuag

Abstract Currently, 8 out of the 10 most populous megacities in the world are vulnerable to severe earthquake damage, while 6 out of 10 are at risk of being severely affected by tsunami. To mitigate ground shaking and tsunami risks for coastal communities, reliable tools for assessing the effects of these hazards on coastal structures are needed. Methods for assessing the seismic performance of buildings and infrastructure are well established, allowing for seismic risk assessments to be performed with some degree of confidence. In the case of tsunami, structural assessment methodologies are much less developed. This stems partly from a general lack of understanding of tsunami inundation processes and flow interaction with the built environment. This chapter brings together novel numerical and experimental work being carried out at UCL EPICentre and highlights advances

---

T. Rossetto (✉)
EPICentre, Department of Civil, Environmental and Geomatic Engineering, University College London, London, UK
e-mail: t.rossetto@ucl.ac.uk

C. Petrone
Earthquake Research Analyst, Willis Towers Watson, London, UK
e-mail: Crescenzo.Petrone@WillisTowersWatson.com

I. Eames
EPICentre, Department of Mechanical Engineering, University College London, London, UK
e-mail: i.eames@ucl.ac.uk

C. De La Barra
EPICentre, Department of Civil, Environmental and Geomatic Engineering, University College London, London, UK
e-mail: camilo.bustamante.16@ucl.ac.uk

A. Foster
School of Mechanical, Aerospace and Civil Engineering, University of Manchester, Manchester, UK
e-mail: andrew.foster-3@manchester.ac.uk

J. Macabuag
NatCat and Research and Development Analyst SCOR, London, UK

made in defining tsunami loads for use in structural analysis, and in the assessment of buildings for tsunami loads. The results of this work, however, demonstrate a conflict in the design targets for seismic versus tsunami-resistant structures, which raise questions on how to provide appropriate building resilience in coastal areas subjected to both these hazards. The Chapter therefore concludes by summarizing studies carried out to assess building response under successive earthquakes and tsunami that are starting to address this question.

23.1 Introduction

Currently, 8 out of the 10 most populous megacities in the world are vulnerable to severe earthquake damage, while 6 out of 10 are at risk of being severely affected by tsunami, (Sundermann et al. 2014). In order to mitigate ground shaking and tsunami risks for coastal communities, there is first a need to understand and quantify these risks. As a significant portion of the economic and life losses sustained in natural hazards stem directly or indirectly from damage to the built environment, two fundamental components of risk assessment are the characterization of hazard-induced actions on buildings and their response to these actions.

Seismic hazard analysis is an established field of study with many tools widely available for both probabilistic and scenario strong ground motion assessments, which allow the actions on structures from earthquakes to be evaluated with some confidence. The literature also presents significant advances in the modelling of earthquake-triggered tsunami hazards (e.g. Goda et al. 2017; Suppasri et al. 2016), with tsunami transformations into coastal margins being well-modelled by existing numerical codes (e.g. MOST by Titov and Synolakis 1998 and FUNWAVE by Grilli et al. 2007, amongst others). These numerical models are able to simulate offshore wave characteristics of tsunami wave forms, however modelling of the flow inundation depths and velocities as the tsunami travels onshore remains a challenge. The latter is highly complex and requires the use of very high bathymetric and topographic resolutions for the numerical models to provide a realistic simulation of the flow (e.g. as seen in Mader 2004). Furthermore, the computational expense required to explicitly model the presence of coastal buildings on the tsunami inundation means that it is almost never done in practice; the effect of the built environment on the tsunami flow more commonly modelled through the use of an increased onshore bed roughness. This means that the vast majority of existing tsunami onshore inundation numerical models and simulations are unable to provide a direct evaluation of tsunami forces on buildings. Instead, these have to be calculated from empirical or semi-empirical equations that relate tsunami force to the flow characteristics that can be predicted by these models, e.g. from the inundation depth, h, velocity, u. Such force equations can be found in current and past tsunami design guidance documents (e.g. Okada et al. 2006; FEMA 2008; ASCE 7-16 2017), but show limited consensus. Due to limited observational data on tsunami, the empirical closures of the presented force equations are based either on expert opinion or on experiments; the latter being very limited in their representation of realistic tsunami, (see Sect. 23.2).

Similarly to the case of seismic hazards, methods for assessing the seismic performance of both individual and classes of buildings are well established. Several approaches for the numerical analysis of structural response to earthquakes exist, which range in computational expense from more burdensome non-linear response history analyses of complex structural models to rapid non-linear static-based analyses of highly simplified structural models. Consequently, seismic fragility functions exist for buildings that are based on numerical analysis (e.g. see compendium presented in Yepes-Estrada et al. 2016 as an example). A fragility function relates the probability that a building (or building class) will reach or exceed a number of pre-defined damage states when subjected to increasing hazard actions. Fragility functions provide a concise overview of structural performance under the natural hazard and hence are commonly used in natural catastrophe risk modelling.

In the case of tsunami, structural assessment methodologies are much less developed, with very few analysis-based fragility functions existing in the literature. The lack of literature in this field relates partly to the aforementioned general lack of understanding of tsunami inundation processes and flow interaction with the built environment. However, recent advances in the physical modelling of tsunami in the laboratory (see Rossetto et al. 2011) are helping to shed light on these issues and opening opportunities to significantly progress the field of structural analysis for tsunami.

This Chapter presents a concise summary of the journey the authors have taken over the last years to start to answer some of the structural engineering questions that still pose a significant challenge in the study of building response to tsunami:

1. What are the tsunami forces on buildings?
2. How do we analyse buildings for tsunami loads?
3. What is the role of ductility on structural response to tsunami?
4. How can we best analyse a structure under sequential earthquake and tsunami loads?

Most of the work presented in this chapter is part of a larger programme of research being conducted by University College London and HR Wallingford in the European Research Council funded URBANWAVES project, which has also significantly advanced the understanding of tsunami onshore flows through innovative large scale experiments and computational fluid dynamics.

23.2 What Are the Tsunami Forces on Buildings?

The first author began to look at the issue of tsunami forces on buildings after returning from a field reconnaissance of the 2004 Indian Ocean Tsunami (Rossetto et al. 2007), with the idea that tsunami inundation loads on buildings have a strong horizontal component, and that if this loading could be appropriately characterized,

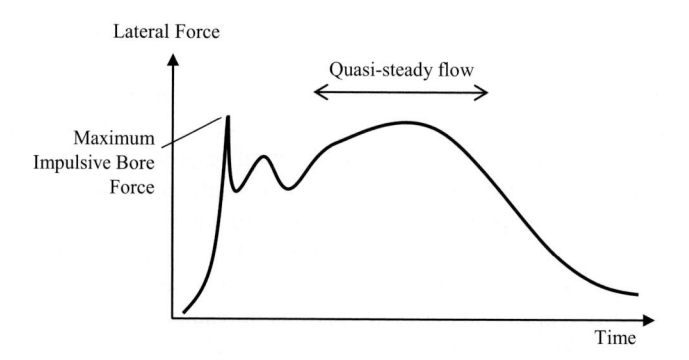

Fig. 23.1 Illustration of general tsunami loading history on an onshore structure

similar techniques to those used in earthquake engineering could be used to assess structural response to tsunami.

Guidance documents for the design of tsunami-resistant structures at the time, treated tsunami loading as a severe flood. Most adopted a tsunami force formulation composed of three components (e.g. FEMA 2005): hydrostatic force (dependent on flow depth, d), hydrodynamic force (dependent on onshore flow velocity, v) and impulse load (initial overshoot associated with the impact of the leading tongue of a wave or surge). Only peak forces were designed for, with the time dependence of tsunami forces on the buildings being completely ignored.

Despite knowledge that loads from tsunami inundation can be of long duration due to the typical wave periods of tsunami (20–40 min), see Fig. 23.1, the time-dependent nature of tsunami loading is also ignored in the later published FEMA P-646 guidelines (FEMA 2008) and current ASCE-16 (2017) code. One of the reasons for this is that all these guidelines are based on experiments that use relatively short-period waves and highly idealized waveforms to represent tsunami.

Most worldwide facilities adopt piston wave-makers, which simply do not have the stroke length to generate realistic tsunami wavelengths at the scale necessary to reproduce its physical processes. Such facilities are typically limited to the generation of solitary waves with periods <10 s (i.e. prototype tsunami wave period of approximately 2 min, assuming scale of 1:50), and have great difficulty in reproducing stable trough-led waves that can characterise tsunami. Through a collaboration between UCL and HR Wallingford, in 2008 a new type of pneumatic tsunami generator was developed (Rossetto et al. 2011), which underwent several improvements over the following years as described in Allsop et al. (2014) and Chandler et al. (2016). This new tsunami generation system, when operated in a long flume equipped with a sloping bathymetry, allowed, for the first time, the study of tsunami forces on onshore buildings subjected to extremely long waves of different waveform (i.e. both elevated and trough-led waves).

Foster et al. (2017) presents the results of one such series of experiments wherein impermeable rectangular model buildings of different widths, b, (representing different blockage ratios, b/w, with respect to the flume width, w), are subjected to both

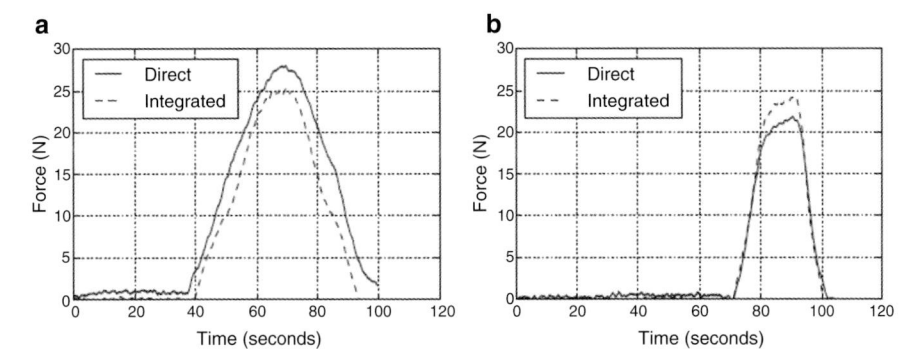

23 Advances in the Assessment of Buildings Subjected to Earthquakes and Tsunami

Fig. 23.2 Sample tsunami inundation force time histories measured at the model building, (blockage ratio 0.6), by Foster et al. (2017) for (**a**) an elevated wave with 80 s period and (**b**) trough-led wave with 80s period. The full lines show the measured force obtained from the load cells and the dashed line, the force obtained by integrating the pressures measured at the pressure transducers

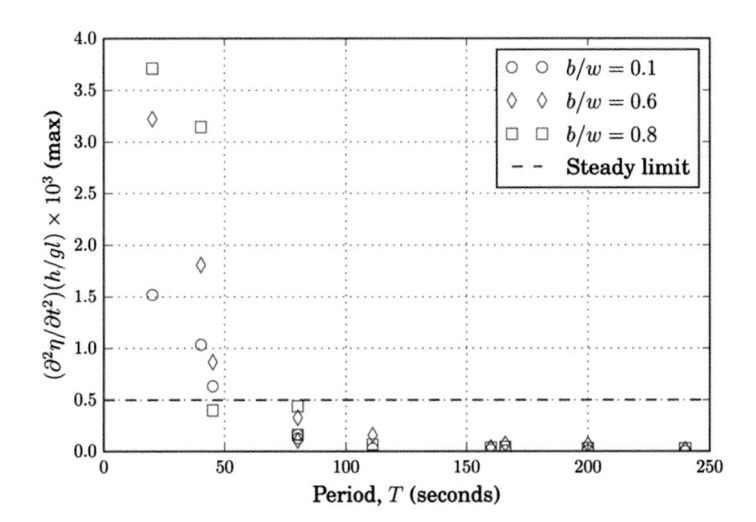

Fig. 23.3 Plot showing the degree of steadiness with respect to the wave period, T. The flow is considered unsteady when $\frac{\partial^2 h}{\partial t^2}$ is greater than 0.05% of the quantity (h/gl), which corresponds to a term that can adequately non-dimensionalise the water depth time derivative whilst also encompassing the influence of the length, l, of the obstacle in the flow. Figure modified from Foster et al. (2017)

crest- and trough-led waves of periods ranging between 20–240 s at 1:50 scale (i.e. up to 20 min tsunami prototype). The following key observations are made in this study:

1. For the long waves adopted, for blockage ratios less than 1.0, no initial impulse loading is detected (see Fig. 23.2).

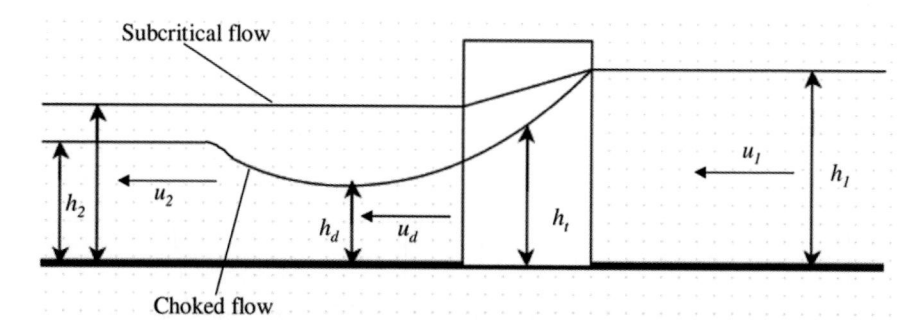

Fig. 23.4 Schematic diagram of the subcritical and choked flow conditions around a rectangular body, redrawn from Qi et al. 2014. In the figure the subscript 1 refers to the incident flow parameters, subscripts d and 2 to the flow parameters immediately and far downstream of the rectangular body, respectively

2. Tsunami waves with periods exceeding 80s produce quasi-steady flows (see Fig. 23.3).
3. There is a local influence, due to the presence of the obstacle in the flow, which results in higher blocking ratios introducing greater unsteadiness (see Fig. 23.3).

These observations resulted in the authors proposing a modification to the steady-flow force equations proposed in Qi et al. (2014) for representing the measured tsunami loads on buildings.

From tests conducted in a small scale laboratory at UCL, Qi et al. (2014) shows that steady flows around rectangular bodies are subcritical for low incident flow Froude numbers (Fr_1), inducing drag-dominated forces on the structure. However, when the incident Froude number reaches or exceeds a critical value (Fr_c), the flow transitions to a choked state, where hydrostatic forces dominate (Fig. 23.4). They also show that the value of Fr_c is affected by the blocking ratio of the body with respect to the flume (b/w). Finally, by proposing a relationship between the incident and downstream Froude numbers, Qi et al. (2014) propose simple equations for the estimation of the overall force on the rectangular body that can be evaluated solely from knowledge of the incident flow depth (h_1), velocity (u_1) and the blockage ratio.

In the context of tsunami onshore flow numerical modelling, where as stated in Sect. 23.1, the presence of the buildings is not explicitly modelled, the force formulations of Qi et al. (2014) forms a viable empirical closure for force calculation that, as opposed to other equations, accounts for the state of the flow around the building. Hence, in Foster et al. (2017) the Qi et al. (2014) formulae are updated and modified to better fit the large scale tsunami experimental data, and to account for tsunami forces from unsteady inundation flows (associated with the shorter tsunami waves). These result in Eqs. 23.1 and 23.2 for steady flows, i.e. when $\left(\frac{\delta^2 h}{\delta t^2}\right)\left(\frac{h}{gl}\right) < 0.0005$, where l is the length of the building:

$$F_T = \frac{1}{2}C_D\rho b u_1^2 h_1 \quad \text{for} \quad F_{R1} < F_{Rc} \quad \text{(subcritical conditions)} \quad (23.1)$$

$$F_T = \lambda_s b g^{1/3} u_1^{4/3} h_1^{4/3} \quad \text{for} \quad F_{R1} \geq F_{Rc} \quad \text{(choked conditions)} \quad (23.2)$$

$$\text{where,} \quad \lambda_s = 0.73 + 1.2(b/w) + 1.1(b/w)^2 \quad (23.3)$$

Instead Eqs. 23.1 and 23.4 result for unsteady flows, i.e. when $\left(\frac{\delta^2 h}{\delta t^2}\right)\left(\frac{h}{gt}\right) \geq 0.0005$:

$$F_T = \lambda b g^{1/3} u_1^{4/3} h_1^{4/3} \quad \text{for} \quad F_{R1} \geq F_{Rc} \quad \text{(choked conditions)} \quad (23.4)$$

$$\text{where,} \quad \lambda = 1.37 - 1.35(b/w) + 1.37(b/w)^2 \quad (23.5)$$

A further important observation made is that irrespective of the flow conditions, the pressure distribution along the front of the structure follows a triangular (hydrostatic) distribution.

23.3 How Do We Analyse Buildings for Tsunami Loads?

Current guidelines for the design and assessment of buildings under tsunami actions do not contain specific guidance as to how to apply the tsunami loads to the building for the structural analysis nor which analysis methods to use for the structural response assessment. Having developed a means to better evaluate both the tsunami loading on buildings and the imposed pressure distribution, we began to investigate how best to analyse coastal buildings for tsunami loading. A summary of this investigation to date is presented in Petrone et al. (2017). In the latter paper, a case study structure is used to compare the tsunami response parameters predicted by three different analysis approaches, and collapse fragility functions are built.

The case study structure used is a Japanese 10-storey reinforced concrete moment resisting frame vertical evacuation structure. The structure is designed to resist both earthquake and tsunami actions, and the reader is referred to the paper for full details of the model structure. The building is evaluated using the existing non-linear constant-height pushover approach (CHPO) used in Attary et al. (2017), amongst others. For the first time, the paper also assesses the building using tsunami non-linear response history analysis (TDY) and the newly proposed variable height pushover analysis (VHPO).

TDY follows the same principles as a seismic response history analysis, except that, in this case, the tsunami force time history ($F_T(t)$) is applied to the structure using a triangular load distribution up to the tsunami inundation height at the relevant time step, and the resulting structural deformations are measured. For their application Petrone et al. (2017) adopt the steady state force formulations of Qi et al. (2014) to derive over 800 $F_T(t)$ from tsunami inundation height ($h(t)$) and velocity ($u(t)$) time histories calculated by Goda et al. (2015) for the 2011 Tohoku earthquake and tsunami.

The CHPO method is an approach that is similar to a conventional earthquake pushover analysis, but is modified to account for the characteristics of tsunami loading. In CHPO a constant inundation depth (h) is considered, and a displacement-controlled analysis is carried applying the lateral load to the structure according to a hydrostatic-type distribution. As the inundation height is constant, the force is increased by increasing the velocity of the inundation flow, thus changing the Froude number (Fr) as the analysis progresses.

The non-linear variable height pushover analysis (VHPO) differs from CHPO as it applies lateral loads to the structure according to a hydrostatic-type distribution, however it linearly increases the inundation depth up to a target value, h_{max}, whilst maintaining a constant Froude number. VHPO is a force-controlled procedure, and its disadvantage is that it is unable to capture the post-peak behaviour in the pushover curve.

The three procedures are used to assess the collapse of the structure. For the pushover analyses, the structure is assumed to be failed when the tsunami peak force exceeds the structural strength; the structural strength is assessed as the peak force in the pushover curve. This definition of collapse implicitly assumes that ductility does not play a role in the structural assessment. In the case of TDY collapse is predicted to occur when the structure exhibits an inter-storey drift ratio (IDR) equal to that which occurs when the structure reaches peak strength in the pushover analysis. Such a collapse criterion is consistent with that defined for the pushover.

Comparison of the structure response and collapse prediction obtained using TDY, CHPO and VHPO results in the following main observations:

1. The tsunami response of the structure is sensitive to the load discretization used to apply the lateral loads. Ideally the tsunami load should be distributed along the height of the structure with five or more load application points per storey.
2. VHPO approximates well the engineering demand parameters and collapse fragility curves obtained from TDY for a wide range of tsunami time-histories.
3. The VHPO outperforms CHPO in predicting the maximum IDR and column shear at the ground storey (see Fig. 23.5).
4. Neither pushover-based approach can predict the structural response if there is a strong second peak in the tsunami inundation time history (see Fig. 23.5).
5. The tsunami peak force is better correlated to the maximum IDR than flow velocity and inundation depth, and results in fragility curves with lower dispersion values.

The tsunami applies significant shear forces and concentrated deformations at the bottom storeys of the building. To the collapse point, the static pushover approaches (particularly VHPO) provide a very good estimate of the structure response under TDY, suggesting that the loading does not significantly excite the structure's dynamic properties. However, in the described study an explicit decision was made to ignore the contribution of ductility to the structure response. In the case of the Japanese evacuation building this assumption is justified by the fact that, despite being designed for earthquake and tsunami actions, column shear failure precipitates collapse before the maximum strength of the structure is reached. However, if this

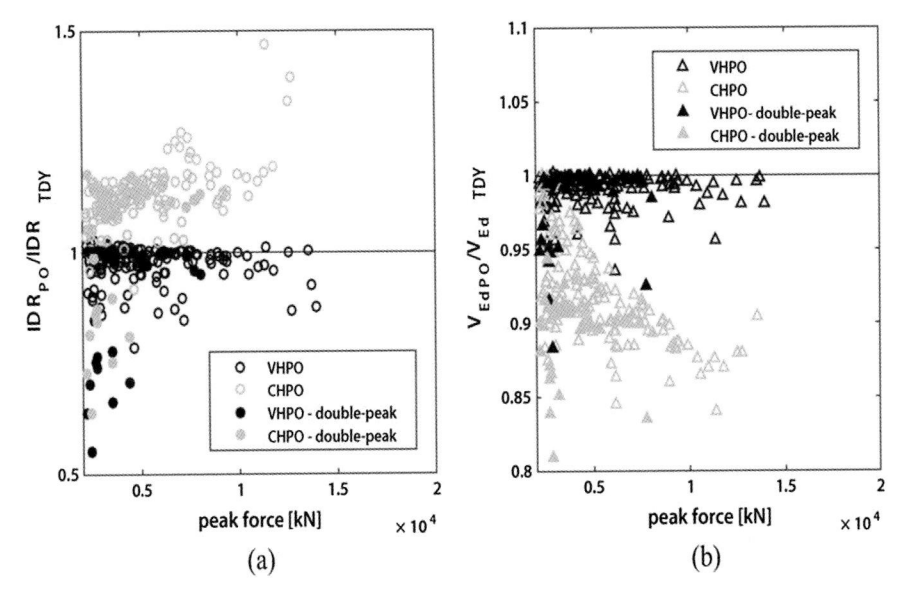

Fig. 23.5 Comparison of IDR and shear demand (V_{Ed}) for CHPO and VHPO versus TDY. Double-peak wave cases are shown with filled markers (Adapted from Petrone et al. 2017)

shear failure was avoided, what role would the building ductility play in the structural response to tsunami?

23.4 What Is the Role of Ductility on Structural Response to Tsunami?

To answer this question we went back to fundamental structural dynamics concepts. As a structure deforms under a time-dependent load, it develops inertia forces, damping forces and internal (spring) forces. By solving the equation of motion, it can be shown analytically (see Rossetto et al. 2018b) for an elastically perfectly plastic (EPP) single degree of freedom system of mass m, and stiffness k, that the plastic displacement ductility demand (μ_p) is a function of the applied load, the yield force, Fy, and the natural period of the structure, T ($2\pi\sqrt{m/k}$). In Rossetto et al. (2018b) the duration over which a structure can sustain a load greater than its yield load (t_p) is calculated for a two idealized tsunami load histories: a triangular and parabolic load history. The analytical results are then verified numerically, before extending the analysis to consider structural damping, strain-hardening and realistic tsunami loading profiles.

Figure 23.6 presents the results for the EPP SDoF subjected to the triangular loading history. The figure shows the relationship between the calculated maximum

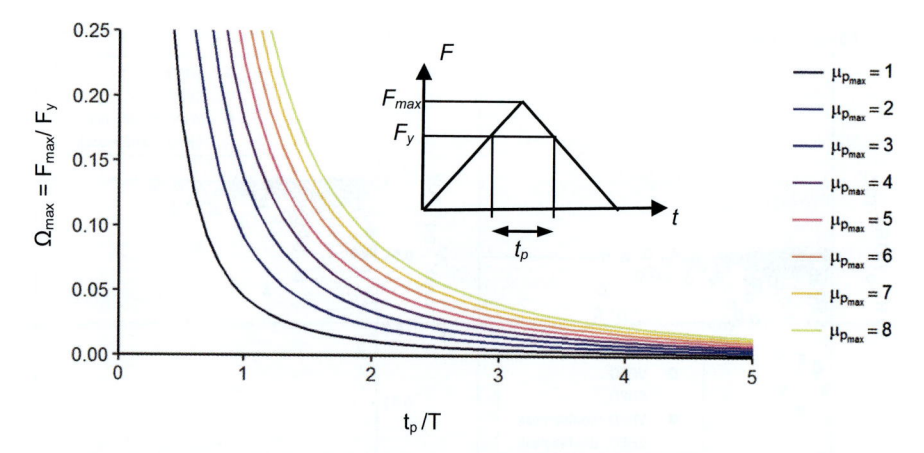

Fig. 23.6 Relationship between the calculated maximum overstrength (Ω_{max}) and t_p/T for different ductility values (Modified from Rossetto et al. 2018b)

overstrength (Ω_{max}) and t_p/T. It is observed that an increase in structural ductility corresponds to a higher achievable overstrength, but that overstrength values greater than 5% can only be achieved if the time over which the force applied to the structure exceeds yield is less than 1 to 2.5 times the structure natural period. Translating this into practical numbers, if we consider a 10 storey reinforced concrete moment resisting frame with a natural period of around 1 s and ductility of 4.0, this would achieve negligible overstrength for any duration of plastic loading (t_p) exceeding 3.5 s. If we consider that strong tsunami inundations can exceed minutes in duration, it becomes clear that structural ductility cannot be relied upon to allow the structure to sustain loads exceeding its yield capacity. Rossetto et al. (2018b) show that the inclusion of strain hardening in the SDoF can improve the achievable overstrength. However, again, for realistic values of strain hardening this increase in overstrength is negligible.

The results of this study indicate that in order to achieve tsunami resistance, a structure needs to be designed to resist the full tsunami loading elastically. This can be extremely expensive if the structure does not allow the tsunami inundation to flow through it (relieving pressures on external walls and structural elements). As the structural strength, rather than its deformation capacity, governs the tsunami resistance of the building, and given the steady nature of tsunami inundation flows, this study reinforces the suitability of pushover-based analysis methods for the assessment of buildings under tsunami.

23.5 How Can We Best Analyse a Structure Under Sequential Earthquake and Tsunami Loads?

As maximum strength is found to govern the tsunami design of buildings (Sect. 23.4), in areas at risk from both seismic excitation and tsunami inundation, the effect on the tsunami response of damage to the structure in a preceding earthquake should be considered.

Only two existing studies have been found that have looked into the tsunami response of structures previously damaged by earthquake ground shaking through numerical techniques. Park et al. (2012) and Latcharote and Kai (2014) both adopt non-linear response history analysis for assessing their structural models under the earthquake loading, and follow this with a constant height pushover (CHPO) for the tsunami response assessment. Both adopt a coarse discretisation of the applied tsunami loading along the height of the building for the CHPO, which is shown by Petrone et al. (2017) to significantly affect the reliability of the tsunami pushover analysis results. CHPO is also shown to provide an over-prediction of the tsunami induced shear forces and displacement response at the building's lower storeys. Neither of the existing studies have compared their overall structural response against an earthquake and tsunami non-linear response history analysis. Equally, none have attempted to look at the possibility of simplifying the earthquake analysis phase, through use of an earthquake pushover (PO).

In Rossetto et al. (2018a), we systematically change the analysis approach used in each of the three phases involved in the assessment of structural behavior under sequential earthquakes and tsunami, namely the earthquake loading phase, unloading of the structure until at-rest condition and the tsunami loading phase. In the earthquake loading phase, non-linear response history analysis (DY) was considered as well as a static nonlinear pushover (PO) with a typical lateral load distribution following the shape of the first mode response of the structure (e.g. FEMA 2000). Two types of unloading analyses were considered. However, these were seen to have little effect on the final earthquake and tsunami response of the building, and hence are not reported here. In the tsunami loading phase, tsunami nonlinear response history analysis (TDY), constant height pushover (CHPO) and variable height pushover (VHPO) were considered. For both pushover analyses, the performance point (P.P. in Fig. 23.7) is determined at the point of intersection between the tsunami pushover curve and a horizontal line representing the tsunami force demand, F_T. As in Petrone et al. (2017), the structure is assumed to be collapsed if the tsunami demand is larger than the structural lateral load capacity. This definition of collapse relies on the fact that ductility has been proven to not to play a significant role in the tsunami performance of buildings (Rossetto et al. 2018b).

All combinations of earthquake, unloading and tsunami analysis approaches were implemented for the response assessment of the same Japanese evacuation building used in Petrone et al. (2017) and previously discussed in Sect. 23.3. The results of the earthquake and tsunami non-linear response history analysis combination

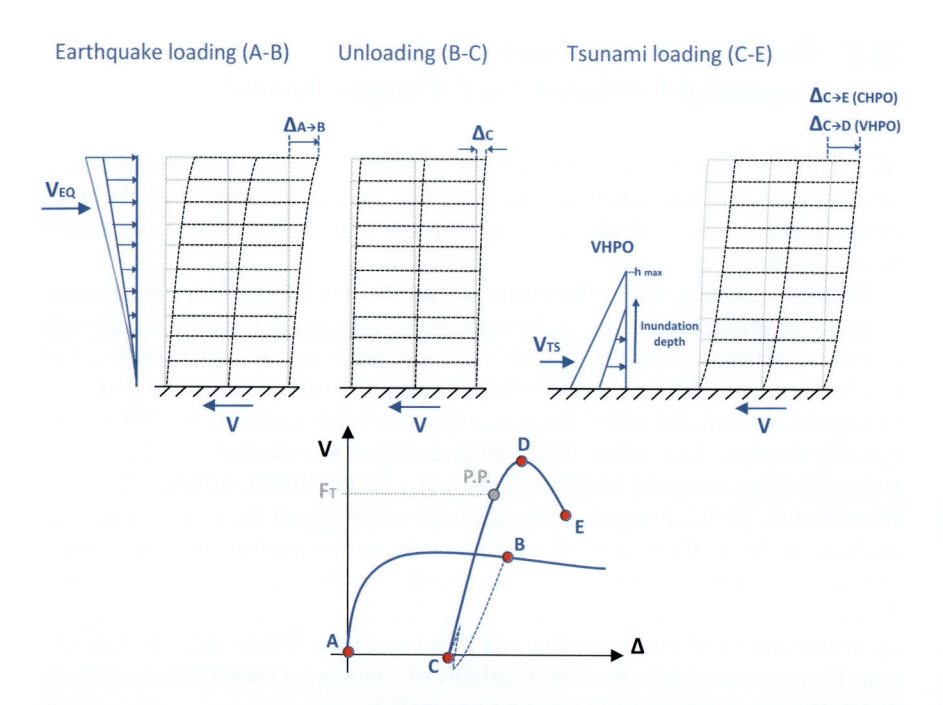

Fig. 23.7 Schematic representation of the double pushover methodology for the PO-VHPO case

(DY-TDY) were used as the reference against which to measure the reliability and accuracy of the other earthquake and tsunami analysis method combinations. Sixteen of the 800+ simulated earthquake ground motion and tsunami inundation time-history pairs of Goda et al. (2015) for the 2011 Tohoku earthquake and tsunami were selected to run the reference case. These were chosen to cover a range of earthquake and tsunami intensities.

For brevity, in this section only the main results of the comparison of DY-TDY with DY-VHPO and PO-VHPO are presented. The reader is referred to Rossetto et al. (2018a) for the complete comparison of approaches and the sensitivity analyses performed. It is highlighted that the double pushover approach (PO-VHPO), illustrated schematically in Fig. 23.7, presents significant computational savings as compared to DY-TDY. PO-CHPO is still faster to run, however, this is seen to come at the significant expense of accuracy, and this analysis combination is not recommended for use in sequential earthquake and tsunami analysis.

Several engineering response parameters (ERP) such as inter-storey drift ratios at each floor (IDR), top displacements, base shear, internal forces and floor accelerations can be measured across the analyses. However, in the case of the earthquake pushover, the analysis needs to be stopped at a desired point (i.e. point B in Fig. 23.7) before the subsequent unloading and tsunami loading is applied. The analyst can choose where to stop the earthquake PO analysis, for example, when a specific damage state is achieved in the structure (associated with the occurrence of a specific value for the measured ERPs). Else, a capacity spectrum based method, (for instance, FRACAS as described in Rossetto et al. 2016), can be used to determine the structure performance point under a given earthquake ground motion or spectrum.

However, in order to eliminate the effect of estimation errors in the earthquake performance point arising from, for instance, the use of a capacity spectrum based assessment, Rossetto et al. (2018a) stop the PO analysis when the maximum inter-storey drift ratio experienced by the structure in the corresponding DY analysis ($IDR_{max,\ DY}$) is achieved in a matching floor within the structure subjected to the PO analysis. It is acknowledged that this matching procedure can lead to different estimates of the overall damage distribution on the structure (with the exception of the matched floor), as the response under dynamic earthquake excitation differs from the one under static pushover.

Figure 23.8 presents a comparison of the global and storey-level response of the case study structure when analysed using combinations DY-TDY, DY-VHPO and PO-VHPO for two selected earthquake-tsunami pairs of moderately high earthquake intensity and varying tsunami intensities. It is clear that, when compared with the reference DY-TDY case, DY-VHPO shows an excellent agreement in terms of the global behaviour and IDR distribution. On the other hand, the change in the analysis type for the earthquake phase from DY to PO, yields a worse estimate of the global displacements (Fig. 23.8c) and IDR distribution (Fig. 23.8d).

The reason for this behaviour is that for the second earthquake-tsunami pair (Fig. 23.8c–d), under the earthquake loading phase for the PO analysis, the structure is pushed to a large value of IDR_{max} (0.670%). This results in a significant residual displacement after the structure unloading phase. This in turn, results in the observable shift along the x-axis of the reloading tsunami VHPO curve in Fig. 23.8c. Such a large residual displacement is not observed in the corresponding DY-VHPO case. Furthermore, in this particular case, as the tsunami intensity is relatively low, the tsunami pushover does not significantly modify the residual IDR profile resulting from the earthquake PO analysis. Hence, the final IDR profile in the PO-VHPO analysis is significantly different from the reference DY-TDY case. This observation is repeated for all the cases where the earthquake pushover has induced significant plasticity in the structure, and the subsequent tsunami has a relatively low intensity.

Despite the observed differences in displacement and IDR response when the earthquake PO is used instead of the DY analysis, PO-VHPO predicts quite well the shear internal force in the most critical column of the structure, (at the tsunami performance point, Fig. 23.9) for the 16 cases assessed. This is explained as the shear internal force at the ground storey of the building is driven by the tsunami loading; observation which is sustained by the lack of difference in shear internal force values when the earthquake analysis type is changed, whilst a visible difference in shear internal force results when the tsunami analysis is changed from CHPO to VHPO.

These results suggest that DY-VHPO could be used as an alternative to DY-TDY for the analysis of structures. This approach provides good accuracy, a reduced computational time, and, as it adopts a pushover-based tsunami assessment, the same analysis can be used to assess the earthquake damaged structure under numerous subsequent tsunami events.

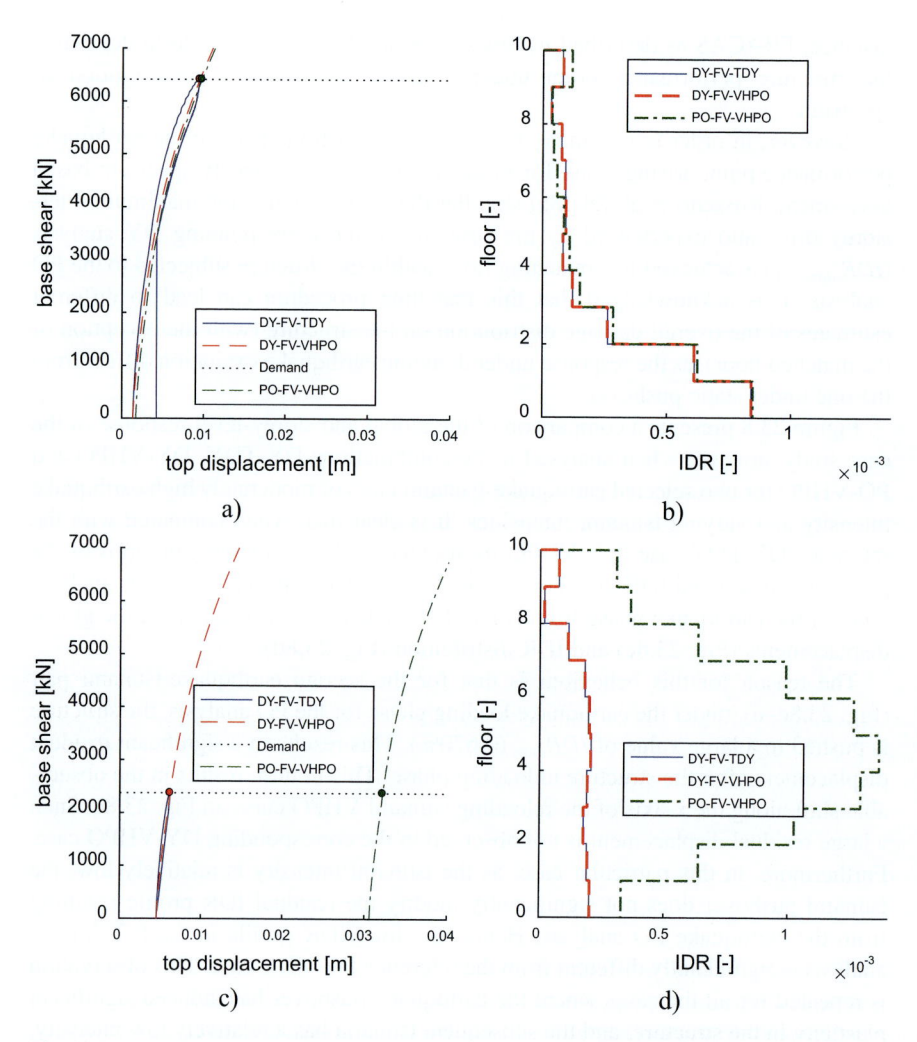

Fig. 23.8 Comparison of tsunami global response (**a** and **c**) and IDR distribution (**b** and **d**) for two different earthquake-tsunami pairs

Furthermore, despite the observed discrepancies in the IDR and displacement responses predicted by DY-TDY and PO-VHPO, the latter double pushover approach does provide reasonable estimates of the shear internal forces. The performance of this approach could also be improved if applied to buildings that are predominantly dominated by a first mode response, as it is expected that the difference in the structure deformed shape between DY and PO for these would be less. Due to the fact that in Sect. 23.4 we see that the tsunami response of a structure is inherently strength-based (with the ductility playing a secondary role), the double pushover method might be suitable for use in fragility assessments of populations of buildings, where a significant saving in computational expense might justify the loss

Fig. 23.9 Error in shear force estimation in the most critical column at the end of all three loading phases

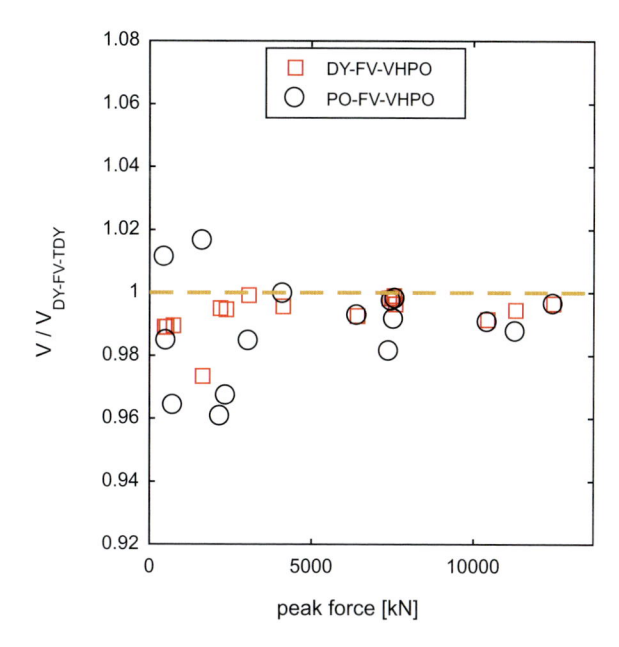

in accuracy of response prediction. For such cases, the use of a capacity spectrum based approach for the estimation of the structure performance under the earthquake loading phase would provide acceptable results.

23.6 Conclusions

Over the last years experimental capabilities for simulating tsunami have evolved and are allowing the study of ever more realistic tsunami in the laboratory. This is providing the vital information needed to better characterize the forces imparted by tsunami inundation flows on coastal infrastructure, and has allowed the inclusion of the flow state in tsunami force equations presented in Sect. 23.2.

The time is now ripe for transferring this new knowledge on the physics of tsunami inundation flows into the field of structural engineering, so as to develop new approaches and guidance for the design and assessment of coastal infrastructure for this hazard. Towards this goal, a new variable height pushover approach (VHPO) for the assessment of building response under tsunami inundation flows has been proposed and has been found to be highly promising.

It is shown that due to the long duration of tsunami inundation flows, large lateral loads can be applied to the structure for a relatively long duration. Due to the long duration of the loading and the limited ductility of structures, when the tsunami loads exceed the lateral strength of the structure it is likely to collapse. In tsunami design this observation would translate into the elastic design of structures for tsunami loading. However, this design concept conflicts directly with the use of ductile design in the seismic load case.

How do we reconcile these in cases when a building is at risk from both earthquakes and tsunami?

Fortunately, as seen in Sect. 23.5, due to the different loading characteristics of earthquakes and tsunami, major differences in the lateral load resistance of a structure can be achieved under the two load cases. For example, in the case study building, a much higher lateral strength can be achieved under the tsunami loading than under the earthquake loading. Sect. 23.3 suggests that a particularly high lateral strength might be achieved in the structure if its vertical elements are designed such that they can resist the high shear loads imposed by a tsunami at the building's lower storeys. For this case, seismic detailing for shear can help, but might have to be extended to the entire member rather than solely at element critical sections defined through earthquake loading considerations. However, this recommendation needs to be informed by further studies.

But what is the effect of preceding earthquake damage on the tsunami strength of buildings? According to the analyses run by Rossetto et al. (2018a) on the Japanese tsunami evacuation structure, the effect of the earthquake damage on the tsunami strength is very limited unless the earthquake damage is extensive or has induced partial collapse. This is encouraging for the design of buildings under earthquake and tsunami loading. However, these findings are based on a limited number of analyses carried out on a special structure, and throughout the Chapter gross assumptions have been made in neglecting tsunami inundation forces associated with buoyancy and debris impact. Hence, significant work remains to be carried out to advance the field of tsunami engineering.

Acknowledgements The research presented in this Chapter has been predominantly funded by the European Research Council under the European Union's Seventh Framework Programme (FP7/2007–2013)/ERC grant agreement number 336084 'URBANWAVES', awarded to Professor Tiziana Rossetto. The authors are grateful to Dr. David McGovern, Dr. Ian Chandler, Professor William Allsop, Dr. Tristan Robinson and Dr. Christian Klettner for their assistance in this research.

References

Allsop W, Chandler I, Zaccaria M (2014) Improvements in the physical modelling of tsunamis and their effects. In: Coastlab14 (5th International conference on the application of physical modelling in coastal and port engineering and science), Varna, Bulgaria, 29 September–2 October 2014

ASCE 7-16 (2017). ASCE 7 standard minimum design loads and associated criteria for buildings 488 and other structures

Attary N, van de Lindt JW, Unnikrishnan VU, Barbosa AR, Cox DT (2017) Methodology for development of physics-based tsunami fragilities. J Struct Eng 143(5)

Chandler, I., Allsop, W., Barranco Granged, I., McGovern, D. (2016) Understanding wave generation in pneumatic tsunami simulators. In: Coastlab16 (6th International conference on the application of physical modelling in coastal and port engineering and science), Ottawa, Canada, 10–13 May 2016

FEMA (2000) FEMA 356. Prestandard and Commentary for the Seismic Rehabilitation of Buildings, Washington

FEMA (2005) Coastal construction manual. In: FEMA 55, 3rd edn. Washington, DC, Federal Emergency Management Agency

FEMA (2008) Guidelines for Design of Structures for vertical evacuation from tsunamis. In: Technical report P646. Washington, DC, Federal Emergency Management Agency

Foster ASJ, Rossetto T, Allsop W (2017) An experimentally validated approach for evaluating tsunami inundation forces on rectangular buildings. Coast Eng 128(November 2016):44–57. https://doi.org/10.1016/j.coastaleng.2017.07.006

Goda K, Yasuda T, Mori N, Mai M (2015) Variability of tsunami inundation footprints considering stochastic scenarios based on a single rupture model: application to the 2011 Tohoku earthquake. J Geophys Res Oceans 120(6):4552–4575

Goda, K., Petrone, C., De Risi, R., Rossetto, T. (2017) 'Stochastic coupled simulation of strong motion and tsunami for the 2011 Tohoku, Japan earthquake', Stochastic environmental research and risk assessment. Springer Berlin Heidelberg, (November), pp 1–19. https://doi.org/10.1007/s00477-016-1352-1

Grilli ST, Ioualalen M, Asavanant J, Shi F, Kirby JT, Watts P (2007) Source constraints and model simulation of the December 26, 2004, Indian Ocean tsunami. J Waterw Port Coast Ocean Eng 133:414–428

Latcharote P, Kai Y (2014) Nonlinear structural analysis of reinforced concrete buildings suffering damage from earthquake and subsequent tsunami. In 10th U.S. National Conference on Earthquake Engineering – Frontiers of Earthquake Engineering

Mader CL (2004) Numerical modeling of water waves, 2nd edn. CRC Press, London

Okada T, Ishikawa T, Tateno T, Sugano T, Takai S (2006) Tsunami loads and structural design of tsunami refuge buildings. Technical report, The Building Centre of Japan

Park S, van de Lindt JW, Cox D, Gupta R, Aguiniga F (2012) Successive earthquake-tsunami analysis to develop collapse fragilities. J Earthq Eng 16(6):851–863

Petrone C, Rossetto T, Goda K (2017) Fragility assessment of a RC structure under tsunami actions via nonlinear static and dynamic analyses. Eng Struct 136:36–53. https://doi.org/10.1016/j.engstruct.2017.01.013

Qi ZX, Eames I, Johnson ER (2014) Force acting on a square cylinder fixed in a free-surface channel flow. J Fluid Mech 756:716–727. https://doi.org/10.1017/jfm.2014.455

Rossetto T, Peiris N, Pomonis A, Wilkinson SM, Del Re D, Koo R, Gallocher S (2007) The Indian Ocean tsunami of December 26, 2004: observations in Sri Lanka and Thailand. Nat Hazards 42(1):105–124

Rossetto T, Allsop W, Charvet I, Robinson DI (2011) Physical modelling of tsunami using a new pneumatic wave generator. Coast Eng 58(6):517–527

Rossetto T, Gehl P, Minas S, Galasso C, Duffour P, Douglas J, Cook O (2016, October 15). FRACAS: a capacity spectrum approach for seismic fragility assessment including record-to-record variability. Eng Struct ;125:337–348.

Rossetto T, De La Barra C, Petrone C, De La Llera JC (2018a) Comparison of nonlinear static and dynamic analysis methods for assessing structural response under earthquake and tsunami in sequence. Earthquake Engineering and Structural Dynamics (submitted)

Rossetto T, Macabuag J, Petrone C, Eames I (2018b) Does ductility play a significant role in the response of structures to tsunami Loading?. Eng struct (submitted)

Sundermann L, Schelske O, Hausmann P (2014) Mind the risk – a global ranking of cities under threat from natural disaster'. Swiss Re. Report No. 1505715_13_en12/14. pp 39

Suppasri A, Latcharote P, Bricker JD, Leelawat N, Hayashi A, Yamashita K, Makinoshima F, Roeber V, Imamura F (2016) Improvement of tsunami countermeasures based on lessons from the 2011 great East Japan earthquake and tsunami—situation after five years. Coast Eng J 58(4):1640011. https://doi.org/10.1142/S0578563416400118

Titov VV, Synolakis CE (1998) Numerical modeling of tidal wave runup. J Waterw Port Coast Ocean Eng 124:157–171

Yepes-Estrada C, Silva V, Rossetto T, D'Ayala D, Ioannou I, Meslem A, Crowley H (2016) The global earthquake model physical vulnerability database. Earthquake Spectra 32(4):2567–2585

Chapter 24
Seismic Vulnerability of Classical Monuments

Ioannis N. Psycharis

Abstract Classical monuments are articulated structures consisting of multi-drum columns made of discrete stone blocks that are placed one on top of the other without mortar. Despite the lack of any lateral load resisting mechanism except friction, classical monuments are, in general, earthquake resistant, as proven from the fact that they have survived several strong earthquakes over the centuries. However, in their current condition, they present many different types of damage that affect significantly their stability. This chapter presents the results of theoretical and experimental research on the earthquake resisting features and the assessment of the vulnerability of these structures, which is not straightforward due to the high nonlinearity and the sensitivity of the response. Recent trends towards a performance-based philosophy for the seismic risk assessment of these structures, based on conditional limit-state probabilities and seismic fragility surfaces, are also discussed.

24.1 Introduction

Classical monuments are made of structural elements (drums in case of columns), which lie one on top of the other without mortar. Columns are connected to each other with architraves (also called "epistyles") consisting of stone beams, usually made of marble. A characteristic example is shown in Fig. 24.1 from the Olympieion of Athens, Greece.

Architrave beams are usually connected to each other with iron clamps and dowels. However, in most cases no structural connections are provided between the drums of the columns. Only in few cases, iron shear connectors (dowels) are provided at the joints, which restrict, up to their yielding, sliding but do not affect rocking. The wooden dowels that were usually placed at the joints among the drum of the columns were aiming at

---

I. N. Psycharis (✉)
Laboratory for Earthquake Engineering, School of Civil Engineering, National Technical University of Athens, Athens, Greece
e-mail: ipsych@central.ntua.gr

centring the stones during construction and, practically, do not have any effect on their seismic response.

Despite their articulated construction and the fact that many of them are located in seismically active regions, quite a few classical monuments are standing for more than 2000 years, although with damages in most cases. Of course, many others have collapsed while, in early literature, there are frequent references to extensive repair of these structures because of earthquake damage, whereupon the opportunity to introduce changes in their design and construction that enhanced their earthquake resistance was taken. Thus, while not all classical monuments are intrinsically earthquake-resistant, they are in fact more resistant to earthquakes than might be expected (Psycharis et al. 2000, 2003).

Due to their spinal construction, columns and walls of ancient monuments respond to strong earthquakes with intense rocking and sliding. As a result, the dynamic analysis of ancient monuments and the assessment of their vulnerability to earthquakes is a difficult problem to treat, since their seismic response is nonlinear, complicated and very sensitive to even trivial changes of the parameters of the system or the excitation.

Several investigators have examined the seismic response of classical monuments analytically, numerically or experimentally, mostly using two-dimensional models (e.g. Allen et al. 1986; Sinopoli 1989; Psycharis 1990; Winkler et al. 1995; Psycharis et al. 2000; Konstantinidis and Makris 2005; Papaloizou and Komodromos 2009 among others) and lesser three-dimensional ones (e.g. Papantonopoulos et al. 2002; Mouzakis et al. 2002; Psycharis et al. 2003, 2013; Dasiou et al. 2009a, b). These studies have shown that such structures do not possess natural modes in the classical

Fig. 24.1 Olympieion of Athens, Greece: Left: Columns and architraves at a corner; Right: A free-standing and a fallen column showing their multi-drum construction

24.2 Main Features of the Rocking Response of Rigid Blocks

The earthquake response of classical monuments is dominated by the rocking that occurs at the joints of the stone elements, following the dynamics of rocking rigid blocks. For this reason, the main features of the rocking response of a single, free-standing block are presented in this section.

The rocking response of a free-standing rigid block, despite its apparent simplicity, is a difficult problem to treat because it is nonlinear and extremely sensitive. Thus, although the problem has been observed since late 19th century (Milne 1885, Milne and Omori 1893, Kirkpatrick 1927) and the first systematic analysis was presented by Housner in 1963, this problem continues to attract the interest of several investigators.

The nonlinear feature of the rocking response of rigid blocks is illustrated in Fig. 24.2, in which the time history of the rocking angle of an orthogonal block with dimensions: base width $b = 0.50$ m and total height $2h = 1.50$ m is shown for the El Centro (1940) earthquake amplified to four different values of the peak ground acceleration, namely, $pga = 0.60$ g, 0.70 g, 0.80 g and 0.90 g. In all cases, the coefficient of restitution, which counts for the dissipation of energy during the

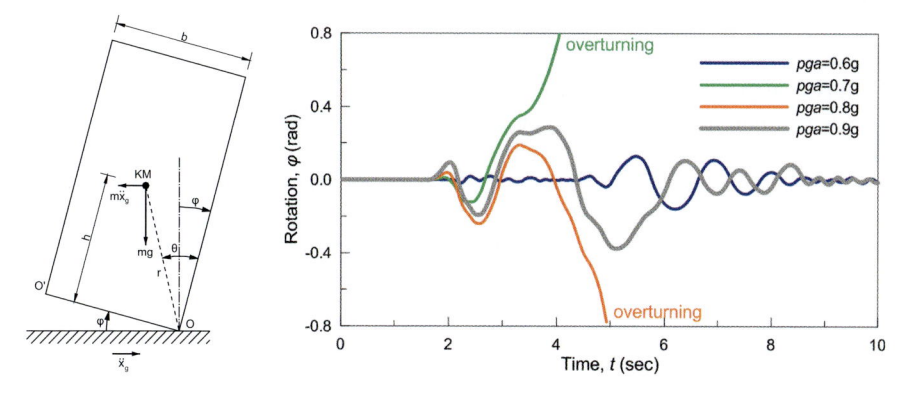

Fig. 24.2 Rocking response of an orthogonal block of dimensions $b = 0.50$ m, $2h = 1.5$ m for the El Centro (1940) earthquake amplified to several values of pga ($\varepsilon = 0.85$)

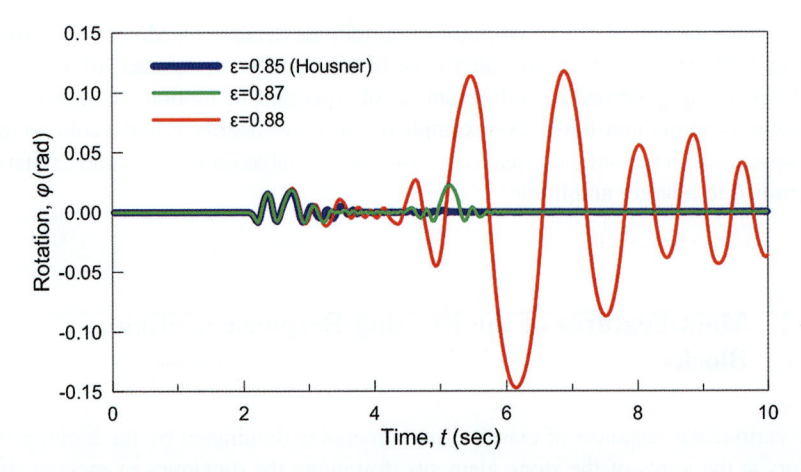

Fig. 24.3 Rocking response of an orthogonal block of dimensions $b = 0.50$ m, $2h = 1.5$ m for the El Centro earthquake amplified to $pga = 0.50$ g for various values of the coefficient of restitution ε

impacts of the block with the base, was set to $\varepsilon = 0.85$, which corresponds to the Housner's theoretical value (Housner 1963). It is seen that the response of the block is stable for $pga = 0.60$ g (blue line) while the block overturns in the direction of positive rotations for $pga = 0.70$ g (green line). If the base excitation is increased to $pga = 0.80$ g the block overturns in the opposite direction (negative rotations). However, if the base motion is amplified even more to $pga = 0.90$ g the response is stable again and overturning does not occur (grey line).

Apart from the nonlinearity, another characteristic of the rocking response is its sensitivity to even trivial changes of the parameters. This sensitivity has been proven by the non-repeatability of the same experiment (Yim and Chopra 1984; Mouzakis et al. 2002). In Fig. 24.3, the sensitivity of the response of the above-mentioned block to the value of the coefficient of restitution ε is shown. In this plot, the response of the block is shown for the El Centro record amplified to $pga = 0.50$ g and for three values of the coefficient of restitution: $\varepsilon = 0.85$ (Housner's value), $\varepsilon = 0.87$ and $\varepsilon = 0.88$.

It is seen that the response for $\varepsilon = 0.87$ (green line) is very similar with the one for $\varepsilon = 0.85$ (blue line), except for an additional small rocking response of the block around $t = 5$ s, which does not occur for $\varepsilon = 0.85$. However, if we slightly increase the coefficient of restitution to $\varepsilon = 0.88$, intense rocking occurs after $t = 4$ s with significantly larger amplitude than the amplitude in the time interval $2.0 < t < 3.5$ s when all the rocking response takes place for $\varepsilon = 0.85$. It is interesting to notice that this intense rocking for $\varepsilon = 0.88$ occurs after the strong motion of the ground excitation.

It is worthmentioning that, although, in general, a decrease in the value of ε leads to smaller rocking amplitude, due to the larger dissipation of energy during impact, it is also possible that a smaller coefficient of restitution produces larger rocking response (Aslam et al. 1980). This counter-intuitive phenomenon is attributed to the nonlinearity of the response. Note that the appropriate value of the coefficient of

Fig. 24.4 Animated rocking response under the same impulse base excitation of two similar orthogonal blocks of the same slenderness (tanθ = 0.5) and different size: base width $b = 0.50$ m for the left block and $b = 1.50$ m for the right block

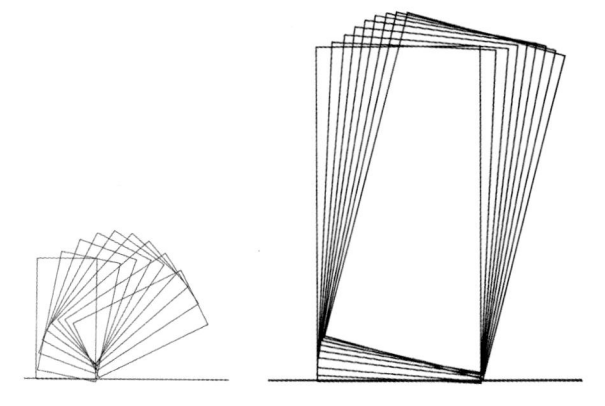

restitution is not easy to define, since experimental investigation (e.g., Priestley et al. 1978; Aslam et al. 1980) showed that the actual value of ε might be significantly different than the theoretical one of Housner, depending on the materials of the block and the base.

Concerning the parameters that affect the rocking response, it has been proven that the normalized response under harmonic excitation can be expressed solely by four dimensionless terms (Zhang and Makris 2001; Dimitrakopoulos and DeJong 2012), namely:

- The ratio ω_g/p, where $p = \sqrt{mgr/I_O}$ is the characteristic frequency parameter of the system (r is the half diagonal and I_O is the moment of inertia around point O, refer to Fig. 24.2) and ω_g is the frequency of the harmonic excitation. This ratio increases with the frequency of the excitations and the size of the block (measured through r);
- The ratio $a_g/(g\tan\theta)$, with θ being the slenderness angle (refer to Fig. 24.2) and a_g being the amplitude of the harmonic excitation. This ratio measures the strength of the excitation compared to the critical acceleration $g\tan\theta$ required for the initiation of rocking;
- The slenderness of the block, which is measured with the angle θ; and
- The coefficient of restitution ε.

Assuming that ε is known and constant, the dimensionless analysis reveals that:

- For a given base excitation (given a_g and ω_g), the response depends on the slenderness θ and the characteristic frequency p. The latter decreases inversely with the size of the block, measured with the half-diagonal r, therefore, for the same slenderness there is an important size effect on the response. Actually, among two blocks with the same slenderness θ but different size, the smaller one will experience more intense rocking than the larger one. This is shown in Fig. 24.4, in which the response of two blocks with tanθ = 0.5 but different size ($b = 0.50$ m for the left block and $b = 1.5$ m for the right one) is shown for the

same impulse base excitation. It is seen that the small block overturns while the large one does not.
- For a given block (given θ and p), the rocking response and the overturning risk greatly depend on the predominant period of the base excitation. In general, the required normalized amplitude of the base acceleration, $a_g/(g\tan\theta)$, to cause overturning decreases as the excitation period T_g increases (Zhang and Makris 2001; Dimitrakopoulos and DeJong 2012). In other words, the block is more vulnerable to long-period earthquakes than to high-frequency ones.

It should be noted that, if φ is the angle of rotation, the inequality $\varphi > \theta$ is a necessary but not a sufficient condition for overturning to occur, since it is possible that the rocking angle attains temporarily values larger than θ (i.e. $\varphi_{max} > \theta$) without overturning. Of course such cases are exceptional, since for $\varphi > \theta$ the weight of the block produces an overturning moment instead of a restoring one; thus the block will not topple only if at the same time a quite large restoring inertial force develops due to the ground motion, capable to reverse this situation and bring the block back to stable state.

The above-mentioned conclusions on the response and toppling of rigid blocks, although they have been derived for harmonic base excitations, apply qualititevely to earthquake ground motions as well, at least near-faults ones containing strong directivity pulses (Fragiadakis et al. 2016a). In this case, the frequency of the ground motion should be set equal to the frequency of the predominant pulse.

24.3 Seismic Response of Classical Monuments

The earthquake response of classical monuments is governed by the motion of the stones they are constructed of, which can rock and slide individually or in groups. In case of columns, wobbling also occurs during rocking due to the cylindrical shape of the drums. Since rocking is dominant in the dynamic behaviour, the earthquake response of classical monuments is characterized by the strong nonlinearity and the sensitivity discussed in the previous section.

A typical example of the seismic response of multi-drum columns is shown in Fig. 24.5, in which snapshots of the response of two columns of the Olympieion of Athens at two different time instances during intense ground shaking are shown. It is evident that rocking dominates the response; however, the response of each column is different, as it is significantly affected by the geometry. In particular, the height of the drums varies, while the left column has 14 drums and the right one has 15 drums.

In general, there are many 'modes' of response in which multi-block systems can respond during an earthquake and the system continuously moves from one 'mode' to another. For example, there are four 'modes' of vibration for two-block assemblies (Psycharis 1990), depicted on Fig. 24.6. For systems with many blocks, the number of 'modes' increases exponentially with the number of blocks.

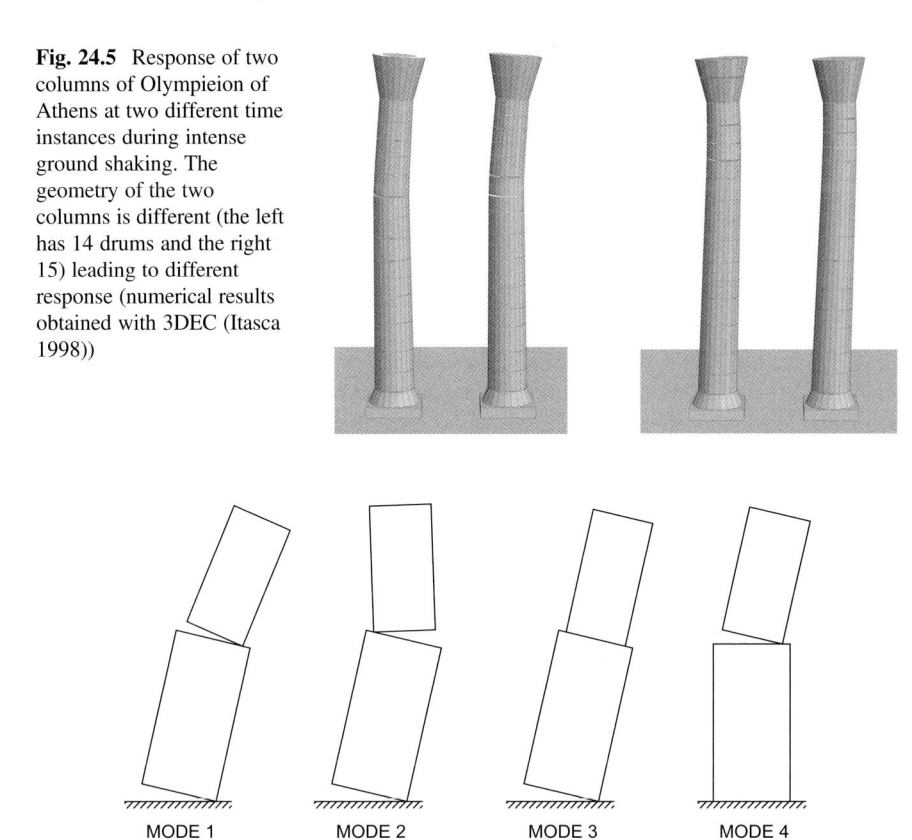

Fig. 24.5 Response of two columns of Olympieion of Athens at two different time instances during intense ground shaking. The geometry of the two columns is different (the left has 14 drums and the right 15) leading to different response (numerical results obtained with 3DEC (Itasca 1998))

Fig. 24.6 The four rocking 'modes' of vibration of a two-block assembly (Psycharis 1990)

24.4 Vulnerability to Earthquakes

As mentioned above, classical monuments can generally sustain large earthquakes without collapse in their intact condition; however, they are not earthquake proof for all seismic motions. Also, if damage is present, their vulnerability decreases significantly.

The assessment of the seismic reliability of a monument is a prerequisite for the correct decision making during a restoration process. The seismic vulnerability of the column, not only in what concerns the collapse risk, but also the magnitude of the expected maximum and residual displacements of the drums, is vital information that can help the authorities decide the necessary interventions. This assessment is not straightforward, not only because fully accurate analyses for the near-collapse state are practically impossible due to the difficulty in modelling accurately the existing imperfections and the sensitivity of the response to even small changes in the geometry, but also because the results highly depend on the ground motions characteristics.

In general, excluding the effect of damage, the vulnerability of ancient monuments depends on two main parameters: the size of the structure and the predominant period of the ground motion (Psycharis et al. 2000). These issues are discussed in the following.

24.4.1 Size Effect

Similarly to the single rocking block, the size of ancient monuments affects their dynamic response and their vulnerability to earthquakes, with larger structures being more stable than smaller ones. This is shown in Fig. 24.7, in which the minimum acceleration amplitude of a harmonic excitation of varying period, required to cause collapse (stability threshold), is shown for two cases: (a) the columns of the temple of Apollo at Bassae, Greece, of height 5.95 m; and (b) the columns of the temple of Zeus at Nemea, Greece, of height 10.33 m (Psycharis et al. 2000). Results are given for the free-standing column and the set of two columns connected with an architrave. It is seen that, for the same period of excitation, significantly larger acceleration is needed to overturn the larger columns of Zeus compared with the smaller columns of Apollo.

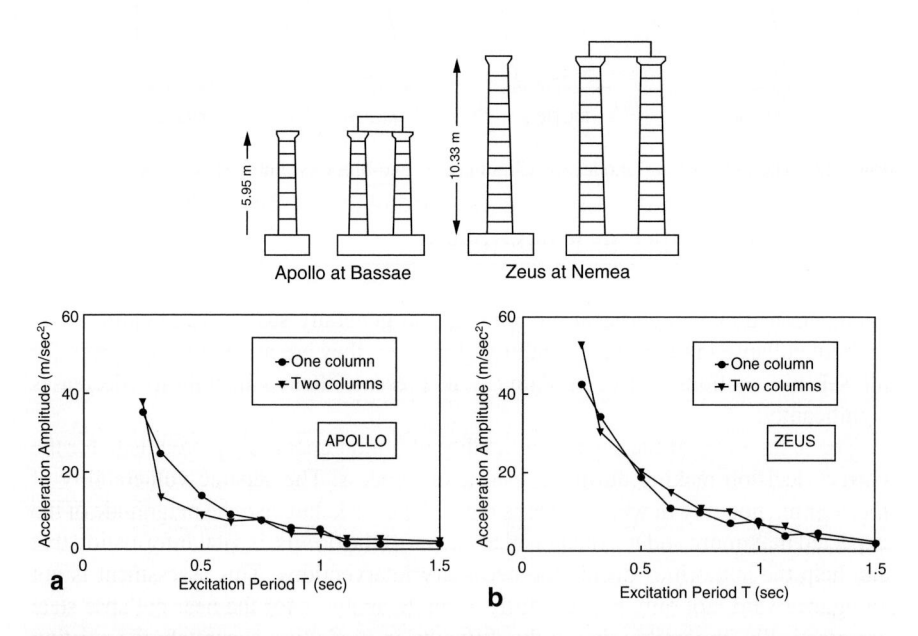

Fig. 24.7 Minimum acceleration amplitude of harmonic excitations required for the collapse of free-standing columns and sets of two columns connected with an architrave: (**a**) columns of the temple of Apollo; (**b**) columns of the temple of Zeus (Psycharis et al. 2000)

The results depicted on Fig. 24.7 also show another interesting observation: the stability threshold of each monument is similar for the free-standing column and the set of two columns. This means that restoration of fallen architraves does not necessarily lead to enhanced stability of the monument against future earthquakes. Figure 24.7 shows that such restoration of the architraves might be favourable or unfavourable depending on the characteristics of the structure and the excitation: in case of Apollo, it was generally unfavourable while in case of Zeus, it was generally favourable.

It should be mentioned that the above observation concerns the in-plane collapse of the columns (2D analyses). However, shaking table tests on sets of columns connected with architraves in line or in corner have shown that the architrave beams are quite vulnerable in the out-of-plane direction, being the first pieces that fall down. The collapse of the architraves endangers the stability of the whole monument, since it is possible that they hit the columns during their fall.

24.4.2 Effect of Predominant Period of Ground Motion

The earthquake response of ancient monuments is dominated by the rocking of the drums of the columns, therefore, it is greatly affected by the predominant period of the excitation with low-frequency earthquakes being much more dangerous than high-frequency ones. In this sense, near field ground motions, which contain long-period directivity pulses, might bring these structures to collapse.

The effect of the period of excitation to the risk of collapse in case of harmonic excitations is shown in Fig. 24.7 for the columns of the Temples of Apollo and Zeus. It is seen that, in all cases examined, the stability threshold decreases exponentially as the period of excitation increases. The same trend is also observed in Fig. 24.8, in which the stability of a free-standing column of the Parthenon of Athens, Greece is examined under near-fault earthquake excitations containing a directivity pulse of frequency f_p. In this plot, the threshold between safe (non-collapse) and unsafe (collapse) regions on the PGA–f_p plane for 3500 near-fault simulated earthquake motions with magnitudes M_w ranging from 5.5 to 7.5 and epicentral distances ranging from 0 to 20 km is shown (Psycharis et al. 2013). It is seen that the minimum required PGA for collapse of the column decreases for smaller f_p (larger predominant period).

In general, previous analyses (Psycharis et al. 2000) have shown that low-frequency earthquakes force the structure to respond with intensive rocking, whereas high-frequency ones produce significant sliding of the drums, especially at the upper part of the columns.

These results show that the choice of the earthquakes that will be used in time-history analyses is very important, as the dynamic response and the risk of collapse are sensitive to the energy and frequency content of the time history of the input ground motion. Apart from the above-mentioned strong effect of the predominant period of the ground motion, the time sequence of the various phases in the record

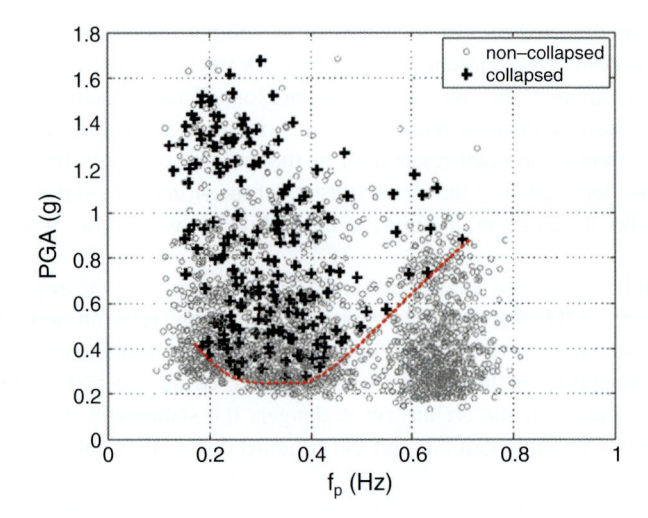

Fig. 24.8 Threshold (red line) between safe and unsafe regions on the *PGA–f*$_p$ plane for a free-standing column of the Parthenon subjected to 3500 near-fault simulated earthquake motions (f_p is the frequency of the predominant pulse contained in each record). (Psycharis et al. 2013)

might also be significant. In this sense, it is essential to constrain the selection of the base excitations to what one may call suitable surrogate ground acceleration time histories that could replicate as closely as possible the time histories of past and anticipated earthquakes.

It is evident, therefore, that the choice of which time-histories to include and which to exclude in order to constrain ground motions is an important decision. There is a balance to be struck between being not restrictive enough in the time histories used, leading to unreliable results and hence predictions due to errors and uncertainties; and being too restrictive, which leads to a too small set of time histories and hence non-conclusive results.

24.4.3 Effect of Existing Damage

Although classical monuments without significant damages are, in general, not vulnerable to usual earthquake motions, collapse can occur much easier if imperfections are present. In their current condition, ruins of ancient structures present many different types of damage (Fig. 24.9). Most common are: missing pieces (cut-offs) that reduce the contact areas, foundation problems resulting in tilting of the columns, dislocated drums from previous earthquakes and cracks in the structural elements that, in some cases, split the blocks in two parts. Such imperfections may endanger the safety of the structure in future earthquakes.

An example of the effect of existing imperfections on the stability of ancient columns is shown in Fig. 24.10 for the free-standing column of the Parthenon of

Fig. 24.9 Classical columns with significant drum dislocations. Left: Columns at Propylaia of the Acropolis of Athens, Greece; Right: Column of the temple of Hera in Samos, Greece

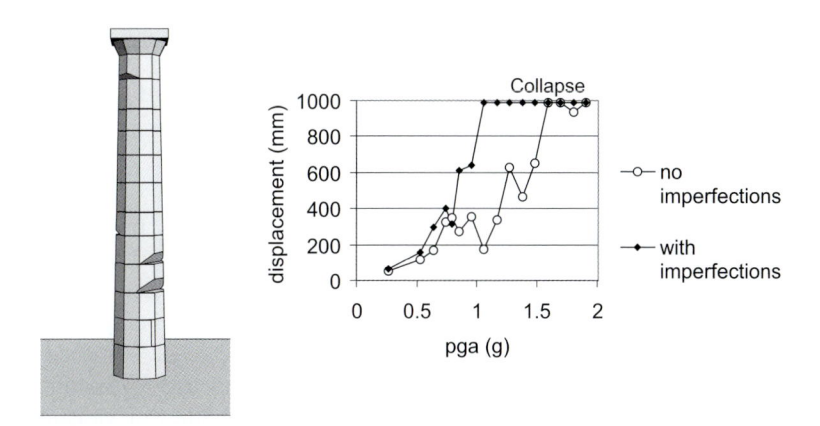

Fig. 24.10 Maximum permanent displacements of a free-standing column of the Parthenon under the Aigion, Greece (1995) earthquake amplified to several values of *PGA* without and with the imperfections shown in the left diagram (Psycharis et al. 2003)

Athens (Psycharis et al. 2003), where the maximum permanent displacement of the column is plotted versus the *PGA* of the ground motion. It is seen that the presence of the imperfections shown in the left drawing of Fig. 24.10 leads to larger displacements and significantly earlier collapse.

Similar results were obtained when the column of the Propylaia of the Acropolis of Athens with the dislocated drums (left photo in Fig. 24.9) was

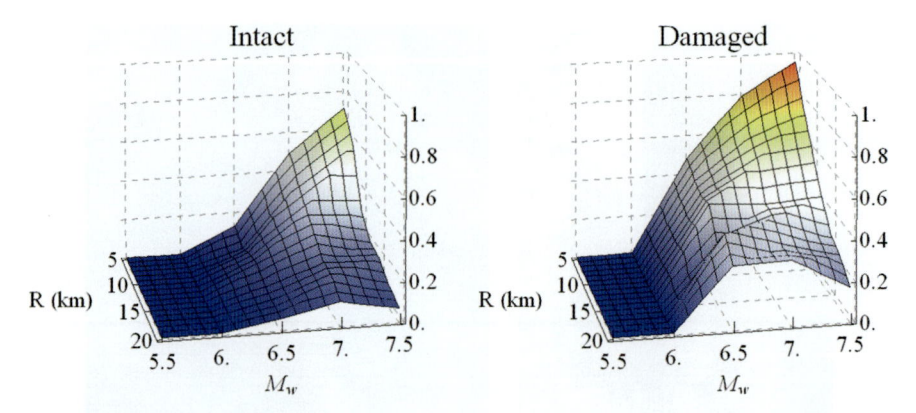

Fig. 24.11 Collapse probabilities for the intact and the damaged column of the Propylaia of the Acropolis of Athens (left photo in Fig. 24.9). The damaged column is evidently more prone to collapse (Fragiadakis et al. 2016b)

subjected to 3500 near-fault simulated earthquake motions with magnitudes M_w ranging from 5.5 to 7.5 and epicentral distances ranging from 0 to 20 km (Fragiadakis et al. 2016b). In Fig. 24.11, the collapse probabilities of the intact and the damaged column are presented as function of earthquake magnitude and distance. Evidently, the damaged column is clearly more prone to collapse compared to the one that is intact.

24.5 Performance-Based Reliability Assessment

Performance-Based Earthquake Engineering (PBEE) and seismic risk assessment combine computational tools and reliability assessment procedures to obtain the system fragility for a wide range of limit states. The seismic risk assessment requires the calculation of the failure probabilities of a pre-set number of performance objectives. According to PBEE, the acceptable level of damage sustained by a structural system depends on the level of ground shaking and its significance. Thus, the target in risk assessment is to obtain the probabilities of violating the stated performance levels, ranging from little or no damage for frequent earthquakes to severe damage for rare events.

Today, these concepts are well understood among earthquake engineers, but when classical monuments are considered the performance-based criteria may differ considerably. For example, to retrofit an ancient column one has to decide what is the 'acceptable level' of damage for a given intensity level. The approach for making such decisions is not straightforward. A consensus among various experts in archaeology and monument preservation is necessary, while a number of non-engineering decisions have to be taken.

In order to assess the risk of a monument, the performance levels of interest and the corresponding levels of capacity of the monument need first to be decided. Demand and capacity should be measured with appropriate parameters at critical locations, in accordance to the different damage (or failure) modes of the structure. Subsequently, this information has to be translated into one or a combination of engineering demand parameters (*EDP*s), e.g., permanent or maximum column deformation, drum dislocation, foundation rotation or maximum axial and shear stresses. For the *EDP*s chosen, appropriate threshold values that define the various performance objectives e.g. light damage, collapse prevention, etc. need to be established.

In case of classical columns, two engineering demand parameters (*EDP*s) have been suggested by Psycharis et al. (2013) for the assessment of their vulnerability, namely: (a) the maximum displacement at the capital normalized by the base diameter; and (b) the relative residual dislocation of adjacent drums normalized by the diameter of the corresponding drums at their interface. The first *EDP* is the maximum of the normalized displacement of the capital (top displacement) over the whole time history and is denoted as u_{top}, i.e. $u_{top} = \max[u(top)]/D_{base}$. This is a parameter that provides a measure of how much a column has been deformed during the ground shaking and also shows how close to collapse the column was brought during the earthquake. Note that the top displacement usually corresponds to the maximum displacement among all drums. The second *EDP* is the residual relative drum dislocations at the end of the seismic motion normalised by the drum diameter at the corresponding joints and is denoted as u_d, i.e. $u_d = \max(resu_i)/D_i$. This parameter provides a measure of how much the geometry of the column has been altered after the earthquake increasing thus the vulnerability of the column to future events.

These *EDP*s have a clear physical meaning and allow to easily identify various damage states and set empirical performance objectives. For example a u_{top} value equal to 0.3 indicates that the maximum displacement was 1/3 of the bottom drum diameter and thus there was no danger of collapse, while values of u_{top} larger than unity imply intense shaking and large deformations of the column, which, however, do not necessarily lead to collapse. It is not easy to assign a specific value of u_{top} that corresponds to collapse, as collapse depends on the 'mode' of deformation, which in turn depends on the ground motion characteristics. For example, for a cylindrical column that responds as a monolithic block with a pivot point at the corner of its base (Fig. 24.12a), collapse is probable to occur for $u_{top} > 1$, as the weight of the column turns to an overturning force from a restoring one when u_{top} becomes larger than unity. But, if the same column responds as a multi-drum one with rocking at all joints (Fig. 24.12b), a larger value of u_{top} can be attained without threatening the overall stability. In fact, the top displacement can be larger than the base diameter without collapse, as long as the weight of each part of the column above an opening joint gives a restoring moment about the pole of rotation of the specific part.

Based on the above defined *EDP*s, the performance criteria of Tables 24.1 and 24.2 have been adopted. For u_{top}, three performance levels were selected (Table 24.1), similarly to the ones that are typically assigned to modern structures.

Fig. 24.12 Top displacement for two extreme modes of rocking: (**a**) as a monolithic block; (**b**) with opening of all joints (displacements are shown exaggerated) (Psycharis et al. 2013)

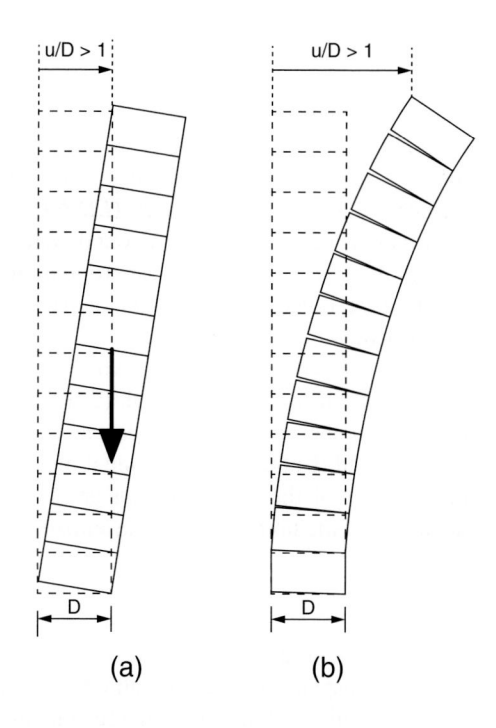

Table 24.1 Performance criteria concerning the risk of collapse (Psycharis et al. 2013)

u_{top}	Performance level	Description
0.15	Damage limitation	No danger for the column. No permanent drum dislocations expected.
0.35	Significant damage	Large opening of the joints with probable damage due to impacts and considerable residual dislocation of the drums. No serious danger of collapse.
1.00	Near collapse	Very large opening of the joints, close to partial or total collapse.

The first level (*damage limitation*) corresponds to weak shaking of the column with very small or no rocking. At this level of shaking, no damage, nor any severe residual deformations are expected. The second level (*significant damage*) corresponds to intense shaking with significant rocking and evident residual deformation of the column after the earthquake; however, the column is not brought close to collapse. The third performance level (*near collapse*) corresponds to very intense shaking with significant rocking and probably sliding of the drums. The column does not collapse at this level, as $u_{top} < 1$, but it is brought close to collapse. In most cases, collapse occurred when this performance level was exceeded. The values of u_{top} that are assigned at every performance level are based on the average assumed risk of collapse.

Three performance levels were also assigned to the normalised residual drum dislocation, u_d (Table 24.2). This *EDP* is not directly related to how close to collapse

Table 24.2 Performance criteria concerning permanent deformation (residual drum dislocations) (Psycharis et al. 2013))

u_d	Performance level	Description
0.005	Limited deformation	Insignificant residual drum dislocations without serious effect to future earthquakes
0.01	Light deformation	Small drum dislocations with probable unfavourable effect to future earthquakes
0.02	Significant deformation	Large residual drum dislocations that increase significantly the danger of collapse during future earthquakes

the column was brought during the earthquake, since residual displacements are caused by wobbling and sliding and are not, practically, affected by the amplitude of the rocking. However, their importance to the response of the column to future earthquakes is significant, as previous damage/dislocation has generally an unfavourable effect to the seismic response to future events (Psycharis 2007).

The first performance level (*limited deformation*) concerns very small residual deformation which is not expected to affect considerably the response of the column to future earthquakes. The second level (*light deformation*) corresponds to considerable drum dislocations that might affect the dynamic behaviour of the column to forthcoming earthquakes, increasing its vulnerability. The third performance level (*significant deformation*) refers to large permanent displacements at the joints that increase considerably the danger of collapse to future strong seismic motions. It must be noted, however, that the threshold values assigned to u_d are not obvious, as the effect of pre-existing damage to the dynamic response of the column varies significantly according to the column properties and the characteristics of the ground motion. The values proposed are based on engineering judgment taking into consideration the size of drum dislocations that have been observed in monuments and also the experience of the authors from previous numerical analyses and experimental tests.

This approach was applied to the free-standing column of the Parthenon of Athens subjected to 3500 near-fault simulated earthquake motions with magnitudes M_w ranging from 5.5 to 7.5 and epicentral distances ranging from 5 to 20 km (Psycharis et al. 2013). The comparison of the two proposed *EDP*s is shown in Fig. 24.13 for all ground motions considered excluding those that caused collapse. Although there is a clear trend showing that, generally, strong ground motions lead to large top displacements u_{top} during the strong shaking and also produce large permanent deformation u_d of the column, there is significant scattering of the results indicating that intense rocking does not necessarily imply large residual dislocations of the drums and also that large drum dislocations can occur for relatively weak shaking of the column. This was also observed during shaking table experiments (Mouzakis et al. 2002) where cases of intense rocking with very small residual drum displacements have been identified.

The proposed fragility assessment methodology can be applied to derive fragility curves or surfaces. For example, for the free-standing column of the Parthenon

Fig. 24.13 Comparison of u_d versus u_{top} for the free-standing column of Parthenon subjected to 3500 synthetic near-fault ground motions, excluding ground motions causing collapse (Psycharis et al. 2013)

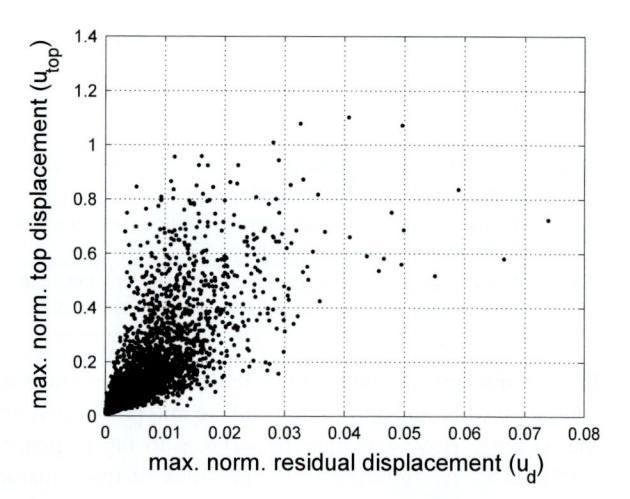

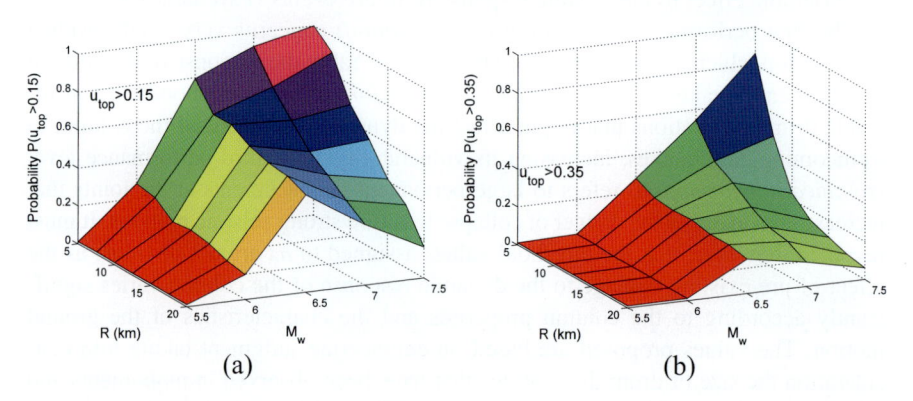

Fig. 24.14 Fragility surfaces of the Parthenon column with respect to the maximum capital displacement u_{top} for the performance levels of Table 24.1: (**a**) $u_{top} > 0.15$; (**b**) $u_{top} > 0.35$ (Psycharis et al. 2013)

Fig. 24.14 shows the fragility surfaces for two performance levels of Table 24.1 corresponding to the above-mentioned 3500 simulated near-fault ground motions. It is seen that for both *damage limitation* and *significant damage*, the exceedance probability generally increases for ground shakings of larger magnitude. However, the exceedance probability decreases with magnitude in the range $M_w = 6.5$–7.5 and $R > 15$ km. This counter-intuitive response, which was verified for real earthquakes as well, is attributed to the saturation of the *PGV* for earthquakes with magnitude larger than $M_{sat} = 7.0$ (e.g. see Rupakhety et al. 2011) while the period of the pulse is increasing exponentially with the magnitude. As a result, the directivity pulses haves small acceleration amplitude for large magnitudes, which is not capable to produce intense rocking.

Figure 24.15 shows the fragility surfaces when the *EDP* is the normalized permanent drum dislocation, u_d, and considering the performance levels of

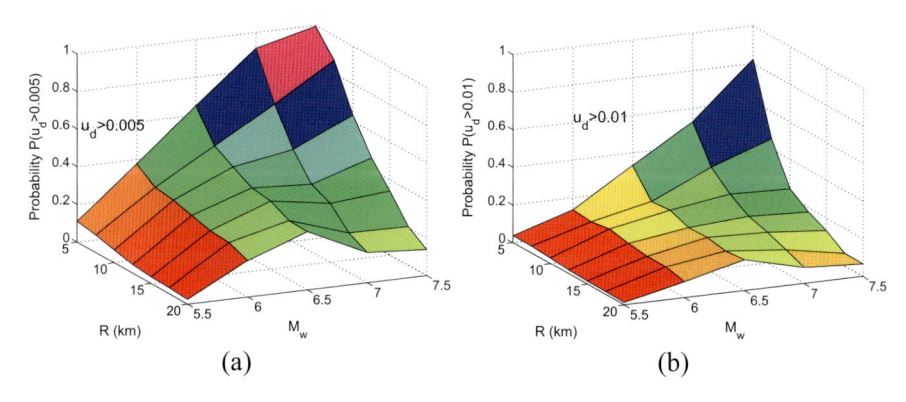

Fig. 24.15 Fragility surfaces of the Parthenon column with respect to the permanent drum dislocations, u_d for the performance levels of Table 24.2: (**a**) $u_d > 0.005$; (**b**) $u_d > 0.01$ (Psycharis et al. 2013)

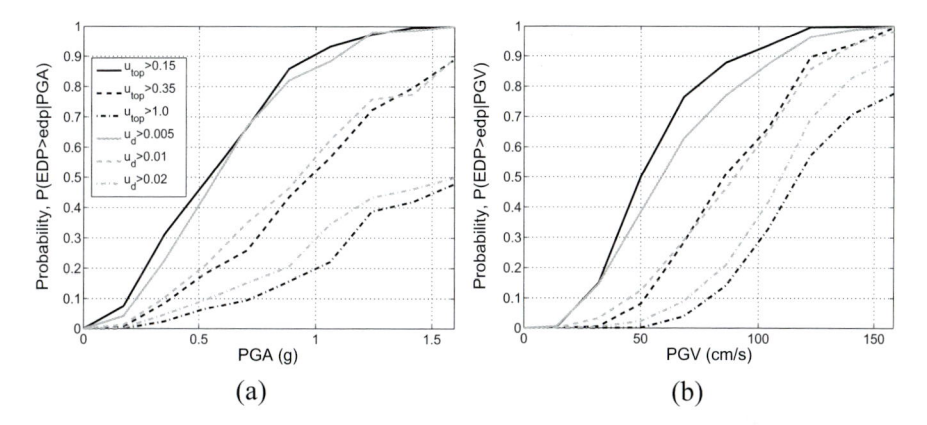

Fig. 24.16 Fragility curves of the Parthenon column using different intensity measures: (**a**) peak ground acceleration; (**b**) peak ground velocity

Table 24.2. For the *limited deformation* limit state ($u_d > 0.005$), probabilities around 0.3 are observed for magnitudes close to 6. Note that, for the column of the Parthenon with an average drum diameter about 1600 mm, $u_d > 0.005$ refers to residual displacements at the joints exceeding 8 mm. The probability of exceedance of the *light deformation* performance criterion ($u_d > 0.01$), which corresponds to residual drum dislocations larger than 16 mm, is less than 0.2 for all earthquake magnitudes examined and for distances from the fault larger than 10 km.

Finally, fragility curves for the *EDP*s thresholds defined in Tables 24.1 and 24.2 and using *PGA* and *PGV* as intensity measures, are shown in Fig. 24.16. It is seen that the probability that a moderate earthquake with $PGA \sim 0.3$ g and $PGV \sim$ 40–50 cm/s has only 10% probability to cause considerable rocking to the column with $u_{top} > 0.35$ and to produce permanent dislocations of the drums that exceed 1% of their diameter.

24.6 Summary

In this chapter the main parameters that affect the vulnerability of classical monuments to earthquakes are presented and discussed. Based on the results of previous studies, the main features of the response can be summarized as follows:

- Owing to rocking and sliding, the response is nonlinear. The nonlinear nature of the response is pronounced even for the simplest case of a rocking single block. In addition, multi-drum columns can rock in various 'modes', which alternate during the response increasing thus the complexity of the problem. The word 'mode' denotes the pattern of rocking motion rather than a natural mode in the classical sense, since rocking structures do not possess such modes and periods of oscillation.
- The dynamic behaviour is sensitive to even trivial changes in the geometry of the structure or the base-motion characteristics. The sensitivity of the response has been verified experimentally, since 'identical' experiments produced significantly different results in some cases. The sensitivity of the response is responsible for the significant out-of-plane motion observed during shaking table experiments for purely planar excitations.
- The vulnerability of the structure greatly depends on the predominant period of the ground motion, with earthquakes containing low-frequency pulses being in general much more dangerous than high-frequency ones. The former force the structure to respond with intensive rocking, whereas the latter produce significant sliding of the drums, especially at the upper part of the columns.
- The size of the structure affects significantly the stability, with bulkier structures being much more stable than smaller ones of the same slenderness.
- Classical monuments are not, in general, vulnerable to earthquakes. However, their stability might have been significantly reduced in the damaged condition that they are found today. Types of damage that might increase their vulnerability to earthquakes include cut-off of drums, displaced drums, inclined columns due to foundation failure, cracks in the stones, etc.
- Two engineering demand parameters (*EDPs*) are adopted for the assessment of the vulnerability of classical columns in terms of PBEE: (a) the maximum displacement at the capital normalized by the base diameter; and (b) the relative residual dislocation of adjacent drums normalized by the diameter of the corresponding drums at their interface. Three performance levels are assigned to each *EDP* and the values of the corresponding thresholds are proposed.

References

Allen RH, Oppenheim IJ, Parker AP, Bielak J (1986) On the dynamic response of rigid body assemblies. Earthq Eng Struct Dyn 14:861–876

Aslam M, Godden WG, Scalise DT (1980) Earthquake rocking response of rigid bodies. J Struct Div ASCE 106(ST2):377–392

Dasiou M-E, Mouzakis HP, Psycharis IN, Papantonopoulos C, Vayas I (2009a) Experimental investigation of the seismic response of parts of ancient temples. Prohitech Conference, Rome, 21–24 June

Dasiou M-E, Psycharis IN, Vayias I (2009b) Verification of numerical models used for the analysis of ancient temples. Prohitech Conference, Rome, 21–24 June

Dimitrakopoulos EG, DeJong MJ (2012) Revisiting the rocking block: closed form solutions and similarity laws. Proc R Soc A 468(2144):2294–2318

Fragiadakis M, Psycharis I, Cao Y, Mavroeidis GP (2016a) Parametric investigation of the dynamic response of rigid blocks subjected to synthetic near-source ground motion records. ECCOMAS Congress, Crete Island, Greece, 5–10 June

Fragiadakis M, Stefanou I, Psycharis I (2016b) Vulnerability assessment of damaged classical multidrum columns. In: Sarhosis V, Bagi K, Lemos JV, Milani G (eds) Computational modeling of masonry structures using the discrete element method. IGI Global, Hershey

Housner GW (1963) The behaviour of inverted pendulum structures during earthquakes. Bull Seismol Soc Am 53:403–417

Itasca Consulting Group (1998). 3DEC – universal distinct element code. Minneapolis, USA

Kirkpatrick P (1927) Seismic measurements by the overthrow of columns. Bull Seismol Soc Am 17(2):95–109

Konstantinidis D, Makris N (2005) Seismic response analysis of multidrum classical columns. Earthq Eng Struct Dyn 34:1243–1270

Milne J (1885) Seismic experiments. Trans Seismol Soc Japan 8:1–82

Milne J, Omori F (1893) On the overturning and fracturing of brick and columns by horizontally applied motion. Seismol J Japan 17:59–86

Mouzakis H, Psycharis IN, Papastamatiou DY, Carydis PG, Papantonopoulos C, Zambas C (2002) Experimental investigation of the earthquake response of a model of a marble classical column. Earthq Eng Struct Dyn 31:1681–1698

Papaloizou L, Komodromos P (2009) Planar investigation of the seismic response of ancient columns and colonnades with epistyles using a custom-made software. Soil Dyn Earthq Eng 29:1437–1454

Papantonopoulos C, Psycharis IN, Papastamatiou DY, Lemos JV, Mouzakis H (2002) Numerical prediction of the earthquake response of classical columns using the distinct element method. Earthq Eng Struct Dyn 31:1699–1717

Priestley MJN, Evison RJ, Carr AJ (1978) Seismic response of structures free to rock on their foundation. Bull N Z Nat Soc Earthq Eng 11(3):141–150

Psycharis IN (1990) Dynamic behaviour of rocking two-block assemblies. Earthq Eng Struct Dyn 19:555–575

Psycharis I (2007) A probe into the seismic history of Athens, Greece from the current state of a classical monument. Earthquake Spectra 23:393–415

Psycharis IN, Papastamatiou DY, Alexandris AP (2000) Parametric investigation of the stability of classical columns under harmonic and earthquake excitations. Earthq Eng Struct Dyn 29:1093–1109

Psycharis IN, Lemos JV, Papastamatiou DY, Zambas C, Papantonopoulos C (2003) Numerical study of the seismic behaviour of a part of the Parthenon Pronaos. Earthq Eng Struct Dyn 32:2063–2084

Psycharis I, Fragiadakis M, Stefanou I (2013) Seismic reliability assessment of classical columns subjected to near-fault ground motions. Earthq Eng Struct Dyn 42:2061–2079

Rupakhety E, Sigurdsson SU, Papageorgiou AS, Sigbjörnsson R (2011) Quantification of ground-motion parameters and response spectra in the near-fault region. Bull Earthq Eng 9:893–930

Sinopoli A (1989) Dynamic analysis of a stone column excited by a sine wave ground motion. Appl Mech Rev Part 2(44):246–255

Winkler T, Meguro K, Yamazaki F (1995) Response of rigid body assemblies to dynamic excitation. Earthq Eng Struct Dyn 24:1389–1408

Yim CS, Chopra AK (1984) Dynamics of structures on two-spring foundation allowed to uplift. J Eng Mech ASCE 110(7):1124–1146

Zhang J, Makris N (2001) Rocking response of free-standing blocks under cycloidal pulses. J EngMech, ASCE 127:473–483

Chapter 25
What Seismic Risk Do We Design for When We Design Buildings?

Iunio Iervolino

Abstract This paper discusses two issues related to the seismic performance of code-conforming structures from the probabilistic standpoint: (i) the risk structures are implicitly exposed to when designed via state-of-the-art codes; (ii) which earthquake scenarios are expected to erode the portion of safety margins determined by elastic seismic actions for these structures. Both issues are addressed using recent research results referring to Italy.

Regarding (i), during the last few years, the Italian earthquake engineering community is putting effort to assess the seismic risk of structures designed according to the code currently enforced in the country, which has extended similarities with Eurocode 8. For the scope of the project, five structural typologies were designed according to standard practice at five sites, spanning a wide range of seismic hazard levels. The seismic risk assessment follows the principles of performance-based earthquake engineering, integrating probabilistic hazard and vulnerability, to get the annual failure rates. Results, although not fully consolidated yet, show risk increasing with hazard and uneven seismic reliability across typologies.

With regard to (ii) it is discussed that, in the case of elastic design actions based on probabilistic hazard analysis (i.e., uniform hazard spectra), exceedance of spectral ordinates can be likely-to-very-likely to happen in the epicentral area of earthquakes, which occur relatively frequently over a country such as Italy. Although this can be intuitive, it means that design spectra, by definition, do not necessarily determine (elastic) design actions that are conservative for earthquakes occurring close to the construction site. In other words, for these scenarios protection is essentially warranted by the rarity with which it is expected they occur close to the structure

---

This manuscript is largely based on the papers by Iervolino et al. (2017) and Iervolino and Giorgio (2017)

I. Iervolino (✉)
Dipartimento di Strutture per l'Ingegneria e l'Architettura, Università degli Studi di Napoli Federico II, Naples, Italy
e-mail: iunio.iervolino@unina.it

© Springer International Publishing AG, part of Springer Nature 2018
K. Pitilakis (ed.), *Recent Advances in Earthquake Engineering in Europe*,
Geotechnical, Geological and Earthquake Engineering 46,
https://doi.org/10.1007/978-3-319-75741-4_25

and further safety margins implicit to earthquake-resistant design (i.e., those discussed in the first part).

25.1 Introduction

In the current state-of-the-art seismic codes (e.g., the Italian building code, CS.LL. PP. 2008, NTC08 hereafter, similar to Eurocode 8 or EC8, CEN 2004) structural performance, with respect to violation of given limit states (*failure* hereafter), must be verified for levels of ground motions associated with specific exceedance return period (T_r) at the building site. In case of ordinary structures, for example, safety verifications for *life-safety* and *collapse-prevention* limit states are required against ground motion levels that are exceeded on average once every 475 and 975 years (probabilities of exceedance of 10% and 5% in 50 years), respectively.[1] In such a design practice, if failure were to always occur for intensities larger than those considered during design, and never did occur for intensities lower than the design one, then the risk of failure (i.e., the seismic risk) would be equal to the exceedance rate of the design intensity, that is the reciprocal of the return period. However, thanks to code requirements, it is generally expected that the seismic risk of failure is smaller than that of exceedance of the design ground motion. On the other hand, these further safety margins are neither explicitly controlled nor quantified, which means that the resulting seismic risk, that is the rate of earthquakes causing failure of code-conforming structures, is implicit to structural design.[2]

When modern codes are concerned, a rational safety goal might be that designing two different structures for the seismic intensity with the same exceedance return period brings comparable seismic risk. For example, two structures belonging to the same structural typology, with the same use, designed in different sites, or different structural typologies designed for the same limit state at the same site. However, because there's no probabilistic control beyond exceedance of elastic design actions, it is not granted neither that the same exceedance probability determines the seismic risk nor that such a risk is necessarily acceptable.

The research work discussed herein intends to shed some light on what is the seismic risk of (Italian) code-conforming structures designed for seismic actions based on probabilistic seismic hazard analysis and, given that failure is allowed by state-of-the-art codes, in which earthquakes it is more likely. To this aim, the rest of the manuscript is divided in two parts. In the first one, the results of a large research project attempting to assess the implicit-by-design seismic risk of standard code-conforming buildings, is discussed. For the scope of the project, five structural

---

[1] In EC8 the same actions are used for the limit states identified as *significant damage* and *near collapse*, respectively.

[2] Other quantities such as material design characteristics or design loads originate from probabilistic considerations, yet their reflection of the global safety margins of the structure is structure-specific and is not explicitly controlled.

typologies were designed according to the most recent seismic code and standard practice at five sites, spanning a wide range of design hazard from low- to high-seismicity. These structures are also finely modeled to capture dynamically their three-dimensional non-linear behavior during earthquakes. The assessment of their risk follows the principles of performance-based earthquake engineering integrating probabilistic hazard and vulnerability, to get the annual failure rates.

Once the rate of earthquakes causing violation of the performance levels of interest is assessed, it may be interesting to change the perspective and to look, from the seismic hazard side, what are the earthquake scenarios (i.e., magnitude and location with respect to the construction site) for which design intensity is exceeded with high probability. To this aim, it is discussed that, in the case of elastic actions based on probabilistic hazard analysis (i.e., uniform hazard spectra), exceedance of spectral ordinates can be likely-to-very-likely to happen in the epicentral area of earthquakes, which are not necessarily of extreme magnitude. Although this can be intuitive, it means that design spectra do not necessarily determine (elastic) design actions conservative for moderate-to-high magnitude earthquakes (i.e., those occurring every few years over Italy) in case they occur close to the construction site. In other words, for these scenarios protection is basically warranted by the rarity with which it is expected they occur close to the structure or, in case of occurrence, by further safety margins implicit to earthquake-resistant design (i.e., those not explicitly controlled that are discussed in the first part).

25.2 The RINTC 2015–2017 Project

To quantitatively address the seismic risk code-conforming design implicitly exposes structures to, a large research project is ongoing in Italy. This project, named *Rischio Implicito di strutture progettate secondo le Norme Tecniche per le Costruzioni* (RINTC),[3] has been developed by a joint working group formed between *Rete dei Laboratori Universitari di Ingegneria Sismica* (ReLUIS) and *Centro Europeo di Ricerca e Formazione in Ingegneria Sismica* (EUCENTRE), with the funding of *Presidenza del Consiglio dei Ministri – Dipartimento della Protezione Civile* (see, RINTC Workgroup 2017).

In the RINTC project, structures, belonging to a variety of typologies and configurations, were designed according to the current Italian code provisions in a number of sites at different hazard levels (Milan or MI, Caltanissetta or CL, Rome or RM, Naples or NA, and L'Aquila or AQ) and local site conditions (A and C according to EC8 classification). In Fig. 25.1 the considered sites are shown on the official Italian map of peak ground acceleration (PGA) with 475 years exceedance

---

[3]Project website's URL: http://www.reluis.it/index.php?option=com_content&view=article&id=549&Itemid=198&lang=it

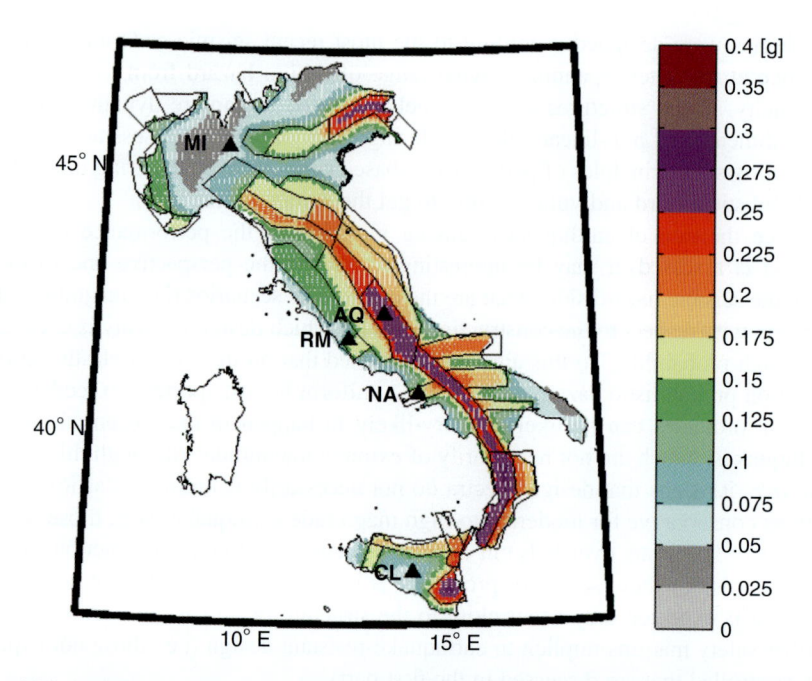

Fig. 25.1 Location of construction sites considered in the RINTC project on the official seismic hazard map used for design (Stucchi et al. 2011) in terms of peak ground acceleration (PGA) with $T_r = 475$ years on A-type EC8 soil site class

return period, which is the basis for design actions in NTC08. The map refers to A-type soil site class (see Stucchi et al. 2011 for details).

The structures have been designed for two code-defined limit states, that is, (a) *damage-control* and (b) *life-safety*.[4] The buildings are considered as *ordinary*, that is, the reference design actions at the construction site are those with return period equal to 50 years and 475 years for the former and latter limit states respectively. Elastic design action is represented by the *uniform hazard spectrum*, or UHS, for the site; i.e., the spectrum whose ordinates have all the return period of exceedance of interest (see also Sect. 25.8).[5]

The final results of the project are represented by the annual failure rates of the considered code-conforming structures. Failure is herein understood as the violation of two different performance levels: *convenient-to-repair damage* to non-structural elements (e.g., infills) and *global collapse* (i.e., life-safety-threatening structural failure). The risk is quantified in a state-of-the-art approach referring to performance-based earthquake engineering (PBEE; Cornell and Krawinkler 2000). In fact, for all the structures the failure rates are obtained by integrating probabilistic

---

[4]Base-isolated structures (to follow) are designed for *collapse-prevention*.

[5]In fact, in the Italian code spectra are UHS' approximated by via a simplified EC8-type functional form.

seismic structural vulnerability (i.e., *fragility*) and seismic hazard for the sites where the structures are located. To compute the failure rates, Eq. (25.1) is employed.

$$\lambda_f = \int_0^{+\infty} P[failure|IM = im] \cdot |d\lambda_{IM}(im)| \qquad (25.1)$$

In the equation, IM indicates a ground motion intensity measure, while $|d\lambda_{IM}(im)| = -[d\lambda_{IM}(im)/d(im)] \cdot d(im)$ is the differential of $\lambda_{IM}(im)$, or the *hazard curve*. It is the function providing the annual rate of earthquakes causing the exceedance of an IM threshold, indicated as *im*. The hazard curve is obtained from probabilistic seismic hazard analysis or PSHA (Cornell 1968); to follow. $P[failure|IM = im]$ is the *fragility* function of the structure under analysis. It provides the probability of failure for $IM = im$; i.e., for any arbitrary value of the ground motion intensity measure.

25.3 Structures and Modeling

The five structural types of buildings refer to standard modern constructions and are widely representative of residential or industrial structures in Italy. Design procedures refer as much as possible to common professional engineering practice. The considered cases as of the end of 2017 (the project is still ongoing) are:

1. cast-in-place reinforced concrete (RC) regular 3-, 6-, and 9-story residential moment-resisting-frame structures, designed via modal analysis (Camata et al. 2017), with the following configurations:

    (a) bare-frames (BF);
    (b) pilotis-frames (PF);
    (c) infilled-frames (IF);
    (d) (9-storey) with concrete structural walls (SW);

2. un-reinforced masonry (URM) 2- and 3-story residential buildings, with four different geometries, designed with the *simple building* and *linear* or *non-linear static analysis* approaches (Camilletti et al. 2017), with the following configurations

    (a) regular;
    (b) irregular;

3. pre-cast reinforced concrete (PRC) 1-story industrial buildings with two different plan geometries and two different heights (Ercolino et al. 2017), with the following configurations:

    (a) without cladding panels;
    (b) with cladding panels;

4. steel (S) 1-story industrial buildings (Scozzese et al. 2017) with two different plan geometries and two different heights, in analogy of configurations with respect to PRC:

    (a) without cladding panels;
    (b) with cladding panels;

5. base-isolated (BI) 6-story reinforced concrete residential buildings (Ponzo et al. 2017) with base isolation system made of:

    (a) rubber bearings (HDRB);
    (b) double-curvature friction pendulums (DCFP);
    (c) hybrid (HDRB and sliders).

In the computation of failure rates, record-to-record variability of seismic response is the primary source of uncertainty; i.e., structural models are generally deterministic. However, for selected cases (indicated as ModUnc) of each typology, the uncertainty in structural modeling and in design has been accounted for following the approach described in Franchin et al. (2017); however, the effect of factors, such as quality of construction or design errors, was always neglected. Moreover, one selected case of 9-story RC with structural walls, includes explicit modeling of soil-structure-interaction (SSI).

Table 25.1 summarizes the case studies at the end of 2017. Note that, to reduce the effort, not all structures have been designed for all five sites, although most of

Table 25.1 Designed and analyzed structures at the end of 2017

Type	Soil	MI	NA	AQ
RC	A	–	–	9-story (BF/PF/IF)
	C	3/6/9-story (BF/PF/IF)	3/6/9-story (BF/PF/IF)	3/6/9-story (BF/PF/IF)
		9-story SW	ModUnc	ModUnc
			9-story SW (also w/ SSI)	9-story SW
URM	A	2/3-story, regular/irregular	2/3-story, regular/irregular	2/3-story, regular
				ModUnc
	C	2/3-story, regular/irregular	2/3-story, regular/irregular	2/3-story, regular/irregular
PRC	A	1-story, 4 geometries	1-story, 4 geometries	1-story, 4 geometries
	C	1-story, 4 geometries	1-story, 4 geometries	1-story, 4 geometries
S	A	1-story, 4 geometries w/ and w/o panels	1-story, 4 geometries w/ and w/o panels	1-story, geometry 1/2/3/4 w/ and w/o panels
	C	1-story, 4 geometries w/ and w/o panels	1-story, 4 geometries w/ and w/o panels	1-story, 4 geometries w/ and w/o panels
BI	A	–	–	–
	C	–	6-story, HDRB/HDRB+slider	6-story, HDRB/HDRB w/ slider/DCFP (11 configurations)
				ModUnc

them have been designed for at least three sites reflecting low, moderate and high hazard levels (Milan, Naples, and L'Aquila, respectively)

25.4 Hazard

Equation (25.1) requires hazard curves to compute failure rates. The results of the probabilistic hazard study at the basis of NTC08 are available at http://esse1-gis.mi.ingv.it/. They are given in terms of hazard curves for 5%-damped (pseudo) spectral acceleration on A-type soil site class for eleven oscillation periods (*T*) ranging from 0 s (PGA) to 2 s, computed for a grid featuring more than ten-thousands locations that covers the entire country. The curves are discretized at nine return periods, between 30 years and 2475 years. In the RINTC project the spectral (pseudo) acceleration, *Sa*, at the fundamental period of each structure (T_1) is chosen as the ground motion intensity measure. Therefore, due to limitations in the soil type, oscillation periods for which *Sa* hazard is available, and return periods at which hazard is computed in the official study, the hazard curves at the sites of interest had to be re-computed for the scope of the RINTC project.

The hazard curves were calculated according to Eq. (25.2), where ν_i, $i = \{1, 2, \ldots, s\}$, is the rate of earthquakes above a minimum magnitude for each of the *s* seismic sources affecting the site of interest. The term $f_{M,R,i}(m,r)$ is the joint probability density function of magnitude (*M*) and source-to-site distance (*R*) for the i-th source, and $P[IM > im|M = m, R = r]$ is the probability of exceeding the IM threshold conditional to $\{M, R\}$, provided by a ground motion prediction equation (GMPE).

$$\lambda_{IM}(im) = \sum_{i=1}^{s} \nu_i \cdot \iint_{M,R} P[IM > im|M = m, R = r] \cdot f_{M,R,i}(m,r) \cdot dm \cdot dr \quad (25.2)$$

If the calculation of Eq. (25.2) is repeated for all possible values of *im* within an interval of interest, one obtains the hazard curve providing $\lambda_{IM}(im)$ as a function of *im*. As an illustration, Fig. 25.2 shows PGA hazard curves (for soil site class B according to EC8) for the five locations of the project. The curves were computed considering the *branch 921* of the logic tree described in Stucchi et al. (2011), which features the source characterization described in Meletti et al. (2008) and the GMPE of Ambraseys et al. (1996).[6] These models that constitute the core of the hazard model developed to produce the official seismic hazard map used for design in Italy, and were also used in the RINTC to determine the seismic risk according to Eq. (25.1).

---

[6]Note that assessing the performance of some structures required considering spectral accelerations at vibration periods not contemplated by the GMPE of Ambraseys et al. (1996), for these cases that of Akkar and Bommer (2010) was employed.

Fig. 25.2 Example of hazard curves for the considered sites in terms of PGA on siite class B (EC8 classification)

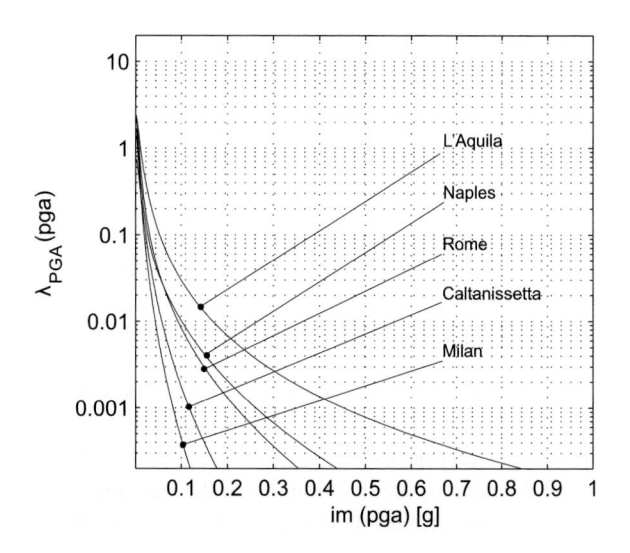

In addition, *disaggregation* of seismic hazard (e.g., Bazzurro and Cornell 1999), was carried out. It was required to perform hazard-consistent record selection required to run the non-linear dynamic analyses forming the basis of the risk assessment.

25.5 Fragility

Three-dimensional computer models were developed for all the designed structures with the aim of evaluating their seismic performance via non-linear dynamic analysis. The structural response measure or EDP (*engineering demand parameter*) considered was the maximum (in the two horizontal directions of the structure) demand over capacity ratio, expressed in terms of interstory drift angle or roof-drift angle. The main failure criterion for the assessment of global collapse was the drift corresponding to the 50% drop in base shear from the static push-over analysis (see RINTC Workgroup 2017). However, for some structural configurations some additional failure criteria were needed, for example, PRC required control of failure of connections, while base-isolated buildings required specific criteria for failure of the isolation system. For damage to non-structural elements, multiple failure criteria considering the extent of damage over the building, were considered.

All models are lumped-plasticity, except for the industrial steel building cases that were modeled using distributed plasticity elements. All structures are analyzed with OPENSEES (Mazzoni et al. 2006) apart from the masonry structures that are analyzed using TREMURI (Lagomarsino et al. 2013).

It was mentioned that $P[failure|IM = im]$ as a function of *im* is the fragility function of the structure. In this study, for each considered structure, the fragility curve was computed via non-linear dynamic analysis using Eq. (25.3). To this aim, the domain of IM, that is $Sa(T_1)$, has been discretized to ten values, corresponding to the following return periods from the hazard curve of interest: $T_r = \{10, 50, 100, 250, 500, 1000, 2500, 5000, 10000, 100000\}$ years.

$$P[failure|IM = im_i]$$
$$= \left\{1 - \Phi\left[\frac{\log(edp_f) - \mu_{\log(EDP|IM=im_i)}}{\sigma_{\log(EDP|IM=im_i)}}\right]\right\} \cdot \left(1 - \frac{N_{col,IM=im_i}}{N_{tot,IM=im_i}}\right) + \frac{N_{col,IM=im_i}}{N_{tot,IM=im_i}}$$
(25.3)

In the equation, edp_f indicates structural seismic capacity for the performance level of interest; $\{\mu_{\log(EDP|IM=im_i)}, \sigma_{\log(EDP|IM=im_i)}\}$ are the mean and standard deviation of the logs of EDP when $IM = im_i$, $i = \{1, ..., 10\}$; $\Phi(\cdot)$ is the cumulative Gaussian distribution function; $N_{col,IM=im_i}$ is the number of collapse cases (i.e., those reaching global instability according to the terminology in Shome and Cornell 2000); and $N_{tot,IM=im_i}$ is the number of ground motion records with $IM = im_i$.

The method to probabilistically evaluate structural response, and then fragility, was the *multi-stripe* nonlinear dynamic analysis (e.g., Jalayer 2003). To select the ground motion records to be used as input for dynamic analysis, the hazard-consistent *conditional spectrum* (CS) approach (e.g., Lin et al. 2013), has been considered. Ground motion record sets selected for each CS are consistent with the earthquake scenarios (expressed in terms of magnitude and source-to-site distance) that contributed the most to $Sa(T_1) = im_i$ according to disaggregation of site hazard. Because the scenarios controlling the hazard, in general, change with the specific value of $Sa(T_1)$ considered, different sets of records were selected for each hazard level (see Iervolino et al. 2017, for some details). All the analyses neglected, so far, the vertical components of ground motion as specific analyses show no need to take them into account.

25.6 Trend of Failure Rates

Equation (25.1) can be used for the computation of failure rates only for the values of $\lambda_{IM}(im)$ provided by hazard analysis. The latter, has a limit at $1/T_r = 10^{-5}$ [event/year]; in fact, no $Sa(T_1)$ values for return periods longer than $T_r = 10^5$ years were calculated, to avoid large hazard extrapolations. Therefore, it has been conservatively assumed that ground motions with an IM larger than that corresponding to $T_r = 10^5$ years, cause failure with certainty. This means that the failure rate has been approximated in excess by Eq. (25.4).

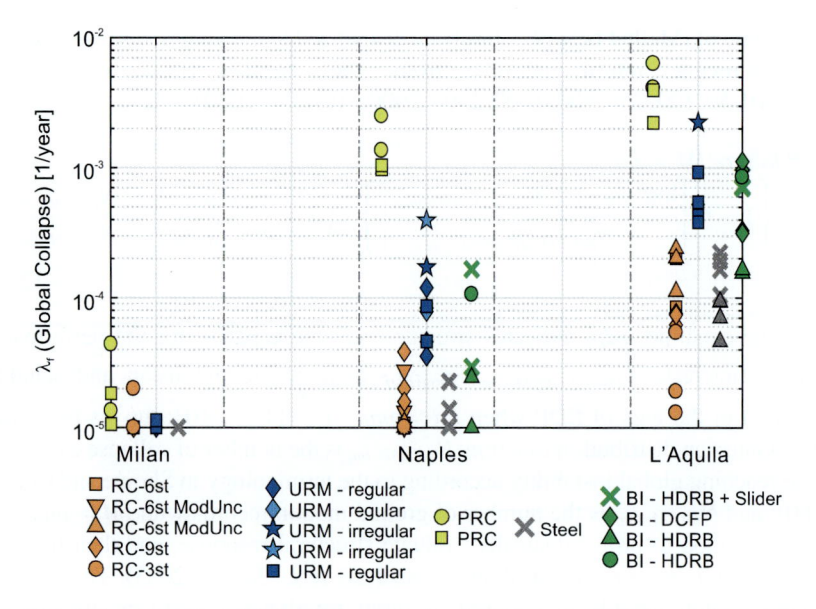

Fig. 25.3 Failure rates for global collapse (soil site class C) at three of the considered sites in ascending order of hazard. (Figure adapted from Iervolino et al. 2017)

$$\lambda_f = \int_0^{im_{T_r=10^5}} P[failure|IM = im] \cdot |d\lambda_{IM}(im)| + 10^{-5} \quad (25.4)$$

In the equation, $im_{T_r=10^5}$ indicates the last available *im*-value for which a return period of exceedance has been calculated. Consequently, in those cases when the first part of the integral is negligible with respect to 10^{-5}, then Eq. (25.4) only allows to state that the annual failure rate is lower than 10^{-5}.

Figures 25.3 reports the global collapse failure rates, as of the end of 2016, for soil site class C at three of the considered sites. These results, although not final, indicate the following:

1. as a general trend, the collapse failure rates generally tend to increase with the site hazard, independent of the structural type considered (likely due to over-strength imposed at moderate-to-low hazard sites by, for example, minimum design requirements; see also Suzuki et al. 2017);
2. the failure rates tend not to be uniform among different structural types designed for the same site hazard;
3. in some cases, the collapse failure rates are so low that only an upper bound to the actual failure rate can be provided; i.e., $\lambda_f \leq 10^{-5}$; however, in other cases it is comparable to (or larger than) the annual rate of exceedance of the design seismic intensity; e.g., $1/475 = 0.0021$.

Although these general trends clearly emerge from the last three years of the RINTC project, it is emphasized that it is still ongoing and several of these results (and other not shown here) are undergoing verification and investigation towards consolidation. Caution should be applied in using all results presented. For example, the critically-high risk exhibited by some PRC structures are likely due to some design and modeling options of beam-to-column connections and, therefore, are not definitive. Similarly, the comparatively high collapse failure rates of base-isolated strctures seem due to their more controlled behavior during design and the lower margin of safety with respect to collapse beyond the maximum design displacement; conversely, base-isolated structures show comparatively lower rates of onset of damage to non-structural elements (not shown here).

25.7 The Nature of Uniform-Hazard Design Spectra

It has been discussed with what frequency (annual rate) failure is expected for code-conforming structures; in this part of the paper it is analyzed which earthquakes erode the safety margins that depend on the elastic design seismic actions (Iervolino and Giorgio 2017).

The latest version of the Italian earthquake catalogue (CPTI15; http://emidius.mi.ingv.it/CPTI15-DBMI15/), assigns moment magnitude larger than six to thirteen earthquakes in the 1915–2014 period, which translates to an average of one event every eight years in the past century. During the last decade, among the main (i.e., severely damaging) seismic sequences for the country, one counts that of L'Aquila (2009), that of Emilia (2012) and that of central Italy 2016–2017 (the latter not included in CPTI15), whose largest (moment) magnitude earthquakes were 6.3, 6.1 and 6.5, respectively. In the same period of these events, NTC08 went into effect, which, as mentioned, prescribe seismic design actions determined on a probabilistic basis by means of the uniform hazard spectra or UHS' that were also used for design in the RINTC project.

The recent seismic events provided an unprecedented level of instrumental recordings, for the country (see for example Luzi et al. 2017). These data allow a comparison of actually-observed seismic actions with their code-prescribed counterparts used for designing new structures. Said comparison has repeatedly shown registered seismic actions, in the epicentral areas, systematically exceeding design spectra, which, in turn, vamped a debate on whether the design actions were incorrectly evaluated. The objective herein is to demonstrate that such exceedance is well expected based on the nature of UHS'. Observed exceedances cannot be considered sufficient to claim that the code-mandated seismic actions underestimate the seismic hazard. On the contrary, they are a foreseeable consequence of the philosophy that underlies definition of seismic actions in the code, when it is based on probabilistic seismic hazard. Consequently, it is also shown that UHS represent design action likely exceeded in the epicentral areas of earthquakes occurring relatively frequently in Italy. Thus, in these events, safety is mostly entrusted to the safety margins beyond the elastic design spectrum.

To prove the proposition, the starting point is discussing the seismic actions observed during the 2009 L'Aquila earthquake (e.g., Chioccarelli et al. 2009).

Fig. 25.4 Response spectra of horizontal ground motion recorded at L'Aquila (AQV station) during the 6.3 moment-magnitude earthquake of 2009 and code spectrum for $T_r = 475$ years (soil site class is B according to the EC8 classification, as reported for AQV the ITalian Accelerometric Archive; http://itaca.mi.ingv.it/)

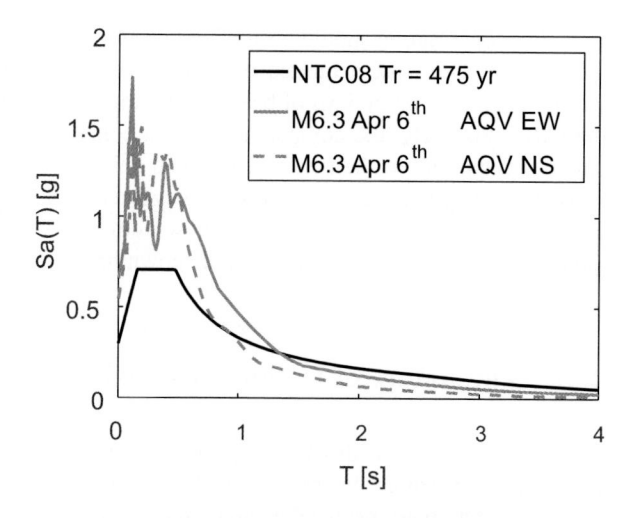

Figure 25.4 shows the spectra of the horizontal components of seismic ground motion recorded in L'Aquila (AQV monitoring station of the *Rete Accelerometrica Nazionale* managed by the Italian *Dipartimento della Protezione Civile*; http://ran.protezionecivile.it) during the considered earthquake. The same figure also shows the NTC08 spectrum to be used for life-safety limit-state design of ordinary construction at the AQV site (i.e., the UHS with 475 years return period of exceedance), this is the same as the one used in RINTC except for soil site class, which matches the one of the recording station, for comparison purposes. One notices that the design actions have been greatly exceeded over a relatively wide interval of natural vibration periods, including the $Sa(T = 1s)$ ordinate, which will be analyzed in the following. It will be shown forthwith that this is neither strange nor an indicator of deficiency of the code spectra themselves, but it is instead a predictable consequence of the nature of code spectra.

25.8 Scenario Contributions to Design Hazard

As already mentioned, the elastic design spectra of the Italian code are uniform hazard spectra. Such spectra are computed, for a given construction site, by means of probabilistic seismic hazard analysis described by Eq. (25.2) above.

It is now useful, for the purposes of this study, to rewrite the hazard integral as in Eq. (25.5), considering, for example, the $Sa(T = 1s)$ as the IM.

$$\lambda_{Sa(T=1s)}(sa) = \iint_{M,R} P[Sa(T=1s) > sa | M = m, R = r] \cdot \nu_{M=m, R=r} \cdot dm \cdot dr$$

(25.5)

In the equation, $\nu_{M=m,R=r} \cdot dm \cdot dr = \sum_{i=1}^{s} \nu_i \cdot f_{M,R,i}(m,r) \cdot dm \cdot dr$ represents the rate of earthquakes of magnitude $(m, m + dm)$ that originate at a distance equal $(r, r + dr)$ (accounting for all considered seismic sources).

The hazard integral can be further compacted as in Eq. (25.6), where $\lambda_{Sa(T=1s) > sa, M=m, R=r} \cdot dm \cdot dr = P[Sa(T=1s) > sa|M=m, R=r] \cdot \nu_{M=m, R=r} \cdot dm \cdot dr$ is the rate of earthquakes of magnitude $(m, m + dm)$ that originate at a distance $(r, r + dr)$, and cause exceedance of the intensity threshold, sa.

$$\lambda_{Sa(T=1s)}(sa) = \iint_{M,R} \lambda_{Sa(T=1s)>sa, M=m, R=r} \cdot dm \cdot dr \qquad (25.6)$$

$\lambda_{Sa(T=1s) > sa, M=m, R=r} \cdot dm \cdot dr$ is the contribution to hazard of the earthquake scenario with magnitude $(m, m + dm)$ at distance $(r, r + dr)$ from the site. (In the following, for computation/representation purposes, dm and dr are replaced by small, yet finite, magnitude and distance bins, Δm and Δr, respectively.)

Focusing, for example, attention on $Sa(T=1s)$ with $\lambda_{Sa(T=1s)} = 0.0021$ (i.e., the $Sa(T=1s)$ value with $T_r = 475$ years at the site), let us call this value $sa_{T_r=475}$, it is evident from Eq. (25.6) that none of the aforementioned single contributions may exceed this value; i.e., $\lambda_{Sa(T=1s) > sa, M=m, R=r} \cdot \Delta m \cdot \Delta r \leq 0.0021, \forall \{m, r\}$. Because the lowest magnitude earthquakes are, typically, more frequent than higher magnitudes, this limitation is met in a way that when the former events occur, they have a low probability of exceeding the acceleration threshold that corresponds to that rate. Conversely, the latter events, having low recurrence rate can have high exceedance probability; i.e., $P[Sa(T=1s) > sa|M=m, R=r]$ can approach one.

To better illustrate the point, Fig. 25.5 provides a discretized representation, in terms of magnitude-distance bins, of the individual contributions of magnitude and distance pairs to $sa_{T_r=475}$ at L'Aquila. Such a representation is obtained using the same hazard component models described in Sect. 25.4 and used for the RINTC project, which yielded $sa_{T_r=475} = 0.27g$ on EC8 soil site class B.

Figure 25.5 (bottom) provides the rates $\lambda_{Sa(T=1s) > sa, M=m, R=r} \cdot \Delta m \cdot \Delta r$ [1/year], that is, for each $\{M, R\}$ bin (i.e., a scenario), the average number of earthquakes per year causing exceedance of $sa_{T_r=475} = 0.27g$ (by definition, the sum of these rates over all the bins is equal to 0.0021).[7] Fig. 25.5 (top-left) gives the rates of occurrence of earthquakes corresponding to each bin; i.e., $\nu_{M=m, R=r} \cdot \Delta m \cdot \Delta r$ [1/year]. Finally, Fig. 25.5 (top-right) provides the probability that earthquakes corresponding to each magnitude-distance scenario cause exceedance of 0.27 g; i.e., $P[Sa(T=1s) > sa_{T_r=475}|M=m, R=r]$. The product of values in Fig. 25.5 (top-left) and Fig. 25.5 (top-right), corresponding to the same $\{M, R\}$ bin, provides the value in Fig. 25.5 (bottom) for that bin; see Eq. (25.5).

The rates of occurrence in Fig. 25.5 (top), rapidly decrease with increasing magnitude, independently of distance, as expected. Looking at the dependence of $\nu_{M=m, R=r} \cdot \Delta m \cdot \Delta r$ on source-to-site distance, it appears that the rates tend to

---

[7]Note that the table factually represents the distribution of magnitude and distance one obtains from hazard disaggregation multiplied by 0.0021.

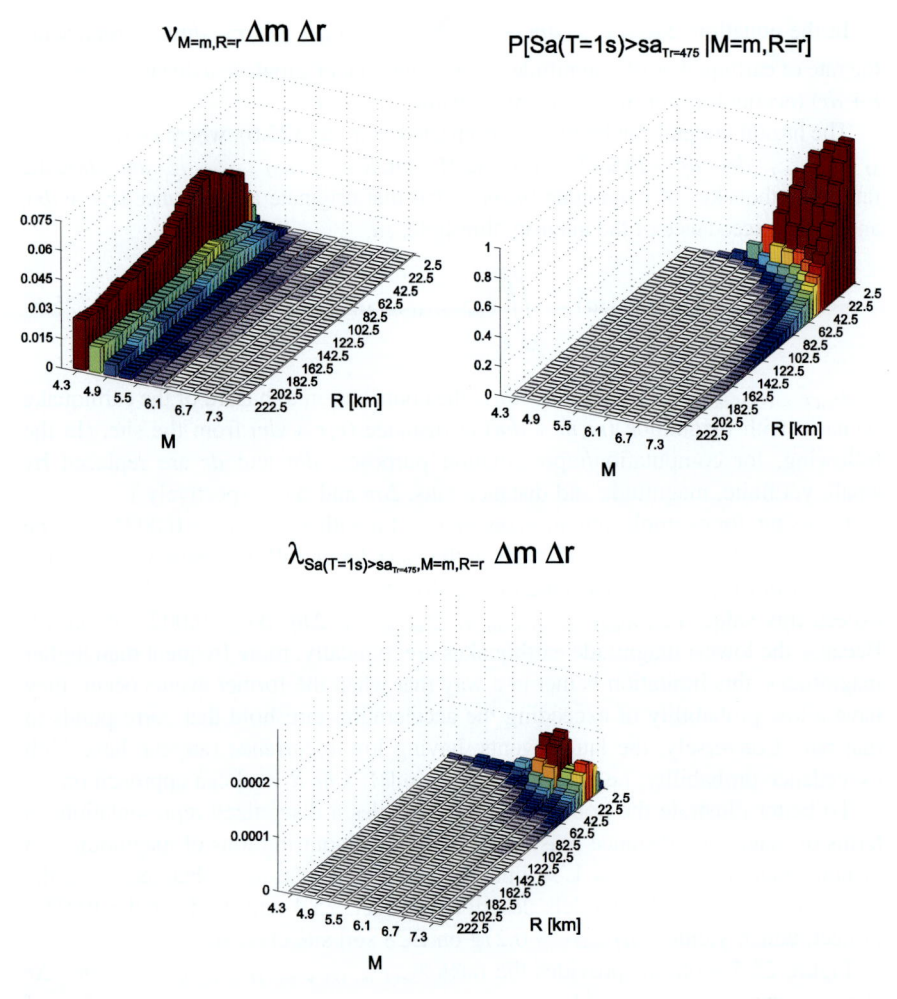

Fig. 25.5 Scenario representation of the hazard at L'Aquila in terms of $Sa(T = 1s)$ with exceedance return period equal to 475 years. The top-left panel reports the rates of occurrence [1/year] of magnitude-distance bins; the top-right panel provides the probability of exceedance, of $sa_{T_r=475}$, for each bin; the bottom is panel the product of the previous two bin-by-bin. Summing-up bars in the bottom panel provides $0.0021 = 1/475$ and corresponds to the integral of Eq. (25.6).

decrease with distance approaching to zero, because relatively smaller portions of the source zones fall in a circle/ring with center in the site of interest and smaller radius. Consequently, there is a large number of bins with small-to-very-small occurrence rate (i.e., white areas in the figure), they correspond to large magnitudes at all distances, or moderate-to-large magnitudes at small distance.

Figure 25.5 (center) provides the effect of the magnitude-distance scenarios in terms of probability of exceeding $sa_{T_r=475}$. Because such a probability increases with increasing magnitude and decreasing distance as indicated by GMPEs, it can be seen that several bins with low rate in the top panel can have very large exceedance probability, if they actually occur, that approaches one in some cases. In particular, the exceedance probability, conditional to $\{M, R\}$, starts to be significant for $M \geq 6$, $R \leq 5$km, but it is large up to $R \approx 60$km for magnitudes larger than seven.

As mentioned, Fig. 25.5 (bottom) reports the hazard contributions of magnitude-distance scenarios, weighing the exceedance probability of each scenario by its occurrence rate: $\lambda_{Sa(T\,=\,1s)\,>\,sa,\,M\,=\,m,\,R\,=\,r} = P[Sa(T=1s) > sa | M = m, R = r] \cdot \nu_{M\,=\,m,\,R\,=\,r}$. It appears from the figure that the $\{M, R\}$ pairs giving the largest contributions are, in general, close earthquakes, because of their large $P[Sa(T=1s) > sa | M = m, R = r]$. However, the largest values of $\lambda_{Sa(T\,=\,1s)\,>\,sa,\,M\,=\,m,\,R\,=\,r} \cdot \Delta m \cdot \Delta r$ do not correspond to the largest magnitude occurring at the smallest distance, because these are very rare events; i.e., with comparatively small $\nu_{M\,=\,m,\,R\,=\,r} \cdot \Delta m \cdot \Delta r$.

In fact, among the close-by earthquakes the most significant exceedance rates are given by smaller magnitudes. These events are more frequently occurring close to the site than extreme magnitudes, yet the probability of exceeding the threshold for some of them is not small at all, as demonstrated in the next section.

25.9 Close Earthquakes

As expected, close-by earthquakes give the largest contribution to hazard. In fact, summing up the values of the bars from $\{M, R\}$ bins up to $R \leq 50$km in Fig. 25.5 (bottom) one obtains 0.019, which means that the earthquakes occurring within this distance account for 90% of the contributions to $Sa(T=1s)$ hazard with exceedance return period $T_r = 475$ years in L'Aquila. This is a common situation when the site is within a seismic source zone that dominates the hazard (see Iervolino et al. 2011).

Focusing exclusively on the contributions of the earthquakes occurring within 50 km, it can be seen that there are different magnitudes with similar contributions. For example, an earthquake of magnitude $M \in (5.05, 5.35)$ at a distance between 0km $\leq R <$ 5km has $\lambda_{Sa(T\,=\,1s)\,>\,sa,\,M\,=\,m,\,R\,=\,r} \cdot \Delta m \cdot \Delta r$ equal to about 1.2E-5 [1/year], which is about the same of an earthquake of magnitude $M \in (6.85, 7.15)$ at distance 0km $\leq R <$ 5km. However, this equivalent contribution to hazard arises from very different occurrence rates and conditional exceedance probabilities, as it can be seen in Table 25.2, where the values from the three panels Fig. 25.5 are given for the two scenarios.

Analyzing Table 25.2 is crucial in demonstrating the initial proposition of this part of the paper. Despite the same threshold exceedance rate, the two scenarios are very different in rarity, as expected. With the lower magnitude being about fifteen times more frequent than the larger. Conversely, when an earthquake $M \approx 5.2$ occurs close to the site, it has 6% probability of exceeding the $Sa(T=1s)$ ordinate of the $T_r = 475$ years UHS for L'Aquila; i.e., in case of occurrence there is 0.94% chance that the UHS is not exceeded at the $T=1s$ ordinate. Conversely, if L'Aquila were close to an $M \approx 7$ event, then the probability of exceeding the threshold would be larger than 90%. It immediately follows from this reasoning that the UHS is hard to

Table 25.2 Two scenarios with comparable contributions to hazard, but very different frequency of occurrence and probability of exceeding the design $Sa(T = 1s)$ in L'Aquila in case of occurrence

	$5.05 \leq M < 5.35$, $0 \leq R < 5$	$6.85 \leq M < 5.15$, $0 \leq R < 5$
$\nu_{R=r, M=m} \cdot \Delta m \cdot \Delta r [1/\text{year}]$	2.0E-4	1.3E-5
$P[Sa(T=1s) > sa_{T_r=475} \mid M=m, R=r]$	6E-2	9E-1
$\lambda_{Sa(T=1s) > sa_{T_r=475}, M=m, R=r} \cdot \Delta m \cdot \Delta r$ [1/year]	1.2E-5	1.2E-5

be exceeded only by distant earthquakes or by the relatively more frequent among close earthquakes. On the other hand, it is very likely to almost certain, depending on the magnitude (see Fig. 25.5, center), that it is going to be exceeded by the more rare among close-by earthquakes. Therefore, the UHS may not represent a high threshold in the case of occurrence of this kind of earthquakes.

It must be underlined that this reasoning does not question that the ordinates of the UHS for the site are exceeded, as intended, on average once every 475 years. However, this exceedance return period, for the rarest earthquakes, is warranted by the fact that their occurrence close to the site is unlikely. On the other hand, when such earthquakes do occur near a given site, the exceedance of design actions can be probable-to-very-probable, depending on the considered scenario.

Similar reasoning can be applied to any other spectral ordinate and/or return period, although the range of scenarios to which it applies is expected to change in the very same way disaggregation depends on the spectral ordinate or return period under consideration.

25.10 What to Expect for a Magnitude 6.3 Event

In the light of all that was shown above, one may now return to examine what happened at AQV during the mainshock of the L'Aquila 2009 earthquake.

AQV was at close distance to the epicenter (zero in terms of Joyner and Boore 1981, distance, which is the distance from the surface projection of the earthquake rupture), then it was somewhat likely to observe exceedance of the $T_r = 475$ years UHS at that site. In fact, according to the GMPE of Ambraseys et al. (1996), for an earthquake of magnitude 6.3 at 0 km, $P[Sa(T=1s) > 0.27g \mid M=6.3, R=0] > 0.6$ on soil site class B (see Fig. 25.5, center). Therefore, first, exceedance of the $T_r = 475$ years $Sa(T = 1s)$, observed in Fig. 25.4, is in accord with the models underlying hazard analysis. Second, the earthquakes exceeding the UHS with very high probability are not necessarily of especially high magnitude (see Fig. 25.5 center).

As a matter of fact, this discussion could be extended for the entire epicentral area of the earthquake. In order to understand which are the locations where exceedance of design actions ought to have been avoided, one should examine Fig. 25.6. First thing shown on that figure is the surface projection of the rupture that caused

25 What Seismic Risk Do We Design for When We Design Buildings? 599

Fig. 25.6 Design Sa ($T=1s$) (i.e., those from Fig. 25.1 with 475 years exceedance return period) for the area hit by the April 6th 2009 (moment magnitude 6.3) L'Aquila earthquake and equal probability contours for their exceedance due to an earthquake with the same magnitude and localization as the one that actually occurred

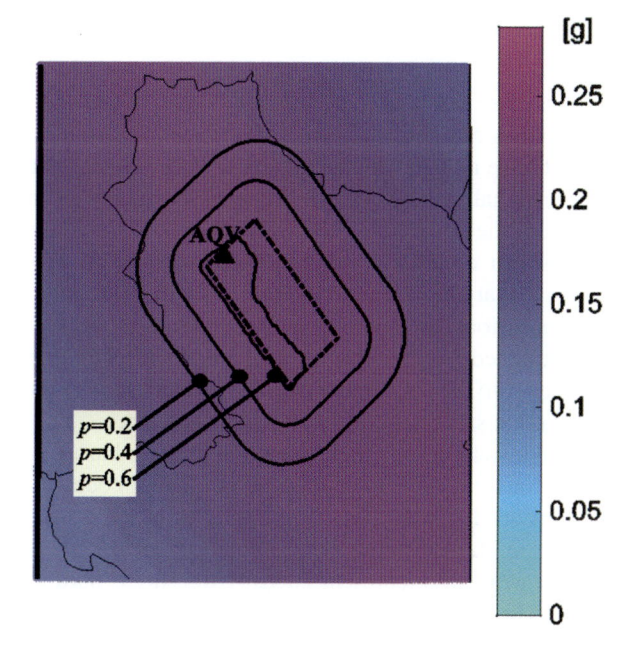

the earthquake (dash-dot line) and the administrative limits of the area's municipalities (thin black lines). On the same figure, the $sa_{Tr\,=\,475}$ values for the area, from the analysis described in Stucchi et al. 2011,[8] are shown as colored contours. The black iso-probability delimit areas exhibiting various probabilities, $P[Sa(T=1s) > sa_{T_r=475} | M=m, R=r]$, of observing the exceedance of the code-mandated design actions depicted in the underlying colored contours. The probabilities were calculated using the GMPE of Ambraseys et al. (1996), the same employed when determining the code design actions of the underlying colored map, so that the calculations are consistent.

One notes that, in a relatively wide area on/around the source, the exceedance of design $Sa(T=1s)$ was likely (e.g., larger than 50%) for an earthquake of the magnitude and location as those occurred on April 6th 2009. As argued for the individual case of AQV, this is by no means contradicting the hazard map but it is rather an intrinsic characteristic thereof. On the other hand, the probability of exceeding the seismic actions rapidly decreases as one moves farther away from the rupture.

---

[8]Consistent with Figure 25.1, the design $Sa(T=1s)$ map refers to type A EC8 site class.

25.11 Conclusions

This article presented some arguments about when and where damages are expected for code-conforming structures. The developed studies refer to Italy, however they might have an international appeal as Italy is at the state of the art of seismic codes internationally and its design norms have extensive similarities with Eurocode 8.

From the description of the RINTC project, a large research effort towards the assessment of seismic risk for different structural typologies, it seems to mainly emerge that the seismic risk for structures designed with nominally equivalent design actions is increasing with the hazard at the site; thus, more risky structures are designed for the most hazardous sites, although design actions at the different sites refer to the same return periods. Thus, design based on the same hazard does not lead to the same risk for the designed structures.

From a closer look to the nature of uniform hazard spectra (on the basis of which design actions are determined in the Italian code) it emerges that design elastic actions are likely-to-very-likely going to be exceeded in the case of moderate-to-high magnitude earthquakes, were they to occur close to the site. In other words, the UHS', generally, represent intensity thresholds hard to surpass by far-away earthquakes or by those of lower magnitude among those close-by. Conversely, they do not represent conservative thresholds for earthquakes relatively rare in occurrence and near the site. This is well expected, and does not represent a reason to blame the way that probabilistic spectra are determined. However, the conclusion for code-conforming constructions is that safety against violation of the design limit states in the epicentral area of earthquakes, which are observed relatively frequently all-over a country such as Italy, is entrusted to safety margins beyond the elastic design spectrum.

Acknowledgments The study was developed between 2015 and 2017 in the framework the *Rete dei Laboratori Universitari di Ingegneria Sismica* activities funded by *Presidenza del Consiglio dei Ministri – Dipartimento della Protezione Civile* (2014–2018 program). The help from Andrea Spillatura (IUSS, Pavia, Italy) and Pasquale Cito (University of Naples Federico II, Italy) is also gratefully acknowledged.

References

Akkar S, Bommer JJ (2010) Empirical equations for the prediction of PGA, PGV, and spectral accelerations in Europe, the Mediterranean region, and the Middle East. Seismol Res Lett 81:195–206

Ambraseys NN, Simpson KU, Bommer JJ (1996) Prediction of horizontal response spectra in Europe. Earthq Eng Struct Dyn 25:371–400

Bazzurro P, Cornell CA (1999) Disaggregation of seismic hazard. Bull Seismol Soc Am 89:501–520

Camata G, Celano F, De Risi MT et al (2017) RINTC project: nonlinear dynamic analyses of Italian code-conforming reinforced concrete buildings for risk of collapse assessment. In Papadrakakis M, Fragiadakis M (eds) Proceedings of COMPDYN 2017 – 6th ECCOMAS

thematic conference on computational methods in structural dynamics and earth-quake engineering, Rhodes (GR) 2017
Camilletti D, Cattari S, Lagomarsino S et al (2017) RINTC project: Nonlinear dynamic analyses of Italian code-conforming URM buildings for collapse risk assessment. In Papadrakakis M, Fragiadakis M (eds) Proceedings of COMPDYN 2017 – 6th ECCOMAS thematic conference on computational methods in structural dynamics and earth-quake engineering, Rhodes (GR) 2017
CEN, European Committee for Standardisation (2004) Eurocode 8: design provisions for earthquake resistance of structures, Part 1.1: general rules, seismic actions and rules for buildings, EN 1998-1
Chioccarelli E, De Luca F, Iervolino I (2009) Preliminary study on L'Aquila earthquake ground motion records V5.2. Available at http://www.reluis.it/. Last accessed December 2017
Cornell CA (1968) Engineering seismic risk analysis. Bull Seismol Soc Am 58:1583–1606
Cornell CA, Krawinkler H (2000) Progress and challenges in seismic performance assessment. PEER Center News 3:1–3
CS.LL.PP (2008) DM 14 gennaio, Norme tecniche per le costruzioni Gazzetta Ufficiale della Repubblica Italiana 29 (in Italian)
Ercolino M, Cimmino M, Magliulo G et al (2017) RINTC project: nonlinear analyses of Italian code conforming precast R/C industrial buildings for risk of collapse assessment. In Papadrakakis M, Fragiadakis M (eds) Proceedings of COMPDYN 2017 – 6th ECCOMAS thematic conference on computational methods in structural dynamics and earth-quake engineering, Rhodes (GR) 2017
Franchin P, Mollaioli F, Noto F (2017) RINTC project: influence of structure-related uncertainties on the risk of collapse of Italian code-conforming reinforced concrete buildings. In Papadrakakis M, Fragiadakis M (eds) Proceedings of COMPDYN 2017 – 6th ECCOMAS thematic conference on computational methods in structural dynamics and earth-quake engineering, Rhodes (GR) 2017
Iervolino I, Giorgio M (2017) Regional exceedance frequency of uniform hazard spectra (Which earthquakes do design spectra protect against?) (In review)
Iervolino I, Chioccarelli E, Convertito V (2011) Engineering design earthquakes from multimodal hazard disaggregation. Soil Dyn Earthq Eng 31:1212–1231
Iervolino I, Spillatura A, Bazzurro P (2017) RINTC project: assessing the (implicit) seismic risk of code-conforming structures in Italy. In Papadrakakis M, Fragiadakis M (eds) Proceedings of COMPDYN 2017 – 6th ECCOMAS thematic conference on computational methods in structural dynamics and earth-quake engineering, Rhodes (GR) 2017
Jalayer F (2003) Direct probabilistic seismic analysis: implementing non-linear dynamic assessments. Thesis, Stanford University
Joyner WB, Boore DM (1981) Peak horizontal acceleration and velocity from strongmotion records including records from the 1979 Imperial Valley, California, Earthquake. Bull Seismol Soc Am 71:2011–2038
Lagomarsino S, Penna A, Galasco A et al (2013) TREMURI program: an equivalent frame model for the nonlinear seismic analysis of masonry buildings. Eng Struct 56:1787–1799
Lin TG, Haselton CB, Baker JW (2013) Conditional spectrum-based ground motion selection. Part I: hazard consistency for risk-based assessments. Earthq Eng Struct Dyn 42:1847–1865
Luzi L, Pacor F, Puglia R et al (2017) The Central Italy seismic sequence between august and December 2016: analysis of strong-motion observations. Seismol Res Lett 88:1219–1231
Mazzoni S, McKenna F, Scott MH et al (2006) OpenSees command language manual. Available at http://opensees.berkeley.edu/. Last accessed December 2017
Meletti C, Galadini F, Valensise G et al (2008) A seismic source zone model for the seismic hazard assessment of the Italian territory. Tectonophysics 450:85–108
Ponzo FC, Cardone D, Dall'Asta A et al (2017) RINTC project: nonlinear analyses of Italian code-conforming base-isolated buildings for risk of collapse assessment. In Papadrakakis M, Fragiadakis M (eds) Proceedings of COMPDYN 2017 – 6th ECCOMAS thematic conference

on computational methods in structural dynamics and earth-quake engineering, Rhodes (GR) 2017

RINTC Workgroup (2017) Results of the 2015–2016 RINTC project. Available at http://www.reluis.it/. Last accessed December 2017

Scozzese F, Terracciano G, Zona A et al (2017) RINTC project: nonlinear dynamic analyses of Italian code-conforming steel single-story buildings for collapse risk assessment. In Papadrakakis M, Fragiadakis M (eds) Proceedings of COMPDYN 2017 – 6th ECCOMAS thematic conference on computational methods in structural dynamics and earth-quake engineering, Rhodes (GR) 2017

Shome N, Cornell CA (2000) Structural seismic demand analysis: consideration of "collapse". In PMC2000 – 8th ASCE Specialty Conference on Probabilistic Mechanics and Structural Reliability. University of Notre Dame, South Bend, Indiana, 24–26 July 2000

Stucchi M, Meletti C, Montaldo V et al (2011) Seismic hazard assessment (2003–2009) for the Italian building code. Bull Seismol Soc Am 101:1885–1911

Suzuki A, Baltzopoulos G, Iervolino I et al (2017) A look at the seismic risk of Italian code-conforming RC buildings. In: Proceedings of 16th Europeassn conference on earthquake engineering, Thessaloniki (GR), 18–21 June 2018

Chapter 26
The 2016–2017 Central Apennines Seismic Sequence: Analogies and Differences with Recent Italian Earthquakes

Mauro Dolce and Daniela Di Bucci

Abstract On August 24th, 2016, a severe, very long seismic sequence started in Central Italy. It was characterized by nine major shocks M5+, two of which with moment magnitude Mw 6.0 (August 24th, 2016) and 6.5 (October 30th, 2016). A complex seismogenic fault system was activated, with the rupture of several segments. The affected area, which develops in NNW-SSE direction along the Apennines, was very large, due to both the large magnitude values and the distance among the epicenters of the nine major shocks. The maximum observed (cumulated) intensity was XI in both MCS and EMS scales. After 1 year, 78,500 seismic events had been recorded by the National Institute of Geophysics and Volcanology national seismic network. 299 people lost their life, all due to the first main shock. Devastating damage was experienced by buildings, cultural heritage, roads and other lifelines, resulting in huge economical direct losses.

The emergency response was coordinated, according to the Law 225/1992, by the Italian National Department of Civil Protection. The main scientific features of the sequence and the main technical emergency activities are shown, discussed and, when possible, compared to the main recent Italian earthquakes, i.e., 1997 Umbria-Marche, the 2009 Abruzzo and 2012 Emilia earthquakes, pointing out analogies and differences.

26.1 Introduction

The earthquake sequence that has occurred in Central Apennines (Central Italy) since August 24th, 2016, includes nine seismic events of Mw $\geq$ 5.0, two of which of Mw 6.0 and 6.5, in 5 months (Table 26.1). It caused a total of 299 fatalities and 412 hospitalized injured people. Observed intensities attained the degree XI on the MCS scale.

M. Dolce (✉) · D. Di Bucci
Department of Civil Protection, Presidency of the Council of Ministers, Rome, Italy
e-mail: mauro.dolce@protezionecivile.it; daniela.dibucci@protezionecivile.it

Table 26.1 2016-17 Central Apennines seismic sequence: Mw ≥ 5.0 earthquakes

yyyy-mm-dd	Time CET	Mw	Zone	Depth (km)	Latitude	Longitude
2016-08-24	03:36:32	6.0	1 km W Accumoli (RI)	8	42.70	13.23
2016-08-24	04:33:28	5.3	5 km E Norcia (PG)	8	42.79	13.15
2016-10-26	19:10:36	5.4	3 km SW Castelsantangelo sul Nera (MC)	9	42.88	13.13
2016-10-26	21:18:05	5.9	3 km NW Castelsantangelo sul Nera (MC)	8	42.91	13.13
2016-10-30	07:40:17	6.5	5 km NE Norcia (PG)	9	42.83	13.11
2017-01-18	10:25:40	5.1	3 km NW Capitignano (AQ)	10	42.55	13.28
2017-01-18	11:14:09	5.5	2 km NW Capitignano (AQ)	10	42.53	13.28
2017-01-18	11:25:23	5.4	3 km SW Capitignano (AQ)	9	42.50	13.28
2017-01-18	14:33:36	5.0	2 km N Barete (AQ)	10	42.47	13.28

Data from INGV; http://cnt.rm.ingv.it/

Although partially overlapping the 1997 Umbria-Marche and 2009 Abruzzo earthquakes area, it determined a seismic emergency quite different from the emergencies previously managed by the Italian National Service of Civil Protection (SNPC) in the past 30 years (Dolce and Di Bucci 2017). Once again, partly unexpected complexities had to be dealt with, and the civil protection system did it by balancing well-established procedures and a sufficient flexibility to adapt them to this specific case. The emergency response was coordinated, according to Law 225/1992, by the Department of Civil Protection (DPC).

The sequence started on August 24th, 2016, at 3:36 a.m. CET, when a strong earthquake (Ml 6.0, Mw 6.0, depth 8 km) occurred with epicenter in Accumoli, a small village of the Lazio Region located in the Central Apennines, causing 299 fatalities and 390 hospitalized injured people. Observed intensities attained the degree X-XI on the MCS scale and X on the EMS scale (Galli et al. 2016; Tertulliani and Azzaro 2016a). Three small municipalities, namely Amatrice, Accumoli and Arquata, experienced severe disruption. No foreshocks preceded this main shock, whereas one aftershock, which occurred about 1 h after it, exceeded magnitude 5, reaching the value of Mw 5.3. Until October 26th, 2016, in addition to the main shock and the strongest aftershock, the seismic sequence was formed by more than 18,000 events, 15 of which with $4.0 \leq Ml < 5.0$, and 250 with $3.0 \leq Ml < 4.0$. At that moment, the sequence was distributed over a length of more than 40 km with a NNW-SSE strike.

The area affected by this first main shock was located at the boundaries among four Regions, namely Abruzzo, Lazio, Marche and Umbria, and involved 7 provinces: Ascoli Piceno, Fermo, L'Aquila, Macerata, Perugia, Rieti, Teramo. This territory is prevailingly mountainous, with more than 70% of the topographic surface exceeding 900 m elevation. The population density was quite low, with a mean value in the order of 15 people per square kilometer (about 200 is the mean national value; ISTAT 2016), distributed over small municipalities, each formed by a large number of localities. Just as an example, about 2600 people lived in the Amatrice municipality, distributed over 47 localities (including the village of Amatrice) and a number of sparse houses.

The mean income per person was less than € 10,000 per year (the national equivalent is € 12,790). About 65% people worked in the same municipalities where they lived, a value higher than the national mean value (54%). This indicates the existence of a local labor market, which corresponds to a mainly agricultural economy with a high percentage of farms (especially breeding farms) with respect to the resident people: more than 7 farms per 100 inhabitants (national mean value 2.7).

Tourism was an important economic activity before the earthquake. It was mainly related to the environmental context of that territory, which is located in an area of great natural interest and low urbanization. Therefore, the accommodation availability was higher than the national average (more than 300 beds per 1000 inhabitants vs. 80), with a large number of B&B and holiday farms. It has to be noticed that many tourists were house-owners living in the surrounding cities, in Central Italy, and were used to spend holidays in their houses, just in those villages located in the epicentral area: these houses are a relevant part of those damaged by the earthquake.

On October 26th, 2016, at 20:18 CET, a Mw 5.9 (Ml 5.9) earthquake occurred with epicenter in Visso, a nearby small village in the Marche Region, preceded by a Mw 5.4 seismic event at 19:10. These events and the following seismic sequence occurred to the NNW of the previous seismic activity. The Mw 5.9 earthquake occurred 25 km to the NNW of the Mw 6.0 epicenter. Fortunately, no further casualties occurred, because most of the local population had been yet arranged in safe, temporary lodging after the August 24th main shock.

On October 30th, at 7:40 CET, the strongest seismic event of this sequence occurred in an area located between the zones hit by the previous two main shocks, with epicenter in Norcia, in the Umbria Region. This earthquake occurred 18 km to the NNW of the first main shock epicenter. The local magnitude Ml was 6.1, whereas the moment magnitude Mw was 6.5. This is the highest magnitude observed in Italy since the Mw 6.8, 1980 Irpinia earthquake (I_0 X MCS). Also in this case, no casualties occurred, even because of the time when the event occurred. The ongoing surveys and activities carried out by firefighters and technical experts usually started at 8:00 o'clock, and therefore none of them was inside any building when the earthquake occurred.

After the latter two events, the affected area became much wider, as it will be shown later, and included larger towns and zones hosting well-developed industrial activities.

On January 18th, 2017, four seismic events with magnitude 5+ hit the southernmost part of the area already interested by the ongoing seismic sequence. The first three events (Ml 5.3-Mw 5.1, Ml 5.4-Mw 5.5, Ml 5.3-Mw 5.4) occurred in 1 h, between 10:25 and 11:25 CET, whereas the fourth one (Ml 5.1-Mw 5.0) occurred ca. 3 h later (14:33 CET; Table 26.1). No casualties occurred, due to the earthquakes. However, an extreme weather event contemporarily hit the same area. A long and heavy snowfall, whose intensity can be referred to a return period in the order of hundred years, had started 2 days before just in the same part of the Central

Apennines. This determined enormous difficulties for at least 10 days, affecting the electricity power distribution and transports, and causing uncomfortable out of home accommodation for people with damaged houses. To give an idea of the difficulties encountered in the assistance to the population activities, it is worthwhile considering that the first self-protection action in case of earthquake is to go outside the buildings, whereas the first one in case of heavy snowfall is to stay inside them. Moreover, an avalanche at about 40 km to the East of the January 18th epicenters (at Rigopiano, in the Abruzzo Region), caused the death of 34 people in a resort hotel. Two emergencies coming from two independent yet concomitant perils (earthquake and exceptional snowfall+avalanche) had to be managed. In addition, extensive landslides were caused by the following melting snow, perhaps along with some effects due to shaking.

After January 2017, although no earthquakes with magnitude 5 and more had occurred anymore, the seismic sequence still went on. On October 31st, 2017, it was formed by more than 78,500 events (data INGV; http://cnt.rm.ingv.it/), among which:

- 2 with $Mw \geq 6.0$
- 7 with $5.0 \leq Mw < 6.0$
- 62 with $4.0 \leq Ml < 5.0$
- 1102 with $3.0 \leq Ml < 4.0$.

The distance between the northernmost and southernmost Mw5+ events was about 50 km along a NNW-SSE strike.

In the present paper, the main scientific features of the sequence and the main technical emergency activities are described, discussed and, when possible, compared to the main recent Italian earthquakes, i.e., the 1997 Umbria-Marche, 2009 Abruzzo and 2012 Emilia earthquakes, pointing out analogies and differences.

26.2 Geological/Seismotectonic Setting, Seismic Hazard and Historical Seismicity

From a tectonic point of view, the epicentral area is part of the Apennines fold-and-thrust belt, an orogenic chain that formed in Cenozoic times with a general NW-directed motion towards the Adriatic foreland (Bally et al. 1986; Mostardini and Merlini 1986; Barchi et al. 1998). In Quaternary times, this compressional tectonic phase was replaced by an extensional tectonic phase, which is still ongoing and currently affecting the region with a SW-NE–striking extension (e.g., Boncio and Lavecchia 2000). The numerous active normal faults present in this part of the Italian territory, as well as the related seismicity, respond to this extensional tectonic regime.

The seismogenic faults responsible for the main shocks of the considered seismic sequence are coherent with this extensional stress field, being segments of a

NNW-SSE-striking, WSW-dipping normal fault system, with lengths of ca. 15–20 km and a dip angle in the order of 40–50°. The activation of antithetic or low angle, pre-existing fault planes has been suggested as well (Bonini et al. 2016; Cheloni et al. 2017).

A glance at the historical seismicity observed in the considered zone shows that the 2016–2017 Central Apennines seismic sequence occurred in an area well known for having been yet affected, in the past, by other moderate-to-high seismic events (catalogue CPTI15, updated to 2015: http://emidius.mi.ingv.it/CPTI15-DBMI15/; Rovida et al. 2016). Among these earthquakes, the 1627 Accumoli event (Io 7–8 MCS, Mw 5.3), known for the damage of few important buildings (Monachesi and Castelli 1992), was followed, a dozen years later, by the 1639 Amatrice event (Io 9–10 MCS, Mw 6.2). This earthquake (October 8th, 0:35 GMT), in particular, can be regarded as a possible twin earthquake of the August 24th, 2016, Amatrice event, both for its epicentral location and for the damage (and therefore the equivalent magnitude). In particular, according to CFTI4Med catalogue (http://storing.ingv.it/cfti4med/; Guidoboni et al. 2007), this earthquake caused about 500 fatalities. Differently from the 2016 Amatrice earthquake, the 1639 event had been forerun in the preceding half an hour by two foreshocks that alerted the population. Severe damages are reported, also to the livestock activities. For this reason, as well as for the continuously ongoing aftershocks, people abandoned the epicentral area and migrated to nearby cities like Ascoli Piceno and Rome. A strong aftershock occurred on October 15th, 1639, causing further collapses and damage in Amatrice and the quasi-total destruction of some small surrounding villages. Moreover, two moderate earthquakes occurred in the same area in 1646 (Laga Mounts, Io 9 MCS, Mw 5.9) and 1672 (Amatrice, Io 7–8 MCS, Mw 5.3). Finally, the zone was affected by one of the two catastrophic events that hit the Central Apennines in 1703 (the Valnerina earthquake, Io 11 MCS, Mw 6.9).

The seismic hazard of the region mirrors this well-known historical seismicity. According to the Seismic Hazard Map of Italy (MPS Working Group 2004), with exceedance probability 10% in 50 years, the epicentral area is characterized by PGA values in the range of 0.250–0.275 g (where g = acceleration of gravity, 9.8 m/s^2). In Italy, these values are among the highest seismic hazard levels.

26.3 Emergency Management

In the first hour after the August 24th, 2016 earthquake, once the epicentral coordinates and Richter magnitude were made available by the National Institute of Geophysics and Volcanology (INGV), an earliest picture of its possible consequences was estimated by the Department of Civil Protection by elaborating a damage scenario that will be described later in detail. This returned a severe potential impact, which required an emergency management at national level.

The scenario reliability was checked, in the first hours after the event, against the reports directly coming from the epicentral area to the Department Operation Room.

Further checks were made on the basis of the data collected by the RAN-DPC strong-motion network and of the shake maps provided by INGV (http://shakemap.rm.ingv.it). These pieces of information allowed the earthquake effects simulated by the damage scenario to be tuned, in order to calibrate the emergency intervention on the actual needs.

The entire Italian SNPC was immediately activated, its mandate being the safeguarding of human life and health, goods, national heritage, human settlements and environment from all natural or man-made disasters, under the DPC coordination. In particular, the operational structures of the Ministry of Interior—i.e., National Fire Brigades, Police, Prefectures—and of the Ministry of Defense—i.e., Army, Navy, Air Force and Carabinieri—contribute to SNPC actions, together with the Financial Police. Companies of road and railway transportation, electricity and telecommunication, as well as Volunteers Associations are part of the system. Finally, an important strength of the system is represented by its link with the scientific community, which enables, through the DPC Competence Centers, timely translation of up-to-date scientific knowledge into operability and decision-making.

On August 24th, 2016, while the first assessment activities were under way, the following actions were undertaken at national level for the search-and-rescue (SAR) and assistance to the population. The Operational Committee started its activities at the DPC headquarters at 4:00 a.m. CET, and went on until August 28th, ensuring a unified direction and coordination of the emergency management.

The first actions after the event can be summarized as follows:

3.36 a.m. CET	Earthquake
4.00	Operational Committee meeting
4.30	Provincial coordination center established in Rieti (the capital of one of the affected provinces)
5.00	Assessment and Coordination Teams deployed on site by DPC
6.30	First press conference at DPC
10.00	Survey of the Head of the Department of Civil Protection and Minister of Infrastructures and Transports in the affected areas
Early afternoon	Prime Minister's visit to the affected areas

On August 25th, 2016, the Council of Ministers acknowledged the severity of the earthquake and declared the "state of emergency" for a period of 180 days (later extended), referred to seven provinces in four Regions, namely Abruzzo, Lazio, Marche and Umbria, under the coordination of the Head of DPC.

On August 28th, 2016, the Direction of Command and Control (Di.Coma.C.), i.e., the national coordination system on site, was established in Rieti at 12:00 CET. It was aimed at better coordinating the components and operational structures of the Italian SNPC involved in the SAR, assistance and provisional activities. It operated up to April 7th, 2017, date that corresponds to the transfer to the Regions of many activities of emergency management coordination.

Through the Di.Coma.C., different activities were coordinated, for instance:

- assisting the population in temporary accommodation areas and structures, like tent camps and hotels;
- assigning and monitoring the self-lodging financial support to families whose houses had been partially or totally destroyed, or evacuated;
- preparing, managing and closing temporary accommodation areas and structures;
- assessing damage and usability of buildings, including schools;
- managing the procedures for the requests of urgent provisional interventions;
- providing authorizations for the expenses needed for the various activities;
- facilitating link and cooperation among local and central coordination centers.

Dealing with the first assistance to the population, 400 people were hosted in tent camps in the first 24 h after the first main shock, and 2500 in the first 48 h. The maximum peak in those first days was of 4807 people. At the end of August 2016, 43 tent camps had been set up, while other temporary solutions were operating in safe sports arenas and gyms available in the damaged area. On October 25th, 2016, assisted people were 1136 (http://www.protezionecivile.gov.it/jcms/it/assistenza_alla_popolazio.wp). The forces operating during that period reached a maximum of 6.806 people in the days following August 24th, and then progressively decreased to 2.617 (http://www.protezionecivile.gov.it/jcms/it/forze_in_campo.wp). These figures have been largely exceeded after the October 26th and 30th main shocks.

The general strategy envisaged for the rescue and the assistance to the population in this disaster management is summarized in Table 26.2.

It has to be noticed that on September 9th, 2016, a Government Commissioner for the Reconstruction was nominated through a Prime Minister's Decree. He was in charge of all the activities related to the reconstruction of the area affected by the earthquake, whereas the emergency management coordination remained in charge of the Head of DPC. These two authorities were in charge of different but temporarily overlapping and interfingering activities, which required collaboration and continuous interaction.

After the two main shocks occurred at the end of October, 2016, the population directly assisted by the civil protection system reached a new maximum peak of 31.763 people. The forces operating after these two earthquakes reached a maximum of 6.916 people in the first days, and then progressively decreased to 4.292 during the first half of January 2017 (http://www.protezionecivile.gov.it/jcms/it/forze_in_campo.wp).

Table 26.2 General strategy for the assistance to the population in the short and long run

Activity	Time frame
Search-and-rescue/health response	Hours/days
Tents and other short term solutions	Weeks
Hotels, CAS, MAPRE, usable houses, containers	Weeks/months
SAE	Months/years
Buildings repaired or rebuilt (reconstruction process)	Years

CAS self-lodging financial support, MAPRE post-emergency rural temporary housing modules, SAE post-emergency long term temporary housing

Although the earthquake sequence had started at the end of summer, these new main shocks occurred while the winter was approaching, and, what is more, in a mountainous region. Therefore, the choice was not to install tent camps again, but to favor people displacement in hotels along the coast of the Adriatic Sea and of the Trasimeno Lake in Umbria, both at safe - although relatively short - distance from the epicentral area. One day before the upsurge of the seismic sequence on January 18th, 2017, the number of assisted people was 10.076, and it became 15.107 the day after. The forces operating in the days after January 18th reached a maximum of 7.482 people, also because of the concurrent exceptional snowfall, and then progressively decreased to 3.423 by the end of March (http://www.protezionecivile.gov.it/jcms/it/forze_in_campo.wp).

On August 2017, 1 year after the first main shock, the assisted people were 7.559 (http://www.protezionecivile.gov.it/jcms/it/assistenza_alla_popolazio.wp).

The considered previous earthquakes had different numbers of assisted people, consistently with the higher population density of their epicentral area and the earthquakes characteristics. For instance, the maximum number of people needing assistance soon after the 2009 Abruzzo earthquake was of ca. 67,000, due to the epicenter located in correspondence with city of L'Aquila, with more than 70,000 inhabitants. After the 2012 Emilia earthquake, instead, the assisted people reached the number of about 16,000. In this case, the type of accommodation in the first period consisted mainly of tents and hotels, backed up by the contribution for autonomous arrangement.

26.4 Technical Activities

In addition to the SAR and assistance to the population, many technical activities have been carried out to support the civil protection management since the first emergency phase. Some of them are reported in Table 26.3, which shows how some activities have to be completed in the first hours after the main shock, whereas others need a much longer period to be properly accomplished.

It is worthwhile noticing that for many of the activities reported in Table 26.3 the timing had to be restarted each time a new main shock occurred. This restart was especially burdensome for the macroseismic and coseismic surveys and for the damage assessment.

Many of the above activities were carried out by academic and research institutions, in their role of Competence Centers, to support civil protection needs under the coordination of DPC at Di.Coma.C.. The nature and role of DPC Competence Centers in the general framework of the national warning system is defined as follows: *Competence Centers* (Centers for Technological and Scientific services, development and transfer) are institutions that provide services, information, data, elaborations, technical and scientific contributions for specific topics, to share the best practices in risk assessment and management. Competence Centers for seismic risk are research institutes and academic consortia (Table 26.4).

26 The 2016–2017 Central Apennines Seismic Sequence: Analogies and...

Table 26.3 Timetable of the main post-event technical activities

Time interval	Technical activity	Description
2′ → 5′–30′	Epicenter and magnitude evaluation	Collecting and processing instrumental data acquired by the INGV seismic network
10′ → 30′–60′	Simulation of damage scenarios and data processing from monitoring systems	Simulating the earthquake impact on constructions through a DPC scenario software
		Collecting and processing field and building accelerometric data acquired by the DPC monitoring networks
6 h → 7–14 days	Site surveys for macroseismic and coseismic effects	Macroseismic surveys for the assessment of MCS intensities at built localities
		Geological surveys of rock-falls and landslides, surface faulting and fracturing, soil liquefaction and other hydrological effects, aimed at evaluating safety conditions of exposed buildings and infrastructures
6 h → 6–12 months	Temporary monitoring of field and structures	Installing temporary seismic and accelerometric networks, and structure monitoring systems
24 h → 6–12 months	Post – Earthquake damage and safety assessment	Conducting building inspections for damage and usability assessment for private and public buildings
		Carrying out damage and usability assessment of cultural heritage, as well as other activities aimed at reducing the residual risk of built heritage and movable cultural heritage
		Carrying out technical, geological and engineering evaluations for the choice of sites for temporary houses and settlements

Representatives of the Competence Centers were present in the Di.Coma.C. in order to directly interact with DPC officers. Their activities were organized as follows:

- INGV was responsible for the seismic monitoring and leaded the geological field surveys concerning tectonic features;
- ISPRA leaded the geological field surveys concerning landslides, in particular those affecting roads and buildings;
- CNR leaded the geological field surveys concerning microzonation;
- ENEA was especially in charge with the rubble management;
- EUCENTRE supported the tent camps management and contributed to the damage assessment;
- ReLUIS contributed to the damage assessment, both in general and with a specific focus on peculiar structures and problems, as well as on cultural heritage, for which it also coordinated the technical-scientific aspects.

Table 26.4 Competence Centers for seismic risk and other institutions

Competence center	Main activities
INGV	Seismic surveillance, applied research projects on seismology and earthquake geology; emergency scientific-technical support
ReLUIS	Applied research projects on earthquake engineering; emergency scientific-technical support
EUCENTRE	Applied research projects on earthquake engineering; emergency scientific-technical support
CNR (IGAG, IRPI, IREA)	Seismic microzonation, landslides surveys, satellite interferometry; emergency scientific-technical support
ISPRA	Geological mapping, induced geological effects; emergency scientific-technical support
ENEA	Rubble management; emergency scientific-technical support
ASI	Satellite data provision

Researchers from each institution, as well from academy, contributed to each different topic under the coordination of the leading Competence Center.

In the following sections, the attention will be especially focused on:

- the first assessment of damage and losses through computer simulated scenarios;
- the acquisition and elaboration of the accelerometric data by the National Accelerometric Network and the Seismic Observatory of the Structures, owned and operated by DPC;
- the macroseismic and geological surveys;
- the organization and outcomes of the inspections for damage and post-earthquake usability assessment of public and private buildings, with priority for schools, cultural heritage and other structures;
- the actions aimed at assuring the continuity of the school activities and reducing the residual risk of built heritage and movable cultural heritage.

All these aspects will be briefly presented from a civil protection perspective.

26.5 First Damage Simulation Scenario

As anticipated in the previous section, in the first hour after the August 24th, 2016, earthquake, an earliest picture of the possible consequences was obtained starting from the epicentral coordinates and Richter magnitude provided by INGV. Based on these parameters, a damage scenario was immediately developed by DPC officers through the SIGE software (Bramerini and Lucantoni 2001) and it returned an estimate of the earthquake consequences in terms of number of people involved in collapsed buildings (between 38 and 1724) and homeless (between 6135 and 115,912), as well as of collapsed or unusable buildings (between 5561 and 56,630). An expected IX degree MCS epicentral macroseismic intensity was also estimated.

Matching these figures with the real data, one can observe that the scenario had given a good representation of the event, thus allowing the first emergency response to be properly calibrated. The sum of the numbers of fatalities and hospitalized injured, 299 and 390 respectively, are in the order of the corresponding estimates of people involved in collapsed buildings. For the total number of collapsed or unusable buildings, a corresponding real figure is not available, because the subsequent main shocks of October 26th and 30th changed the damage scenario when the building inspections were only at 37% of completion with respect to the total number of requests, equal to 77,000. Since about 45% inspected buildings turned out to be unusable, a simple projection of the unusable or collapsed buildings can be made to the entire population of buildings for which an inspection had been requested. If one considers the 45% of 77,000, a result of about 35,000 can be obtained, and this figure as well is in agreement with the estimated interval 5561–56,630.

The total number of homeless, i.e., people whose house was no more usable because of its damage or collapse, cannot be compared with the maximum number of assisted people, 4807, but rather with the number of the unusable or collapsed buildings multiplied by the average number of inhabitants per building. This can be assumed of the order of 1, due to the small size of the buildings, their use other than residential and the high number of second houses. With this assumption, the number of homeless is again of the order of 35,000, which well fits the estimated interval 6135–115,912.

The maximum observed intensities exceeded the estimated value of IX in few (10) localities in the municipalities of Amatrice, Accumoli and Arquata del Tronto.

Obviously, no other scenarios were elaborated for the following main shocks, because SIGE software refers, like most of the similar simulation softwares, to a territory in ordinary conditions in terms of vulnerability and exposure, that was not the case after the August 24th earthquake.

In the two previous important earthquakes to which it is possible to make a comparison, the SIGE scenario simulations provided opposite results with respect to reality. In the case of the 2009 Abruzzo earthquake, the outcomes of SIGE were quite consistent with the real figures in terms of fatalities, injured people, homeless and unusable buildings (Dolce 2010). In the case of the 2012 Emilia earthquake, SIGE returned largely overestimated earthquake consequences (Dolce and Di Bucci 2014). Indeed an expected VIII-IX degree MCS epicentral macroseismic intensity was estimated vs. an actual VII-VIII MCS intensity. This was probably due to the kinematics of the seismogenic faults, that for the Emilia earthquake consisted of thrusts activated within a compressional tectonic regime, and to the subsurface geological setting of the Po Plain, filled in by kilometers thick terrigenous deposits. The observed discrepancies can be explained by the fact that SIGE results are calibrated on the most frequent Italian earthquakes, which typically occur along the peninsula, especially in the Apennines mountain chain, and are caused by normal fault systems activated within an extensional tectonic regime.

26.6 Seismic Monitoring of Ground and Structures

The strong motion accelerometric records provide a clear picture of the local intensities of the earthquakes, which can help explain the amount and areal distribution of the damage. It is, then, worthwhile examining the main features of the available records. It has also to be emphasized how the prompt availability of these records, in few minutes after each earthquake, contribute to get a general idea of its severity and possible consequences, so as to size the emergency activities in the first moments after each main shock.

The RAN-DPC strong-motion network, owned and managed by the National Department of Civil Protection, is made up of more than 560 permanent digital stations, whose data are tele-transmitted to the DPC monitoring center. This network is able to guarantee a dense cover of all the zones of the national territory characterized by high seismic hazard, with an instrumental density proportional to the hazard level. Since the Central Apennines, where the main shock epicenters were located, is a high to medium-high hazard zone, the number of stations was quite high and a great amount of data was made available.

All the data recorded by RAN-DPC are published on the website http://ran.protezionecivile.it/, from where those here presented have been drawn. The same recordings, although with different filtering, are also reported in the ITACA - Italian Accelerometric Archive web site (http://itaca.mi.ingv.it/).

In Tables 26.5, 26.6 and 26.7, the main features of the most important epicentral records relevant to the August 24th Mw 6.0 (Accumoli), October 26th Mw 5.9 (Visso) and October 30th Mw 6.5 (Norcia) main shocks are reported. Within a maximum distance of 200 km, the number of stations triggered by the above three main shocks were, respectively, 120, 121 and 125. All the related recordings are archived in the above said website.

Concerning the first main shock (Table 26.5), the measured peak ground acceleration (PGA) in the epicentral area, at the Amatrice site (ca. 10 km from the epicenter), reached 916 cm/s^2 on the E-W horizontal component and 400 cm/s^2 on the vertical one. The same E-W component also provided the maximum values of peak ground velocity, PGV = 44 cm/s, spectral acceleration at 0.3 s, PSA03 = 1787 cm/s^2, and Arias intensity, 171 cm/s. It has to be noticed, however, that the maximum values of peak ground displacement, PGD = 8.2 cm, spectral acceleration at 1 s, PSA10 = 411 cm/s^2, and Housner Intensity, 138 cm, occurred at the NOR (Norcia) station, 14 km from the epicenter, which is characterized by a soil profile "C", according to EC8 classification.

Concerning the second main shock (October 26th, 2016; Table 26.6), the maximum values of all the strong motion parameters herein considered occurred at the same station CMI, located in Campi, a small village in the municipality of Norcia, 7 km from the epicenter, and all in the E-W horizontal direction. The measured PGA reached 684 cm/s^2. The other parameters were of the same order as the maximum ones recorded for the first main shock, and some of them even greater. In particular, the spectral acceleration at 0.3 s, PSA03 = 1779 cm/s^2, was practically equal, while

Table 26.5 Main characteristics of the recorded accelerograms of the August 24th, 2016, main shock, in the seven epicenter nearest stations

Epicenter (from INGV) Mw 6.0, 2016-08-24 01:36:32 (UTC) – Accumoli, Rieti

Sta	Can	Lat (°)	Lon (°)	Site Name	Dist km	PGA cm/s^2	PGV cm/s	PGD cm	PSA03 cm/s^2	PSA10 cm/s^2	Arias cm/s	Housner cm	EC8
AMT	E-W	426.325	132.866	Amatrice	9.58	**915.97**	44.25	2.96	**1,786.88**	199.93	**171.23**	130.40	B*
AMT	N-S	426.325	132.866	Amatrice	9.58	445.59	39.11	7.03	566.87	356.08	65.80	135.27	B*
AMT	Vert.	426.325	132.866	Amatrice	9.58	**399.94**	27.45	4.46	414.57	328.56	51.82	94.64	B*
RQT	E-W	428.130	133.110	Arquata del tronto	13.91	447.87	13.85	2.05	938.23	75.56	136.68	47.05	B*
RQT	Vert.	428.130	133.110	Arquata del tronto	13.91	396.54	9.16	1.92	411.45	42.19	88.59	34.37	B*
NOR	E-W	427.924	130.924	Norcia	14.25	192.12	31.06	**8.20**	306.03	**411.44**	50.90	**137.85**	C*
NOR	N-S	427.924	130.924	Norcia	14.25	165.66	15.21	4.33	442.27	242.97	31.25	80.39	C*
NOR	Vert.	427.924	130.924	Norcia	14.25	258.33	14.68	2.82	279.99	120.86	28.24	47.81	C*
NRC	E-W	427.925	130.964	Norcia	14.25	331.61	29.20	6.25	711.12	237.14	94.72	108.96	B
NRC	N-S	427.925	130.964	Norcia	14.25	376.96	19.16	5.67	631.13	193.98	75.39	84.24	B
NRC	Vert.	427.925	130.964	Norcia	14.25	208.60	8.74	2.27	563.85	100.20	34.71	42.87	B
CSC	E-W	427.190	130.122	Cascia	17.45	104.40	5.46	0.90	196.47	74.73	7.20	24.19	B
CSC	N-S	427.190	130.122	Cascia	17.45	91.91	5.47	1.11	197.07	51.42	8.15	19.01	B
CSC	Vert.	427.190	130.122	Cascia	17.45	64.32	2.27	0.67	94.74	39.26	3.74	10.74	B
PCB	E-W	425.580	133.380	Poggio_Cancelli	18.91	190.70	10.64	1.33	372.79	110.44	18.35	33.32	B*
PCB	N-S	425.580	133.380	Poggio_Cancelli	18.91	287.02	10.67	1.73	528.22	148.39	29.35	44.59	B*
PCB	Vert.	425.580	133.380	Poggio_Cancelli	18.91	80.89	5.43	1.09	218.86	103.23	3.62	26.71	B*
MTR	E-W	425.240	132.448	Montereale	20.15	88.90	9.35	2.22	141.88	120.78	5.26	39.69	B*
MTR	N-S	425.240	132.448	Montereale	20.15	69.30	6.82	2.30	156.79	96.64	5.61	27.88	B*
MTR	Vert.	425.240	132.448	Montereale	20.15	39.16	5.75	1.77	74.61	75.13	2.24	18.63	B*

Most significant figures in bold

Table 26.6 Main characteristics of the recorded accelerograms of the October 26th, 2016, main shock, in the seven epicenter nearest stations Epicenter (from INGV) Mw 5.9, 2016-10-26 19:18:05 (UTC) – Ussita, Macerata

Sta	Can	Lat (°)	Lon (°)	Site Name	Dist km	PGA cm/s^2	PGV cm/s	PGD cm	PSA03 cm/s^2	PSA10 cm/s^2	Arias cm/s	Housner cm	EC8
CNE	E-W	428.944	131.528	Castel Santangelo S.N.	3.11	553.54	23.36	3.02	1,288.22	232.33	107.17	94.15	C*
CNE	N-S	428.944	131.528	Castel Santangelo S.N.	3.11	420.07	30.53	5.28	865.69	373.01	93.42	116.97	C*
CNE	Vert.	428.944	131.528	Castel Santangelo S.N.	3.11	489.29	15.16	2.83	531.47	116.68	93.74	51.71	C*
CMI	E-W	428.504	130.928	Campi	7.62	**684.48**	**48.66**	**6.92**	**1,778.56**	**507.98**	**207.48**	**174.65**	C*
CMI	N-S	428.504	130.928	Campi	7.62	349.39	23.51	2.95	798.03	224.95	90.75	78.02	C*
CMI	Vert.	428.504	130.928	Campi	7.62	494.80	11.84	1.73	335.12	113.66	56.53	41.94	C*
PRE	E-W	428.793	130.334	Preci	8.80	282.13	7.66	1.13	343.69	61.49	29.07	30.23	B*
PRE	N-S	428.793	130.334	Preci	8.80	239.33	9.61	1.69	365.35	49.51	34.80	27.26	B*
PRE	Vert.	428.793	130.334	Preci	8.80	174.52	5.22	1.40	188.20	51.47	16.67	27.19	B*
CLO	E-W	428.294	132.060	Castelluccio	11.43	165.73	13.27	1.42	338.02	148.60	22.56	43.96	A*
CLO	N-S	428.294	132.060	Castelluccio	11.43	168.56	15.08	2.36	479.79	184.38	25.32	59.04	A*
CLO	Vert.	428.294	132.060	Castelluccio	11.43	214.27	8.60	1.09	176.82	94.20	20.93	32.03	A*
MCV	E-W	429.934	130.013	Monte Cavallo	13.55	408.58	11.22	1.29	390.40	74.56	80.42	35.38	B*
MCV	N-S	429.934	130.013	Monte Cavallo	13.55	559.74	14.32	1.76	954.87	62.46	111.35	47.70	B*
MCV	Vert.	429.934	130.013	Monte Cavallo	13.55	**529.49**	7.43	0.84	192.98	61.33	66.78	22.97	B*
NOR	E-W	427.924	130.924	Norcia	13.91	222.45	16.47	2.58	425.23	248.26	25.09	66.18	C*
NOR	N-S	427.924	130.924	Norcia	13.91	137.55	9.92	2.58	235.95	154.52	16.74	51.76	C*
NOR	Vert.	427.924	130.924	Norcia	13.91	96.90	7.09	1.43	147.12	105.09	10.87	34.97	C*
NRC	E-W	427.925	130.964	Norcia	13.91	242.27	18.99	2.00	357.20	174.55	27.60	60.02	B
NRC	N-S	427.925	130.964	Norcia	13.91	346.67	19.95	1.74	397.90	105.77	49.79	63.57	B
NRC	Vert.	427.925	130.964	Norcia	13.91	211.33	8.74	1.09	275.35	63.95	17.12	32.93	B

Most significant figures in bold

Table 26.7 Main characteristics of the recorded accelerograms of the October 30th, 2016, main shock, in the seven epicenter nearest stations

Epicenter (from INGV) Mw 6.5, 2016-10-30 06:40:17 (UTC) – Norcia, Perugia

Sta	Can	Lat (°)	Lon (°)	Site Name	Dist km	PGA cm/s²	PGV cm/s	PGD cm	PSA03 cm/s²	PSA10 cm/s²	Arias cm/s	Housner cm	EC8
NOR	E-W	427.924	130.924	Norcia	4.92	361.15	56.37	10.59	784.71	**874.41**	260.11	**268.81**	C*
NOR	N-S	427.924	130.924	Norcia	4.92	282.42	43.41	9.91	568.03	853.24	129.10	188.26	c*
NOR	Vert.	427.924	130.924	Norcia	4.92	275.50	17.34	4.60	473.40	200.35	111.65	83.56	c*
NRC	E-W	427.925	130.964	Norcia	5.39	477.19	47.05	10.22	**1,894.83**	806.57	**327.02**	226.69	B
NRC	N-S	427.925	130.964	Norcia	5.39	326.71	38.81	8.40	1,130.05	535.24	218.22	165.70	B
NRC	Vert.	427.925	130.964	Norcia	5.39	378.34	18.38	4.88	682.85	273.60	164.08	85.38	B
ONE	E-W	428.944	131.528	Castel Santangelo S.N.	6.95	454.66	39.54	5.28	1,064.25	516.19	161.13	150.45	C*
ONE	N-S	428.944	131.528	Castel Santangelo S.N.	6.95	343.25	27.87	3.86	699.35	308.12	90.27	104.40	C*
ONE	Vert.	428.944	131.528	Castel Santangelo S.N.	6.95	595.87	16.83	3.09	985.74	186.41	228.54	74.71	C*
PRE	E-W	428.793	130.334	Preci	7.93	260.91	10.89	1.83	478.65	122.67	72.65	50.22	B*
PRE	N-S	428.793	130.334	Preci	7.93	315.42	14.05	2.42	421.45	128.12	84.85	53.71	B*
PRE	Vert.	428.793	130.334	Preci	7.93	202.52	6.72	1.61	173.17	89.09	52.21	30.83	B*
CLO	E-W	428.294	132.060	Castelluccio	8.23	478.45	66.10	**14.25**	1,126.09	862.01	222.18	250.87	A*
CLO	N-S	428.294	132.060	Castelluccio	8.23	**634.00**	54.36	9.31	1,568.22	1,017.45	381.90	250.95	A*
CLO	Vert.	428.294	132.060	Castelluccio	8.23	649.74	53.99	18.13	1,452.43	607.55	430.28	178.78	A*
CSC	E-W	427.190	130.122	Cascia	15.86	150.74	10.99	3.54	331.04	120.12	27.72	41.95	B
CSC	N-S	427.190	130.122	Cascia	15.86	172.34	12.37	3.16	303.00	124.69	33.34	40.82	B
CSC	Vert.	427.190	130.122	Cascia	15.86	168.53	5.64	1.74	196.90	62.42	21.97	26.65	B
MMO	E-W	428.993	133.268	Montemonaco	18.91	203.06	10.23	2.93	349.73	95.97	49.12	40.26	A*
MMO	N-S	428.993	133.268	Montemonaco	18.91	191.65	12.06	3.32	620.74	105.87	52.91	40.84	A*
MMO	Vert.	428.993	133.268	Montemonaco	18.91	144.28	7.66	2.32	383.03	83.66	35.05	37.00	A*

Most significant figures in bold

PSA10 = 508 cm/s^2, Arias intensity = 207 cm/s and Housner Intensity = 175 cm were quite higher. Also the CMI station is characterized by a soil profile classified as "C", according to EC8.

Concerning the third main shock (October 30th, 2016; Table 26.7), the most energetic one, the maximum values of horizontal PGA = 634 cm/s^2, PGV = 66 cm/s and PGD = 14 cm occurred at the same station CLO (Castelluccio di Norcia), another small village in the municipality of Norcia, 8 km from the epicenter. The first one was recorded in the N-S direction and the other two in the E-W direction. At the same station, also the maximum vertical PGA, equal to 649 cm/s^2, occurred. The maximum values of the other parameters were recorded, instead, by the stations located at Norcia, at about 5 km from the epicenter. PSA03 = 1895 cm/s^2 and Arias Intensity = 327 cm/s were recorded at NRC station in the E-W direction, while PSA10 = 874 cm/s^2 and Housner Intensity = 269 cm were recorded at NOR station. Apart from the peak ground acceleration, all the highest values of the other parameters were recorded during the third, Mw 6.5, main shock, as one could expect. They resulted to be up to 75% greater than the previously recorded maximum values. Although this could also depend on the distance from the epicenter, the directivity and other near fault effects, it is worth noticing that, in terms of destructiveness potential, the records of the October 30th in the two stations of Norcia were the most dangerous ones.

A first comparison between the intensity parameters for the seismic design of structures in the epicentral area, according to the current official Seismic Hazard Map of Italy (MPS Working Group 2004), and the PGA actually recorded can be useful to understand the destructive potential of the earthquakes. If one considers the hazard map referred to a 475 years return period, which represents the target for the design of dwelling buildings referred to the ultimate life safety state, then the maximum value of PGA on stiff horizontal soil is as high as 0.26 g, while the maximum value all over Italy is 0.28 g. Even considering magnifying factors to account for site amplification effects, which typically have a value lower than 2, it appears clear that the maximum recorded values of PGA and of spectral accelerations at 0.3 s exceed the corresponding ones in the current seismic hazard model for a 475 years return period, which is used for verification of the life safety limit state and are better compared with higher return period values, corresponding to collapse limit states. It can be easily concluded that, in many areas affected by the sequence, the seismic actions on the constructions were probably higher than those prescribed by the current code for the life safety verification, for which severe damage has to be expected. No matter about the assumption of the current code, what is important to underline is that, in spite of this, many buildings, and not only those designed according to the current design standards, survived with minor, or even without, damage. However, and not surprisingly, many old buildings not complying with any seismic code reported severe damage and even collapsed under the actions of such destructive earthquakes.

A comparison with the 2009 and 2012 earthquakes emphasizes the much higher values of PGA and of other intensity quantities recorded in the 2016–2017 sequence. As a matter of fact, the maximum recorded PGA value in the 2009 Abruzzo earthquake, at AQV station, was 664 cm/s^2, while the maximum PGV and PGD values, recorded at station AQK near the centre of L'Aquila city, were 38.6 cm/s and 11.9 cm, respectively (Dolce and Di Bucci 2017).

For the 2012 Emilia earthquake, the maximum horizontal PGA value, recorded 6 km far from the epicentre during the second main shock, was 290 cm/s^2, while the maximum PSA03 and PSA10 were 700 cm/s^2 and 370 cm/s^2, respectively, the maximum PGV was 57 cm/s and the maximum PGD was 18 cm. The maximum vertical acceleration was 900 cm/s^2 (Dolce and Di Bucci 2017). In this case, the peculiar characteristics of the alluvial and terrigenous infill of the Po Plain affected significantly the values of the various intensity quantities. It limited the horizontal acceleration, which resulted to be three times lower than the ones recorded in 2016, but determined high velocity and displacement values, which were of the same order as those recorded in 2016.

The OSS-DPC is a national permanent network that monitors the seismic response of more than 150 structures, including schools, hospitals, town halls, bridges, and a dam, mainly using accelerometric sensors suitably distributed both over the height of the structure and at its foundation (Dolce et al. 2015). By processing the acquired signals, the OSS-DPC allows a remote estimation of the damage suffered by the monitored structures to be made in few minutes after an earthquake. The level of damage is evaluated on the basis of the maximum interstory drift (ID) experienced by the building while shaking, according to the threshold values reported in Table 26.8. An idea of the damage possibly suffered by similar structures and, more in general, of the destructive potential of the shaking near around is therefore obtained.

In Table 26.9, a summary of the most significant behavioral parameters, namely the maximum peak ground acceleration (PGA) at the foundation of the structure and the maximum interstory drift (ID) along the height of the building are reported for three buildings strongly shaken by the three main shocks (http://www.mot1.it/ossdownload/index_it.php).

During the August 24th earthquake, the nearest monitored structure for which the records were available in few minutes was a hospital building at Norcia, 14 km from the epicenter, while a total of 37 monitoring systems were triggered within a 200 km distance from the epicenter.

During the October 26th earthquake, the nearest monitored structure was a masonry school at Visso, 4 km far from the epicenter, which resulted strongly

Table 26.8 OSS-DPC thresholds for the estimation of damage suffered by the monitored structures after an earthquake

Damage level	R.C. buildings	Masonry buildings
No damage	0.00% ≤ ID <0.50%	0.00% ≤ ID <0.20%
Slight damage	0.50% ≤ ID <0.90%	0.20% ≤ ID <0.45%
Moderate damage	0.90% ≤ ID <1.50%	0.45% ≤ ID <0.80%
Heavy damage	1.50% ≤ ID	0.80% ≤ ID

The level of damage is evaluated on the basis of the maximum interstory drift (ID)

damaged, whereas the other buildings were classified as not damaged. A total of 59 monitoring systems were triggered within a 200 km distance from the epicenter.

During the October 30th earthquake, the nearest monitored structure was a reinforced concrete school building at Norcia, 5 km from the epicenter, which displayed a light damage. A total of 60 monitoring systems were triggered within a 200 km distance from the epicenter.

As can be seen in Table 26.9, PGA values between 0.19 g and 0.57 g were experienced by the three above mentioned buildings. According to the threshold values of the interstory drift shown in Table 26.9, only the school masonry building of Visso experienced a very heavy damage (ID 1.6% during the October 26th event, vs. 0.80% threshold), and actually a near collapse condition was found when the school was visually inspected afterword. Damage to non-structural elements was instead found in those R.C. school buildings whose drift values were greater than the slight damage threshold value.

With respect to the previous recent earthquake sequences occurred in 2009 and 2012, a larger number and a better quality of records has been obtained by the Seismic Observatory of Structures. Actually, this is the first time, in Italy, that some complete instrumental records are available for buildings subjected to strong motions that have reached near collapse or significant damage conditions. Their exploitation in scientific studies can provide important contributions to the understanding of the behavior of masonry and R.C. buildings.

26.7 Macroseismic Survey

Immediately after the first main shock occurrence, macroseismic teams including DPC officers, CNR-IGAG and INGV researchers carried out a number of field surveys to assign a macroseismic intensity MCS to each municipality and locality of the epicentral area. This quick evaluation of the damage in the epicentral area is aimed at providing a first picture of the zones where the damage is from moderate to high.

Two days after the earthquake, 44 localities had been surveyed, while 5 days after the event a first report was produced including 116 localities. Results dated September 21st, 2016, show values up to intensity X-XI MCS, that means from "completely destructive" to "catastrophic", in three localities of two municipalities, namely Amatrice and Arquata del Tronto. More than 291 localities in 76 municipalities were surveyed and the related intensities assigned, ranging from V to X-XI MCS (Galli et al. 2016; Fig. 26.1a).

The first surveys aimed at re-evaluating the macroseismic field in MCS scale, after the October 26th Mw 5.9 second main shock, had just started when, on October 30th, 2016, the Mw 6.5 strongest seismic event of the sequence occurred. The second round of surveys returned a damaged area much larger than the previous one (Fig. 26.1b). More than 450 localities in 115 municipalities were surveyed. The maximum observed (cumulated) intensity is XI in the MCS scale, reached in five

Table 26.9 Most significant behavioral parameters recorded for three buildings strongly shaken by the three main shocks

Id.code	Use type	Municipality	Structural type	Max PGA 24.08.16	Max PGA 26.10.16	Max PGA 30.10.16	Max Interstory Drift 24.08.16	Max Interstory Drift 26.10.16	Max Interstory Drift 30.10.16
EA080	Hospital	Norcia	R.C. (**)	0.23 g	0.19 g	0.32 g	0.04%	0.37%	0.08%
15SNO	School	Norcia	R.C. (*)	0.52 g	0.48 g	0.57 g	0.23%	0.20%	0.56%
BC037	School	Visso	Masonry	0.33 g	0.48 g	0.30 g	0.61%	1.60%	1.10%

(*) The R.C. structure had been retrofitted, before the earthquakes, using steel braces with energy dissipation devices embedded
(**) Equipped with a simplified monitoring system. The interstory drift is evaluated on the total height of the building

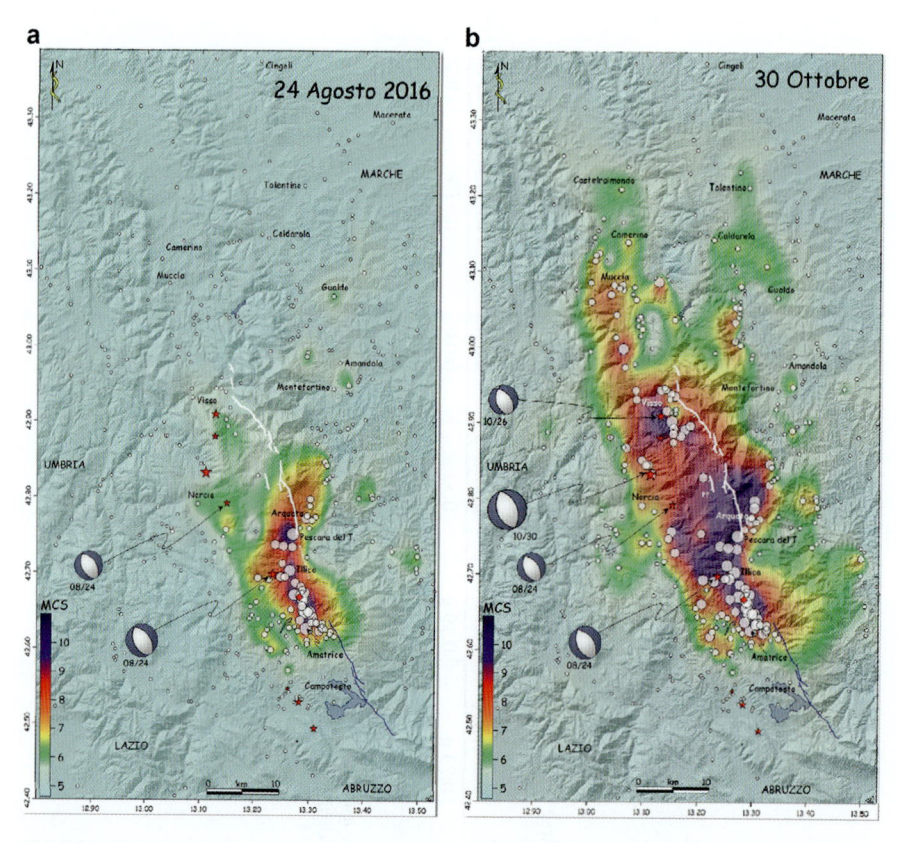

Fig. 26.1 Macroseismic survey of the 2016–2017 Central Italy seismic sequence in MCS scale: (**a**) after the August 24th, 2016, first main shock, and (**b**) after the October 30th, 2016, strongest main shock (Maps from Galli et al. 2017)

localities of three municipalities, namely Amatrice, Accumoli and Arquata del Tronto (Galli et al. 2017). Moreover, 31 localities in 7 municipalities reported damage corresponding to an intensity range between IX-X and X-XI MCS.

Finally, surveys carried out after the occurrence of four seismic events with Mw ≥ 5.0, on January 18th, 2017, allowed a further increase of the observed damage, in particular in the southern part of the macroseismic field.

In the meanwhile, researchers from INGV, ENEA and EUCENTRE, working within the INGV-QUEST Team, carried out a macroseismic survey using the European Macroseismic Scale EMS98 (Tertulliani and Azzaro 2016a, b; 2017. Fig. 26.2). This survey as well was repeated after the occurrence of the October 2016 and January 2017 major seismic events visiting, respectively, more than 240 and 70 localities, partly overlapping. In a range from V to XI EMS, the maximum intensity values have been assigned to Amatrice and Arquata del Tronto that, therefore, result to be the most damaged towns in both MCS and EMS scale.

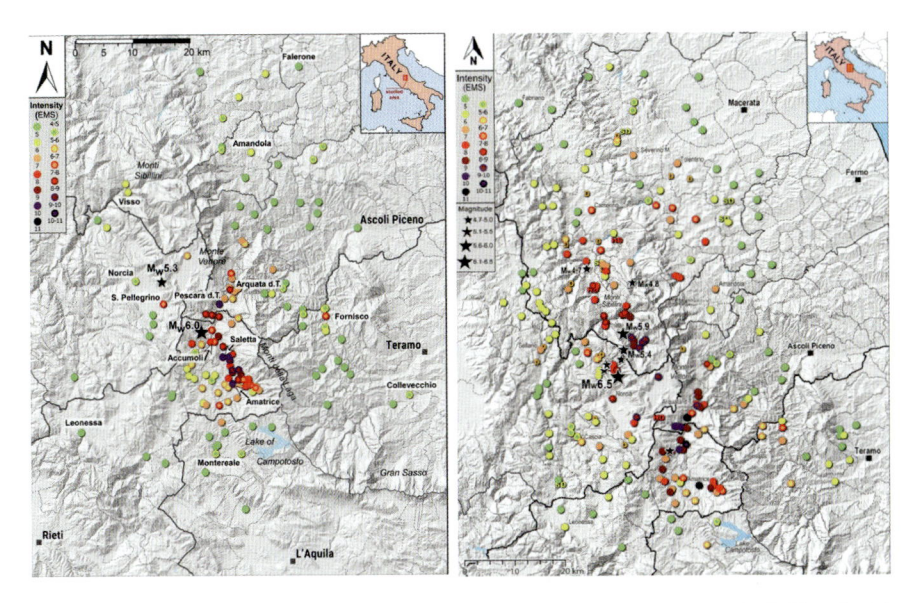

Fig. 26.2 Macroseismic survey of the 2016–2017 Central Italy seismic sequence in EMS scale: (**a**) after the August 24th, 2016, first main shock, and (**b**) after the October 30th, 2016, strongest main shock (Maps from: https://ingvterremoti.wordpress.com/2017/08/24/i-rilievi-macrosismici-dei-terremoti-del-2016-2017-in-italia-centrale/)

Moreover, 13 localities in six municipalities reported damage corresponding to an intensity range between IX-X and X-XI EMS.

In general, both the macroseismic surveys show, as one can expect, a widening of the damaged area, mainly after the October 2016 and, partly, also after the January 2017 events. This can be explained: by the areal distribution of the main shock epicenters, which spans over a length of 50 km, as already pointed out in the Introduction; by the severity of the October 30th main shock; and, finally, by the progressive increase of the building stock vulnerability due to previous damage.

If one compares the described macroseismic field in MCS intensity scale with the equivalent for the strongest Italian earthquakes occurred in the previous 30 years (1997 Umbria-Marche, 2009 Abruzzo and 2012 Emilia, all displaying Ml 5.9; Dolce and Di Bucci 2017), one can notice that the latter are characterized by lower maximum intensities (Umbria-Marche IX–X; Abruzzo IX–X; Emilia VII–VIII MCS). This difference also corresponds to a difference in the size of the damaged areas. In the 2016–2017 case, the macroseismic field of cumulated intensities $I_{MCS} \geq$ VII was about 70 km long and 30 km wide. In the Emilia earthquake, the equivalent area was about 20 km long and 10 km wide, in the Abruzzo earthquake it was 55 km long and 15 km wide, while in the Umbria-Marche sequence it was about 45 km long and 20 km wide. These figures reflect that the 2016–2017 Central Italy seismic sequence was, both for number of seismic events and for magnitudes, much more energetic than the previous ones.

What all of the cases share is, instead, the occurrence of more than one main shock in each of them. In particular, the 1997 Umbria–Marche and 2009 Abruzzo seismic sequences have been characterized by three main shocks, whereas the 2012 Emilia by two. As known by the literature (e.g., Di Bucci et al. 2010; Dolce and Di Bucci 2017) this behavior of the seismic sequences is quite frequent in Italy and characterizes all the tectonic environments (extensional and compressional, but also strike-slip).

26.8 Geological Coseismic Effects

The long and energetic seismic sequence, with three main shocks and a total of nine seismic events with $Mw \geq 5.0$ distributed over a large area, caused many geological coseismic effects, which have been surveyed in detail by researchers operating within the Di.Coma.C.. These effects encompass surface fracturing and faulting processes, which mainly re-activated preexisting faults already known in the literature (e.g., Boncio et al. 2004; Pizzi and Galadini 2009; Valensise et al. 2016), surveyed under INGV coordination (Emergeo Working Group 2017), and many rock-falls and landslides, as it always happens when moderate-to-strong earthquakes hit the Apennines chain. The latter phenomena were surveyed, in particular, by geologists from ISPRA and CNR, with special regard to those cases potentially or really affecting emergency management activities (e.g., Amanti and Galluzzo eds 2016). Landslides and rock-falls were numerous, affected the entire region and were in part responsible for disruptions of the transportation system. Apart from some severe landslides, continuous rock-falls also induced by the aftershocks following the October 30th seismic event made the circulation on many provincial and local roads very difficult, also for the SAR teams.

At a wider scale, coseismic surface deformation was detected by CNR-IREA and INGV investigators by analyzing a large geodetic data set of interferometric synthetic aperture radar (InSAR) and GPS measurements (Gruppo di lavoro IREA-CNR and INGV 2016; Cheloni et al. 2017). The interferometric technique was applied to radar data coming from ALOS2 and Sentinel-1 satellite constellations yet in the first days after the main shock, and then to Cosmo-SkyMed data provided by the National Space Agency (ASI), which allowed a much more detailed analysis, also of the slope instabilities due to the shaking.

The surface coseismic deformation related to the seismogenic fault motion of the August 24th event corresponds to a depressed, roughly elliptical area, as usually happens when the fault kinematics is extensional. The maximum value of subsidence was ca. 20 cm, in correspondence with the Accumoli village. GPS data obtained from the integrated elaboration of measures coming from several permanent networks present in the area are coherent with the described deformation.

Also the October 26th major event was generated by a WSW-dipping normal fault, which is part of the same fault system which includes the August 24th main shock seismogenic fault. The fault motion caused a subsidence of the epicentral area with a maximum displacement in the order of 20 cm.

Due to the higher magnitude characterizing the October 30th earthquake, thus to the larger seismogenic source, the ground displacement detected through interferometric investigations on satellite radar imagery for this event depicted instead a wider area of deformation. East-West and vertical displacement components were accurately measured all over the epicentral area (Cheloni et al. 2017). A maximum subsidence of ~90 cm was detected in the surroundings of Castelluccio, whereas a slight uplift of ~15 cm was observed in correspondence of Norcia. E-W displacement were also observed, with a westward movement of the Norcia area, an eastward one of the footwall of the main normal fault system, and localized westward motions close to Castelluccio (up to ~60 cm).

Dealing with the four earthquakes occurred on January 18th, 2017, in the considered seismotectonic environment their magnitudes are usually associated to limited surface deformation. Moreover, the thick snow cover present at that time affected the signal correlation. Nevertheless, some elaborations by CNR-IREA on ALOS-2 and Sentinel_1 images returned a subsidence of few centimeters.

If the coseismic effects just described are compared with those characterizing the 1997 Umbria-Marche, 2009 Abruzzo and 2012 Emilia earthquakes, it can be seen that the geographic and seismotectonic location of the first two events is essentially the same as of the 2016–2017 seismic sequence. All of them occurred at the core of the Apennines, a mountain chain characterized by high peaks, which made the rockfalls a distinctive coseismic effect. Moreover, they developed in an extensional tectonic regime, as also shown by active faults at surface, and their seismogenic faults, at depth, display normal kinematics. As a consequence, the overall surface deformation essentially corresponds to a general subsidence, whose values depend on the earthquake magnitude, hypocenter depth and dip angle of the causative fault. Therefore, also in this case the severe magnitude Mw 6.5 of the October 30th earthquake made the difference. The 2012 Emilia seismic sequence remains a different case: it occurred in the Po Plain, i.e., the largest alluvial plain of Italy, and the related most important coseismic effects were, therefore, widespread liquefaction phenomena. Moreover, due to the compressional tectonic regime characterizing that region, the thrust kinematics of the seismogenic faults and the limited earthquakes magnitude, surface faulting was totally lacking (Dolce and Di Bucci 2017, and references therein).

26.9 Post-earthquake Damage/Usability Assessment of Buildings

Post-earthquake usability evaluation is a quick assessment, based on expert judgement of specially trained technical teams, on visual screening and easily collected data. It is aimed at detecting whether, during an ongoing seismic crisis, damaged buildings can be still used, being reasonably safeguarded human life.

The damage and usability assessment of buildings is a fundamental activity in a post-earthquake emergency management, as its outcome allows a number of issues to find a correct solution:

1. the population to safely stay in or re-enter their homes;
2. properly scaling shelter and temporary housing needs, both in the emergency (tent camps, hotels, self-lodging financial support) and in the post-emergency (temporary housing) phases;
3. restarting social and economic activities as soon as possible;
4. defining economic needs for reconstruction;
5. identifying priorities and funding criteria for the repair/reconstruction of each building.

In Italy, the damage and usability assessment of ordinary buildings is executed using well established procedures, based on the experience of surveys carried out in Italy in the aftermaths of earthquakes since 1997 (Dolce and Di Bucci 2017). The Decree issued by the Italian Prime Minister on May 5th, 2011, later updated by the Decrees of July 8th, 2014 and of January 14th, 2015, enforced such procedures at national level, including the AeDES (Agibilità e Danno nell'Emergenza Sismica - usability and damage in seismic emergency) inspection form and the related manual (issued in 1997 for the first time; Baggio et al. 2000). Moreover, a greater impulse was given to the damage and usability assessment organization after the two most recent strong Italian earthquakes (2009 Abruzzo, 2012 Emilia) by:

- enforcing rules for the recruitment of experts, by making a large number of training courses of at least 60 h, with a final test,
- implementing a new inspection form, and the relevant compilation manual, for large span and prefabricated buildings, called GL-AeDES.

On August 24th, 2016, about 6000 experts, professionals and public administrations employees, had followed the training courses and were then in principle available to be recruited, on a voluntary basis, for inspections in the affected areas.

According to the AeDES form, building usability is classified into 6 categories: A - Usable building, B - Building usable only after short term countermeasures, C - Partially usable building, D - Building to be re-inspected, requiring a specific, yet visual, investigation, E - Unusable building, due to high structural risk, high non-structural risk or high geotechnical risk, F - Unusable building because of an external risk. For category F buildings, also the usability outcome corresponding to the damage state of the inspected building has to be made explicit in the inspection form. More details on the above definitions can be found in Baggio et al. (2000) and in Dolce and Di Bucci (2014).

The inspection activities started just a few days after the first main shock of August 24th. The AeDES assessment was performed by experts coming from different Regions and from the National Fire Brigades, by researchers from the DPC Competence Centres (namely, ReLUIS and EUCENTRE), and by professionals - engineers, architects and surveyors - coordinated through the related national professional Councils. All the inspectors had to fulfil the requirement of either having passed the final test of the above mentioned training courses, or having a considerable experience from past earthquake AeDES surveys.

Almost 30,000 inspections had been carried out until October 26th, 2016, and the corresponding AeDES inspection forms filled in. During the period of maximum activity, the damage and usability assessment involved about 160 teams per day, each team being formed by 2–3 experts. The maximum number of inspections per day ranged around 1000. On the one hand, as in past earthquake emergencies, the DPC officers coordinating these activities underwent a high pressure exerted by the local authorities and the population, needing a fast response especially to the requests of those citizens living in slightly or non-damaged buildings. On the other hand, assessing the usability of a building implies awareness and responsibility, issues that have to be managed and carefully considered even in an emergency hurry.

The outcomes of the survey until October 26th, 2016, provided the following figures: the amount of usable buildings (category A) corresponded to the 50% of the inspected buildings, while the amount of unusable buildings for their damage state, for external risk or for both (categories E, EF and AF) corresponded to the 32%. The remaining 18% was relevant to categories B, C, D, BF, CF and DF. Until October 26th, the citizens' requests for inspections were 77,000, i.e., approximately the total number of inspections made after the 2009 Abruzzo earthquake.

Soon after the October 26th and 30th main shocks, the number of requests raised quickly, due to the widening of the damaged area. Moreover, most of the buildings already inspected had to be re-evaluated. It was easy to foresee that a number of requests as high as 200,000, and even more, would have been attained. Based on this, it was also very clear that a different strategy had to be set up, as the number of available teams, always on a voluntary basis, was not adequate to deal with such a huge number of requests in a reasonable time lapse. A more rapid procedure for the usability assessment of private buildings, called FAST (scheda per il rilevamento sui Fabbricati per l'Agibilità Sintetica post Terremoto – form for the inspection on buildings for rapid post-earthquake usability), was therefore set up. It was especially devised to quickly define the number of temporary dwelling solutions needed, whereas schools and other public buildings continued to be inspected following the AeDES procedure. The main features of the FAST procedure were:

- Long training courses were no longer required, allowing new different teams to be composed;
- «Usable» or «Not Usable» were the only possible inspection outcomes;
- No internal inspections were needed to assign the «Not Usable» evaluation, as the damage had not to be quantified;
- The «Usable» outcome, instead, still required internal inspection, and it was assumed as the final outcome, equivalent to AeDES «Usable» outcome;
- The «Not Usable» outcome required a further AeDES inspection to assign the specific category from B to E;
- Inspections that were not completed, needing an evaluation from inside, required a second inspection;

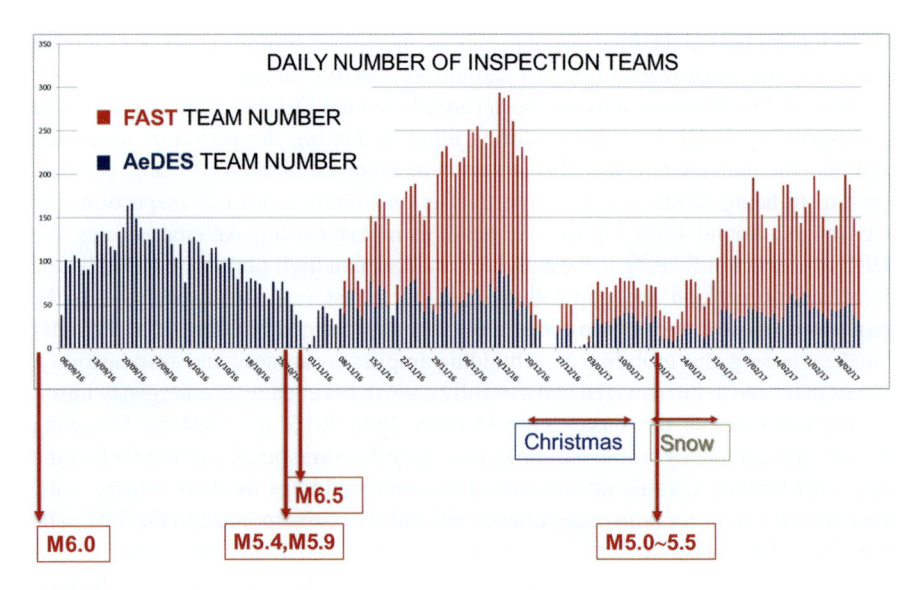

Fig. 26.3 Number of inspection teams from the start of the usability assessment activity until the end of February 2017

- The potential production was expected to be of about 10 FAST inspections/day/team, doubling the about 5 AeDES inspections/day/team.

The diagram in Fig. 26.3 shows how the number of AeDES and FAST inspection teams was distributed through time, from the second week of November 2016, i.e., the start of the usability assessment activity, until the end of February 2017. As it can be seen, the introduction of the FAST procedure considerably increased the number of teams per day, which at mid-December reached the figure of almost 300, 75% of which being FAST teams. In the meanwhile, AeDES teams continued to operate for specific cases, in a consistent framework. Looking at the diagram, it can also be noticed the increasing trend after each main shock or the cluster of Mw5+ seismic events (August 24th, October 26th and 30th, January 18th), which lasted one or 1 month and half, followed by a decreasing trend. Moreover, a quite long period of reduced activity corresponded to the Christmas holidays and to the exceptional snowfall in the second half of January 2017.

At the end of October 2017, just 1 year after the Mw 6.5 main shock, the total number of inspections carried out was almost 210,000, while the residual inspections to be still carried out were ca. 6500, half of which of AeDES kind. Most of the remaining inspections (70%) were in the Marche Region. The total number of AeDES forms filled in was about 75,000, including ca. 2600 school buildings and almost 4000 other public buildings. The total number of FAST inspections, carried out only on private buildings, was ca. 134,000, 79% of which with an assigned outcome (i.e., no further evaluations needed). 57% of the FAST inspections had a positive («Usable») outcome.

It has to be emphasized the use, for the first time after an earthquake, of the GL-AeDES inspection form, which had been officialized by a Prime Minister's Decree on January 14th, 2015. It was particularly useful for the inspection of sheds for cattle farms and food-processing industry that were present in the affected area, whose economy is based in part on agricultural and breeding activities, as said in the Introduction.

In the light of what just described, issues numbered 1 to 4 in the list at the beginning of this section can be considered as fulfilled, whereas the fifth one was not completely satisfied. This is because, according to the funding scheme adopted by the Government Commissioner for the Reconstruction, the compilation of the AeDES form is in principle required to decide the grant provided by the State to repair and strengthen the damaged private buildings. In order to operate in a consistent framework, when only the FAST form was available from the inspection activities coordinated by the Department of Civil Protection and the outcome was "Unusable", the Commissioner established that the process of damage and usability assessment had to be completed through the compilation of the AeDES form by a professional appointed by the owner.

Once again, the post-earthquake damage and usability assessment of ordinary buildings has turned out to be a fundamental activity not only for the emergency, but also for the reconstruction management. Many differences with respect to the 1997, 2009 and 2012 earthquake sequences, however, have to be pointed out, which depend on some conditioning factors characterizing each of the three emergencies. They are:

1. the space, time and intensity characteristics of each sequence: while in the previous sequences the main shocks occurred within 10 days maximum, the time lapse between the main shocks of this sequence required the restart of the inspection activities and change of procedure. Some similitudes can be found with the 1997 Umbria-Marche sequence, when a third main shock occurred after almost 20 days, but its consequences have had a relatively lower impact. Moreover, the distance among the main shock epicenters of the 2016–2017 sequence has determined a widespread distribution of heavy damage, which, for instance, did not occur in the 2009 case. Here, the low population density, the short distance between the epicenters and their vicinity to L'Aquila caused, instead, the concentration in this city of most of the damage and, therefore, of the buildings needing inspection;
2. the administrative complexity: as said before, the area affected by the 2016–2017 sequence involves four regions. This circumstance, in addition to the extension of the affected territory, required the decentralization of the inspections management. In 1997 the regions were only two, while in 2009 and 2012 most of the inspections were managed at central level in the Di.Coma.C.;
3. the urgency of completing inspections, due to their importance in both the emergency and reconstruction phases: this is a condition that recurs at each earthquake, but it is even more boosted in case of a vast territory interested by earthquakes, like in the present case;

4. the number of building inspections to be made, that was in the order of 80,000 in 2009 and 40,000 in 2012, while in the case of the 2016–2017 sequence it was much larger, in the order of 210,000;
5. the conditions of the road infrastructure after the earthquakes and the areal distribution of the building damage can determine logistic difficulties that have to be managed by the inspection teams and the survey organization. These conditions were much worse for the 2016-2017 sequence with respect to the previous earthquakes, as intensity MCS 7+ was observed in a vast territory, as seen in the "Macroseismic survey" section, and many road interruptions occurred due to damaged bridges, landslides, rock-falls, building rubble;
6. the season when the earthquake and the following emergency period occur, on the one hand, influences the urgency of the completion of the inspections, on the other hand, conditions the commuting of the inspection teams and the daylight hours. The 2016–2017 sequence essentially developed through autumn and winter, thus in the worst conditions. Differently, the 2009 and 2012 earthquakes occurred during spring, and the emergency was managed during spring, summer and autumn. Again, some similitudes can be found instead with the 1997 sequence, whose first two main shocks occurred on September 26th;
7. the preparedness that is needed to deal with the different situations related to damage/usability assessment: in 1997 there were no trained inspectors, as the AeDES form was just released in its first version, while in 2009 and 2012 there were a limited number of AeDES trained inspectors. Therefore, very short training courses had to be organized in the first day of the period of engagement of the voluntary teams (typically 1 week). In 2016, the quality of the AeDES inspections had been improved, but the intervened rules limited the recruitment of AeDES teams only among those experts already trained before the emergency, thus reducing the experts offer in terms of quantity.

All the above conditioning factors required different strategies and procedures to be followed and, in case, set up to deal with unforeseen situations with respect to the previous experience. The urgency of the completion of inspections was, obviously, one major factor that drove some choices. The most important difference with respect to the previous earthquakes has been the set up and adoption, for private building inspections, of the FAST procedure in parallel with the AeDES one, postponing the compilation of most of the AeDES forms to the repair/reconstruction phase.

At the present state of advancement of this inspection activity, it is not possible to directly compare the outcomes of the inspections to private buildings of this sequence with those of the previous earthquakes. Looking separately AeDES and FAST outcomes, and also considering that the provided figures include duplications, one can say that out of ca. 70,000 AeDES outcomes, about 42% are "Usable", while out of ca. 110,000 FAST outcomes, about 57% are "Usable". For the 2012 earthquake, out of 40,000 AeDES forms, 37% gave "Usable" outcome (Dolce and Di Bucci 2014), while for the 2009 earthquake, out of about 75,000 AeDES forms, 52% provided "Usable" outcome (Dolce and Goretti 2015). These different percentages of "usable" outcomes strongly depend on the adopted procedures. For instance, the 37% "Usable" outcomes for the 2012 Emilia earthquake may appear surprising with respect to the other earthquakes, also considering the lower moment magnitude of the main shocks. However, it can be easily explained taking into account that, in that case, a preliminary screening was carried out before starting the AeDES inspections, in order to rapidly identify the "undoubtedly" usable buildings that did not need any further AeDES inspection.

26.10 Schools

Due to the period of the year when the August 24th main shock occurred, i.e., in late summer, the main problems with schools were fortunately not related to the presence of students inside them, but rather to the need to regularly restart the new scholastic year at mid-September. Scholastic activities are, indeed, fundamental for the normal life recovery. Therefore, the capacity of school buildings to resist earthquakes with no or minor damage, not compromising their use, is a fundamental element for the resilience of the population. The urgency of a solution for this problem was further increased by the depopulation trend in those mountainous areas.

In the damage and usability assessment survey, started few days after the first main shock (see the previous section) priority was given, therefore, to school buildings, in order to immediately find the most suitable alternative solutions in case of heavy damage, or to carry out fast repair interventions in buildings where slight damage had occurred. For school buildings, as well as for public buildings in general, all the inspections were carried out with the AeDES procedure in all the emergency phases.

Due to the low population density in the affected area, the school emergency management after August 24th was not particularly difficult, considering the quite limited number of heavily damaged school buildings. Indeed, until October 25th, 2016, 942 inspections had been carried out on the same number of school buildings, 76% of which were usable (category A in the AeDES form) and only about 5.5% (47 buildings) were unusable for heavy damage or collapse (category E; Di Ludovico et al. 2017a).

After October 30th, inspections had to be completely remade, also reconsidering the already inspected school buildings that had not been heavily damaged or did not collapse after the first main shock. Until January 18th, 2017, 1980 inspections were carried out, including the repeated inspections, resulting in 72.4% "usable" (A) and 5.5% "unusable for heavy damage" (E) outcomes. Inspections have been continued, and until October 30th, 2017, 2633 AeDES forms had been filled in, including several repetitions, resulting in 66% "usable" (A) and 6% "unusable for heavy damage" (E) outcomes.

The behavior of school buildings has been examined with respect to the estimated earthquake intensity experienced, drawn from the INGV PGA shake maps, as reported by Di Ludovico et al. (2017a, b, c). According to these investigators, after the most energetic part of the seismic sequence, the percentage of usable school buildings, among the inspected ones, ranged from 68% structures that experienced PGA lower than 0.10 g, down to 14% structures affected by PGA greater than 0.30 g. R.C. buildings resulted to be the most common structural type, with damage mainly to non-structural members (usability outcome B/C), while masonry buildings experienced more frequently severe damage to structural members (usability outcome E). The damage level of the school buildings was clearly affected by plan and elevation structural regularity or irregularity, construction age, number of stories, pre-existing damage, construction quality, and previous strengthening interventions, according to Dolce (2004).

The strategy to deal with the unavailability of school buildings until their repair or reconstruction was initially based on the temporary arrangement of students in other usable school buildings in the same municipality or in nearby municipalities, also resorting to double shift teaching. Some large tents were adopted as schools for four municipalities in the first phase of the emergency. In the meanwhile, making use of donations, temporary school modules were being arranged in about 20 municipalities to overcome the most critical situations. Generally speaking, different solutions were adopted for different situations, as several factors (e.g., population remained or not in the village or town, availability of other schools or other adaptable undamaged buildings, etc.) concurred to define the "optimal" one, in terms of costs and benefits.

The Government Commissioner for the Reconstruction too considered as primary the problem of restarting school activities, firstly in acceptable and then in comfortable conditions, and he immediately began a recognition to identify the damaged school buildings needing (for several reasons) a complete reconstruction. In this activity, the DPC Competence Center ReLUIS was involved, and a new, more detailed recognition was made on 88 heavily damaged school buildings characterized by an "E" outcome in the AeDES inspections, relevant to 72 schools. A comparison of estimated reconstruction costs vs. repair+retrofit costs was carried out, assuming that the latter costs depend on vulnerability and damage level, in order to select schools to be reconstructed.

The main problems relevant to school emergency management are similar to those experienced for the 2009 and 2012 earthquakes, due to the time of occurrence of the first main shocks: all of them occurred during the night, resulting in no casualties in the schools. Even those previous earthquakes damaged many school buildings (e.g., Di Ludovico et al. 2009), thus requiring a big effort to allow students to continue their scholastic activities at the best. Differences between the 2009 and 2012, and the 2016–2017 earthquakes are, instead, mainly related to the season of occurrence and, therefore, to the possibility of benefiting from favorable climatic conditions and summer holidays for the school management in the short-medium term. Earthquakes occurred in springtime, like in the 2009 and 2012 cases, allowed both the emergency and/or the reconstruction management to focus efforts on setting up temporary or final solutions having some months at disposal. Therefore, temporary school modules, repair and strengthening interventions of damaged buildings, as well as the construction of new school buildings (using fast construction technologies) could have been implemented in order to start the new scholastic year as normally programmed and in quite comfortable conditions.

26.11 Cultural Heritage

Cultural heritage had already been heavily damaged by the August 24th earthquake, but the October 26th and 30th following main shocks dramatically widened and increased the level of damage, up to the partial or total collapse of many churches in the epicentral areas. Thousands of churches, palaces and heritage manufacts (towers, walls, etc.) were heavily damaged. Luckily, mainly due to the day of the week and the time of day when the main shocks occurred, no fatalities were found inside heritage buildings.

Actually, the size of the damage area referred to built heritage (especially churches), due to its higher vulnerability, is much wider than that referred to ordinary buildings. Damaged heritage buildings can be found as far as 100 km from the main shock epicenters. Therefore, there is a considerable overlap between the area affected by this seismic sequence and the areas affected by the 1997 Umbria-Marche and 2009 Abruzzo earthquakes. This concerns both the increase of previous damage, in case of buildings not repaired and strengthened yet after the previous earthquakes, and the re-occurrence of the damage, even worse than before, up to collapse, in case of simply repaired and weakly strengthened buildings.

In order to deal with the emergency management of cultural heritage, a strong collaboration was established within the Di.Coma.C. among the Ministry of Cultural Heritage, the operational structures of the civil protection system (Firefighter Corps, Carabinieri, Army), the Competence Centers and the DPC officers.

The main activities carried out on cultural heritage are:

1. assessment of damage and usability of churches, historical palaces and other heritage manufacts;
2. evaluation of safety conditions and execution of safety countermeasures;
3. protection of mobile heritage using temporary coverage;
4. displacement and sheltering of mobile heritage (artworks) at risk;
5. selection, displacement and sheltering of «architectural elements» from rubble, in case of collapsed buildings;
6. cost analysis of damage.

The emergency management, especially for what concerns points 1 and 4, was essentially based on a ministerial Directive on the cultural heritage issued on April 23rd, 2015 (Minister of Cultural Heritage 2015). It includes a number of inspection forms, with different levels of detail, for the description of the construction and of the movable heritage content. Among them, also the inspection forms for churches and palaces can be found. A further inspection form for other manufacts (towers, walls, fountains, etc.) had to be set up during the emergency, in order to properly conclude the survey and get a complete picture of the damage to all types of cultural heritage.

The assessment of damage and usability of churches and historical palaces was carried out with the technical support of ReLUIS for the structural damage evaluation and usability assessment. The outcome of this latter consists of a classification similar to that used for ordinary buildings (from A to F). Teams were formed by ReLUIS structural experts and Cultural Heritage Ministry experts.

Due to the evolution of the sequence, the survey had to be restarted several times. Focusing only on the detailed inspections, they started soon after August 24th. Before October 30th, 995 inspections had been carried out. After the October 30th main shock, the inspections were temporarily stopped for the safety of the operators. They then restarted and continued until Christmas at a quite slow rate, reaching a total number of 451 inspected buildings. With the new year, when the sequence was apparently slowing down, a new inspection campaign was organized, aimed at evaluating damage and usability of all the cultural heritage manufacts, including many that had already been inspected, for which one or more requests had been received until January 10th, 2017. This was deemed necessary because of the increase of damage determined not only by the five M5+ seismic events, but also by the 50 M4+ and about one thousand M3+ events that had occurred until the end of 2016. Also in this case, the survey management was centralized in the DiComaC., and a renewed effort was asked to both the experts of ReLUIS and of the Cultural Heritage Ministry. The new inspection campaign started on January 16th and ended in 2 months. It involved 20–30 teams per day, on average. Each team, made up of 3–4 experts, was able to carry out 3–5 inspection per day, depending on the complexity of the manufact to inspect and on the logistic conditions. This new campaign was just started when the four M5+ quakes of January 18th, 2017 occurred, along with the intense snowfall mentioned above. Both events determined some further difficulties in the surveys, which required a re-modulation of the inspection program, concentrating the inspections, during the first weeks, in a working area far from the old and new epicenters and, even more, from the area most interested by the snowfall.

The inspection requests received until January 10th, 2017, were 3680 (543 of which in Abruzzo, 1832 in Marche, 342 in Lazio and 963 in Umbria Regions). Further 449 requests arrived afterwards, summing up to a total of more than 4000.

Until April 7th, 2017, about 5.000 detailed inspections, including also those repeated, had been carried out. As for churches, in particular, about 4000 inspections had been carried out, ca. 1000 of which replied. Out of 3100 inspected churches, 23.5% were judged usable, 48% unusable, 21.5% usable after short term interventions, and 7% partially usable, unusable for external risk or temporarily unusable.

The execution of temporary safety countermeasures was carried out by Firefighters, in connection with the Ministry, or by private owners, under the supervision of the heritage Superintendent. A total number of ca. 500 interventions had been completed by April 7th, 2017.

Generally speaking, this activity requires a long time to be completed, given the large number of buildings that need safety countermeasures. In many cases of slight or moderate damage it has to be evaluated, however, whether it is more convenient to directly proceed with the final repair and strengthening intervention, without any intermediate provisional works.

The activities related to the displacement and sheltering of mobile heritage at risk were carried out with the involvement of the Firefighters Corps, the specialized Command of Carabinieri for the protection of cultural heritage, the Army, as well as specialized volunteer organizations of civil protection, besides Cultural Heritage Ministry officials. By April 7th, 2017, more than 15,000 movable artworks had been recovered, including statues, pictures and other types of movable cultural heritage, from more than 400 buildings, mainly churches. Moreover, almost 3 km of archives and 7000 volumes were recovered.

A quantitatively important and complex activity, which is requiring a very long time, is the recovery of valuable architectural elements from rubble of collapsed buildings and their displacement into safe and secure places. Rubble is classified into three types: A – of protected heritage, B – of historical buildings, C – of buildings of no cultural interest. Type A rubble should, as far as possible, be preserved in situ, while B and C rubble is transported in temporary deposit sites, where valuable architectural elements are selected by Ministry experts, with the help of voluntary associations of civil protection.

The main criticalities during the emergency management were related to the time of execution of all the above listed activities, but especially those concerning: the usability assessment of not severely damaged churches; the displacement and sheltering of mobile heritage (artworks) in heavily damaged or collapsed churches; the safety countermeasures; the on-site temporary coverage of artworks under collapsed churches, also because of the snowfall and the consequent melting of the snow; the operations to secure precious "heritage rubble". Faster procedures and a big effort of the civil protection system components were requested and put in place after the October 30th main shock.

26.12 Conclusion

The long earthquake sequence of Central Italy that started on August 24th, 2016, with nine Mw 5+ earthquakes, three of which were main shocks with moment magnitude Mw 6.0, 5.9 and 6.5 in a 65 days time frame and a subsequence of four M5+ after 5 months, has represented a big challenge in the emergency management for the entire SNPC of Italy. A renewed greater effort after each main shock was required, along with some change of strategy and procedure to deal with the different emergency management problems. Moreover, the Mw 6.5 main shock was the strongest earthquake after the November 23rd, 1980, Irpinia earthquake that, however, occurred in a totally different situation, where neither the SNPC nor the DPC existed yet.

In the present paper, after a description of the main features of the earthquake sequence in terms of intensities and coseismic effects, only some significant emergency management problems and the related adopted solutions have been described, focusing the attention especially on the technical and scientific activities finalized at the emergency management and carried out within the DPC coordination. Many of these activities, indeed, are based on the scientific and technical information provided by the DPC technical offices and by its Competence Centers. Among them, the ones relevant to the damage assessment of constructions, both private and public buildings, including schools as well as cultural heritage, have deserved a specific focus, along with some of the related activities aimed at reducing the residual risk and finding temporary solutions.

There are activities that have not been described here but are, however, fundamental for the emergency management. They are not characterized by a full involvement of the technical-scientific components of the civil protection system and are relevant, for instance, to the SAR operations in the first days, to the recovery of the transport infrastructures, with urgent interventions in the first days and final permanent rehabilitation in the long run, to the management of first solutions for homeless and the realization of long term temporary housing, and to the management of the enormous amount of rubble. Further details on these activities and, more in general, on the emergency management are reported in the DPC website: http://www.protezionecivile.gov.it/.

A few days after the first main shock, therefore during the first emergency phase, the reconstruction process virtually started with the endorsement of the Government Commissioner for the Reconstruction. The responsibilities and duties for the emergency management coordination and the reconstruction process, however, were kept well separated. Even the reconstruction process had to adapt its strategy to the upsurges of the seismic sequence. Three reconstruction decree-laws, then converted in laws, were enacted after each main shock or the late strong earthquake subsequence, i.e., after August 24th, 2016, October 30th, 2016 and January 18th, 2017.

Important lessons can be drawn from the experience derived from both the emergency management and the complete build up of the reconstruction process. The long duration of the seismic sequence and the high magnitude of the strongest main shock posed unprecedented problems to the modern organization of the Italian civil protection, many of which had to find a solution during the emergency management.

Acknowledgments The Authors are responsible for the contents of this work, which do not necessarily reflect the position and official policy of the Italian Department of Civil Protection.

References

Amanti M, Galluzzo F (eds) (2016) Report riassuntivo ISPRA sopralluoghi viabilità 24–31/08/2016. ISPRA, attività connessa all'Emergenza terremoto del 24 agosto 2016, p 15

Baggio C, Bernardini A, Colozza R, Corazza L, Della Bella M, Di Pasquale G, Dolce M, Goretti A, Martinelli A, Orsini G, Papa F, Zuccaro G (2000) Manuale per la Compilazione della Scheda di I Livello di Rilevamento Danno, Pronto Intervento e Agibilità per Edifici Ordinari nell'Emergenza Post-sismica (Manual for the Compilation of the 1st Level Forms for the Assessment of the Damage, the Provisional Interventions and the Usability of Ordinary Buildings in the Post-Earthquake Emergency). Servizio Sismico Nazionale e Gruppo Nazionale per la Difesa dai Terremoti, p 112

Bally A, Burbi L, Cooper C, Ghelardoni R (1986) Balanced sections and seismic reflection profiles across the central Apennines. Mem Soc Geol It 35:257–310

Barchi M, Minelli G, Pialli G (1998) The crop 03 profile: a synthesis of result on deep structures of the northern Apennines. Mem Soc Geol It 52:383–400

Boncio P, Lavecchia G (2000) A structural model for active extension in Central Italy. J Geodyn 29:233–244

Boncio P, Lavecchia G, Pace B (2004) Defining a model of 3D seismogenic sources for seismic hazard assessment applications: the case of central Apennines (Italy). JoSe 8:407–425

Bonini L, Maesano FE, Basili R, Burrato P, Carafa MMC, Fracassi U, Kastelic V, Tarabusi G, Tiberti MM, Vannoli P, Valensise G (2016) Imaging the tectonic framework of the 24 August 2016, Amatrice (central Italy) earthquake sequence: new roles for old players? Ann Geophys 59 (Fast Track 5). https://doi.org/10.4401/AG-7229

Bramerini F, Lucantoni A (2001) Simulazioni di scenari di danno per la pianificazione dell'emergenza postterremoto (Simulation of damage scenarios for the post-earthquake emergency planning). Ingegneria Sismica 18(1):37–46

Cheloni D, De Novellis V, Albano M, Antonioli A, Anzidei M, Atzori S, Avallone A, Bignami C, Bonano M, Calcaterra S, Castaldo R, Casu F, Cecere G, De Luca C, Devoti R, Di Bucci D, Esposito A, Galvani A, Gambino P, Giuliani R, Lanari R, Manunta M, Manzo M, Mattone M, Montuori A, Pepe A, Pepe S, Pezzo G, Pietrantonio G, Polcari M, Riguzzi F, Salvi S, Sepe V, Serpelloni E, Solaro G, Stramondo S, Tizzani P, Tolomei C, Trasatti E, Valerio E, Zinno I, Doglioni C (2017) Geodetic model of the 2016 Central Italy earthquake sequence inferred from SAR and GPS data. Geophys Res Lett. https://doi.org/10.1002/2017GL073580

Di Bucci D, Burrato P, Vannoli P, Valensise G (2010) Tectonic evidence for the ongoing Africa-Eurasia convergence in the Central Mediterranean: a journey among large earthquakes, slip reversals and elusive fault motions. J Geophys Res 115:B12404. https://doi.org/10.1029/2009JB006480

Di Ludovico M, Di Pasquale G, Dolce M, Manfredi G, Moroni C, Prota A (2009) Behaviour of scholastic buildings after L'Aquila earthquake. Progettazione Sismica 3:155–157. Special issue: L'Aquila, April 6th 2009, 3:32 am (English version)

Di Ludovico M, Santoro A, De Martino G, Prota A, Dolce M, Manfredi G (2017a) Analisi di correlazione tra danno e caratteristiche tipologiche degli edifici scolastici colpiti dalla sequenza sismica del centro Italia nel 2016. XVII Convegno ANIDIS "l'Ingegneria Sismica in Italia", Pistoia

Di Ludovico M, De Martino G, Santoro A, Moroni C, Prota A, Dolce M, Manfredi G (2017b) Cumulative damage on school buildings following the 2016 central Italy earthquake sequence. GNGTS 2017, Trieste

Di Ludovico M, Digrisolo A, Moroni C, Graziotti F, Manfredi V, Prota A, Dolce M, Manfredi G (2017c) Remarks on damage and response of school buildings during the Central Italy earthquake sequence. Submitted to BEE

Dolce M (2004) Seismic safety of schools in Italy. Proceeding. of the OECD ad hoc experts' group meeting on earthquake safety in schools, 9–11 February 2004, Paris, France

Dolce M (2010) Emergency and post-emergency management of the Abruzzi Earthquake, theme-leader lecture. In: Garevski M, Ansal A (eds) Earthquake Engineering in Europe. Springer, Dordrecht. https://doi.org/10.1007/978-90-481-9544-2_19

Dolce M, Di Bucci D (2014) National civil protection organization and technical activities in the 2012 Emilia earthquakes (Italy). Bull Earthq Eng 12(5):2231–2253. Special issue on the Emilia earthquake. https://doi.org/10.1007/s10518-014-9597-x

Dolce M, Di Bucci D (2017) Comparing recent Italian earthquakes. Bull Earthq Eng 15(2):497–533. Special issue: post-earthquake assessment. https://doi.org/10.1007/s10518-015-9773-7

Dolce M, Goretti A (2015) Building damage assessment after the 2009 Abruzzi earthquake. Bull Earthq Eng 13:2241. https://doi.org/10.1007/s10518-015-9723-4

Dolce M, Nicoletti M, De Sortis A, Marchesini S, Spina D, Talanas F (2015) Osservatorio Sismico delle Strutture: the Italian structural seismic monitoring network. Bull Earthq Eng 15:621. https://doi.org/10.1007/s10518-015-9738-x

Emergeo Working Group (2017) Photographic collection of the coseismic geological effects originated by the 24th August 2016, Amatrice (Central Italy) seismic sequence. Misc INGV 34:1–114. http://www.ingv.it/editoria/miscellanea/2017/miscellanea34/

Galli P, Peronace E, Tertulliani A (eds) (2016) Rapporto sugli effetti macrosismici del terremoto del 24 Agosto 2016 di Amatrice in scala MCS. Roma, rapporto congiunto DPC, CNR-IGAG, INGV, p 15. https://doi.org/10.5281/zenodo.161323

Galli P, Castenetto S, Peronace E (2017) Rapporto sugli effetti macrosismici del terremoto del 30 Ottobre 2016 (Monti Sibillini) in scala MCS. Roma, rapporto congiunto DPC, CNR-IGAG, p 17+8 http://www.protezionecivile.gov.it/resources/cms/documents/Rapporto_effetti_macrosismici_terremoto_30_Ottobre_2016_18_genn__.pdf

Gruppo di lavoro IREA-CNR and INGV (2016) Sequenza sismica di Amatrice: risultati iniziali delle analisi interferometriche satellitari. https://doi.org/10.5281/zenodo.60938

Guidoboni E, Ferrari G, Mariotti D, Comastri A, Tarabusi G, Valensise G (2007) CFTI4Med, catalogue of strong earthquakes in Italy (461 B.C.-1997) and Mediterranean area (760 B.C.-1500). INGV-SGA

ISTAT (2016) Caratteristiche dei territori colpiti dal sisma del 24 agosto 2016. http://www.istat.it/it/archivio/190370

Minister of Cultural Heritage (2015) Directive cultural heritage of 23 April, 2015

Monachesi G, Castelli V (eds) (1992) Sismicità dell'area aquilano-teramana dalla "analisi attraverso i cataloghi". Rapporto tecnico per la Regione Abruzzo, Osservatorio Geofisico Sperimentale, Macerata, p 245. https://www.emidius.eu/AHEAD/study/MONCA992

Mostardini F, Merlini S (1986) Appennino centro-meridionale. Sezioni geologiche e proposta di modello strutturale. Mem Soc Geol It 35:177–202

MPS Working Group (2004) Redazione della mappa di pericolosità sismica prevista dall'Ordinanza PCM 3274 del 20 marzo 2003. Rapporto Conclusivo per il Dipartimento della Protezione Civile, INGV, Milano-Roma, aprile 2004, p 65, 5 appendici. http://zonesismiche.mi.ingv.it/

Pizzi A, Galadini F (2009) Pre-existing cross-structures and active fault segmentation in the northern-central Apennines (Italy). Tectonophys 476(1–2):304. https://doi.org/10.1016/j.tecto.2009.03.018

Tertulliani A, Azzaro R (eds) (2016a) QUEST – Rilievo macrosismico in EMS98 per il terremoto di Amatrice del 24 agosto 2016. https://doi.org/10.5281/zenodo.160707

Tertulliani A, Azzaro R (eds) (2016b) QUEST – Rilievo macrosismico per i terremoti nell'Italia centrale. Aggiornamento dopo le scosse del 26 e 30 ottobre 2016. Aggiornamento al 21 dicembre 2016. https://doi.org/10.5281/zenodo.238778

Tertulliani A, Azzaro R (eds) (2017) QUEST – Rilievo macrosismico in EMS98 per la sequenza sismica in Italia Centrale: aggiornamento dopo il 18 gennaio 2017, Rapporto interno INGV https://doi.org/10.5281/zenodo.556929

Rovida A, Locati M, Camassi R, Lolli B, Gasperini P (eds) (2016) CPTI15, the 2015 version of the parametric catalogue of Italian earthquakes. Istituto Nazionale di Geofisica e Vulcanologia. https://doi.org/10.6092/INGV.IT-CPTI15

Valensise G, Vannoli P, Basili R, Bonini L, Burrato P, Carafa MMC, Fracassi U, Kastelic V, Maesano FE, Tiberti MM, Tarabusi G (2016) Fossil landscapes and youthful seismogenic sources in the central Apennines: excerpts from the 24 August 2016, Amatrice earthquake and seismic hazard implications. Ann Geophys 59(Fast Track 5). https://doi.org/10.4401/Ag-7215

Chapter 27
Eurocode 8. Evolution or Revolution?

Philippe Bisch

Abstract The development of the 2nd generation of Eurocodes is under way, under a mandate of the European Commission to CEN. The history of the development of the Eurocodes since 1975 until their release in 2005 as European standards EN (1st generation) is put into perspective. It is pointed out that the evolution of the texts during successive updates of Eurocode 8 leaves open discussions on certain topics for the development of the 2nd generation. After having explained how the work in progress is organised, the most important topics for the current or future work are discussed, with the solutions that are proposed.

27.1 Introduction

In 2012, following an exchange with CEN/TC250, responsible for the development of Eurocodes (EC), the European Commission entrusted CEN with the mandate M/515, which consists of developing certain themes not covered by the first generation of ECs. Two topics are common to all Eurocodes: reducing the number of nationally determined parameters (NDP) and improving ease of use.

As far as Eurocode 8 (EC8) is concerned, work has been carried out to reorganise all the parts, in order to improve their readability, to avoid repetitions and even contradictions. On the menu of the work to be carried out on part 1: the revision of the soils classification and of the definition of the seismic action, the development of displacement methods, the development of a new chapter on structures equipped with dissipating devices, revision of the ductility classes, a thorough review of the chapters on materials and the addition of provisions concerning aluminium, better treatment of infills and claddings. For the other parts: redesign of part 3 on

---

P. Bisch (✉)
Ecole des Ponts et Chaussées, Marne-la-Vallée, France

Egis Industries, Montreuil, France
e-mail: philippe.bisch@egis.fr

© Springer International Publishing AG, part of Springer Nature 2018
K. Pitilakis (ed.), *Recent Advances in Earthquake Engineering in Europe*,
Geotechnical, Geological and Earthquake Engineering 46,
https://doi.org/10.1007/978-3-319-75741-4_27

re-evaluation with the addition of bridges, development of soil-structure interaction in part 5, etc.

This important work should make it possible to deliver a deeply renewed second-generation EC8, without the fundamental principles being altered.

This article describes the progress made at the end of 2017. Given the remaining development work, this report is necessarily incomplete and some of the options presented may be subject to changes at later stages of the program, the texts not being fixed until the formal vote, which has not yet been scheduled.

27.2 The Programme for a Second Generation of Eurocodes

27.2.1 A Bit of History

27.2.1.1 The Origins

In 1975, the Commission of the European Community adopted a programme of action in the field of construction, on the basis of Article 95 of the Treaty of Rome. The objective of the programme was to remove barriers to trade and harmonize technical specifications. As part of this action programme, the European Commission took the initiative to establish a set of harmonised technical rules for the design of structures, to replace the national rules in force in the Member States. For 15 years, the Commission piloted the development of the Eurocodes, which led to a first version of texts distributed within the member states, but not applied in practice.

The objective of the ECs assigned by the Commission is to provide common structural design rules for everyday use, for the calculation of complete structures and component products of a traditional or innovative nature.

In 1989, the Commission and the Member States of the European Union and EFTA decided, on the basis of an agreement between the Commission and CEN, to transfer the preparation and publication of the ECs to CEN with a view to give them the status of European Norms (EN).

Nevertheless, the European Commission has remained very involved in the process of drawing up the Eurocodes, participating in its financing, on the one hand, and intervening directly in the process of controlling texts, on the other.

A first set of "experimental" standards (ENV statutes) was therefore drawn up and disseminated, most of them between 1994 and 1997. These standards were not meant for a "compulsory" application in the Member States, but only to full-scale tests in projects, in order to prepare texts with EN status.

… 27 Eurocode 8. Evolution or Revolution?

27.2.1.2 Conversion to EN

The different standards established as ENVs have been the subject of a review among the Member States in order to give an opinion on the changeover to the European standard EN (which has been acquired) and to gather the comments to be taken into account for the elaboration of the ENs. These comments have proved to be very numerous and sometimes fundamental, so that a work which was originally to be a mere refinement following experimentation has in fact proved to be a fairly profound revision of the texts.

The work of converting the ENV experimental version into the European EN standard was undertaken in 1998 and the different parts were published between 2005 and 2007, in the form of "packages", the first of which consisted of all parts of the Eurocodes for the design of buildings. The conversion took place in continuity with the development of the ENVs, since the last ENV parts were not finalised when this work began.

27.2.2 The Eurocode 8 Case

27.2.2.1 The ENVs

The first "experimental" series of EC8 was established from 1990 onwards and published gradually from 1995 to 1998. For the design of structures in seismic regions, the provisions of EC8 complement the provisions of EN 1990 to EN 1997 and EN 1999. There are therefore many references to the provisions of the other ECs, which compels the development schedule of the different parts of EC8, which can only be finalised after the standards they reference.

The program consisted of eight standards, listed below (with indication of the effective date of publication):

- ENV 1998–1-1: 1994: General rules.
- ENV 1998–1-2: 1994: Buildings.
- ENV 1998–1-3: 1995: Specific rules for materials.
- ENV 1998–1-4: 1996: Strengthening and repair.
- ENV 1998–2: 1995: Bridges.
- ENV 1998–3: 1996: Towers, masts and chimneys.
- ENV 1998–4: 1998: Silos, tanks and pipes.
- ENV 1998–5: 1994: Foundations and retaining structures.

27.2.2.2 The First Generation EN of Eurocode 8

It was decided during the conversion work to merge the ENV parts 1–1, 1–2 and 1–3 into EN 1998–1 and to give an integer number to the reinforcement part, which found itself in 1998–3, rejecting old part 3 relating to towers, masts and chimneys as EN 1998–6. This is the current numbering of the first generation EN of the EC8.

Significant developments have been introduced between ENV and EN, notably:

- The use of maximum acceleration on rock, a_{gR}, instead of nominal acceleration and the introduction of two standard spectral shapes associated with a revision of the soil classification. Moreover, the exponents of the analytical expressions of the design spectra are taken equal to those of the elastic spectra for flexible structures.
- The development of displacement methods, allowing their use for new structures. In particular, elastic displacement spectra have been introduced in an annex. It should be noted, however, that, although the analysis method has been quite sufficiently described (demand side), all the criteria necessary for justification have not been developed (capacity side).
- The distinction between two types of reinforced concrete walls: ductile walls and long walls with little ductility.
- Development of rules for reinforced concrete foundations.
- The introduction of provisions for reinforced concrete prefabricated elements.
- The introduction of a chapter for composite steel-concrete structures.
- The introduction of a chapter on base isolation in Part 1 as Section 10.
- An in-depth revision of the standard dedicated to existing building to take account of recent developments on this subject and, in particular, to add informative annexes on materials.
- The development in part 5 of rules for the verification of spread foundations.

It was very seriously considered to remove the ductility class H and to improve the class L with additional provisions in order to return to two ductility classes, but this option was eventually dropped.

27.3 Review Process for the Second Generation

27.3.1 The Origin of the Process

Following an exchange with TC250, the European Commission, by means of a mandate M/466 of May 2010, officially submitted to CEN a request for the preparation of a mandate relating to the evolution of the Eurocodes, including new developments:

- The development of Eurocodes parts relating to new materials (structural glass, FRP, very high strength concrete) and a new structural type (membranes).
- Development of Eurocodes parts for the assessment and reinforcement of existing structures.
- Strengthening the provisions on structural robustness.

The mandate to be established would also include developments and deep review of existing parts, including:

- The reduction of the number of Nationally Determined Parameters (NDP).
- Consideration of recent research and innovation.
- Simplifying rules by focusing on the most commonly identified situations.

The aim was to define in detail the work to be undertaken and to establish the development programme. Note that this programme does not include "normal" maintenance of existing texts. CEN responded to the Commission under this mandate after consultation with stakeholders, including TC250 and its subcommittees.

After reviewing the proposals, the Commission requested CEN, by mandate M/515 of December 2012, to detail the parts of the programme of development of ECs it accepts (detailed in the appendix to the mandate) and a new Eurocode dedicated to structural glass. CEN responded to this mandate with a final report in May 2013. This document contains a precise description of the program to be undertaken for a second generation of Eurocodes and the organisation to be set up, such as described below. It also contains a detailed description of the 77 tasks to be undertaken.

As the Commission accepted CEN's proposals in response to the mandate, the development of the second generation of Eurocodes could start with an official start-up on 1st January 2015 and for an expected duration of 78 months, with all texts issued by TC250 in June 2021.

27.3.2 The Organisation

The organisation set up to draft and edit the second-generation Eurocodes is similar to the one that was used for the first generation. Indeed, the main structure constituted by TC250 and its subcommittees (SC) has remained active over time, with reduced maintenance activity. It was simply a matter of adapting the working rhythm of the organisation to the production of the new codes. In the original configuration, each SC is in charge of the editing of a Eurocode, with the same number (SC8 for Eurocode 8). The development of EN1990 was directly supported by TC250.

The organisation of the CEN technical committees also makes it possible to set up working groups (WG) made up of experts appointed by the Members. These WGs are decided either by TC250 or by its SCs, which assign them a delimited working area. It is also possible to create "task groups" with a very precise scope and "ad hoc groups" with a more ephemeral existence, generally responsible for writing specific reports.

Some WGs had been created during the maintenance period, with one notable example: WG7, which was responsible for considering the evolution of EN1990, which had been transformed into SC10 at the beginning of the mandate. Similarly, SC11 was created on the basis of WG3 to develop a new Eurocode dedicated to structural glass.

Most WGs have been set up to assist SCs in their production work. As regards SC8, WGs have been formed to cover all materials: WG1 for masonry, WG2 for steel, aluminium and composite steel-concrete structures, WG3 for timber and WG5 for concrete. A WG4 was also formed to advise SC8 on site classification and spectral shapes; its scope was then extended to geotechnics in relation to Part 5. In addition, a WG6 was formed on the basis of a TG2 to advise on all developments related to bridges (new chapter of Part 3 and Part 2). Finally, a TG1 is in charge of cogitating and advising SC8 on the reorganisation of Eurocode 8.

For the effective drafting of texts or parts of texts, TC250 designates project teams (PT) selected from a call for experts according to predetermined criteria, the main one being the scientific competence in the area of the PT under consideration. Members of a PT have a contract financed by the grant awarded by the European Commission to fulfil the mandate. Depending on their field of activity, the work of the PTs is supervised either by TC250 or by the concerned SC. With respect to SC8, six PTs have been planned and their scope is defined in the mandate.

The drafting work is spread over four phases, SC8 being concerned by the first three, at the rate of two PTs per phase. The work of each PT is spread over 3 years, the phases being expected to be shifted by 1 year (in reality the shift has been greater between the first and the second phase). The PTs are responsible for drafting the pieces of text defined in their contracts and when they have completed their work, the SC takes responsibility for the completion of the standards and their editing.

At the same time, a Systematic Review was launched according to CEN rules, aimed at obtaining the opinion of the Members on the first-generation standards. The comments collected at the end of each review are examined by the SC concerned and, if accepted, integrated by the PT in charge of the corresponding part.

The production schedule envisaged a work end on the texts in 2021, but, given the lag in the second phase, a new target at 2022 is more reasonable. Given CEN's production time, translation and diffusion, in the current perspective, the texts should be available according to a schedule that remains to be defined, the largest number of parts from 2022. It will then remain for the Members, for each EC part, to prepare the National Annex. Note that since Eurocode 8 is complementary to other Eurocodes, its schedule is necessarily dependent on other EC schedules.

27.3.3 Reduction of Nationally Determined Parameters

This is the first task assigned to PTs under the mandate, at the request of the European Commission. The ultimate goal is clearly the harmonisation of standards in the Union. This is a historical trend since already the possibilities of national variations had been singularly reduced from the ENV stage, with National Application Documents (NAD), to the EN stage, with simply NDPs.

Nevertheless, some NDPs of the first generation were more the result of not reaching consensus on procedures or values than safety choices. By hypothesis, additional work of analysis and discussion should make it possible to improve the situation. TC250 was therefore concerned with clarifying the overall situation and defining a reduction policy, resulting in a report to be implemented by the PTs.

"Legitimate" NDPs are primarily those related to safety, such as partial factors. They are therefore not meant to disappear, even if all Members choose the same value. Conversely, all the parameters entering into a physical or mechanical model are of an intrinsic nature and are not intended to give rise to NDPs, the consensus then having to relate to the limit of the models used. But between these two extremes there remains a "grey" area, not least because the local economy must also be taken into consideration. The relevant parameters must be discussed and if the differences are not reduced, then they may eventually give rise to NDP.

For SC8, a very particular case arises: the definition of the seismic action as input hazard. In principle, the only safety choice is that of the return period to be considered for the limit states, or alternatively performance factors, therefore an NDP table. Nevertheless, although the hazard is determined from physical models, it is not possible to define it completely in the standard, because it depends on the local seismicity and especially on the geology of the sites. In this case, although an effort has been made to reduce the number of NDPs, there are still some parameters to be defined by the Member States.

To carry out this reduction work, PTs can use the database developed by the Joint Research Centre (JRC) in Ispra (Italy), containing the national annexes provided by the Member States and the corresponding NDPs.

27.3.4 Ease of Use and Development Method

Ease of Use is the second task assigned to PTs. The aim is to improve the ease of use of the codes and to improve the navigational fluidity between the different parts of the Eurocodes, which requires very good coordination. As in the case of NDP reduction, TC250 did some preparatory work to clarify how this issue should be addressed.

It was first necessary to determine who is the main user of the Eurocodes. Indeed, the codes can be useful to several families of stakeholders, among which the Authorities, the owners, the experts, the young graduates, the teachers, the software editors, the other TCs, the producers and construction companies... It has been established that the main users are the well-trained designers of structures and that the Eurocodes must therefore be developed primarily as standards for the design of the works, without forgetting the other users.

The first objective is to have a clear and precise drafting, with an organization of texts such that the imbrication of procedures is easily apprehended by the user. The number of methods proposed (for example for the structural analysis) must be limited and in principle there must not be two methods with the same domain of application. In addition, the text should first introduce simplified methods with a generally smaller domain of validity, before exposing the more general methods.

The second consideration is that, in order to obtain an easy-to-read text, that is to say not cluttered with considerations for dealing with complex cases, it is necessary to exclude the latter and cover only the usual cases constituting the bulk of the market. This is why the objective has been set (without being binding) in 80% of cases. This requires on the one hand to be very precise on what is covered or not, on the other hand to give requirements of a general nature to frame other cases. This can be done by carefully choosing the principles.

The third criterion is the ease of circulation through the different parts of Eurocodes, which implies close coordination in the development of Eurocodes. This is particularly true for Eurocode 8 which complements the others.

Coherence with ENs other than Eurocodes, including materials, construction products and execution, needs to be improved, for example by ensuring that the properties necessary for design are effectively defined in the other ENs and that they contain the appropriate test methods.

In general, the development of the new Eurocodes should be based on the best available knowledge, with methods which use is based on commonly accepted research results and which has been validated by sufficient practical experience.

In addition, the second generation of ECs must be an evolution, avoiding fundamental changes in design approach and text structure, unless duly justified. This is partly the case of EC8 where the general philosophy of seismic design is retained, but the organisation of the different parts is quite profoundly modified.

Finally, the CEN editorial rules are strict and must be respected. This concerns in particular the table of contents, which must be common; for EC8, this rule is followed as far as possible, i. e. for the first chapters, because the structure of the code cannot be the same as for materials ECs. An important point for drafting concerns the use of verbs, which determines the status of each requirement: "Shall" describes a principle that suffers no alternative; "Should" is an application rule which may, subject to justification, suffer an alternative if the situation is outside the domain of application of the rule; "May" is a possible procedure, which may be used to justify compliance with a stronger status rule. Again, these rules (common to all ECs) are likely to evolve and it will be necessary to refer to the final drafts.

27.4 Revision of EN1998

27.4.1 EC8 PTs

Part 1 is covered by two PTs: the first one (PT1) deals with all the material-independent chapters, it intervenes in the first phase. The second one (PT2) is responsible for all the chapters relating to materials (chapters 5–9 in the 1st generation); he intervenes in the second phase.

Part 3 is covered by PT3 during the first phase, thus at the same time as PT1. Experience has shown that the works are very closely linked and that some of the clauses developed by PT3 could eventually be located in Part 1.

Part 5 (geotechnics) is developed in phase 2 by PT4.

Parts 4 and 6 are covered simultaneously by PT5, with a view to a possible merger between the two parts.

Finally, PT6 is responsible for the development of Part 2 (bridges) and for the final coherence of all parts. PTs 5 and 6 are active in phase 3.

The following paragraphs describe the framework of the PTs resulting from M/515 and the systematic review.

27.4.2 Terms of Mandate M/515 for EC8

27.4.2.1 Part 1

The mandate covers all the important additions and modifications concerned by Part 1. It includes:

- European seismic map and definition of seismic action. It was originally proposed to include a European zoning map from the European SHARE project. Nevertheless, the maps established by this project have been challenged by the SC8 members concerned by low or moderate seismicity and the work has been reoriented towards a better harmonization of the seismicity maps, the classification of sites and the definition of the spectra.
- Clarification of the required performance levels and associated limit states by adding a level of operability.
- The development of displacement methods, including static nonlinear analysis (pushover). The description of the method was incomplete in terms of demand and developed only in Part 3 for capacity, with deformation criteria.
- The development of rules for the use of distributed dissipative systems.
- The review of ductility classes to relate them to the level of seismicity, especially to cover areas of low and moderate seismicity with simpler rules.
- Development of rules for concrete primary flat slabs.
- Updated chapters on steel, composite and wood structures, to reflect recent technological and knowledge developments.
- The revision of the "masonry" chapter to integrate a similar approach to that of other materials, and revision of the clauses for simple masonry buildings, considered not sufficiently secure.
- The development of clauses relating to infills and claddings.
- The development of clauses for aluminium structures.

Reducing the number of NDPs is a real issue for Part 1, where there are many, the other parts containing much less. In particular, the material chapters include 39 NDPs, of which 15 for masonry. Reviewing these NDPs in the light of TC250 criteria would lead to only 4, which is indeed a difficult goal to achieve.

27.4.2.2 Part 3

The mandate for this part comes immediately after that of PT1, because it was dealt with during the first phase. It includes:

- Updating the text to take into account the evolution of knowledge on this relatively new subject.
- Complete covering of the displacement based approach (capacity side).
- The development of clauses for bridges.

27.4.2.3 Part 5

The main extension according to the mandate is a comprehensive treatment of soil-structure interaction; but the common theme of ease of use, in relation to other parts and related to SC8 discussions, has led to create a larger scope for the PT.

27.4.2.4 Parts 2, 4 and 6

The requests included in the mandate are less important for these parts, which explains why only two PTs were set up. Nevertheless, the theme of seismic isolation was identified, especially for bridges. Also, the modifications made previously to the other parts would affect these texts sufficiently, as well as the updating according to the evolution of knowledge, to define a work plan for PT 5 and 6.

27.4.3 *Systematic Reviews*

PTs should in principle incorporate comments accepted by the SC into their text. Nevertheless, some of these comments may overlap with the PTs' main task with respect to their terms of reference, and, in these cases, it was asked to PTs to advise SC8 on how to consider these comments, depending on the evolution they propose. As a result, few comments from members were initially rejected. In addition, many of the comments were specific and related to clarification requests, so there was good reason to approve them for ease of use.

For part 1, 374 comments were received from the systematic review. SC8 decided, in the light of these comments, that ease of use should take precedence over the stability of the text. The most important comments on the general chapters are:

- The limit states must be standardised between the different parts.
- Review the definition of spectra, including anchor parameters.
- Review the soil classification to better cover the diversity of sites.

27 Eurocode 8. Evolution or Revolution?

- Have rules adapted to areas with low or moderate seismicity. This must be done, also with the review of ductility classes.
- Review the application of analysis methods: the linear analysis with a behaviour factor, which must be kept for simplicity, and non-linear methods that must be developed, but cannot be mandatory.
- Review the rules regarding torsion, including the application of accidental torsion.

SC8 finally decided not to develop rules on modelling and to stick to modelling rules adapted to analysis methods. Most of the comments made and accepted were consistent and close to the terms of reference and could therefore be taken into account by the PT.

With regard to the material chapters, the comments are very diverse and only few important ones are given below:

- Restructuring ductility classes. Revise ductility H, little used because too complex, especially in concrete.
- The redefinition of types of concrete structures, including walls.
- Define more precisely the structures sensitive to torsion.
- Develop clauses relating to transfer floors and primary flat slabs with or without flat beams.
- For steel, complete the clauses concerning material properties and connections, complete the types of structures.
- For timber and masonry, develop the verification criteria associated with non-linear methods. The clauses are considered obsolete for both materials and must be completely revised.
- The simple rules for small masonry buildings are considered in-sufficient or even unsafe.

For Part 3 (83 comments), detailed remarks are numerous. In particular, SC8 notes that the concrete annex is generally acceptable, but that the annexes on steel and masonry are obsolete and therefore need to be rewritten. Consequently, it must be able to pass them from an informative annex to a normative chapter.

For Part 5, there are 93 comments, SC8 has decided that most are to be considered. The most important ones relate to consistency with Eurocode 7, for example the properties of materials and verification methods. It is also requested to complete the clauses on the foundations, which led SC8 to request a better coverage of the actions on buried structures.

There are 160 comments on Part 2, many on lack of clarity and inconsistencies, particularly with Part 1 and EN 15129 for clauses related to isolation. New developments are required, mainly to cover integral bridges or other types of bridges; only the first are retained by SC8. In addition, it is requested to improve the definition of actions on abutments and culverts, which will eventually be dealt with by PT4.

Finally, parts 4 and 6 receive respectively 148 and 48 comments. Most of them concern an improvement of the clarity of clauses, which obviously meets the ease of use requirement. It is especially considered that many clauses are obsolete and should take into account the more recent advances, such as those implemented for example in the New Zealand guides.

27.5 Main Developments of Eurocode 8

The last part of this presentation develops some topics that have already been dealt with in the first phase, although not final, and topics considered for the second phase.

27.5.1 Structuring Part 1 and Other Parts

The different parts of Eurocode 8 have been developed successively, the first being Parts 1 and 5. In principle, Part 1 contains the general provisions applicable to all parts, but taking into account the development sequence of ENs, needs appeared for the other parts, different from what had already been developed in part 1. This led to some unnecessary repetitions, even contradictions. It soon became apparent that the structure of the code should be revised, first affecting part 1.

In the general chapters of Part 1 are the general requirements, the definition of seismic action, associated with the classification of soils to define the design spectra; chapters 4–9 are devoted to buildings.

In the general part, the way the topics were treated was reviewed and updated, as explained thereafter. It was decided to include in this part of the text all the general modelling principles and analysis methods applicable in the other parts, keeping in the latter only what is specific to each type of structure. In addition to the advantage of avoiding repetitions and contradictions, this significantly reduces the volume of parts dedicated to structures.

Likewise, this general part contains the general provisions relating to isolation and distributed dissipation devices. It is even envisaged to transfer the verification rules specific to materials, common to all structures. The final content of this general part will be decided later in the development of the programme, to have a better visibility on all parts. Finally, new annexes complete the general provisions.

Part 1 also contains provisions relating to new buildings which, because of the development of the general part, are reduced in volume. It is conceivable that the provisions relating to buildings may be disconnected from the general rules to form a separate part.

27.5.2 Limit States

Among the inconsistencies between parts of Eurocode 8, the one concerning the limit states to be checked quickly appeared to SC8 to be absolutely reduced. The ultimate limit state to be verified in Part 1 corresponds to the "Significant Damage" state of Part 3 and it was considered useful to consider in this Part an additional limit state "Near Collapse ", intended primarily for life safety. In both texts, it is also considered a limit state of damage limitation specific to Eurocode 8. It appeared that the definitions given in Part 3 could serve as a starting point for a common definition of limit states. In addition, at the request of certain delegations, a limit state of operability has been added.

Beyond the considerations internal to Eurocode 8, there is the question of the coherence of the Eurocodes, particularly with regard to EN1990 which defines the general philosophy and in particular the categories of limit states. From this point of view, the limit state of damage limitation is in a particular situation since it is not qualified as a serviceability limit state, although it strongly resembles it. In addition, some Member States had chosen not to verify it, which is not an option open by the present standard. So, a certain vagueness reigned over the nature of this limit state, which had to be clarified for its status and use.

Finally, the limit states were classified into:

(a) ELU, intended in the philosophy of Eurocodes to ensure the safety of people and assets:

 – Near Collapse (NC). The structure is very damaged, people's lives are saved.
 – Significant Damage (SD). The structure is damaged but can be repaired in most cases. Non-structural elements can be very damaged.

(b) ELS:

 – Damage Limitation (DL). The structural and non-structural elements are repairable at reasonable cost.
 – Operability (OP). This limit state is to ensure the functionality of the structure, which mainly entails the functioning of the related equipment. The precise description of the criteria can only be defined case by case, since they depend on the functions to be saved. This affects, for example, certain hospital services.

This new classification does not lead to a modification of the ULS to be verified. For new structures, verification of the SD remains mandatory. It allows on the one hand to save lives with sufficient margin, but also to recover most structures, which is important to limit the social impact of the earthquake.

For existing structures, verification of the NC is advised, but it is not possible by the q-factor method, except in low seismicity areas.

On the other hand, the decision to verify the ELSs is at the initiative of the Member States for types of structures to be defined, or for the owners of these structures, in the absence of a decision by the Authorities.

27.5.3 Importance Classes

One of the issues incompletely resolved in the current text is the consistency between the reliability objectives associated with the consequence classes of EN1990 and those underlying the definition of the importance classes of Eurocode 8. A note indicates a parallel between the consequence classes and the importance classes, but the two concepts remained rather distant. It was therefore preferable to better connect the two notions, especially to help users to more easily deduce the importance class from the consequence class, while keeping a ranking option more appropriate to the seismic situation.

A necessary first step was to bring the definitions of classes closer together, which seemed quite possible in looking at the current definitions. A second action was to make the classes coincide. This was achieved by a redefinition of consequence classes in EN1990, where CC2 corresponds to current structures of importance classes II and CC3 can be subdivided into a. and b. to correspond to importance classes III and IV.

Finally, as this comparison was made, SC8 decided to replace the term "importance class" by that of "consequence class ", while keeping a description of the subclasses CC3a and CC3b.

27.5.4 Definition of Seismic Action

Mandate M/515 explicitly mentions the revision of Chapter 3 on the definition of the seismic action. In addition, the definition of the spectrum and the classification of soils were the subject of many remarks during the systematic review, which had to be answered.

In the ENV, the elastic spectrum on rock was anchored to a zero-period nominal acceleration defined by multiplying the maximum ground acceleration by 0,7. At the EN stage, the nominal acceleration has been replaced by the maximum acceleration on rock, a_{gR}, the main argument being that the nominal acceleration is usually defined by dividing the spectral accelerations in an interval where they are maximum (corresponding to the plateau) by 2,5. Taking a_{gR} as the anchoring parameter and 2,5 at the plateau for the spectrum on rock would be equivalent. Nevertheless, this choice was the subject of discussions during the elaboration of the text, so that it was even envisaged to introduce a correction factor of between 0,7 and 1 to cover the regional situations where the maximum soil acceleration would not be sufficiently representative. But this possibility was ultimately not retained.

The determination of the value of a_{gR} for a given return period, in each seismic zone, had to be carried out by National Authorities from the average value (at 50%) of the maximum acceleration provided for by the attenuation relations, without integration along the scatter.

27 Eurocode 8. Evolution or Revolution?

As regards the shape of the spectrum, two types have been proposed in the current text (Type 1 and Type 2), depending only on the magnitude M_s of the surface waves of the earthquakes that may affect the region in question. For M_s less than 5,5, it was recommended that the Type 2 spectrum be adopted, the magnitude to be considered corresponding to a realistic earthquake and not to an extreme conservative earthquake. This distinction had been introduced to avoid excessive conservatism at large periods in the case of moderate magnitudes, thus leading to a narrower plateau for type 2. In the new version, given the new parameters of the elastic spectrum, it was not considered useful to maintain the two spectral shapes.

It was decided to define the new shape of the elastic spectrum by anchoring it on spectral acceleration values different from a_{gR}. Indeed, the spectrum is meant to be used for the analysis of structures and a zero-period acceleration is not appropriate for this purpose. This is why it was chosen to anchor the spectrum on accelerations at the plateau, representative of stiff structures, and to 1 s, representative of flexible structures. In principle, this second point is on the branch at constant spectral velocity, but there may be deep basin conditions (for example in Romania) where 1 s is still located on the plateau. This is why flexibility in the definition of this second point was introduced when T_C is greater than 1 s.

The general shape of the elastic spectrum has been preserved, nevertheless, the T_A period below which the acceleration is constant has been reintroduced so as not to penalize very stiff structures and equipment. This results in the rock spectrum at 5% damping as in Fig. 27.1a for acceleration and Fig. 27.1b for displacement.

PGA is nevertheless useful for geotechnical applications. It is therefore defined in a conventional way by dividing the acceleration at the plateau by a factor F_0, taken equal to 2,5 unless a specific study gives a different value.

27.5.5 Site Classification

The question of taking into account the deep geology is not recent, since this subject had been debated during the conversion to EN. This is one of the topics discussed to develop the new classification. But it was also necessary to take into account the

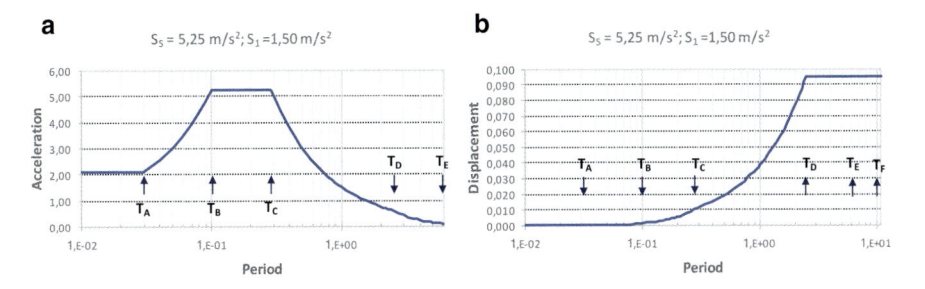

Fig. 27.1 Elastic spectrum on rock (**a**) acceleration, (**b**) displacement

comments made by the Members during the systematic review and to take into account the progress of knowledge of the European seismicity.

The first issue to be addressed was the choice of "objective" parameters allowing the description of the site response. The main parameter used in the current version is the shear waves velocity over 30 m, $v_{s,30}$, considered appropriate but insufficient; it was therefore necessary to improve the model. One possibility was to choose the predominant frequency f_0 of the site, which is relatively easy to measure, but whose use is not widespread. It was finally preferred to add to $v_{s,30}$, which remains the parameter most used in the available studies, the height H_{800} of soil to the substratum, which is defined by a velocity at least equal to 800 m/s.

A second question was that the soil classification of the current version causes significant discontinuities of spectrum at classes boundary. It has therefore been proposed by PT1 to offer an analytical formulation of the two site amplification parameters F_s and F_1 at the two anchoring periods of the spectrum, as a function of v_s, and of H_{800}, the celerity being taken at height H_{800} if it is less than 30 m.

The choice of the number of classes and their limits is an open subject that must win the consensus of the Members. PT1 proposed to keep 5 classes of sites A to F, for which the spectral parameters are fully defined (they are not defined for classes E and F in the current version). For now, SC8 has accepted the proposed definitions of these classes.

Nevertheless, the two parameters v_s and H_{800} are not available in all projects and it was therefore necessary to find possible alternatives. The first alternative is to use conventional geotechnical measurements, as in the current version. The second is to use the frequency f_0, identified mainly from the H/V spectral acceleration ratios. The third is to use geological knowledge of soil layers on a descriptive basis. These various methods make it possible to cover the possibilities offered in the projects according to their importance and the possibilities of measures that they can use.

It remains possible to establish site spectra from a specific study whose rules are given in a normative annex and this is even specified in the case of importance classes III or IV on sites D or F.

27.5.6 Good Design Criteria

The current version includes criteria for good design, but these criteria cannot be prescriptive because they are sometimes incompatible with the functionalities of the structure or the architectural arrangements. They have therefore been reported in an informative annex.

27.5.7 Torsion of Buildings

It is recognised that torsion around the vertical axis is a potentially very damaging phenomenon and should therefore be strictly controlled. This leads in particular to take special precautions for buildings flexible in torsion, i. e. those which have a small torsion radius compared to dimensions in plan. Such situations at risk arise especially when the first mode of the structure is strongly influenced by torsion. It is therefore necessary to maintain provisions to control the torsion, even if the rules evolve.

A first action carried out by the PT was to clarify the way to take into account torsion, which was not always well perceived by the users, because disseminated in the text.

A question arises then: should we keep the accidental eccentricity, added to the natural eccentricity, which can be calculated from the drawings? The purpose of accidental eccentricity is to take into account in a inclusive manner certain variability of the input parameters which are not taken into account in the analysis, in particular: (i) a torsional movement at the base level, due to the non-uniformity of the motion on all the foundations; (ii) the non-simultaneous formation of the plastic zones in the bracing elements (which is not reflected in the linear analysis); (iii) the non-uniformity of distribution of the variable masses in plan. But taking it into account introduces an additional complexity into the calculations, which is weak in most cases, but practically insurmountable in the case of time history analysis. So, the question of maintaining this accidental eccentricity is posed in an objective of "ease of use".

In cases where the torsional stiffness and/or the natural eccentricity are sufficiently large, this variability does not have a great influence and accidental eccentricity may not be taken into account. On the other hand, when either of the two quantities is small, the variability of the parameters can significantly influence the first mode and the analysis can underestimate the effect of torsion.

A way to solve the problem, if the accidental eccentricity is removed, is to impose a minimum value of eccentricity that would replace the natural eccentricity if it is too low. This is the current focus of SC8, with particular attention to buildings with low natural eccentricity or torsion-sensitive.

27.5.8 Regularity of Buildings

The two types of regularity, in plan and in elevation, are separated and intervene in different ways in the analysis process. The provisions related to regularity had been widely debated during the elaboration of the current EN, in particular the torsional radius r for plan regularity.

The main purpose of plan regularity criteria is to evaluate the influence of torsion to allow analysis in two vertical planes. The torsional radius, defined in a given direction as the ratio between the torsional stiffness and the stiffness related to the overall flexion, is an intuitive notion and qualitatively easy to understand, but quantitatively difficult to establish, since it depends on the type of bracing and the architectural configuration of the building. Also, there is no formula in the current EN to calculate the torsional radius from geometric and mechanical data. Such formulas can be added for simple situations, but can hardly be generalised.

It must be taken into account that these criteria were introduced at a time when it was necessary to simplify the analysis, but great progress has been made in the finite element models, the algorithms for resolution and the power of the computers. Therefore, the limitation of the models in size has lost a lot of interest from these points of view; most designers are now using 3D models that take into account the natural torsion. Finally, it was decided to move the regularity in plan criteria to an informative annex as guidance for good design. The model is a designer's choice, however, the use of 2D models remains limited.

The criteria for regularity in elevation play a more important role, in particular for the value of the behaviour factor in relation to the overall ductility. Nevertheless, it was decided to simplify these criteria by keeping only the more important ones.

27.5.9 Analysis Methods

All analysis methods of the current version are retained because their use is related to different situations.

The application of the lateral forces method is extended as it is no longer necessary to meet the criteria for regularity in elevation. In return, the simplified formulas for calculating the period and the linear distribution of acceleration cannot be used in general. This is why the Rayleigh approximation method is preferred.

For nonlinear analysis, the pushover analysis method is supplemented to take into account the effects of torsion and the influence of higher modes. For buildings, in order to keep a sufficiently simple procedure, a single acceleration profile is applied and the above-mentioned effects are taken into account by applying to the displacement demand factors calibrated on the results of the linear analysis.

27.5.10 Behaviour Factor q

The values of the behaviour factor have been an object of debate since the beginning of the Eurocode 8 project. This is a delicate subject because its economic implications can be important. It had already been discussed in the conversion phase where several Member States considered that they were not high enough compared to other codes. It remains an open topic since there have been comments on this subject during the systematic review. It is also about establishing a good balance between materials, so as not to introduce competitive advantage (or disadvantage) on a basis that is not purely physical.

One of the important points to be clarified is the role of overstrength in the value of q. This overstrength may have various origins, for example the margin between the resistance taken into account in the verification and the ultimate resistance and the favourable effect of redundancy. The participation of overstrength in the current version is recognised, but it is not separated from what is due to ductility. It is this that justifies taking a value of q equal to 1,5 for ductility class structures "L", which in reality is not ductile as it corresponds to a quasi-elastic behaviour. On the other hand, the role of redundancy (α_u/α_1) appears explicitly in the expression of q for some structures, but not in a generalised way. This is why it was decided to distinguish these components by decomposing q into three terms:

$$q = q_R \, q_S \, q_D$$

q_R is the part of overstrength due to redundancy and is specified as a ratio α_u/α_1 for some structures, q_S is the part due to other causes of overstrength. q_D allows to represent ductility for classes M and H.

27.5.11 Base Isolation

Base isolation has been developed in the current EN version, in Part 1 for general rules and application to buildings, limiting to total isolation. In Part 2, for bridges, more diversified devices are envisaged, in particular rigid connection devices.

The initial clause for recentering in Part 2 has been found to be far too restrictive as it eliminates much of the European devices. For this reason, it was later modified taking into account the results of researches carried out under the LESSLOSS project. For its part, TC340 had developed an energy based formulation for EN 15129, different from that finally adopted by Eurocode 8, based on displacements. Nevertheless, point-of-view exchanges showed that both formulations were roughly equivalent.

However, there is still a coordination problem between SC8 and TC340, the product standard addressing structural issues (such as recentering) and Eurocode 8 Part 2 addressing product specifications, for example in Annex K. It is important to clearly define the scope of the two standards: EC8 for structural design and EN 15129 for devices design and testing; this is one of the objectives of the new generation.

27.5.12 Structures with Distributed Energy Dissipation

A new topic is introduced in Part 1, that of distributed energy dissipation, which can be obtained by very dissipative structural elements or by specific devices such as dampers. In fact, the corresponding clauses have been introduced in the general part, in particular the analysis methods that can be applied to all types of structures, but also in a chapter specific to buildings. The energy dissipation considered may depend on displacement or velocity and both possibilities are therefore considered.

The difficulty in calculating the response of the structure is the simultaneous consideration of the energy dissipation and of the post-elastic behaviour of the main structure. Two approaches are proposed: one by a nonlinear multimodal analysis, the second by energy balance.

In the first approach, well adapted to velocity dependent devices, the calculation of the response of the first mode is similar to a pushover analysis and is based on the response of an equivalent oscillator, but with an effective period and an effective damping adapted to this type of situation. The other modes are calculated in the elastic domain, but taking into account a suitable damping. The combination of modes is then effected as for the conventional multimodal method. This necessitates the introduction of a correction formula for high values of damping, the current formula remaining valid up to 28%.

In the second approach, adapted to displacement dependent systems, the energy that can be absorbed by the structure comes from the main structure and the dissipative system. The different energy contributions can be calculated from formulas given in the text. The overall energy demand is calculated from the spectrum and must remain below the absorption capacity of the structure. Checks are also to be done at each level of the building.

27.5.13 Ductility Classes

The definition of ductility classes and their number have since the beginning been a subject of debate within SC8. It had been envisaged during the conversion to improve ductility L in order to keep only two classes. This option was discussed again for the next generation, but eventually SC8 decided to keep three classes. Several arguments lead to a fairly thorough review of how classes are defined and constructed:

– Ductility class H is often too complicated and ultimately not used, especially in concrete structures, so there is no point in keeping it as it is.
– The clauses have been developed without any clear relation to the seismicity level. Most of the European territory has a low or moderate seismicity where simplified rules can suffice.

- The current standard recommends using ductility L only in low seismicity areas, but on the one hand this limit is a NDP and, on the other hand, some countries with high seismicity areas have not limited its use.

SC8 is therefore working towards a redefinition of ductility classes (possibly with different names), more suited to the conditions of seismicity, without there being an automatic and imposed relationship between the two notions. In principle, these classes should provide about equivalent safety and the choice would be that of the designer on economic criteria related to the nature of his project.

27.5.14 Reinforced Concrete Walls

At the ENV stage, only one type of reinforced concrete wall was considered: the cantilever walls took up a large part of the seismic forces, and can develop a plastic hinge at their base. This concept, developed in particular by Prof. Paulay in New Zealand, consists, as for columns, in organizing the ductility of the plastic hinge by connecting the ultimate plastic rotation to the level of ductility used and ensuring the quasi-elastic behaviour of the non-critical zones of the wall by capacity design. The existence of a plastic hinge at the base assumes that the wall is anchored in a rigid foundation and cannot lift. This ductile wall concept is well suited to a type of structure where there are few walls and where vertical loads are carried in good proportion by columns.

But this architecture is not unique and, in some countries, the use is that the walls are proportionally numerous and carry the bulk of the vertical loads. In this case, the walls are interconnected and constitute a low ductile but very resistant structural arrangement.

It was decided at the EN stage to develop a concept more appropriate for this type of walls, starting from what had been developed in the French standard on an experimental basis. In this concept, because of the lack of ductility, there is no plastic hinge; the non-linearity is essentially geometrical: it is related to the length of the walls and the shift of the neutral axis during the movement, which causes the masses to rise. This phenomenon related to cracking is similar to that involved in the uplift of foundations, where the potential energy of weight is involved; this energy serves as an accumulator of the energy injected into the structure by the earthquake and is clearly preponderant with respect to the strain energy, as soon as the lifting of the masses is substantial. In order not to create a new difficulty, this non-linearity has been taken into account by means of the behaviour factor q.

On the other hand, walls with openings (dual walls) are mentioned in the text and recognised as more ductile due to lintel yielding, but the rules for distinguishing them from other walls are not detailed.

It is therefore necessary to clearly define these different types of walls, which correspond to different architectures and physical behaviours. This work remains to be done.

27.5.15 Masonry

Masonry presents a particular difficulty, since the mode of construction and the properties of the materials used, particularly the units, depend strongly on local uses and production. Since it is necessary to respect the needs of all CEN members, the text cannot introduce provisions that effectively eliminate a significant part of local production, without there being any substantiated physical reason for that. Moreover, since this production is very different from a region to another, it is not possible to cover all the particular cases. This has led to a large number of NDPs, contrary to the current philosophy of reduction.

Moreover, the structure itself of the masonry chapter differs from that of other materials, which has led to provisions considered less restrictive for masonry than for other materials, for example the values of the behaviour factors.

Therefore, the chapter needs to be reviewed with more general physical considerations that can be adapted to different local productions and thus avoid so many NDPs.

27.6 Conclusion

The work of developing the second generation of Eurocode 8 is underway. The evolution of the text is important, in particular because the organisation of the different parts is substantially modified, in order to improve the readability and the overall consistency. Nevertheless, the physical principles that are those of Earthquake Engineering remain at the base of the standard and the designers trained to the use of the first generation should not have difficulty to adapt to the second. Many topics are added or modified to provide better help to the designer, especially for routine projects.

Some aspects have been developed here, among the most important, but the work is far from complete and a more complete assessment will be possible at the next ECEE.

References

CEN/TC250 PUBLIC INFORMATION. https://standards.cen.eu/dyn/www/f?p=204:7:0::::FSP_ORG_ID:6231&cs=1475B1C21B51CE51CCD000F68519ABE9C

MANDATE M/515. http://ec.europa.eu/growth/tools-databases/mandates/index.cfm?fuseaction=search.detail&id=523

Chapter 28
Research Needs Towards a Resilient Community

Vulnerability Reduction, Infrastructural Systems Model, Loss Assessment, Resilience-Based Design and Emergency Management

Paolo Franchin

Abstract Most of the literature on resilience is devoted to its assessment. It seems time to move from analysis to design, to develop the tools needed to enhance resilience. Resilience enhancement, a close relative of the less fashionable risk mitigation, adds to the latter, at least in the general perception, a systemic dimension. Resilience is often paired with community, and the latter is a system. This chapter therefore discusses strategies to enhance resilience, endorses one of prevention rather than cure, and focuses in the remainder on the role played by systemic analysis, i.e. the analysis of the built environment modelled beyond a simple collection of physical assets, with due care to the associated interdependencies. Research needs are identified and include challenges in network modelling, the replacement of generic fragility curves for components, how to deal with evolving state of information.

28.1 Increasing Resilience

28.1.1 Resilience Definition and Quantification

The term resilience originated in Mechanics, was adopted in Ecology and has recently seen increasing use, alongside the term sustainability, in many other fields. Growing population and urbanization, increase in complexity of infrastructural systems, and of interconnection in general, translate into increasing and longer-lasting impacts from natural and man-made hazards and are probably behind this

---

P. Franchin (✉)
Department of Structural and Geotechnical Engineering,
Sapienza University of Rome, Rome, Italy
e-mail: paolo.franchin@uniroma1.it

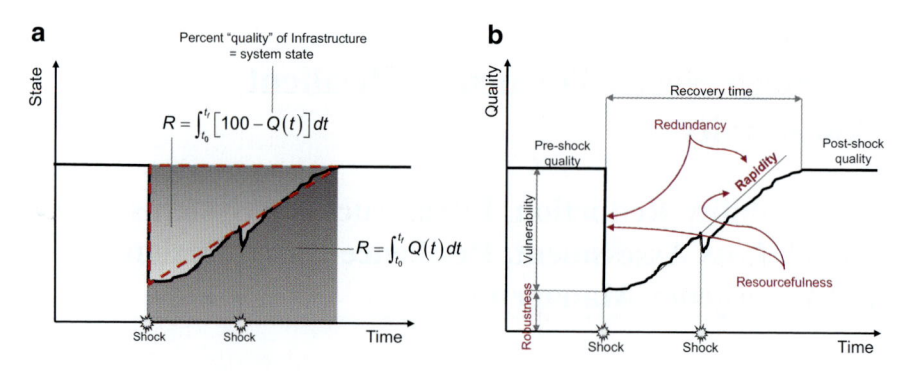

Fig. 28.1 The so-called "engineering" definition of resilience (**a**) and the "4R's" (**b**)

revival of the concept of resilience. Its widespread use in many disciplines has led to a host of different definitions, adapted to the needs or points of view in each particular field. Many papers are devoted just to the collection of these definitions. Herein the definition of resilience evolved from that first given in (Bruneau et al. 2003) is adopted. This definition, graphically illustrated in Fig. 28.1a, is by now well accepted in the civil engineering field, even though interesting alternative proposals have been advanced (Sun et al. 2015).

Resilience is a system property defined in terms of the evolution of system state over time in the presence of a disturbance. If a meaningful global variable describing state at the system level is found, called quality in (Bruneau et al. 2003), then resilience can be defined with respect to the variation (fall and recovery) of Q over time, due to one or more shocks. R can be related to either the area above the recovery path (sometimes referred to as the "resilience triangle", in dashed red in the figure), or below. Figure 28.1b introduces a number of properties related to Q and R: the fall in Q due to the shock, measure of the immediate impact of the disturbance, is called vulnerability; conversely, its complement is the system "Robustness"; the average slope of the recovery path is the "Rapidity", whose increase reduces the recovery time; "Redundancy" in the system, as well as "Resources" invested in the system or in recovery actions, decrease vulnerability and increase rapidity. Robustness, rapidity, redundancy and resources are also evocatively known as "the 4Rs".

Within this context, it can be observed that the main problem to be solved in order to assess resilience in a quantitative manner, is the formulation of an effective, computable and descriptive measure of system state, i.e. the variable Q, along with the development of tools to describe in a reliable manner its evolution over time. Resilience then follows. The issue is not trivial to solve and much work has been devoted to it, e.g. by Bruneau and co-workers (Cimellaro et al. 2010), as well as others (Bocchini et al. 2014). Just to give two examples, difficulties stem from prediction of the sequence of shocks (in the context of earthquake engineering, seismic sequences), which are random in both intensity, position of each shock and occurrence time and require refined tools not yet mature (see e.g., Iervolino et al. 2014), or the analysis of cumulative damage to system components (e.g., among many others, Franchin and Pinto 2009).

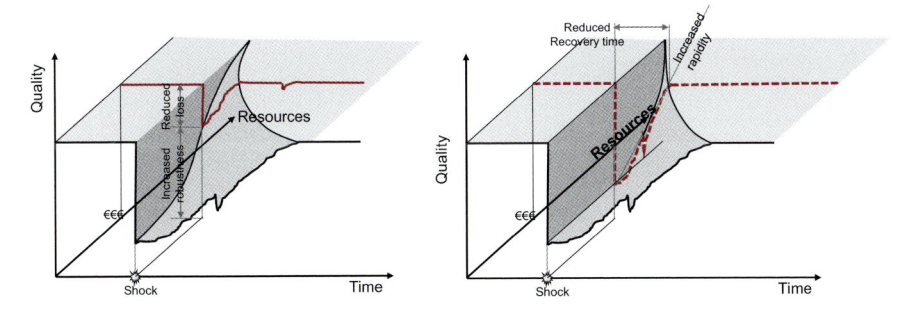

Fig. 28.2 Resilience enhancement: investment of resources in vulnerability reduction (**a**) versus recovery actions (**b**)

28.1.2 Resilience Enhancement

Leaving aside for now the problems related to the definition and quantification of Q and hence R, one can examine, still in abstract terms, what are the strategies to increase the resilience of a system.

Figure 28.2 illustrates the two extreme strategies, which are not mutually exclusive. Panel (a) shows investment of resources into increasing robustness or, conversely, reducing vulnerability. This is pure prevention. With increasing resources put into enhancement of the existing system components and in higher performance of the new ones, the initial impact goes progressively to zero. Decreased damage obviously translates in faster recovery, as well as in overall lower loss. Panel (b) shows investment of resources into recovery actions. The latter leaves the initial impact unaltered but increases rapidity reducing recovery time and loss, even though, most likely, not to the extent achievable by prevention. As an old advertisement said, it appears that prevention is better than cure. Why a society, or system administrator would want to incur damage in the first instance?

The answer to the question, as well as more arguments in favour of prevention, require consideration of one aspect of paramount importance: uncertainty. Figures 28.1 and 28.2 show qualitative diagrams of Q that represent one possible sequence of shocks, with one possible initial impact, followed by one recovery path. The reality is, just focusing on the cause, that the number, intensity, position and time of occurrence of earthquake in a seismic sequence cannot be deterministically predicted. Moving on to the effects, components' damage given intensity is also uncertain. Even more uncertain is the path of recovery, which is affected not just by physical (cumulative) damage but also by the recovery strategy. One could speculate about the very possibility of predicting the latter. Recent examples have shown unpredictable political decisions can lead to completely unforeseeable outcomes (Calvi and Spaziante 2009). Therefore, prediction of recovery path and time are affected by large uncertainty. This is all shown in Fig. 28.3a. Further, the new stable state to which the recovery curve tends is almost never equal to the pre-shock

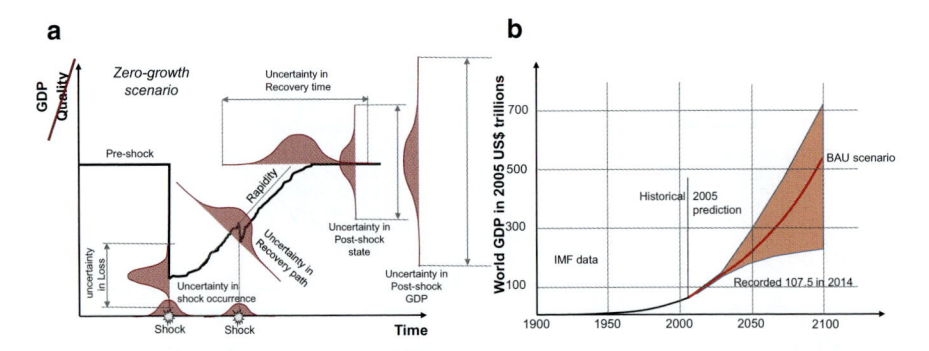

Fig. 28.3 Uncertainty in impact, recovery path and time and post-shock state (**a**); Uncertainty in world GDP up to 2100 AD

one, as simplistically shown in Fig. 28.3a as well as in both Figs. 28.1 and 28.2. Thus, further uncertainty characterizes the post-shock state.

All of the above is further exacerbated when the system is a "community" or "society". While it is probably still manageable to constrain uncertainty in post-shock state for a "simpler" physical system such as a transportation or water distribution network, when the system has also social and economic dimensions, like a community does, the issue complicates considerably. What is quality for a community? The very simple answer, which avoids the debate and adopts the default choice of an economic measure, is to measure quality by the gross domestic product (GDP), whole, per capita, purchasing power parity, etc.[1] The associated uncertainty of prediction of the post shock state is much larger, as indicated in Fig. 28.3a. The reason for this increased uncertainty is that the very trend line, i.e. the evolution over time of GDP in the absence of shocks, is highly uncertain, as shown by Fig. 28.3b, which reports a 2005 prediction from International Monetary Fund (IMF). Interestingly enough, the latter figure shows that, according to all predictions, GDP will keep increasing (which is why Fig. 28.3a, where for simplicity the GDP baseline is flat, is denoted as a "zero-growth" scenario). Certainly these predictions reflect the dominant line of thought among economists of our time, and one wonders how much sustainability enters into these considerations (Meadows et al. 2004).

Going now back to the initial question of prevention versus cure, one can say that, due to the large uncertainty in occurrence time and place of earthquakes, as well as on the associated damage, and considering the stark difference in time horizon between this class of natural hazards and the next political election, it may appear more convenient to put resources in more pressing matters than in a possibility of future loss. This view is supported by observation of different approaches to the problem of prevention in different countries. Quite invariably and expectedly the countries that invest more are those where frequency and intensity of earthquakes are larger. Indeed, to cite the Italian case, only due to a number of seismic events closely

---

[1] The reader is warned that what follow are amateurish economic considerations.

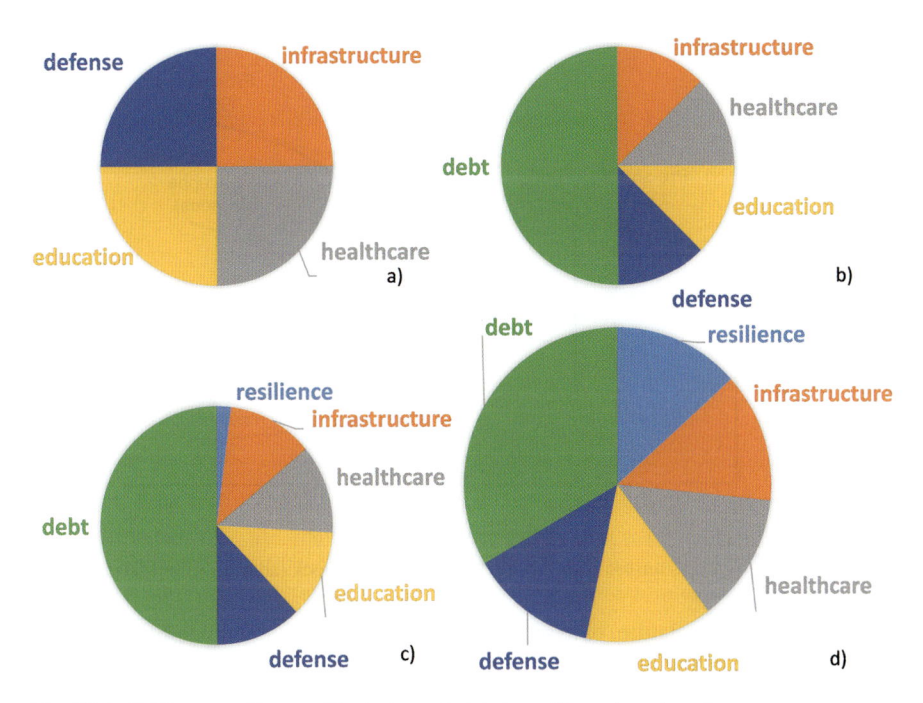

Fig. 28.4 Public expenditure and budget constraints to resilience investment

spaced in time (L'Aquila 2009; Emilia 2012 and Central Italy 2016–2017, (Dolce 2018)), measures towards prevention through state-subsidized strengthening have started being adopted, e.g. the "Sismabonus" scheme (Dolce 2017). The problem, however, remains. Economic boundary conditions are not favourable, as qualitatively illustrated in Fig. 28.4. In ideal conditions (panel (a)) public expenditure covers a number of functions (e.g. education, defence, healthcare, infrastructure). In real conditions, public expenditure is partly absorbed by interests on debt (panel (b)). Investment in resilience, a long term goal which requires vision, competes with running expenses associated with strong social interests. In a way, the situation with resilience is similar to that of education. In both cases the return on the investment is postponed and requires more political will and long-term vision. One possibility is to integrate resilience-oriented actions, like seismic strengthening, into ordinary maintenance programs of infrastructures (panel (c)). Of course a generous GDP growth (panel (d)) would allow larger investments, but, once again, current economic boundary conditions are different. It is obviously easier to divert budget to resilience (emergency management plus repair/reconstruction for higher performance) at the time an earthquake strikes, when public attention is high and opposition could be easily pointed as morally unjustified in view of the disaster victims.

The argument of high uncertainty can actually support an opposed point of view, a strong position in favour of prevention, since all the prediction problems related to the post-shock state (the entire recovery path ending with the new state) are greatly reduced by limiting initial damage, or simply eliminated together with it. The issue is

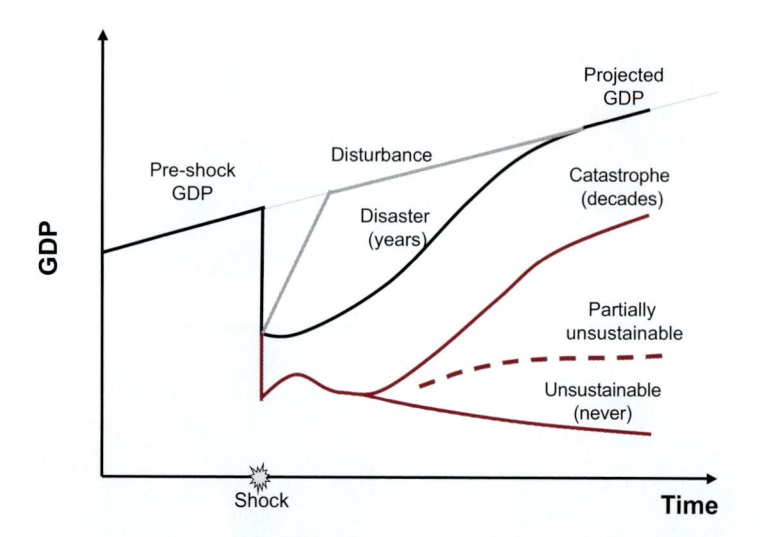

Fig. 28.5 Sustainable vs unsustainable outcomes following a shock (adapted from Davis (2015))

not one of academic nature, of avoiding a difficult (impossible?) modelling problem, but, rather, one of the utmost practical relevance. As shown in Fig. 28.5, recovery paths, as they are called, could well be characterized by no recovery at all. Impact may be so large that it takes decades to attain a new state that is below the previous trend line (a catastrophe), or even large enough to be partially or totally unsustainable, with a community being hit too hard to ever recover resulting in a large-scale outmigration. Think of Pompeii (79 AD), or, more recently, of the situation after Tohoku earthquake (2011) in some regions of Japan or in Christchurch after the 2010 Canterbury earthquake sequence (Newell et al. 2012).

In sum, informed decision-making requires reliable predictions. When uncertainties are of the order of magnitude illustrated, and contributed by sources other than physical, requiring specialist knowledge from different disciplines, including softer ones like economy or sociology, it can be argued from the limited perspective of an engineer that vulnerability reduction is the safer, more reliable way towards resilience enhancement. This is especially true when the range of possible outcomes includes extreme unsustainable ones as shown in Fig. 28.5 (which is not always the case, for instance in regions of medium seismicity).

28.1.3 Resilience-Based Design, or Performance-Based Design with Resilience-Based Targets

If this "engineering take" on resilience enhancement is adopted, the next question arising is how to relate a global system-level objective like resilience (which, as shown in Fig. 28.1, is computed in terms of the system state) to individual safety levels for the elements of the built environment. In other words, what seismic risk should we be designing new structures or retrofitting existing ones for, in order for our communities to meet predefined resilience targets? This theme is attracting some interest, ranging from attempts to answer the preliminary question of what is the risk implied by our current design procedures (Iervolino 2018), to attempts to link component safety to system safety through classical reliability methods (Lin et al. 2016). Actually, the most interesting approach to this problem is the one put forward in (Mieler et al. 2013, 2015).

Mieler et al. (2013, 2015) start by analysing the regulatory framework within which critical systems like nuclear power plants (NPPs) are designed, and end up comparing it with the regulatory framework for the system of interest in our case, i.e. a Community.[2] They draw an analogy between NPPs and communities, highlighting differences. The main difference, however, does not require to enter into the details. It is in the design philosophy. NPPs are designed top-down, starting from a clear explicit statement of the system-level goal, or undesired outcome to avoid. For NPPs these are core damage and radioactivity release, as shown on top of Fig. 28.6a. Next, functions vital for the plant and associated failure modes related to the undesired outcome are described through event tress and linked to the performance, or damage, of each of the physical system in the system of systems that represents the NPP (these are divided in primary and secondary, or support systems). On the contrary, the design philosophy behind communities is obviously bottom-up, with a fragmented and often non-consistent regulatory framework. Systems can also be subdivided into primary, essentially the buildings housing the vital community functions (VCFs) identified in analogy with those of a NPP, and secondary, i.e. the lifelines satisfying the demands for energy, transportation, communication, goods and services. These systems, however, are designed according to codes and guidelines that are drafted and maintained by non-coordinated bodies. Often, with reference to the specific problem of resilience to earthquakes, the seismic action is not considered at all in some systems or considered in non-consistent manner across different systems. As a result, the probability of occurrence of the undesired outcome, i.e. an unsustainable outmigration, is not a target but the end result of an uncontrolled process.

---

[2]A system is a dynamic entity comprising a collection of interacting components assembled to perform an intended function. As such, a community can be described as a system, albeit an incredibly large and multi-faceted one. It is a complex dynamic system of people and organizations with relationships and interactions. Most of these relationships and interactions are physically supported by the community's built environment, which plays a crucial role in enabling a community to successfully function: it provides the physical foundations for much of the economic and social activities that characterize a modern society. Natural and man-made hazards can damage the built environment, thus disrupting the security, economy, safety, health, and welfare of the public. In response, regulatory frameworks were developed and implemented to ensure minimum levels of performance for individual parts of the built environment.

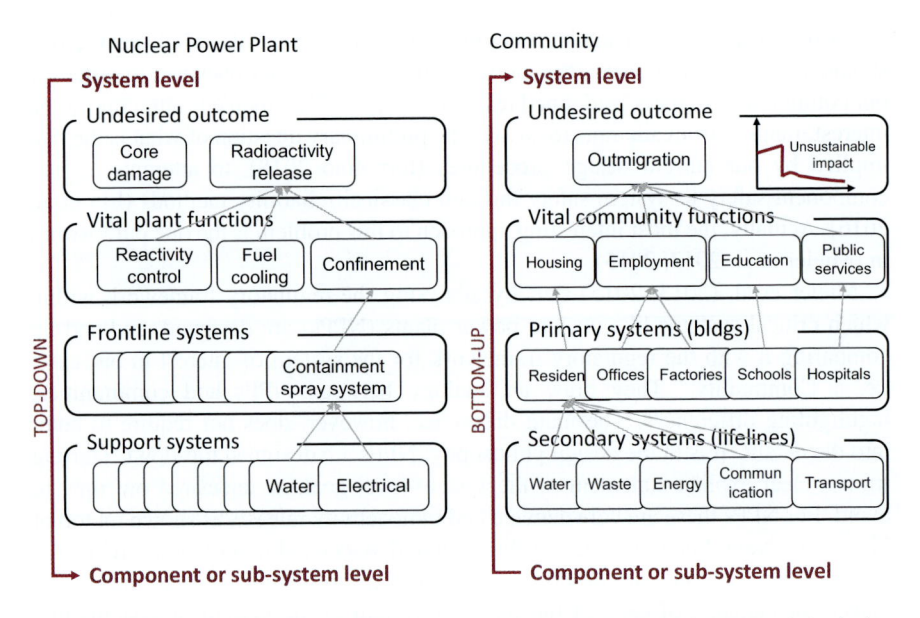

Fig. 28.6 The analogy between a NPP and a community and the opposite approach in design between the two systems

It would be sensible to adapt the NPPs regulatory framework to communities, but a number of differences pose challenges to this operation:

1. Physical scale. Communities occupy much larger geographic areas than NPPs, thus subsystems and components such as lifelines are spatially distributed over a potentially large area. As a result, it becomes necessary to account for partial failures of these subsystems and components. For example, an earthquake may cause damage to portions of an electric power grid, resulting in service disruptions to particular neighbourhoods or city blocks only. The evaluation of NPPs does not account for partial failures: components and sub-systems are either functional or non-functional.
2. External boundaries of the system. Most components and subsystems in a NPP reside within the well-defined physical boundaries of the plant. A community, on the other hand, can rely on components and subsystems that fall outside its jurisdictional boundaries. An electric power grid can draw electricity from a generating station far away and events disrupting the functionality of the station may cause service disruptions in the community, even though its power grid is not directly affected by the event. This issue of where to draw the boundary of the system is a difficult one, and in one case this led to modelling at very large scale (the entire US in (Karaca 2005)).
3. Time scale. A community's built environment is constructed over time, over decades or even centuries, especially in Europe. Individual components have likely been designed and constructed using substantially different specifications

and standards (the problem of existing buildings and other physical assets), meaning that the expected performance of similar components (e.g., residential buildings or highway bridges) within a community can vary drastically. In comparison, NPPs are built over a relatively short period of time.

This said, the overall approach conceptually set out in Mieler et al. (2013, 2015) is rational and can be considered possibly the more solid base for developing the next generation design guidelines. This is indeed the basis for resilience-based design (RBD), which, in the view of the author, is nothing else than performance-based design (PBD) with resilience-based performance targets. Of course, while concepts of PBD still struggle to make their way into codes (Fardis 2018), and, when they do, they really are watered down versions of PBD, devoid of any explicit consideration of probability (Vamvatsikos 2017), PBD methods are being developed, extended, improved (Vamvatsikos et al. 2015) (Franchin et al. 2017) and at some point they will be ready for practical application. At that time the missing link will be proper resilience-based performance targets. Research is needed in this direction because the framework put forward by Mieler et al. is not operational.

28.1.3.1 Setting Resilience-Based Performance Targets for Individual Physical Assets

In this section an idea is presented on how to operationalize Mieler et al. framework to establish the performance target for an individual structure. This is one of the research needs mentioned in the title and this section identifies specific aspects that need focused research efforts.

Figure 28.7 illustrates the flow chart of the procedure from the community resilience goal to the performance targets for the vital functions. Step 1 requires defining an undesired community-level outcome (outmigration, or any other). Step 2 requires establishing an accepted threshold value for the probability (e.g. annual, or mean annual frequency) of the undesired outcome, P_{max}. The undesired outcome and the associated P_{max} jointly represent the community resilience goal. Step 3 involves establishing VCFs event trees and associated (tentative) target performances. In the figure for illustration purposes, only two VCFs are considered, Housing and Public services, with three possible events (partial functioning states) defined in terms of an appropriate tracking variable. Probabilities are assigned to each state j of each function i (these are desired targets), p_{ij}.

Step 4 involves combing the VCFs trees into the Community tree. In the example the probabilities of each sequence resulting from the combination of VCF events are computed by simple multiplication assuming statistical independence, like the 42% probability of the percent residents displaced, R, and the percent capacity disrupted, C, being lower than their lowest respective thresholds. This is beyond doubt an aspect where substantial improvement is needed, since obviously capacity disruption in services and percentage of residents displaced are not statistically independent, being caused by damage to common systems due to a common cause.

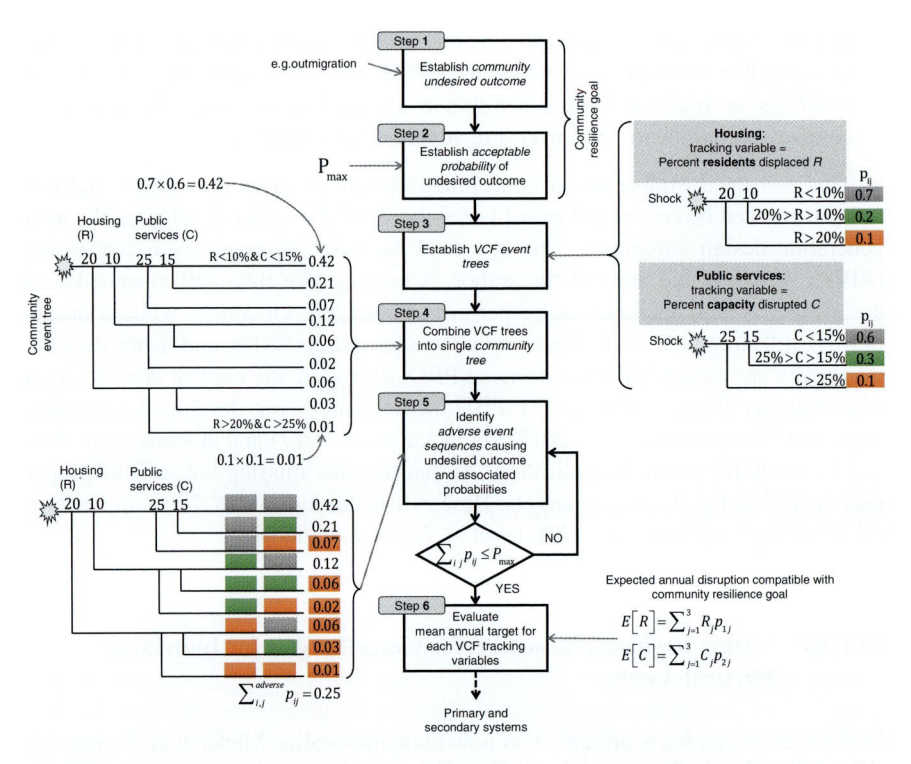

Fig. 28.7 How to establish the target annual disruption in vital community functions compatible with predefined community resilience goals, according to the framework by Mieler et al. (2013, 2015)

Step 5 requires the identification of adverse event sequences leading to the undesired outcome, as well as of the associated probabilities. Identification of adverse sequences is done in the figure according to the simplified rule used in Mieler et al., i.e. a sequence leads to the undesired outcome when at least one VCF "is in the red", or two VCFs "are in the orange". This is obviously one more aspect that needs to be formalized and tackled in a more robust manner for operationalization. The sum over all adverse event sequences provides the total undesired outcome probability, which must be lower or equal than the target. Once, after iteration, this condition is positively verified, the target mean annual value of the tracking variable of each VCF is obtained (Step 6).

Even considering the problem of statistical dependence in Step 4 or that of identifying adverse sequences in Step 5, the procedure in Fig. 28.7 is well defined and provides a link from the community resilience goal, to the VCF performance targets. The next step, however, is not trivial. Herein, one idea is put forward on how this last link to the performance of an individual physical asset can be established. The idea is illustrated with reference to the hypothetical problem of determining the target performance for a new hospital, in terms for instance of ratio of post-event to pre-event beds or operating theatres available. The hospital is one of the primary systems upon which the health-care service depends. The latter is part of the "Public Services" vital

28 Research Needs Towards a Resilient Community

	Police	Health-care	Food
Public services	0.3	0.3	0.4

Service importance matrix (1×n): $\mathbf{I}_{service}$

	Police stations	Hospitals	Stores	Water	Electric power	Fuel (oil)	Roads
Police	0.4				0.2	0.2	0.2
Health-care		0.5		0.1	0.2	0.1	0.1
Food			0.5	0.1	0.1	0.1	0.2

System importance matrix (n×m): $\mathbf{I}_{system}$ — Primary | Secondary

Fig. 28.8 Matrices linking functions to services (**a**) and services to systems (**b**)

function. A relation is needed between the "Public Services", for which a target disruption value compatible with community resilience goal, E[C] in Fig. 28.7, is now available, and each service, as well as between the latter services and the primary and secondary systems. According to the proposed framework these relationships can be expressed with service and systems importance matrices, denoted in Fig. 28.8 by $\mathbf{I}_{service}$ and $\mathbf{I}_{system}$, respectively. The "Public Services" VCF is over-simplified in the figure for the sake of illustration and is reduced to the police, health-care and food services. Also, the list of supporting systems, primary and secondary, is also reduced for the sake of the example, and the actual numbers in the matrices are made up, their sound selection process representing one more important aspect that requires focused research efforts. With reference to the health-care service, the figure shows that it strongly depends on the service level of hospitals (a weight of 0.5) and to a minor extent on the service level of lifelines such as the potable water, electric power and road networks (weights between 0.1 and 0.2).

With the above matrices in place, the total disruption in public services D_{tot} can be linked to the disruption in the primary and secondary systems. Further, by partitioning the matrices, the total disruption can be divided into a contribution coming from the new hospital to be designed (system 1) and a contribution coming from all the surrounding systems:

$$D_{tot} = \underbrace{\mathbf{I}_{service}}_{1 \times n} \underbrace{\mathbf{I}_{system}}_{n \times m} \underbrace{\mathbf{D}_{system}}_{m \times 1} = \mathbf{I}_{service} \begin{bmatrix} \mathbf{I}_{system,1} & \mathbf{I}_{system,2} \end{bmatrix} \begin{bmatrix} \mathbf{D}_{system,1} \\ \mathbf{D}_{system,2} \end{bmatrix} =$$

$$= \underbrace{\underbrace{\mathbf{I}_{service}}_{1 \times n} \underbrace{\mathbf{I}_{system,1}}_{n \times 1} \underbrace{D_{system,1}}_{1 \times 1}}_{1 \times 1} + \underbrace{\underbrace{\mathbf{I}_{service}}_{1 \times n} \underbrace{\mathbf{I}_{system,2}}_{n \times (m-1)} \underbrace{\mathbf{D}_{system,2}}_{(m-1) \times 1}}_{1 \times 1}$$

(28.1)

By equating this disruption to the annual percent capacity disruption compatible with the community resilience goal, E[C] in Fig. 28.7, one could obtain the target unknown maximum disruption for the new hospital $D_{system,1}$:

$$D_{system,1} = \frac{E[C] - \mathbf{I}_{service}\mathbf{I}_{system,2}\mathbf{D}_{system,2}}{\mathbf{I}_{service}\mathbf{I}_{system,1}} \quad (28.2)$$

The missing item in the above equation is the disruption due to seismic events to all the other systems of the built environment supporting the community, $\mathbf{D}_{system,2}$. The latter requires a systemic analysis of the entire built environment, i.e. an analysis to assess the impact of an earthquake event at the regional or urban scale. While this is one further aspect where research is needed, unlike the aspects previously mentioned, systemic analysis has already attracted considerable attention in the last decade at least, and several frameworks or partial models exist. The next main section is thus devoted to this topic and to identify some of its research gaps.

28.2 Systemic Analysis of the Built Environment

28.2.1 Existing Frameworks

Among the many available, the definition of infrastructure given in (PCCIP 1997) as the "network of distinct man-made systems and processes that function collaboratively and synergistically to produce and distribute a continuous flow of essential goods and services", is adopted. From a system-theoretic point of view, the infrastructure is thus a system of systems (SOS) (Rinaldi 2004), a super-system containing all other systems (buildings, lifelines, critical facilities, etc.) and constitutes the physical layer supporting the life of our society, the built environment, i.e. the two bottom layers in Fig. 28.9a. The term infrastructural system is thus used to indicate any of the component systems in Fig. 28.9a. Analysis of the impact of an earthquake on infrastructural systems has started with the analysis of single systems. Most of the research in earthquake engineering still focuses on the characterization of the single-site point assets in these systems, like buildings or bridges, which are themselves (structural) systems but, at the scale of interest herein, are just components. This component-oriented point of view dominates the scene and this is reflected, just to give an example, in the HAZUS collection of components' fragility curves (NIBS 1999). Of the much smaller proportion of research that looks into the spatially distributed portion of these systems, most work focused initially on road networks, e.g. (Shinozuka et al. 2003), with fewer works devoted to other lifelines, like power networks (Vanzi 1996) or water networks (Wang et al. 2010).[3] Studies dealing with two or more systems are even scarcer and typically referring to the power and another dependent network, e.g. (Dueñas-Osorio et al. 2007) (Poljanšek et al. 2012), while those aiming at modelling consequences beyond simple physical damage are rare (Cho et al. 2001) (Karaca 2005). To the knowledge of the author, the

---

[3]This chapter is not a state-of-the-art on either resilience or the assessment of infrastructural systems, but, rather, a point of view on some research gaps in the field. For this reason, only a subjective, partial selection of examples is given here, before focusing from the next section on the framework developed by the author and co-workers.

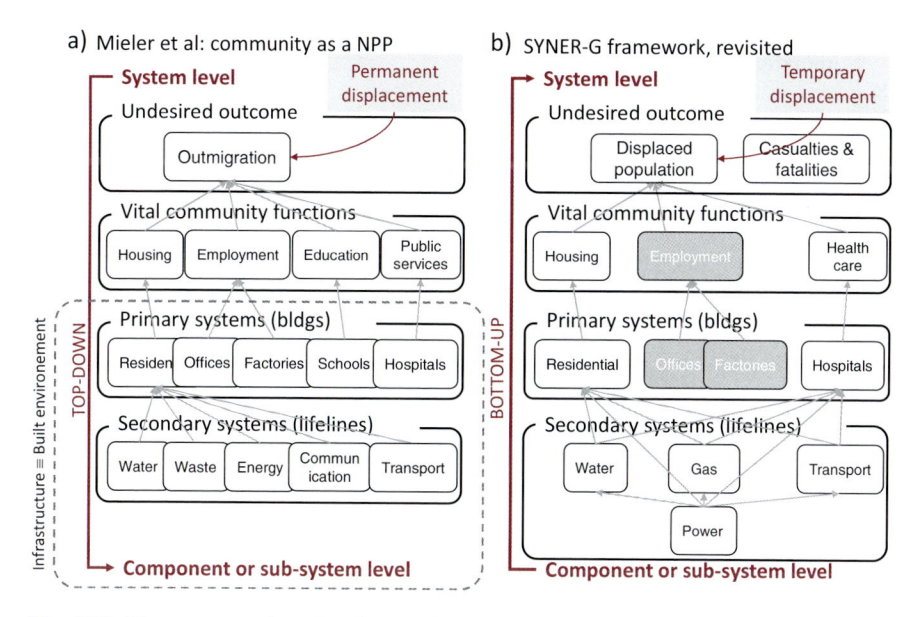

Fig. 28.9 The representation of a Community as a NPP by Mieler et al. (2013, 2015), with identification of the supporting system of systems (primary+secondary) at its base (**a**) and the analogy with the SYNER-G framework (grey fill indicate the considered systems, implemented in OOFIMS)

first notable large scale effort to model the problem of the impact of an earthquake at urban or regional scale from a systemic point of view originated on the US East Coast under the umbrella of the MCEER and later of the MAE. This is the research the led to seminal works like that by Bruneau et al. (2003).

More recently, in Europe, the author and co-workers contributed to the development of a framework for the analysis of interdependent infrastructural systems, within the context of the SYNER-G project (2009–2013). This project is described in a number of papers and in two dedicated books (Pitilakis et al. 2014a, b). In particular, the systemic framework and the general object-oriented model developed to support it are described in (Franchin 2014). The framework presents high similarity with that put forward in (Mieler et al. 2013) and can be described in the same way, as shown in Fig. 28.9b. The main difference is in the perspective, a top-down design one versus a more traditional bottom-up, assessment perspective in SYNER-G. On the other hand, the SYNER-G framework, implemented in an open-source software, namely Object-Oriented Framework for Infrastructure Modelling and Simulation (OOFIMS), is fully operationalized and considers already an important subset of the primary and secondary systems, with their interactions. In this respect, this model, or other similar in capabilities, are good candidates for the evaluation of systemic impact needed in Eqs. (28.1) and (28.2).

Figure 28.10 illustrates qualitatively the main features of this systemic model. It is multi-layered, with some layers in the physical space (collectively denoted as

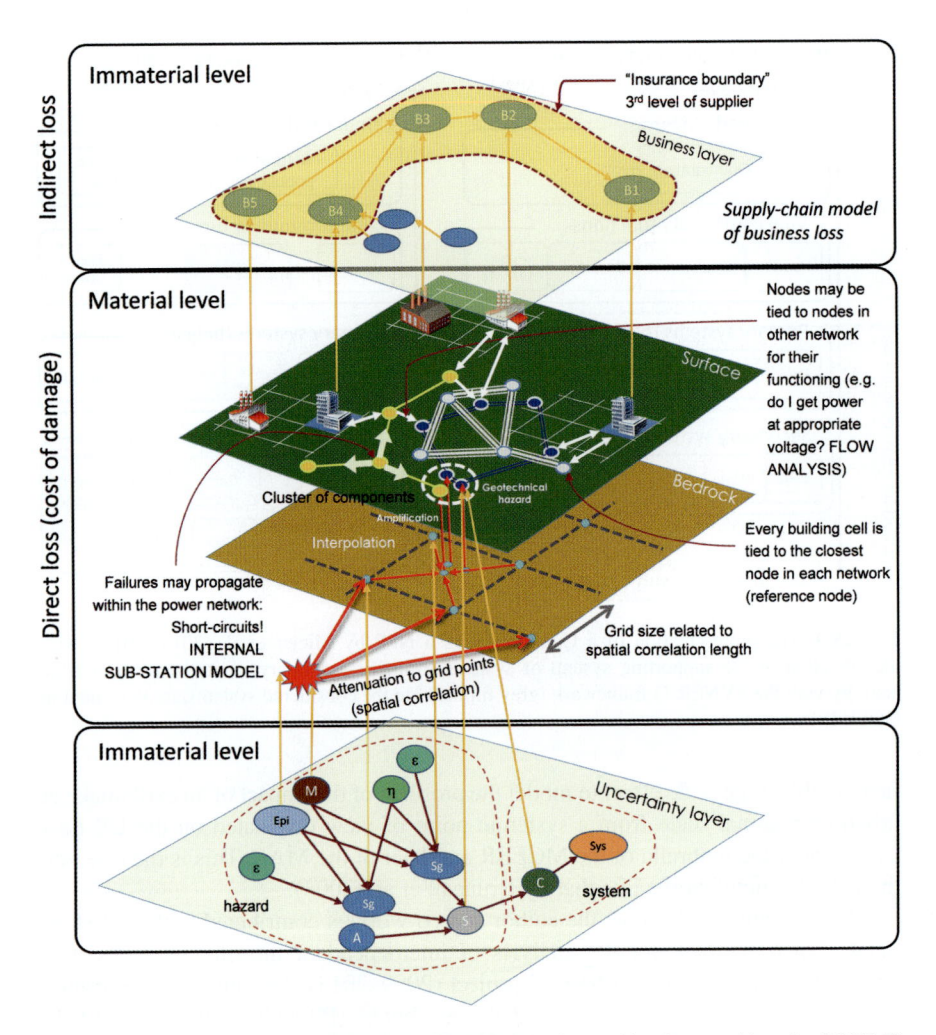

Fig. 28.10 The model developed within the SYNER-G project and implemented into the OOFIMS software, and its possible extension to consider indirect loss due to business interruption

"material level"), while others are non-material. Thus, for each physical quantity needed at the material level, there is a corresponding random variable in the Uncertainty layer (links in this non-physical network denote statistical dependence). At the bottom lies a probabilistic model of spatially distributed seismic hazard, described in detail in (Weatherill et al. 2014), which allows prediction of consistent seismic intensity fields (or maps). The separation of inter- and intra-event errors in the prediction of simultaneous intensity at different sites (Bommer and Crowley 2006), and of spatial correlation (Jayaram and Baker 2009), by now a consolidated acquisition, wasn't immediately recognized in its importance. For quite some time most so-called regional studies, which generally where portfolio loss assessments with no consideration of interactions whatsoever, used design maps. The model in Fig. 28.10 generates scenarios starting from magnitude and location, it then predicts local intensities S_g on a regular grid (considering inter-event η and correlated intra-event ε errors), from which values S at the location of individual components are interpolated. When spatially consistent values at the bedrock under each component are determined, surface values are obtained via site-dependent random amplification (A), and values of other intensity measures, needed for components of different types clustering at the same site, are also obtained in a probabilistically consistent manner.

The innovative portion of the model begins where the hazard ends, i.e. in the system portion where the components states (C) are propagated into the system states (Sys). The focus in its development was on the capability to include interactions, on the possibility of integrating different system models with non-consistent granularity of the input data, and on refinement of each system internal model. Thus, just to give an example, the primary systems (buildings) are automatically tied to the closest node in each of the secondary, establishing a two-way relationship with each of them (assembly of demand from the buildings to the system, loss of service from the system to the buildings). Other building-system interactions include, e.g., obstruction of road segments due to building damage. Ties exist also between nodes of the secondary systems, so that loss of power due to a failure in the electric power network cascades into failure of the connected systems. But the main important difference with other models is in the refinement of the individual network system models. All models are analysed in terms of flows, rather than simple connectivity. This aspect is one that has been shown to change the results of the assessments (Cavalieri et al. 2012, 2014) (Franchin and Cavalieri 2015), and is considered in the next section.

Before moving on, however, let's consider once more Fig. 28.10. The top portion of the figure shows a layer, denoted as "business layer", where another non-physical network lies. Nodes B_i in this network are businesses, and the network of links connecting them represent the chains of supply and demand from prime sources to the final customer. This portion of the model is just an idea and, to the best knowledge of the author, a complete full-blown systemic view of this type (i.e. with a consistent level of completeness at all levels, from hazard, to physical behaviour of the interconnected systems, to the organizational aspects of business) is still missing and probably beyond the current modelling capabilities. Nonetheless, it shows one possibility to go beyond the simple summation of direct loss and venture into that of indirect loss armed with the capability of modelling business interruption due to a range of causes, many of which non-local. This kind of studies may never reach the point where they provide accurate assessments, but will serve the purpose of exploring complex patterns. It may be of interest to the insurance industry (for instance to choose where to trace their boundary for liability, the figure shows, as an example, the boundary of business B_1 stopping at third-order suppliers).

28.2.2 Challenges in Network Modelling

The production and distribution of essential goods and services mentioned before requires a number of specialised networked systems, consisting of production, exchange and consumption sites, connected by links. From a mathematical point of view, they can all be regarded as graphs consisting of nodes or vertices, and edges. At a basic level, where most physical differences among networks are disregarded and the focus is only on connectivity, this graph theoretic point of view, and the associated mathematical apparatus, is all that is needed to assess the impact of damage. Figure 28.11a shows a simple network, with $n_V = 4$ vertices and $n_E = 5$ edges connecting them. Whatever is conveyed by the network, can travel along edges in both directions, depending on what drives or directs the flows. This situation arises, e.g., in power networks, where flow can occur in both directions depending on the voltage at the edge ends. The corresponding graph is called undirected, and its mathematical representation can be through a symmetric adjacency matrix **A**. Alternatively, an incidence matrix **I** can be used, where each row indicates which nodes are connected by the corresponding edge. When edges cannot be necessarily travelled in both directions, like e.g. with one-way roads in transportation networks, the graph is directed and **A** is not symmetric any more. Damage is modelled at the network level, by an update of either **A** or **I**, as shown for instance by removal of the third and fifth row of **I** in Fig. 28.11c. At this basic level, the physical differences between networks enter only into the way damage to components is predicted: for each system a different set of fragility curves is considered.

The limits of this basic modelling level can be easily exposed, both at the global and the local level. Figure 28.12a shows the same network as Fig. 28.11a, but now the four nodes are divided into $n_S = 2$ sources (v1 and v2) and $n_D = 2$ demand nodes (v3 and v4). An often used connectivity-based global measure of network

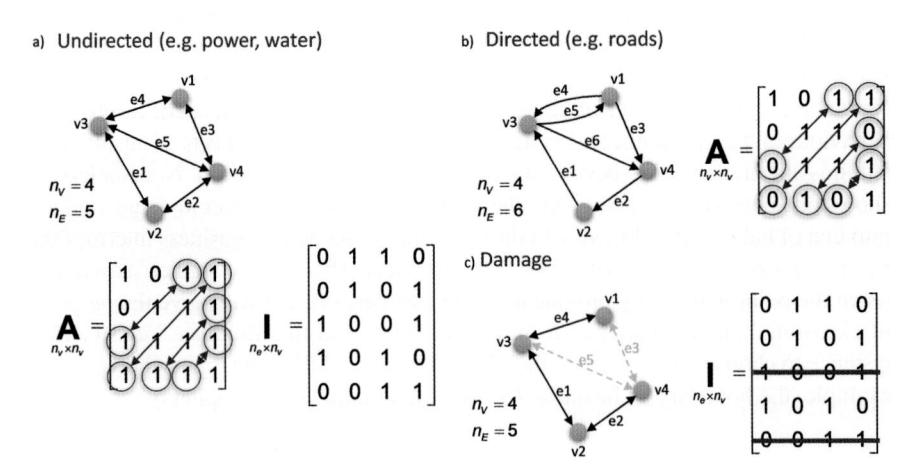

Fig. 28.11 Graph representation of networks: undirected (**a**) and directed (**b**). Damaged network (**c**)

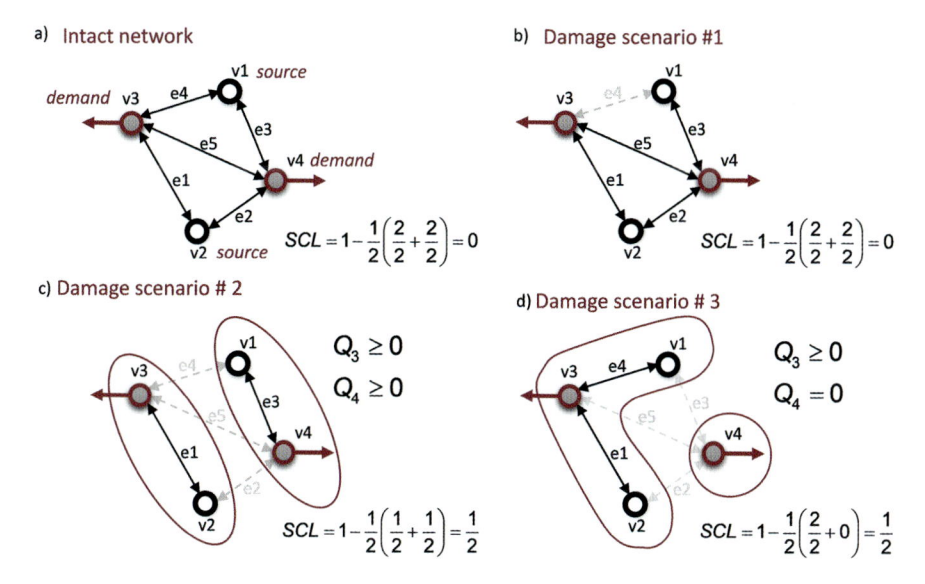

Fig. 28.12 Sources and demand nodes in the network of Fig. 28.11(**a**). Different damage scenarios (**b**) to (**d**)

performance is the so-called simple connectivity loss (SCL), defined as (Franchin and Cavalieri 2013):

$$SCL = 1 - \frac{1}{n_D} \sum_{i=1}^{n_D} \frac{N_{Si,0}}{N_{Si,s}} \qquad (28.3)$$

where N_{Si} is the number of sources connected to the i-th demand node, and the subscript "0" and "s" denote the "zero", undamaged, reference conditions, and the "seismic", damaged one. It clearly takes upon the value 0 in the initial condition where, for any demand node, $N_{Si,\,s} = N_{Si,\,0}$. Figure 28.11b shows a damage scenario where edge 4 is removed from the network. Even though not directly, demand node v3 is still connected through node v4 to the source v1. As a result, SCL = 0, as for the intact conditions. According to the connectivity approach this damage goes undetected. Two more damage scenario, shown in Fig. 28.11c, d, are also assigned the same value of SCL of 0.5, even though in case (c) both demand nodes are still connected to one source, while in case (d) demand node v4 is disconnected and therefore its demand cannot be satisfied ($Q_4 = 0$).

The situation in Fig. 28.11b is worth considering again. It should be stressed that being still connected to a source is a necessary but not sufficient condition to satisfy a node demand. What matters is the actual flow delivered. For instance, a number of pipes may connect a source with a demand node, but leaks along the path may reduce pressure to the point that the water head is below the height of the buildings served by the demand node. No water would get out of the tap, especially at higher floors, without any broken pipe in between. Similarly, electric apparatuses are not very tolerant to voltage, and power provided with voltage lower than 90% the regular working condition would make them unusable. These are good reasons to go for flow modelling, one more is given later on (flow can be measured after the event). Against this approach is the elegant and efficient reliability methods available for connectivity-only models. Solving the real problem, however, remains for now a matter of brute force and Monte Carlo-like simulations. In the view of the author, efforts to devise more efficient, affordable schemes to reduce the burden of Monte Carlo simulations, like e.g. the reduction of seismic scenarios to be analysed (Chang et al. 2000) (Jayaram and baker 2010), are more appreciable than those trying to push the limits of connectivity-based methods. It is probably wise, however, to keep pursuing both, given the stark difference in computational effort. For instance, while it has been shown that simple connectivity models cannot predict the correct retrofit priorities for power-network components, enhanced connectivity approaches like hierarchical decomposition come closer to the flow-based results (Cavalieri et al. 2014).

Table 28.1 reports, for the five of the many networked systems implemented so far in OOFIMS (Cavalieri et al. 2012, 2014a, b, 2016, Esposito et al. 2014 and Cavalieri 2017), the topology, i.e. whether edges form closed loops as in grid-like systems or they don't, as in tree-like ones, and which portion of each network has one topological structure or the other. The table then provides information on damage, and finally on flow. For what concerns damage, in order to describe it fully, models for damage and its consequence should be formulated and implemented for both nodes and edges. For nodes, damage at the sources (S) and in intermediate junction nodes (J, in general, denoted TD/D for power, where voltage transformation can occur, or Re/ReMe for gas, where pressure reduction can happen) should be modelled. Junction in road networks are more complex. They can be at-grade, in which case they are basically not vulnerable, or interchanges, made up of ramps and bridges, tunnels, in which case they resemble those in power or gas networks, where they are themselves systems. In OOFIMS node damageability, with the exception of TD/D and Re/ReMe substations, is not modelled, as indicated by the italic in the table. On the other hand, damage to edges is modestly modelled. It can be continuous (with progressive reduction of flow capacity), as in all system with pipes (water, gas, etc), or discrete, binary (fail,safe), as for power lines, or multi-level as in road networks, for lanes that can be closed to limit vertical load on damaged structures. Furthermore, damage can be direct, for all components, or indirect, as when a power line (especially an overhead one that, contrarily to buried one is almost insensitive to seismic motion) is damaged due to overcharge. The consequences of damage are for all networks a decrease in flow for direct damage (consequence on the damaged edge) and a possible increase in flow for indirect damage (consequence of damage to other edges, causing change of flow patterns).

Coming finally to the lower portion of Table 28.1, dedicated to flow, the table first reports the quantity whose gradient drives flows through the network, and the quantity flowing. In all cases, with the exception of road networks, the flow equations express flow of a physical quantity under physical constraints. Road networks are complicated by the fact that drivers are (still) human, and elements of behaviour modelling are included. Without entering into the details, several

Table 28.1 Networked systems: topology, flow and damage type and consequences

System		Power	Gas	Freshwater	Stormwater/ Wastewater	Roads
Topology	Grid-like	Transmission/ distribution	Transmission/ distribution	Transmission/ distribution	N/A	Yes
	Tree-like	Distribution (LV)	Distribution (LP)	Distribution	Collection	Yes
Damage	Nodes	S, TD&D	S, Re, ReMe	S, J	J	No
	Edges	Binary.Buried, direct. *Overhead, indirect.*	Continuous. Pipes, direct (leaks/ breaks).	Continuous. Pipes, direct (leaks/ breaks).	Continuous. Pipes/chan- nels, direct (leaks/ breaks).	Discrete. Bridges/ embank- ments/ trenches/ tunnels, direct.
	Consequence (direct)	Zero current	Flow decrease	Flow decrease	Flow decrease	Flow decrease
	Consequence (indirect)	Current increase (overcharge)	Flow increase	Flow increase	Flow increase	Flow increase (traffic congestion)
Flow	Driver	Voltage V	Pressure P	Hydraulic head $h = z+P/\gamma_w$	Head	Trips (O/D matrix)
	Quantity	Power (complex)	Gas (compressible)	Liquid	Liquid	Vehicles
	Equations(line loss)	Multiple voltage ranges	Multiple pressure ranges	Single pressure range	Free surface gravity flow	Single speed range
	Equations(line capacity)	*Yes*	No	No	*Yes*	Yes

S = Source node; D/TD = Distribution or Transformation-distribution sub-station (power); Re/ReMe = pressure Reduction or Reduction-Measurement sub-station (gas); J = generic Junction (where more edges converge)

algorithms exist to determine flow, looking for an optimum either from the user or the system perspective. Perhaps in the future, with the advent of autonomous and coordinated vehicles the predictability of these networks will improve considerably. Herein, however, what is relevant is that the demand on the road network is the end result of another analysis whose output is the so-called origin-destination (O/D) matrix, which expresses the amount of vehicles leaving each node (traffic analysis zone) for each possible destination. Most studies in the field of seismic assessment of road networks used and still use so-called static O/D matrices, i.e. use pre-earthquake demand on post-earthquake damaged network. Attempts to use so-called dynamic O/D matrices, linking the change in traffic demand to damage in residential buildings and economic activities, are scarce and seem not to have drawn enough attention (Cho et al. 2004). This is one more aspect that requires considerable improvement and seems to have defied researchers' efforts so far. On the other hand, perhaps because flow-based assessments and multi-system studies are still a minority in lifelines research, it may have escaped the attention that the same problem afflicts also power, water or gas systems. All these systems have demand proportional to population needs, and if the population is displaced and economic activity pattern is altered after the event, the demand pattern should reflect it. In this respect the automatic link between buildings and demand nodes in the SYNER-G framework represents a solution, at least for these systems (no progress has been made instead on the post-earthquake travel demand front).

The last portion of the table provides information on the actual flow equations. In light of the differences in the conveyed quantities, equations are much simpler where water is involved, than in the case of electricity or gas. Flow models include two types of equations, balance equations expressing flow continuity at the nodes, and resistance equations expressing line loss. For freshwater networks these are written as:

$$\begin{cases} \mathbf{I}_D^{*T}\mathbf{q} - \mathbf{Q} = \mathbf{0} \\ \Delta\mathbf{h} - \mathbf{r}(\mathbf{q}) = \left(\mathbf{I}_S^*\mathbf{h}_S + \mathbf{I}_D^*\mathbf{h}_D\right) - \mathbf{r}(\mathbf{q}) = \mathbf{0} \\ \mathbf{r}(\mathbf{q}) = \mathbf{R}\mathbf{q} \circ |\mathbf{q}| \end{cases} \qquad (28.4)$$

where $\Delta\mathbf{h}$, $\mathbf{r}$ and $\mathbf{q}$ are the $n_E \times 1$ vectors of node head difference, resistance (function of the edge flow) and edge flows, $\mathbf{Q}$ is the $n_D \times 1$ vector of node demands (zero if the node is a junction) and $\mathbf{I}_S^*$ and $\mathbf{I}_D^*$ are the partitions of the incidence matrix related to source nodes and demand nodes, respectively. With reference to the simple network in Fig. 28.11a, the incidence matrix is expressed as:

$$\mathbf{I}^*_{n_e \times n_v} = \begin{bmatrix} 0 & -1 & 1 & 0 \\ 0 & -1 & 0 & 1 \\ -1 & 0 & 0 & 1 \\ -1 & 0 & 1 & 0 \\ 0 & 0 & -1 & 1 \end{bmatrix} = \begin{bmatrix} \mathbf{I}_S^* & \mathbf{I}_D^* \\ {\scriptstyle n_e \times n_S} & {\scriptstyle n_e \times n_D} \end{bmatrix} \qquad (28.5)$$

It is important to note that within the flow analysis, although the water network is undirected, the incidence matrix must reflect the actual edge directions as specified in the network connectivity matrix, and thus it has 0, +1 and -1 entries.

For gas networks, pressure ranges are large enough that a single set of equations does not suffice (line loss changes its proportionality), so that equations for multiple levels of pressure, including the transformation from one level to the other are needed:

$$\begin{cases} \mathbf{I}_D^{*T}\mathbf{q} - \mathbf{Q} = \mathbf{0} \\ \Delta \mathbf{p} - \mathbf{r}(\mathbf{q}) = (\mathbf{I}_S^*\mathbf{p}_S + \mathbf{I}_D^*\mathbf{P}_R\mathbf{p}_D) - \mathbf{r}(\mathbf{q}) = \mathbf{0} \end{cases}$$

$$\mathbf{r}(\mathbf{q}) = \begin{cases} r_{ij}(q_{ij}) = p_i - p_j = K_L q_{ij}^2 & \text{low−pressure} \\ r_{ij}(q_{ij}) = p_i^2 - p_j^2 = K_M q_{ij}^2 & \text{medium(high)−pressure} \end{cases}$$

(28.6)

where the pressure vector $\mathbf{p}$ replaces the head difference vector $\Delta\mathbf{h}$. The difficulty arises when one realizes that the above equations are design equations, not assessment equations. They are routinely used in design of new systems where demand satisfaction is a requirement. This is a simplification akin to that of modelling structures linearly, or with simple bilinear models neglecting post-peak behaviour, since no such extreme behaviour would be acceptable for a new structure (Fardis 2018). But in a damaged network system loss can be so large that demands are not met at many nodes. Head-driven (or pressure-driven, etc), rather than demand-driven equations are thus needed, such as:

$$\begin{cases} \mathbf{I}_D^{*T}\mathbf{q} - \mathbf{Q}(\mathbf{h}_D) - \mathbf{Q}_{seismic}(\mathbf{h}_D) = \mathbf{0} \\ (\mathbf{I}_S^*\mathbf{h}_S + \mathbf{I}_D^*\mathbf{h}_D) - \mathbf{r}(\mathbf{q}) = \mathbf{0} \end{cases}$$

(28.7)

for water networks, or:

$$\begin{cases} \mathbf{I}_D^{*T}\mathbf{q} - \mathbf{Q}(\mathbf{p}_D) - \mathbf{Q}_{seismic}(\mathbf{p}_D) = \mathbf{0} \\ (\mathbf{I}_S^*\mathbf{p}_S + \mathbf{I}_D^*\mathbf{P}_R\mathbf{p}_D) - \mathbf{r}(\mathbf{q}) = \mathbf{0} \end{cases}$$

(28.8)

where $\mathbf{P}_R$ is the pressure reduction vector (Cavalieri 2017). In the above equations, node demands are reduced as a function of driving pressure, and additional "seismic" demands are added to nodes, lumping line loss from each pipe segment in its end nodes and summing up. This is the type of equations implemented into OOFIMS. Alternatives include the use of a third-party software in an iterative manner, adjusting water head and demands until satisfaction, like done for instance by GIRAFFE with EPANET. It could then be added that the above equations are still only stationary ones, assuming that the state of damage, supply and demand are fixed in time. Just to give an example, and not considering multiple shock sequences causing incremental damage, in the presence of variable head sources, with important network damage and associated loss, the problem of computing the sustainability of demand in the absence of repair and during a dry season become of very high relevance (studies like these have been performed for instance for Wellington, in New Zealand).

The situation for power networks is even more complicated, since, to the best knowledge of the author, available flow equations are not still demand-driven. But there is more than that, the problem being that since power cannot (could not?) be easily stored, power networks are operated with continuous adjustments to balance demand and capacity, trying to solve the so called Alternate Current Optimal Power Flow (ACOPF) problem, which is a nonlinear (non-convex) constrained optimization problem, the constraints being posed by limited line capacity or generators power limits. The problem is so tough that, formulated in 1962 (Carpentier 1962), it still awaits to date a robust and fast solution. It is solved on a yearly, monthly, daily and hourly basis for the needs of the operators and to adjust market prices (it is also indeed called the Security-Constrained Economic Dispatch or SCED), and depending on the case, solution of increasing efficiency (speed) and simplification are adopted. The alternatives include using the Economic Dispatch, i.e. solving the optimization problem without constraints, or solving nonlinear quadratic AC equations that yield mathematically but not necessarily physically feasible solutions, nor optimal ones. The latter are those used already in the (pioneering?) study in (Vanzi 1996) and implemented in OOFIMS. Most risk assessment studies have not even tried this, and when flow was considered, it was in the simplified linearized direct current (DC) variant, only recently moving on (Li et al. 2017).

In conclusion, difficulties in flow-based analysis are not related exclusively to the increased computational burden but also to the improvement and adaptation of flow models to the case of assessment of damaged networks. This is one more research gap towards the capability of designing resilience of communities in the real world.

28.2.3 Components' Fragility: Beyond Generic Fragility Models

The previous section discussed the merits and difficulties of flow-based analysis at the network level to determine functional consequences of physical damage. This section goes back to the previous step of evaluation of the components' state of damage. Damage is usually determined sampling from a damage state distribution, which is obtained at the local intensity level in each scenario event from the set of the component fragility curves. In the simplest of cases, with so-called binary components, a single curve is used, separating functional from non-functional state. In general, at least two curves are considered, as shown for instance in Fig. 28.13, with three resulting state: intact, damaged and collapsed. This allows modelling intermediate states of partial functioning. Systemic studies and simpler portfolio loss assessments make extensive use of so-called generic fragility curves, the main source of which is the already recalled HAZUS collection (NIBS 1999). The applicability of these literature fragility is seldom if ever questioned, the aim of the

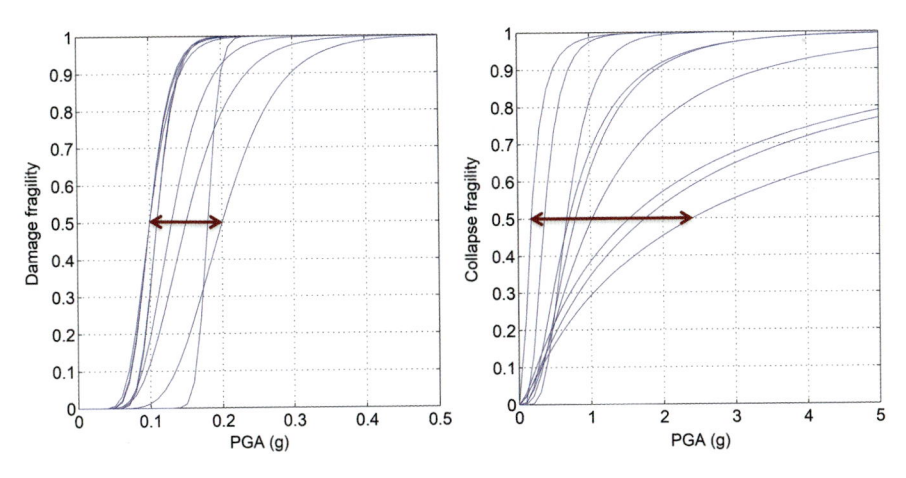

Fig. 28.13 Damage and collapse fragility for a set of bridges belonging to the same class (Adapted from Borzi et al. 2015)

analysis being testing the systemic part, rather than realism in the results. Sometimes, applicability issues are considered and project specific fragilities, which may include literature ones, are collected/derived, as done for instance during the SYNER-G project (Pitilakis et al. 2014a) or for cases where previous experience on structural behaviour is not available (Crowley et al. 2017). The problem, however, with generic fragility curves, is that they are conceptually questionable.

Fragility is well-known to be site-specific (Veneziano et al. 1983), meaning that the same structure at different sites will have a different fragility curve. This is because the fragility is not a property of the structure. This by itself would be enough to say that generic fragility curves are a nonsense, but there is more to add. During a nation-wide assessment of 485 bridges in Italy, with refined inelastic response history analyses of 3D models carried out in a consistent manner across the bridge stock, damage and collapse fragility curves where obtained (Borzi et al. 2015). The important finding of interest here is that these curves allowed to challenge the usual classifications of bridges used for the purpose of generic fragility developments. Figure 28.13 shows the fragility curves of a set of nine bridges that were obtained querying the 485 bridges data base with the criteria: multiple simply supported spans, single-stem hollow-core piers of height between 5 m and 30 m, rubber bearings. The HAZUS, Turkish and Greek typological classification would attach to all bridges the same fragility curve. The results in the figure speak for themselves. Variation only in the median collapse intensity is more than 2 g.

In sum, fragility is only an intermediate step in the evaluation of risk (or damage probability), it does depend on the site and, most importantly, on the structure properties in a stronger manner than assumed by lumping "similar" structures into the same bin for the purpose of assigning a fragility curve. What is shown in Fig. 28.13 for girder bridges can be easily extended to buildings. Can we expect that all 5-storeys regular RC buildings have the same fragility curve? There is a strong need for improved tools for damage prediction. Generic fragility curves should be replaced by more flexible fragility models that yield structure and site specific fragility curves. A first attempt in this direction, more of a demonstration study, is the Bayesian-network-based fragility model for girder bridges developed based on the data from the above mentioned 485 bridges (Franchin et al. 2016), capable of predicting a structure and site-specific fragility, even when values are assigned just to a portion of its random variables (a considerable advantage of using Bayesian Networks over alternative methods). Something along these lines, however, is needed for all types of components, if a higher level of reliability of results of systemic analysis is sought. This is an entire area of work that needs fresh concerted efforts.

28.2.4 Evolving State of Information and Emergency Management

Approaching the conclusion of this chapter, one last issue is worth mentioning. A systemic model, like the one illustrated so far, can also be used beyond planning for resilience enhancement (or risk mitigation, as it used to be called). Such a model, if a sufficient level of realism was attained (e.g. by limiting its use to a subsystem that can be more reliably predicted), could be used to inform the construction of a decision support system (DSS) for emergency management purposes. Research in this direction is ongoing, and efficient management of the post-event phase is another action towards higher resilience.

A promising tool for building DSSs is represented by Bayesian Networks. These are network of random variables used to represent in an efficient manner the uncertainty in a problem and for which efficient methods have been developed to perform Bayesian inference (Nielsen and Jensen 2009). The latter, i.e. the capability of quickly updating probabilities in response to changes in the state of knowledge, makes BNs a natural candidate for DSSs. Work in this direction in the field of seismic risk of distributed civil systems has mainly looked into connectivity-based modelling of systems, e.g. (Bensi et al. 2013). After developing a BN for the hazard portion of the problem (compare the uncertainty layer in Fig. 28.10 with Fig. 28.14a), Bensi et al. worked in order to improve the system part. The easiest way to describe the latter is the converging structure where all components are linked to the system, as in Fig. 28.14b. This structure is straightforward to obtain and for this reason is called Naïve formulation. Without entering into the theoretical details, the computational burden associated with Bayesian inference on a BN increases, among other things, with the number of incoming links to a variable, making the Naïve formulation intractable with any realistic system. This reason led to the formulation of alternative system descriptions, like the minimum link set formulation (MLS) shown in Fig. 28.14c, where components are linked to the MLS they belong, and these in turn are linked to the system, or the Efficient MLS formulation,

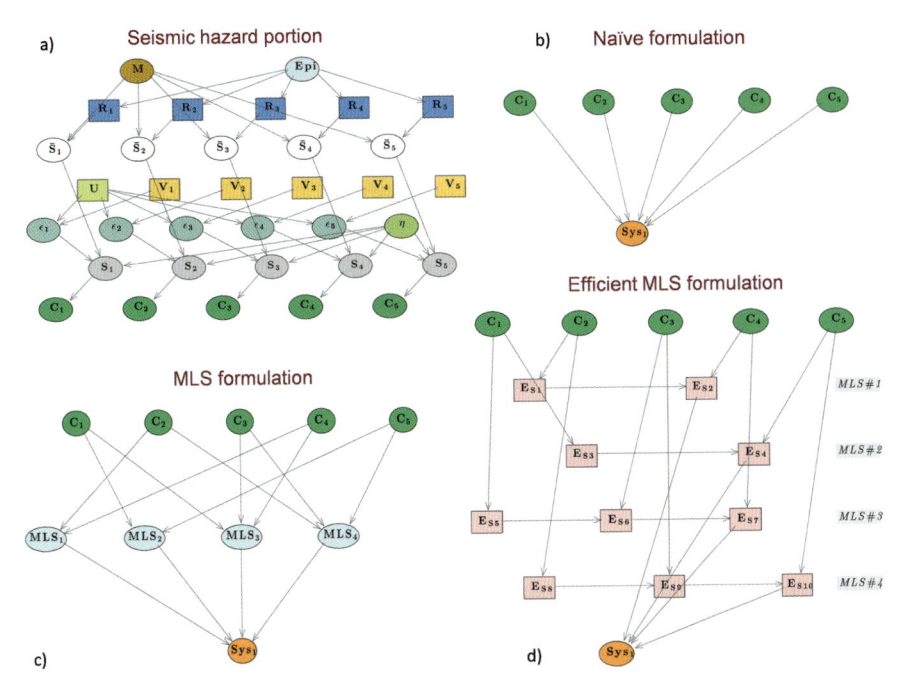

Fig. 28.14 BN for the hazard (**a**) acting on a 5-components system, and alternative formulations for the system portion of the BN (**b**) to (**d**) (Modified from Cavalieri et al. 2017)

where the number of incoming links to each variable is further reducing introducing survival path sequences (Fig. 28.14d).

The above efficient formulations, however, cannot be extended to handle the case of systems where flows are of interest, as shown in (Cavalieri et al. 2017). On the other hand, the naïve formulation, in its simplicity does not impose limitations on the type of performance metrics used to describe the system's and components' states, allowing treatment of the flow problem. For this reason, an improved Naïve formulation has been devised in (Cavalieri et al. 2017), whereby the number of incoming links is kept below a manageable threshold by eliminating edges corresponding to low correlation between component and systems states, as illustrated in Fig. 28.15a. The key to this operation is a pre-event systemic analysis to establish a set of component-system state vectors to establish these correlations. There are several advantages to this approach, besides the fact that it can solve the actual problem, rather than its connectivity simplification. First of all, as it can be seen by comparing Fig. 28.15a, b, the BN is much simpler than that obtained with the efficient MLS formulation (the figure refers to a connectivity case, where both approaches can be applied, and where inference results of the improved Naïve model have been compared to the exact MLS-based results showing excellent accuracy). This simplicity makes it also possible for the BN to be set up automatically based on the systemic simulation results, a welcome feature when dealing with real size complicated systems. Further, the BN size scales up linearly with the system size, while the

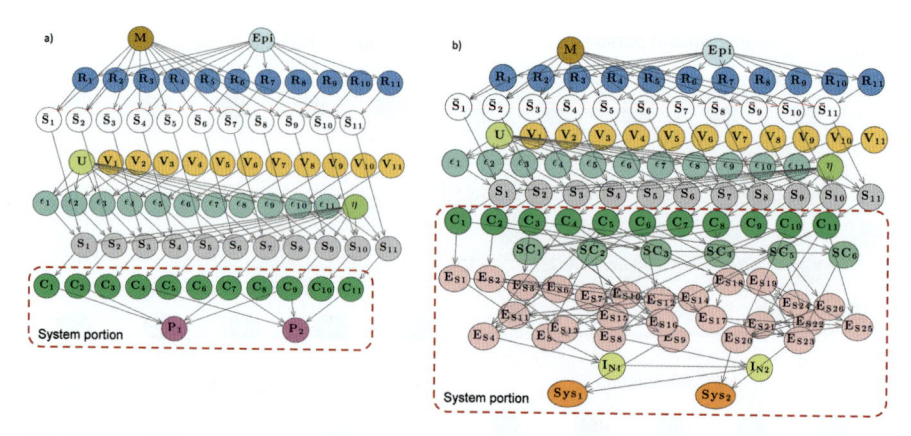

Fig. 28.15 BN for an 11-components system, thrifty-Naïve formulation (**a**) versus Efficient MLS formulation (**b**) (Modified from Cavalieri et al. 2017)

MLS-based formulations grow exponentially. This makes the improved (or thrifty, as it has been called) Naïve formulation tractable even for large systems (Gehl et al. 2017). Last but not least, since the goal is to perform Bayesian inference on the BN, it is of high practical relevance to choose a system model where performance is described in measurable terms. Flow is such a quantity: water coming out of a tap can be measured locally, without knowledge of the state of damage of the entire system. On the contrary, disconnection from or connection to a source, are information that require an analysis of the system and knowledge of the damage in all components, which is exactly what one wants to infer from a limited set of measurements.

Once again, this an instance of the problem of choosing between elegance and efficiency of the reliability method, and the solution of a realistic problem. If the latter is chosen, the only alternative seems to be trying to improve the efficiency of lower-level methods (before it was Monte Carlo, now it is the Naïve system representation), which in their simplicity exhibit the necessary flexibility to accommodate the more accurate description of system performance. This is the last of the research directions for improvement of systemic analysis towards resilience identified in this chapter.

28.3 Conclusions

Based on the previous discussion, the author's personal opinions on some of the research efforts needed towards resilience are the following:

- Investing in resilience before the event, by reducing vulnerability of existing infrastructure and designing new assets for higher performance, is the best way to

enhance resilience, associated with less uncertainty and therefore higher confidence in the end result, possibly avoiding unsustainable outcomes.
- Non-technical obstacles against this action are the lack of political will and, of course, but to a lesser extent, the unfavourable economic conditions. After all, when money is short, it should be spent wisely. Public regulators, instead, tend to reward efficiency rather than redundancy/robustness, thus fostering a more fragile infrastructure. The problem is that investment in resilience can only be made concurrently with other current public expenses and the temporal horizon of politicians (the next election) is not as long as recurrence period of natural hazards. A healthy growth of the economy would help, but sustainability issues cannot be disregarded and it may be time to start thinking to solutions that do not necessarily imply or rely on an indefinite growth of the economy.
- On the technical side, much work is needed to improve systemic analysis, so that performance targets for each asset, to be retrofitted or newly built, can be set in a rigorous manner to attain community resilience goals. It is not anticipated that this will ever become a mainstream tool, but it is important that the idea informs decision-maker at a higher level, to move on the discussion on the acceptable seismic risk levels of structures.
- In order to improve systemic analysis, network modelling should be flow-based, or at least new smart connectivity-based methods that mimic flow-based solutions should be devised. Flow is better because survival of connection does not guarantee a satisfactory level of service at demand nodes, the importance of components is ranked differently based on connectivity or flow, and, last but not least, flow can be measured and used as evidence input to BN-based decision-support systems for emergency management, while connectivity cannot.
- The obstacles against this generalized adoption of flow-based modelling are that: flow on damaged networks cannot be computed as in undamaged ones; in many cases, appropriate flow equations still need to be developed; flow forces to use Monte Carlo-like simulation, the least efficient among reliability methods.
- Fragility of components is only an intermediate step in the evaluation of damage. It does depend on the site and, most importantly, on the structure properties in a stronger manner than usually assumed. There is a strong need for improved tools for damage prediction. Generic fragility curves should be replaced by more flexible fragility models that yield structure and site specific fragility curves. This is an entire area of work that needs fresh concerted efforts.
- Improvements in systemic analysis will also benefit the other side of resilience, i.e. the post-event management phase, possibly leading to useful decision support systems informed based on comprehensive pre-event simulations.

Finally, a remark is due that is not based on what has been presented so far, but it arises naturally in every discussion about these type of analyses, especially when the background of the researchers is a "hard" one, like e.g. structural mechanics. For systemic analysis to provide useful input to all of the above, there is a strong need for validation studies and sensitivities to plug-in models. The latter is relatively easier to obtain, since it requires only running analyses with different models for each sub-system and component, assessing the sensitivity of the results to each. The former represents instead the greatest of challenges. The problem is one of scale, spatial and temporal. It is easy to calibrate a model for confined strength of concrete, by crushing concrete samples. It is probably impossible to calibrate a system of systems model at regional scale. The issue is one of spatial scale, as much as it is one of temporal scale, since even if one could measure "everything" in a region of interest during a single event, the recurrence time between damaging events is such that acquiring a reliable (multi-event) dataset would be practically impossible. The only way out at present is to accept this impossibility and be content with partial validations, i.e. validations of the intermediate models, like the ground motion prediction equations or the magnitude recurrence laws used in predicting the seismic intensity, or the fragility models for the components, and relying on the rationality of the framework used to combine them. In this respect, the sensitivity studies to the input models are essential to determine their importance and better focus validation on those that are most relevant. Designing good validations is the last important step needed in this research field.

Acknowledgements Research from the author and co-workers is cited in this contribution extensively, to an extent that obviously does not reflect its relative weight in the field, but the intent of this contribution is to put forward some thoughts on research needs, rather than providing an exhaustive and balanced state of the art. This research was developed over a number of years with financial support of the European Commission, through the SYNER-G research project (grant number 244061), and the Italian Department of Civil Protection, through the RELUIS consortium (Special project RS6). This support is gratefully acknowledged. The author wishes also to especially acknowledge the long-lasting and fruitful collaboration with Dr. Francesco Cavalieri, who was, among other things, the main developer of the OOFIMS implementation of the systemic analysis framework. Finally, the views expressed in this chapter are those of the author, and do not necessarily reflect those of the funding agencies or of the collaborators.

References

Bensi M, Der Kiureghian A, Straub D (2013) Efficient Bayesian network modelling of systems. Reliab Eng Syst Saf 112:200–213

Bocchini P, Frangopol DM, Ummenhofer T, Zinke T (2014) Resilience and sustainability of civil infrastructure: toward a unified approach. J Infrastruct Syst 20(2)

Bommer JJ, Crowley H (2006) The influence of ground-motion variability in earthquake loss modelling. Bull Earthq Eng 4(3):231–248

Borzi B, Ceresa P, Franchin P, Noto F, Calvi GM, Pinto PE (2015) Seismic vulnerability of the Italian roadway bridge stock. Earthquake Spectra 31(4):2137–2161

Bruneau M, Chang SE, Eguchi RT, Lee GC, O'Rourke TD, Reinhorn AM et al (2003) A framework to quantitatively assess and enhance the seismic resilience of communities. Earthquake Spectra 19(4):733–752

Calvi GM, Spaziante V (2009) Reconstruction between temporary and definitive: the CASE project. Progettazione Sismica, 03/English, 2009, pp 221–250

Carpentier J (1962) Contribution á l'étude du dispatching économique. Bulletin de la Société Française des Électriciens 3(8):431–447

Cavalieri F (2017) Steady-state flow computation in gas distribution networks with multiple pressure levels. Energy 121:781–791

Cavalieri F, Franchin P, Buriticá Cortés JA, Tesfamariam S (2014a) Models for seismic vulnerability analysis of power networks: comparative assessment. Comput Aided Civ Inf Eng 29(8):590–607

Cavalieri F, Franchin P, Pinto PE (2014) Chapter: Application to selected transportation and electric networks in Italy. In: Pitilakis K, Franchin P, Khazai B, Wenzel H (eds) SYNER-G: systemic seismic vulnerability and risk assessment of complex urban, utility, lifeline systems and critical facilities, vol 31. Springer, Dordrecht, pp 331–346. https://doi.org/10.1007/978-94-017-8835-9_11. Online ISBN: 978-94-017-8835-9

Cavalieri F, Franchin P, Gehl P, Khazai B (2012) Quantitative assessment of social losses based on physical damage and interaction with infrastructural systems. Earthq Eng Struct Dyn 41(11):1569–1589

Cavalieri F, Franchin P, Giovinazzi S (2016) Earthquake-altered flooding hazard induced by damage to storm water systems. Sustain Resilient Infrastruct 1(1–2):14–31

Cavalieri F, Franchin P, Gehl P, D'Ayala D (2017) Bayesian networks and infrastructure systems: computational and methodological challenges. In: Risk and reliability analysis: theory and applications. Springer, Cham, pp 385–415

Chang SE, Shinozuka M, Moore JE II (2000) Probabilistic earthquake scenarios: extending risk analysis methodologies to spatially distributed systems. Earthquake Spectra 16(3):557–572

Cho S, Gordon P, Moore JE II, Richardson HW, Shinozuka M, Chang S (2001) Integrating transportation network and regional economic models to estimate the cost of a large urban earthquake. J Reg Sci 41(1):39–65

Cho S, Murachi Y, Fan Y, Shinozuka M (2004) Transportation network simulation for dynamic origin-destination matrix under earthquake damage (Paper 1697). In: Proceedings of 13th world conference on earthquake engineering, Vancouver, BC, Canada

Cimellaro GP, Reinhorn AM, Bruneau M (2010) Framework for analytical quantification of disaster resilience. Eng Struct 32(11):3639–3649

Crowley H, Polidoro B, Pinho R, van Elk J (2017) Framework for developing fragility and consequence models for local personal risk. Earthquake Spectra 33(4):1325–1345

Davis C (2015) Overview on the multi-hazard earthquake-flooding project, and activities of the ASCE Infrastructure Resilience Division. Presentation at the 3rd UC Lifeline Week, Rome, Italy. https://sites.google.com/site/resilientinfrastructures/events/3rd-uc-lifeline-week/presentations/day-2

Dolce M (2017) Reduction of vulnerability (and risk), development of guidelines - optimised use of resources for the reduction of seismic risk while improving energy performance. Kick off meeting Italian Department of Civil Protection and EC Joint Research Centre, Rome, Italy, June 1st 2017

Dolce M (2018) The 2016–2017 Central Apennines seismic sequence: analogies and differences with recent Italian earthquakes. In: Pitilakis K (ed) Recent advances in earthquake engineering in Europe. Springer, Cham, pp 603–639

Dueñas-Osorio L, Craig JI, Goodno BJ (2007) Seismic response of critical interdependent networks. Earthq Eng Struct Dyn 36(2):285–306

Esposito E, Iervolino I, d'Onofrio A, Santo A, Cavalieri F, Franchin P (2014) Simulation-based seismic risk assessment of gas distribution networks. Comput Aided Civ Inf Eng 30(7):508–523. https://doi.org/10.1111/mice.12105

Fardis MN (2018) Practical modelling of RC structures for displacement-based evaluation: toward the second generation of EN-Eurocode 8 and beyond. In: Pitilakis K (ed) Recent advances in earthquake engineering in Europe. Springer, Cham, pp 101–123

Franchin P (2014) A computational framework for systemic seismic risk analysis of civil infrastructural systems. In: Pitilakis K, Franchin P, Khazai B (eds) SYNER-G: systemic seismic vulnerability and risk assessment of complex urban, utility, lifeline systems and critical facilities, vol 31. Springer, Dordrecht, pp 23–56. https://doi.org/10.1007/978-94-017-8835-9_2

Franchin P, Cavalieri F (2013) Seismic vulnerability analysis of a complex interconnected civil infrastructure. In: Tesfamariam S, Goda K (eds) Handbook of seismic risk analysis and Management of civil infrastructure systems. Woodhead Publishing Ltd, Cambridge, pp 465–513. ISBN 0-85709-268-5

Franchin P, Cavalieri F (2015) Probabilistic assessment of civil infrastructure resilience to earthquakes. Comput Aided Civ Inf Eng 30(7):583–600

Franchin P, Lupoi A, Noto F, Tesfamariam S (2016) Seismic fragility of reinforced concrete girder bridges using Bayesian belief network. Earthq Eng Struct Dyn 45(1):29–44

Franchin P, Petrini F, Mollaioli F (2017) Improved risk-targeted performance-based seismic design of reinforced concrete frame structures. Earthq Eng Struct Dyn. https://doi.org/10.1002/eqe.2936

Franchin P, Pinto PE (2009) Allowing traffic over mainshock-damaged bridges. J Earthq Eng 13(5):585–599

Gehl P, Cavalieri F, Franchin P, Negulescu C (2017) Robustness of a hybrid simulation-based/Bayesian approach for the risk assessment of a real-world road network. In: Proceedings of the 12th international conference on structural safety and reliability

Iervolino I (2018) What seismic risk we do design for when we design buildings? In: Pitilakis K (ed) Recent advances in earthquake engineering in Europe. Springer, Cham, pp 583–602

Iervolino I, Giorgio M, Polidoro B (2014) Sequence-based probabilistic seismic hazard analysis. Bull Seismol Soc Am 104(2):1006–1012

Jayaram N, Baker JW (2009) Correlation model for spatially distributed ground-motion intensities. Earthq Eng Struct Dyn 38(15):1687–1708

Jayaram N, Baker JW (2010) Efficient sampling and data reduction techniques for probabilistic seismic lifeline risk assessment. Earthq Eng Struct Dyn 39(10):1109–1131

Nielsen TD, Jensen FV (2009) Bayesian networks and decision graphs. Springer Science & Business Media, New York

Karaca E (2005) Regional earthquake loss estimation: role of transportation network, sensitivity and uncertainty, and risk mitigation. Doctoral dissertation, Massachusetts Institute of Technology

Li J, Dueñas-Osorio L, Chen C, Shi C (2017) AC power flow importance measures considering multi-element failures. Reliab Eng Syst Saf, accepted. https://doi.org/10.1016/j.ress.2016.11.010

Lin P, Wang N, Ellingwood BR (2016) A risk de-aggregation framework that relates community resilience goals to building performance objectives. Sustain Resilient Infrastruct 1(1–2):1–13

Meadows D, Randers J, Meadows D (2004) Limits to growth: the 30-year update. Chelsea Green Publishing, White River Junction

Mieler MW, Stojadinovic B, Budnitz RJ, Mahin SA, Comerio MC (2013) Toward resilient communities: a performance-based engineering framework for design and evaluation of the built environment (PEER Report 2013/19). Pacific Earthquake Engineering Research Center, Berkeley

Mieler MW, Stojadinovic B, Budnitz R, Comerio M, Mahin S (2015) A framework for linking community-resilience goals to specific performance targets for the built environment. Earthquake Spectra 31(3):1267–1283

Newell J, Beaven S, Johnston DM (2012) Population movements following the 2010–2011 Canterbury earthquakes: summary of research workshops November 2011 and current evidence, GNS Miscellaneous Series 44. 23 p + Appendix C

NIBS (National Institute of Building Sciences) (1999) Earthquake loss estimation methodology: HAZUS99 (SR2). Technical manual, Federal Emergency Management Agency, Washington, DC. http://www.fema.gov/hazus

OOFIMS Object-oriented framework for infrastructure modelling and simulation. Aavailable at https://sites.google.com/a/uniroma1.it/oofims/

PCCIP (1997) Critical foundations: protecting America's infrastructures. Report of the President's Commission on Critical Infrastructure Protection. Available from http://www.fas.org/sgp/library/pccip.pdf . Accessed 7 Sept 2011

Pitilakis K, Crowley H, Kaynia AM (2014a) SYNER-G: typology definition and fragility functions for physical elements at seismic risk, Geotechnical, Geological and Earthquake Engineering, 27. Springer, Dordrecht

Pitilakis K, Franchin P, Khazai B, Wenzel H (eds) (2014b) SYNER-G: systemic seismic vulnerability and risk assessment of complex urban, utility, lifeline systems and critical facilities: methodology and applications, vol 31. Springer, Dordrech

Poljanšek K, Bono F, Gutiérrez E (2012) Seismic risk assessment of interdependent critical infrastructure systems: the case of European gas and electricity networks. Earthq Eng Struct Dyn 41(1):61–79

Rinaldi SM (2004) Modelling and simulating critical infrastructures and their interdependencies. In: Proceedings of the thirty-seventh annual Hawaii international conference on system sciences. https://doi.org/10.1109/HICSS.2004.1265180

Shinozuka M, Murachi Y, Dong X, Zhou Y, Orlikowski MJ (2003) Effect of seismic retrofit of bridges on transportation networks. Earthq Eng Eng Vib 2(2):169–179

Sun L, Didier M, Delé E, Stojadinovic B (2015) Probabilistic demand and supply resilience model for electric power supply system under seismic hazard. In: 12th international conference on applications of statistics and probability in civil engineering, ICASP12 Vancouver, Canada, July 12–15, 2015

Vamvatsikos D (2017) Performance-based seismic design in real life: the good, the bad and the ugly. Atti del XVII Convegno ANIDIS L'ingegneria Sismica in Italia, 17–24

Vamvatsikos D, Kazantzi AK, Aschheim MA (2015) Performance-based seismic design: avant-garde and code-compatible approaches. ASCE-ASME J Risk Uncertain Eng Syst Part A Civil Eng 2(2):C4015008

Vanzi I (1996) Seismic reliability of electric power networks: methodology and application. Struct Saf 18:311–327

Veneziano D, Casciati F, Faravelli L (1983) Method of seismic fragility for complicated systems. In: Proceeding of the 2nd Committee of Safety of Nuclear Installation (CNSI). Specialist meeting on probabilistic methods in seismic risk assessment for nuclear power plants, Lawrence Livermore Laboratory, CA

Wang Y, Au S-K, Fu Q (2010) Seismic risk assessment and mitigation of water supply systems. Earthquake Spectra 26(1):257–274

Weatherill G, Esposito S, Iervolino I, Franchin P, Cavalieri F (2014) Framework for seismic hazard analysis of spatially distributed systems. In: SYNER-G: systemic seismic vulnerability and risk assessment of complex urban, utility, lifeline systems and critical facilities. Springer, Dordrecht, pp 57–88